MATRIX Book Series 1

Editors
David R. Wood (*Editor-in-Chief*)
Jan de Gier
Cheryl E. Praeger
Terence Tao

MATRIX is Australia's international and residential mathematical research institute. It facilitates new collaborations and mathematical advances through intensive residential research programs, each lasting 1–4 weeks.

More information about this series at http://www.springer.com/series/15890

David R. Wood
Editor-in-Chief

Jan de Gier • Cheryl E. Praeger • Terence Tao
Editors

2016 MATRIX Annals

 Springer

Editors
David R. Wood (*Editor-in-Chief*)
Monash University
Clayton, Australia

Jan de Gier
University of Melbourne
Melbourne, Australia

Cheryl E. Praeger
University of Western Australia
Crawley, Australia

Terence Tao
UCLA
Los Angeles, USA

ISSN 2523-3041   ISSN 2523-305X (electronic)
MATRIX Book Series
ISBN 978-3-030-10183-1   ISBN 978-3-319-72299-3 (eBook)
https://doi.org/10.1007/978-3-319-72299-3

Mathematics Subject Classification (2010): 51-XX, 81-XX, 58-XX, 20-XX, 22-XX, 05-XX, 19-XX, 37-XX, 41-XX, 49-XX

Preface

MATRIX is Australia's first international and residential mathematical research institute. It was established in 2015 and launched in 2016 as a joint partnership between Monash University and the University of Melbourne, with seed funding from the ARC Centre of Excellence for Mathematical and Statistical Frontiers. The purpose of MATRIX is to facilitate new collaborations and mathematical advances through intensive residential research programs, which are currently held in Creswick, a small town nestled in the beautiful forests of the Macedon Ranges, 130 km west of Melbourne.

This book, *2016 MATRIX Annals*, is a scientific record of the five programs held at MATRIX in 2016:

- *Higher Structures in Geometry and Physics*
- *Winter of Disconnectedness*
- *Approximation and Optimisation*
- *Refining C*-Algebraic Invariants for Dynamics Using KK-Theory*
- *Interactions Between Topological Recursion, Modularity, Quantum Invariants and Low-Dimensional Topology*

The MATRIX Scientific Committee selected these programs based on scientific excellence and the participation rate of high-profile international participants. This committee consists of Jan de Gier (Melbourne University, Chair), Ben Andrews (Australian National University), Darren Crowdy (Imperial College London), Hans De Sterck (Monash University), Alison Etheridge (University of Oxford), Gary Froyland (University of New South Wales), Liza Levina (University of Michigan), Kerrie Mengersen (Queensland University of Technology), Arun Ram (University of Melbourne), Joshua Ross (University of Adelaide), Terence Tao (University of California, Los Angeles), Ole Warnaar (University of Queensland), and David Wood (Monash University).

The selected programs involved organisers from a variety of Australian universities, including Federation, Melbourne, Monash, Newcastle, RMIT, Sydney, Swinburne, and Wollongong, along with international organisers and participants. Each program lasted 1–4 weeks and included ample unstructured time to encourage collaborative research. Some of the longer programs had an embedded conference or lecture series. All participants were encouraged to submit articles to the MATRIX Annals.

The articles were grouped into refereed contributions and other contributions. Refereed articles contain original results or reviews on a topic related to the MATRIX program. The other contributions are typically lecture notes based on talks or activities at MATRIX. A guest editor organised appropriate refereeing and ensured the scientific quality of submitted articles arising from each program. The editors (Jan de Gier, Cheryl E. Praeger, Terence Tao, and myself) finally evaluated and approved the papers.

Many thanks to the authors and to the guest editors for their wonderful work.

MATRIX has hosted eight programs in 2017, with more to come in 2018; see www.matrix-inst.org.au. Our goal is to facilitate collaboration between researchers in universities and industry, and increase the international impact of Australian research in the mathematical sciences.

David R. Wood
MATRIX Book Series Editor-in-Chief

Higher Structures in Geometry and Physics

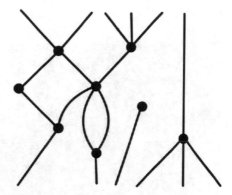

6–17 June 2016

Organisers

Marcy Robertson (Melbourne)
Philip Hackney (Macquarie)

The inaugural program at MATRIX took place on June 6–17, 2016, and was entitled "Higher Structures in Geometry and Physics". It was both a pleasure and a privilege to take part in this first ever program at MATRIX. The excellent working conditions, cosy environment, friendly staff and energetic participants made this time both memorable and productive.

The scientific component of our program was comprised of a workshop with lecture series by several invited speakers, followed in the subsequent week by a conference featuring talks on a range of related topics from speakers from around the globe. The two events were separated by a long weekend which gave the participants free time to discuss and collaborate. Within this volume is a collection of lecture notes and articles reflecting quite faithfully the ideas in the air during these two weeks.

Our title Higher Structures (not unlike the term down under) suggests a certain fixed perspective. For the participants in our program, this perspective comes from the twentieth-century examples of algebraic and categorical constructions associated to topological spaces, possibly with geometric structures and possibly taking motivation from physical examples. From this common frame of reference stems a range of new and rapidly developing directions, activities such as this program at MATRIX play a vital role in weaving these threads into a collective understanding. The excitement of working in this rapidly developing field was felt during our time at MATRIX, and I hope it comes across in this volume as well.

I would like to thank all of the authors who took the time to contribute to this volume. I would also like to thank the MATRIX staff and officials for hosting and facilitating this event and giving us the opportunity to share our work with this volume. Most importantly, I would like to thank the organizers of our program Marcy and Philip for all of their hard work and for giving all of us participants this unique opportunity.

Ben Ward
Guest Editor

Participants

Ramon Abud Alcala (Macquarie), Clark Barwick (Massachusetts Institute of Technology), Alexander Campbell (Macquarie), David Carchedi (George Mason), Gabriel C. Drummond-Cole (IBS Center for Geometry and Physics), Daniela Egas Santander (Freie Universitat Berlin), Nora Ganter (Melbourne), Christian Haesemeyer (Melbourne), Philip Hackney (Osnabrueck), Ralph Kauffman (Purdue), Edoardo Lanari (Macquarie), Martin Markl (Czech Academy of Sciences), Branko Nikolic (Macquarie), Simona Paoli (Leicester), Sophia Raynor (Aberdeen), Emily Riehl (Johns Hopkins), David Roberts (Adelaide), Marcy Robertson (Melbourne), Chris Rogers (Louisiana), Martina Rovelli (EPFL),

Matthew Spong (Melbourne), Michelle Strumila (Melbourne), TriThang Tran (Melbourne), Victor Turchin (Kansas State), Dominic Verity (Macquarie), Raymond Vozzo (Adelaide), Ben Ward (Stony Brook), Mark Weber (Macquarie), Felix Wierstra (Stockholm), Sinan Yalin (Copenhagen), Jun Yoshida (Tokyo), Dimitri Zaganidis (EPFL)

Winter of Disconnectedness

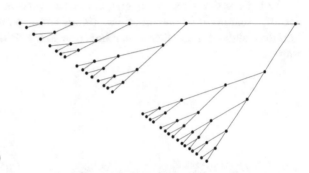

27 June–8 July 2016

Organisers

Murray Elder (Newcastle)
Jacqui Ramagge (Sydney)
Colin Reid (Newcastle)
Anne Thomas (Sydney)
George Willis (Newcastle)

Our understanding of totally disconnected locally compact (t.d.l.c.) groups has been growing rapidly in recent years. These groups are of interest for general theoretical reasons, because half the task of describing the structure of general locally compact groups falls into the totally disconnected case, and also for purposes of specific applications, because of the significance that various classes of t.d.l.c. groups have in geometry, number theory and algebra. The workshop held at Creswick from 27 June to 8 July 2016 was the first part of a program in which leading researchers presented the most recent advances by giving short courses and individual lectures. Time was also set aside for collaboration between established researchers and students. Both general techniques and results relating to particular classes of t.d.l.c. groups were covered in the lectures.

Four courses of five lectures each were delivered at Creswick, as follows:

- Helge Glöckner (Paderborn): *Endomorphisms of Lie groups over local fields*
- George Willis (Newcastle): *The scale, tidy subgroups and flat groups*
- Anne Thomas (Sydney): *Automorphism groups of combinatorial structures*
- Phillip Wesolek (Binghamton): *A survey of elementary totally disconnected locally compact groups*

A second workshop was held in Newcastle at the end of July 2016 with the same structure.

Notes from the four courses just listed and from the courses

- Adrien Le Boudec (Louvain Le Neuve): *Groups of automorphisms and almost automorphisms of trees: subgroups and dynamics*
- Colin Reid (Newcastle): *Normal subgroup structure of totally disconnected locally compact groups*

delivered at the Newcastle workshop are published here. Although the distinction is not absolute, the notes by Reid, Wesolek and Willis cover general methods and those by Glöckner, Le Boudec and Thomas treat specific classes of t.d.l.c. groups. In several cases, they are based on notes taken by listeners at the workshops, and we are grateful for their assistance. Moreover, the notes by Glöckner, which have been refereed, are an expanded version of what was delivered in lectures and contain calculations and proofs of some results that have not previously been published.

We believe that these notes give the 2016 overview of the state of knowledge and of research directions on t.d.l.c. groups and hope that they will also serve to introduce students and other researchers new to the field to this rapidly developing subject.

Dave Robertson
Guest Editor

Participants

Benjamin Brawn (Newcastle), Timothy Bywaters (Sydney), Wee Chaimanowong (Melbourne), Murray Elder (Newcastle), Helge Glöckner (Paderborn), John Harrison

(Baylor), Waltraud Lederle (ETH Zurich), Rupert McCallum (Tübingen), Sidney Morris (Federation), Uri Onn (ANU), C.R.E. Raja (Indian Statistical Institute, Bangalore), Jacqui Ramagge (Sydney), Colin Reid (Newcastle), David Robertson (Newcastle), Anurag Singh (Utah), Simon Smith (City University of New York), George Willis (Newcastle), Thomas Taylor (Newcastle), Anne Thomas (Sydney), Stephan Tornier (ETH Zurich), Tian Tsang (RMIT), Phillip Wesolek (Universite Catholique de Louvain)

Approximation and Optimisation

10–16 July 2016

Organisers

Vera Roshchina (RMIT)
Nadezda Sukhorukova
(Swinburne)
Julien Ugon (Federation)
Aris Daniilidis (Chile)
Andrew Eberhard (RMIT)
Alex Kruger (Federation)
Zahra Roshanzamir
(Swinburne)

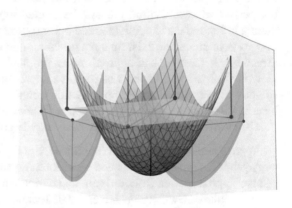

There are many open problems in the field of approximation theory where tools from variational analysis and nonsmooth optimisation show promise. One of them is the Chebyshev (also known as uniform) approximation problem. Chebyshev's work is generally considered seminal in approximation theory and remains influential to this day in this field of mathematics. His work also has many significant implications in optimisation, where uniform approximation of functions is considered an early example of an optimisation problem where the objective function is not differentiable. In fact, the problem of best polynomial approximation can be reformulated as an optimisation problem which provides very nice textbook examples for convex analysis.

The joint ancestry and connections between optimisation and approximation are still very apparent today. Many problems in approximation can be reformulated as optimisation problems. On the other hand, many optimisation methods require set and/or function approximation to work efficiently.

These connections were explored in the 1950s, 1960s and 1970s. This was also the period when the area of nonsmooth analysis emerged. Convex and nonsmooth analysis techniques can be applied to obtain theoretical results and algorithms for solving approximation problems with nonsmooth objectives. These problems include Chebyshev approximation: univariate (polynomial and fixed-knots polynomial spline) approximation and multivariate polynomial approximation. In 1972, P.-J. Laurent published his book, where he demonstrated interconnections between approximation and optimisation. In particular, he showed that many difficult (Chebyshev) approximation problems can be solved using optimisation techniques.

Despite this early work, historically the fields of optimisation and approximation have remained separate, each (re)developing their own methodology and terminology. There is a clear pattern though: advances in optimisation result in significant breakthroughs in approximation and, on the other hand, new approximation techniques and approaches advance the development of new optimisation methods and improve the performance of existing ones. Therefore, the lift-off theme of the program was multivariate and polynomial spline approximation from the perspective of optimisation theory. We invited top researchers in approximation theory, polynomial and semialgebraic optimisation and variational analysis to find new ways to attack the existing open problems and to establish major new research directions. A fresh look at approximation problems from the optimisation point of view is vital, since it enables us to solve approximation problems that cannot be solved without very advanced optimisation techniques (and vice versa) and discover beautiful interconnections between approximation and optimisation.

The morning sessions of the program consisted of lectures:

- Approximation of set-valued functions (Nira Dyn, Tel Aviv University)
- Algebraic, convex analysis and semi-infinite programming approach to Chebyshev approximation (Julien Ugon, Federation University Australia, and Nadezda Sukhorukova, Swinburne University of Technology)
- The sparse grid combination technique and optimisation (Markus Hegland, Australian National University)
- Quasi-relative interior and optimisation (Constantin Zalinescu, Al. I. Cuza University)

The afternoons were dedicated to smaller group discussions. The following discussions were vital for the research papers submitted to this volume:

1. **Compact convex sets with prescribed facial dimensions, by Vera Roshchina, Tian Sang and David Yost**

    The rich soup of research ideas that was stirred up during the workshop helped us with a breakthrough in a seemingly unrelated research direction. During the workshop, David Yost came up with a neat inductive idea that finished the proof of the dimensional sequence theorem that he and Vera Roshchina have been working on for a while. Later during the workshop the fractal ideas introduced by Markus Hegland motivated us to consider convex sets with fractal facial structure. Even more surprisingly, we discovered that a beautiful example of such a set can be obtained from the spherical gasket studied by Tian Sang in her prior research on infinite Coxeter groups.

2. **Chebyshev multivariate polynomial approximation: alternance interpretation, by Nadezda Sukhorukova, Julien Ugon and David Yost**

The notion of alternating sequence (alternance) is central for univariate Chebyshev approximation problems. How can we extend this notion to the case of multivariate approximation, where the sets are not totally ordered? In one of the papers, namely, "Chebyshev multivariate polynomial approximation: alternance interpretation" by Sukhorukova, Ugon and Yost, the authors work on this issue and propose possible solutions, in particular a very elegant formulation for necessary and sufficient optimality conditions for multivariate Chebyshev approximation.

Julien Ugon and Nadezda Sukhorukova
Guest Editors

Participants

Alia Al nuaimat (Federation), Fusheng Bai (Chongqing Normal University), Yi Chen (Federation), Jeffrey Christiansen (RMIT), Brian Dandurand (RMIT), Reinier Diaz Millan (Federal Institute of Goias), Nira Dyn (Tel Aviv), Andrew Eberhard (RMIT), Gabriele Eichfelder (Technische Universitat Ilmenau), Markus Hegland (ANU), Alexander Kruger (Federation), Vivek Laha (Indian Institute of Technology, Patna), Jeffrey Linderoth (Wisconsin- Madison), Prabhu Manyem (Nanchang Institute of Technology), Faricio Oliveira (RMIT), Zahra Roshan Zamir (Swinburne), Vera Roshchina (RMIT), Tian Sang (RMIT), Jonathan Scanlan, Vinay Singh (National Institute of Technology, Mizoram), Nadezda Sukhorukova (Swinburne), Julien Ugon (Federation), Dean Webb (Federation), David Yost (Federation), Constantin Zalinescu (University "Al. I. Cuza" Iasi), Jiapu Zhang (Federation)

Refining C*-Algebraic Invariants for Dynamics Using KK-Theory

18–29 July 2016

Organisers

Magnus Goffeng
(Chalmers / Gothenburg)
Adam Rennie (Wollongong)
Aidan Sims (Wollongong)

THE BOTT PROJECTION

This graduate school and workshop were motivated by intense recent interest and progress in the non-commutative geometry of dynamical systems.

The progress has been of several sorts. The assignment of C^*-algebras to dynamical systems is not new but has become much more sophisticated in recent years. The K-groups of such dynamical C^*-algebras provide invariants of the original dynamical system but are not always fine enough to capture the structural features of greatest interest. In response to this, precise characterisations of the relationships between more detailed K-theoretic invariants and equivalence classes of dynamical systems have recently been sharpened significantly.

The extra ingredient whose potential applications to such problems we hoped to highlight to attendees is recent progress in importing ideas from algebraic topology to dynamical systems theory through the computability of the Kasparov product. The Kasparov product is a far-reaching generalisation of index theory and provides an abstract composition rule for morphisms in the KK-category of C^*-algebras. The KK-category extends and refines the correspondence category of C^*-algebras; and a correspondence can be regarded as a generalised dynamical system and is closely related to the construction of dynamical C^*-algebras.

With this background in mind, we started the first week with three lecture series during the mornings, with informal Q&A and research in the afternoons. The three lecture courses were:

- Robin Deeley: *Groupoids and C*-algebras*
- Bram Mesland: *Kasparov's KK-theory*
- Adam Rennie and Aidan Sims: *Hilbert modules and Cuntz–Pimsner algebras*

Groupoid C^*-algebras and Cuntz–Pimsner algebras are two of the most flexible and best developed frameworks for modelling dynamical systems using C^*-algebras. The basics of Kasparov's KK-theory and the recent advances in the computability of the product proved central to the progress seen during the workshop.

The lecture series by Deeley, Mesland, Rennie, and Sims set the stage for the second week, bringing attendees, particularly the significant student cohort, together around a common language and a joint leitmotif. The talks ranged widely over the core topics, their applications and neighbouring disciplines. The five papers which follow give a good indication of the breadth of the conference.

Deeley offers refinements of Putnam's homology theory for dynamical systems, and Ruiz et al. provide strong invariants for a special class of dynamical systems. Bourne (with Schulz-Baldes) provides an application of KK-theoretic techniques to topological insulators. Goffeng and Mesland provide a detailed account of novel aspects of the non-commutative geometry of the Cuntz algebras, while Arici probes the non-commutative topology and geometry of quantum lens spaces using techniques which are perfectly in tune with the theme of the workshop.

Adam Rennie
Guest Editor

Participants

Zahra Afsar (Wollongong), Francesca Arici (Radboud University Nijmegen), Chris Bourne (Erlangen-Nurnberg), Guo Chuan Thiang (Adelaide), Robin Deeley (Hawaii), Anna Luise Duwenig (Victoria), James Fletcher (Wollongong), Iain Forsyth (Leibniz University Hannover), Elizabeth Anne Gillaspy (Universitet Manster), Magnus Goffeng (Chalmers Technology/Gothenburg), Peter Hochs (Adelaide), Marcelo Laca (Victoria), Lachlan MacDonald (Wollongong), Michael Mampusti (Wollongong), Bram Mesland (Leibniz University Hannover), Alexander Mundey (Wollongong), Adam Rennie (Wollongong), Karen Rught Strung (Polish Academy of Sciences), Efren Ruiz (Hawaii at Hilo), Thomas Scheckter (UNSW), Aidan Sims (Wollongong), Hang Wang (Adelaide), Yasuo Watatani (Kyushu)

Interactions Between Topological Recursion, Modularity, Quantum Invariants and Low-Dimensional Topology

28 November–23 December 2016

Organisers

Motohico Mulase (UC Davis)
Norman Do (Monash)
Neil Hoffman (Oklahoma State)
Craig Hodgson (Melbourne)
Paul Norbury (Melbourne)

$$\left\langle \vcenter{\hbox{}} \right\rangle = (-A^2 - A^{-2})\left\langle \vcenter{\hbox{}} \right\rangle$$

$$= (-A^2 - A^{-2})(A^9 - A^5 - A - 2A^{-3} + A^{-7} - A^{-11})$$

This program contributed to the active international research effort under way at present to connect structures in mathematical physics with those in low-dimensional topology, buoyed by recent theoretical advances and a broad range of applications. It brought together people in cognate areas with common interests, including algebraic geometry, conformal field theory, knot theory, representation theory, quantum invariants and combinatorics.

The program was motivated by recent generalisations of the technique of topological recursion, as well as fundamental conjectures concerning invariants in low dimensional topology. These include the AJ conjecture, the Jones slope conjecture of Garoufalidis and the underlying topological significance of the 3D-index. Algorithmic techniques to compute these and related invariants are also featured in the program.

The program began with a week of short courses, comprising three lectures each, on the following topics:

- *Conformal field theory* (Katrin Wendland, Freiburg)
- *Hyperbolic knot theory* (Jessica Purcell, Monash)
- *Quantum invariants* (Roland van der Veen, Leiden)
- *Topological recursion* (Norman Do, Monash)

The program for the middle week was largely informal and reserved for research collaborations. The final week was dedicated to an international conference, which gathered together leading experts in the areas of mathematical physics, topological recursion, quantum invariants and low-dimensional topology to address recent advances and explore new connections between these fields.

The program attracted a total of 51 attendees. During the conference, there were 33 talks, of which 23 were delivered by international visitors. Among these talks were the following:

- Jørgen Andersen: Verlinde formula for Higgs bundles
- Feng Luo: Discrete uniformization for polyhedral surfaces and its convergence
- Rinat Kashaev: Pachner moves and Hopf algebras
- Scott Morrison: Modular data for Drinfeld doubles
- Hyam Rubinstein and Craig Hodgson: Counting genus two surfaces in 3-manifolds
- Gaëtan Borot: Initial conditions for topological recursion
- Tudor Dimofte: Counting vortices in the 3D index
- George Shabat: Counting Belyi pairs over finite fields
- Leonid Chekhov: Abstract topological recursion and Givental decomposition
- Piotr Sułkowski: Knots and BPS/super-quantum curves

The articles in these proceedings represent different aspects of the program. Kashaev's contribution describes a topological quantum field theory in four dimensions. Licata-Mathews and Spreer-Tillmann describe topological and geometric results for 3-manifolds. Shabat describes first steps towards generalising Belyi maps to finite fields. Roland van der Veen kindly contributed notes from his short course on quantum invariants of knots.

Norman Do, Neil Hoffman, Paul Norbury

Guest Editors

Participants

Jørgen Andersen (Aarhus), Vladimir Bazhanov (ANU), Gaetan Borot (Max Planck), Benjamin Burton (Queensland), Alex Casella (Sydney), Wee Chaimanowong (Melbourne), Abhijit Champanerkar (CUNY), Anupam Chaudhuri (Monash), Leonid

Chekhov (Steklov), Blake Dadd (Melbourne), Tudor Dimofte (California, Davis), Norman Do (Monash), Petr Dunin-Barkovskiy (Moscow), Omar Foda (Melbourne), Evgenii Fominykh, Sophie Ham, Robert Cyrus Haraway III (Sydney), Craig Hodgson (Melbourne), Neil Hoffman (Oklahoma), Joshua Howie (Monash), Adele Jackson (ANU), Max Jolley (Monash), Rinat Kashaev (Geneva), Seonhwa Kim (IBS), Ilya Kofman (CUNY), Reinier Kramer (Amsterdam), Andrew James Kricker (NTU), Priya Kshirsagar (UC Davis), Alice Kwon (CUNY), Tung Le (Monash), Oliver Leigh (British Columbia), Danilo Lewanski (Amsterdam), Joan Elizabeth Licata (ANU), Beibei Liu (UC Davis), Feng Luo (Rutgers), Joseph Lynch (Melbourne), Alessandro Malusa (Aarhus), Clément Maria (Queensland), Daniel Mathews (Monash), Sergei Matveev (Chelyabinsk), Todor Milanov Kavli (IPMU), Scott Morrison (ANU), Motohico Mulase (UC Davis), Paul Norbury (Melbourne), Nicolas Orantin (EPFL), Erik William Pettersson (RMIT), Aleksandr Popolitov (Amsterdam), Jessica Purcell (Monash), Robert Quigley-McBride, Hyam Rubinstein (Melbourne), Axel Saenz Rodriguez (Virginia), Sjabbo Schaveling, Henry Segerman (Oklahoma), Georgy Shabat (Independent), Rafael Marian Siejakowski (NTU Singapore), Ruifang Song (UC Davis), Piotr Sulkowski (Warsaw & Caltech), Dominic James Tate (Sydney), Stephan Tillmann (Sydney), Roland van der Veen (Leiden), Paul Wedrich (Imperial), Katrin Wendland (Friburg), Campbell Wheeler (Melbourne), Adam Wood (Melbourne), Tianyu Yang (Melbourne)

Contents

Refining C* Algebraic Invariants for Dynamics Using KK-Theory

Interactions Between Topological Recursion, Modularity, Quantum Invariants and Low-dimensional Topology

PART II   OTHER CONTRIBUTED ARTICLES

Higher Structures in Geometry and Physics

Winter of Disconnectedness

Interactions Between Topological Recursion, Modularity, Quantum Invariants and Low-dimensional Topology

Part I
Refereed Articles

Homotopical Properties of the Simplicial Maurer–Cartan Functor

Christopher L. Rogers

Abstract We consider the category whose objects are filtered, or complete, L_∞-algebras and whose morphisms are ∞-morphisms which respect the filtrations. We then discuss the homotopical properties of the Getzler–Hinich simplicial Maurer–Cartan functor which associates to each filtered L_∞-algebra a Kan simplicial set, or ∞-groupoid. In previous work with V. Dolgushev, we showed that this functor sends weak equivalences of filtered L_∞-algebras to weak homotopy equivalences of simplicial sets. Here we sketch a proof of the fact that this functor also sends fibrations to Kan fibrations. To the best of our knowledge, only special cases of this result have previously appeared in the literature. As an application, we show how these facts concerning the simplicial Maurer–Cartan functor provide a simple ∞-categorical formulation of the Homotopy Transfer Theorem.

1 Introduction

Over the last few years, there has been increasing interest in the homotopy theory of filtered, or complete, L_∞-algebras[1] and the role these objects play in deformation theory [10, 12], rational homotopy theory [3, 4, 13], and the homotopy theory of homotopy algebras [6, 8, 9]. One important tool used in these applications is the simplicial Maurer–Cartan functor $\mathfrak{MC}_\bullet(-)$ which produces from any filtered L_∞-algebra a Kan simplicial set, or ∞-groupoid. This construction, first appearing in the work of Hinich [12] and Getzler [10], can (roughly) be thought of as a "non-abelian analog" of the Dold–Kan functor from chain complexes to simplicial

---

[1] Throughout this paper, all algebraic structures have underlying $\mathbb{Z}$ graded $\mathbb{k}$-vector spaces with char $\mathbb{k} = 0$.

C.L. Rogers (✉)
Department of Mathematics and Statistics, University of Nevada, Reno, 1664 N. Virginia Street, Reno, NV 89557-0084, USA
e-mail: chrisrogers@unr.edu; chris.rogers.math@gmail.com

© Springer International Publishing AG, part of Springer Nature 2018
D.R. Wood et al. (eds.), *2016 MATRIX Annals*, MATRIX Book Series 1,
https://doi.org/10.1007/978-3-319-72299-3_1

3

vector spaces. In deformation theory, these ∞-groupoids give higher analogs of the Deligne groupoid. In rational homotopy theory, this functor generalizes the Sullivan realization functor, and has been used to study rational models of mapping spaces.

A convenient presentation of the homotopy theory of filtered L_∞-algebras has yet to appear in the literature. But based on applications, there are good candidates for what the weak equivalences and fibrations should be between such objects. One would also hope that the simplicial Maurer–Cartan functor sends these morphisms to weak homotopy equivalences and Kan fibrations, respectively. For various special cases, which are recalled in Sect. 3, it is known that this is indeed true. In joint work with Dolgushev [5], we showed that, in general, $\mathfrak{MC}_\bullet(-)$ maps any weak equivalence of filtered L_∞-algebras to a weak equivalence of Kan complexes. This can be thought of as the natural L_∞ generalization of the Goldman–Millson theorem in deformation theory.

The purpose of this note is to sketch a proof of the analogous result for fibrations (Theorem 2 in Sect. 3 below): The simplicial Maurer–Cartan functor maps any fibration between any filtered L_∞-algebras to a fibration between their corresponding Kan complexes. Our proof is not a simple generalization of the special cases already found in the literature, nor does it follow directly from general abstract homotopy theory. It requires some technical calculations involving Maurer–Cartan elements, similar to those found in our previous work [5].

As an application, we show in Sect. 4 that "∞-categorical" analogs of the existence and uniqueness statements that comprise the Homotopy Transfer Theorem [1, 2, 14, 15] follow as a corollary of our Theorem 2. In more detail, suppose we are given a cochain complex A, a homotopy algebra B of some particular type (e.g., an A_∞, L_∞, or C_∞-algebra) and a quasi-isomorphism of complexes $\phi: A \to B$. Then, using the simplicial Maurer–Cartan functor, we can naturally produce an ∞-groupoid $\mathfrak{F}$ whose objects correspond to solutions to the "homotopy transfer problem". By a solution, we mean a pair consisting of a homotopy algebra structure on A, and a lift of ϕ to a ∞-quasi-isomorphism of homotopy algebras $A \xrightarrow{\sim} B$. The fact that $\mathfrak{MC}_\bullet(-)$ preserves both weak equivalences and fibrations allows us to conclude that: (1) The ∞-groupoid $\mathfrak{F}$ is non-empty, and (2) it is contractible. In other words, a homotopy equivalent transferred structure always exists, and this structure is unique in the strongest possible sense.

2 Preliminaries

2.1 Filtered L_∞-Algebras

In order to match conventions in our previous work [5], we define an L_∞-**algebra** to be a cochain complex (L, ∂) for which the reduced cocommutative coalgebra $\underline{S}(L)$ is equipped with a degree 1 coderivation Q such that $Q(x) = \partial x$ for all $x \in L$ and $Q^2 = 0$. This structure is equivalent to specifying a sequence of *degree 1* multi-brackets

$$\{\,,\,,\ldots,\,\}_m : S^m(L) \to L \quad m \geq 2 \tag{1}$$

satisfying compatibility conditions with the differential ∂ and higher-order Jacobi-like identities. (See Eq. 2.5 in [5].) More precisely, if $\mathrm{pr}_L : \underline{S}(L) \to L$ denotes the usual projection, then

$$\{x_1, x_2, \ldots, x_m\}_m = \mathrm{pr}_L Q(x_1 x_2 \ldots x_m), \qquad \forall x_j \in L.$$

This definition of L_∞-algebra is a "shifted version" of the original definition of L_∞-algebra. A shifted L_∞-structure on L is equivalent to a traditional L_∞-structure on sL, the suspension of L.

A **morphism** (or ∞-morphism) Φ from an L_∞-algebra (L, Q) to an L_∞-algebra $(\tilde{L}, \tilde{Q})$ is a dg coalgebra morphism

$$\Phi : \big(\underline{S}(L), Q\big) \to \big(\underline{S}(\tilde{L}), \tilde{Q}\big). \tag{2}$$

Such a morphism Φ is uniquely determined by its composition with the projection to $\tilde{L}$:

$$\Phi' := \mathrm{pr}_{\tilde{L}} \Phi.$$

Every such dg coalgebra morphism induces a map of cochain complexes, e.g., the linear term of Φ:

$$\phi := \mathrm{pr}_{\tilde{L}} \Phi|_L : (L, \partial) \to (\tilde{L}, \tilde{\partial}), \tag{3}$$

and we say Φ is **strict** iff it consists only of a linear term, i.e.

$$\Phi'(x) = \phi(x) \qquad \Phi'(x_1, \ldots, x_m) = 0 \quad \forall m \geq 2 \tag{4}$$

A morphism $\Phi : (L, Q) \to (\tilde{L}, \tilde{Q})$ of L_∞-algebras is an ∞-**quasi-isomorphism** iff $\phi : (L, \partial) \to (\tilde{L}, \tilde{\partial})$ is a quasi-isomorphism of cochain complexes.

We say an L_∞-algebra (L, Q) is a **filtered L_∞-algebra** iff the underlying cochain complex (L, ∂) is equipped with a complete descending filtration,

$$L = \mathscr{F}_1 L \supset \mathscr{F}_2 L \supset \mathscr{F}_3 L \cdots \tag{5}$$

$$L = \varprojlim_k L/\mathscr{F}_k L, \tag{6}$$

which is compatible with the brackets, i.e.

$$\left\{ \mathscr{F}_{i_1}L, \mathscr{F}_{i_2}L, \ldots, \mathscr{F}_{i_m}L \right\}_m \subseteq \mathscr{F}_{i_1+i_2+\cdots+i_m}L \quad \forall \; m > 1.$$

A filtered L_∞-algebra in our sense is a shifted analog of a "complete" L_∞-algebra, in the sense of Berglund [3, Def. 5.1].

Remark 1 Due to its compatibility with the filtration, the L_∞-structure on L induces a filtered L_∞-structure on the quotient $L/\mathscr{F}_nL$. In particular, $L/\mathscr{F}_nL$ is a **nilpotent L_∞-algebra** [3, Def. 2.1], [10, Def. 4.2]. Moreover, when the induced L_∞-structure is restricted to the sub-cochain complex

$$\mathscr{F}_{n-1}L/\mathscr{F}_nL \subseteq L/\mathscr{F}_nL$$

all brackets of arity ≥ 2 vanish. Hence, the nilpotent L_∞-algebra $\mathscr{F}_{n-1}L/\mathscr{F}_nL$ is an **abelian L_∞-algebra**.

Definition 1 We denote by $\widehat{\mathsf{Lie}}_\infty$ the category whose objects are filtered L_∞-algebras and whose morphisms are ∞-morphisms $\Phi\colon (L,Q) \to (\tilde{L}, \tilde{Q})$ which are compatible with the filtrations:

$$\Phi'(\mathscr{F}_{i_1}L \otimes \mathscr{F}_{i_2}L \otimes \cdots \otimes \mathscr{F}_{i_m}L) \subset \mathscr{F}_{i_1+i_2+\cdots+i_m}\tilde{L}, \tag{7}$$

Definition 2 Let $\Phi\colon (L,Q) \to (\tilde{L}, \tilde{Q})$ be a morphism in $\widehat{\mathsf{Lie}}_\infty$.

1. We say Φ is a **weak equivalence** iff its linear term $\phi\colon (L, \partial) \to (\tilde{L}, \tilde{\partial})$ induces a quasi-isomorphism of cochain complexes

$$\phi|_{\mathscr{F}_nL}\colon (\mathscr{F}_nL, \partial) \to (\mathscr{F}_n\tilde{L}, \tilde{\partial}) \qquad \forall n \geq 1.$$

2. We say Φ is a **fibration** iff its linear term $\phi\colon (L, \partial) \to (\tilde{L}, \tilde{\partial})$ induces a surjective map of cochain complexes

$$\phi|_{\mathscr{F}_nL}\colon (\mathscr{F}_nL, \partial) \to (\mathscr{F}_n\tilde{L}, \tilde{\partial}) \qquad \forall n \geq 1.$$

3. We say Φ is an **acyclic fibration** iff Φ is both a weak equivalence and a fibration.

Remark 2 If (L, Q) is a filtered L_∞-algebra, then for each $n \geq 1$, we have the obvious short exact sequence of cochain complexes

$$0 \to \mathscr{F}_{n-1}L/\mathscr{F}_nL \xrightarrow{i_{n-1}} L/\mathscr{F}_nL \xrightarrow{p_n} L/\mathscr{F}_{n-1}L \to 0. \tag{8}$$

It is easy to see that (8) lifts to a sequence of filtered L_∞-algebras, in which all of the algebras in the sequence are nilpotent L_∞-algebras (see Remark 1), and in which all of the morphisms in the sequence are strict. In particular, the morphism $L/\mathscr{F}_n L \xrightarrow{p_n} L/\mathscr{F}_{n-1}L$ is a fibration.

2.2 Maurer–Cartan Elements

Our reference for this section is Section 2 of [6]. We refer the reader there for details. Let L be a filtered L_∞-algebra. Since $L = \mathscr{F}_1 L$, the compatibility of the multi-brackets with the filtrations gives us well defined map of sets $\mathsf{curv}\colon L^0 \to L^1$:

$$\mathsf{curv}(\alpha) = \partial\alpha + \sum_{m \geq 1} \frac{1}{m!}\{\alpha^{\otimes m}\}_m. \tag{9}$$

Elements of the set

$$\mathrm{MC}(L) := \{\alpha \in L^0 \mid \mathsf{curv}(\alpha) = 0\}$$

are called the **Maurer–Cartan (MC) elements** of L. Note that MC elements of L are elements of degree 0. Furthermore, if $\Phi\colon (L, Q) \to (\tilde{L}, \tilde{Q})$ is a morphism in $\widehat{\mathsf{Lie}}_\infty$ then the compatibility of Φ with the filtrations allows us to define a map of sets

$$\Phi_*\colon \mathrm{MC}(L) \to \mathrm{MC}(\tilde{L})$$

$$\Phi_*(\alpha) := \sum_{m \geq 2} \frac{1}{m!}\Phi'(\alpha^{\otimes m}). \tag{10}$$

The fact that $\mathsf{curv}(\Phi(\alpha)) = 0$ is proved in [6, Prop. 2.2].

Given an MC element $\alpha \in \mathrm{MC}(L)$, we can "twist" the L_∞-structure on L, to obtain a new filtered L_∞-algebra L^α. As a graded vector space with a filtration, $L^\alpha = L$; the differential ∂^α and the multi-brackets $\{\,,\ldots,\,\}_m^\alpha$ on L^α are defined by the formulas

$$\partial^\alpha(v) := \partial(v) + \sum_{k=1}^{\infty} \frac{1}{k!}\{\alpha,\ldots,\alpha,v\}_{k+1}, \tag{11}$$

$$\{v_1, v_2, \cdots, v_m\}_m^\alpha := \sum_{k=0}^{\infty} \frac{1}{k!}\{\alpha,\ldots,\alpha,v_1,v_2,\cdots,v_m\}_{k+m}. \tag{12}$$

2.3 Getzler–Hinich Construction

The MC elements of (L, Q) are in fact the vertices of a simplicial set. Let Ω_n denote the de Rham-Sullivan algebra of polynomial differential forms on the geometric simplex Δ^n with coefficients in $\Bbbk$. The simplicial set $\mathfrak{MC}_\bullet(L)$ is defined as

$$\mathfrak{MC}_n(L) := \mathrm{MC}(L \widehat{\otimes} \Omega_n) \tag{13}$$

where $L \widehat{\otimes} \Omega_n$ is the filtered L_∞-algebra defined as the projective limit of nilpotent L_∞-algebras

$$L \widehat{\otimes} \Omega_n := \varprojlim_k \big((L/\mathscr{F}_k L) \otimes \Omega_n \big).$$

Recall that the L_∞-structure on the tensor product of chain complexes $(L/\mathscr{F}_k L) \otimes \Omega_n$ is induced by the structure on $L/\mathscr{F}_k L$, and is well-defined since Ω_n is a commutative algebra. For example:

$$\{\bar{x}_1 \otimes \omega_1, \bar{x}_2 \otimes \omega_2, \dots, \bar{x}_l \otimes \omega_l\} := \pm\{\bar{x}_1, \bar{x}_2, \dots, \bar{x}_l\} \otimes \omega_1 \omega_2 \cdots \omega_l.$$

Proposition 4.1 of [6] implies that the simplicial set $\mathfrak{MC}_\bullet(L)$ is a Kan complex, or ∞-groupoid, which is sometimes referred to as the "Deligne–Getzler–Hinich" ∞-groupoid of L.

Any morphism $\Phi: L \to \tilde{L}$ in $\widehat{\mathsf{Lie}}_\infty$ induces a morphism $\Phi^{(n)}: L \widehat{\otimes} \Omega_n \to \tilde{L} \widehat{\otimes} \Omega_n$ for each $n \geq 0$ in the obvious way:

$$\Phi^{(n)}(x_1 \otimes \theta_1, x_2 \otimes \theta_2, \dots, x_m \otimes \theta_m) := \pm\Phi(x_1, x_2, \dots, x_m) \otimes \theta_1 \theta_2 \cdots \theta_m. \tag{14}$$

This then gives us a map of MC sets $\Phi_*^{(n)}: \mathrm{MC}(L \widehat{\otimes} \Omega_n) \to \mathrm{MC}(\tilde{L} \widehat{\otimes} \Omega_n)$ defined via Eq. (10). It is easy to see that $\Phi^{(n)}$ is compatible with the face and degeneracy maps, which leads us to the **simplicial Maurer–Cartan functor**

$$\mathfrak{MC}_\bullet: \widehat{\mathsf{Lie}}_\infty \to \mathsf{Kan}$$
$$\mathfrak{MC}_\bullet\big(L \xrightarrow{\Phi} \tilde{L}\big) := \mathfrak{MC}_\bullet(L) \xrightarrow{\Phi_*^\bullet} \mathfrak{MC}_\bullet(\tilde{L}) \tag{15}$$

3 The Functor $\mathfrak{MC}_\bullet(-)$ Preserves Weak Equivalences and Fibrations

Our first observation concerning the simplicial Maurer–Cartan functor is that it sends a weak equivalence $\Phi: L \xrightarrow{\sim} \tilde{L}$ in $\widehat{\mathsf{Lie}}_\infty$ to a weak homotopy equivalence. For the special case in which Φ is a *strict* quasi-isomorphism between (shifted) *dg Lie algebras*, Hinich [12] showed that $\mathfrak{MC}_\bullet(\Phi)$ is a weak equivalence. If Φ happens to be a *strict* quasi-isomorphism between *nilpotent* L_∞-algebras, then Getzler [10] showed that $\mathfrak{MC}_\bullet(\Phi)$ is a weak equivalence. The result for the general case of ∞-quasi-isomorphisms between filtered L_∞-algebras was proved in our previous work with V. Dolgushev.

Theorem 1 ([5, Thm. 1.1]) *If $\Phi: (L, Q) \xrightarrow{\sim} (\tilde{L}, \tilde{Q})$ is a weak equivalence of filtered L_∞-algebras, then*

$$\mathfrak{MC}_\bullet(\Phi): \mathfrak{MC}_\bullet(L) \to \mathfrak{MC}_\bullet(\tilde{L})$$

is a homotopy equivalence of simplicial sets.

It is interesting that the most subtle part of the proof of the above theorem involves establishing the bijection between $\pi_0(\mathfrak{MC}_\bullet(L))$ and $\pi_0(\mathfrak{MC}_\bullet(\tilde{L}))$.

The second noteworthy observation is that if $\Phi: L \twoheadrightarrow \tilde{L}$ is a fibration then $\mathfrak{MC}_\bullet(\Phi)$ is a Kan fibration. To the best of our knowledge, this result, at this level of generality, is new.

Theorem 2 *If $\Phi: (L, Q) \to (\tilde{L}, \tilde{Q})$ is a fibration of filtered L_∞-algebras, then*

$$\mathfrak{MC}_\bullet(\Phi): \mathfrak{MC}_\bullet(L) \to \mathfrak{MC}_\bullet(\tilde{L})$$

is a fibration of simplicial sets.

Two special cases of Theorem 2 already exist in the literature. If Φ happens to be a *strict* fibration between *nilpotent* L_∞-algebras, then the result is again due to Getzler [10, Prop. 4.7]. If Φ is a *strict* fibration between *profinite* filtered L_∞-algebra, then Yalin showed [16, Thm. 4.2(1)] that $\mathfrak{MC}_\bullet(\Phi)$ is a fibration.

The proof of Theorem 2 is technical and will appear in elsewhere in full detail. We give a sketch here.

Suppose $\Phi: L \to \tilde{L}$ is a fibration. This induces a morphism between towers of nilpotent L_∞-algebras

$$\begin{array}{ccccccccc} \cdots \longrightarrow\!\!\!\!\rightarrow L/\mathscr{F}_{n+1}L & \xrightarrow{p_{n+1}} & L/\mathscr{F}_n L & \xrightarrow{p_n} & L/\mathscr{F}_{n-1}L & \longrightarrow\!\!\!\!\rightarrow & \cdots & \longrightarrow\!\!\!\!\rightarrow & 0 \\ {\scriptstyle\overline{\Phi}}\downarrow & & {\scriptstyle\overline{\Phi}}\downarrow & & {\scriptstyle\overline{\Phi}}\downarrow & & & & \downarrow \\ \cdots \longrightarrow\!\!\!\!\rightarrow \tilde{L}/\mathscr{F}_{n+1}\tilde{L} & \xrightarrow{\tilde{p}_{n+1}} & \tilde{L}/\mathscr{F}_n\tilde{L} & \xrightarrow{\tilde{p}_n} & \tilde{L}/\mathscr{F}_{n-1}\tilde{L} & \longrightarrow\!\!\!\!\rightarrow & \cdots & \longrightarrow\!\!\!\!\rightarrow & 0 \end{array}$$

$$(16)$$

which gives us a morphism between towers of Kan complexes:

$$
\begin{array}{ccc}
\vdots & & \vdots \\
\downarrow & & \downarrow \\
\mathfrak{MC}_\bullet\!\left(L/\mathscr{F}_{n+1}L\right) & \xrightarrow{\ \overline{\Phi}_*^{(\bullet)}\ } & \mathfrak{MC}_\bullet\!\left(\tilde{L}/\mathscr{F}_{n+1}\tilde{L}\right) \\
\ \downarrow{\scriptstyle p_{n+1*}^{(\bullet)}} & & \ \downarrow{\scriptstyle \tilde{p}_{n+1*}^{(\bullet)}} \\
\mathfrak{MC}_\bullet\!\left(L/\mathscr{F}_{n}L\right) & \xrightarrow{\ \overline{\Phi}_*^{(\bullet)}\ } & \mathfrak{MC}_\bullet\!\left(\tilde{L}/\mathscr{F}_{n}\tilde{L}\right) \\
\ \downarrow{\scriptstyle p_{n*}^{(\bullet)}} & & \ \downarrow{\scriptstyle \tilde{p}_{n*}^{(\bullet)}} \\
\mathfrak{MC}_\bullet\!\left(L/\mathscr{F}_{n-1}L\right) & \xrightarrow{\ \overline{\Phi}_*^{(\bullet)}\ } & \mathfrak{MC}_\bullet\!\left(\tilde{L}/\mathscr{F}_{n-1}\tilde{L}\right) \\
\downarrow & & \downarrow \\
\vdots & & \vdots \\
\downarrow & & \downarrow \\
* & \longrightarrow\!\!\!\!\rightarrow & *
\end{array}
\tag{17}
$$

The morphisms p_n and $\tilde{p}_n$ are strict fibrations between nilpotent L_∞-algebras. Hence, Prop. 4.7 of [10] implies that their images under $\mathfrak{MC}_\bullet(-)$ are fibrations of simplicial sets. The inverse limit $\varprojlim: \mathsf{tow}(\mathsf{sSet}) \to \mathsf{sSet}$ of this morphism of towers is $\mathfrak{MC}_\bullet(\Phi): \mathfrak{MC}_\bullet(L) \to \mathfrak{MC}_\bullet(\tilde{L})$. The functor $\varprojlim$ is right Quillen [11, Ch. VI, Def. 1.7]. Hence, to show $\mathfrak{MC}_\bullet(\Phi)$ is a fibration, it is sufficient to show that the morphism of towers (17) is a fibration. By definition, this means we must show, for each $n > 1$, that the morphism induced by the universal property in the pullback diagram:

$$
\begin{array}{ccc}
\mathfrak{MC}_\bullet(L/\mathscr{F}_n L) & & \\
& \searrow^{\ p_{n*}^{(\bullet)}} & \\
\ \downarrow{\scriptstyle (\overline{\Phi}_*^{(\bullet)},\, p_{n*}^{(\bullet)})} & & \\
\mathfrak{MC}_\bullet(\tilde{L}/\mathscr{F}_n\tilde{L}) \times_{\mathfrak{MC}_\bullet(\tilde{L}/\mathscr{F}_{n-1}\tilde{L})} \mathfrak{MC}_\bullet(L/\mathscr{F}_{n-1}L) & \longrightarrow & \mathfrak{MC}_\bullet(L/\mathscr{F}_{n-1}L) \\
\ \downarrow & & \ \downarrow{\scriptstyle \overline{\Phi}_*^{(\bullet)}} \\
{\scriptstyle \overline{\Phi}_*^{(\bullet)}} \quad \mathfrak{MC}_\bullet(\tilde{L}/\mathscr{F}_n\tilde{L}) & \xrightarrow[\ \tilde{p}_{n*}^{(\bullet)}\ ]{} & \mathfrak{MC}_\bullet(\tilde{L}/\mathscr{F}_{n-1}\tilde{L})
\end{array}
$$

is a fibration of simplicial sets [11, Ch. VI, Def. 1.1]. So suppose we are given a horn $\gamma\colon \Lambda_k^m \to \mathfrak{MC}_\bullet(L/\mathscr{F}_n L)$ and commuting diagrams:

$$
\begin{array}{ccc}
\Lambda_k^m & \xrightarrow{\ \gamma\ } & \mathfrak{MC}_\bullet(L/\mathscr{F}_n L) \\
\downarrow & & \downarrow{\overline{\Phi}_*^{(\bullet)}} \\
\Delta^m & \xrightarrow{\ \tilde{\beta}\ } & \mathfrak{MC}_\bullet(\tilde{L}/\mathscr{F}_n \tilde{L})
\end{array}
\qquad
\begin{array}{ccc}
\Lambda_k^m & \xrightarrow{\ \gamma\ } & \mathfrak{MC}_\bullet(L/\mathscr{F}_n L) \\
\downarrow & & \downarrow{p_{n*}^{(\bullet)}} \\
\Delta^m & \xrightarrow{\ \beta\ } & \mathfrak{MC}_\bullet(L/\mathscr{F}_{n-1} L)
\end{array}
$$

$$
\begin{array}{ccc}
\Delta^m & \xrightarrow{\ \ \beta\ \ } & \mathfrak{MC}_\bullet(L/\mathscr{F}_{n-1} L) \\
\tilde{\beta}\ \downarrow & & \downarrow{\overline{\Phi}_*^{(\bullet)}} \\
\mathfrak{MC}_\bullet(\tilde{L}/\mathscr{F}_n \tilde{L}) & \xrightarrow{\ \tilde{p}_{n*}^{(\bullet)}\ } & \mathfrak{MC}_\bullet(\tilde{L}/\mathscr{F}_{n-1} \tilde{L})
\end{array}
$$

We need to produce an m-simplex $\alpha\colon \Delta^m \to \mathfrak{MC}_\bullet(L/\mathscr{F}_n L)$ which fills the horn γ and satisfies $\overline{\Phi}_*^{(\bullet)}\alpha = \tilde{\beta}$ and $p_{n*}^{(\bullet)}\alpha = \beta$. Since $p_{n*}^{(\bullet)}$ is a fibration, there exists an m-simplex θ lifting β:

$$
\begin{array}{ccc}
\Lambda_k^m & \xrightarrow{\ \gamma\ } & \mathfrak{MC}_\bullet(L/\mathscr{F}_n L) \\
\downarrow & \nearrow{\theta} & \downarrow{p_{n*}^{(\bullet)}} \\
\Delta^m & \xrightarrow{\ \beta\ } & \mathfrak{MC}_\bullet(L/\mathscr{F}_{n-1} L)
\end{array}
$$

but there is no guarantee that $\overline{\Phi}_*^{(\bullet)}(\theta) = \tilde{\beta}$. However, note that the m-simplex

$$
\eta := \overline{\Phi}_*^{(\bullet)}(\theta) - \tilde{\beta} \tag{18}
$$

of the simplicial vector space $\tilde{L}/\mathscr{F}_n \tilde{L} \otimes \Omega_\bullet$ lies in the kernel of the linear map $\tilde{p}_n^{(m)}$. We now observe that the fibration $\Phi\colon L \to \tilde{L}$ induces a map between the short exact sequences (8) of nilpotent L_∞-algebras:

$$
\begin{array}{ccccc}
\mathscr{F}_{n-1}L/\mathscr{F}_n L & \longrightarrow & L/\mathscr{F}_n L & \xrightarrow{\ p_n\ } & L/\mathscr{F}_{n-1} L \\
\overline{\mathscr{F}_{n-1}\Phi}\ \downarrow & & \overline{\Phi}\ \downarrow & & \downarrow{\overline{\Phi}} \\
\mathscr{F}_{n-1}\tilde{L}/\mathscr{F}_n \tilde{L} & \longrightarrow & \tilde{L}/\mathscr{F}_n \tilde{L} & \xrightarrow{\ \tilde{p}_n\ } & \tilde{L}/\mathscr{F}_{n-1} \tilde{L}
\end{array}
\tag{19}
$$

It follows from the compatibility of Φ with the filtrations, that $\overline{\mathscr{F}_{n-1}\Phi}$ above is simply the linear term of the morphism $\overline{\Phi}$ restricted to the subspace $\mathscr{F}_{n-1}L/\mathscr{F}_n L$. Moreover, since Φ is a fibration, $\overline{\mathscr{F}_{n-1}\Phi}$ is surjective. Hence, $\overline{\mathscr{F}_{n-1}\Phi}$ is a strict fibration between abelian L_∞-algebras, and so Prop. 4.7 of [10] implies that the corresponding map in the diagram of simplicial sets below is a fibration:

$$\begin{array}{ccc}
\mathfrak{MC}_\bullet\left(\mathscr{F}_{n-1}L/\mathscr{F}_nL\right) \longrightarrow \mathfrak{MC}_\bullet\left(L/\mathscr{F}_nL\right) \xrightarrow{p_{n*}^{(\bullet)}} \mathfrak{MC}_\bullet\left(L/\mathscr{F}_{n-1}L\right) \\
\overline{\mathscr{F}_{n-1}\Phi}_*^{(\bullet)}\Big\downarrow \qquad\qquad \overline{\Phi}_*^{(\bullet)}\Big\downarrow \qquad\qquad \Big\downarrow \overline{\Phi}_*^{(\bullet)} \\
\mathfrak{MC}_\bullet\left(\mathscr{F}_{n-1}\tilde{L}/\mathscr{F}_n\tilde{L}\right) \longrightarrow \mathfrak{MC}_\bullet\left(\tilde{L}/\mathscr{F}_n\tilde{L}\right) \xrightarrow{\tilde{p}_{n*}^{(\bullet)}} \mathfrak{MC}_\bullet\left(\tilde{L}/\mathscr{F}_{n-1}\tilde{L}\right)
\end{array} \tag{20}$$

A straightforward calculation shows that the vector η (18) is in fact a m-simplex of $\mathfrak{MC}_\bullet\left(\mathscr{F}_{n-1}\tilde{L}/\mathscr{F}_n\tilde{L}\right)$, whose restriction to the horn Λ_k^m vanishes. Hence, there exists a lift $\lambda: \Delta^m \to \mathfrak{MC}_\bullet\left(\mathscr{F}_{n-1}L/\mathscr{F}_nL\right)$ of η through $\overline{\mathscr{F}_{n-1}\Phi}_*^{(\bullet)}$.

One can then show via a series of technical lemmas that $\alpha = \lambda + \theta$ is a m-simplex of $\mathfrak{MC}_\bullet(L/\mathscr{F}_nL)$ which fills the horn γ and satisfies both $\overline{\Phi}_*^{(\bullet)}\alpha = \tilde{\beta}$ and $p_{n*}^{(\bullet)}\alpha = \beta$. Hence, the morphism of towers (17) is a fibration in tow(sSet), and we conclude that $\mathfrak{MC}_\bullet(\Phi): \mathfrak{MC}_\bullet(L) \to \mathfrak{MC}_\bullet(\tilde{L})$ is a fibration of simplicial sets.

4 Homotopy Transfer Theorem

For this section, we follow the conventions presented in Sections 1 and 2 of [8]. We refer the reader there for further background on dg operads and homotopy algebras. Let $\mathscr{C}$ be a dg cooperad with a co-augmentation $\overline{\mathscr{C}}$ that is equipped with a compatible cocomplete ascending filtration:

$$0 = \mathscr{F}^0\overline{\mathscr{C}} \subset \mathscr{F}^1\overline{\mathscr{C}} \subset \mathscr{F}^2\overline{\mathscr{C}} \subset \mathscr{F}^3\overline{\mathscr{C}} \subset \dots \tag{21}$$

Any co-augmented cooperad satisfying $\mathscr{C}(0) = 0$, $\mathscr{C}(1) = \Bbbk$, for example, admits such a filtration (by arity). Cobar($\mathscr{C}$) algebra structures on a cochain complex (A, ∂_A) are in one-to-one correspondence with codifferentials Q on the cofree coalgebra $\mathscr{C}(A) = \bigoplus_{n\geq 0}\left(\mathscr{C}(n) \otimes A^{\otimes n}\right)_{S_n}$ which satisfy $Q|_A = \partial$. Homotopy algebras such as L_∞, A_∞, and C_∞ algebras are all examples of Cobar($\mathscr{C}$) algebras of this kind. A morphism (or more precisely "∞-morphism") $F: (A, Q_A) \to (B, Q_B)$ between Cobar($\mathscr{C}$) algebras is morphism between the corresponding dg coalgebras $F: (\mathscr{C}(A), \partial_A + Q_A) \to (\mathscr{C}(B), \partial_B + Q_B)$. Such a morphism is an ∞-**quasi-isomorphism** iff its linear term $\mathrm{pr}_B F|_A: (A, \partial_A) \to (B, \partial_B)$ is a quasi-isomorphism of chain complexes.

Given a cochain complex (A, ∂_A), one can construct a dg Lie algebra Conv($\mathscr{C}$, End$_A$) whose Maurer–Cartan elements are in one-to-one correspondence with Cobar($\mathscr{C}$) structures on (A, ∂_A). The underlying complex of Conv($\mathscr{C}$, End$_A$) can be identified with the complex of linear maps Hom($\overline{\mathscr{C}}(A), A$). The filtration (21) induces a complete descending filtration on Conv($\mathscr{C}$, End$_A$) which is compatible with the dg Lie structure. Hence, the desuspension $\mathbf{s}^{-1} \operatorname{Conv}(\mathscr{C}, \operatorname{End}_A)$ is a filtered L_∞-algebra in our sense.

We now indulge in some minor pedantry by presenting the well-known Homotopy Transfer Theorem in the following way. Let (B, Q_B) be a Cobar($\mathscr{C}$)-algebra, (A, ∂) a cochain complex, and $\phi: A \to B$ a quasi-isomorphism of cochain complexes. One asks whether the structure on B can be transferred through ϕ to a homotopy equivalent structure on A. A solution to the **homotopy transfer problem** is a Cobar($\mathscr{C}$)-structure Q_A on A, and a ∞-quasi-isomorphism $F: (A, Q_A) \overset{\sim}{\to} (B, Q_B)$ of Cobar($\mathscr{C}$)-algebras such that $\operatorname{pr}_B F|_A = \phi$.

Solutions to the homotopy transfer problem correspond to certain MC elements of a filtered L_∞-algebra. The cochain complex

$$\operatorname{Cyl}(\mathscr{C}, A, B) := \mathbf{s}^{-1} \operatorname{Hom}(\bar{\mathscr{C}}(A), A) \oplus \operatorname{Hom}(\bar{\mathscr{C}}(A), B) \oplus \mathbf{s}^{-1} \operatorname{Hom}(\bar{\mathscr{C}}(B), B) \quad (22)$$

can be equipped with a (shifted) L_∞-structure induced by: (1) the convolution Lie brackets on $\operatorname{Hom}(\bar{\mathscr{C}}(A), A)$ and $\operatorname{Hom}(\bar{\mathscr{C}}(B), B)$, and (2) pre and post composition of elements of $\operatorname{Hom}(\bar{\mathscr{C}}(A), B)$ with elements of $\operatorname{Hom}(\bar{\mathscr{C}}(A), A)$ and $\operatorname{Hom}(\bar{\mathscr{C}}(B), B)$, respectively. (See Sec. 3.1 in [7] for the details.)

As shown in Sec. 3.2 of [7], the L_∞-structure on $\operatorname{Cyl}(\mathscr{C}, A, B)$ is such that its MC elements are triples (Q_A, F, Q_B), where Q_A and Q_B are Cobar($\mathscr{C}$) structures on A and B, respectively, and F is a ∞-morphism between them. In particular, if $\phi: A \to B$ is a chain map, then $\alpha_\phi = (0, \phi, 0)$ is a MC element in $\operatorname{Cyl}(\mathscr{C}, A, B)$, where "0" denotes the trivial Cobar($\mathscr{C}$) structure.

We can therefore twist, as described in Sect. 2.2, by the MC element α_ϕ to obtain a new L_∞-algebra $\operatorname{Cyl}(\mathscr{C}, A, B)^{\alpha_\phi}$. The graded subspace

$$\overline{\operatorname{Cyl}}(\mathscr{C}, A, B)^{\alpha_\phi} := \mathbf{s}^{-1} \operatorname{Hom}(\bar{\mathscr{C}}(A), A) \oplus \operatorname{Hom}(\bar{\mathscr{C}}(A), B) \oplus \mathbf{s}^{-1} \operatorname{Hom}(\bar{\mathscr{C}}(B), B) \quad (23)$$

is equipped with a filtration induced by the filtration on $\mathscr{C}$. Restricting the L_∞ structure on $\operatorname{Cyl}(\mathscr{C}, A, B)^{\alpha_\phi}$ to $\overline{\operatorname{Cyl}}(\mathscr{C}, A, B)^{\alpha_\phi}$ makes the latter into a filtered L_∞-algebra. The MC elements of $\overline{\operatorname{Cyl}}(\mathscr{C}, A, B)^{\alpha_\phi}$ are those MC elements (Q_A, F, Q_B) of $\operatorname{Cyl}(\mathscr{C}, A, B)$ such that $\operatorname{pr}_B F|_A = \phi$.

We have the following proposition. (See Prop. 3.2 in [7]).

Proposition 1 *The canonical projection of cochain complexes*

$$\pi_B: \mathbf{s}^{-1} \operatorname{Hom}(\bar{\mathscr{C}}(A), A) \oplus \operatorname{Hom}(\bar{\mathscr{C}}(A), B) \oplus \mathbf{s}^{-1} \operatorname{Hom}(\bar{\mathscr{C}}(B), B) \to \mathbf{s}^{-1} \operatorname{Hom}(\bar{\mathscr{C}}(B), B) \quad (24)$$

lifts to a (strict) acyclic fibration of filtered L_∞-algebras:

$$\pi_B: \overline{\operatorname{Cyl}}(\mathscr{C}, A, B)^{\alpha_\phi} \overset{\sim}{\twoheadrightarrow} \mathbf{s}^{-1} \operatorname{Conv}(\mathscr{C}, \operatorname{End}_B)$$

We can now express the homotopy transfer theorem as a simple corollary:

Corollary 1 (Homotopy Transfer Theorem) *Let (B, Q_B) be a* Cobar($\mathscr{C}$)-*algebra,* (A, ∂) *a cochain complex, and* $\phi \colon A \to B$ *a quasi-isomorphism of cochain complexes. The solutions to the corresponding homotopy transfer problem are in one-to-one correspondence with the objects of a sub ∞-groupoid*

$$\mathfrak{F}_{Q_B} \subseteq \mathfrak{MC}_\bullet \left(\overline{\mathrm{Cyl}}(\mathscr{C}, A, B)^{\alpha_\phi} \right).$$

Furthermore,

1. *(existence)* $\mathfrak{F}_{Q_B}$ *is non-empty, and*
2. *(uniqueness)* $\mathfrak{F}_{Q_B}$ *is contractible.*

Proof All statements follow from Theorems 1 and 2, which imply that

$$\mathfrak{MC}_\bullet(\pi_B) \colon \mathfrak{MC}_\bullet \left(\overline{\mathrm{Cyl}}(\mathscr{C}, A, B)^{\alpha_\phi} \right) \xrightarrow{\sim\!\!\!\!\twoheadrightarrow} \mathfrak{MC}_\bullet \left(\mathrm{Conv}(\bar{\mathscr{C}}, \mathrm{End}_B) \right) \qquad (25)$$

is an acyclic fibration of Kan complexes. Indeed, we define $\mathfrak{F}_{Q_B}$ as the fiber of $\mathfrak{MC}_\bullet(\pi_B)$ over the object $Q_B \in \mathfrak{MC}_0 \left(\mathrm{Conv}(\bar{\mathscr{C}}, \mathrm{End}_B) \right)$. Since $\mathfrak{MC}_\bullet(\pi_B)$ is a Kan fibration, $\mathfrak{F}_{Q_B}$ is a ∞-groupoid. Objects of $\mathfrak{F}_{Q_B}$ are those MC elements of $\overline{\mathrm{Cyl}}(\mathscr{C}, A, B)^{\alpha_\phi}$ which are of the form (Q_A, F, Q_B), and hence are solutions to the homotopy transfer problem.

Since $\mathfrak{MC}_\bullet(\pi_B)$ is an acyclic fibration, it satisfies the right lifting property with respect to the inclusion $\emptyset = \partial \Delta^0 \subseteq \Delta^0$. Hence, $\mathfrak{MC}_\bullet(\pi_B)$ is surjective on objects. This proves statement (1). Statement (2) follows from the long exact sequence of homotopy groups.

Let us conclude by mentioning the difference between the above formulation of the Homotopy Transfer Theorem and the one given in Section 5 of our previous work [8] with Dolgushev. There we only had Theorem 1 to use, and not Theorem 2. Hence, we proved a slight variant of the transfer theorem [8, Thm. 5.1]. We defined a solution to the homotopy transfer problem as a triple $(Q_A, F, \tilde{Q}_B)$, where Q_A is a Cobar($\mathscr{C}$) algebra structure on A, $\tilde{Q}_B$ is a Cobar($\mathscr{C}$) algebra structure on B homotopy equivalent to the original structure Q_B, and $F \colon (A, Q_A) \to (B, \tilde{Q}_B)$ is a ∞-quasi-isomorphism whose linear term is ϕ. We used the fact that $\mathfrak{MC}_\bullet(\pi_B)$ is a weak equivalence, and therefore gives a bijection

$$\pi_0 \left(\mathfrak{MC}_\bullet \left(\overline{\mathrm{Cyl}}(\mathscr{C}, A, B)^{\alpha_\phi} \right) \right) \cong \pi_0 \left(\mathfrak{MC}_\bullet \left(\mathrm{Conv}(\bar{\mathscr{C}}, \mathrm{End}_B) \right) \right),$$

to conclude that such a solution $(Q_A, F, \tilde{Q}_B)$ exists. It is easy to see that objects of the *homotopy fiber* of $\mathfrak{MC}_\bullet(\pi_B)$ over the vertex Q_B are pairs consisting of a solution $(Q_A, F, \tilde{Q}_B)$ to this variant of the transfer problem, and an equivalence from $\tilde{Q}_B$ to Q_B.

Acknowledgements I would like to acknowledge support by an AMS-Simons Travel Grant, and I thank Vasily Dolgushev and Bruno Vallette for helpful discussions regarding this work. I would also like to thank the organizers of the MATRIX Institute program "Higher Structures in Geometry and Physics" for an excellent workshop and conference.

References

1. Berger, C., Moerdijk, I.: Axiomatic homotopy theory for operads. Comment. Math. Helv. **78**(4), 805–831 (2003)
2. Berglund, A.: Homological perturbation theory for algebras over operads. Algebr. Geom. Topol. **14**(5), 2511–2548 (2014)
3. Berglund, A.: Rational homotopy theory of mapping spaces via Lie theory for L_∞-algebras. Homol. Homotopy Appl. **17**(2), 343–369 (2015)
4. Buijs, U., Félix, Y., Murillo, A.: L_∞ models of based mapping spaces. J. Math. Soc. Jpn. **63**(2), 503–524 (2011)
5. Dolgushev, V.A., Rogers, C.L.: A version of the Goldman-Millson theorem for filtered L_∞-algebras. J. Algebra **430**, 260–302 (2015)
6. Dolgushev, V.A., Rogers, C.L.: On an enhancement of the category of shifted L_∞-algebras. Appl. Categ. Struct. **25**(4), 489–503 (2017)
7. Dolgushev, V., Willwacher, T.: The deformation complex is a homotopy invariant of a homotopy algebra. In: Developments and Retrospectives in Lie Theory. Developments in Mathematics, vol. 38, pp. 137–158. Springer, Cham (2014)
8. Dolgushev, V.A., Hoffnung, A.E., Rogers, C.L.: What do homotopy algebras form? Adv. Math. **274**, 562–605 (2015)
9. Dotsenko, V., Poncin, N.: A tale of three homotopies. Appl. Categ. Struct. **24**(6), 845–873 (2016)
10. Getzler, E.: Lie theory for nilpotent L_∞-algebras. Ann. Math. (2) **170**(1), 271–301 (2009)
11. Goerss, P.G., Jardine, J.F.: Simplicial Homotopy Theory. Progress in Mathematics, vol. 174. Birkhäuser, Basel (1999)
12. Hinich, V.: Descent of Deligne groupoids. Int. Math. Res. Not. **1997**(5), 223–239 (1997)
13. Lazarev, A.: Maurer-Cartan moduli and models for function spaces. Adv. Math. **235**, 296–320 (2013)
14. Loday, J.L., Vallette, B.: Algebraic Operads. Grundlehren der Mathematischen Wissenschaften [Fundamental Principles of Mathematical Sciences], vol. 346. Springer, Heidelberg (2012)
15. Markl, M.: Homotopy algebras are homotopy algebras. Forum Math. **16**(1), 129–160 (2004)
16. Yalin, S.: Maurer–Cartan spaces of filtered L_∞-algebras. J. Homotopy Relat. Struct. **11**, 375–407 (2016)

Fibrations in ∞-Category Theory

Clark Barwick and Jay Shah

Abstract In this short expository note, we discuss, with plenty of examples, the bestiary of fibrations in quasicategory theory. We underscore the simplicity and clarity of the constructions these fibrations make available to end-users of higher category theory.

1 Introduction

The theory of ∞-categories—as formalized in the model of quasicategories—offers two ways of specifying homotopy theories and functors between them.

First, we may describe a homotopy theory via a homotopy-coherent universal property; this is a widely appreciated advantage, and it's a feature that any sufficiently well-developed model of ∞-categories would have. For example, the ∞-category **Top** of spaces is the free ∞-category generated under (homotopy) colimits by a single object [10, Th. 5.1.5.6].

The second way of specifying ∞-categories seems to be less well-loved: this is the ability to perform completely explicit constructions with excellent formal properties. This allows one to avoid the intricate workarounds that many of us beleaguered homotopy theorists have been forced to deploy in order solve infinite hierarchies of homotopy coherence problems. This feature seems to be peculiar to the model of quasicategories, and the main instrument that makes these explicit constructions possible is the theory of *fibrations* of various sorts. In this étude, we study eight sorts of fibrations of quasicategories in use today—left, right, Kan, inner, iso (AKA categorical), cocartesian, cartesian, and flat—and we discuss the beautifully explicit constructions they provide.

In the end, ∞-category theory as practiced today combines these two assets, and the result is a powerful amalgam of universal characterizations and crashingly

C. Barwick • J. Shah (✉)
Department of Mathematics, Massachusetts Institute of Technology, 77 Massachusetts Avenue, Cambridge, MA 02139-4307, USA
e-mail: clarkbar@math.mit.edu; jshah@math.mit.edu

© Springer International Publishing AG, part of Springer Nature 2018
D.R. Wood et al. (eds.), *2016 MATRIX Annals*, MATRIX Book Series 1,
https://doi.org/10.1007/978-3-319-72299-3_2

explicit constructions. We will here focus on the underappreciated latter feature, which provides incredibly concrete constructions to which we would not otherwise have access. We have thus written this under the assumption that readers are more or less familiar with the content of the first chapter of Lurie's book [10].

2   Left, Right, and Kan Fibrations

The universal property of the ∞-category **Top** we offered above certainly characterizes it up to a contractible choice, but it doesn't provide any simple way to specify a functor *into* **Top**.

At first blush, this looks like very bad news: after all, even if C is an ordinary category, to specify a functor of ∞-categories $F: C \longrightarrow$ **Top**, one has to specify an extraordinary amount of information: one has to give, for every object $a \in C$, a space $F(a)$; for every morphism $f: a \longrightarrow b$, a map $F(f): F(a) \longrightarrow F(b)$; for every pair of composable morphisms $f: a \longrightarrow b$ and $g: b \longrightarrow c$, a homotopy $F(gf) \simeq F(g)F(f)$; for every triple of composable morphisms, a homotopy of homotopies; etc., *ad infinitum*.

However, ordinary category theory suggests a way out: Suppose $F: C \longrightarrow$ **Set** a functor. One of the basic tricks of the trade in category theory is to build a category $\mathrm{Tot}\,F$, sometimes called the *category of elements* of F. The objects of $\mathrm{Tot}\,F$ are pairs (a, x), where $a \in C$ is an object and $x \in F(a)$ is an element; a morphism $(a, x) \longrightarrow (b, y)$ of $\mathrm{Tot}\,F$ is a morphism $f: a \longrightarrow b$ such that $F(f)(x) = y$.

The category $\mathrm{Tot}\,F$, along with the projection $p: \mathrm{Tot}\,F \longrightarrow C$, is extremely useful for studying the functor F. For example, the set of sections of p is a limit of F, and the set $\pi_0(\mathrm{Tot}\,F)$ of connected components is a colimit of F. In fact, the assignment $F \rightsquigarrow \mathrm{Tot}\,F$ is an equivalence of categories between the category of functors $C \longrightarrow$ **Set** and those functors $X \longrightarrow C$ such that for any morphism $f: a \longrightarrow b$ of C and for any object $x \in X$ with $p(x) = a$, there exists a unique morphism $\phi: x \longrightarrow y$ with $p(\phi) = f$. (These functors are sometimes called *discrete opfibrations*.) In other words, functors $C \longrightarrow$ **Set** correspond to functors $X \longrightarrow C$ such that for any solid arrow commutative square

$$
\begin{array}{ccc}
\Lambda_0^1 & \longrightarrow & NX \\
\downarrow & \nearrow & \downarrow{\scriptstyle p} \\
\Delta^1 & \longrightarrow & NC,
\end{array}
$$

there exists a unique dotted lift.

We may therefore hope that, instead of working with functors from C to **Top**, one might work with suitable ∞-categories *over* C instead. To make this work, we need to formulate the ∞-categorical version of this condition. To this end, we adopt the same attitude that permits us to arrive at the definition of an ∞-category: instead of demanding a single unique horn filler, we demand a whole hierarchy of horn fillers, none of which we require to be unique. The hierarchy ensures that the filler at any stage is unique up to a homotopy that is unique up to a homotopy, etc., *ad infinitum*.

Definition 2.1 A *left fibration* is a map $p: X \longrightarrow S$ of simplicial sets such that for any integer $n \geq 1$ and any $0 \leq k < n$ and any solid arrow commutative square

$$
\begin{array}{ccc}
\Lambda^n_k & \longrightarrow & X \\
\downarrow & \nearrow & \downarrow p \\
\Delta^n & \longrightarrow & S,
\end{array}
$$

there exists a dotted lift.

Dually, a *right fibration* is a map $p: X \longrightarrow S$ of simplicial sets such that for any integer $n \geq 1$ and any $0 < k \leq n$ and any solid arrow commutative square

$$
\begin{array}{ccc}
\Lambda^n_k & \longrightarrow & X \\
\downarrow & \nearrow & \downarrow p \\
\Delta^n & \longrightarrow & S,
\end{array}
$$

there exists a dotted lift.

Of course, a *Kan fibration* is a map of simplicial sets that is both a left and a right fibration.

To understand these notions, we should begin with some special cases.

Example 2.2 For any simplicial set X, the unique map $X \longrightarrow \Delta^0$ is a left fibration if and only if X is an ∞-groupoid (i.e., a Kan complex). Indeed, we see immediately that X is an ∞-category, so to conclude that X is an ∞-groupoid, it suffices to observe that the homotopy category hX is a groupoid; this follows readily from the lifting condition for the horn inclusion $\Lambda^2_0 \hookrightarrow \Delta^2$.

Since pullbacks of left fibrations are again left fibrations, we conclude immediately that the fibers of a left fibration are ∞-groupoids.

Example 2.3 ([10, Cor. 2.1.2.2]) If C is an ∞-category and $x \in C_0$ is an object, then recall that one can form the *undercategory* $C_{x/}$ uniquely via the following functorial bijection:

$$
\mathrm{Mor}(K, C_{x/}) \cong \mathrm{Mor}(\Delta^0 \star K, C) \times_{\mathrm{Mor}(\Delta^0, C)} \{x\}.
$$

The inclusion $K \hookrightarrow \Delta^0 \star K$ induces a forgetful functor $p: C_{x/} \longrightarrow C$. The key fact (due to Joyal) is that p is a left fibration; in particular, $C_{x/}$ is an ∞-category [10, Cor. 2.1.2.2].

The fiber of p over a vertex $y \in C_0$ is the ∞-groupoid whose n-simplices are maps $f: \Delta^{n+1} \longrightarrow C$ such that $f(\Delta^{\{0\}}) = x$ and $f|_{\Delta^{\{1,\ldots,n+1\}}}$ is the constant map at y. In other words, it is $\mathrm{Hom}^L_C(x, y)$ in the notation of [10, Rk. 1.2.2.5].

Dually, we can define the *overcategory* $C_{/x}$ via

$$\mathrm{Mor}(K, C_{x/}) \cong \mathrm{Mor}(K \star \Delta^0, C) \times_{\mathrm{Mor}(\Delta^0, C)} \{x\},$$

and the inclusion $K \hookrightarrow K \star \Delta^0$ induces a right fibration $q \colon C_{/x} \longrightarrow C$. The fiber of q over a vertex $y \in C_0$ is then $\mathrm{Hom}_C^R(y, x)$.

Subexample 2.3.1 *In the introduction, we mentioned that* **Top** *is the ∞-category that is freely generated under colimits by a single object $*$. This generator turns out to be the terminal object in* **Top**. *Let us write* **Top**$_*$ *for the overcategory* **Top**$_{*/}$. *The forgetful functor* **Top**$_* \longrightarrow$ **Top** *is a left fibration.*

The fiber over a vertex $X \in$ **Top**$_0$ *is the ∞-groupoid* $\mathrm{Hom}_{\mathbf{Top}}^L(*, X)$, *which we will want to think of as a model for X itself.*

Here is the theorem that is going to make that possible:

Theorem 2.4 (Joyal) *Suppose C an ∞-category (or more generally, any simplicial set). For any functor $F \colon C \longrightarrow$* **Top**, *we may consider the left fibration*

$$\mathbf{Top}_* \times_{\mathbf{Top}, F} C \longrightarrow C.$$

This defines an equivalence of ∞-categories

$$\mathrm{Fun}(C, \mathbf{Top}) \xrightarrow{\sim} \mathbf{LFib}(C),$$

where **LFib**(C) *is the simplicial nerve of the full simplicial subcategory of $s\mathbf{Set}_{/C}$ spanned by the left fibrations.*

Dually, for any functor $G \colon C \longrightarrow$ **Top**^{op}, *we may consider the right fibration*

$$\mathbf{Top}_*^{op} \times_{\mathbf{Top}^{op}, G} C \longrightarrow C.$$

This defines an equivalence of ∞-categories

$$\mathrm{Fun}(C^{op}, \mathbf{Top}) \xrightarrow{\sim} \mathbf{RFib}(C),$$

where **RFib**(C) *is the simplicial nerve of the full simplicial subcategory of $s\mathbf{Set}_{/C}$ spanned by the right fibrations.*

A left fibration $p \colon X \longrightarrow C$ is said to be *classified by* $F \colon C \longrightarrow$ **Top** just in case it is equivalent to the left fibration

$$\mathbf{Top}_* \times_{\mathbf{Top}, F} C \longrightarrow C.$$

Dually, a right fibration $q \colon Y \longrightarrow C$ is said to be *classified by* $G \colon C^{op} \longrightarrow$ **Top** just in case it is equivalent to the right fibration

$$\mathbf{Top}_*^{op} \times_{\mathbf{Top}^{op}, G} C \longrightarrow C.$$

The proofs of Joyal's theorem[1] are all relatively involved, and they involve breaking this assertion up into several constituent parts. But rather than get distracted by these details (beautiful though they be!), let us instead swim in the waters of appreciation for this result as end-users.

Example 2.5 Even for $C = \Delta^0$, this theorem is nontrivial: it provides an equivalence **Top** $\xrightarrow{\sim}$ **Gpd**$_\infty$, where of course **Gpd**$_\infty$ is the simplicial nerve of the full simplicial subcategory of s**Set** spanned by the Kan complexes. In other words, this result provides a concrete model for the ∞-category that was might have only been known through its universal property.

But the deeper point here is that with the universal characterization of the introduction, it's completely unclear how to specify a functor from an ∞-category C *into* **Top**. Even with the equivalence **Top** $\simeq$ **Gpd**$_\infty$, we would still have to specify an infinite hierarchy of data to check this. However, with Joyal's result in hand, our task becomes to construct a left fibration $X \longrightarrow C$. In practice, these are the constructions which are tractable, because one trades the explicit specification of coherence data for horn-filling conditions.

Example 2.6 One may use Joyal's theorem to find that if $p: X \longrightarrow C$ is a left fibration classified by a functor $F: C \longrightarrow$ **Top**, then the colimit of F is weakly homotopy equivalent (i.e., equivalent in the Quillen model structure) to X, and the the limit of F is weakly homotopy equivalent to the space $\mathrm{Map}_C(C, X)$ of sections of p.

Example 2.7 One attitude toward ∞-categories is that they are meant to be categories "weakly enriched" in spaces. Whatever this means, it should at least entail corepresentable and representable functors

$$h^x: C \longrightarrow \textbf{Top} \quad \text{and} \quad h_x: C^{op} \longrightarrow \textbf{Top}$$

for $x \in C_0$. But thanks to Joyal's theorem, we already have these: the former is given by the left fibration $C_{x/} \longrightarrow C$, and the latter is given by the right fibration $C_{/x} \longrightarrow C$.

This also provides a recognition principle: a left fibration $X \longrightarrow C$ corresponds to a corepresentable functor if and only if X admits an initial object; in this case, we call the left fibration itself *corepresentable*. Dually, a right fibration $X \longrightarrow C$ corresponds to a representable functor if and only if X admits a terminal object; in this case, we call the right fibration itself *representable*.

Example 2.8 In the same vein, we expect to have a functor

$$\mathrm{Map}_C: C^{op} \times C \longrightarrow \textbf{Top}$$

---

[1]We know (with the referee's help) five proofs: the original one due to Joyal, a modification thereof due to Lurie [10], a simplification due to Dugger and Spivak [5], a more conceptual and self-contained version due to Stevenson [14], and a recent simplification due to Heuts and Moerdijk [7, 8].

for any ∞-category C. From Joyal's Theorem, our job becomes to construct a left fibration $(s, t)\colon \widetilde{\mathscr{O}}(C) \longrightarrow C^{op} \times C$. It turns out that this isn't so difficult: define $\widetilde{\mathscr{O}}(C)$ via the formula

$$\widetilde{\mathscr{O}}(C)_n := \mathrm{Mor}(\Delta^{n,op} \star \Delta^n, C),$$

and s and t are induced by the inclusions $\Delta^{n,op} \hookrightarrow \Delta^{n,op} \star \Delta^n$ and $\Delta^n \hookrightarrow \Delta^{n,op} \star \Delta^n$, respectively.

So the claim is that (s, t) is a left fibration. This isn't a completely trivial matter, but there is a proof in [12], and another, slightly simpler, proof in [2]. The key point is to study the behavior of the left adjoint of the functor $\widetilde{\mathscr{O}}$ on certain "left anodyne" monomorphisms.

These two examples illustrate nicely a general principle about working "vertically"—i.e., with left and right fibrations—versus working "horizontally"— i.e., with functors to **Top**. It is easy to write down a left or right fibration, but it may not be easy to see that it *is* a left or right fibration. On the other hand, it is quite difficult even to write down a suitable functor to **Top**. So working vertically rather than horizontally relocates the difficulty in higher category theory from a struggle to make good definitions to a struggle to prove good properties.

Example 2.9 Suppose C an ∞-category. A Kan fibration to C is simultaneously a left fibration and a right fibration. So a Kan fibration must correspond to a both a covariant functor and a contravariant functor to **Top**, and one sees that the "pushforward" maps must be homotopy inverse to the "pullback" maps. That is, the following are equivalent for a map $p\colon X \longrightarrow C$ of simplicial sets:

- p is a Kan fibration;
- p is a left fibration, and the functor $C \longrightarrow$ **Top** that classifies it carries any morphism of C to an equivalence;
- p is a right fibration, and the functor $C^{op} \longrightarrow$ **Top** that classifies it carries any morphism of C to an equivalence.

From this point view, we see that when C is an ∞-groupoid, Kan fibrations $X \longrightarrow C$ are "essentially the same thing" as functors $C \longrightarrow$ **Top**, which are in turn indistinguishable from functors $C^{op} \longrightarrow$ **Top**.

3 Inner Fibrations and Isofibrations

Inner fibrations are tricky to motivate from a 1-categorical standpoint, because the nerve of any functor is automatically an inner fibration. We will discuss here a reasonable way of thinking about inner fibrations, but we do not know a reference for complete proofs, yet.

Definition 3.1 An *inner fibration* is a map $p: X \longrightarrow S$ of simplicial sets such that for any integer $n \geq 2$ and any $0 < k < n$ and any solid arrow commutative square

$$
\begin{array}{ccc}
\Lambda_k^n & \longrightarrow & X \\
\downarrow & \nearrow & \downarrow p \\
\Delta^n & \longrightarrow & S,
\end{array}
$$

there exists a dotted lift.

Example 3.2 Of course a simplicial set X is an ∞-category just in case the canonical map $X \longrightarrow \Delta^0$ is an inner fibration. Consequently, any fiber of an inner fibration is an ∞-category.

Example 3.3 If X is an ∞-category and D is an ordinary category, then it's easy to see that any map $X \longrightarrow ND$ is an inner fibration.

On the other hand, a map $p: X \longrightarrow S$ is an inner fibration if and only if, for any n-simplex $\sigma \in S_n$, the pullback

$$X \times_{S,\sigma} \Delta^n \longrightarrow \Delta^n$$

is an inner fibration. Consequently, we see that p is an inner fibration if and only if, for any n-simplex $\sigma \in S_n$, the pullback $X \times_{S,\sigma} \Delta^n$ is an ∞-category.

So in a strong sense, we'll understand the "meaning" of inner fibrations one we understand the "meaning" of functors from ∞-categories to Δ^n.

Example 3.4 When $n = 1$, we have the following. For any ∞-categories C_0 and C_1, there is an equivalence of ∞-categories

$$\{C_0\} \times_{\mathbf{Cat}_{\infty/\Delta\{0\}}} \mathbf{Cat}_{\infty/\Delta^1} \times_{\mathbf{Cat}_{\infty/\Delta\{1\}}} \{C_1\} \overset{\sim}{\longrightarrow} \mathrm{Fun}(C_0^{op} \times C_1, \mathbf{Top}).$$

A proof of this fact doesn't seem to be contained in the literature yet, but we will nevertheless take it as given; it would be a consequence of Proposition 5.1, which we expect to be proven in a future work of P. Haine.

Now the ∞-category on the right of this equivalence can also be identified with the ∞-category

$$\mathrm{Fun}^L(P(C_1), P(C_0))$$

of colimit-preserving functors between $P(C_1) = \mathrm{Fun}(C_1^{op}, \mathbf{Top})$ and $P(C_0) = \mathrm{Fun}(C_0^{op}, \mathbf{Top})$. Such a colimit-preserving functor is sometimes called a *profunctor*.

Example 3.5 When $n = 2$, if C is an ∞-category, and $C \longrightarrow \Delta^2$ is a functor, then we have three fibers C_0, C_1, and C_2 and three colimit-preserving functors $F: P(C_2) \longrightarrow P(C_1)$, $G: P(C_1) \longrightarrow P(C_0)$, and $H: P(C_2) \longrightarrow P(C_0)$. Furthermore, there is natural transformation $\alpha: G \circ F \longrightarrow H$.

In the general case, a functor $C \longrightarrow \Delta^n$ amounts to the choice of ∞-categories $C_0, C_1, \ldots, C_n$ and a lax-commutative diagram of colimit-preserving functors among the various ∞-categories $P(C_i)$.

What we would like to say now is that the ∞-category of inner fibrations $X \longrightarrow S$ is equivalent to the ∞-category of normal (i.e. identity-preserving) lax functors from S^{op} to a suitable "double ∞-category" of ∞-categories and profunctors, generalizing a classical result of Bénabou [15]. We do not know, however, how to make such an assertion precise.

In any case, we could ask for more restrictive hypotheses. We could, for example, ask for fibrations $X \longrightarrow S$ that are classified by functors from S to an ∞-category of profunctors (so that all the 2-morphisms that appear are equivalences); this is covered by the notion of *flatness*, which will discuss in the section after next. More restrictively, we can ask for fibrations $X \longrightarrow S$ that are classified by functors from S to $\mathbf{Cat}_\infty$ itself; these are *cocartesian fibrations*, which we will discuss in the next section.

For future reference, let's specify an extremely well-behaved class of inner fibrations.

Definition 3.6 Suppose C an ∞-category. Then an *isofibration* (AKA a *categorical fibration*[2]) $p \colon X \longrightarrow C$ is an inner fibration such that for any object $x \in X_0$ and any equivalence $f \colon p(x) \longrightarrow b$ of C, there exists an equivalence $\phi \colon x \longrightarrow y$ of X such that $p(\phi) = f$.

We shall revisit this notion in greater detail in a moment, but for now, let us simply comment that an isofibration $X \longrightarrow C$ is an inner fibration whose fibers vary functorially in the equivalences of C.

For more general bases, this definition won't do, of course, but we won't have any use for isofibrations whose targets are not ∞-categories. The model-theoretically inclined reader should note that isofibrations are exactly the fibrations with target an ∞-category for the Joyal model structure; consequently, any functor of ∞-categories can be replaced by an isofibration.

4 Cocartesian and Cartesian Fibrations

If $F \colon C \longrightarrow \mathbf{Cat}$ is an (honest) diagram of ordinary categories, then one can generalize the category of elements construction as follows: form the category X whose objects are pairs (c, x) consisting of an object $c \in C$ and an object $x \in F(c)$,

---

[2]Emily Riehl makes the clearly compelling case that "isofibration" is preferable terminology, because it actually suggests what kind of lifting property it will have, whereas the word "categorical" is unhelpful in this regard. She also tells us that "isofibration" is a standard term in 1-category theory, and that the nerve of functor is an isofibration iff the functor is an isofibration. We join her in her view that "isofibration" is better.

in which a morphism $(f, \phi): (d, y) \longrightarrow (c, x)$ is a morphism $f: d \longrightarrow c$ of C and a morphism

$$\phi: X(f)(y) \longrightarrow x$$

of $F(c)$. This is called the *Grothendieck construction*, and there is an obvious forgetful functor $p: X \longrightarrow C$.

One may now attempt to reverse-engineer the Grothendieck construction by trying to extract the salient features of the forgetful functor p. What we may notice is that for any morphism $f: d \longrightarrow c$ of C and any object $y \in F(d)$ there is a special morphism

$$\Phi = (f, \phi): (d, y) \longrightarrow (c, X(f)(y))$$

of X in which

$$\phi: F(f)(y) \longrightarrow F(f)(y)$$

is simply the identity morphism. This morphism is *initial* among all the morphisms Ψ of X such that $p(\Psi) = f$; that is, for any morphism Ψ of X such that $p(\Psi) = f$, there exists a morphism I of X such that $p(I) = id_c$ such that $\Psi = I \circ \Phi$.

We call morphisms of X that are initial in this sense *p-cocartesian*. Since a p-cocartesian edge lying over a morphism $d \longrightarrow c$ is defined by a universal property, it is uniquely specified up to a unique isomorphism lying over id_c. The key condition that we are looking for is then that *for any morphism of C and any lift of its source, there is a p-cocartesian morphism with that source lying over it*. A functor p satisfying this condition is called a *Grothendieck opfibration*.

Now for any Grothendieck opfibration $p: X \longrightarrow C$, let us attempt to extract a functor $F: C \longrightarrow \mathbf{Cat}$ that gives rise to it in this way. We proceed in the following manner. To any object $a \in C$ assign the fiber X_a of p over a. To any morphism $f: a \longrightarrow b$ assign a functor $F(f): X_a \longrightarrow X_b$ that carries any object $x \in X_a$ to the target $F(f)(x) \in X_b$ of "the" q-cocartesian edge lying over f.

Right away, we have a problem: q-cocartesian edges are only unique up to isomorphism. So these functors cannot be strictly compatible with composition; rather, one will obtain natural isomorphisms

$$F(g \circ f) \simeq F(g) \circ F(f)$$

that will satisfy a secondary layer of coherences that make F into a *pseudofunctor*. Fortunately, one can rectify this pseudofunctor to an equivalent honest functor, which in turn gives rise to p, up to equivalence.

As we have seen in our discussion of left and right fibrations, there are genuine advantages in homotopy theory to working with fibrations instead of functors. Consequently, we define a class of fibrations that is a natural generalization of the class of Grothendieck opfibrations.

Definition 4.1 If $p: X \longrightarrow S$ is an inner fibration of simplicial sets, then an edge $f: x \longrightarrow y$ of X is p-*cocartesian* just in case, for each integer $n \geq 2$, any extension

$$\Delta^{\{0,1\}} \xrightarrow{\ f\ } X,$$
$$\Lambda_0^n \overset{F}{\nearrow}$$

and any solid arrow commutative diagram

$$
\begin{array}{ccc}
\Lambda_0^n & \xrightarrow{\ F\ } & X \\
\downarrow & \overset{\overline{F}}{\nearrow} & \downarrow{\scriptstyle p} \\
\Delta^n & \longrightarrow & S,
\end{array}
$$

a dotted lift exists. Equivalently, if S is an ∞-category, f is p-cocartesian if for any object $z \in X$ the commutative square of mapping spaces

$$
\begin{array}{ccc}
\mathrm{Map}_X(y, z) & \xrightarrow{\ f^*\ } & \mathrm{Map}_X(x, z) \\
\downarrow{\scriptstyle p} & & \downarrow{\scriptstyle p} \\
\mathrm{Map}_S(py, pz) & \xrightarrow[p(f)^*]{} & \mathrm{Map}_S(px, pz)
\end{array}
$$

is a homotopy pullback square.

We say that p is a *cocartesian fibration* if, for any edge $\eta: s \longrightarrow t$ of S and for every vertex $x \in X_0$ such that $p(x) = s$, there exists a p-cocartesian edge $f: x \longrightarrow y$ such that $\eta = p(f)$.

Cartesian edges and *cartesian fibrations* are defined dually, so that an edge of X is p-cartesian just in case the corresponding edge of X^{op} is cocartesian for the inner fibration $p^{op}: X^{op} \longrightarrow S^{op}$, and p is a cartesian fibration just in case p^{op} is a cocartesian fibration.

Example 4.2 ([10, Rk 2.4.2.2]) A functor $p: D \longrightarrow C$ between ordinary categories is a Grothendieck opfibration if and only if the induced functor $N(p): ND \longrightarrow NC$ on nerves is a cocartesian fibration.

Example 4.3 Any left fibration is a cocartesian fibration, and a cocartesian fibration is a left fibration just in case its fibers are ∞-groupoids.

Dually, of course, the class of right fibrations coincides with the class of cartesian fibrations whose fibers are ∞-groupoids.

Example 4.4 Suppose C an ∞-category and $p: X \longrightarrow C$ an inner fibration. Then for any morphism η of X, the following are equivalent.

- η is an equivalence of X;
- η is p-cocartesian, and $p(\eta)$ is an equivalence of C;
- η is p-cartesian, and $p(\eta)$ is an equivalence of C.

It follows readily that if p is a cocartesian or cartesian fibration, then it is an isofibration.

Conversely, if C is an ∞-groupoid, then the following are equivalent.

- p is an isofibration;
- p is a cocartesian fibration;
- p is a cartesian fibration.

Example 4.5 ([10, Cor. 2.4.7.12]) For any ∞-category C, we write $\mathscr{O}(C) :=$ $\mathrm{Fun}(\Delta^1, C)$. Evaluation at 0 defines a cartesian fibration $s\colon \mathscr{O}(C) \longrightarrow C$, and evaluation at 1 defines a cocartesian fibration $t\colon \mathscr{O}(C) \longrightarrow C$.

One can ask whether the functor $s\colon \mathscr{O}(C) \longrightarrow C$ is also a *cocartesian* fibration. One may observe [10, Lm. 6.1.1.1] that an edge $\Delta^1 \longrightarrow \mathscr{O}(C)$ is s-cocartesian just in case the corresponding diagram $(\Lambda_0^2)^\rhd \cong \Delta^1 \times \Delta^1 \longrightarrow C$ is a pushout square.

In the following, we will denote by $\mathbf{Cat}_\infty$ the simplicial nerve of the (fibrant) simplicial category whose objects are ∞-categories, in which $\mathrm{Map}(C, D)$ is the maximal ∞-groupoid contained in $\mathrm{Fun}(C, D)$. Similarly, for any ∞-category C, we will denote by $\mathbf{Cocart}(C)$ (respectively, $\mathbf{Cart}(C)$) the simplicial nerve of the (fibrant) simplicial category whose objects are cocartesian (resp., cartesian) fibrations $X \longrightarrow C$, in which $\mathrm{Map}(X, Y)$ is the ∞-groupoid whose n-simplices are functors $X \times \Delta^n \longrightarrow Y$ over C that carry any edge (f, τ) in which f is cocartesian (resp. cartesian) to a cocartesian edge (resp., a cartesian edge).

Example 4.6 Consider the full subcategory $\mathbf{RFib}^{rep} \subset \mathscr{O}(\mathbf{Cat}_\infty)$ spanned by the representable right fibrations. The restriction of the functor $t\colon \mathscr{O}(\mathbf{Cat}_\infty) \longrightarrow \mathbf{Cat}_\infty$ to $\mathbf{RFib}^{rep}$ is again a cocartesian fibration.

Theorem 4.7 *Suppose C an ∞-category. For any functor $F\colon C \longrightarrow \mathbf{Cat}_\infty$, we may consider the cocartesian fibration*

$$\mathbf{RFib}^{rep} \times_{\mathbf{Cat}_\infty, F} C \longrightarrow C.$$

This defines an equivalence of categories

$$\mathrm{Fun}(C, \mathbf{Cat}_\infty) \overset{\sim}{\longrightarrow} \mathbf{Cocart}(C).$$

Dually, for any functor $G\colon C \longrightarrow \mathbf{Cat}_\infty^{op}$, we may consider the cartesian fibration

$$\mathbf{RFib}^{rep,op} \times_{\mathbf{Cat}_\infty^{op}, G} C \longrightarrow C.$$

This defines an equivalence of categories

$$\mathrm{Fun}(C^{op}, \mathbf{Cat}_\infty) \overset{\sim}{\longrightarrow} \mathbf{Cart}(C).$$

A cocartesian fibration $p\colon X \longrightarrow C$ is said to be *classified* by F just in case it is equivalent to the cocartesian fibration

$$\mathbf{RFib}^{rep} \times_{\mathbf{Cat}_\infty, F} C \longrightarrow C.$$

Dually, a cartesian fibration $q: Y \longrightarrow C$ is said to be *classified* by F just in case it is equivalent to the cocartesian fibration

$$\mathbf{RFib}^{rep,op} \times_{\mathbf{Cat}_\infty^{op},G} C \longrightarrow C.$$

Example 4.8 Suppose C an ∞-category, and suppose $X \longrightarrow C$ an isofibration. If $\iota C \subseteq C$ is the largest ∞-groupoid contained in C, then the pulled back isofibration

$$X \times_C \iota C \longrightarrow \iota C$$

is both cocartesian and cartesian, and so it corresponds to a functor

$$\iota C \simeq \iota C^{op} \longrightarrow \mathbf{Cat}_\infty.$$

This is the sense in which the fibers of an isofibration vary functorially in equivalences if C.

Example 4.9 For any ∞-category C, the functor $C^{op} \longrightarrow \mathbf{Cat}_\infty$ that classifies the cartesian fibration $s: \mathscr{O}(C) \longrightarrow C$ is the functor that carries any object a of C to the undercategory $C_{a/}$ and any morphism $f: a \longrightarrow b$ to the forgetful functor $f^\star: C_{b/} \longrightarrow C_{a/}$.

If C admits all pushouts, then the cocartesian fibration $s: \mathscr{O}(C) \longrightarrow C$ is classified by a functor $C \longrightarrow \mathbf{Cat}_\infty$ that carries any object a of C to the undercategory $C_{a/}$ and any morphism $f: a \longrightarrow b$ to the functor $f_!: C_{a/} \longrightarrow C_{b/}$ that is given by pushout along f.

One particularly powerful construction with cartesian and cocartesian fibrations comes from [10, §3.2.2]. We've come to call this the *cartesian workhorse*.

Example 4.10 Suppose $p: X \longrightarrow B^{op}$ a cartesian fibration and $q: Y \longrightarrow B^{op}$ a cocartesian fibration. Suppose $F: B \longrightarrow \mathbf{Cat}_\infty$ a functor that classifies p and $G: B^{op} \longrightarrow \mathbf{Cat}_\infty$ a functor that classifies q. Clearly one may define a functor

$$\mathrm{Fun}(F, G): B^{op} \longrightarrow \mathbf{Cat}_\infty$$

that carries a vertex s of B^{op} to the ∞-category $\mathrm{Fun}(F(s), G(s))$ and an edge $\eta: s \longrightarrow t$ of B^{op} to the functor

$$\mathrm{Fun}(F(s), G(s)) \longrightarrow \mathrm{Fun}(F(t), G(t))$$

given by the assignment $F \rightsquigarrow G(\eta) \circ F \circ F(\eta)$.

If one wishes to work instead with the fibrations directly (avoiding straightening and unstraightening), the following construction provides an elegant way of writing explicitly the cocartesian fibration classified by the functor $\mathrm{Fun}(F, G)$.

Suppose $p: X \longrightarrow B^{op}$ is a cartesian fibration classified by a functor

$$F: B \longrightarrow \mathbf{Cat}_\infty,$$

and suppose $q: Y \longrightarrow B^{op}$ is a cocartesian fibration classified by a functor

$$G: B^{op} \longrightarrow \mathbf{Cat}_\infty.$$

One defines a simplicial set $\widetilde{\mathrm{Fun}}_B(X, Y)$ and a map $r: \widetilde{\mathrm{Fun}}_B(X, Y) \longrightarrow B^{op}$ defined by the following universal property: for any map $\sigma: K \longrightarrow B^{op}$, one has a bijection

$$\mathrm{Mor}_{/B^{op}}(K, \widetilde{\mathrm{Fun}}_B(X, Y)) \cong \mathrm{Mor}_{/B^{op}}(X \times_{B^{op}} K, Y),$$

functorial in σ.

It is then shown in [10, Cor. 3.2.2.13] (but see Example 7.3 below for a proof which places this result in a broader context) that r is a cocartesian fibration, and an edge

$$g: \Delta^1 \longrightarrow \widetilde{\mathrm{Fun}}_B(X, Y)$$

is r-cocartesian just in case the induced map $X \times_{B^{op}} \Delta^1 \longrightarrow Y$ carries p-cartesian edges to q-cocartesian edges. The fiber of the map $\widetilde{\mathrm{Fun}}_B(X, Y) \longrightarrow S$ over a vertex s is the ∞-category $\mathrm{Fun}(X_s, Y_s)$, and for any edge $\eta: s \longrightarrow t$ of B^{op}, the functor $\eta_!: T_s \longrightarrow T_t$ induced by η is equivalent to the functor $F \rightsquigarrow G(\eta) \circ F \circ F(\eta)$ described above.

Warning 4.11 *We do not know of a proof that the functor associated to this fibration via straightening is actually the expected one (In [6], this is verified in the case where the functor G is constant).*

Finally, let us mention a weakening of the notion of a cocartesian fibration to that of a *locally cocartesian fibration*, which is an inner fibration $p : X \longrightarrow S$ such that for every edge $f : \Delta^1 \longrightarrow S$, the pullback p_f is a cocartesian fibration. Since p is cocartesian when restricted to live over any 1-simplex, the equivalence of Theorem 4.7 produces functors $f_! : X_s \longrightarrow X_t$ for every edge $f : s \longrightarrow t$; however, one loses the equivalence $(d_1\sigma)_! \simeq (d_0\sigma)_! \circ (d_2\sigma)_!$ for any 2-simplex σ in S. Instead, the analogue of Theorem 4.7 (proven in [11]) states that locally cocartesian fibrations $X \longrightarrow S$ are equivalent to normal lax functors from S to $\mathbf{Cat}_\infty$ considered as an ∞ double category.

5 Flat Inner Fibrations

We have observed that if C is an ∞-category, and if $C \longrightarrow \Delta^1$ is any functor with fibers C_0 and C_1, then there is a corresponding profunctor from C_1 to C_0, i.e., a colimit-preserving functor $P(C_1) \longrightarrow P(C_0)$. Furthermore, the passage from ∞-categories over Δ^1 to profunctors is even in some sense an equivalence.

So to make this precise, let **Prof** denote the full subcategory of the ∞-category $\mathbf{Pr}^L$ of presentable ∞-categories and left adjoints spanned by those ∞-categories of the form $P(C)$. We can *almost*—but not quite—construct an equivalence between the ∞-categories $\mathbf{Cat}_{\infty/\Delta^1}$ and the ∞-category $\mathrm{Fun}(\Delta^1, \mathbf{Prof})$.

The trouble here is that there are strictly more equivalences in $\mathbf{Prof}$ than there are in $\mathbf{Cat}_\infty$; two ∞-categories are equivalent in $\mathbf{Prof}$ if and only if they have equivalent idempotent completions. So that suggests the fix for this problem: we employ a pullback that will retain the data of the ∞-categories that are the source and target of our profunctor.

Proposition 5.1 *There is an equivalence of ∞-categories*

$$\mathbf{Cat}_{\infty/\Delta^1} \simeq \mathrm{Fun}(\Delta^1, \mathbf{Prof}) \times_{\mathrm{Fun}(\partial\Delta^1, \mathbf{Prof})} \mathrm{Fun}(\partial\Delta^1, \mathbf{Cat}_\infty).$$

We do not know of a reference for this result, yet, but we expect this to appear in a future work of P. Haine.

When we pass to ∞-categories over Δ^2, we have a more complicated problem: a functor $C \longrightarrow \Delta^2$ only specifies a lax commutative diagram of profunctors: three fibers C_0, C_1, and C_2; three colimit-preserving functors $F: P(C_0) \longrightarrow P(C_1)$, $G: P(C_1) \longrightarrow P(C_2)$, and $H: P(C_0) \longrightarrow P(C_2)$; and a natural transformation $\alpha: G \circ F \longrightarrow H$. In order to ensure that α be a natural equivalence, we need a condition on our fibration. This is where flatness comes in.

Definition 5.2 An inner fibration $p: X \longrightarrow S$ is said to be *flat* just in case, for any inner anodyne map $K \hookrightarrow L$ and any map $L \longrightarrow S$, the pullback

$$X \times_S K \longrightarrow X \times_S L$$

is a categorical equivalence.

We will focus mostly on flat *iso*fibrations. Let us see right away that some familiar examples and constructions yield flat inner fibrations.

Example 5.3 (Lurie, [13, Ex. B.3.11]) Cocartesian and cartesian fibrations are flat isofibrations; more precisely, the combination of the locally cocartesian and flat conditions on an isofibration exactly yield the class of cocartesian fibrations [4, Prp. 1.5]. In particular, if C is an ∞-groupoid, then any isofibration $X \longrightarrow C$ is flat.

Example 5.4 (Lurie, [13, Pr. B.3.13]) If $p: X \longrightarrow S$ is a flat inner fibration, then for any vertex $x \in X_0$, the inner fibrations $X_{x/} \longrightarrow S_{p(x)/}$ and $X_{/x} \longrightarrow S_{/p(x)}$ are flat as well.

It is not necessary to test flatness with *all* inner anodyne maps; in fact, one can make do with the inner horn of a 2-simplex:

Proposition 5.5 (Lurie, [13, Pr. B.3.14]) *An inner fibration $p: X \longrightarrow S$ is flat if and only if, for any 2-simplex $\sigma \in S_2$, the pullback*

$$X \times_{p,S,(\sigma|\Lambda_1^2)} \Lambda_1^2 \longrightarrow X \times_{p,S,\sigma} \Delta^2$$

is a categorical equivalence.

Proposition 5.6 (Lurie, [13, Pr. B.3.2, Rk. B.3.9]) *An inner fibration $X \longrightarrow S$ is flat just in case, for any 2-simplex*

of S any for any edge $x \longrightarrow y$ lying over f, the simplicial set $X_{x/}{}_{/y} \times_S \{v\}$ is weakly contractible.

5.7 *Note that the condition of the previous result is vacuous if the 2-simplex is degenerate. Consequently, if S is 1-skeletal, then any inner fibration $X \longrightarrow S$ is flat.*

Proposition 5.8 *Suppose C an ∞-category. Then there is an equivalence*

$$\mathbf{Flat}(C) \xrightarrow{\sim} \mathrm{Fun}(C^{op}, \mathbf{Prof}) \times_{\mathrm{Fun}(\iota C^{op}, \mathbf{Prof})} \mathrm{Fun}(\iota C^{op}, \mathbf{Cat}_\infty),$$

where $\mathbf{Flat}(C)$ is an ∞-category of flat isofibrations $X \longrightarrow C$.

Once again, we do not know a reference for this in the literature yet, but we expect that this will be shown in future work of P. Haine.

6   Marked Simplicial Sets and Categorical Patterns

In order to model the ∞-category **Cocart**(C) of cocartesian fibrations at the level of simplicial sets (and thereby implement the known proof of Theorem 4.7 via Lurie's straightening and unstraightening functors), it turns out that one must remember the data of the cocartesian edges. Moreover, it is often useful to remember still more data on the fibration side in order to reflect special features of the corresponding functors. This leads us to the notions of marked simplicial sets and categorical patterns.

Definition 6.1 A *marked simplicial set* consists of a simplicial set S together with the additional datum of a set $M \subset S_1$ of *marked edges* that contains all the degenerate edges. A *categorical pattern* on a simplicial set S is a triple (M, T, P) where (S, M) constitutes a marked simplicial set and one additionally has:

- a set $T \subset S_2$ of *scaled 2-simplices* that contains all the degenerate 2-simplices, and
- a set P of maps $f_\alpha : K_\alpha^\lhd \longrightarrow S$ such that $f_\alpha((K_\alpha^\lhd)_1) \subset M$ and $f_\alpha((K_\alpha^\lhd)_2) \subset T$.

Let $s\mathbf{Set}^+$ be the category of marked simplicial sets.

Example 6.2 There exists a *marked model structure* on $s\mathbf{Set}^+$ such that the fibrant objects are ∞-categories with the equivalences marked and the cofibrations are the monomorphisms. We have the Quillen equivalence

$$(-)^\flat : s\mathbf{Set}_{\mathrm{Joyal}} \rightleftarrows s\mathbf{Set}^+ : U$$

where $(-)^\flat$ marks the degenerate edges and U forgets the marking; also note that U is left adjoint to the functor $(-)^\sharp$ which marks all edges.

As a demonstration of the utility of $s\mathbf{Set}^+$ over $s\mathbf{Set}_{\mathrm{Joyal}}$, observe that $s\mathbf{Set}^+$ is a *simplicial* model category (unlike $s\mathbf{Set}_{\mathrm{Joyal}}$) via the Quillen bifunctor

$$\mathrm{Map}^\sharp(-,-) : (s\mathbf{Set}^+)^{op} \times s\mathbf{Set}^+ \longrightarrow s\mathbf{Set}$$

defined by $\mathrm{Hom}_{s\mathbf{Set}}(K, \mathrm{Map}^\sharp((X,E),(Y,F))) = \mathrm{Hom}_{s\mathbf{Set}^+}(K^\sharp \times (X,E),(Y,F))$. Consequently, we may define $\mathbf{Cat}_\infty$ to be the homotopy coherent nerve of the simplicial subcategory $(s\mathbf{Set}^+)^\circ$ of the fibrant objects in $s\mathbf{Set}^+$, and indeed this is the definition adopted in [10]. In addition, $s\mathbf{Set}^+$ is also enriched over $s\mathbf{Set}_{\mathrm{Joyal}}$ via

$$\mathrm{Map}^\flat(-,-) : (s\mathbf{Set}^+)^{op} \times s\mathbf{Set}^+ \longrightarrow s\mathbf{Set}$$

defined by $\mathrm{Hom}_{s\mathbf{Set}}(K, \mathrm{Map}^\flat((X,E),(Y,F))) = \mathrm{Hom}_{s\mathbf{Set}^+}(K^\flat \times (X,E),(Y,F))$, and over itself via the internal hom.

We now pass to a relative situation and introduce a class of model structures on $s\mathbf{Set}^+_{/(S,M)}$ which serve to model (variants of) ∞-categories of $\mathbf{Cat}_\infty$-valued functors with domain S.

Notation 6.3 *For the purposes of this section, if $p\colon X \longrightarrow S$ and $f\colon K \longrightarrow S$ are maps of simplicial sets, then let us write*

$$p_f\colon X \times_S K \longrightarrow K$$

for the pullback of p along f.

Definition 6.4 Suppose (M,T,P) a categorical pattern on a simplicial set S. Then a marked map $p\colon (X,E) \longrightarrow (S,M)$ is said to be (M,T,P)-*fibered* if the following conditions obtain.

- The map $p\colon X \longrightarrow S$ is an inner fibration.
- For every marked edge η of S, the pullback p_η is a cocartesian fibration.
- An edge ϵ of X is marked just in case $p(\epsilon)$ is marked and ϵ is $p_{p(\epsilon)}$-cocartesian.
- For any commutative square

$$
\begin{array}{ccc}
\Delta^{\{0,1\}} & \xrightarrow{\ \epsilon\ } & X \\
\downarrow & & \downarrow{\scriptstyle p} \\
\Delta^2 & \xrightarrow[\ \sigma\ ]{} & S
\end{array}
$$

in which σ is scaled, if ϵ is marked, then it is p_σ-cocartesian.
- For every element $f_\alpha\colon K_\alpha^\triangleleft \longrightarrow S$ of P, the cocartesian fibration

$$p_{f_\alpha}\colon X \times_S K_\alpha^\triangleleft \longrightarrow K_\alpha^\triangleleft$$

is classified by a limit diagram $K_\alpha^\triangleleft \longrightarrow \mathbf{Cat}_\infty$.

- For every element $f_\alpha \colon K_\alpha^\lhd \longrightarrow S$ of P, any cocartesian section σ of p_{f_α} is a p-limit diagram in X [10, Dfn. 4.3.1.1]. If X and S are ∞-categories, then this is equivalent to requiring that for any $x \in X$, the homotopy commutative square of mapping spaces

$$
\begin{array}{ccc}
\mathrm{Map}_X(x, \sigma(v)) & \longrightarrow & \lim_{k \in K_\alpha} \mathrm{Map}_X(x, \sigma(k)) \\
\downarrow{\scriptstyle p} & & \downarrow{\scriptstyle p} \\
\mathrm{Map}_S(p(x), p\sigma(v)) & \longrightarrow & \lim_{k \in K_\alpha} \mathrm{Map}_S(px, p\sigma(k))
\end{array}
$$

is a homotopy pullback square (where $v \in K_\alpha^\lhd$ is the cone point).

Theorem 6.5 (Lurie, [13, Th. B.0.20]) *Suppose (M, T, P) a categorical pattern on a simplicial set S. Then there exists a left proper, combinatorial, simplicial model structure on $s\mathbf{Set}^+_{/(S,M)}$ in which the cofibrations are monomorphisms, and a marked map $p \colon (X, E) \longrightarrow (S, M)$ is fibrant just in case it is (M, T, P)-fibered.*
We will denote this model category $s\mathbf{Set}^+_{/(S,M,T,P)}$.

Example 6.6 For any simplicial set S, the $(S_1, S_2, \varnothing)$-fibered maps $(X, E) \longrightarrow S^\sharp$ are precisely those maps of the form $\natural X \longrightarrow S^\sharp$, where the underlying map of simplicial sets $X \longrightarrow S$ is a cocartesian fibration and $\natural X$ denotes X with the cocartesian edges marked. We are therefore entitled to call the model structure of Theorem 6.5 on $s\mathbf{Set}^+_{/S^\sharp}$ given by $(S_1, S_2, \varnothing)$ the *cocartesian* model structure. Abusing notation, the underlying category is also denoted as $s\mathbf{Set}^+_{/S}$. In [10, §3.2], Lurie proves an equivalence $\mathrm{Fun}(S, \mathbf{Cat}_\infty) \simeq N((s\mathbf{Set}^+_{/S})^\circ)$.

We have a convenient characterization of the fibrations between fibrant objects in the model structures of Theorem 6.5:

Proposition 6.7 (Lurie, [13, Th. B.2.7]) *Suppose S an ∞-category, and suppose (M, T, P) a categorical pattern on S such that every equivalence of S is marked, and every 2-simplex $\Delta^2 \longrightarrow S$ whose restriction to $\Delta^{\{0,1\}}$ is an equivalence is scaled. If (Y, F) is fibrant in $s\mathbf{Set}^+_{/(S,M,T,P)}$, then a map $f \colon (X, E) \longrightarrow (Y, F)$ over (S, M) is a fibration in $s\mathbf{Set}^+_{/(S,M,T,P)}$ if and only if (X, E) is fibrant, and the underlying map $f \colon X \longrightarrow Y$ is an isofibration.*

Example 6.8 For any marked simplicial set (S, M), we may as well call the model structure on $s\mathbf{Set}^+_{/(S,M)}$ supplied by the categorical pattern $(M, S_2, \emptyset)$ the cocartesian model structure. Directly from the definition, we see that the fibrant objects are those inner fibrations $C \longrightarrow S$ that possess cocartesian edges lifting the marked edges in the base (such that those are the edges marked in C). Moreover, in light of the previous proposition, we have another description of the cocartesian model structure on $s\mathbf{Set}^+_{/S}$ in the case of $p \colon S \longrightarrow T$ a cocartesian fibration, as created by the forgetful functor to the cocartesian model structure $s\mathbf{Set}^+_{/T}$—i.e., by regarding $s\mathbf{Set}^+_{/S}$ as $(s\mathbf{Set}^+_{/T})/p$. (Note that a model structure is uniquely characterized by its cofibrations and fibrant objects.) For example, if S is an ∞-category and M is the set of equivalences in S, then the cocartesian model structure on $sSet^+_{/(S,M)}$ is Quillen equivalent to $(sSet_{Joyal})_{/S}$.

Example 6.9 Specifying the set of scaled 2-simplices allows us to model lax functoriality. For example, in the extreme case where we take the categorical pattern $(S_1, (S_2)_{degen}, \emptyset)$ on S, then the resulting model structure on $sSet^+_{/S^\sharp}$ has fibrant objects the locally cocartesian fibrations over S.

Let us now consider examples where the set P of maps in the definition of a categorical pattern comes into play.

Example 6.10 ([13, Prp. 2.1.4.6] and [13, Var. 2.1.4.13]) Let $S = \mathbf{F}_*$ be the category of finite pointed sets, with objects $\langle n \rangle = \{1, \ldots, n, *\}$. We call a map $\alpha \colon \langle n \rangle \longrightarrow \langle m \rangle$ *inert* if for every non basepoint element $i \in \langle m \rangle$, the preimage $\alpha^{-1}(i)$ consists of exactly one element. Then α determines an injection $\langle m \rangle^\circ \longrightarrow \langle n \rangle^\circ$, where $\langle n \rangle^\circ = \langle n \rangle \setminus \{*\}$. Let P be the set of maps $\Lambda^2_0 \longrightarrow \mathbf{F}_*$ given by all

$$\langle p \rangle \longleftarrow \langle n \rangle \longrightarrow \langle q \rangle$$

where the maps are inert and determine a bijection $\langle p \rangle^\circ \sqcup \langle q \rangle^\circ \longrightarrow \langle n \rangle^\circ$. Then the categorical pattern $((\mathbf{F}_*)_1, (\mathbf{F}_*)_2, P)$ on $\mathbf{F}_*$ yields a model structure on $sSet^+_{/(\mathbf{F}_*)^\sharp}$ with fibrant objects symmetric monoidal ∞-categories, and where maps between fibrant objects are (strong) symmetric monoidal functors. To see this, observe that the chosen set P implies that for a fibrant object $C \longrightarrow \mathbf{F}_*$, the fiber $C_{\langle n \rangle}$ decomposes as $\prod_{1 \le i \le n} C_{\langle 1 \rangle}$, with the n comparison maps induced by the n inert morphisms ρ^i : $\langle n \rangle \longrightarrow \langle 1 \rangle$ which send i to 1 and the other elements to the basepoint. Then the *active* maps, that is the maps in the image of the functor $(-)_+ : \mathbf{F} \longrightarrow \mathbf{F}_*$ which adds a basepoint, induce the multiplication maps which endow $C_{\langle 1 \rangle}$ with the structure of a symmetric monoidal ∞-category.

If instead we took the categorical pattern $((\mathbf{F}_*)_{inert}, (\mathbf{F}_*)_2, P)$ on $\mathbf{F}_*$, then we would obtain a model structure on $sSet^+_{/(\mathbf{F}_*, (\mathbf{F}_*)_{inert})}$ whose fibrant objects are ∞-operads. Note that the last condition in the definition of (M, T, P)-fibered object is now essential to obtain the correct description of the mapping spaces in the total category of an ∞-operad.

The remaining two examples require an acquaintance with the theory of perfect operator categories [1, §6], and generalize the examples of ∞-operads and symmetric monoidal ∞-categories, respectively.

Example 6.11 Suppose Φ a perfect operator category, and consider the following categorical pattern

$$(\mathrm{Ne}, N(\Lambda(\Phi))_2, P)$$

on the nerve $N\Lambda(\Phi)$ of the Leinster category [1, §7]. Here the class $\mathrm{Ne} \subset N(\Lambda(\Phi))_1$ consists of all the inert morphisms of $N\Lambda(\Phi)$. The class P is the set of maps $\Lambda^2_0 \longrightarrow N\Lambda(\Phi)$ given by diagrams $I \longleftarrow J \longrightarrow I'$ of $\Lambda(\Phi)$ in which both $J \longrightarrow I$ and $J \longrightarrow I'$ are inert, and

$$|J| = |J \times_{TI} I| \sqcup |J \times_{TI'} I'| \cong |I| \sqcup |I'|.$$

Then a marked map $(X, E) \longrightarrow (N\Lambda(\Phi), N\Lambda^\dagger(\Phi))$ is $(\mathrm{Ne}, N(\Lambda(\Phi))_2, P)$-fibered just in case the underlying map of simplicial sets $X \longrightarrow N\Lambda(\Phi)$ is a ∞-operad over Φ, and E is the collection of cocartesian edges over the inert edges [1, §8].

In particular, when $\Phi = \mathbf{F}$, the category of finite sets, this recovers the collection of ∞-operads.

Example 6.12 If Φ is a perfect operator category, we can contemplate another categorical pattern

$$(N(\Lambda(\Phi))_1, N(\Lambda(\Phi))_2, P)$$

on $\Lambda(\Phi)$. Here P is as in the previous example.

Now a marked map $(X, E) \longrightarrow (N\Lambda(\Phi), N\Lambda^\dagger(\Phi))$ is $(N(\Lambda(\Phi))_1, N(\Lambda(\Phi))_2, P)$-fibered just in case the underlying map of simplicial sets $X \longrightarrow N\Lambda(\Phi)$ is a Φ-monoidal ∞-category, and E is the collection of all cocartesian edges.

7 Constructing Fibrations via Additional Functoriality

The theory of flat isofibrations can be used to describe some extra functoriality of these categorical pattern model structures in S. We will use this extra direction of functoriality to perform some beautifully explicit constructions of fibrations of various kinds. The main result is a tad involved, but the constructive power it offers is worth it in the end.

Notation 7.1 *Let us begin with the observation that, given a map of marked simplicial sets* $\pi\colon (S, M) \longrightarrow (T, N)$, *there is a string of adjoints*

$$\pi_! \dashv \pi^\star \dashv \pi_\star,$$

where the far left adjoint

$$\pi_!\colon s\mathbf{Set}^+_{/(S,M)} \longrightarrow s\mathbf{Set}^+_{/(T,N)}$$

is simply composition with π; *the middle functor*

$$\pi^\star\colon s\mathbf{Set}^+_{/(T,N)} \longrightarrow s\mathbf{Set}^+_{/(S,M)}$$

is given by the assignment $(Y, F) \rightsquigarrow (Y, F) \times_{(T,N)} (S, M)$; *and the far right adjoint*

$$\pi_\star\colon s\mathbf{Set}^+_{/(S,M)} \longrightarrow s\mathbf{Set}^+_{/(T,N)}$$

is given by a "space of sections." That is, an n-simplex of $\pi_\star(X, E)$ is a pair (σ, f) consisting of an n-simplex $\sigma \in T_n$ along with a marked map

$$(\Delta^n)^\flat \times_{(T,N)} (S, M) \longrightarrow (X, E)$$

over (S, M); an edge (η, f) of $\pi_\star(X, E)$ is marked just in case η is marked, and f carries any edge of $\Delta^1 \times_T S$ that projects to a marked edge of S to a marked edge of X.

Theorem 7.2 (Lurie, [13, Th. B.4.2]) *Suppose S, S', and X three ∞-categories; suppose (M, T, P) a categorical pattern on S; suppose (M', T', P') a categorical pattern on S'; and suppose E a collection of marked edges on X. Assume these data satisfy the following conditions.*

- *Any equivalence of either S or X is marked.*
- *The marked edges in S and X are each closed under composition.*
- *Every 2-simplex $\Delta^2 \longrightarrow S$ whose restriction to $\Delta^{\{0,1\}}$ is an equivalence is scaled.*

Suppose

$$\pi : (X, E) \longrightarrow (S, M)$$

a map of marked simplicial sets satisfying the following conditions.

- *The functor $\pi : X \longrightarrow S$ is a flat isofibration.*
- *For any marked edge η of S, the pullback*

$$\pi_\eta : X \times_S \Delta^1 \longrightarrow \Delta^1$$

  is a cartesian fibration.
- *For any element $f_\alpha : K_\alpha^\triangleleft \longrightarrow S$ of P, the simplicial set K_α is an ∞-category, and the pullback*

$$\pi_{f_\alpha} : X \times_S K_\alpha^\triangleleft \longrightarrow K_\alpha^\triangleleft$$

  is a cocartesian fibration.
- *Suppose*

a 2-simplex of X in which $\pi(\phi)$ is an equivalence, ψ is locally π-cartesian, and $\pi(\psi)$ is marked. Then the edge ϕ is marked just in case χ is.

- *Suppose $f_\alpha\colon K_\alpha^\triangleleft \longrightarrow S$ an element of P, and suppose*

a 2-simplex of $X \times_S K_\alpha^\triangleleft$ in which ϕ is π_{f_α}-cocartesian and $\pi_{f_\alpha}(\psi)$ is an equivalence. Then the image of ψ in X is marked if and only if the image of χ is.

Finally, suppose

$$\rho\colon (X, E) \longrightarrow (S', M')$$

a map of marked simplicial sets satisfying the following conditions.

- *Any 2-simplex of X that lies over a scaled 2-simplex of S also lies over a scaled 2-simplex of S'.*
- *For any element $f_\alpha\colon K_\alpha^\triangleleft \longrightarrow S$ of P and any cocartesian section s of π_{f_α}, the composite*

$$K_\alpha^\triangleleft \xrightarrow{\ s\ } X \times_S K_\alpha^\triangleleft \longrightarrow X \xrightarrow{\ \rho\ } S'$$

  lies in P'.

Then the adjunction

$$\rho_! \circ \pi^\star\colon s\mathbf{Set}^+_{/(S,M,T,P)} \rightleftarrows s\mathbf{Set}^+_{/(S',M',T',P')}\colon\pi_\star \circ \rho^\star$$

is a Quillen adjunction.

Before applying this theorem, let us comment on the relevance of some of the various hypotheses (sans the ones dealing with the class P in the definition of a categorical pattern, which we only include for completeness).

- Flatness: We introduced flat inner fibrations in connection with composing profunctors. Composition of profunctors involves inner anodyne maps such as $\Lambda_1^2 \longrightarrow \Delta^2$ being sent to categorical equivalences upon pulling back, and the added property of being an isofibration ensures that the larger class of categorical equivalences is sent to itself by pullback [13, Cor. B.3.15]. Indeed, if $\pi : X \longrightarrow S$ is a flat isofibration of ∞-categories then $\pi^* : (s\mathbf{Set}_{\text{Joyal}})_{/S} \longrightarrow (s\mathbf{Set}_{\text{Joyal}})_{/X}$ is left Quillen [13, Prp. B.4.5]. Flatness is thus the basic necessary condition which precedes additional considerations stemming from the categorical patterns in play.
- Cartesian conditions: Cartesian fibrations are a subclass of the *cosmooth* maps, which are those maps $X \longrightarrow S$ such that for any right cofinal (i.e. initial) map

$K \longrightarrow L, X \times_S K \longrightarrow X \times_S L$ is again right cofinal; dually, cocartesian fibrations are a subclass of the *smooth* maps ([10, Dfn. 4.1.2.9] and [10, Prp. 4.1.2.15]). This fact is clear in view of the total category X of a cocartesian fibration $X \longrightarrow S$ being a model for the lax colimit of the associated functor. Moreover, a map $K^\sharp \longrightarrow L^\sharp$ is an equivalence in the cocartesian model structure on $s\mathbf{Set}^+_{/S^\sharp}$ if and only if $K \longrightarrow L$ is right cofinal, because this is equivalent to $\mathrm{Map}^\sharp_S(-, {}_\natural C)$ yielding an equivalence for any cocartesian fibration $C \longrightarrow S$, and the ∞-category of cocartesian sections is a model for the limit of the associated functor. Thus, one sees the relevance of $\pi : X \longrightarrow S$ being a cartesian fibration, for π_* to be right Quillen with respect to the cocartesian model structures; to be precise about this, one then needs to handle the added complications caused by markings.

Let us now put Theorem 7.2 to work. We first discard some of the baggage of categorical patterns which we will not need.

Corollary 7.2.1 *Suppose $\pi\colon X \longrightarrow S$ and $\rho\colon X \longrightarrow T$ two functors of ∞-categories. Suppose $E \subset X_1$ a collection of marked edges on X, and assume the following.*

- *Any equivalence of X is marked, and the marked edges are closed under composition.*
- *The functor $\pi\colon X \longrightarrow S$ is a flat locally cartesian fibration.*
- *Suppose*

a 2-simplex of X in which $\pi(\phi)$ is an equivalence, and ψ is locally π-cartesian. Then the edge ϕ is marked just in case χ is.

Then the functor $\pi_\star \circ \rho^\star$ is a right Quillen functor

$$s\mathbf{Set}^+_{/T} \longrightarrow s\mathbf{Set}^+_{/S}$$

for the cocartesian model structure. In particular, if $Y \longrightarrow T$ is a cocartesian fibration, then the map $r\colon Z \longrightarrow S$ given by the universal property

$$\mathrm{Mor}_S(K, Z) \cong \mathrm{Mor}_T(K^\flat \times_{S^\sharp} (X, E), {}_\natural Y)$$

is a cocartesian fibration, and an edge of Z is r-cocartesian just in case the map $\Delta^1 \times_S X \longrightarrow Y$ carries any edge whose projection to X is marked to a q-cocartesian edge.

We will speak of applying Corollary 7.2.1 to the span

$$S \xleftarrow{\;\pi\;} (X, E) \xrightarrow{\;\rho\;} T$$

to obtain a Quillen adjunction

$$\rho_! \circ \pi^\star : s\mathbf{Set}^+_{/S} \rightleftarrows s\mathbf{Set}^+_{/T} : \pi_\star \circ \rho^\star.$$

Example 7.3 Suppose $p: X \longrightarrow S$ is a cartesian fibration, and suppose $q: Y \longrightarrow S$ is a cocartesian fibration. The construction of Example 4.10 gives a simplicial set $\widetilde{\mathrm{Fun}}_S(X, Y)$ over S given by the universal property

$$\mathrm{Mor}_S(K, \widetilde{\mathrm{Fun}}_S(X, Y)) \cong \mathrm{Mor}_S(K \times_S X, Y).$$

By the previous result, the functor $\widetilde{\mathrm{Fun}}_S(X, Y) \longrightarrow S$ is a cocartesian fibration. Consequently, one finds that [10, Cor. 3.2.2.13] is a very elementary, very special case of the previous result.

Another compelling example can be obtained as follows. Suppose $\phi: A \longrightarrow B$ is a functor of ∞-categories. Of course if one has a cocartesian fibration $Y \longrightarrow B$ classified by a functor $\mathbf{Y}$, then one may pull back to obtain the cocartesian fibration $A \times_B Y \longrightarrow A$ that is classified by the functor $\mathbf{Y} \circ \phi$. In the other direction, however, if one has a cocartesian fibration $X \longrightarrow A$ that is classified by a functor $\mathbf{X}$, then how might one write explicitly the cocartesian fibration $Z \longrightarrow B$ that is classified by the right Kan extension? It turns out that this is just the sort of situation that Corollary 7.2.1 can handle gracefully. We'll need a spot of notation for comma ∞-categories:

Notation 7.4 *If $f: M \longrightarrow S$ and $g: N \longrightarrow S$ are two maps of simplicial sets, then let us write*

$$M \downarrow_S N := M \underset{\mathrm{Fun}(\Delta^{\{0\}}, S)}{\times} \mathrm{Fun}(\Delta^1, S) \underset{\mathrm{Fun}(\Delta^{\{1\}}, S)}{\times} N.$$

A vertex of $M \downarrow_S N$ is thus a vertex $x \in M_0$, a vertex $y \in N_0$, and an edge $f(x) \longrightarrow g(y)$ in S_1. When M and N are ∞-categories, the simplicial set $M \downarrow_S N$ is a model for the lax pullback of f along g.

Proposition 7.5 *If $\phi: A \longrightarrow B$ is a functor of ∞-categories, and if $p: X \longrightarrow A$ is a cocartesian fibration, then form the simplicial set Y over B given by the following universal property: for any map $\eta: K \longrightarrow Z$, we demand a bijection*

$$\mathrm{Mor}_{/B}(K, Y) \cong \mathrm{Mor}_{/A}(K \downarrow_B A, X),$$

functorial in η. Then the projection $q: Y \longrightarrow B$ is a cocartesian fibration, and if p is classified by a functor $\mathbf{X}$, then q is classified by the right Kan extension of $\mathbf{X}$ along ϕ.

Proof As we have seen, we need conditions on ϕ to ensure that the adjunction

$$\phi^* : s\mathbf{Set}^+_{/B} \rightleftarrows s\mathbf{Set}^+_{/A} : \phi_*$$

is Quillen. However, we can rectify this situation by applying Corollary 7.2.1 to the span

$$B \xleftarrow{\text{ev}_0 \, \circ \, \text{pr}_{\mathscr{O}(B)}} (\mathscr{O}(B) \times_{\text{ev}_1, B, \phi} A)^{\sharp} \xrightarrow{\text{pr}_A} A$$

in view of the following two facts, which we leave to the reader to verify (or see [10, Cor. 2.4.7.12] and [3, Lem. 9.8] for proofs):

- the functor $\text{ev}_0 \circ \text{pr}_{\mathscr{O}(B)}$ is a cartesian fibration;
- for all $C \longrightarrow B$ fibrant in $s\mathbf{Set}^+_{/B}$, the identity section $B \longrightarrow \mathscr{O}(B)$ induces a functor

$$C \times_B A^{\sharp} \longrightarrow C \times_B \mathscr{O}(B)^{\sharp} \times_B A^{\sharp}$$

which is a cocartesian equivalence in $s\mathbf{Set}^+_{/A}$.

We thereby obtain a Quillen adjunction

$$(\text{pr}_A)_! \circ (\text{ev}_0 \circ \text{pr}_{\mathscr{O}(B)})^* \colon s\mathbf{Set}^+_{/B} \rightleftarrows s\mathbf{Set}^+_{/A} \colon (\text{ev}_0 \circ \text{pr}_{\mathscr{O}(B)})_* \circ (\text{pr}_A)^*$$

which models the adjunction of ∞-categories

$$\phi^* \colon \text{Fun}(B, \mathbf{Cat}_\infty) \rightleftarrows \text{Fun}(A, \mathbf{Cat}_\infty) \colon \phi_*,$$

as desired.

This result illustrates the basic principle that replacing strict pullbacks by lax pullbacks allows one to make homotopically meaningful constructions. Of course, any suitable theory of ∞-categories would allow one to produce the right adjoint ϕ_* less explicitly, by recourse to the adjoint functor theorem. This example rather illustrates, yet again, how the model of quasicategories affords more explicit descriptions of standard categorical constructions.

Let us end our study with some simple but gratifying observations concerning the interaction of Theorem 7.2 with compositions and homotopy equivalences of spans. First the compositions:

Lemma 7.6 *Suppose we have spans of marked simplicial sets*

$$C_0 \xleftarrow{\pi_0} D_0 \xrightarrow{\rho_0} C_1$$

and

$$C_1 \xleftarrow{\pi_1} D_1 \xrightarrow{\rho_1} C_2$$

and categorical patterns $(M_i, T_i, \emptyset)$ *on* C_i *such that the hypotheses of Theorem 7.2 are satisfied for each span. Then the hypotheses of Theorem 7.2 are satisfied for the span*

$$D_0 \xleftarrow{\text{pr}_0} D_0 \times_{C_1} D_1 \xrightarrow{\text{pr}_1} D_1$$

with the categorical patterns $(M_{D_i}, \pi_i^{-1}(T_i), \varnothing)$ on each D_i. Consequently, we obtain a Quillen adjunction

$$(\rho_1 \circ \mathrm{pr}_1)_! \circ (\pi_0 \circ \mathrm{pr}_0)^\star : s\mathbf{Set}^+_{/(C_0, M_0, T_0, \varnothing)} \rightleftharpoons s\mathbf{Set}^+_{/(C_2, M_2, T_2, \varnothing)} : (\pi_0 \circ \mathrm{pr}_0)_\star \circ (\rho_1 \circ \mathrm{pr}_1)^\star,$$

which is the composite of the Quillen adjunction from $s\mathbf{Set}^+_{/(C_0, M_0, T_0, \varnothing)}$ to $s\mathbf{Set}^+_{/(C_1, M_1, T_1, \varnothing)}$ with the one from $s\mathbf{Set}^+_{/(C_1, M_1, T_1, \varnothing)}$ to $s\mathbf{Set}^+_{/(C_2, M_2, T_2, \varnothing)}$.

Proof The proof is by inspection. However, one should beware that the "long" span

$$C_0 \longleftarrow D_0 \times_{C_1} D_1 \longrightarrow D_1$$

can fail to satisfy the hypotheses of Theorem 7.2, because the composition of locally cartesian fibrations may fail to be again be locally cartesian; this explains the roundabout formulation of the statement. Finally, observe that if we employ the base-change isomorphism $\rho_0^* \pi_{1,*} \cong \mathrm{pr}_{0,*} \circ \mathrm{pr}_1^*$, then we obtain our Quillen adjunction as the composite of the two given Quillen adjunctions.

Now let us see that Theorem 7.2 is compatible with homotopy equivalences of spans:

Lemma 7.7 *Suppose a morphism of spans of marked simplicial sets*

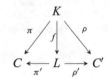

where $\rho_! \pi^*$ and $(\rho')_! (\pi')^*$ are left Quillen with respect to the model structures given by categorical patterns $\mathfrak{P}_C$ and $\mathfrak{P}_{C'}$ on C and C'. Suppose moreover that f is a homotopy equivalence in $s\mathbf{Set}^+_{/\mathfrak{P}_{C'}}$—i.e., suppose that there exists a homotopy inverse g and homotopies

$$h: \mathrm{id} \simeq g \circ f \quad and \quad k: \mathrm{id} \simeq f \circ g.$$

Then the natural transformation $\rho_! \pi^* \longrightarrow (\rho')_! (\pi')^*$ induced by f is a weak equivalence on all objects, and, consequently, the adjoint natural transformation $(\pi')_* (\rho')^* \longrightarrow \pi_* \rho^*$ is a weak equivalence on all fibrant objects.

Proof The homotopies h and k pull back to show that for all $X \longrightarrow C$, the map

$$\mathrm{id}_X \times_C f: X \times_C K \longrightarrow X \times_C L$$

is a homotopy equivalence with inverse $\mathrm{id}_X \times_C g$. The last statement now follows from [9, 1.4.4(b)].

Acknowledgements We are thankful to Bob Bruner, Emily Riehl, and the anonymous referee for their helpful comments on this paper.

References

1. Barwick, C.: From operator categories to topological operads (2013). Preprint. arXiv:1302.5756
2. Barwick, C., Glasman, S.: On the fibrewise effective Burnside ∞-category (2016). Preprint available from arXiv:1607.02786
3. Barwick, C., Dotto, E., Glasman, S., Nardin, D., Shah, J.: Elements of parametrized higher category theory (2016). Preprint available from arXiv:1608.03657.
4. Barwick, C., Glasman, S., Shah, J.: Spectral Mackey functors and equivariant algebraic K-theory (II) (2015). Preprint. arXiv:1505.03098
5. Dugger, D., Spivak, D.I.: Rigidification of quasi-categories. Algebr. Geom. Topol. **11**(1), 225–261 (2011)
6. Gepner, D., Haugseng, R., Nikolaus, Th.: Lax colimits and free fibrations in ∞-categories (2015). Preprint. arXiv:1501.02161
7. Heuts, G., Moerdijk, I.: Left fibrations and homotopy colimits (2016). Preprint available from arXiv:1308.0704
8. Heuts, G., Moerdijk, I.: Left fibrations and homotopy colimits II (2016). Preprint available from arXiv:1602.01274
9. Hovey, M.: Model Categories, vol. 63 American Mathematical Society, Providence, RI (1999)
10. Lurie, J.: Higher Topos Theory. Annals of Mathematics Studies, vol. 170. Princeton University Press, Princeton, NJ (2009). MR 2522659 (2010j:18001)
11. Lurie, J.: $(\infty, 2)$-categories and the Goodwillie calculus I. Preprint from the web page of the author (2009)
12. Lurie, J.: Derived algebraic geometry X. Formal moduli problems. Preprint from the web page of the author (2011)
13. Lurie, J.: Higher algebra. Preprint from the web page of the author (2012)
14. Stevenson, D.: Covariant model structures and simplicial localization (2016). Preprint. arXiv:1512.04815
15. Street, R.: Powerful functors. Preprint from the web page of the author (2001)

The Smooth Hom-Stack of an Orbifold

Check for
updates

David Michael Roberts and Raymond F. Vozzo

Abstract For a compact manifold M and a differentiable stack $\mathfrak{X}$ presented by a Lie groupoid X, we show the Hom-stack $\mathcal{H}om(M, \mathfrak{X})$ is presented by a Fréchet–Lie groupoid $\mathrm{Map}(M, X)$ and so is an infinite-dimensional differentiable stack. We further show that if $\mathfrak{X}$ is an orbifold, presented by a proper étale Lie groupoid, then $\mathrm{Map}(M, X)$ is proper étale and so presents an infinite-dimensional orbifold.

This note serves to announce a generalisation of the authors' work [8], which showed that the smooth loop stack of a differentiable stack is an infinite-dimensional differentiable stack, to more general mapping stacks where the source stack is a compact manifold (or more generally a compact manifold with corners). We apply this construction to differentiable stacks that are smooth orbifolds, that is, they can be presented by proper étale Lie groupoids (see Definition 9).

Existing work on mapping spaces of orbifolds has been considered in the case of C^k maps [4], of Sobolev maps [10] and smooth maps [3]; in the latter case several different notions of smooth orbifold maps are considered, from the point of view of orbifolds described by orbifold charts. In all these cases, some sort of orbifold structure has been found (for instance, Banach or Fréchet orbifolds).

Noohi [6] solved the problem of constructing a topological mapping stack between more general *topological* stacks, when the source stack has a presentation by a compact topological groupoid. See [8] for further references and discussion.

We take as given the definition of Lie groupoid in what follows, using finite-dimensional manifolds unless otherwise specified. Manifolds will be considered as trivial groupoids without comment. We pause only to note that in the infinite-dimensional setting, the source and target maps of Fréchet–Lie groupoids must be submersions between Fréchet manifolds, which is a stronger hypothesis than asking the derivative is surjective (or even split) everywhere, as in the finite-dimensional or Banach case.

D.M. Roberts (✉) • R.F. Vozzo
School of Mathematical Sciences, University of Adelaide, Adelaide, SA 5005, Australia
e-mail: david.roberts@adelaide.edu.au; raymond.vozzo@adelaide.edu.au

© Springer International Publishing AG, part of Springer Nature 2018
D.R. Wood et al. (eds.), *2016 MATRIX Annals*, MATRIX Book Series 1,
https://doi.org/10.1007/978-3-319-72299-3_3

We will also consider groupoids in diffeological spaces. Diffeological spaces (see e.g. [1]) contain Fréchet manifolds as a full subcategory and admit all pullbacks (in fact all finite limits) and form a cartesian closed category such that for K and M smooth manifolds with K compact, the diffeological mapping space M^K is isomorphic to the Fréchet manifold of smooth maps $K \to M$.

Differentiable stacks are, for us, stacks of groupoids on the site $\mathcal{M}$ of finite-dimensional smooth manifolds with the open cover topology that admit a presentation by a Lie groupoid [2]. We can also consider the more general notion of stacks that admit a presentation by a diffeological or Fréchet–Lie groupoid.

Definition 1 Let $\mathcal{X}, \mathcal{Y}$ be stacks on $\mathcal{M}$. The *Hom-stack* $\mathcal{H}om(\mathcal{Y}, \mathcal{X})$ is defined by taking the value on the manifold N to be $\mathbf{Stack}_{\mathcal{M}}(\mathcal{Y} \times N, \mathcal{X})$.

Thus we have a Hom-stack for any pair of stacks on $\mathcal{M}$. The case we are interested in is where we have a differentiable stack $\mathcal{X}$ associated to a Lie groupoid X, e.g. an orbifold, and the resulting Hom-stack $\mathcal{H}om(M, \mathcal{X})$ for M a compact manifold.

We define a *minimal cover* of a manifold M to be a cover by regular closed sets V_i such that the interiors V_i° form an open cover of M, and every V_i° contains a point not in any other V_j°. We also ask that finite intersections $V_i \cap \ldots \cap V_k$ are also regular closed. Denote the collection of minimal covers of a manifold M by $C(M)_{min}$, and note that such covers are cofinal in open covers. Recall that a cover V of a manifold defines a diffeological groupoid $\check{C}(V)$ with objects $\coprod_i V_i$ and arrows $\coprod_{i,j} V_i \cap V_j$.[1]
We are particularly interested in the case when we take the closure $\{\overline{U_i}\}$ of $\{U_i\}$, a good open cover, minimal in the above sense.

We denote the arrow groupoid of a Lie groupoid X by X^2—it is again a Lie groupoid and comes with functors $S, T: X^2 \to X$, with object components source and target, resp. Let M be a compact manifold with corners and X a Lie groupoid. Define the *mapping groupoid* $\mathrm{Map}(M, X)$ to be the following diffeological groupoid. The object space $\mathrm{Map}(M, X)_0$ is the disjoint union over minimal covers V of the spaces $X^{\check{C}(V)}$ of functors $\check{C}(V) \to X$. The arrow space $\mathrm{Map}(M, X)_1$ is

$$\coprod_{V_1, V_2 \in C(M)_{min}} X^{\check{C}(V_1)} \times_{X^{\check{C}(V_{12})}} (X^2)^{\check{C}(V_{12})} \times_{X^{\check{C}(V_{12})}} X^{\check{C}(V_2)}$$

where the chosen minimal refinement $V_{12} \subset V_1 \times_M V_2$ is defined using the boolean product on the algebra of regular closed sets. The maps

$$S, T: (X^2)^{\check{C}(V_{12})} \to X^{\check{C}(V_{12})} \qquad \text{and} \qquad X^{\check{C}(V_i)} \to X^{\check{C}(V_{12})} \quad (i = 1, 2) \tag{1}$$

give us a pullback and the two projections

$$X^{\check{C}(V_1)} \times_{X^{\check{C}(V_{12})}} (X^2)^{\check{C}(V_{12})} \times_{X^{\check{C}(V_{12})}} X^{\check{C}(V_2)} \longrightarrow X^{\check{C}(V_i)}, \tag{2}$$

---

[1] We can in what follows safely ignore the issue of intersections of boundaries.

induce, for $i = 1, 2$, the source and target maps for our groupoid resp. Composition in the groupoid is subtle, but is an adaptation of the composition of transformations of anafunctors given in [7]. The proof of the following theorem works exactly as in Theorem 4.2 in [8].

Theorem 2 *For X a Lie groupoid and M a compact manifold the Hom-stack $\mathcal{H}om(M, X)$ is presented by the diffeological groupoid $\mathrm{Map}(M, X)$.* □

We need some results that ensure the above constructions give Fréchet manifolds.

Proposition 3 *For M a compact smooth Riemannian manifold (possibly with corners), K a compact regular closed Lipschitz subset of M and N a smooth manifold, the induced restriction map $N^M \to N^K$ is a submersion of Fréchet manifolds.*

Proof Recall that a submersion of Fréchet manifolds is a smooth map that is locally, for suitable choices of charts, a projection out of a direct summand. This means we have to work locally in charts and show that we have a split surjection of Fréchet spaces. We can reduce this to the case that $N = \mathbb{R}^n$, since the charts are given by spaces of sections of certain vector bundles, and we can consider these spaces locally and patch them together, and thence to $N = \mathbb{R}$. The proof then uses [5, Theorem 3.15], as we can work in charts bi-Lipschitz to flat $\mathbb{R}^n$, hence reduce to the case of $K \subset \bar{B} \subset \mathbb{R}^n$, for B some large open ball. □

In particular this is true for sets K that are closures of open geodesically convex sets, and even more specifically such open sets that are the finite intersections of geodesically convex charts in a good open cover. We also use a special case of Stacey's theorem [9, Corollary 5.2]; smooth manifolds with corners are smoothly $\mathcal{T}$-compact spaces in Stacey's sense.

Theorem 4 (Stacey) *Let $N_1 \to N_2$ be a submersion of finite-dimensional manifolds and K a compact manifold, possibly with corners. Then the induced map of Fréchet manifolds $N_1^K \to N_2^K$ is a submersion.* □

The following proposition is the main technical tool in proving the mapping stack is an infinite-dimensional differentiable stack.

Proposition 5 *The diffeological space $X^{\check{C}(V)}$ is a Fréchet manifold.*

Proof First, the diffeological space of functors is isomorphic to the space of simplicial maps $N\check{C}(V) \to NX$ between the nerves of the groupoids. Then, since the subspaces of degenerate simplicies in $N\check{C}(V)$ are disjoint summands, we can remove those, and consider semi-simplicial maps between semisimplicial diffeological spaces instead. Then, since inverses in $\check{C}(V)$ are also disjoint, we can remove those as well, and consider the diffeological space of semisimplicial maps from the 'nerve' of the smooth irreflexive partial order $\check{C}^<(V)$ to the nerve of X, considered as a semisimplicial space (where we have chosen an arbitrary total ordering on the finite minimal cover V). *This* diffeological space is what we show is a Fréchet manifold, by carefully writing the limit as an iterated pullback of diagrams involving maps that are guaranteed to be submersions by Proposition 3 and Theorem 4, and using the fact that X is appropriately coskeletal, i.e. $(NX)_n = X_1 \times_{X_0} \ldots \times_{X_0} X_1$. The original space of functors is then a diffeological space isomorphic to this Fréchet manifold, hence is a Fréchet manifold.   □

Lemma 6 *Let $X \to Y$ be a functor between Lie groupoids with object and arrow components submersions, and $V_1 \to V_2$ a refinement of minimal covers. Then the induced map $X^{\check{C}(V_2)} \to Y^{\check{C}(V_1)}$ is a submersion between Fréchet manifolds.* □

We will consider the special cases that the functor $X \to Y$ is the identity, and also that the refinement $V_1 \to V_2$ is the identity.

Theorem 7 *For a Lie groupoid X and compact manifold M, $\mathrm{Map}(M, X)$ is a Fréchet–Lie groupoid.*

Proof The object space $\mathrm{Map}(M, X)_0$ is a manifold by Proposition 5. The arrow space $\mathrm{Map}(M, X)_1$ is a manifold since it is given by a pullback diagram built with the maps (1), which are submersions by Lemma 6. The identity map is smooth, as it is a smooth map between diffeological spaces that happen to be manifolds, and so is composition.   □

The following theorem is the first main result of the paper. The proof uses the technique of [6, Theorem 4.2] as adapted in [8], where it is shown that all of the constructions remain smooth.

Theorem 8 *For a Lie groupoid X and compact manifold M, the stack $\underline{\mathcal{H}\mathrm{om}}(M, \mathcal{X})$ is weakly presented by the Fréchet–Lie groupoid $\mathrm{Map}(M, X)$.* □

A *weak* presentation means that the pullback of the map $\mathrm{Map}(M, X)_0 \to \underline{\mathcal{H}\mathrm{om}}(M, \mathcal{X})$ against itself gives a stack representable by the Fréchet manifold $\mathrm{Map}(M, X)_1$, and the two projections are submersions. For the site of diffeological spaces, a weak presentation is an ordinary presentation. This is also the case if we allow non-Hausdorff manifolds [2, Proposition 2.2], so we either have to pay the price of a weak presentation or working over a site of non-Hausdorff manifolds. If the groupoid $\mathrm{Map}(M, X)$ is *proper*, as in Theorem 10 below, then we can upgrade this weak presentation to an ordinary one over Hausdorff manifolds.

Definition 9 A (Fréchet–)Lie groupoid Z is *proper* if the map $(s, t)\colon Z_1 \to Z_0 \times Z_0$ is a proper map (i.e. closed with compact fibres), *étale* if the source and target maps are local diffeomorphisms, and an *orbifold groupoid* if it is a proper and étale.

It is a theorem of Moerdijk–Pronk that orbifold groupoids are equivalent to orbifolds defined in terms of orbifold charts—in finite dimensions—if we ignore issues of effectivity (which we do, for now). For infinite dimensional orbifolds this is not yet known, but may be possible for Fréchet–Lie groupoids with local additions on their object and arrow manifolds, of which our construction is an example.

Our second main result is then:

Theorem 10 *If X is an étale Lie groupoid, then $\mathrm{Map}(M, X)$ is étale. If X is an orbifold groupoid, then $\mathrm{Map}(M, X)$ is an orbifold groupoid.*

Proof Stability of local diffeomorphisms under pullback mean that we only need to show that the smooth maps $S^{\check{C}(V)}, T^{\check{C}(V)}\colon (X^2)^{\check{C}(V)} \to X^{\check{C}(V)}$, for any minimal cover V, are local diffeomorphisms. If X is an étale Lie groupoid then the fibres of its source and target maps are discrete, and one can show that $S^{\check{C}(V)}$, $T^{\check{C}(V)}$ have discrete diffeological spaces as fibres. But these maps are submersions of Fréchet manifolds, hence are local diffeomorphisms.

Properness follows if we can show that (s, t) for the mapping groupoid is closed and every object has a finite automorphism group. This reduces to showing that $(S, T)^{\check{C}(V)}$ is closed and its fibres are finite. We can show the latter by again working in the diffeological category and showing that the fibres of $(S, T)^{\check{C}(V)}$ are discrete, and also a subspace of a finite diffeological space. As all the spaces involved are metrisable, we use a sequential characterisation of closedness together with the local structure of the proper étale groupoid X, and find an appropriate convergent subsequence in the required space of natural transformations. □

Acknowledgements This research was supported under the Australian Research Council's Discovery Projects funding scheme (project numbers DP120100106 and DP130102578).

References

1. Baez, J.C., Hoffnung, A.: Convenient categories of smooth spaces. Trans. Am. Math. Soc. **363**(11), 5789–5825 (2011). arXiv:0807.1704
2. Behrend, K., Xu, P.: Differentiable stacks and gerbes. J. Symplectic Geom. **9**(3), 285–341 (2011). arXiv:math/0605694
3. Borzellino, J.E., Brunsden, V.: The stratified structure of spaces of smooth orbifold mappings. Commun. Contemp. Math. **15**(5), 1350018, 37 (2013). arXiv:0810.1070
4. Chen, W.: On a notion of maps between orbifolds I. Function spaces. Commun. Contemp. Math. **8**(5), 569–620 (2006). arXiv:math/0603671
5. Frerick, L.: Extension operators for spaces of infinite differentiable Whitney jets. J. Reine Angew. Math. **602**, 123–154 (2007). https://doi.org/10.1515/CRELLE.2007.005
6. Noohi, B.: Mapping stacks of topological stacks. J. Reine Angew. Math. **646**, 117–133 (2010). arXiv:0809.2373
7. Roberts, D.M.: Internal categories, anafunctors and localisation. Theory Appl. Categ. **26**(29), 788–829 (2012). arXiv:1101.2363
8. Roberts, D.M., Vozzo, R.F.: Smooth loop stacks of differentiable stacks and gerbes (2016). Preprint. arXiv:1602.07973
9. Stacey, A.: Yet more smooth mapping spaces and their smoothly local properties (2013). Preprint. arXiv:1301.5493
10. Weinmann, T.: Orbifolds in the framework of Lie groupoids. Ph.D. thesis, ETH Zürich (2007). https://doi.org/10.3929/ethz-a-005540169

Complicial Sets, an Overture

Emily Riehl

Abstract The aim of these notes is to introduce the intuition motivating the notion of a *complicial set*, a simplicial set with certain marked "thin" simplices that witness a composition relation between the simplices on their boundary. By varying the marking conventions, complicial sets can be used to model (∞, n)-categories for each $n \geq 0$, including $n = \infty$. For this reason, complicial sets present a fertile setting for thinking about weak infinite dimensional categories in varying dimensions. This overture is presented in three acts: the first introducing simplicial models of higher categories; the second defining the Street nerve, which embeds strict ω-categories as *strict* complicial sets; and the third exploring an important saturation condition on the marked simplices in a complicial set and presenting a variety of model structures that capture their basic homotopy theory. Scattered throughout are suggested exercises for the reader who wants to engage more deeply with these notions.

1 Introduction

As the objects that mathematicians study increase in sophistication, so do their natural habitats. On account of this trend, it is increasingly desirable to replace mere 1-categories of objects and the morphisms between them, with infinite-dimensional categories containing 2-morphisms between 1-morphisms, 3-morphisms between 2-morphisms, and so on. The principle challenge in working with infinite-dimensional categories is that the naturally occurring examples are *weak* rather than *strict*, with composition of n-morphisms only associative and unital up to an $n + 1$-morphism that is an "equivalence" in some sense. The complexity is somewhat reduced in the case of (∞, n)-*categories*, in which all k-morphisms are weakly invertible for

E. Riehl (✉)
Johns Hopkins University, 3400 N Charles Street, Baltimore, MD, USA
e-mail: eriehl@math.jhu.edu

© Springer International Publishing AG, part of Springer Nature 2018
D.R. Wood et al. (eds.), *2016 MATRIX Annals*, MATRIX Book Series 1,
https://doi.org/10.1007/978-3-319-72299-3_4

$k > n$, but even in this case, explicit models of these schematically defined (∞, n)-categories can be extremely complicated.

Complicial sets provide a relatively parsimonious model of infinite-dimensional categories, with special cases modeling $(\infty, 0)$-categories (also called ∞-*groupoids*), $(\infty, 1)$-categories (the ubiquitous ∞-*categories*), indeed (∞, n)-categories for any n, and also including the general case of (∞, ∞)-categories. Unlike other models of infinite-dimensional categories, the definition of a complicial set is extremely simple to state: it is a simplicial set with a specified collection of marked "thin" simplices, in which certain elementary anodyne extensions exist. These anodyne extensions provide witnesses for a weak composition law and guarantee that the thin simplices are equivalences in a sense defined by this weak composition.

This overture is dividing into three acts, each comprising one part of the 3-h mini course that generated these lecture notes. In the first, we explore how a simplicial set can be used to model the weak composition of an $(\infty, 1)$-category and consider the extra structure required to extend these ideas to provide a simplicial model of $(\infty, 2)$-categories. This line of inquiry leads naturally to the definition of a a complicial set as a *stratified* (read "marked") simplicial set in which composable simplices admit composites.

In the second part, we delve into the historical motivations for this model for higher categories based on stratified simplicial sets. John Roberts proposed the original definition of *strict* complicial sets, which admit unique extensions along the elementary anodyne inclusions, as a conjectural model for strict ω-categories [6]. Ross Street defined a nerve functor from ω-categories into simplicial sets [7], and Dominic Verity proved that it defines a full and faithful embedding into the category of stratified simplicial sets whose essential image is precisely the strict complicial sets [9]. While we do not have the space to dive into proof of this result here, we nonetheless describe the Street nerve in some detail as it is an important source of examples of both strict and also weak complicial sets, as is explained in part three.

In the final act, we turn our attention to those complicial sets that most accurately model (∞, n)-categories. Their markings are *saturated*, in the sense that every simplex that behaves structurally like an equivalence, is marked. We present a variety of model structures, due to Verity, that encode the basic homotopy theory of complicial sets of various flavors, including those that are *n-trivial*, with every simplex above dimension n marked, and saturated. The saturation condition is essential for a conjectural equivalence between the complicial sets models of (∞, n)-categories and other models known to satisfy the axiomatization of Barwick–Schommer–Pries [1], which passes through a complicial nerve functor due to Verity. This result will appear in a future paper.

2 Introducing Complicial Sets

Infinite dimensional categories have morphisms in each dimension that satisfy a
weak composition law, which is associative and unital up to higher-dimensional
morphisms rather than on the nose. There is no universally satisfactory definition of
"weak composition"; instead a variety of models of infinite-dimensional categories
provide settings to work with this notion.

A *complicial set*, nee. *weak complicial set*, is a *stratified* simplicial set, with
a designed subset of "thin" marked simplices marked, that admits extensions
along certain maps. Complicial sets model weak infinite-dimensional categories,
sometimes called (∞, ∞)-*categories*. By requiring all simplices above a fixed
dimension to be thin, they can also model (∞, n)-categories for all $n \in [0, \infty]$.

Strict complicial sets were first defined by Roberts [6] with the intention of
constructing a simplicial model of strict ω-categories. He conjectured that it should
be possible to extend the classical nerve to define an equivalence from the category
of strict ω-categories to the category of strict complicial sets. Street defined this
nerve [7], providing a fully precise statement of what is known as the Street–Roberts
conjecture, appearing as Theorem 3.1. Verity proved the Street–Roberts conjecture
[9] and then subsequently defined and developed the theory of the weak variety of
complicial sets [8, 10] that is the focus here.

We begin in Sect. 2.1 by revisiting how a quasi-category (an unmarked simplicial
set) models an $(\infty, 1)$-category. This discussion enables us to explore what would be
needed to model an $(\infty, 2)$-category as a simplicial set in Sect. 2.2. These excursions
motivate the definition of stratified simplicial sets in Sect. 2.3 and then complicial
sets in Sect. 2.4. We conclude in Sect. 2.5 by defining *n-trivial* complicial sets,
which, like (∞, n)-categories, have non-invertible simplices concentrated in low
dimensions.

We assume the reader has some basic familiarity with the combinatorics of
simplicial sets and adopt relatively standard notations, e.g., $\Delta[n]$ for the standard
n-simplex and $\Lambda^k[n]$ for the horn formed by those faces that contain the kth vertex.

2.1 Quasi-Categories as $(\infty, 1)$-Categories

The most popular model for $(\infty, 1)$-categories were first introduced by Michael
Boardman and Rainer Vogt under the name *weak Kan complexes* [2].

Definition 2.1 A **quasi-category** is a simplicial set A so that every inner horn
admits a filler

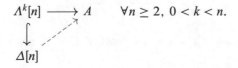

$$\Lambda^k[n] \longrightarrow A \qquad \forall n \geq 2, \ 0 < k < n.$$

This presents an $(\infty, 1)$-category with:

- A_0 as the set of objects;
- A_1 as the set of 1-cells with sources and targets determined by the face maps

$$A_1 \underset{d_0 = \text{target}}{\overset{d_1 = \text{source}}{\rightrightarrows}} A_0$$

and degenerate 1-simplices serving as identities;
- A_2 as the set of 2-cells;
- A_3 as the set of 3-cells, and so on.

The weak 1-category structure arises as follows. A 2-simplex

$$
\begin{array}{c}
1 \\
{}^{f}\nearrow \quad \searrow^{g} \\
\alpha \\
0 \xrightarrow{\quad h \quad} 2
\end{array}
\tag{1}
$$

provides a witness that $h \simeq gf$.

Notation 2.2 *We adopt the convention throughout of always labeling the vertices of an n-simplex by $0, \ldots, n$ to help orient each picture. This notation does not assert that the vertices are necessarily distinct.*

A 3-simplex then

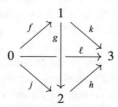

provides witnesses that $h(gf) \simeq hj \simeq \ell \simeq kf \simeq (hg)f$.

The **homotopy category** of a quasi-category A is the category whose objects are vertices and whose morphisms are a quotient of A_1 modulo the relation $f \simeq g$ that identifies a pair of parallel edges if and only if there exists a 2-simplex of the following form:

$$
\begin{array}{c}
1 \\
{}^{f}\nearrow \quad \searrow \\
\alpha \\
0 \xrightarrow{\quad g \quad} 2
\end{array}
\tag{2}
$$

Notation 2.3 *Here and elsewhere the notation "=" is used for degenerate simplices.*

Exercise 2.4 Formulate alternate versions of the relation $f \simeq g$ and prove that in a quasi-category each of these relations defines an equivalence relation and furthermore that these relations are all equivalent.

The composition operation witnessed by 2-simplices is not unique on the nose but it is unique up to the notion of homotopy just introduced.

Exercise 2.5 Prove that the homotopy category is a strict 1-category.

A quasi-category is understood as presenting an $(\infty, 1)$-category rather than an (∞, ∞)-category because each 2-simplex is invertible up to a 3-simplex, and each 3-simplex is invertible up to a 4-simplex, and so on, in a sense we now illustrate. First consider a 2-simplex as in (2). This data can be used to define a horn $\Lambda^1[3] \to A$ whose other two faces are degenerate

which can be filled to define a "right inverse" β in the sense that this pair of 2-cells bound a 3-simplex with other faces degenerate. Similarly, there is a "special outer horn"[1] $\Lambda^3[3] \to A$

which can be filled to define a "left inverse" γ. In this sense, α is an *equivalence* up to 3-simplices, admitting left and right inverses along the boundary of a pair of three simplices.

This demonstrates that 2-simplices with a degenerate outer edge admit left and right inverses, but what if α has the form (1)? In this case, we can define a horn $\Lambda^1[3] \to A$ horn

---

[1] Special outer horns $\Lambda^0[n] \to A$ and $\Lambda^n[n] \to A$ have first or last edges mapping to 1-*equivalences* (such as degeneracies) in A, as introduced in Definition 4.1 below.

whose 3rd face is constructed by filling a horn $\Lambda^1[2] \to A$. In this sense, any 2-simplex is equivalent to one with last (or dually first) edge degenerate.

Exercise 2.6 Generalize this argument to show that the higher-dimensional simplices in a quasi-category are also weakly invertible.

2.2 Towards a Simplicial Model of $(\infty, 2)$-Categories

Having seen how a simplicial set may be used to model an $(\infty, 1)$-category, it is natural to ask how a simplicial set might model an $(\infty, 2)$-category. A reasonable idea would be interpret the 2-simplices as inhabited by not necessarily invertible 2-cells pointing in a consistent direction. The problem with this is that the 2-simplices need to play a dual role: they must also witness composition of 1-simplices, in which case it does not make sense to think of them as inhabited by non-invertible cells. The idea is to mark as "thin" the witnesses for composition and then demand that these marked 2-simplices behave as 2-dimensional equivalences in a sense that can be intuited from the preceding three diagrams.

Then 3-simplices can be thought of as witnesses for composition of not-necessarily thin 2-simplices. For instance, given a pair of 2-simplices α and β with boundary as displayed below, the idea is to build a $\Lambda^2[3]$-horn

whose 0th face is a thin filler of the $\Lambda^1[2]$-horn formed by g and k. The 2nd face, defined by filling the horn $\Lambda^2[3]$-horn, defines a composite 2-simplex, which is witnessed by the (thin) 3-simplex. Note that because the 0th face is thin, its 1st edge is interpreted as a composite kg of g and k, which is needed so that the boundary of the new 2-cell agrees with the boundary of the pasted composite of β and α. Since the 3-simplex should be thought of as a witness to a composition relation involving the 2-simplices that make up its boundary, the three simplex should also be regarded as "thin."

A similar $\Lambda^1[3]$-horn can be used to define composites where the domain of α is the last, rather than the first, edge of the codomain of β. It is in this way that simplicial sets with certain marked simplices are used to model $(\infty, 2)$-categories or indeed (∞, n)-categories for any $n \in [0, \infty]$. We now formally introduce *stratified simplicial sets* before stating the axioms that define these *complicial sets*.

2.3   Stratified Simplicial Sets

We have seen that for a simplicial set to model an infinite-dimensional category with non-invertible morphisms in each dimension, it should have a distinguished set of "thin" n-simplices witnessing composition of $(n-1)$-simplices. Degenerate simplices are always thin in this sense. Furthermore, the intuition that the "thin" simplices are the equivalences, in a sense that is made precise in Sect. 4, suggests that certain 1-simplices might also be marked as thin. This motivates the following definition:

Definition 2.7   A **stratified simplicial set** is a simplicial set with a designated subset of **marked** or **thin** positive-dimensional simplices that includes all degenerate simplices. A map of stratified simplicial sets is a simplicial map that preserves thinness.

Notation 2.8   *The symbol "$\simeq$" is used throughout to decorate thin simplices.*
There are left and right adjoints

$$\mathsf{Strat} \; \underset{\underset{(-)^{\sharp}}{\overset{(-)^{\flat}}{\underset{\longleftarrow}{\overset{\longleftarrow}{\underset{\bot}{\overset{\bot}{-U\to}}}}}}{} \; \mathsf{sSet}$$

to the forgetful functor from stratified simplicial sets to ordinary simplicial sets, both of which are full and faithful. The left adjoint assigns a simplicial set its **minimal stratification**, with only degenerate simplices marked, while the right adjoint assigns the *maximal stratification*, marking all simplices. When a simplicial set is regarded as a stratified simplicial set, the default convention is to assign the minimal stratification, with the notation "$(-)^{\flat}$" typically omitted.

Definition 2.9   An inclusion $U \hookrightarrow V$ of stratified simplicial sets is:

- **regular**, denoted $U \hookrightarrow_r V$, if thin simplices in U are created in V (a simplex is thin in U if and only if its image in V is thin); and
- **entire**, denoted $U \hookrightarrow_e V$, if the map is the identity on underlying simplicial sets (in which case the only difference between U and V is that more simplices are marked in V).

A standard inductive argument, left to the reader, proves:

Proposition 2.10   *The monomorphisms in* Strat *are generated under pushout and transfinite composition by*

$$\{\partial\Delta[n] \hookrightarrow_r \Delta[n] \mid n \geq 0\} \cup \{\Delta[n] \hookrightarrow_e \Delta[n]_t \mid n \geq 1\},$$

where the top-dimensional n-simplex in $\Delta[n]_t$ *is thin.*

Exercise 2.11   Prove this.

2.4 Complicial Sets

A stratified simplicial set is a simplicial set with enough structure to talk about composition of simplices. A complicial set is a stratified simplicial set in which composites exist and in which thin witnesses to composition compose to thin simplices, an associativity condition that will also play a role in establishing their equivalence-like nature. The following form of the definition of a (weak) *complicial set*, due to Verity [10], modifies an earlier equivalent presentation due to Street [7]. Verity's modification focuses on a particular set of *k-admissible n-simplices*, thin *n*-simplices that witness that the *k*th face is a composite of the $(k + 1)$th and $(k - 1)$th simplices.

Definition 2.12 (*k*-**Admissible** *n*-**Simplex**) The *k*-**admissible** *n*-**simplex** is the entire superset of the standard *n*-simplex with certain additional faces marked thin: a non-degenerate *m*-simplex in $\Delta^k[n]$ is thin if and only if it contains all of the vertices $\{k - 1, k, k + 1\} \cap [n]$. Thin faces include:

- the top dimensional *n*-simplex
- all codimension-one faces except for the $(k - 1)$th, *k*th, and $(k + 1)$th
- the 2-simplex spanned by $[k - 1, k, k + 1]$ when $0 \leq k \leq n$ or the edge spanned by $[k - 1, k, k + 1] \cap [n]$ when $k = 0$ or $k = n$.

Definition 2.13 A **complicial set** is a stratified simplicial set that admits extensions along the **elementary anodyne extensions**, which are generated under pushout and transfinite composition by the following two sets of maps:

i. The **complicial horn extensions**

$$\Lambda^k[n] \hookrightarrow_r \Delta^k[n] \quad \text{for} \quad n \geq 1, \ 0 \leq k \leq n$$

are regular inclusions of *k*-**admissible** *n*-**horns**. An inner admissible *n*-horn parametrizes "admissible composition" of a pair of $(n - 1)$-simplices. The extension defines a composite $(n - 1)$-simplex together with a thin *n*-simplex witness.

$$
\begin{array}{ccc}
\Lambda^k[n] & \longrightarrow & A \\
\downarrow & \nearrow & \\
\Delta^k[n] & &
\end{array}
\tag{3}
$$

ii. The **complicial thinness extensions**

$$\Delta^k[n]' \hookrightarrow_e \Delta^k[n]'' \quad \text{for} \quad n \geq 1, \ 0 \leq k \leq n,$$

are entire inclusions of two entire supersets of $\Delta^k[n]$. The stratified simplicial set $\Delta^k[n]'$ is obtained from $\Delta^k[n]$ by also marking the $(k - 1)$th and $(k + 1)$th faces, while $\Delta^k[n]''$ has all codimension-one faces marked. This extension problem

$$
\begin{array}{ccc}
\Delta^k[n]' & \longrightarrow & A \\
\downarrow & \nearrow & \\
\Delta^k[n]'' & &
\end{array}
$$

(4)

demands that whenever the composable pair of simplices in an admissible horn are thin, then so is any composite.

Definition 2.14 A **strict complicial set** is a stratified simplicial set that admits unique extensions along the elementary anodyne extensions (3) and (4).

Example 2.15 (Complicial Horn Extensions) To gain familiarity with the elementary anodyne extensions, let us draw the complicial horn extensions in low dimensions, using red to depict simplices present in the codomain but not the domain and "$\simeq$" to decorate thin simplices. The labels on the simplices are used to suggest the interpretation of certain data as composites of other data, but recall that in a (non-strict) complicial set there is no single simplex designated as *the* composite of an admissible pair of simplices. Rather, the fillers for the complicial horn extensions provide *a* composite and a witness to that relation.

- $\Lambda^1[2] \hookrightarrow_r \Delta^1[2]$

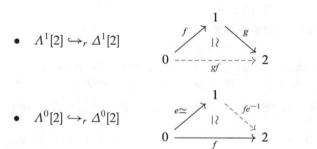

- $\Lambda^0[2] \hookrightarrow_r \Delta^0[2]$

- $\Lambda^2[3] \hookrightarrow_r \Delta^2[3]$

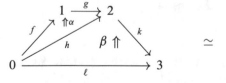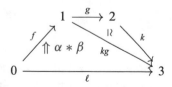

- $\Lambda^0[3] \hookrightarrow_r \Delta^0[3]$

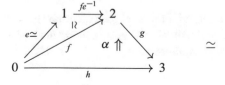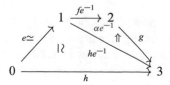

- For $\Lambda^2[4] \hookrightarrow_r \Delta^2[4]$ the non-thin codimension-one faces in the horn define the two 3-simplices with a common face displayed on the left, while their composite is a 3-simplex as displayed on the right.

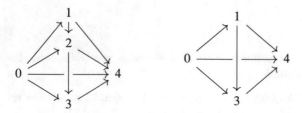

It makes sense to interpret the right hand simplex, the 2nd face of the 2-admissible 4-simplex, as a composite of the 3rd and 1st faces because the 2-simplex

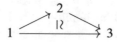

is thin.

2.5 n-Trivialization and the n-Core

We now introduce the complicial analog of the condition that an (∞, ∞)-category is actually an (∞, n)-category, in which each r-cell with $r > n$ is weakly invertible.

Definition 2.16 A stratified simplicial set X is n-**trivial** if all r-simplices are marked for $r > n$.

The full subcategory of n-trivial stratified simplicial sets is reflective and coreflective

$$\mathsf{Strat}_{n\text{-}tr} \xleftrightarrow[\;\;\;\;\;\;\;]{\;\;\;\;\;\;\;} \mathsf{Strat}$$

in the category of stratified simplicial sets. That is n-**trivialization** defines an idempotent monad on Strat with unit the entire inclusion

$$X \hookrightarrow_e \mathrm{tr}_n X$$

of a stratified simplicial set X into the stratified simplicial set $\mathrm{tr}_n X$ with the same marked simplices in dimensions $1, \ldots, n$, and with all higher simplices "made thin." A complicial set is n-**trivial** if this map is an isomorphism.

The n-**core** $\mathrm{core}_n X$, defined by restricting to those simplices whose faces above dimension n are all thin in X, defines an idempotent comonad with counit the regular inclusion

$$\mathrm{core}_n X \hookrightarrow_r X.$$

Again, a complicial set is n-trivial just when this map is an equivalence. As is always the case for a monad-comonad pair arising in this way, these functors are adjoints: $\mathrm{tr}_n \dashv \mathrm{core}_n$.

The subcategories of n-trivial stratified simplicial sets assemble to define a string of inclusions with adjoints

$$
\mathrm{sSet} \xrightarrow[\cong]{(-)^\sharp} \mathrm{Strat}_{0\text{-tr}} \underset{\underset{\mathrm{core}_1}{\xleftarrow{\quad\bot\quad}}}{\overset{\overset{\mathrm{tr}_1}{\xleftarrow{\quad}}}{\xrightarrow{\ \bot\ }}} \mathrm{Strat}_{1\text{-tr}} \ \cdots \ \mathrm{Strat}_{(n-1)\text{-tr}} \underset{\underset{\mathrm{core}_{n-1}}{\xleftarrow{\quad\bot\quad}}}{\overset{\overset{\mathrm{tr}_{n-1}}{\xleftarrow{\quad}}}{\xrightarrow{\ \bot\ }}} \mathrm{Strat}_{n\text{-tr}} \cdots \mathrm{Strat}
$$

that filter the inclusion of simplicial sets, considered as maximally marked stratified simplicial sets, into the category of all stratified simplicial sets.

Exercise 2.17 Show that the two right adjoints restrict to complicial sets to define functors that model the inclusion of $(\infty, n - 1)$-categories into (∞, n)-categories and its right adjoint, which takes an (∞, n)-category to the "groupoid core," an $(\infty, n - 1)$-category.

Remark 2.18 By contrast, the left adjoint, which just marks things arbitrarily, does not preserve complicial structure; this construction is too naive to define the "freely invert n-arrows" functor from (∞, n)-categories to $(\infty, n - 1)$-categories.[2]

Exercise 2.19 Show that a 0-trivial complicial set is exactly a Kan complex with the maximal "$(-)^\sharp$" marking.

Exercise 2.20 Prove that the underlying simplicial set of any 1-trivial complicial set is a quasi-category.

Conversely, any quasi-category admits a stratification making it a complicial set. The markings on the 1-simplices cannot be arbitrarily assigned. At minimum, certain automorphisms (endo-simplices that are homotopic to identities) must be marked. More to the point, each edge that is marked necessarily defines an equivalence in the quasi-category. But it is not necessary to mark all of the equivalences.

---

[2]For instance, if A is a naturally marked quasi-category, that is 1-trivial, then its zero trivialization is not a Kan complex (because we have not changed the underlying simplicial set) but its groupoid core is (by a theorem of Joyal).

Example 2.21 Strict n-categories define n-trivial *strict* complicial sets, with unique fillers for the admissible horns, via the Street nerve, which is the subject of the next section.

In the third part of these notes, we argue that the complicial sets that most closely model (∞, n)-categories are the n-trivial *saturated* complicial sets, in which all *equivalences* are marked. In the case of an n-trivial stratification, the equivalences are canonically determined by the structure of the simplicial set. One bit of evidence for the importance of the notion of saturation discussed below is the fact that the category of quasi-categories is isomorphic to the category of saturated 1-trivial complicial sets (Example 4.19).

3 The Street Nerve of an ω-Category

The *Street nerve* is a functor

$$N: \omega\text{-Cat} \to \mathsf{sSet}$$

from strict ω-categories to simplicial sets. As is always the case for nerve constructions, the Street nerve is determined by a functor

$$\mathscr{O}: \Delta \to \omega\text{-Cat}.$$

In this case, the image of $[n] \in \Delta$ is the nth *oriental* $\mathscr{O}_n$, a strict n-category defined by Street [7]. The nerve of a strict ω-category C is then defined to be the simplicial set whose n-simplices

$$NC_n := \hom(\mathscr{O}_n, C)$$

are ω-functors $\mathscr{O}_n \to C$. There are various ways to define a stratification on the nerve of an ω-category, defining a lift of the Street nerve to a functor valued in stratified simplicial sets. One of these marking conventions turns Street nerves of strict ω-categories into strict complicial sets, and indeed all strict complicial sets arise in this way. This is the content of the Street–Roberts conjecture, proven by Verity, which motivated the definition of strict complicial sets.

Theorem 3.1 (Verity) *The Street nerve defines a fully faithful embedding*

$$\omega\text{-Cat} \xhookrightarrow{\ N\ } \mathsf{Strat}$$

of ω-categories into stratified simplicial sets, where an n-simplex $\mathscr{O}_n \to C$ in NC is marked if and only if it carries the top dimensional n-cell on $\mathscr{O}_n$ to an identity in C. The essential image is the category of strict complicial sets.

In Sect. 3.1, we introduce strict ω-categories, and then in Sect. 3.2 we introduce the orientals. In Sect. 3.3, we then define the Street nerve and revisit the Street–Roberts conjecture, though we leave the details of its proof to [9]. At the conclusion of this section, we look ahead to Sect. 4.1, which explores other marking conventions for Street nerves of strict n-categories. In this way, the Street nerve provides an important source of examples of weak, as well as strict, complicial sets. These are obtained by marking the equivalences and not just the identities in NC, the consideration of which leads naturally to the notion of saturation in a complicial set, which is a main topic for the final section of these notes.

3.1 ω-Categories

Street's "The algebra of oriented simplexes" [7] gives a single-sorted definition of a (strict) n-category in all dimensions $n = 1, \ldots, \omega$. In the single-sorted definition of a 1-category, an object is identified with its identity morphism, and these 0-cells are recognized among the set of 1-cells as the fixed points for the source and target maps.

Definition 3.2 A 1-**category** $(C, s, t, *)$ consists of

- a set C of **cells**
- functions $s, t: C \rightrightarrows C$ so that $ss = ts = s$ and $tt = st = t$ (a target or source has itself as its target and its source).
- a function $*: C \times_C C \to C$ from the pullback of s along t to C so that $s(a * b) = s(b)$ and $t(a * b) = t(a)$ (the source of a composite is the source of its first cell and the target is the target of the second cell).

and so that

- $s(a) = t(v) = v$ implies $a * v = a$ (right identity)
- $u = s(u) = t(a)$ implies $u * a = a$ (left identity)
- $s(a) = t(b)$ and $s(b) = t(c)$ imply $a * (b * c) = (a * b) * c$ (associativity).

The **objects** or 0-**cells** are the fixed points for s and then also for t and conversely.

Definition 3.3 A 2-**category** $(C, s_0, t_0, *_0, s_1, t_1, *_1)$ consists of two 1-categories

$$(C, s_0, t_0, *_0) \quad \text{and} \quad (C, s_1, t_1, *_1)$$

so that

- $s_1 s_0 = s_0 = s_0 s_1 = s_0 t_1, t_0 = t_0 s_1 = t_0 t_1$ (globularity plus 1-sources and 1-targets of points are points)
- $s_0(a) = t_0(b)$ implies $s_1(a *_0 b) = s_1(a) *_0 s_1(b)$ and $t_1(a *_0 b) = t_1(a) *_0 t_1(b)$ (1-cell boundaries of horizontal composites are composites).

- $s_1(a) = t_1(b)$ and $s_1(a') = t_1(b')$ and $s_0(a) = t_0(a')$ imply that

$$(a *_1 b) *_0 (a' *_1 b') = (a *_0 a') *_1 (b *_0 b')$$

(middle four interchange).

Identities for $*_0$ are 0-**cells** and identities for $*_1$ are 1-**cells**.

Definition 3.4 An ω^+-**category**[3] consists of 1-categories $(C, s_n, t_n, *_n)$ for each $n \in \omega$ so that $(C, s_m, t_m, *_m, s_n, t_n, *_n)$ is a 2-category for each $m < n$. The identities for $*_n$ are n-**cells**. An ω^+-**functor** is a function that preserves sources, targets, and composition for each n.

An ω-**category** is an ω^+-category in which every element is a **cell**, an n-cell for some n. Every ω^+-category has a maximal sub ω-category of cells and all of the constructions described here restrict to ω-categories.

An n-**category** is an ω-category comprised of only n-cells. This means that the 1-category structures $(C, s_m, t_m, *_m)$ for $m > n$ are all discrete.

Example 3.5 The underlying set functor ω^+-Cat $\to$ Set is represented by the **free** ω^+-**category** 2_ω **on one generator**,[4] whose underlying set is

$$(2 \times \omega) \cup \{\omega\}.$$

The element ω is the unique non-cell, while the objects $(0, n)$ and $(1, n)$ are n-cells, respectively the n-source and n-target of ω:

$$s_n(\omega) = (0, n) \quad \text{and} \quad t_n(\omega) = (1, n).$$

An m-cell is necessarily its own n-source and n-target for $m \leq n$; thus:

$$s_n(\epsilon, m) = t_n(\epsilon, m) = (\epsilon, m) \quad \text{for } m \leq n,$$

while:

$$s_n(\epsilon, m) = (0, n) \quad \text{and} \quad t_n(\epsilon, m) = (1, n) \quad \text{for } n < m.$$

The identity laws dictate all of the composition relations, e.g.:

$$\omega *_n (0, n) = \omega = (1, n) *_n \omega.$$

Using 2_ω one can define the **functor** ω^+-**category** $[A, B]$ for two ω^+-categories A and B: elements are ω^+-functors $A \times 2_\omega \to B$.

---

[3] Street called these "ω-categories" but we reserve this term for something else.
[4] In personal communication, Ross suggests that there may be something wrong with this example, but I do not see what it is.

Exercise 3.6 Work out the rest of the definition of the ω^+-category $[A, B]$ and prove that ω^+-Cat is cartesian closed.

Theorem 3.7 (Street) *There is an equivalence of categories*

$$(\omega^+\text{-Cat})\text{-Cat} \xrightarrow{\simeq} \omega^+\text{-Cat}$$

which restricts to define an equivalence

$$(n\text{-Cat})\text{-Cat} \xrightarrow{\simeq} (1 + n)\text{-Cat}$$

for each $n \in [0, \omega]$.[5]

Proof The construction of this functor is extends the construction of a 2-category from a Cat-enriched category. Let $\mathscr{C}$ be a category enriched in ω^+-categories. Define an ω^+-category C whose underlying set is

$$C := \sqcup_{u,v \in \mathrm{ob}\,\mathscr{C}}\,\mathscr{C}(u, v).$$

The 0-source and 0-target of an element $a \in \mathscr{C}(u, v)$ are u and v, respectively, and 0-composition is defined using the enriched category composition. The n-source,n-target, and n-composition are defined using the $(n - 1)$-category structure of the ω^+-category $\mathscr{C}(u, v)$.

Conversely, given an ω^+-category C, the associated ω^+-category enriched category $\mathscr{C}$ can be defined by taking the 0-cells of C as the objects of $\mathscr{C}$, defining $\mathscr{C}(u, v)$ to be the collection of elements with 0-source u and 0-target v, using the operations $(s_n, t_n, *_n)$ for $n > 0$ to define the ω^+-category structure on $\mathscr{C}(u, v)$.

3.2 Orientals

The nth *oriental* $\mathcal{O}_n$ is a strict n-category with a single n-cell whose source is the pasted composite of $(n - 1)$-cells, one for each of the odd faces of the simplex $\Delta[n]$, and whose target is a pasted composite of $(n-1)$-cells, one for each of the even faces of the simplex $\Delta[n]$. The orientals $\mathcal{O}_n$ can be recognized as full sub ω-categories of an ω-category $\mathcal{O}_\omega$, the free ω-category on the ω-simplex $\Delta[\omega]$, spanned by the objects that correspond to the vertices of $\Delta[n]$. The precise combinatorial definition of $\mathcal{O}_n$ is rather subtle to state, making use of Street's notion of *parity complex*, which we decline to introduce in general. Before defining the orientals as special cases of parity complexes, we first describe the low-dimensional cases.

---

[5]Recall that in ordinal arithmetic $1 + \omega = \omega$.

The orientals $\mathscr{O}_0, \mathscr{O}_1, \mathscr{O}_2, \ldots$ are ω-categories, where each $\mathscr{O}_n$ is an n-category. In low dimensions:

($n = 0$) $\mathscr{O}_0$ is the ω-category with a single 0-cell:

$$0$$

($n = 1$) $\mathscr{O}_1$ is the ω-category with two 0-cells 0, 1 and a 1-cell:

$$0 \longrightarrow 1$$

($n = 2$) $\mathscr{O}_2$ is the ω-category with three 0-cells $0, 1, 2$ and four 1-cells as displayed:

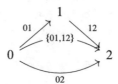

Note that only two of these are composable, with their composite the 1-cell denoted by $\{01, 12\}$. The underlying 1-category of $\mathscr{O}_2$ is the non-commutative triangle, the free 1-category generated by the ordinal [2].
There is a unique 2-cell

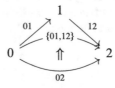

whose 0-source is 0 and whose 0-target is 2, and whose 1-source is 02 and whose 1-target is $\{01, 12\}$. We can simplify our pictures by declining to draw the free composites that are present in $\mathscr{O}_2$, as they must be in any ω-category. Under this simplifying convention, $\mathscr{O}_2$ is depicted as:

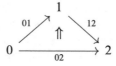

($n = 3$)   Similarly $\mathscr{O}_3$ has four 0-cells, abbreviated 0,1,2,3; has the free category on the graph [3] as its underlying 1-category, with six atomic 1-cells and five free composites; has four atomic 2-cells plus two composites; and has a 3-cell from one of these composites to the other. Under the simplifying conventions established above, $\mathscr{O}_3$ can be drawn as:

Definition 3.8 (The nth Oriental, Informally) The nth **oriental** is the strict n-category $\mathscr{O}_n$ whose atomic k-cells corresponding to the k-dimensional faces of $\Delta[n]$ (the non-degenerate k-simplices, which can be identified with $(k+1)$-element subsets of $[n]$). The codimension-one source of a k-cell is a pasted composite of the odd faces of the $\Delta[k]$-simplex, while the codimension-one target is a pasted composite of the even faces of the k-simplex.

If S is a subset of faces of $\Delta[n]$ write S^- for the union of the odd faces of simplices in S and write S^+ for the union of even faces of simplices in S. Write S_k for the k-dimensional elements of S and $|S|_k$ for the elements of dimension at most k.

Definition 3.9 (The nth Oriental, Precisely) The k-cells of the n-category $\mathscr{O}_n$ are pairs (M, P) where M and P are non-empty, **well-formed**, finite subsets of faces of $\Delta[n]$ of dimension at most k so that M and P both **move M to P**. Here a subset S of faces of $\Delta[n]$ is **well-formed** if it contains at most one vertex and if for any distinct elements $x \neq y$, x and y have no common sources and no common targets. A subset S **moves M to P** if

$$M = (P \cup S^-)\backslash S^+ \quad \text{and} \quad P = (M \cup S^+)\backslash S^-.$$

If (M, P) is a m-cell, the axioms imply that $M_m = P_m$. The k-source and k-target are given by

$$s_k(M, P) := (|M|_k, M_k \cup |P|_{k-1})$$
$$t_k(M, P) := (|M|_{k-1} \cup P_k, |P|_k)$$

and composition is defined by

$$(M, P) *_k (N, Q) := (M \cup (N\backslash N_k), (P\backslash P_k) \cup Q).$$

Example 3.10 The oriental $\mathscr{O}_4$ has a unique 4-cell given by the pair

$$M = \{01234, 0124, 0234, 012, 023, 034, 04, 0\}$$
$$P = \{01234, 0123, 0134, 1234, 124, 234, 014, 01, 12, 23, 34, 4\}.$$

Exercise 3.11 Identify the source and target of the unique 4-cell in $\mathscr{O}_4$.

Exercise 3.12 Show that $\mathscr{O}_n$ has a unique n-cell.

The orientals satisfy the universal property of being freely generated by the faces of the simplex, in the sense of the following definition of free generation for an ω-category.

Definition 3.13 For an ω-category C, write $|C|_n$ for its n-categorical truncation, discarding all higher-dimensional cells. The ω-category C is **freely generated** by a subset $G \subset C$ if for each ω-category X, $n \in \omega$, n-functor $|C|_n \to X$, and map $G \cap |C|_{n+1} \to X$, compatible with n-sources and targets there exists a unique extension to an $(n+1)$-functor $|C|_{n+1} \to X$.

$$
\begin{array}{ccc}
G \cap |C|_{n+1} & \longrightarrow & X \\
\cap & \overset{\exists!}{\dashrightarrow} & \\
|C|_{n+1} & s_n \Big\Vert t_n & \\
{\scriptstyle s_n} \Big\downarrow\Big\downarrow {\scriptstyle t_n} & & \\
|C|_n & \longrightarrow & X
\end{array}
$$

Theorem 3.14 (Street) *The category $\mathcal{O}_n$ is freely generated by the faces of $\Delta[n]$.*

Exercise 3.15 Use this universal property to show that the orientals define a cosimplicial object in ω-categories

$$\mathcal{O} \colon \Delta \to \omega\text{-}\mathsf{Cat}.$$

This cosimplicial object gives rise to the Street nerve, to which we now turn.

3.3 The Street Nerve as a Strict Complicial Set

Definition 3.16 The **Street nerve** of an ω-category C, is the simplicial set NC whose n-simplices are ω-functors $\mathcal{O}_n \to C$.

Example 3.17 (Street Nerves of Low-Dimensional Categories)

 i. The Street nerve of a 1-category is its usual nerve.
ii. The Street nerve of a 2-category has 0-simplices the objects, 1-simplices the 1-cells, and 2-simplices the 2-cells $\alpha \colon h \Rightarrow gf$ whose target is a specified composite

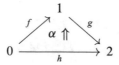

The 3-simplices record equations between pasted composites of 2-cells of the form

This simplicial set is 3-coskeletal, with a unique filler for all spheres in higher dimensions.

In general:

Theorem 3.18 (Street) *The nerve of an n-category is $(n + 1)$-coskeletal.*

The Street nerve can be lifted along $U\colon \mathsf{Strat} \to \mathsf{sSet}$ by choosing a stratification for the simplicial set NC.

Definition 3.19 In the **identity stratification** of the Street nerve of an ω-category C, an n-simplex in NC is marked if and only if the corresponding ω-functor $\mathcal{O}_n \to C$ carries the n-cell in $\mathcal{O}_n$ to a cell of lower dimension in C. That is, in the identity stratification of NC, only those n-simplices corresponding to identities are marked.

The identity stratification defines a functor $\omega\text{-}\mathsf{Cat} \to \mathsf{Strat}$. This terminology allows us to restate the Street–Roberts conjecture more concisely:

Theorem 3.20 (Verity) *The Street nerve with the identity stratification defines a fully faithful embedding*

$$\omega\text{-}\mathsf{Cat} \overset{N}{\hookrightarrow} \mathsf{Strat}$$

of ω-categories into stratified simplicial sets, with essential image the category of strict complicial sets.

Example 3.21

i. If C is a 1-category, the identity stratification turns NC into a 2-trivial strict complicial set with only the identity (i.e., degenerate) 1-simplices marked.
ii. If C is a 2-category, the identity stratification turns NC into a 3-trivial strict complicial set with only the degenerate 1-simplices marked and with a 2-simplex marked if and only if it is inhabited by an identity 2-cell, whether or not there are degenerate edges, e.g.,:

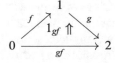

An interesting feature of the complicial sets model of higher categories is that strict ω-categories can also be a source of *weak* rather than *strict* complicial sets, simply by choosing a more expansive marking convention. We begin the next section by exploring this possibility.

4 Saturated Complicial Sets

In the previous section, we defined the Street nerve of an ω-category C, a simplicial set NC whose n-simplices are diagrams $\mathcal{O}_n \to C$ indexed by the nth oriental. We observed that this simplicial set becomes a strict complicial set if we mark precisely those diagrams $\mathcal{O}_n \to C$ that carry the n-cell of $\mathcal{O}_n$ to a cell of dimension less than n in C (i.e., to an identity).

One of the virtues of the complicial sets model of weak higher categories is the possibility of changing the stratification on a given simplicial set if one desires a more generous or more refined notion of thinness, corresponding to a tighter or looser definition of composition. The identity stratification of NC is the smallest stratification that makes this simplicial set into a weak complicial set, but we will soon meet other larger stratifications that are more categorically natural.

In Sect. 4.1, we begin by looking in low dimensions for limitations on which simplices can be marked in a complicial set, and discover that any marked 1-simplex is necessarily an 1-equivalence, in a sense that we define. In Sect. 4.2, we introduce the higher-dimensional generalization of these notions. We conclude in Sect. 4.3 by summarizing the work of Verity that establishes the basic homotopy theory of complicial sets of various flavors.

To construct weak complicial sets from nerves of strict ω-categories, the stratification on the Street nerve is enlarged, but in other instances refinement of the markings is desired. For example, Verity constructs a Kan complex of simplicial cobordisms between piecewise-linear manifolds. Because the underlying simplicial set is a Kan complex, it becomes a weak complicial set under the 0-trivial stratification where all cobordisms (all positive-dimensional simplices) are marked. Other choices, in increasing order of refinement, are to mark the h-cobordisms (cobordisms for which the negative and positive boundary inclusions are homotopy equivalences), the quasi-invertible cobordisms (the "equivalences"), or merely the trivial cobordisms (meaning the cobordism "collapses" onto its negative and also its positive boundary).

4.1 Weak Complicial Sets from Strict ω-Categories

To explore other potential markings of Street nerves of strict ω-categories, we first ask whether it is possible to mark more than just the degenerate 1-simplices.

If f is a marked edge in any complicial set A, then the $\Lambda^2[2]$-horn with 0th face f and 1st face degenerate is admissible, so f has a right equivalence inverse. A dual construction involving a $\Lambda^0[2]$-horn shows that f has a left equivalence inverse.

$$
\begin{array}{ccc}
 & 1 & \\
\overset{\simeq}{\nearrow} \quad \Big\downarrow \simeq & & \overset{f\simeq}{\searrow} \\
0 =\!\!=\!\!=\!\!=\!\!= 2 & &
\end{array}
\qquad
\begin{array}{ccc}
 & 1 & \\
\overset{f\simeq}{\nearrow} \quad \Big\downarrow \simeq & & \overset{\simeq}{\searrow} \\
0 =\!\!=\!\!=\!\!=\!\!= 2 & &
\end{array}
\qquad (5)
$$

The elementary thinness extensions imply further than these one-sided inverses are also marked, so they admit further inverses of their own.

Definition 4.1 A 1-simplex in a stratified simplicial set is a 1-**equivalence** if there exist a pair of thin 2-simplices as displayed

Note the notion of 1-equivalence is defined relative to the 2-dimensional stratification.

Remark 4.2 There are many equivalent ways to characterize the 1-equivalences in a complicial set A. We choose Definition 4.1 because of its simplicity and naturality, and because this definition provides a homotopically well-behaved type of equivalences in homotopy type theory; see [3, 2.4.10].

The elementary anodyne extensions displayed in (5) prove:

Proposition 4.3 *Any marked 1-simplex in a complicial set is a 1-equivalence.*

This result suggests an alternate stratification for nerves of 1-categories:

Proposition 4.4 *If C is a 1-category then the 1-trivial stratification of NC with the isomorphisms as marked 1-simplices defines a complicial set.*

Depending on the 1-category there may be intermediate stratifications where only some of the isomorphisms are marked (the set of marked edges has to satisfy the 2-of-3 property) but these are somehow less interesting.

Exercise 4.5 Prove Proposition 4.4.

Let us now consider the degenerate edges, the thin edges, and the 1-equivalences as subsets of the set of 1-simplices in a complicial set A. In any stratified simplicial set, the degenerate 1-simplices are necessarily thin. In a complicial set A, Proposition 4.3 proves that the thin 1-simplices are necessarily 1-equivalences, but there is nothing in the complicial set axioms that guarantees that all equivalences are marked. We introduce terminology that characterizes when this is the case:

Definition 4.6 A complicial set A is 1-**saturated** if every 1-equivalence is marked.

If a 1-trivial complicial set is 1-saturated then it is *saturated* in the sense of Definition 4.15 below. From the definitions, it is easy to prove:

Proposition 4.7 *If C is a strict 1-category, there is a unique saturated 1-trivial complicial structure on NC, namely the one in which every isomorphism in C is marked. Moreover, this is the maximal 1-trivial stratification making NC into a complicial set.*

Exercise 4.8 Prove this.

To build intuition for higher dimensional generalizations of these notions, next consider the Street nerve of a strict 2-category as a 2-trivial stratified simplicial set.

As the notion of 1-saturation introduced in Definitions 4.1 and 4.6 depends on the markings of 2-simplices, it makes sense to consider the markings on the 2-simplices first. If only identity 2-simplices are marked, then the 1-saturation of NC is as before: marking all of the 1-cell isomorphisms in the 2-category C. But we might ask again whether a larger stratification is possible at level 2.

In any complicial set, consider a thin 2-simplex α with 0th edge degenerate. From α one can build admissible $\Lambda^1[3]$ and admissible $\Lambda^3[3]$-horns admitting thin fillers:

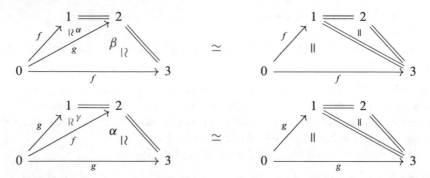

So again we conclude that any thin 2-simplex of this form is necessarily an "equivalence" up to thin 3-simplices, in the sense of the displayed diagrams. Informally, a complicial set is 2-*saturated* if all 2-simplices that are equivalences in this sense are marked. A precise definition of saturation that applies in any dimension appears momentarily as Definition 4.15. It follows that:

Proposition 4.9 *If C is a strict 2-category, there is a unique saturated 2-trivial complicial structure on NC, in which the 2-cell isomorphisms and the 1-cell equivalences are marked. Moreover, this is the maximal 2-trivial stratification making NC into a complicial set.*

Unlike the 1-trivial saturated stratification on the Street nerve of a 1-category described in Proposition 4.7, the 2-trivial saturated stratification on the Street nerve of a 2-category described in Proposition 4.9 describes a *weak* and not a *strict* complicial set.

4.2   Saturation

To define saturation in any dimension, it is convenient to rephrase the definition of 1-saturation as a lifting property. The pair of thin 2-simplices

define the 3rd and 0th faces of an inner admissible $\Lambda^1[3]$- or $\Lambda^2[3]$-horn that fills to define a thin 3-simplex

This 3-simplex defines a map $\Delta[3]_{eq} \to A$, where $\Delta[3]_{eq}$ is the 3-simplex given a 1-trivial stratification with the edges [02] and [13] also marked.

Proposition 4.10 *A complicial set A is 1-saturated if and only if it admits extensions along the entire inclusion of $\Delta[3]_{eq}$ into the maximally marked 3-simplex:*

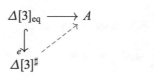

Exercise 4.11 Prove this.

There are similar extension problems that detects saturation in any dimension, which are defined by forming the join of the inclusion $\Delta[3]_{eq} \hookrightarrow_e \Delta[3]^\sharp$ with simplices on one side or the other.

Definition 4.12 (Join and Slice) The ordinal sum on $\mathbb{\Delta}_+$ extends via Day convolution to a bifunctor on the category of augmented simplicial sets called the **join**. Any simplicial set can be regarded as a trivially augmented simplicial set. Under this inclusion, the join restricts to define a bifunctor

$$\mathsf{sSet} \times \mathsf{sSet} \overset{\star}{\to} \mathsf{sSet}$$

so that $\Delta[n] \star \Delta[m] = \Delta[n+m+1]$. More generally, an n-simplex in the join $A \star B$ of two simplicial sets is a pair of simplices $\Delta[k] \to A$ and $\Delta[n-k-1] \to B$ for some $-1 \le k < n$. Here $\Delta[-1]$ is the trivial augmentation of the empty simplicial set, in which case the functors $\Delta[-1] \star -$ and $- \star \Delta[-1]$ are naturally isomorphic to the identity.

The **left** and **right slices** of a simplicial set A over a simplex $\sigma \colon \Delta[n] \to A$ are the simplicial sets $\sigma \backslash A$ and A/σ whose k-simplices correspond to diagrams

$$\Delta[k] \longrightarrow \sigma\backslash A \qquad \longleftrightarrow \qquad \Delta[n] \hookrightarrow \Delta[n] \star \Delta[k] \underset{\sigma}{\longrightarrow} A$$

$$\Delta[k] \longrightarrow A/\sigma \qquad \longleftrightarrow \qquad \Delta[n] \hookrightarrow \Delta[k] \star \Delta[n] \underset{\sigma}{\longrightarrow} A/\sigma \tag{6}$$

See [4] for more.

Definition 4.13 (Stratified Join) The simplicial join lifts to a join bifunctor

$$\text{Strat} \times \text{Strat} \overset{\star}{\to} \text{Strat}$$

in which a simplex $\Delta[n] \to A \star B$, with components $\Delta[k] \to A$ and $\Delta[n-k-1] \to B$, is marked in $A \star B$ if and only if at least one of the simplices in A or B is marked. More details can be found in [9].

Exercise 4.14 Define a stratification on the slices $\sigma \backslash A$ and A/σ over an n-simplex $\sigma \colon \Delta[n] \to A$ so that the correspondence (6) extends to stratified simplicial sets.

Definition 4.15 A complicial set is **saturated** if it admits extensions along the set of entire inclusions

$$\{\Delta[m] \star \Delta[3]_{\text{eq}} \star \Delta[n] \hookrightarrow_e \Delta[m] \star \Delta[3]^{\sharp} \star \Delta[n] \mid n, m \geq -1\}.$$

In fact, it suffices to require only extensions

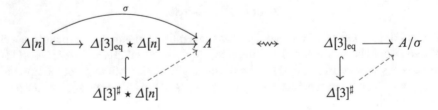

along inclusions of one-sided joins of the inclusion $\Delta[3]_{\text{eq}} \hookrightarrow_e \Delta[3]^{\sharp}$ with an n-simplex for each $n \geq -1$, and as it turns out only the left-handed joins or right-handed joins are needed.

By Proposition 4.10, the $n = -1$ case of Definition 4.15 asserts that every 1-equivalence in A, defined relative to the marked 2-simplices and marked 3-simplices, is marked. By Proposition 4.10 again, the general extension property

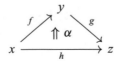

asserts that every 1-equivalence in the slice complicial set A/σ is marked.

At first blush, Definition 4.15 does not seem to be general enough. In the case of a vertex $\sigma \colon \Delta[0] \to A$, 1-equivalences in A/σ define 2-simplices in A whose [01]-edge is a 1-equivalence. In particular, a generic 2-simplex

$$\begin{array}{ccc}
 & \overset{y}{} & \\
 \overset{f}{\nearrow} & \Uparrow \alpha & \overset{g}{\searrow} \\
 x & \underset{h}{\longrightarrow} & z
\end{array}$$

with no 1-equivalence edges along its boundary, does not define a 1-equivalence in any slice complicial set. However, there are admissible 3-horns that can be filled to define the pasted composites of α with 1_f and 1_g, respectively:

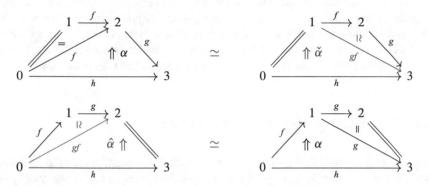

By the complicial thinness extension property, if any of α, $\hat{\alpha}$, or $\check{\alpha}$ are marked, then all of them are.

Exercise 4.16 Generalize this "translation" argument to prove that any n-simplex in a complicial set is connected via a finite sequence of n-simplices to an n-simplex whose first face is degenerate and an n-simplex whose last face is degenerate in such a way that if any one of these simplices is thin, they all are.

Definition 4.17 In an n-trivial complicial set, an n-simplex $\sigma: \Delta[n] \to A$ is an n-**equivalence** if it admits an extension

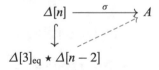

along the map $\Delta[n] \hookrightarrow \Delta[3]_{eq} \star \Delta[n-2]$ whose image includes the edge $[1,2]$ of $\Delta[3]_{eq}$ and all of the vertices of $\Delta[n-2]$.

Remark 4.18 The set of n-equivalences identified by Definition 4.17 depends on the marked $(n+1)$-simplices, which is the reason we have only stated this definition for an n-trivial complicial set. The n-equivalences in a generic complicial set are characterized by an inductive definition, the formulation of which we leave to the reader.

Example 4.19 (Quasi-Categories as Complicial Sets) Expanding on the work of Sect. 2.1, a quasi-category has a unique saturated stratification making it a complicial set: namely the 1-trivial saturation where all of the 1-equivalences are marked. This is the "natural marking" discussed in [5]. Conversely, any 1-trivial saturated complicial set is a quasi-category. So quasi-categories are precisely the 1-trivial saturated complicial sets.

Each simplicial set has a minimum stratification, with only degeneracies marked. Because the definition of saturation is inductive, each simplicial set also has a minimum saturated stratification. Larger saturated stratifications also exist (e.g., the maximal marking of all positive-dimensional simplices). It is more delicate to describe how the process of saturating a given complicial set interacts with the complicial structure: adding new thin simplices adds new admissible horns which need fillers. What is more easily understood are model structures whose fibrant objects are complicial sets of a particular form, a subject to which we now turn.

4.3 Model Categories of Complicial Sets

The category of stratified simplicial sets is cartesian closed, where the cartesian product $\times$ is referred to as the *Gray tensor product* because this is the analogous tensor product in higher category theory.[6] We write "hom" for the internal hom characterized by the 2-variable adjunction

$$\mathsf{Strat}(A \times B, C) \cong \mathsf{Strat}(A, \hom(B, C)) \cong \mathsf{Strat}(B, \hom(A, C)).$$

Let

$$I := \{\partial\Delta[n] \hookrightarrow \Delta[n] \mid n \geq 0\} \cup \{\Delta[n] \hookrightarrow \Delta[n]_t \mid n \geq 0\}$$

denote the generating set of monomorphisms of stratified simplicial sets introduced in Proposition 2.10 and let

$$J := \{\Lambda^k[n] \hookrightarrow_r \Delta^k[n] \mid n \geq 1, k \in [n]\} \cup \{\Delta^k[n]' \hookrightarrow_e \Delta^k[n]'' \mid n \geq 2, k \in [n]\}$$

denote the set of **elementary anodyne extensions** introduced in Definition 2.13, the right lifting property against which characterizes the complicial sets. A combinatorial lemma proves that the pushout product $I \hat{\times} J$ of maps in I with maps in J is an **anodyne extension**: that is, may be expressed as a retract of a transfinite composite of pushouts of coproducts of elements of J (here mere composites of pushouts suffice). As a corollary:

Proposition 4.20 (Verity [10]) *If X is a stratified simplicial set and A is a weak complicial set, then $\hom(X, A)$ is a weak complicial set.*

Verity provides a very general result for constructing model structures whose fibrant objects are defined relative to some set of monomorphisms K containing J.

---

[6]Note that in the theory of bicategories, the cartesian product plays the role of the Gray tensor product in 2-category theory, in the sense that there is a biadjunction between the cartesian product and the hom-bicategory of pseudofunctors, pseudo-natural transformations, and modifications.

Call a stratified simplicial set a K-**complicial set** if it admits extensions along each map in K. Suppose K is a set of monomorphisms of Strat so that

i. every elementary anodyne extension is in K

and moreover each of/all of the following equivalent conditions hold for each $j \in K$:

ii. Each element j of K is a K-**weak equivalence**: i.e., $\mathrm{hom}(j, A)$ is a homotopy equivalence[7] for each K-fibrant stratified set.
iii. $\mathrm{hom}(j, A)$ is a trivial fibration for each K-complicial set.
iv. Each K-complicial set admits extensions along all the maps $i \hat{\times} j$ for all $i \in I$ and $j \in K$.

Call a map that has the right lifting property with respect to the set K a K-**complicial fibration**.

Theorem 4.21 (Verity [10]) *Each set of stratified inclusions K satisfying the conditions (i)–(iv) gives rise to a cofibrantly generated model structure whose:*

- *weak equivalences are the K-weak equivalences,*
- *cofibrations are monomorphisms,*
- *fibrant objects are the K-complicial sets, and*
- *fibrations between fibrant objects are K-complicial fibrations.*

Moreover, such a model structure is monoidal with respect to the Gray tensor product.

Proof Apply Jeff Smith's theorem [10, 125].

Example 4.22 Theorem 4.21 applies to the minimal set of elementary anodyne extensions

$$J := \{\Lambda^k[n] \hookrightarrow_r \Delta^k[n] \mid n \geq 1, k \in [n]\} \cup \{\Delta^k[n]' \hookrightarrow_e \Delta^k[n]'' \mid n \geq 2, k \in [n]\}$$

defining the **model structure for complicial sets**.

Example 4.23 Theorem 4.21 applies to the union of the minimal J with

$$K_n^{\mathrm{tr}} := \{\Delta[r] \hookrightarrow_e \Delta[r]_t \mid r > n\}$$

defining the **model structure for n-trivial complicial sets**.

Example 4.24 Theorem 4.21 applies to the union of the minimal J with

$$K^{\mathrm{s}} := \{\Delta[m] \star \Delta[3]_{\mathrm{eq}} \star \Delta[n] \hookrightarrow \Delta[m] \star \Delta[3]^{\sharp} \star \Delta[n] \mid m, n \geq -1\}$$

---

[7]Two maps $f, g : X \to A$ are **homotopic** if they extend to a map $X \times \Delta[1]^{\sharp} \to A$. If A is a weak complicial set, this "simple homotopy" is an equivalence relation.

defining the **model structure for saturated complicial sets**.

Example 4.25 Theorem 4.21 applies to the union of the minimal J with both K_n^{tr} and K^s defining the **model structure for n-trivial saturated complicial sets**.

By Example 4.19, the $n = 1$ case of this last result gives a new proof of Joyal's model structure for quasi-categories.

Acknowledgements This document evolved from lecture notes written to accompany a 3-h mini course entitled "Weak Complicial Sets" delivered at the Higher Structures in Geometry and Physics workshop at the MATRIX Institute from June 6–7, 2016. The author wishes to thank Marcy Robertson and Philip Hackney, who organized the workshop, the MATRIX Institute for providing her with the opportunity to speak about this topic, and the NSF for financial support through the grant DMS-1551129. In addition, the author is grateful for personal conversations with the two world experts—Dominic Verity and Ross Street—who she consulted while preparing these notes. Finally, thanks are due to an eagle-eyed referee who made several cogent suggestions to improve the readability of this document.

References

1. Barwick, C., Schommer-Pries, C.: On the unicity of the homotopy theory of higher categories (2013). arXiv:1112.0040
2. Boardman, J.M., Vogt, R.M.: Homotopy Invariant Algebraic Structures on Topological Spaces. Lecture Notes in Mathematics, vol. 347. Springer, Berlin (1973)
3. Homotopy Type Theory: Univalent Foundations of Mathematics: The Univalent Foundations Program, Institute for Advanced Study (2013)
4. Joyal, A.: Quasi-categories and Kan complexes. J. Pure Appl. Algebra **175**(1–3), 207–222 (2002)
5. Lurie, J.: Higher Topos Theory. Annals of Mathematical Studies, vol. 170. Princeton University Press, Princeton, NJ (2009)
6. Roberts, J.E.: Complicial sets. Handwritten manuscript (1978)
7. Street, R.H.: The algebra of oriented simplexes. J. Pure Appl. Algebra **49**, 283–335 (1987)
8. Verity, D.: Weak complicial sets II, nerves of complicial Gray-categories. In: Davydov, A. (ed.) Categories in Algebra, Geometry and Mathematical Physics (StreetFest). Contemporary Mathematics, vol. 431. American Mathematical Society, Providence, RI (2007)
9. Verity, D.: Complicial Sets, Characterising the Simplicial Nerves of Strict ω-Categories. Memoirs of the American Mathematical Society, vol. 905. American Mathematical Society, Providence, RI (2008)
10. Verity, D.: Weak complicial sets I, basic homotopy theory. Adv. Math. **219**, 1081–1149 (2008)

A Non-crossing Word Cooperad for Free Homotopy Probability Theory

Gabriel C. Drummond-Cole

Abstract We construct a cooperad which extends the framework of homotopy probability theory to free probability theory. The cooperad constructed, which seems related to the sequence and cactus operads, may be of independent interest.

1 Introduction

The purpose of this paper is to provide a convenient operadic framework for the cumulants of free probability theory. In [4, 5], the author and his collaborators described an operadic framework for classical and Boolean cumulants. This framework involves a choice of governing cooperad, and in both the classical and Boolean cases, the choice is an "obvious" and well-studied algebraic object. Namely, for classical cumulants, the governing cooperad is the cocommutative cooperad, while for Boolean cumulants it is the coassociative cooperad.

Extending this framework to free probability requires the construction of a governing cooperad with certain properties. The main construction of this paper is a cooperad, called the *non-crossing word cooperad*, satisfying these properties. As far as the author can tell, this cooperad is, at least to some degree, new. No well-studied cooperad (such as those in [21]) seems to satisfy the requisite properties. That said, there is clearly some sort of relationship between the newly constructed cooperad and the sequence [1, 10] and cactus [8, 9, 20] operads. This line of thinking is not pursued in this article beyond the remark at the end of Sect. 3. If it turns out that this cactus variant is well-known, that would be delightful—please let us know.

Also, we make no attempt here to axiomatize the properties necessary to interface appropriately with free probability or to prove any uniqueness results. That is to say, there is every likelihood that this is the "wrong" cooperad. First of all, there is the near miss in terms of structure compared to the previously known

G.C. Drummond-Cole (✉)
Center for Geometry and Physics, Institute for Basic Science (IBS), Pohang 37673, Republic of Korea
e-mail: gabriel@ibs.re.kr

© Springer International Publishing AG, part of Springer Nature 2018
D.R. Wood et al. (eds.), *2016 MATRIX Annals*, MATRIX Book Series 1,
https://doi.org/10.1007/978-3-319-72299-3_5

operads. In addition, there are at least two failures of parallelism between the classical and Boolean cases and the new case presented here. See the remark following Theorem 1. One possible explanation for these failures is that the correct framework requires *operator-valued* free cumulants, that is, free cumulants with a not necessarily commutative ground ring. This line of reasoning has been pursued in other work [2]. It would also be exciting to hear about other potential frameworks to bring free cumulants into the framework of this kind of operadic algebra, whether along the same rough lines as in this paper or not.

The remainder of the paper is organized as follows. For convenience, we work with *unbiased* definitions of operads and cooperads, writing them in terms of finite sets and never choosing a particular ordered set. This is not usual in the literature although it should be familiar to experts. The paper begins with a review of this formalism.

Next, we describe the kind of words we will use and construct two cooperads spanned by them. The first, the *word cooperad*, is auxiliary for our purposes although it may have independent interest. We construct the *non-crossing word cooperad* as a quotient of the word cooperad. After a brief review of necessary notions from homotopy probability theory and free probability theory, we apply the non-crossing word cooperad to the motivating question and show that it fits into the framework of homotopy probability theory.

1.1 Conventions

We will use the notation $[n]$ to denote the set $\{1, \ldots, n\}$. We work over a field $\mathbb{K}$ of characteristic zero.

2 Unbiased Operads and Cooperads

We will use an unbiased definition for operads and cooperads, as it significantly reduces the notation necessary to describe our structures at the cost of requiring a few explicit definitions rather than a reference. There are several distinct issues that one faces with cooperadic algebra in full generality, related to issues like conilpotency, 0-ary operations, and the "handedness" of the categories we generally work in. We will make several strong simplifying assumptions to avoid the most obvious pitfalls.

Let **Lin** be either the category of vector spaces, the category of graded vector spaces, or the category of chain complexes over $\mathbb{K}$. We consider vector spaces as graded vector spaces concentrated in degree zero and graded vector spaces as chain complexes with zero differential without further comment.

2.1 Species and Plethysm

Definition 1 A linear *species* is a functor from finite sets and their isomorphisms to **Lin**. A species is *reduced* if it takes value 0 on the empty set.

All species will be linear in this paper.

The *unit species* I has $I(S) = \mathbb{K}$ if $|S| = 1$ and $I(S) = 0$ otherwise, with the identity for every nonzero morphism.

The *coinvariant composition* or *coinvariant plethysm* of two species F and G is the species $F \circ G$ given by

$$(F \circ G)(S) = \operatorname*{colim}_{S \xrightarrow{f} T} \left(F(T) \otimes \bigotimes_{t \in T} G(f^{-1}(t)) \right) .$$

The *invariant composition* or *invariant plethysm* of two species F and G is the species $F \bar{\circ} G$ given by

$$(F \bar{\circ} G)(S) = \lim_{S \xrightarrow{f} T} \left(F(T) \otimes \bigotimes_{t \in T} G(f^{-1}(t)) \right) .$$

In both cases the limits and colimits are taken over the diagram category whose objects are maps out of S and whose morphisms are isomorphisms under S.

Lemma 1 *Let F be a species and let G be a reduced species. Then there is an isomorphism between $F \bar{\circ} G$ and $F \circ G$, defined below.*

Proof For a fixed set S, choose a set of representatives $\{f_i : S \to T_i\}$, one for each isomorphism type of surjection $f : S \to T$ in the diagram category defining both plethysms. This set is a fortiori finite because we have restricted to surjections.

The invariant plethysm projects onto the defining factor

$$(F \circ G)(S)_i := F(T_i) \otimes \bigotimes_{t \in T_i} G(f_i^{-1}(t_i)) .$$

Likewise, the coinvariant plethysm receives a map from $(F \circ G)(S)_i$.

This collection of maps then determines both:

1. a map from the invariant plethysm to the direct product $\prod (F \circ G)(S)_i$ and
2. a map from the direct sum $\bigoplus (F \circ G)(S)_i$ to the coinvariant plethysm.

But since the product is finite, the natural map from the sum to the product is invertible and so we can compose to get a map

$$F \bar{\circ} G \to \prod_i (F \circ G)(S)_i \cong \bigoplus_i (F \circ G)(S)_i \to F \circ G .$$

This overall composition is independent of the choices of representatives. Since G is reduced, this runs over all isomorphism types necessary to define both the invariant and coinvariant plethysm. Moreover, because we are working in characteristic zero, the map, for each fixed isomorphism class, is an isomorphism.   □

There are two points that require care. First of all, we should make sure that when we actually move between the two, that we consistently adhere to the particular choice of isomorphism outlined here. That is, there are two or three different normalizations of this isomorphism present in the literature. The others differ by something like a factor of $|S|!$ or $\frac{1}{|S|!}$ on each component of the product/sum above. Secondly, we do not have such a map when G is not reduced.

Lemma 2 *There are natural isomorphisms making linear species equipped with the unit species and coinvariant plethysm a monoidal category. There are natural isomorphisms making reduced linear species equipped with the unit species and invariant plethysm a monoidal category.*

Proof The left and right unitor isomorphisms can be constructed by direct computation of the (co)limits involved.

Colimits (and essentially finite limits) commute with tensor product. Then (F ∘ G) ∘ H and F ∘ (G ∘ H) are both naturally isomorphic to

$$\operatorname*{colim}_{S \xrightarrow{f} T \xrightarrow{g} U} \left(F(U) \otimes \bigotimes_{u \in U} G(g^{-1}(u)) \otimes \bigotimes_{t \in T} H(f^{-1}(t)) \right) .$$

Verifying that these natural isomorphisms satisfy the triangle and pentagon axioms is straightforward. The case of the invariant plethysm is basically the same.   □

2.2 Operads and Cooperads

Definition 2 An *operad* is a monoid in the monoidal category of linear species with coinvariant plethysm. A (reduced) *cooperad* is a comonoid in the monoidal category of reduced species with invariant plethysm.

The data of an operad $\mathcal{P} = (P, \eta, \mu)$ consists of a species P equipped with maps $\eta : I \to P$ (the *unit*) and $P \circ P \xrightarrow{\mu} P$ (the *composition*). The composition must be associative and the unit must satisfy left and right unit properties.

More explicitly, to specify a composition map out of the defining colimit of $P \circ P$ it suffices to give a map out of each term with the appropriate equivariance. So for a map $f : S \to T$, one can specify a map

$$\mu_f : P(T) \otimes \bigotimes_{t \in T} P(f^{-1}(t)) \to P(S)$$

and then define the composition map as the colimit of μ_f.

Similarly, the data of a cooperad $C = (\mathsf{C}, \varepsilon, \Delta)$ consists of a reduced species C equipped with maps $\varepsilon : \mathsf{C} \to I$ (the *counit*) and $\mathsf{C} \xrightarrow{\Delta} \mathsf{C} \bar{\circ} \mathsf{C}$ (the *decomposition*). The decomposition must be coassociative and the counit must satisfy left and right counit properties.

More explicitly, to specify a decomposition map into the defining limit of $\mathsf{C} \bar{\circ} \mathsf{C}$ it suffices to give a map into each term with the appropriate coequivariance. So for a surjection $f : S \twoheadrightarrow T$, one can specify a map

$$\Delta_f : \mathsf{C}(S) \to \mathsf{C}(T) \otimes \bigotimes_{t \in T} \mathsf{C}(f^{-1}(t))$$

and then define Δ as the limit of Δ_f.

In practice, the (co)equivariance and (co)unital conditions are easy to verify and the main thing to check is (co)associativity.

Remark 1 The expression of operads as monoids in a monoidal category is due to Smirnov [16]; the dual picture was written down in [7]. In general, biased definitions are more common in the literature. Given a (co)operad in this unbiased definition, one can recover the data of a (co)operad under a more standard definition by restricting to the full subcategory containing only the objects $[n]$.

2.3 Examples

We shall use a few simple operads and cooperads. In all of the following,

1. by definition all the species in the examples are reduced, and sets S are assumed to be non-empty.
2. all units and counits are given by the identity map $\mathbb{K} \to \mathbb{K}$ for each singleton set S,
3. it is easy to verify (co)unitality and (co)equivariance, and
4. it is a straightforward (potentially tedious) calculation to verify (co)associativity of the specified (co)composition.

Verifications of (co)unitality, (co)equivariance, and (co)associativity are omitted.

Example 1

1. The unit species I, along with the identity and the canonical isomorphisms $\mathsf{I} \bar{\circ} \mathsf{I} \cong \mathsf{I} \cong \mathsf{I} \circ \mathsf{I}$, has both an operad and cooperad structure. We denote both of these by $\mathcal{I}$.
2. Let Com be the species with $\mathsf{Com}(S) = \mathbb{K}$ for all S (and Com applied to all maps is the identity on $\mathbb{K}$). We give this species an operad structure by specifying

$$\mu_f : \mathbb{K} \otimes \bigotimes_{t \in T} \mathbb{K} \to \mathbb{K}$$

given by the natural identification. This is the *commutative operad* and is denoted
Com.

3. Similarly, we give the data Δ_f for a cooperad with underlying species Com. In
   this case as well,

$$\Delta_f : \mathbb{K} \to \mathbb{K} \otimes \bigotimes_{t \in T} \mathbb{K}$$

is the natural identification. This is the *cocommutative cooperad* and is denoted
coCom.

4. Let Ass be the species such that Ass(S) is the $\mathbb{K}$-linear span of total orders on S:

$$\text{Ord}(S) := \text{Iso}(S, [|S|]) \ .$$

We will specify an operad with underlying species Ass. Given a surjection f :
$S \to T$, there is an embedding $\iota_f : \text{Ord}(T) \times \prod \text{Ord}(f^{-1}(t)) \to \text{Ord}(S)$ given by

$$\iota_f \left(\varrho \times \prod \tau_t \right) (s) = \tau_{f(s)}(s) + \sum_{\varrho(t) < \varrho(f(s))} |f^{-1}(t)| \ .$$

Define the composition map μ_f as the $\mathbb{K}$-linear extension of ι_f. The resulting
operad is the *associative operad*, denoted $\mathcal{A}ss$.

5. Finally, we specify a cooperad with the same underlying species Ass. The
   decomposition map

$$\Delta_f : \mathbb{K}\langle \text{Ord}(S) \rangle \to \mathbb{K}\langle \text{Ord}(T) \rangle \otimes \bigotimes_{t \in T} \mathbb{K}\langle \text{Ord}(f^{-1}(t)) \rangle \ .$$

is again determined by ι by the equation

$$\Delta_f(\sigma) = \sum_{\varrho, \tau_t} \delta_{\sigma, \iota_f(\varrho \times \prod \tau_t)} \left(\varrho \times \prod \tau_t \right) \ .$$

The resulting cooperad is the *coassociative cooperad* and is denoted *co$\mathcal{A}ss$*.

2.4 Algebras and Coalgebras

Now we move on to the discussion of algebras over operads and coalgebras over
cooperads. The category **Lin** embeds into the category of (non-reduced) species as
follows. Let V be an object in **Lin**. Then $\iota(V)$ is the species with $\iota(V)(\emptyset) = V$ and
$\iota(V)(S) = 0$ for nonempty S.

Definition 3 Let F be an species. The *Schur functor* associated to F is a functor **Lin** → **Lin**, defined by

$$V \mapsto (\mathsf{F} \circ \iota(V))(\emptyset) .$$

We will abuse notation and use the notation F∘ for this functor.

The Schur functor I∘ for the unit species I is naturally equivalent to the identity functor. Since the coinvariant plethysm is associative, the iterated Schur functor of two species is naturally isomorphic to the Schur functor of the plethysm:

$$\mathsf{F} \circ (\mathsf{G} \circ (V)) \cong (\mathsf{F} \circ \mathsf{G}) \circ (V) .$$

This implies the following.

Lemma 3 *If the species F is equipped with an operad structure, the unit and composition induce a monad structure on the Schur functor F.*

If the reduced species F is equipped with a cooperad structure, the counit and cocomposition induce a comonad structure on the Schur functor F.

Definition 4 Let $\mathcal{P} = (\mathsf{P}, \eta, \mu)$ be an operad. An *algebra* over $\mathcal{P}$ is an algebra over the monad $\mathcal{P}$. This is the same as a **Lin** object V equipped with a morphism $\mathsf{P} \circ V \to V$ compatible with the monad structure.

Let $C = (\mathsf{C}, \varepsilon, \Delta)$ be a cooperad. A *conilpotent coalgebra* over C is a coalgebra over the comonad C∘. This is the same as a **Lin** object V equipped with a morphism $V \to \mathsf{C} \circ V$ compatible with the comonad structure.

As is general for monads, the forgetful functor from the category of algebras over an operad $\mathcal{P} = (\mathsf{P}, \eta, \mu)$ to **Lin** has a left adjoint, the free $\mathcal{P}$-algebra functor, realized by the Schur functor and the monad structure of P∘. We distinguish between the Schur functor P∘ between **Lin** and itself and the Schur functor $\mathcal{P}$∘ between **Lin** and $\mathcal{P}$-algebras.

Similarly, the forgetful functor U from conilpotent coalgebras over a cooperad $C = (\mathsf{C}, \varepsilon, \Delta)$ to **Lin** has a right adjoint, the cofree conilpotent C-coalgebra functor, realized by the Schur functor and the comonad structure of C∘. Again, we distinguish between the Schur functor C∘ between **Lin** and itself and the Schur functor C∘ between **Lin** and C-coalgebras.

In any event, the adjunction above implies that a morphism of conilpotent C-coalgebras from some coalgebra X into $C \circ V$ may be identified via this adjoint with a **Lin** morphism from the underlying **Lin**-object of X to V.

In general, the adjunction $\mathrm{Hom}_{\mathbf{Lin}}(UX, V) \to \mathrm{Hom}_{C\text{-coalgebras}}(X, C \circ V)$ is realized by taking a **Lin**-morphism f to the composite

$$X \xrightarrow{\text{coalgebraic structure map}} C \circ (UX) \xrightarrow{Cof} C \circ V$$

and the inverse map is given by taking a coalgebra map to its composite with the counit applied to V:

$$UX \to U(C \circ V) \cong C \circ V \to I \circ V \cong V .$$

2.5 *Automorphisms of Cofree Coalgebras*

We record a characterization of automorphisms of cofree coalgebras in terms of this adjunction. We call a cooperad (or a species, by abuse of notation) *strongly coaugmented* if it is reduced and takes value $\mathbb{K}$ on a singleton. A strongly coaugmented cooperad C accepts a map from the cooperad $\mathcal{I}$ which fits into the following diagram

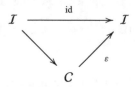

which is necessarily unique.

For a species C, given a **Lin** map $f : \mathsf{C} \circ V \to V$ and a finite set S we let f_S denote the restriction

$$\operatorname*{colim}_{\operatorname{Aut} S} \mathsf{C}(S) \otimes V^{\otimes S} \to \mathsf{C} \circ V \xrightarrow{f} V .$$

Then we have the following.

Lemma 4 *Let* $C = (\mathsf{C}, \varepsilon, \Delta)$ *be a strongly coaugmented cooperad. Let V be an object of* **Lin**. *A morphism* $f : U(C \circ V) \cong \mathsf{C} \circ V \to V$ *is adjoint to a coalgebra automorphism* $C \circ V \to C \circ V$ *if and only if* f_S *is an isomorphism when* $|S| = 1$.

Proof Let $\tilde{f}$ and $\tilde{g}$ be composable morphisms from $C \circ V$ to itself with composite $\tilde{h} = \tilde{g} \circ \tilde{f}$. Write their adjoints from $U(C \circ V)$ to V as f, g, and h. Then by using the above characterization of the adjunction, one can calculate that for S a singleton, we have $h_S = g_S \circ f_S$ (identifying V with $\mathsf{C}(S) \otimes V$). This shows the necessity of the condition.

To show sufficiency, we can proceed by induction on the size of the finite sets in the colimit definining the Schur functor. Let us be a little more explicit for the left inverse to f.

Since we want $\tilde{h}$ to be the identity, we should have $h_S = 0$ for $|S| > 1$. The explicit formula for h_S contains the term $g_S \circ (f_1^{\otimes S})$ plus a sum of terms each of which involves only $f_{S'}$ and $g_{S''}$ for some S'' strictly smaller than S. Then by invertibility of f_1 this suffices to define g_S recursively. A similar procedure defines a right inverse. A priori the formulas defining the right inverse are different but existence of both one-sided inverses forces them to be equal.   □

We conclude the section with a few remarks inessential to the flow of the paper.

Remark 2

1. The proof above explicitly uses the fact that our species are reduced and strongly augmented. In more generality, as long as there is some filtration with good properties (often called weight grading) the same argument works.

2. Algebras over $\mathcal{A}ss$ (respectively $\mathcal{C}om$) are the same thing as associative (associative and commutative) algebra objects in **Lin**, justifying the notation.

3. The reader may have noticed a failure of parallelism, where the coalgebras are conilpotent but the algebras have no dual adjective. This failure of parallelism occurs because we have only used *coinvariant* Schur functor. Even in our restricted setting, the more natural notion for coalgebras over a cooperad would involve an *invariant* Schur functor. As we are interested only in conilpotent coalgebras, the construction here is preferable.

3 Words and Their Cooperads

3.1 Words

This section establishes some basic definitions and lemmas about words.

A word w is a nonempty finite sequence of elements from a set S. In this context, S is called the *alphabet* and elements of S or the sequence w are called letters.

The word w is *pangrammatic* if it contains each letter from the alphabet S.

Definition 5 The word w is *reduced* if it has no subword of the form aa and either is length one or has different first and last letters.

The *reduction* $\overline{w}$ of the word w is the unique minimal length word obtained by repeated reduction by

$$\ldots aa \ldots \mapsto \ldots a \ldots$$

$$a \ldots a \mapsto a \ldots$$

In the second case, a must be the first and last letter of w; this relation is not a "local" move on subwords.

Definition 6 The word w is *non-crossing* if it never contains

$$\ldots a \ldots b \ldots a \ldots b \ldots$$

for distinct a and b in S.

A word is *crossing* unless it is non-crossing.

Remark 3 A map of sets $f : S \to T$ induces a map from words in S to words in T, which will be also denoted by f.

Definition 7 Let w be a word on the alphabet T and let S be a subset of the alphabet T which contains at least one letter of w. Then $w|S$, called *the word restricted to S*, is the word obtained by deleting all letters not in S.

If a word w is pangrammatic then w can be restricted to any nonempty subset of the alphabet and the result is pangrammatic.

The following lemmas about reduction, restriction, and words induced by functions, are immediate.

Lemma 5 *Let w be a word on the alphabet S and let f be a map of sets $S \to T$. Then $\overline{f(\overline{w})} = f(w)$.*

Lemma 6 *Let w be a word on the alphabet T and let S be a subset of T containing at least one letter from w. Then $\overline{w}|S = \overline{w|S}$.*

Lemma 7 *Let w be a word on the alphabet R, let f be a map of sets $R \to S$, and let T be a subset of S containing at least one letter of $f(R)$. Then $f(w|f^{-1}(T)) = f(w)|T$.*
    In general, we do not have $f(w|S) = f(w)|f(S)$ unless $S = f^{-1}f(S)$.

3.2   The Word Cooperad

Now we construct a cooperad spanned by a class of words. In Sect. 3.3, we construct a second, closely related cooperad which will be our main point of interest. As stated in the introduction, there is some relationship between the cooperads constructed here and the sequence and cactus operads. As the relationship is not entirely clear, the following is a self-contained presentation. There is a remark about the connection at the end of Sect. 3.

Definition 8 The *word species* is the species W constructed as follows. To a finite set S, the functor W assigns the $\mathbb{K}$-vector space spanned by pangrammatic reduced words on S. We define the structure necessary to make this species a cooperad, the *word cooperad* $\mathcal{W}$, showing coassociativity in Proposition 1 below.
    The decomposition map $W \to W \bar{\circ} W$ can be specified, as discussed in Sect. 2, by defining Δ_f for each surjection $f : S \twoheadrightarrow T$. We define Δ_f as follows.

$$\Delta_f(w) = \overline{f(w)} \otimes \bigotimes_{t \in T} \overline{w|f^{-1}(t)} .$$

The *counit map* ε, for $|S| = 1$, takes the unique word in $W(S)$ to $1 \in I(S)$.
Checking equivariance with respect to both isomorphisms $S \to S'$ and isomorphisms $T \to T'$ under S is straightforward, so the decomposition map Δ is well-defined.

Example 2 Let $S = \{a_1, a_2, a_3\}$ and let $w = a_1 a_2 a_1 a_3$. Then the limit of interest can be specified in terms of five choices of T and a surjection. $S \to T$. These are:

- the constant map $f_0 : S \to \{b_0\}$,
- the three maps $f_{ij} : S \to T_{ij} = \{b_{ij}, b_k\}$ which take a_i and a_j to b_{ij} and a_k to b_k, and
- the map $f_3 = S \to T_3 = \{b_1, b_2, b_3\}$ which takes a_i to b_i.

Then Δw is (represented by) the sum of Δ_{f_*} over these five choices of f_*. That is:

$$\Delta w = \qquad b_0 \otimes \underbrace{w}_{b_0}$$

$$+ \qquad b_{12}b_3 \otimes \left(\underbrace{a_1 a_2}_{b_{12}} \otimes \underbrace{a_3}_{b_3} \right)$$

$$+ \qquad b_{13}b_2 \otimes \left(\underbrace{a_1 a_3}_{b_{13}} \otimes \underbrace{a_2}_{b_2} \right)$$

$$+ \qquad b_1 b_{23} b_1 b_{23} \otimes \left(\underbrace{a_1}_{b_1} \otimes \underbrace{a_2 a_3}_{b_{23}} \right)$$

$$+ \qquad b_1 b_2 b_3 \otimes \left(\underbrace{a_1}_{b_1} \otimes \underbrace{a_2}_{b_2} \otimes \underbrace{a_3}_{b_3} \right).$$

Proposition 1 *The decomposition map and the counit map give* $\mathcal{W} = (W, \varepsilon, \Delta)$ *the structure of a cooperad.*

Proof It suffices to show coassociativity holds separately on each individual factor in the limit making up $\mathsf{W} \bar{\circ} \mathsf{W} \bar{\circ} \mathsf{W}$. Given a word w in S and surjections $S \xrightarrow{f} T \xrightarrow{g} U$, we have the following two compositions of decompositions:

$$\left(\Delta_g \otimes \mathrm{id} \right) \Delta_f(w) = \overline{g(\overline{f(w)})} \otimes \bigotimes_{u \in U} \overline{\overline{f(w)}|g^{-1}(u)} \otimes \bigotimes_{t \in T} \overline{w|f^{-1}(t)}$$

and

$$\left(\mathrm{id} \otimes \bigotimes_{u \in U} \Delta_{f|_{(gf)^{-1}(u)}} \right) \Delta_{gf}(w)$$

$$= \overline{gf(w)} \otimes \bigotimes_{u \in U} \left(\overline{f(w|(gf)^{-1}(u))} \otimes \bigotimes_{t \in g^{-1}(u)} \overline{w|(gf)^{-1}(u)|f^{-1}(t)} \right)$$

$$= \overline{gf(w)} \otimes \bigotimes_{u \in U} \overline{f(w|(gf)^{-1}(u))} \otimes \bigotimes_{t \in T} \overline{w|(gf)^{-1}(g(t))|f^{-1}(t)}.$$

To show coassociativity, we will show that the terms in the product match up individually. This means that there are three easy verifications to make. First, it is a direct application of Lemma 5 that

$$\overline{g(\overline{f(w)})} = \overline{gf(w)}.$$

Second, using Lemmas 5–7, we see

$$\overline{\overline{f(w)}|g^{-1}(u)} = \overline{f(w)|g^{-1}(u)} = \overline{f(w|(gf)^{-1}(u))} = \overline{f(\overline{w|(gf)^{-1}(u)})} \,.$$

Finally, using Lemma 6 again, we see that

$$\overline{\overline{w|(gf)^{-1}(g(t))}|f^{-1}(t)} = \overline{(w|(gf)^{-1}(g(t)))\,|f^{-1}(t)} = \overline{w|f^{-1}(t)} \,.$$

We omit the verification of counitality.   □

3.3   The Non-crossing Word Cooperad

Definition 9 The *non-crossing species* N assigns to the set S the $\mathbb{K}$-vector space spanned by pangrammatic reduced non-crossing words on S. Similarly, the *crossing species* X assigns to S the span of pangrammatic reduced crossing words on S.

There is a natural inclusion of X into W whose quotient is isomorphic to N.

Proposition 2 *The quotient map* W $\rightarrow$ N *makes the non-crossing species a quotient cooperad of the word cooperad.*

Proof X(1) is zero dimensional so the counit descends to the quotient.

Let w be an arbitrary crossing word in the alphabet S. Then it is only necessary to show that $\Delta(w)$ is in the kernel of the map W $\bar{\circ}$ W $\rightarrow$ N $\bar{\circ}$ N. The word w contains the pattern $\ldots a \ldots b \ldots a \ldots b \ldots$ for distinct a and b in S. Consider $\Delta_f(w)$ for some surjection $f : S \rightarrow T$. If $f(a) \neq f(b)$ then $f(w)$ and hence its reduction $\overline{f(w)}$ is crossing. On the other hand, if $f(a) = f(b)$ then $f|f^{-1}f(a)$ and hence its reduction $\overline{f|f^{-1}f(a)}$ is crossing. Therefore $\Delta(w)$ is contained in X $\bar{\circ}$ W + W $\bar{\circ}$ X.   □

Definition 10 We call $\mathcal{N} = (N, \varepsilon, \Delta)$, where ε and Δ are induced by the quotient map W $\rightarrow$ N, the *non-crossing word cooperad*.

The following is a direct calculation.

Lemma 8 *Let w be a pangrammatic non-crossing word on the alphabet T and let S be a subset of T. Then $w|S$ is non-crossing.*

Corollary 1 *The decomposition map of the non-crossing word cooperad applied to the word w is the limit of $\Delta_f^{nc}(w)$, where $\Delta_f^{nc}(w)$ is equal to $\Delta_f(w)$ if $f(w)$ is non-crossing and 0 if $f(w)$ is crossing.*

Remark 4 Both of the operads constructed here clearly have some relationship to the sequence operad [10] and cactus operad.[8, 20]. This is perhaps easiest to see with the very clean presentation in [6]. There the authors describe two operads whose underlying species differ from those considered here only by allowing words to begin and end with the same letter.

From either a cactus or sequence perspective, the subspecies specified by this additional condition forms a suboperad. For surjections, which are described combinatorially, the condition itself probably gives the best description. For cacti, one can say that it is the suboperad of cellular chains of spineless cacti where the global root coincides with some intersection of lobes.

Based on this, a naive guess might be that the cooperads here are duals of appropriate suboperads of cacti or sequences. However, the decomposition is *not* dual to the composition map of sequences or cacti, at least not in terms of the most straightforward identification of linear basis elements. In fact, a little further thought shows that the straightforward identification of words with themselves could not possibly have been a dual isomorphism. This is because the cacti and sequence operads are graded (in fact differential graded) and so a dual presentation would respect the grading. But it is easy to trace the induced "grading" on the (non-crossing) word cooperad and see that in fact it is only a filtration, not actually a grading because the decomposition maps are not homogeneous with respect to it.

There is still some hope that the word cooperads are dual to (the underlying operads in vector spaces) of some suboperads of cacti or sequences, but this filtration result shows that this could only be possible if the "natural" basis for the cooperads constructed here is actually inhomogeneous with respect to the grading. So the relationship, should it exist, must use some subtler identification. Ben Ward has pointed out that the suboperad of "generic" cacti, where no more than two cactus lobes can meet at a point, is dual to an appropriately defined subcooperad of the non-crossing word cooperad. This corresponds to taking only leading terms in the filtration and constitutes an encouraging sign.

It is also possible that both of these cooperads, along with cacti and sequences, are mutual specializations of some common ancestor, a sort of ur-operad/cooperad of words but do not directly relate to one another without passing through this ancestor.

4   Review of (Homotopy) Probability Theory

This section consists of the glue directly connecting what we have set up to our main application. First we review an operadic framework for homotopy probability theory, and then recall the free cumulants, which govern free independence in non-commutative probability theory.

4.1   Review of Homotopy Probability Theory

We recall in a few words the setup of homotopy probability theory in operadic terms.

Homotopy probability theory was introduced by Park [14] as a simplification of his algebraic model for quantum field theory where Planck's constant plays no role. The most complete reference is Park's monograph [15], which differs in both notation and definitions from this paper but agrees in spirit with what is here.

One of Park's motivations was to generalize and properly axiomatize (algebraic) probability spaces in terms of homotopy algebra. The following is a "classical" definition before generalization (see, for example, [11]).

Definition 11 A *non-commutative probability space* (respectively, a *commutative algebraic probability space*) is a unital associative (unital commutative associative) $\mathbb{K}$-algebra V equipped with a unit-preserving linear map E from V to $\mathbb{K}$. We assume no further compatibility between the linear map and the algebra structure. The elements of V are called *random variables* and the map E is called the *expectation*.

Remark 5 Since commutative algebraic probability spaces most typically arise as measurable functions on a measure space they are often defined to satisfy additional analytic properties that we will ignore here. See e.g., [17].

Two basic ingredients of the motivation to generalize this definition come from physics, where the random variables are the *observables* in a quantum field theory.

First of all, usually a field theory possesses physical symmetries. For symmetries of the classical action, this is an old and well-known part of the BV-BRST formalism that can be dealt with by introducing so-called ghosts. This amounts to replacing the linear space of observables with a chain complex.

There is another kind of symmetry that may come into play, namely symmetry of the expectation. In particular, we only expect closed elements in the complex to be observables, and we expect boundaries in the chain complex to be trivial observables (in well-behaved cases, the converse should also be true, at least morally). This symmetry of the expectation is probably less understood and analyzed in these terms than symmetry of the action. See [15, Section 6] for some discussion of this point.

In the following definition, a unital version of a definition in [4], we stick to the associative framework, but there is clearly a commutative variation.

Definition 12 A *unital associative homotopy probability space* is a unital graded associative $\mathbb{K}$-algebra equipped with a differential which kills the unit and a unit-preserving chain map to the ground field.

A unital associative homotopy probability space concentrated in degree zero is precisely a non-commutative probability space as defined above.

However, this definition cannot capture the full subtlety of the observables in a quantum field theory. Usually, the symmetries of the action are not compatible with the product, so that the product of observables may not be observables (the product of closed elements may not be closed). Instead, the product may need to be "corrected" in some way to be fully defined. Homotopy probability theory can be traced back to Park's observation of this problem and a potential solution for it in [13].

One way to deal with the problem of correcting the classical product is via homotopy algebra, which gathers together these corrections into a coherent package. But this leads naturally to an algebraic generalization where there is not a single product out of which many products can be built, but rather a binary product, an independent trilinear product, and so on. Again, this point of view is espoused at much greater length and in more detail in [15]. Following Park, here we take a broad view and treat this system of corrections as a black box, defining the algebraic structure as minimally as possible.

The following definition defines our spaces of random variables or observables along with mock products, which basically don't need to satisfy any algebraic identities or respect the differential. See Sect. 2.5 for the definition of strong coaugmentation and the notation below.

Definition 13 Let C be a strongly coaugmented species. A C-*correlation algebra* is a chain complex V equipped with a degree zero linear map (not necessarily a chain map) $\varphi_V : C \circ V \to V$ such that, for $|S| = 1$, we have

$$V \cong C_S \circ V \to C \circ V \xrightarrow{\varphi_V} V$$

is the identity.
Next, we encode the expectation.

Definition 14 Let C be a strongly coaugmented species. Fix a C-correlation algebra $\mathbb{A}$. An $\mathbb{A}$-*valued homotopy* C-*probability space* is a C-correlation algebra (V, φ_V) equipped with

1. a map η of chain complexes $\mathbb{A} \to V$, called the *unit*, such that $\varphi_V \circ C\eta = \eta \circ \varphi_{\mathbb{A}}$ and
2. a map E of chain complexes from V to $\mathbb{A}$, called the *expectation*, such that $E \circ \eta = \mathrm{id}_{\mathbb{A}}$.

The conditions on the maps η and E are equivalent to the commutativity of the following diagram.

$$
\begin{array}{ccccc}
C \circ \mathbb{A} & \xrightarrow{\varphi_{\mathbb{A}}} & \mathbb{A} & \xrightarrow{\mathrm{id}_{\mathbb{A}}} & \mathbb{A} \\
\downarrow{\scriptstyle C\eta} & & \downarrow{\scriptstyle \eta} & \nearrow{\scriptstyle E} & \\
C \circ V & \xrightarrow{\varphi_V} & V & &
\end{array}
$$

Remark 6 Definitions 13 and 14 provide definitions for homotopy probability theory over an arbitrary strongly coaugmented species. The case of the species Ass was addressed in [4]; the case of the species Com was addressed in [3, 5]. The specialization of the definition given here to the appropriate cooperads is *not* equivalent to the definitions given there. Rather, the definition here is more general. See Remark 2 of [5]. Park [15] addresses the cocommutative case at a roughly comparable level of generality.

In order to define C-correlation algebras and C-probability spaces as above, the only structure on C is that of a species. From a probabilistic point of view, this structure should be taken as insufficient, because it includes no choice of regime to decide on *independence*. Independence is a critical feature in probability theory. So-called cumulants gather the information of a probability space in a way that facilitates the study of independence; the cumulant of a sum of independent random variables is the sum of the individual cumulants. In order to include a notion of independence in the probability spaces under consideration, we shall endow the species C with additional structure, namely that of a cooperad.

This article is only intended to establish a relationship between the noncrossing word cooperad and free cumulants. It is not intended to establish a full homotopy probability theory in the free setting. Because of this, the recollection below may be too terse for some. Therefore, regardless of any differences in definitions, the interested or puzzled reader is advised to consult the references above (especially the monograph [15]) for more details about homotopy probability theory.

Now let $C = (\mathsf{C}, \varepsilon, \Delta)$ be a strongly coaugmented cooperad and let V be an $\mathbb{A}$-valued homotopy C-probability space. The *C-cumulant morphism* is the C-coalgebra map $\tilde{K}$ (or its adjoint $K : \mathsf{C} \circ V \to \mathbb{A}$) that fits into the following diagram of C-coalgebras (well-defined because $\tilde{\varphi}_{\mathbb{A}}$ is an automorphism by Lemma 4):

$$
\begin{array}{ccc}
C \circ \mathbb{A} & \xrightarrow{\tilde{\varphi}_{\mathbb{A}}} & C \circ \mathbb{A} \\
{\scriptstyle \tilde{K}} \Big\uparrow & & \Big\uparrow {\scriptstyle \tilde{E} = C \circ E} \\
C \circ V & \xrightarrow[\tilde{\varphi}_V]{} & C \circ V .
\end{array}
\tag{1}
$$

Example 3

1. We reinterpret a unital associative homotopy probability space (V, η, E) in our current framework. Since the underlying species of $\mathcal{A}ss$ and $co\mathcal{A}ss$ are the same, the associative algebra structure map $\mathsf{Ass} \circ \mathbb{K} \to \mathbb{K}$ makes $\mathbb{K}$ into a Ass-correlation algebra (and similarly for V).

   Because the unit η is an algebra map and the expectation E respects η, the conditions of Definition 14 are satisfied and we thus have the data of a $\mathbb{K}$-valued homotopy Ass-probability space. The $co\mathcal{A}ss$-cumulant morphism K is made up of the so-called *Boolean cumulants* of the non-commutative (homotopy) probability space. That is, $K_{[n]}$ is the nth Boolean cumulant. This is essentially the main example of [4].

2. Now assume V is as above but also commutative. Then it is a commutative homotopy probability space in the sense of [5]. Again this is supposed to generalize a classical definition. If V is concentrated in degree zero and satisfies two simple inequalities, then it is an algebraic probability space in the sense of [17].

As above, the identification of the underlying species of *Com* and *coCom* gives maps $\varphi_{\mathbb{K}}$ and φ_V which are defined as in the previous example: Com $\circ \mathbb{K} \to \mathbb{K}$ (and likewise for V). Altogether then, this is the data of a $\mathbb{K}$-valued homotopy Com-probability space. The *coCom*-cumulant morphism K encapsulates the so-called *classical cumulants* of the classical algebraic (or homotopy commutative) probability space. This is essentially the main example of [5].

Remark 7

1. The definitions of correlation algebras and probability spaces only required a species, but the cumulant morphism uses the cooperadic structure in a fundamental to extend the correlation algebra structure to a morphism of cofree coalgebras.
2. The cumulants of a probability space (whether classical, Boolean, or free) can be defined combinatorially in terms of Möbius inversion using an appropriate poset of partitions. One can view the encapsulation of the cumulants of a probability space in terms of operadic algebra as a sort of algebraic enrichment of this combinatorial data, where the choice of cooperad corresponds to the choice of appropriate type of partition.

4.2 Review of Free Cumulants

The correct notion for independence in many non-commutative contexts is *free independence*, discovered by Voiculescu [18] (or see the historical survey [19]) and studied by many others since then. We briefly recall free cumulants. See [12] for a quick overview and [11] for a more detailed introduction to free cumulants and their connection to free probability theory in general.

Definition 15 A *non-crossing partition* of N is a surjective map f from $[n]$ to $[k]$ such that:

1. (ordering) if $i < j$ then $\min\left(f^{-1}(i)\right) < \min\left(f^{-1}(j)\right)$ and
2. (non-crossing) $f(1, 2, \ldots N)$ is a non-crossing word in $[k]$.

We call k the *size* of f.

Definition 16 ([11, 11.1]) Let V be a unital $\mathbb{K}$-algebra, let $(\varrho_n)_{n \geq 1}$ be a sequence of functionals $V^{\otimes n} \xrightarrow{\varrho_n} \mathbb{K}$, and let f be a non-crossing partition of N of size k. Then the *multiplicative extension* $\varrho_f : V^{\otimes N} \to \mathbb{K}$ is defined as

$$\varrho_f(a_1 \otimes \cdots \otimes a_n) = \prod_{i=1}^{k} \varrho_{|f^{-1}(i)|}\left(\underline{a_{f^{-1}(i)}}\right) .$$

Here $\underline{a_{f^{-1}(i)}}$ is the tensor product $a_{j_1} \otimes \cdots \otimes a_{j_{|f^{-1}(i)|}}$ where $j_1, \ldots, j_{|f^{-1}(i)|}$ is the restriction $a_1, \ldots, a_n | f^{-1}(i)$.

Definition 17 ([11, 11.4 (3)]) Let (V, E) be a non-commutative probability space. The *free cumulants* of V are the unique functions $\{\kappa_N\}$ whose multiplicative extension satisfies the defining equation

$$E(a_1 \cdots a_N) = \sum_f \kappa_f(a_1 \otimes \cdots \otimes a_N)$$

as f ranges over non-crossing partitions.

5 The Non-crossing Word Cooperad and Free Probability Theory

Finally, we relate non-commutative probability spaces to N-correlation algebras and homotopy N-probability spaces and show that the $\mathcal{N}$-cumulant morphism of a $\mathbb{K}$-valued homotopy N-probability space recovers the free cumulants defined above.

Definition 18 We define a map ψ of species from the non-crossing species N to the underlying species Ass of the associative operad (defined in Example 1). Under the map ψ, a word w in the letters $\{w_1, \ldots, w_{|S|}\}$ goes to the order f_w where $f_w(w_i) = j$ if the subword of w which ends with the first occurrence of w_i in w contains j letters from the alphabet.

Now, as in the first example above, let V be a unital associative homotopy probability space.

We can give both $\mathbb{K}$ and V the structure of N-correlation algebras by composing the map ψ with the structure maps of the associative algebras V and $\mathbb{K}$:

$$N \circ V \xrightarrow{\psi} \text{Ass} \circ V \xrightarrow{\text{structure}} V \,,$$

$$N \circ \mathbb{K} \xrightarrow{\psi} \text{Ass} \circ \mathbb{K} \xrightarrow{\text{structure}} \mathbb{K} \,.$$

As before, since the map E preserves the unit and the unit is a map of associative algebras, they are compatible with this structure and the whole package is then the data of a $\mathbb{K}$-valued homotopy N-probability space.

Now we are ready for the main theorem.

Theorem 1 *Let (V, E) be a non-commutative probability space, viewed as above as a $\mathbb{K}$-valued homotopy N-probability space.*

Then the $\mathcal{N}$-cumulant morphism K recovers the free cumulants of the probability space.

Proof Consider the defining diagram (1) of the cumulant morphism. By adjunction into vector spaces (or chain complexes), we may restrict the right half of the diagram without losing information, as follows.

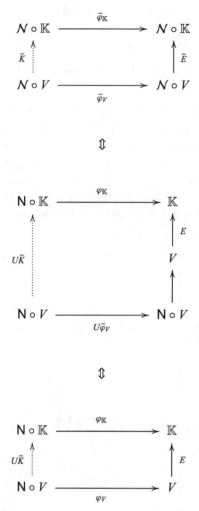

Let w_N be the word $1, \ldots, N$ in the alphabet $[N]$. Define $K_N : V^{\otimes N} \to \mathbb{K}$ in terms of the $\mathcal{N}$-cumulant morphism as

$$K_N(z) = K(w_N \otimes z) .$$

We will show that the map K_N is precisely the Nth free cumulant map.

Apply the maps making up the bottom commutative square to the element of $\mathsf{N} \circ V$ represented by $w_N \otimes (v_1 \otimes \cdots \otimes v_N)$. The map φ_V is just multiplication and so the composition on the bottom and right sides of the square is

$$E(v_1 \cdots v_N) .$$

Recall the vertical map $\tilde{K}$ is defined as the extension of the N-cumulant morphism $K : \mathsf{N} \circ V \to \mathbb{K}$ as follows:

$$
\begin{array}{ccc}
\mathsf{N} \circ V & \xrightarrow{\ U\tilde{K}\ } & \mathsf{N} \circ \mathbb{K} \\[2pt]
{\scriptstyle \Delta_N}\big\downarrow & & \big\uparrow{\scriptstyle \mathsf{N} \circ K} \\[2pt]
(\mathsf{N} \bar{\circ} \mathsf{N}) \circ V & \xrightarrow[\cong]{} & \mathsf{N} \circ (\mathsf{N} \circ V) .
\end{array}
$$

Since $\mathsf{N} \circ (\mathsf{N} \circ V)$ and $\mathsf{N} \circ V$ are defined as colimits (see Sect. 2), in order to evaluate the overall composition $\mathsf{N} \circ V \xrightarrow{U\tilde{K}} \mathsf{N} \circ \mathbb{K} \xrightarrow{\varphi_{\mathbb{K}}} \mathbb{K}$, it suffices to evaluate on a choice of representatives. That is, let S be the (finite) set of surjections f from $[N]$ to $[M]$ such that $i < j$ implies $\min f^{-1}(i) < \min f^{-1}(j)$ (this set exhausts the isomorphism classes of surjections out of $[N]$). Then the following diagram commutes. The diagram may look intimidating but the right hand side is precisely what we are trying to compute while the left hand side just gives a concrete recipe for the calculation.

Using the characterization from Corollary 1, we see that the contribution is 0 for a function f from $[N]$ to $[M]$ if $f(w_N)$ is crossing. Then the subset of functions from $[N]$ to $[M]$ which contribute to the overall composition coincides precisely with the set of functions from $[N]$ to $[M]$ which are non-crossing partitions.

For a given partition f, let us trace the contribution from the f factor in the left side composition. Explicitly, starting with $w_N \otimes V^{\otimes[N]}$, the first vertical map, restricted to the f factor takes this to

$$\overline{f(w_N)} \otimes \bigotimes_{t \in [M]} \overline{w_N | f^{-1}(t)} \otimes V^{\otimes[N]} .$$

The second vertical map is just a change of parenthesization on the factor.

The third vertical map applies K to the factors $\overline{w_N | f^{-1}(t)} \otimes V^{\otimes f^{-1}(t)}$. Because there is no repeated letter in w_N, the reduction is trivial, and we can identify $\overline{w_N | f^{-1}(t)}$ with $w_N | f^{-1}(t)$. Then there is an order-preserving isomorphism between $f^{-1}(t)$ and $[|f^{-1}(t)|]$ which realizes $K(\overline{w_N | f^{-1}(t)} \otimes V^{\otimes f^{-1}(t)})$ as

$$K_N(\underline{w_N | f^{-1}(t) \otimes V^{\otimes f^{-1}(t)}}) .$$

By construction the map ψ takes $f(w_N)$ to the identity order $[M] \rightarrow [M]$ and the final map in the vertical composition is then just the ordered product of the factors corresponding to $f^{-1}(t)$ for t in $[M]$. This product is then

$$\prod_{t=1}^{M} K_N(\underline{w_N | f^{-1}(t) \otimes V^{\otimes f^{-1}(t)}})$$

which is precisely the multiplicative extension of K_f of $(K_1, K_2, \ldots)$.

Thus the overall equation is then

$$E(v_1 \cdots v_N) = \sum_f K_f(v_1 \otimes \cdots \otimes v_N)$$

which demonstrates that K_N satisfy precisely the same defining equations as the free cumulants κ_N.   □

To conclude the paper, we make two caveats about this approach.

Remark 8

1. First of all, this theorem only makes use of the N-cumulant morphism for very special non-crossing words, those of the form $w_N = 1, \ldots, N$. This means that there are many other "cumulants" in this context, not only the free cumulants. For example, applying the same methods with the word $w'_N = 1, 2, \ldots, N - 1, N, N - 1, \ldots, 3, 2$ yields the Boolean cumulants of the same non-commutative probability space. This may be seen either as a feature (flexibility in the method) or a bug (imprecision in the output).

2. More damning is the fact that this method does not seem to work at all in *operator-valued* free probability, where the ground ring is itself non-commutative. In our case, the right hand side of the formula relating expectations and cumulants was a product of individual cumulants κ_n. But in operator-valued free probability, the right hand side includes nested cumulants, like $\kappa_2(a\kappa_1(b) \otimes c)$. This kind of "tree-like" formula does not fit well in this formulism. However, operadic algebra is tailored to describe tree-like compositions and there is a somewhat different and more technical approach using these tools that works in the more general case. This approach is taken in the preprint [2].

Acknowledgements The author gratefully acknowledges useful conversations with Joey Hirsh, John Terilla, Jae-Suk Park, and Ben Ward. An anonymous referee made multiple useful observations.

This work was supported by IBS-R003-D1.

References

1. Berger, C., Fresse, B.: Combinatorial operad actions on cochains. Math. Proc. Camb. Philos. Soc. **137**, 135–174 (2004)
2. Drummond-Cole, G.C.: An operadic approach to operator-valued free cumulants (2016). http://arxiv.org/abs/arxiv:1607.04933
3. Drummond-Cole, G.C., Terilla, J.: Cones in homotopy probability theory (2014). http://arxiv.org/abs/1410.5506
4. Drummond-Cole, G.C., Park, J.S., Terilla, J.: Homotopy probability theory I. J. Homotopy Relat. Struct. **10**, 425–435 (2015). http://link.springer.com/article/10.1007/s40062-013-0067-y
5. Drummond-Cole, G.C., Park, J.S., Terilla, J.: Homotopy probability theory II. J. Homotopy Relat. Struct. **10**, 623–635 (2015). http://link.springer.com/article/10.1007/s40062-014-0078-3
6. Gálvez-Carrillo, I., Lombardi, L., Tonks, A.: An $\mathcal{A}_\infty$ operad in spineless cacti. Mediterr. J. Math. **12**(4), 1215–1226 (2015)
7. Getzler, E., Jones, J.D.S.: Operads, homotopy algebra and iterated integrals for double loop spaces (1994). http://arxiv.org/abs/hep-th/9403055
8. Kaufmann, R.M.: On several varieties of cacti and their relations. Algebr. Geom. Topol. **5**, 237–300 (2005)
9. Kaufmann, R.M.: On spineless cacti, Deligne's conjecture and Connes–Kreimer's Hopf algebra. Topology **46**(1), 39–88 (2007)
10. McClure, J.E., Smith, J.H.: Multivariable cochain operations and little n-cubes. J. Am. Math. Soc. **16**(3), 681–704 (2003)
11. Nica, A., Speicher, R.: Lectures on the Combinatorics of Free Probability. London Mathematical Society Lecture Note Series, vol. 335. Cambridge University Press, Cambridge (2006)
12. Novak, J., Śniady, P.: What is. . . a free cumulant? Not. Am. Math. Soc. **58**(2), 300–301 (2011)
13. Park, J.S.: Flat family of QFTs and quantization of d-algebras (2003). http://arxiv.org/abs/hep-th/0308130
14. Park, J.S.: Einstein chair lecture (2011). City University of New York
15. Park, J.S.: Homotopy theory of probability spaces I: classical independence and homotopy Lie algebras (2015). http://arxiv.org/abs/1510.08289
16. Smirnov, V.: On the cochain complex of topological spaces. Math. USSR Sbornik **43**, 133–144 (1982)

17. Tao, T.: Algebraic probability spaces (2014).   https://terrytao.wordpress.com/2014/06/28/ algebraic-probability-spaces/. Blog post
18. Voiculescu, D.: Symmetries of some reduced free product $C*$-algebras. In: Operator Algebras and Their Connections with Topology and Ergodic Theory: Proceedings of the OATE Conference Held in Bușteni, Romania, Aug 29–Sept 9, 1983. Lecture Notes in Mathematics, vol. 1132, pp. 556–588. Springer, Berlin (1985)
19. Voiculescu, D.: Free probability and the von Neumann algebras of free groups. Rep. Math. Phys. **55**(1), 127–133 (2005)
20. Voronov, A.A.: Notes on universal algebra.   In: Graphs and Patterns in Mathematics and Theoretical Physics. Proceedings of Symposia in Pure Mathematics, vol. 73, pp. 81–103. American Mathematical Society, Providence, RI (2005)
21. Zinbiel, G.W.: Encyclopedia of types of algebras 2010.   In: Operads and Universal Algebra. Nankai Series in Pure Applied Mathematical Theoretical Physics, vol. 9, pp. 217–298. World Scientific Publishing, Singapore (2012)

Endomorphisms of Lie Groups over Local Fields

Helge Glöckner

Abstract Lie groups over totally disconnected local fields furnish prime examples of totally disconnected, locally compact groups. We discuss the scale, tidy subgroups and further subgroups (like contraction subgroups) for analytic endomorphisms of such groups.

1 Introduction

The scale $s(\alpha) \in \mathbb{N}$ of an automorphism (or endomorphism) α of a totally disconnected locally compact group G was introduced in the works of George Willis (see [57, 58, 60]). Following [58] and [60], the scale $s(\alpha)$ can be defined as the minimum of the indices[1]

$$[\alpha(U) : \alpha(U) \cap U],$$

for U ranging through the set $\mathrm{COS}(G)$ of all compact open subgroups of G. Compact open subgroups for which the minimum is attained are called *minimizing*; as shown in [58] and [60], they can be characterized by certain 'tidiness' properties, and therefore coincide with the so-called *tidy subgroups* for α (the definition of which is recalled in Sect. 2).

Besides the tidy subgroups, further subgroups of G have been associated to α which proved to be useful for the study of α, and for the structure theory in general

---

[1]If we wish to emphasize the underlying group G, we write $s_G(\alpha)$ instead of $s(\alpha)$.

H. Glöckner (✉)
Institut für Mathematik, Universität Paderborn, Warburger Str. 100, 33098 Paderborn, Germany

University of Newcastle, Callaghan NSW, Australia
e-mail: glockner@math.upb.de

© Springer International Publishing AG, part of Springer Nature 2018
D.R. Wood et al. (eds.), *2016 MATRIX Annals*, MATRIX Book Series 1,
https://doi.org/10.1007/978-3-319-72299-3_6

(see [1] and [60]). We mention the *contraction group*

$$\overrightarrow{\mathrm{con}}(\alpha) := \left\{ x \in G : \lim_{n \to \infty} \alpha^n(x) = e \right\}$$

and the *parabolic subgroup* $\overrightarrow{\mathrm{par}}(\alpha)$ of all $x \in G$ whose α-orbit $(\alpha^n(x))_{n \in \mathbb{N}_0}$ is *bounded* in the sense that $\{\alpha^n(x) : n \in \mathbb{N}_0\}$ is relatively compact in G. It is also interesting to consider group elements $x \in G$ admitting an *α-regressive trajectory* $(x_{-n})_{n \in \mathbb{N}_0}$ of group elements x_{-n} such that $x_0 = x$ and $\alpha(x_{-n-1}) = x_{-n}$ for all n. Setting $x_n := \alpha^n(x)$ for $n \in \mathbb{N}$, we then obtain a so-called *two-sided α-orbit* $(x_n)_{n \in \mathbb{Z}}$ for x. The *anti-contraction group* $\overleftarrow{\mathrm{con}}(\alpha)$ is defined as the group of all $x \in G$ admitting an α-regressive trajectory $(x_{-n})_{n \in \mathbb{N}_0}$ such that

$$\lim_{n \to \infty} x_{-n} = e;$$

the *anti-parabolic subgroup* $\overleftarrow{\mathrm{par}}(\alpha)$ is the group of all $x \in G$ admitting a bounded α-regressive trajectory. The intersection

$$\mathrm{lev}(\alpha) := \overrightarrow{\mathrm{par}}(\alpha) \cap \overleftarrow{\mathrm{par}}(\alpha)$$

is called the *Levi subgroup* of α; it is the group of all $x \in G$ admitting a bounded two-sided α-orbit (see [1] and [60] for these concepts, which were inspired by terminology in the theory of linear algebraic groups).

In this work, we consider Lie groups over totally disconnected local fields, like the field of p-adic numbers or fields of formal Laurent series over a finite field (see Sects. 2 and 4 for these concepts). The topological group underlying such a Lie group G is a totally disconnected locally compact group, and the analytic endomorphisms $\alpha : G \to G$ we consider are, in particular, continuous endomorphisms of G.

Our goal is twofold: On the one hand, we strive to give an exposition to Lie groups over local fields and their endomorphisms, for readers with varying backgrounds who wish to see examples for the theory of endomorphisms of totally disconnected groups developed in [60]. To this end, we also recall basic concepts concerning Lie groups over totally disconnected local fields, as far as required for the purpose. On the other hand, most of the text can be considered as a research article, as it contains results which are new (or new in the current generality), and which are proved here in full. Compare [21] for a broader (but more sketchy) introduction with a similar thrust, confined to the study of automorphisms. For further information on Lie groups over totally disconnected local fields, see [49] and the references therein, also [48] and [7].[2] Every p-adic Lie group has a compact

---

[2]Contrary to our conventions, the Lie groups in [7] are modelled on Banach spaces which need not be of finite dimension.

open subgroup which is an analytic pro-p-group; see [12, 13], and [48] for the theory of such groups, and Lazard's seminal work [38]. For related studies in positive characteristic, cf. [35] and subsequent studies.

Every group of $\mathbb{K}$-rational points of a linear algebraic group defined over a totally disconnected local field $\mathbb{K}$ can be considered as a Lie group over $\mathbb{K}$ (see [39, Chapter I, Proposition 2.5.2]). We refer to [4, 30, 39], and [52] for further information on such groups, which can be studied with tools from algebraic geometry, and via actions on buildings (see [8] and later work).

The Lie groups we consider need not be algebraic groups, they are merely $\mathbb{K}$-analytic manifolds. Yet, compared to general totally disconnected groups, we have additional structure at our disposal: Every Lie group G over a totally disconnected local field $\mathbb{K}$ has a Lie algebra $L(G)$ (its tangent space $T_e(G)$ at the neutral element $e \in G$), which is a finite-dimensional $\mathbb{K}$-vector space. If $\alpha: G \to G$ is a $\mathbb{K}$-analytic endomorphism, then its tangent map $L(\alpha) := T_e(\alpha)$ at e is a linear endomorphism

$$L(\alpha): L(G) \to L(G)$$

of the $\mathbb{K}$-vector space $L(G)$. It is now natural to ask how the scale and tidy subgroups for α are related to those of $L(\alpha)$. Guided by this question, we describe tidy subgroups and calculate the scale for linear endomorphisms of finite-dimensional $\mathbb{K}$-vector spaces (which also provides a first illustration of the abstract concepts), see Theorem 3.6. For $\alpha: G \to G$ an analytic endomorphism of a Lie group G over a totally disconnected local field $\mathbb{K}$, we shall prove that

$$s(\alpha) = s(L(\alpha)) \tag{1}$$

if and only if the contraction group $\overrightarrow{\mathrm{con}}(\alpha)$ is closed in G (see Theorem 8.13, the main result), which is always the case if char($\mathbb{K}$) = 0 (by Corollary 6.7). If $\overrightarrow{\mathrm{con}}(\alpha)$ is closed, then

$$s(L(\alpha)) = \prod_{\substack{j \in \{1,\dots,m\} \\ \text{s.t. } |\lambda_j|_{\mathbb{K}} \geq 1}} |\lambda_j|_{\mathbb{K}} \tag{2}$$

in terms of the eigenvalues $\lambda_1, \dots, \lambda_m$ of $L(\alpha) \otimes_{\mathbb{K}} \mathrm{id}_{\overline{\mathbb{K}}}$ in an algebraic closure $\overline{\mathbb{K}}$ of $\mathbb{K}$, where $|.|_{\mathbb{K}}$ is the unique extension of the 'natural' absolute value on $\mathbb{K}$ specified in (9) to an absolute value on $\overline{\mathbb{K}}$ (see Theorem 3.6).

The text is organized as follows.

After a preparatory Sect. 2 on background concerning totally disconnected locally compact groups and totally disconnected local fields, we study linear endomorphisms of finite-dimensional $\mathbb{K}$-vector spaces (Sect. 3).

In Sect. 4, we recall elementary definitions and facts concerning $\mathbb{K}$-analytic functions, manifolds, and Lie groups. We then construct well-behaved compact open subgroups in Lie groups over totally disconnected local fields (see Sect. 5).

In Sect. 6, we calculate the scale (and determine tidy subgroups) for α an endomorphism of a p-adic Lie group G. This is simplified by the fact that every p-adic Lie group has an exponential function, which provides a local conjugacy between the dynamical systems $(L(G), L(\alpha))$ and (G, α) around the fixed points 0 (resp., e).

By contrast, analytic endomorphisms $\alpha: G \to G$ cannot be linearized in general if G is a Lie group over a local field of positive characteristic (see [21, 4.8.3] for a counterexample). As a replacement for a linearization, we use (locally) invariant manifolds (viz. local stable, local unstable, and centre manifolds) around the fixed point e of the time-discrete, analytic dynamical system (G, α). As shown in [18] and [19], the latter can be constructed as in the classical real case (cf. [32] and [56]). The necessary definitions and facts are compiled in Sect. 7.

The following section contains the main results, notably a calculation of the scale for analytic endomorphisms $\alpha: G \to G$ of a Lie group G over a totally disconnected local field, if $\overrightarrow{\mathrm{con}}(\alpha)$ is closed (see Theorem 8.13). We also show that if $\overrightarrow{\mathrm{con}}(\alpha)$ is closed, then $\overrightarrow{\mathrm{con}}(\alpha)$, $\mathrm{lev}(\alpha)$, and $\overleftarrow{\mathrm{con}}(\alpha)$ are Lie subgroups of G and the map

$$\overrightarrow{\mathrm{con}}(\alpha) \times \mathrm{lev}(\alpha) \times \overleftarrow{\mathrm{con}}(\alpha) \to \overrightarrow{\mathrm{con}}(\alpha)\,\mathrm{lev}(\alpha)\,\overleftarrow{\mathrm{con}}(\alpha) =: \Omega$$

taking (a, b, c) to abc has open image Ω and is an analytic diffeomorphism (see Theorem 8.15).

The final three sections are devoted to automorphisms with specific properties. An automorphism $\alpha: G \to G$ of a totally disconnected, locally compact group G is called *contractive* if $G = \overrightarrow{\mathrm{con}}(\alpha)$, i.e.,

$$\lim_{n \to \infty} \alpha^n(x) = e \quad \text{for all } x \in G$$

(see [50] and [26]). If

$$\bigcap_{n \in \mathbb{Z}} \alpha^n(V) = \{e\}$$

for some identity neighbourhood $V \subseteq G$, then α is called *expansive* (see [23]), or also *of finite depth* in the case of compact G (see [59]). If

$$e \notin \overline{\{\alpha^n(x): n \in \mathbb{Z}\}}$$

for each $x \in G \setminus \{e\}$, then α is called a *distal* automorphism (cf. [42, 43]). Every contractive automorphism is expansive (see, e.g., [23]).

If G is a Lie group over a totally disconnected local field $\mathbb{K}$ with algebraic closure $\overline{\mathbb{K}}$ and $\alpha: G \to G$ an analytic automorphism, then $\overrightarrow{\mathrm{con}}(\alpha)$ is open in G (resp., α is expansive, resp., α is distal) if and only if

$$|\lambda|_{\overline{\mathbb{K}}} < 1$$

(resp., $|\lambda|_{\mathbb{K}} \neq 1$, resp., $|\lambda|_{\mathbb{K}} = 1$) for each eigenvalue λ of the $\overline{\mathbb{K}}$-linear automorphism $L(\alpha) \otimes_{\mathbb{K}} \mathrm{id}_{\overline{\mathbb{K}}}$ of $L(G) \otimes_{\mathbb{K}} \overline{\mathbb{K}}$ obtained by extension of scalars from $\mathbb{K}$ to $\overline{\mathbb{K}}$ (see Proposition 7.10 for details).

Recall that every continuous homomorphism between p-adic Lie groups is analytic, whence the Lie group structure on a p-adic Lie group is uniquely determined by the underlying topological group (see, e.g., [7]). Lazard [38] characterized p-adic Lie groups within the class of all totally disconnected, locally compact compact groups, and later many further characterizations were found (see [12]).

Recent research showed that p-adic Lie groups are among basic building blocks for general totally disconnected groups in various situations, e.g. in the study of ergodic $\mathbb{Z}^n$-actions on locally compact groups by automorphisms (see [11]) and also in the theory of contraction groups (see [26]). In both cases, Lazard's theory of analytic pro-p-groups was invoked to show that the groups in contention are p-adic Lie groups. Section 9 surveys results concerning contractive automorphisms. We give an alternative, new argument for the appearance of p-adic Lie groups, using the structure theory of locally compact abelian groups (i.e., Pontryagin duality) instead of the theory of analytic pro-p-groups.

Section 10 briefly surveys results concerning expansive automorphisms.

The final section is devoted to distal automorphisms and Lie groups of type R; we prove a criterion for pro-discreteness (Theorem 11.2) which had been announced in [21, Proposition 4.54].

Further papers have been written on the foundation of [60]: Analogues of results from [1, 36], and [58] for endomorphisms of totally disconnected, locally compact groups were developed in [9]; the topological entropy $h_{\text{top}}(\alpha)$ of an endomorphism α of a totally disconnected, locally compact group G was studied in [14]. It was shown there that

$$h_{\text{top}}(\alpha) = \ln s(\alpha) \tag{3}$$

if and only if the so-called nub subgroup $\mathrm{nub}(\alpha)$ of α (as in [60]) is trival (see [14, Corollary 4.11]); the latter holds if and only if $\overrightarrow{\mathrm{con}}(\alpha)$ is closed (as shown in [9, Theorem D]). In the current paper, we can do with the results from [60] and give Lie-theoretic proofs for results which can be generalized further (see [9]), by more involved arguments.[3] The results were obtained before those of [9], and presented in the author's minicourse June 27–July 1, 2016 at the MATRIX workshop and in a talk at the AMSI workshop July 25, 2016.[4] For complementary studies of endomorphisms of pro-finite groups, see [45].

---

[3]Notably, we have the Inverse Function Theorem at our disposal.

[4]Except for results concerning the scale on subgroups and quotients (Proposition 8.27) and the endomorphism case of Lemma 8.20, which were added in 2017. In the talks, I also confined myself to a proof of the equivalence of (a) and (b) in Theorem 8.13 when α is an automorphism, which is easier (while the theorem was stated in full).

Conventions We write $\mathbb{N} := \{1, 2, \ldots\}$ $\mathbb{N}_0 := \mathbb{N} \cup \{0\}$, and $\mathbb{Z} := \mathbb{N}_0 \cup (-\mathbb{N})$. Endomorphisms of topological groups are assumed continuous; automorphisms of topological groups are assumed continuous, with continuous inverse. If we say that a mapping f is an analytic diffeomorphism (or an analytic automorphism), then also f^{-1} is assumed analytic. If E is a vector space over a field $\mathbb{K}$, we write $\mathrm{End}_{\mathbb{K}}(E)$ for the $\mathbb{K}$-algebra of all $\mathbb{K}$-linear endomorphisms of E, and $\mathrm{GL}(E) := \mathrm{End}_{\mathbb{K}}(E)^{\times}$ for its group of invertible elements. If $\mathbb{L}$ is a field extension of $\mathbb{K}$, we let $E_{\mathbb{L}} := E \otimes_{\mathbb{K}} \mathbb{L}$ be the $\mathbb{L}$-vector space obtained by extension of scalars. We identify E with $E \otimes 1 \subseteq E_{\mathbb{L}}$ as usual. Given $\alpha \in \mathrm{End}_{\mathbb{K}}(E)$, we let $\alpha_{\mathbb{L}} := \alpha \otimes \mathrm{id}_{\mathbb{L}}$ be the endomorphism of $E_{\mathbb{L}}$ obtained by extension of scalars. If $\overline{\mathbb{K}}$ is an algebraic closure of $\mathbb{K}$, we shall refer to the eigenvalues $\lambda \in \overline{\mathbb{K}}$ of $\alpha_{\overline{\mathbb{K}}}$ simply as the *eigenvalues of α in $\overline{\mathbb{K}}$*. Given $n \in \mathbb{N}$, we write $M_n(\mathbb{K})$ for the $\mathbb{K}$-algebra of $n \times n$-matrices. If $f: X \to X$ is a self-map of a set X, we say that a subset $Y \subseteq X$ is *α-stable* if $\alpha(Y) = Y$. If $\alpha(Y) \subseteq Y$, then Y is called *α-invariant*. If X is a set, $Y \subseteq X$ a subset, $f: Y \to X$ a map and $x \in Y$, we say that a sequence $(x_{-n})_{n \in \mathbb{N}_0}$ of elements $x_n \in Y$ is an *f-regressive trajectory for x* if $f(x_{-n-1}) = x_{-n}$ for all $n \in \mathbb{N}_0$ and $x_0 = x$. In this situation, we also say that x *admits the f-regressive trajectory* $(x_{-n})_{n \in \mathbb{N}_0}$. If, instead, f is defined on a larger subset of X which contains Y but all x_n are elements of Y, we call $(x_{-n})_{n \in \mathbb{N}_0}$ an *f-regressive trajectory in Y*.

2 Some Basics of Totally Disconnected Groups

In this section, we recall basic definitions and facts concerning totally disconnected locally compact groups and totally disconnected local fields.

The Module of an Automorphism Let G be a locally compact group and $\mathbb{B}(G)$ be the σ-algebra of Borel subsets of G. Let $\lambda_G: \mathbb{B}(G) \to [0, \infty]$ be a Haar measure on G, i.e., a non-zero Radon measure which is left invariant in the sense that $\lambda_G(gA) = \lambda_G(A)$ for all $g \in G$ and $A \in \mathbb{B}(G)$. It is well-known that a Haar measure exists, and that it is unique up to multiplication with a positive real number (cf. [28]). If $\alpha: G \to G$ is an automorphism, then also

$$\mathbb{B}(G) \to [0, \infty], \qquad A \mapsto \lambda_G(\alpha(A))$$

is a left invariant non-zero Radon measure on G and hence a multiple of Haar measure: There exists $\Delta(\alpha) > 0$ such that $\lambda_G(\alpha(A)) = \Delta(\alpha)\lambda_G(A)$ for all $A \in \mathbb{B}(G)$. If $U \subseteq G$ is a relatively compact, open, non-empty subset, then

$$\Delta(\alpha) = \frac{\lambda_G(\alpha(U))}{\lambda_G(U)} \tag{4}$$

(cf. [28, (15.26)], where however the conventions differ). We also write $\Delta_G(\alpha)$ instead of $\Delta(\alpha)$, if we wish to emphasize the underlying group G.

Remark 2.1 Let U be a compact open subgroup of G. If $U \subseteq \alpha(U)$, with index $[\alpha(U) : U] =: n$, we can pick representatives $g_1, \ldots, g_n \in \alpha(U)$ for the left cosets of U in $\alpha(U)$. Exploiting the left invariance of Haar measure, (4) turns into

$$\Delta(\alpha) = \frac{\lambda_G(\alpha(U))}{\lambda_G(U)} = \sum_{j=1}^{n} \frac{\lambda_G(g_j U)}{\lambda_G(U)} = [\alpha(U) : U]. \tag{5}$$

If $\alpha(U) \subseteq U$, applying (5) to α^{-1} instead of α and $\alpha(U)$ instead of U, we obtain

$$\Delta(\alpha^{-1}) = [U : \alpha(U)]. \tag{6}$$

Tidy Subgroups and the Scale If G is a totally disconnected, locally compact group, $\alpha: G \to G$ an endomorphism and U a compact open subgroup of G, following [60] we write

$$U_- := \bigcap_{n \in \mathbb{N}_0} \alpha^{-n}(U) = \{x \in U : (\forall n \in \mathbb{N}_0)\, \alpha^n(x) \in U\},$$

where $\alpha^{-n}(U)$ means the preimage $(\alpha^n)^{-1}(U)$. Let U_+ be the set of all $x \in U$ admitting an α-regressive trajectory $(x_{-n})_{n \in \mathbb{N}_0}$ in U. Then

$$U_+ = \bigcap_{n \in \mathbb{N}_0} U_n \quad \text{with}$$

$$U_0 := U \quad \text{and} \quad U_{n+1} := U \cap \alpha(U_n) \quad \text{for } n \in \mathbb{N}_0; \tag{7}$$

moreover, U_+ and U_- are compact subgroups of G such that

$$\alpha(U_-) \subseteq U_- \quad \text{and} \quad \alpha(U_+) \supseteq U_+$$

(see [60]). The sets

$$U_{--} := \bigcup_{n \in \mathbb{N}_0} \alpha^{-n}(U_-) \quad \text{and} \quad U_{++} := \bigcup_{n \in \mathbb{N}_0} \alpha^n(U_+)$$

are unions of ascending sequences of subgroups, whence they are subgroups of G.

2.2 If we wish to emphasize which endomorphism α is considered, we write $U_{n,\alpha}$, $U_{+,\alpha}$, and $U_{-,\alpha}$ instead of U_n, U_+, and U_-, respectively.
The following definition was given in [60].

Definition 2.3 If $U = U_+ U_-$, then U is called *tidy above* for α. If U_{++} is closed in G and the indices

$$[\alpha^{n+1}(U_+) : \alpha^n(U_+)] \in \mathbb{N}$$

are independent of $n \in \mathbb{N}_0$, then U is called *tidy below* for α. If U is both tidy above and tidy below for α, then U is called *tidy for* α.
The following fact (see [60, Proposition 9]) is useful for our ends:

2.4 *U is tidy for α if and only if U is tidy above and U_{--} is closed in G.*

2.5 As shown in [60], a compact open subgroup U of G is minimizing for α (as defined in Sect. 1) if and only if it is tidy for α, in which case

$$s(\alpha) = [\alpha(U) : \alpha(U) \cap U] = [\alpha(U_+) : U_+].$$

2.6 If α is an automorphism of G, then simply (as in [58])

$$U_+ = \bigcap_{n \in \mathbb{N}_0} \alpha^n(U).$$

Let us consider some easy special cases (which will be useful later).

Lemma 2.7 *Let α be an endomorphism of a totally disconnected, locally compact group G.*

(a) *If $V \subseteq G$ is a compact open subgroup such that $\alpha(V) \subseteq V$, then V is tidy, $V_- = V$ and $s(\alpha) = 1$.*
(b) *If $V \subseteq G$ is a compact open subgroup with $V \subseteq \alpha(V)$, then V is tidy above for α, and $V_+ = V$. If, moreover, V is tidy, then $s(\alpha) = \Delta(\alpha)$.*
(c) *If α is nilpotent (say $\alpha^n = e$) and $U \subseteq G$ a compact open subgroup, then*

$$V := U_- = \bigcap_{k=0}^{\infty} \alpha^{-k}(U) = \bigcap_{k=0}^{n-1} \alpha^{-k}(U) \tag{8}$$

is a compact open subgroup of G with $\alpha(V) \subseteq V$.

Proof

(a) Since $V \subseteq \alpha^{-1}(V)$, we have $V \subseteq \alpha^{-k}(V)$ for all $k \in \mathbb{N}_0$ and thus

$$V_- = \bigcap_{k=0}^{\infty} \alpha^{-k}(V) = V.$$

Hence $V = V_+ V_-$ is tidy above. As the subgroup

$$V_{--} = \bigcup_{k=0}^{\infty} \alpha^{-k}(V_-) = \bigcup_{k=0}^{\infty} \alpha^{-k}(V)$$

contains V, it is open and hence closed. Thus V is tidy for α, by 2.4. Finally $\alpha(V) \subseteq V$ entails that $s(\alpha) = [\alpha(V) : \underbrace{\alpha(V) \cap V}_{=\alpha(V)}] = 1$.

(b) Since $V \subseteq \alpha(V)$, every $x \in V$ has an α-regressive trajectory within V, whence $x \in V_+$. Hence $V = V_+$, and thus $V = V_+ V_-$ is tidy above. If V is tidy, then $s(\alpha) = [\alpha(V) : \alpha(V) \cap V] = [\alpha(V) : V] = \Delta(\alpha)$, using (5).

(c) For integers $k \geq n$, we have $\alpha^k(x) = e \in U$ for all $x \in G$, whence $x \in \alpha^{-k}(U)$ and thus $\alpha^{-k}(U) = G$. This entails the second equality in (8), and so V is compact and open. As $\alpha(U_-) \subseteq U_-$, the final inclusion holds. $\qquad\square$

2.8 If G is a totally disconnected, locally compact group and $g \in G$, let

$$I_g : G \to G, \quad x \mapsto gxg^{-1}$$

be the corresponding inner automorphism of G. Given $g \in G$, abbreviate $s(g) := s(I_g)$. Following [57], the mapping $s : G \to \mathbb{N}$ so obtained is called the *scale function* on G.

Local Fields Basic information on totally disconnected local fields can be found in many books, e.g. [55] and [34].

By a *totally disconnected local field*, we mean a totally disconnected, locally compact, non-discrete topological field $\mathbb{K}$.

Each totally disconnected local field $\mathbb{K}$ admits an *ultrametric absolute value* $|.|$ defining its topology, i.e.,

(a) $|t| \geq 0$ for each $t \in \mathbb{K}$, with equality if and only if $t = 0$;
(b) $|st| = |s| \cdot |t|$ for all $s, t \in \mathbb{K}$;
(c) The *ultrametric inequality* holds, i.e., $|s + t| \leq \max\{|s|, |t|\}$ for all $s, t \in \mathbb{K}$.

An example of an absolute value defining the topology of $\mathbb{K}$ is what we call the *natural absolute value* on $\mathbb{K}$, given by $|0|_{\mathbb{K}} := 0$ and

$$|x|_{\mathbb{K}} := \Delta_{\mathbb{K}}(m_x) \quad \text{for } x \in \mathbb{K} \setminus \{0\} \tag{9}$$

(cf. [55, Chapter II, §2]), where $m_x : \mathbb{K} \to \mathbb{K}$, $y \mapsto xy$ is scalar multiplication by x and $\Delta_{\mathbb{K}}(m_x)$ its module.[5]

It is known that every totally disconnected local field $\mathbb{K}$ either is a field of formal Laurent series over some finite field (if $\mathrm{char}(\mathbb{K}) > 0$), or a finite extension of the field of p-adic numbers for some prime p (if $\mathrm{char}(\mathbb{K}) = 0$). Let us fix our notation concerning these basic examples.

---

[5]Note that if $\mathbb{K}$ is an extension of $\mathbb{Q}_p$ of degree d, then $|p|_{\mathbb{K}} = p^{-d}$ depends on the extension.

Example 2.9 Given a prime number p, the field $\mathbb{Q}_p$ of p-adic numbers is the completion of $\mathbb{Q}$ with respect to the p-adic absolute value,

$$\left| p^k \frac{n}{m} \right|_p := p^{-k} \quad \text{for } k \in \mathbb{Z} \text{ and } n, m \in \mathbb{Z} \setminus p\mathbb{Z}.$$

We use the same notation, $|.|_p$, for the extension of the p-adic absolute value to $\mathbb{Q}_p$. Then the topology coming from $|.|_p$ makes $\mathbb{Q}_p$ a totally disconnected local field, and $|.|_p$ is the natural absolute value on $\mathbb{Q}_p$. Every non-zero element x in $\mathbb{Q}_p$ can be written uniquely in the form

$$x = \sum_{k=n}^{\infty} a_k p^k$$

with $n \in \mathbb{Z}$, $a_k \in \{0, 1, \ldots, p-1\}$ and $a_n \neq 0$. Then $|x|_p = p^{-n}$. The elements of the form $\sum_{k=0}^{\infty} a_k p^k$ form the subring $\mathbb{Z}_p = \{x \in \mathbb{Q}_p : |x|_p \leq 1\}$ of $\mathbb{Q}_p$, which is open and also compact, as it is homeomorphic to $\{0, 1, \ldots, p-1\}^{\mathbb{N}_0}$ via $\sum_{k=0}^{\infty} a_k p^k \mapsto (a_k)_{k \in \mathbb{N}_0}$.

Example 2.10 Given a finite field $\mathbb{F}$ (with q elements), we let $\mathbb{F}((X)) \subseteq \mathbb{F}^{\mathbb{Z}}$ be the field of formal Laurent series $\sum_{k=n}^{\infty} a_k X^k$ with $a_k \in \mathbb{F}$ and $n \in \mathbb{Z}$. Here addition is pointwise, and multiplication is given by the Cauchy product. We endow $\mathbb{F}((X))$ with the topology arising from the ultrametric absolute value

$$\left| \sum_{k=n}^{\infty} a_k X^k \right| := q^{-n} \quad \text{if } a_n \neq 0. \tag{10}$$

Then the set $\mathbb{F}[[X]]$ of formal power series $\sum_{k=0}^{\infty} a_k X^k$ is a compact and open subring of $\mathbb{F}((X))$, and thus $\mathbb{F}((X))$ is a totally disconnected local field. Its natural absolute value is given by (10).

Beyond local fields, we also consider some *ultrametric fields* $(\mathbb{K}, |.|)$. Thus $\mathbb{K}$ is a field and $|.|$ an ultrametric absolute value on $\mathbb{K}$ which defines a non-discrete topology on $\mathbb{K}$. For example, we shall repeatedly use an algebraic closure $\overline{\mathbb{K}}$ of a totally disconnected local field $\mathbb{K}$ and exploit that an ultrametric absolute value $|.|$ on $\mathbb{K}$ extends uniquely to an ultrametric absolute value on $\overline{\mathbb{K}}$ (see, e.g., [47], Theorem 16.1]). The same notation, $|.|$, will be used for the extended absolute value. An ultrametric field $(\mathbb{K}, |.|)$ is called *complete* if $\mathbb{K}$ is a complete metric space with respect to the metric given by $d(x, y) := |x - y|$.

Ultrametric Norms and Balls Let $(\mathbb{K}, |.|)$ be an ultrametric field and $(E, \|.\|)$ be a normed $\mathbb{K}$-vector space whose norm is ultrametric in the sense that $\|x + y\| \leq \max\{\|x\|, \|y\|\}$ for all $x, y \in E$. Since $\|x\| = \|x + y - y\| \leq \max\{\|x + y\|, \|y\|\}$, it follows that $\|x + y\| \geq \|x\|$ if $\|y\| < \|x\|$ and hence

$$\|x + y\| = \|x\| \quad \text{for all } x, y \in E \text{ such that } \|y\| < \|x\|. \tag{11}$$

We shall use the notations

$$B_r^E(x) := \{y \in E : \|y - x\| < r\} \quad \text{and}$$

$$\overline{B}_r^E(x) := \{y \in E : \|y - x\| \leq r\}$$

for balls in E (with $x \in E$, $r \in]0, \infty[$). The ultrametric inequality entails that $B_r^E(0)$ and $\overline{B}_r^E(0)$ are *subgroups* of $(E, +)$ with non-empty interior (and hence both open and closed). Specializing to $E = \mathbb{K}$, we see that

$$\mathbb{O} := \{z \in \mathbb{K} : |z| \leq 1\} \tag{12}$$

is an open subring of $\mathbb{K}$, its so-called *valuation ring*. If $\mathbb{K}$ is a totally disconnected local field, then $\mathbb{O}$ is a compact subring of $\mathbb{K}$ (which is maximal and independent of the choice of absolute value). In this case, also the unit group

$$\mathbb{O}^\times = \{z \in \mathbb{O} : |z| = 1\}$$

of all invertible elements is compact, as it is closed in $\mathbb{O}$.

An *ultrametric Banach space* over a complete ultrametric field is a normed space $(E, \|.\|)$ over $\mathbb{K}$, with ultrametric norm $\|.\|$, such that every Cauchy sequence in E is convergent. We shall always endow a finite-dimensional vector space E over a complete ultrametric field $(\mathbb{K}, |.|)$ with the unique Hausdorff topology making it a topological $\mathbb{K}$-vector space (see Theorem 2 in [6, Chapter I, §2, no. 3]). Then $E \cong \mathbb{K}^m$ (carrying the product topology) as a topological $\mathbb{K}$-vector space, with $m := \dim_{\mathbb{K}}(E)$, entailing that there exists a norm $\|.\|$ on E (corresponding to the maximum norm on $\mathbb{K}^m$) which defines its topology and makes it an ultrametric Banach space. If $(E. \|.\|_E)$ and $(F, \|.\|_F)$ are finite-dimensional normed spaces over a complete ultrametric field, then every linear map $\alpha : E \to F$ is continuous (see Corollary 2 in [6, Chapter I, §2, no. 3]); as usual, we write

$$\|\alpha\|_{\mathrm{op}} := \sup \left\{ \frac{\|\alpha(x)\|_F}{\|x\|_E} : x \in E \setminus \{0\} \right\} \in [0, \infty[$$

for its operator norm. Then $\|\alpha(x)\|_F \leq \|\alpha\|_{\mathrm{op}} \|x\|_E$ and, if α is invertible and $E \neq \{0\}$, then

$$\|\alpha(x)\|_F \geq \frac{1}{\|\alpha^{-1}\|_{\mathrm{op}}} \|x\|_E \quad \text{for all } x \in E.$$

Module of a Linear Automorphism We recall a formula for the module of a linear automorphism.

Lemma 2.11 *Let E be a finite-dimensional vector space over a totally disconnected local field $\mathbb{K}$ and $\alpha \in \mathrm{GL}(E)$. Then*

$$\Delta_E(\alpha) = |\det \alpha|_{\mathbb{K}} = \prod_{i=1}^{n} |\lambda_i|_{\mathbb{K}}, \tag{13}$$

where $\lambda_1, \ldots, \lambda_n$ are the eigenvalues of α in an algebraic closure $\overline{\mathbb{K}}$ of $\mathbb{K}$.

Proof See [7, Proposition 55 in Chapter III, §3, no. 16] for the first equality in (13). The second equality in (13) is clear if all eigenvalues lie in $\mathbb{K}$. For the general case, pick a finite extension $\mathbb{L}$ of $\mathbb{K}$ containing the eigenvalues, and let $d := [\mathbb{L} : \mathbb{K}]$ be the degree of the field extension. Then $\Delta_{E_{\mathbb{L}}}(\alpha_{\mathbb{L}}) = (\Delta_E(\alpha))^d$. Since the extended absolute value is given by

$$|x|_{\mathbb{K}} = \sqrt[d]{\Delta_{\mathbb{L}}(m_x)} \quad \text{for } x \in \mathbb{L} \setminus \{0\}$$

(see [34, Chapter 9, Theorem 9.8] or [47, Exercise 15.E]), the desired equality follows from the special case already treated (applied now to $\mathbb{L}$).   □

3 Endomorphisms of $\mathbb{K}$-Vector Spaces

Linear endomorphisms of vector spaces over totally disconnected local fields provide first examples of endomorphisms of totally disconnected locally compact groups, and their understanding is essential also for our discussion of endomorphisms of Lie groups.

Throughout this section, $\mathbb{K}$ is a totally disconnected local field, E a finite-dimensional $\mathbb{K}$-vector space and $\alpha \colon E \to E$ a $\mathbb{K}$-linear endomorphism. We shall calculate the scale, determine the parabolic, Levi and contraction subgroups for α, and find tidy subgroups.

Our starting point are ideas from [39, Chapter II, §1] concerning iteration of linear endomorphisms. Following [39], we shall decompose E into certain characteristic subspaces, which help us to understand the dynamics of α.

3.1 If the characteristic polynomial p_α of $\alpha \in \mathrm{End}_{\mathbb{K}}(E)$ splits into linear factors in the polynomial ring $\mathbb{K}[X]$, then E is the direct sum of the generalized eigenspaces for α. For $\rho \in [0, \infty[$, let

$$E_\rho$$

be the sum of all generalized eigenspaces $\{v \in E : (\exists n \in \mathbb{N})(\lambda \, id_E - \alpha)^n(v) = 0\}$ for eigenvalues $\lambda \in \mathbb{K}$ with $|\lambda|_{\mathbb{K}} = \rho$; we call E_ρ the *characteristic subspace* for ρ. By construction,

$$E = \bigoplus_{\rho \geq 0} E_\rho. \tag{14}$$

3.2 If $\alpha \in \mathrm{End}_{\mathbb{K}}(E)$ is arbitrary, we choose a finite extension field $\mathbb{L}$ of $\mathbb{K}$ such that p_α splits into linear factors in $\mathbb{L}[X]$. By 3.1, we have a decomposition

$$E_{\mathbb{L}} = \bigoplus_{\rho \geq 0} (E_{\mathbb{L}})_\rho$$

into characteristic subspaces for $\alpha_{\mathbb{L}}$. We call

$$E_\rho := (E_{\mathbb{L}})_\rho \cap E$$

the *characteristic subspace* of E for ρ. If $E_\rho \neq \{0\}$, then ρ is called a *characteristic value* of α. Using the Galois Criterion, it can be shown that each $(E_{\mathbb{L}})_\rho$ is defined over $\mathbb{K}$, i.e.,

$$(E_{\mathbb{L}})_\rho = (E_\rho)_{\mathbb{L}}$$

(see [39, Chapter II, (1.0)]). As a consequence, again (14) holds.[6]

Remark 3.3

(a) By construction, $\alpha(E_\rho) \subseteq E_\rho$ for each $\rho \geq 0$.
(b) $E_0 = \bigcup_{n \in \mathbb{N}_0} \ker(\alpha^n)$ is the generalized eigenspace for the eigenvalue 0 (also known as the "Fitting 0-component"), and thus $\alpha|_{E_0}$ is a nilpotent endomorphism.
(c) For each $\rho > 0$, the restriction $\alpha|_{E_\rho} : E_\rho \to E_\rho$ is an injective endomorphism of a finite-dimensional vector space and hence an automorphism.
(d) The restriction of α to the "Fitting 1-component" $E_{>0} := \bigoplus_{\rho > 0} E_\rho$ is an automorphism, and

$$E = E_0 \oplus E_{>0}.$$

Thus

$$s(\alpha) = s(\alpha|_{E_0})s(\alpha|_{E_{>0}}). \tag{15}$$

Since $s(\alpha|_{E_0}) = 1$ by Lemma 2.7 (c) and (a), we deduce from (15) that $s(\alpha) = s(\alpha|_{E_{>0}})$.

Proposition 3.4 *The scale $s(\alpha)$ of $\alpha \in \mathrm{End}_{\mathbb{K}}(E)$ coincides with the scale $s(\alpha|_{E_{>0}})$ of the automorphism of the Fitting 1-component induced by α.* □
For endomorphisms of p-adic vector spaces, this was already observed in [44].

---

[6]As $E_\rho \neq \{0\}$ for only finitely many $\rho \geq 0$, we can identify the direct sum $E = \bigoplus_{\rho \geq 0} E_\rho$ with the direct product $\prod_{\rho \geq 0} E_\rho$ whenever this is convenient.

We now recall from [19, Proposition 2.4] the existence of norms which are well-adapted to an endomorphism (see already [21, Proposition 4.29] for automorphisms; cf. [39, Chapter II, Lemma 1.1] for a similar, weaker result, also valid for $\mathbb{R}$ and $\mathbb{C}$).

Lemma 3.5 *There exists an ultrametric norm* $\|.\|$ *on E which is* adapted *to α in the following sense:*

(a) $\|.\|$ *is a maximum norm with respect to the decomposition (14) of E into the characteristic subspaces for α;*
(b) $\|\alpha|_{E_0}\|_{\mathrm{op}} < 1$; *and*
(c) *For all $\rho > 0$ and $v \in E_\rho$, we have $\|\alpha(v)\| = \rho\|v\|$.*

If $\varepsilon \in \,]0, 1]$ is given, then $\|.\|$ can be chosen such that $\|\alpha|_{E_0}\|_{\mathrm{op}} < \varepsilon$. $\qquad\square$
As before, in the following theorem we write E_ρ for the characteristic subspace for $\rho > 0$ with respect to α.

Theorem 3.6 *Let $\alpha \in \mathrm{End}_{\mathbb{K}}(E)$ be an endomorphism of a finite-dimensional vector space $E \cong \mathbb{K}^m$ over a local field $\mathbb{K}$. Let $\|.\|$ be a norm on E which is adapted to α. Then the following holds:*

(a) *The ball $B_r^E(0)$ is a compact open subgroup of $(E, +)$ which is tidy for α, for each $r \in \,]0, \infty[$.*
(b) *We have*

$$\overrightarrow{\mathrm{con}}\,(\alpha) = E_{<1} := \bigoplus_{\rho<1} E_\rho, \quad \overleftarrow{\mathrm{con}}\,(\alpha) = E_{>1} := \bigoplus_{\rho>1} E_\rho,$$

$$\overrightarrow{\mathrm{par}}\,(\alpha) = E_{\leq1} := \bigoplus_{\rho\leq1} E_\rho, \quad \overleftarrow{\mathrm{par}}\,(\alpha) = E_{\geq1} := \bigoplus_{\rho\geq1} E_\rho,$$

and $\mathrm{lev}(\alpha) = E_1$.
(c) *The scale of α is given by*

$$s(\alpha) = \prod_{\substack{j\in\{1,\ldots,m\} \\ \text{s.t. } |\lambda_j|_{\mathbb{K}}\geq1}} |\lambda_j|_{\mathbb{K}}, \tag{16}$$

where $\lambda_1, \ldots, \lambda_m$ are the eigenvalues of $\alpha_{\overline{\mathbb{K}}}$ in an algebraic closure $\overline{\mathbb{K}}$ of $\mathbb{K}$, with repetitions according to algebraic multiplicities.
(d) $s(\alpha) = s(\alpha|_{E_{>0}}) = s(\alpha|_{E_{\geq1}}) = s(\alpha|_{E_{>1}}) = \Delta(\alpha|_{E_{\geq1}}) = \Delta(\alpha|_{E_{>1}})$.

Proof We endow vector subspaces of E with the induced norm.

(a) Since E admits the Fitting decomposition $E = E_0 \oplus E_{>0}$ into E_0 and $E_{>0}$ which are α-invariant vector subspaces and

$$B_r^E(0) = B_r^{E_0}(0) \times B_r^{E_{>0}}(0),$$

we need only check that $B_r^{E_0}(0)$ is tidy for $\alpha|_{E_0}$ and $B_r^{E_{>0}}(0)$ is tidy for $\alpha|_{E_{>0}}$. The first property holds by Lemma 2.7, since $\alpha(B_r^{E_0}(0)) \subseteq B_r^{E_0}(0)$ by Lemma 3.5 (b). To check the second property, after replacing α with $\alpha|_{E_{>0}}$ we may assume that α is an automorphism. Thus, let us consider $\alpha \in GL(E)$ and verify that

$$U := B_r^E(0) = \prod_{\rho > 0} B_r^{E_\rho}(0)$$

is tidy for α. For each $k \in \mathbb{Z}$, have

$$\alpha^k(B_r^E(0)) = \prod_{\rho > 0} B_{\rho^k r}^{E_\rho}(0),$$

using Lemma 3.5 (c). Hence

$$U_+ = \bigcap_{k=0}^\infty \alpha^k(U) = \prod_{\rho \geq 1} B_r^{E_\rho}(0) \quad \text{and} \quad U_- = \bigcap_{k=0}^\infty \alpha^{-k}(U) = \prod_{0 < \rho \leq 1} B_r^{E_\rho}(0) \qquad (17)$$

(where we used 2.6), entailing that $U = U_+ + U_-$ is tidy above for α. Since

$$U_{--} := \bigcup_{k=0}^\infty \alpha^{-k}(U_-) = \left(\prod_{0 < \rho < 1} E_\rho \right) \times B_r^{E_1}(0) = E_{<1} \times B_r^{E_1}(0)$$

is closed in E, we deduce with 2.4 that U is tidy.

(b) is obvious from Lemma 3.5 (a), (b), and (c).

(c) Since $U_+ = \prod_{\rho \geq 1} B_r^{E_\rho}(0) = B_r^{E_{\geq 1}}(0)$ is a compact open subgroup of $(E_{\geq 1}, +)$ such that $U_+ \subseteq \alpha(U_+)$, using 2.5 and (5) we obtain

$$s(\alpha) = [\alpha(U_+) : U_+] = \Delta(\alpha|_{E_{\geq 1}}).$$

As the λ_j with $|\lambda_j|_\mathbb{K} \geq 1$ are exactly the eigenvalues of $\alpha|_{E_{\geq 1}}$ in $\overline{\mathbb{K}}$, Lemma 2.11 yields the desired formula.

(d) Eigenvalues λ_j with $|\lambda_j|_\mathbb{K} = 1$ are irrelevant for the product in (c). Using Lemma 2.11, we deduce that also $s(\alpha) = \Delta(\alpha|_{E_{>1}})$. The first equality in (d) holds by Proposition 3.4. Note that $B_r^{E_{\geq 1}}(0)$ and $B_r^{E_{>1}}(0)$ are tidy for $\alpha|_{E_{\geq 1}}$ and $\alpha|_{E_{>1}}$, respectively, and are inflated by the latter. The third and fourth scales in the formula therefore coincide with the corresponding modules, by Lemma 2.7 (b). $\square$

Corollary 3.7 *Let α be a linear endomorphism of a finite-dimensional vector space E over a totally disconnected local field $\mathbb{K}$. Let F be an α-invariant vector* *subspace of E and*

$$\overline{\alpha}: E/F \to E/F, \quad x + F \mapsto \alpha(x) + F$$

be the induced linear endomorphism of the quotient space E/F. Then

$$s_E(\alpha) = s_F(\alpha|_F)s_{E/F}(\overline{\alpha}). \tag{18}$$

Proof The eigenvalues of α in an algebraic closure $\overline{\mathbb{K}}$ of $\mathbb{K}$ are exactly the eigenvalues of $\alpha|_F$, together with those of $\overline{\alpha}$. Hence (18) follows from Theorem 3.6 (c).   □
For basic concepts concerning Lie algebras (which we always assume of finite dimension),[7] see [7, 31], and [49].

Lemma 3.8 *If $\mathfrak{g}$ is a Lie algebra over a totally disconnected local field $\mathbb{K}$ and $\alpha: \mathfrak{g} \to \mathfrak{g}$ a Lie algebra endomorphism, then $\overrightarrow{\mathrm{con}}(\alpha)$, $\overleftarrow{\mathrm{con}}(\alpha)$, $\overrightarrow{\mathrm{par}}(\alpha)$, $\overleftarrow{\mathrm{par}}(\alpha)$, and $\mathrm{lev}(\alpha) = \overrightarrow{\mathrm{con}}(\alpha) \cap \overleftarrow{\mathrm{con}}(\alpha)$ are Lie subalgebras of $\mathfrak{g}$.*

Proof If x and y are elements of $\overrightarrow{\mathrm{con}}(\alpha)$ (resp., of $\overrightarrow{\mathrm{par}}(\alpha)$), then $\alpha^n(x)$ and $\alpha^n(y)$ tend to 0 as $n \to \infty$ (resp., the elements form bounded sequences), entailing that also

$$\alpha^n([x, y]) = [\alpha^n(x), \alpha^n(y)]$$

tends to 0 (resp., is bounded). Hence $[x, y] \in \overrightarrow{\mathrm{con}}(\alpha)$ (resp., $[x, y] \in \overrightarrow{\mathrm{par}}(\alpha)$). If x and y are elements of $\overleftarrow{\mathrm{con}}(\alpha)$ (resp., of $\overleftarrow{\mathrm{par}}(\alpha)$), then we find an α-regressive trajectory $(x_{-n})_{n\in\mathbb{N}_0}$ for x and an α-regressive trajectory $(y_{-n})_{n\in\mathbb{N}_0}$ for y such that $x_{-n} \to 0$ and $y_{-n} \to 0$ as $n \to \infty$ (resp., $(x_{-n})_{n\in\mathbb{N}_0}$ and $(y_{-n})_{n\in\mathbb{N}_0}$ are bounded sequences). Then $([x_{-n}, y_{-n}])_{n\in\mathbb{N}_0}$ is an α-regressive trajectory for $[x, y]$, since

$$\alpha([x_{-n-1}, y_{-n-1}]) = [\alpha(x_{-n-1}), \alpha(y_{-n-1})] = [x_{-n}, y_{-n}] \quad \text{for all } n \in \mathbb{N}_0.$$

Moreover, $[x_{-n}, y_{-n}] \to 0$ as $n \to \infty$ (resp., the sequence $([x_{-n}, y_{-n}])_{n\in\mathbb{N}_0}$ is bounded), showing that $[x, y] \in \overleftarrow{\mathrm{con}}(\alpha)$ (resp., $[x, y] \in \overleftarrow{\mathrm{par}}(\alpha)$).   □
The following lemma will be used in Sect. 11.

Lemma 3.9 *Let E be a finite-dimensional vector space over a totally disconnected local field $\mathbb{K}$ and $\alpha \in \mathrm{GL}(E)$ be an automorphism such that $|\lambda|_{\mathbb{K}} = 1$ for all eigenvalues of α in an algebraic closure $\overline{\mathbb{K}}$ of $\mathbb{K}$. Then the subgroup $\langle \alpha \rangle$ generated by α is relatively compact in $\mathrm{GL}(E)$.*

---

[7]Except for the Lie algebras of analytic vector fields mentioned in Sect. 4.

Proof If $\mathbb{K}$ has characteristic $p > 0$, then it suffices to show that α^{p^n} generates a relatively compact subgroup for some $n \in \mathbb{N}_0$, since $\langle \alpha \rangle$ is contained in the finite union

$$\bigcup_{j=0}^{p^n-1} \alpha^j \circ K$$

of cosets of the compact group $K := \overline{\langle \alpha^{p^n} \rangle}$. We may therefore assume that the characteristic polynomial $p_\alpha \in \mathbb{K}[X]$ of α is separable over $\mathbb{K}$. Let $\mathbb{L} \subseteq \overline{\mathbb{K}}$ be a finite field extension of $\mathbb{K}$ which is Galois and such that p_α splits into linear factors in $\mathbb{L}[X]$. Then α has a unique multiplicative Jordan decomposition

$$\alpha = \alpha_h \circ \alpha_u = \alpha_u \circ \alpha_h$$

such that $(\alpha_h)_\mathbb{L} \in \mathrm{GL}(E_\mathbb{L})$ is diagonalizable and $(\alpha_u)_\mathbb{L} \in \mathrm{GL}(E_\mathbb{L})$ is unipotent (see [4, Theorem I.4.4]). Let $\mathbb{O}$ be the valuation ring of $\mathbb{L}$ and $\mathbb{O}^\times$ be its compact group of invertible elements. Since $|\lambda|_\mathbb{K} = 1$ for all eigenvalues $\lambda \in \mathbb{L} \subseteq \overline{\mathbb{K}}$ of α_h (which coincide with those of α), we have $\lambda \in \mathbb{O}^\times$ and deduce that $(\alpha_h)_\mathbb{L}$ generates a relatively compact subgroup L of $\mathrm{GL}(E_\mathbb{L})$. Identify $\mathrm{GL}(E)$ with the closed subgroup $\{\beta \in \mathrm{GL}(E_\mathbb{L}): \beta(E) \subseteq E\}$ of $\mathrm{GL}(E_\mathbb{L})$. Then $\langle \alpha_h \rangle$ is contained in the compact subgroup $\overline{L} \cap \mathrm{GL}(E)$ and hence relatively compact in $\mathrm{GL}(E)$. Now $(\alpha_u)_\mathbb{L}$ generates a relatively compact subgroup of $\mathrm{GL}(E_\mathbb{L})$, by [15, Lemma 4.1]. Hence α_u generates a relatively compact subgroup of $\mathrm{GL}(E)$, by the preceding argument. Since

$$\langle \alpha \rangle \subseteq \overline{\langle \alpha_h \rangle} \circ \overline{\langle \alpha_u \rangle},$$

we see that $\langle \alpha \rangle$ is relatively compact. $\qquad \square$

4 Analytic Functions, Manifolds and Lie Groups

The section compiles definitions and elementary facts concerning analytic functions, manifolds, and Lie groups over totally disconnected local fields, which we shall use without further explanation. The section ends with two versions of the Inverse Function Theorem, which will be essential in the following.

Analytic Manifolds and Lie Groups Given a totally disconnected local field $(\mathbb{K}, |.|)$ and $n \in \mathbb{N}$, we endow $\mathbb{K}^n$ with an ultrametric norm $\|.\|$ (the choice of norm does not really matter because all norms are equivalent; see [47, Theorem 13.3]). If $\alpha \in \mathbb{N}_0^n$ is a multi-index, we write $|\alpha| := \alpha_1 + \cdots + \alpha_n$. Confusion with the absolute value $|.|$ is unlikely; the intended meaning of $|.|$ will always be clear from the context. If $\alpha \in \mathbb{N}_0^n$ and $y = (y_1, \ldots, y_n) \in \mathbb{K}^n$, we abbreviate $y^\alpha := y_1^{\alpha_1} \cdots y_n^{\alpha_n}$, as usual. Compare [49] for the following concepts.

Definition 4.1 Given an open subset $U \subseteq \mathbb{K}^n$, a map $f \colon U \to \mathbb{K}^m$ is called *analytic*[8] (or $\mathbb{K}$-*analytic*, if we wish to emphasize the ground field) if it is given locally by a convergent power series around each point $x \in U$, i.e.,

$$f(x + y) = \sum_{\alpha \in \mathbb{N}_0^n} a_\alpha y^\alpha \quad \text{for all } y \in B_r^{\mathbb{K}^n}(0),$$

with $a_\alpha \in \mathbb{K}^m$ and some $r > 0$ such that $B_r^{\mathbb{K}^n}(x) \subseteq U$ and

$$\sum_{\alpha \in \mathbb{N}_0^n} \|a_\alpha\| \, r^{|\alpha|} < \infty.$$

Compositions of analytic functions are analytic [49, Theorem, p. 70]. We can therefore define an *m-dimensional analytic manifold* M over a totally disconnected local field $\mathbb{K}$ in the usual way, as a Hausdorff topological space M, equipped with a maximal set $\mathscr{A}$ of homeomorphisms $\phi \colon U_\phi \to V_\phi$ from open subsets $U_\phi \subseteq M$ onto open subsets $V_\phi \subseteq \mathbb{K}^m$ such that the transition map $\psi \circ \phi^{-1}$ is analytic, for all $\phi, \psi \in \mathscr{A}$.

In the preceding situation, the homeomorphisms $\phi \in \mathscr{A}$ are called *charts* for M, and $\mathscr{A}$ is called an *atlas*.

A map $f \colon M \to N$ between analytic manifolds is called *analytic* if it is continuous and $\phi \circ f \circ \psi^{-1}$ (which is a map between open subsets of $\mathbb{K}^m$ and $\mathbb{K}^n$) is analytic, for all charts $\phi \colon U_\phi \to V_\phi \subseteq \mathbb{K}^n$ of N and charts $\psi \colon U_\psi \to V_\psi \subseteq \mathbb{K}^m$ of M.

If $(M, \mathscr{A}_M)$ and $(N, \mathscr{A}_N)$ are analytic manifolds of dimension m and n, respectively, then $M \times N$ with the product topology is an $(m + n)$-dimensional analytic manifold, with the atlas containing $\{\phi \times \psi \colon \phi \in \mathscr{A}_M, \psi \in \mathscr{A}_N\}$.

Every open subset U of a finite-dimensional $\mathbb{K}$-vector space E can be considered as an analytic manifold, endowed with the maximal atlas containing the global chart $\mathrm{id}_U \colon U \to U$. Notably, we can speak about analytic functions

$$f \colon U \to V$$

if U and V are open subsets of finite-dimensional normed $\mathbb{K}$-vector spaces E and F, respectively. Any such function is totally differentiable at each $x \in U$, and we write

$$f'(x) \colon E \to F \tag{19}$$

for its total differential. Deviating from (19), we write $f'(x) = \frac{d}{dt}\big|_{t=0} f(x + t)$ if $E = \mathbb{K}$, as usual (which is $f'(x)(1)$ in the notation of (19)).

---

[8]In other parts of the literature related to rigid analytic geometry, such functions are called *locally analytic* to distinguish them from functions which are globally given by a power series.

A *Lie group* over a totally disconnected local field $\mathbb{K}$ is a group G, equipped with an analytic manifold structure which turns the group multiplication

$$\mu_G \colon G \times G \to G, \quad (x, y) \mapsto xy$$

and the group inversion $\eta_G \colon G \to G, x \mapsto x^{-1}$ into analytic mappings.

Lie groups over $\mathbb{Q}_p$ are also called *p-adic Lie groups*. Besides the additive groups of finite-dimensional $\mathbb{K}$-vector spaces, the most obvious examples of $\mathbb{K}$-analytic Lie groups are general linear groups.

Example 4.2 $\mathrm{GL}_n(\mathbb{K}) = \det^{-1}(\mathbb{K}^\times)$ is an open subset of the space $M_n(\mathbb{K}) \cong \mathbb{K}^{n^2}$ of $n \times n$-matrices and hence is an n^2-dimensional $\mathbb{K}$-analytic manifold. The group operations are rational maps and hence analytic.

More generally, one can show (cf. [39, Chapter I, Proposition 2.5.2]):

Example 4.3 Every (group of $\mathbb{K}$-rational points of a) linear algebraic group defined over $\mathbb{K}$ is a $\mathbb{K}$-analytic Lie group, viz. every subgroup $G \le \mathrm{GL}_n(\mathbb{K})$ which is the set of joint zeros of a set of polynomial functions $M_n(\mathbb{K}) \to \mathbb{K}$. For instance, $\mathrm{SL}_n(\mathbb{K}) = \{A \in \mathrm{GL}_n(\mathbb{K}) \colon \det(A) = 1\}$ is a $\mathbb{K}$-analytic Lie group.

Remark 4.4 See Example 8.26 (first mentioned in [21, Remark 9.7]) for a Lie group G over $\mathbb{K} = \mathbb{F}_p((X))$ which is not a linear Lie group, i.e., which does not admit a faithful, continuous linear representation $G \to \mathrm{GL}_n(\mathbb{K})$ for any n. We shall also encounter a p-adic Lie group which is not isomorphic to a closed subgroup of $\mathrm{GI}_n(\mathbb{Q}_p)$ for any $n \in \mathbb{N}$ (Example 10.3).

Remark 4.5 The analytic manifolds and Lie groups we consider need not be second countable topological spaces. Notably, arbitrary discrete groups (countable or not) can be considered as (0-dimensional) p-adic Lie groups, which is natural from the point of view of topological groups.

All the Lie groups and manifolds considered in these notes are analytic and finite-dimensional. For smooth Lie groups modelled on (not necessarily finite-dimensional) topological vector spaces over a topological field, see [2, 20] and the references therein.

Tangent Vectors, Tangent Spaces, and Tangent Maps Tangent vectors can be defined in many ways. We choose a description which corresponds to the so-called "geometric" tangent vectors in the real case.[9] If M is an m-dimensional analytic manifold over a totally disconnected local field $\mathbb{K}$ and $p \in M$, let us say that two analytic mappings

$$\gamma \colon B_\varepsilon^{\mathbb{K}}(0) \to M \quad \text{and} \quad \eta \colon B_\delta^{\mathbb{K}}(0) \to M$$

---

[9]Compare [5, 7, 49], also [2] for the following facts (although in different formulations).

with $\gamma(0) = \eta(0) = p$ are equivalent (and write $\gamma \sim_p \eta$) if

$$(\phi \circ \gamma)'(0) = (\phi \circ \eta)'(0) \tag{20}$$

for some chart $\phi: U \to V$ of M around p (i.e., with $p \in U$); here $\varepsilon, \delta > 0$. Then (20) holds for all charts around p (by the Chain Rule), and we easily deduce that $\sim_p$ is an equivalence relation. The equivalence classes $[\gamma]$ with respect to $\sim_p$ are called *tangent vectors* for M at p. The set $T_p(M)$ of all tangent vectors at p is called the *tangent space* of M at p. We endow it with the unique vector space structure making the bijection

$$T_p(M) \to \mathbb{K}^m, \quad [\gamma] \mapsto (\phi \circ \gamma)'(0)$$

a vector space isomorphism for some (and hence every) chart ϕ of M around p. The union

$$T(M) := \bigcup_{p \in M} T_p(M)$$

is disjoint and is called the *tangent bundle* of M. If $f: M \to N$ is an analytic map between analytic manifolds, we obtain a linear map

$$T_p(f): T_p(M) \to T_{f(p)}(N), \quad [\gamma] \mapsto [f \circ \gamma]$$

called the *tangent map of f at p*. The map $T(f): T(M) \to T(N)$ taking $v \in T_p(M)$ to $T_p(f)(v)$ is called the *tangent map* of f. If also K is an analytic manifold over $\mathbb{K}$ and $g: K \to M$ an analytic mapping, then

$$T(f \circ g) = T(f) \circ T(g) \tag{21}$$

as both mappings take a tangent vector $[\gamma] \in T(K)$ to $[f \circ g \circ \gamma]$. If U is an open subset of a finite-dimensional vector space E, we identify $T(U)$ with $U \times E$ using the bijection

$$T(U) \to U \times E, \quad [\gamma] \mapsto (\gamma(0), \gamma'(0)). \tag{22}$$

If U is as before, M an analytic manifold and $f: M \to U$ an analytic map, we write df for the second component of the map

$$T(f): T(M) \to T(U) = U \times E,$$

using the identification from (22). Thus

$$df([\gamma]) = (f \circ \gamma)'(0).$$

If U and V are open subsets of finite-dimensional $\mathbb{K}$-vector spaces E and F, respectively, and $f: U \to V$ is an analytic map, then $T(f): T(U) \to T(V)$ is the mapping

$$U \times E \to V \times F, \quad (x, y) \mapsto (f(x), df(x, y))$$

with $df(x, y) = f'(x)(y)$.

Submanifolds and Lie Subgroups Let M be an m-dimensional analytic manifold over a totally disconnected local field $\mathbb{K}$ and $n \in \{0, 1, \ldots, m\}$. A subset $N \subseteq M$ is called an n-dimensional *submanifold* of M if, for each $p \in N$, there exists a chart $\phi: U \to V \subseteq \mathbb{K}^m$ of M around p such that

$$\phi(U \cap N) = V \cap (\mathbb{K}^n \times \{0\}).$$

Identifying $\mathbb{K}^n \times \{0\} \subseteq \mathbb{K}^m$ with $\mathbb{K}^n$ via $(x, 0) \mapsto x$, we get a homeomorphism

$$\phi_N := \phi|_{U \cap N}: U \cap N \to V \cap (\mathbb{K}^n \times \{0\}) \subseteq \mathbb{K}^n.$$

Then N is an n-dimensional analytic manifold in a natural way, using the topology induced by M and the maximal atlas containing all of the maps ϕ_N. Using this manifold structure, the inclusion $j: N \to M$ is analytic. For each $p \in N$, the tangent map $T_p(j): T_p(N) \to T_p(M)$ is injective, and will be used to identify $T_p(N)$ with the image of $T_p(j)$ in $T_p(M)$. Moreover, for each analytic manifold K, a mapping $f: K \to N$ is analytic if and only if $j \circ f: K \to M$ is analytic. We say that a subgroup H of a Lie group G over $\mathbb{K}$ is a *Lie subgroup* if it is a submanifold. By the preceding fact, the submanifold structure then turns the group operations on H into analytic mappings and thus makes H a Lie group.

Lemma 4.6 *Let G be a Lie group over a totally disconnected local field $\mathbb{K}$, of dimension m. A subgroup $H \subseteq G$ is a Lie subgroup of dimension n if and only if there exists a chart $\phi: U \to V \subseteq \mathbb{K}^m$ of G around e such that $\phi(U \cap H) = V \cap (\mathbb{K}^n \times \{0\})$.*

Proof The necessity is clear. Sufficiency: For each $h \in H$, the mapping $\phi_h: hU \to V, x \mapsto \phi(h^{-1}x)$ is a chart for G such that $\phi_h(hU \cap H) = \phi(U \cap H) = V \cap (\mathbb{K}^n \times \{0\})$.   □

The Lie Algebra Functor An *analytic vector field* on an m-dimensional $\mathbb{K}$-analytic manifold M is a mapping $X: M \to T(M)$ with $X(p) \in T_p(M)$ for all $p \in M$, which is analytic in the sense that its local representative

$$X_\phi := d\phi \circ X \circ \phi^{-1}: V \to \mathbb{K}^m$$

is an analytic function for each chart $\phi: U \to V \subseteq \mathbb{K}^m$ of M. The set $\mathscr{V}^\omega(M)$ of all analytic vector fields on M is a $\mathbb{K}$-vector space, with pointwise addition and scalar multiplication. Given $X, Y \in \mathscr{V}^\omega(M)$, there is a unique vector field $[X, Y] \in \mathscr{V}^\omega(M)$ such that

$$[X, Y]_\phi = dY_\phi \circ (\mathrm{id}_V, X_\phi) - dX_\phi \circ (\mathrm{id}_V, Y_\phi)$$

for all charts $\phi: U \to V$ of M, and $[.,.]$ makes $\mathscr{V}^\omega(M)$ a Lie algebra.

If G is a $\mathbb{K}$-analytic Lie group, then its tangent space $L(G) := T_e(G)$ at the identity element can be made a Lie algebra via the identification of $v \in L(G)$ with the corresponding left invariant vector field v_ℓ on G given by $v_\ell(g) := T\lambda_g(v)$ for $g \in G$ with left translation $\lambda_g: G \to G, x \mapsto gx$ (noting that the left invariant vector fields form a Lie subalgebra of $\mathscr{V}^\omega(G)$). Thus

$$[v, w] := [v_\ell, w_\ell](e) \quad \text{for } v, w \in L(G).$$

If $\alpha: G \to H$ is an analytic group homomorphism between $\mathbb{K}$-analytic Lie groups, then the tangent map $L(\alpha) := T_e(\alpha): L(G) \to L(H)$ is a linear map and actually a Lie algebra homomorphism (cf. [7, Chapter III, §3, no. 8] and Lemma 5.1 on p. 129 in [49, Part II, Chapter V.1]). An *analytic automorphism* of a Lie group G is an invertible group homomorphism $\alpha: G \to G$ such that both α and α^{-1} are analytic. For example, each inner automorphism I_g of G is analytic. As usual, we abbreviate $\mathrm{Ad}_g := L(I_g)$.

Since $I_g \circ I_h = I_{gh}$ for $g, h \in G$, we have $\mathrm{Ad}_{gh} = \mathrm{Ad}_g \circ \mathrm{Ad}_h$ by (21). In Sect. 11, we shall use the continuity of the adjoint representation of G on its Lie algebra $\mathfrak{g} := L(G)$. Even more is true (see Definition 8 in [7, Chapter III, §3, no. 12] and the lines preceding it):

4.7 The map $\mathrm{Ad}: G \to \mathrm{Aut}(\mathfrak{g}) \subseteq \mathrm{GL}(\mathfrak{g}), g \mapsto \mathrm{Ad}_g$ is analytic.

Ultrametric Inverse Function Theorems Since small perturbations do not change the size of a given non-zero vector in the ultrametric case (see (11)), the ultrametric inverse function theorem has a nicer form than its classical real counterpart. Around a point p with invertible differential, an analytic map f behaves like an affine-linear map (its linearization). If the differential at p is an isometry, then also f is isometric on a p-neighbourhood.

In the following two lemmas, we let $\mathbb{K}$ be a totally disconnected local field and $|.|$ be an absolute value on $\mathbb{K}$ defining its topology. We fix an ultrametric norm $\|.\|$ on a finite-dimensional $\mathbb{K}$-vector space E and write

$$\mathrm{Iso}(E, \|.\|) := \{\alpha \in \mathrm{GL}(E): (\forall x \in E) \, \|\alpha(x)\| = \|x\|\}$$

for the group of linear isometries. It is well-known that $\mathrm{Iso}(E, \|.\|)$ is open in $\mathrm{GL}(E)$ (see, e.g., [16, Lemma 7.2]), but we shall not use this fact. Given $x \in E$ and $r > 0$, we abbreviate $B_r(x) := B_r^E(x)$. The total differential of f at x is denoted by $f'(x)$. The ultrametric inverse function theorem (for analytic functions) subsumes the following[10]:

Lemma 4.8 *Let $f: U \to E$ be an analytic map on an open subset $U \subseteq E$ and $x \in U$ such that $f'(x) \in \mathrm{GL}(E)$. Then there exists $r > 0$ such that $B_r(x) \subseteq U$,*

$$f(B_t(y)) = f(y) + f'(x)B_t(0) \quad \text{for all } y \in B_r(x) \text{ and } t \in \,]0, r], \tag{23}$$

and $f|_{B_r(x)}: B_r(x) \to f(B_r(x))$ is an analytic diffeomorphism. If $f'(x) \in \mathrm{Iso}(E, \|.\|)$, then r can be chosen such that $B_r(x) \subseteq U$,

$$f(B_t(y)) = B_t(f(y)) \quad \text{for all } y \in B_r(x) \text{ and } t \in \,]0, r], \tag{24}$$

and $f|_{B_r(x)}: B_r(x) \to B_r(f(x))$ is an isometric, analytic diffeomorphism. □

It is useful that r can be chosen uniformly in the presence of parameters. As a special case of [16, Theorem 7.4 (b)′], an 'ultrametric inverse function theorem with parameters' is available[11]:

Lemma 4.9 *Let F be a finite-dimensional $\mathbb{K}$-vector space, $P \subseteq F$ and $U \subseteq E$ be open, $f: P \times U \to E$ be a $\mathbb{K}$-analytic map, $p \in P$ and $x \in U$ such that $f'_p(x) \in \mathrm{Iso}(E, \|.\|)$, where $f_p := f(p, \bullet): U \to E$. Then there exists an open neighbourhood $Q \subseteq P$ of p and $r > 0$ such that $B_r(x) \subseteq U$,*

$$f_q(B_t(y)) = f_q(y) + B_t(0) \tag{25}$$

and $f_q|_{B_t(y)}$ is an isometry, for all $q \in Q$, $y \in B_r(x)$ and $t \in \,]0, r]$. □

5 Construction of Small Open Subgroups

It is essential for our following discussions that Lie groups over totally disconnected local fields have a basis of identity neighbourhoods consisting of compact open subgroups which correspond to balls in the Lie algebra. In this section, we explain how these compact open subgroups can be constructed.

Let G be a Lie group over a totally disconnected local field $\mathbb{K}$ and $|.|$ be an absolute value on $\mathbb{K}$ defining its topology. Fix an ultrametric norm $\|.\|$ on $\mathfrak{g} := L(G)$ and abbreviate $B_t(x) := B_t^{\mathfrak{g}}(x)$ for $x \in \mathfrak{g}$ and $t > 0$. Let

$$\phi: U \to V$$

---

[10]A proof is obtained, e.g., by combining [16, Proposition 7.1 (a)′ and (b)′] with the inverse function theorem for analytic maps from [49, p. 73], recalling that analytic maps are strictly differentiable at each point (in the sense of [5, 1.2.2]), by [5, 4.2.3 and 3.2.4].

[11]To achieve that $f_q|_{B_r(x)}$ is an isometry for all $q \in Q$, note that [16, Lemma 6.1 (b)] applies to all of these functions by [16, p. 239, lines 7–8].

by an analytic diffeomorphism from an open identity neighbourhood $U \subseteq G$ onto an open 0-neighbourhood $V \subseteq \mathfrak{g}$, such that $\phi(e) = 0$ and

$$d\phi|_{\mathfrak{g}} = \mathrm{id}_{\mathfrak{g}} . \tag{26}$$

5.1 After shrinking U (and V), we may assume that U is a compact open subgroup of G. Then

$$\mu_V \colon V \times V \to V, \ (x, y) \mapsto x * y := \phi(\phi^{-1}(x)\phi^{-1}(y))$$

is a group multiplication on V with neutral element 0 which turns V into an analytic Lie group and ϕ into an isomorphism of Lie groups. It is easy to see that the first order Taylor expansions of multiplication and inversion in $(V, *)$ at $(0, 0)$ and 0, respectively, are given by

$$x * y = x + y + \cdots \tag{27}$$

and

$$x^{-1} = -x + \cdots \tag{28}$$

(compare [49, p. 113]). Applying the Ultrametric Inverse Function Theorem with Parameters (Lemma 4.9) to the maps $(x, y) \mapsto x * y$ and $(x, y) \mapsto y * x$ around $(0, 0)$, we find $R > 0$ with $B_R(0) \subseteq V$ such that

$$x * B_t(0) = x + B_t(0) = B_t(0) * x \tag{29}$$

for all $x \in B_R(0)$ and $t \in \,]0, R]$ (exploiting that both relevant partial differentials are $\mathrm{id}_{\mathfrak{g}}$ and hence an isometry, by (27)). Notably, (29) entails that

$$B_t(0) * B_t(0) = B_t(0) \ \text{ for each } t \in \,]0, R],$$

whence $y^{-1} \in B_t(0)$ for each $t \in \,]0, R]$ and $y \in B_t(0)$.
Summing up (with $B_t := B_t(0)$):

Lemma 5.2 $(B_t, *)$ *is a group for each* $t \in \,]0, R]$ *and hence* $B_t^{\phi} := \phi^{-1}(B_t)$ *is a compact open subgroup of* G, *for each* $t \in \,]0, R]$. *Moreover,* B_t *is a normal subgroup of* $(B_R, *)$, *whence* B_t^{ϕ} *is normal in* B_R^{ϕ}. $\qquad \square$
Thus small balls in $\mathfrak{g}$ correspond to compact open subgroups in G.

Remark 5.3 (29) entails that the indices of B_t in $(B_R, +)$ and $(B_R, *)$ coincide (as the cosets coincide), for all $t \in \,]0, R]$.

5.4 Now consider an analytic endomorphism $\alpha \colon G \to G$, or, more generally, an analytic homomorphism

$$\alpha \colon G_0 \to G$$

defined on an open subgroup $G_0 \subseteq G$. For the domain U of ϕ, assume that $U \subseteq G_0$. After shrinking R (if necessary) we may assume that

$$\alpha(B_R^\phi) \subseteq U, \tag{30}$$

whence an analytic homomorphism

$$\beta := \phi \circ \alpha|_{B_R^\phi} \circ \phi^{-1}|_{B_R} : B_R \to V \tag{31}$$

can be defined such that

$$\beta \circ \phi|_{B_R^\phi} = \phi \circ \alpha|_{B_R^\phi}. \tag{32}$$

As a consequence of (26), we have

$$\beta'(0) = L(\alpha). \tag{33}$$

For $\alpha : G \to G$ an analytic automorphism and $\|.\|$ adapted to $L(\alpha)$, we shall see in Sect. 8 that the groups $B_t^\phi := \phi^{-1}(B_t)$ are tidy for α and $t \in {]0, R]}$ close to 0, as long as $\overrightarrow{\mathrm{con}}\,(\alpha)$ is closed (and also the case of $\alpha : G_0 \to G$ will be used). This motivates us to calculate the displacement indices for the compact open subgroups $B_t^\phi \subseteq G$.

Lemma 5.5 *Let G be a Lie group over a totally disconnected local field, G_0 be an open subgroup of G and $\alpha : G_0 \to G$ an analytic homomorphism which is an analytic diffeomorphism onto an open subgroup $\alpha(G_0)$ of G. Let ϕ be as before, $\|.\|$ be adapted to $L(\alpha)$, and R be as in 5.4. Then there exists $t_0 \in {]0, R]}$ such that*

$$[\alpha(B_t^\phi) : \alpha(B_t^\phi) \cap B_t^\phi] = s(L(\alpha)) \ \textit{for all } t \in {]0, t_0]}.$$

Proof Let β be as in (31). By (33) and the Ultrametric Inverse Function Theorem (Lemma 4.8), there is $t_0 \in {]0, R]}$ with $L(\alpha)(B_{t_0}) \subseteq B_R$ such that

$$(\phi \circ \alpha \circ \phi^{-1})(B_t) = L(\alpha)(B_t)$$

for all $t \in {]0, t_0]}$ and hence

$$\alpha(B_t^\phi) = \phi^{-1}(L(\alpha)(B_t)). \tag{34}$$

Given $t \in {]0, t_0]}$, there exists $\theta \in {]0, t]}$ such that $B_\theta \subseteq L(\alpha)(B_t)$. Then

$$s(L(\alpha)) = [L(\alpha)(B_t) : L(\alpha)(B_t) \cap B_t]$$

$$= \frac{[L(\alpha)(B_t) : B_\theta]}{[L(\alpha)(B_t) \cap B_t : B_\theta]} \quad \text{in } (B_R, +)$$

$$= \frac{[L(\alpha)(B_t) : B_\theta]}{[L(\alpha)(B_t) \cap B_t : B_\theta]} \quad \text{in } (B_R, *)$$

$$= \frac{[\alpha(B_t^\phi) : B_\theta^\phi]}{[\alpha(B_t^\phi) \cap B_t^\phi : B_\theta^\phi]} = [\alpha(B_t^\phi) : \alpha(B_t^\phi) \cap B_t^\phi]$$

using Remark 5.3 for the third equality; to obtain the final equality, (34) was used and the fact that $\phi: B_R^\phi \to (B_R, *)$ is an isomorphism.   □

The following lemma shows that different choices of ϕ do not affect the B_t^ϕ for small t (as long as the norm is unchanged).

Lemma 5.6 *Let M be an analytic manifold over a totally disconnected local field $\mathbb{K}$, E be a finite-dimensional $\mathbb{K}$-vector space, and $\|.\|$ be an ultrametric norm on E. Let $p \in M$ and $\phi_j: U_j \to V_j$, for $j \in \{1, 2\}$, be an analytic diffeomorphism from an open neighbourhood U_j of p in M onto an open 0-neighbourhood $V_j \subseteq E$ such that $\phi_j(p) = 0$. If $d\phi_1|_{T_p(M)} = d\phi_2|_{T_p(M)}$, then there exists $\varepsilon > 0$ with $B_\varepsilon^E(0) \subseteq V_1 \cap V_2$ such that*

$$\phi_1^{-1}(B_t^E(0)) = \phi_2^{-1}(B_t^E(0)) \quad \text{for all } t \in \,]0, \varepsilon].$$

Proof The map $h := \phi_2 \circ \phi_1^{-1}: \phi_1(U_1 \cap U_2) \to \phi_2(U_1 \cap U_2)$ is an analytic diffeomorphism between open 0-neighbourhoods in E. Since $T_0(h) = \mathrm{id}_{T_0(E)}$, we have $h'(0) = \mathrm{id}_E$, which is an isometry. Thus, the Ultrametric Inverse Function Theorem provide $\varepsilon > 0$ with $B_\varepsilon^E(0) \subseteq \phi_1(U_1 \cap U_2)$ such that $h(B_t^E(0)) = B_t^E(0)$ for all $t \in \,]0, \varepsilon]$. Notably, $B_t^E(0) \subseteq \phi_1(U_1 \cap U_2)$ and $\phi_1^{-1}(B_t^E(0)) = \phi_2^{-1}(\phi_2(\phi_1^{-1}(B_t^E(0)))) = \phi_2^{-1}(B_t^E(0))$.   □

6 Endomorphisms of p-Adic Lie Groups

In this section, we first recall general facts concerning p-adic Lie groups which go beyond the properties of Lie groups over general local fields already described. In particular, we recall that every p-adic Lie group has an exponential function, and show that contraction groups of endomorphisms of p-adic Lie groups are always closed. We then calculate the scale and describe tidy subgroups for endomorphisms of p-adic Lie groups.

Basic Facts Concerning p-Adic Lie Groups For each $n \in \mathbb{N}$, the exponential series $\sum_{k=0}^\infty \frac{1}{k!} A^k$ converges for matrices A in some 0-neighbourhood V in the algebra $M_n(\mathbb{Q}_p)$ of $n \times n$-matrices and defines an analytic mapping $\exp: V \to GL_n(\mathbb{Q}_p)$. More generally, every analytic Lie group G over $\mathbb{Q}_p$ has an exponential function (see Definition 1 and the following lines in [7, Chapter III, §4, no. 3]):

6.1 An analytic map $\exp_G: V \to G$ on an open $\mathbb{Z}_p$-submodule $V \subseteq \mathfrak{g} := L(G)$ is called an *exponential function* if $\exp_G(0) = e$, $T_0(\exp_G) = \mathrm{id}_\mathfrak{g}$ (identifying

$T_0(\mathfrak{g}) = \{0\} \times \mathfrak{g}$ with $\mathfrak{g}$ via $(0, v) \mapsto v)$ and

$$\exp_G((s + t)x) = \exp_G(sx) \exp_G(tx)$$

for all $x \in U$ and $s, t \in \mathbb{Z}_p$.

6.2 Since $T_0(\exp_G) = \mathrm{id}_\mathfrak{g}$, after shrinking V one can assume that $\exp_G(V)$ is open in G and $\exp_G$ is a diffeomorphism onto its image (by the Inverse Function Theorem). After shrinking V further if necessary, we may assume that $\exp_G(V)$ is a subgroup of G (cf. Lemma 5.2). Hence also V can be considered as a Lie group. The Taylor expansion of multiplication with respect to the logarithmic chart $\exp_G^{-1}$ is given by the Baker-Campbell-Hausdorff (BCH-) series

$$x * y = x + y + \frac{1}{2}[x, y] + \cdots \tag{35}$$

(all terms of which are nested Lie brackets with rational coefficients), and hence $x * y$ is given by this series for small V (see Proposition 5 in [7, Chapter III, §4, no. 3] and proof of Proposition 3 in [7, Chapter III, §7, no. 2], also [49]). If $*$ is given on all of $V \times V$ by the BCH-series, we call $\exp_G(V)$ a *BCH-subgroup* of G.
Next, let us consider homomorphisms between p-adic Lie groups.

6.3 If $\alpha \colon G \to H$ is an analytic homomorphism between p-adic Lie groups, we can choose exponential functions $\exp_G \colon V_G \to G$ and $\exp_H \colon V_H \to H$ such that $L(\alpha).V_G \subseteq V_H$ and

$$\exp_H \circ L(\alpha)|_{V_G} = \alpha \circ \exp_G \tag{36}$$

(see Proposition 8 in [7, Chapter III, §4, no. 4], also [49]).
The following classical fact (see Theorem 1 in [7, Chapter III, §8, no. 1], also [49]) is important:

6.4 Every continuous homomorphism between p-adic Lie groups is analytic.

As a consequence, there is at most one p-adic Lie group structure on a given topological group. As usual, we say that a topological group is a p-adic Lie group if it admits a p-adic Lie group structure. Closed subgroups of p-adic Lie groups are Lie subgroups (see Theorem 2 in [7, Chapter III, §8, no. 2] or [49]), finite direct products and Hausdorff quotient groups of p-adic Lie groups are p-adic Lie groups (see Proposition 11 in [7, Chapter III, §1, no. 6], also [49]).

Closedness of Ascending Unions and Contraction Groups Another fact is vital:

Lemma 6.5 *Every p-adic Lie group G has an open subgroup which satisfies the ascending chain condition on closed subgroups. As a consequence, $\bigcup_{n \in \mathbb{N}} H_n$ is closed for each ascending sequence $H_1 \subseteq H_2 \subseteq \cdots$ of closed subgroups of G.*

Proof See, e.g. [21, Propositions 4.19 and 4.20]; cf. also step 1 of the proof of [54, Theorem 3.5]. □

Two important applications are now described.

Corollary 6.6 *Let α be an endomorphism of a p-adic Lie group G and V be a compact open subgroup of G. If V is tidy above for α, then V is tidy.*

Proof The subgroup $V_{--} = \bigcup_{n \in \mathbb{N}_0} \alpha^{-n}(V_-)$ is an ascending union of closed subgroups of G and hence closed, by Lemma 6.5. Thus V is tidy, by 2.4. $\square$

The second application of Lemma 6.5 concerns contraction groups. For automorphisms, see already [54, Theorem 3.5 (ii)].

Corollary 6.7 *Let G be a p-adic Lie group. Then the contraction group $\overrightarrow{\mathrm{con}}\,(\alpha)$ is closed in G, for each endomorphism $\alpha \colon G \to G$.*

Proof Let $V_1 \supseteq V_2 \supseteq \cdots$ be a sequence of compact open subgroups of G which form a basis of identity neighbourhoods (cf. Lemma 5.2). Then an element $x \in G$ belongs to $\overrightarrow{\mathrm{con}}\,(\alpha)$ if and only if

$$(\forall n \in \mathbb{N})\,(\exists m \in \mathbb{N})\,(\forall k \geq m)\ \alpha^k(x) \in V_n.$$

Since $\alpha^k(x) \in V_n$ if and only if $x \in \alpha^{-k}(V_n)$, we deduce that

$$\overrightarrow{\mathrm{con}}\,(\alpha) = \bigcap_{n \in \mathbb{N}} \bigcup_{m \in \mathbb{N}} \bigcap_{k \geq m} \alpha^{-k}(V_n).$$

Note that $W_n := \bigcup_{m \in \mathbb{N}} \bigcap_{k \geq m} \alpha^{-k}(V_n)$ is an ascending union of closed subgroups of G and hence closed, by Proposition 6.5. Consequently, $\overrightarrow{\mathrm{con}}\,(\alpha) = \bigcap_{n \in \mathbb{N}} W_n$ is closed. $\square$

Remark 6.8 We shall see later that also $\overleftarrow{\mathrm{con}}\,(\alpha)$ is always closed in the situation of Corollary 6.7 (see Theorem 8.15). Alternatively, this follows from the general structure theory (see [9, Proposition 10.4]).

Scale and Tidy Subgroups The following lemma prepares the construction of tidy subgroups in *p*-adic Lie groups, and can also be re-used later when we turn to Lie groups over general local fields. As two endomorphisms are discussed simultaneously in the lemma, we use notation as in 2.2.

Lemma 6.9 *Let G and H be totally disconnected, locally compact topological groups, $\alpha \colon G \to G$ and $\beta \colon H \to H$ be endomorphisms, $U \subseteq G$ and $V \subseteq H$ be subsets and $\psi \colon V \to U$ be a bijection. Assume that there exists a compact open subgroup $B \subseteq H$ such that $B \subseteq V$, $\beta(B) \subseteq V$, the image $W := \psi(B)$ is a compact open subgroup of G, and*

$$\alpha \circ \psi|_B = \psi \circ \beta|_B. \tag{37}$$

Write $B_+ := B_{+,\beta}$, $B_- := B_{-,\beta}$, $W_+ := W_{+,\alpha}$ and $W_- := W_{-,\alpha}$. Then

$$\psi(B_+) = W_+, \quad \psi(B_-) = W_- \quad and \quad \psi(\beta(B_+)) = \alpha(W_+).$$

Proof We define $B_n := B_{n,\beta}$ and $W_n := W_{n,\alpha}$ for $n \in \mathbb{N}_0$ as in 2.2. Then $B_n \subseteq B$ for each $n \in \mathbb{N}_0$, by construction. We show that

$$\psi(B_n) = W_n \tag{38}$$

for all $n \in \mathbb{N}_0$, by induction. The case $n = 0$ is clear: we have $\psi(B_0) = \psi(B) = W = W_0$. Now assume that (38) holds for some n. Since $B_n \subseteq B$, we have $\beta(B_n) \subseteq V$. Using that ψ is injective, (37), and the inductive hypothesis, we see that

$$\psi(B_{n+1}) = \psi(\beta(B_n) \cap B) = \psi(\beta(B_n)) \cap \psi(B) = \alpha(\psi(B_n)) \cap W$$
$$= \alpha(W_n) \cap W = W_{n+1}.$$

Thus (38) holds for all $n \in \mathbb{N}_0$. Since ψ is injective, we deduce that

$$\psi(B_+) = \psi\left(\bigcap_{n\in\mathbb{N}_0} B_n\right) = \bigcap_{n\in\mathbb{N}_0} \psi(B_n) = \bigcap_{n\in\mathbb{N}_0} W_n = W_+.$$

As $B_+ \subseteq B$, using (37) also $\psi(\beta(B_+)) = \alpha(\psi(B_+)) = \alpha(W_+)$ follows. Finally, for $n \in \mathbb{N}_0$ let B_{-n} be the set of all $x \in B$ such that $\beta^k(x) \in B$ for all $k \in \{0, 1, \ldots, n\}$, and W_{-n} be the set of all $w \in W$ such that $\alpha^k(w) \in W$ for all $k \in \{0, 1, \ldots, n\}$. We claim that

$$\psi(B_{-n}) = W_{-n} \quad \text{for all } n \in \mathbb{N}_0. \tag{39}$$

Since $B_- = \bigcap_{n\in\mathbb{N}_0} B_{-n}$ with $B_{-n} \subseteq B \subseteq V$ for all $n \in \mathbb{N}_0$, using the injectivity of ψ we then get

$$\psi(B_-) = \psi\left(\bigcap_{n\in\mathbb{N}_0} B_{-n}\right) = \bigcap_{n\in\mathbb{N}_0} \psi(B_{-n}) = \bigcap_{n\in\mathbb{N}_0} W_{-n} = W_-.$$

It only remains to prove the claim. It suffices to show that

$$\psi(B_{-n}) \subseteq W_{-n} \tag{40}$$

for all $n \in \mathbb{N}_0$, as the arguments can also be applied to $G, \alpha, H, \beta, \psi^{-1}, W$, and B in place of $H, \beta, G, \alpha, \psi, B$, and W, respectively. In fact, (37) implies that $\alpha(W) \subseteq U$, enabling us to compose the functions in (37) with ψ^{-1} on the left. Composing also with $(\psi|_B^W)^{-1}$ on the right, we find that

$$\psi^{-1} \circ \alpha|_W = \beta \circ \psi^{-1}|_W. \tag{41}$$

We now prove (40) by induction, starting with the observation that $\psi(B_0) = \psi(B) = W = W_0$. If (39) holds for some $n \in \mathbb{N}_0$, let $x \in B_{-(n+1)}$. Then $\psi(x) \in \psi(B) = W$ and $\beta^j(\beta(x)) = \beta^{j+1}(x) \in B$ for $j \in \{0, 1, \ldots, n\}$ shows that $\beta(x) \in B_{-n}$, whence $\alpha(\psi(x)) = \psi(\beta(x)) \in W_{-n}$ by induction. Hence $\psi(x) \in \{w \in W : \alpha(w) \in W_{-n}\} = W_{-(n+1)}$.   □

We are now ready to calculate the scale and find tidy subgroups for endomorphisms of p-adic Lie groups. It is illuminating to look at this easier case first, before we turn to endomorphisms of Lie groups over general local fields. Of course, the p-adic case is subsumed by the later discussion, but the latter is more technical as techniques from dynamical systems (local invariant manifolds) will be used as a replacement for the exponential function, which provides a local conjugacy between the linear dynamical system $(L(G), L(\alpha))$ and (G, α) in the case of an endomorphism α of a p-adic Lie group G, and thus enables a more elementary reasoning.

Preparations If G is a p-adic Lie group and $\alpha : G \to G$ an endomorphism, then there exists an open subgroup V of $(L(G), +)$ which is a BCH-Lie group with BCH-multiplication $*$, and an exponential function $\exp_G : V \to U$ which is an isomorphism from the Lie group $(V, *)$ onto a compact open subgroup U of G, as recalled above. Fix a norm $\|.\|$ on $\mathfrak{g} := L(G)$ which is adapted to $L(\alpha)$; after shrinking V, we may assume that

$$V = B_R^{\mathfrak{g}}(0) \tag{42}$$

for some $R > 0$. Abbreviate $B_t := B_t^{\mathfrak{g}} := B_t^{\mathfrak{g}}(0)$ for $t > 0$. Applying 6.3 and Lemma 5.2 to $\phi := (\exp_G)^{-1} : U \to V$, we find $r \in]0, R]$ such that $B_t^{\phi} := \phi^{-1}(B_t) = \exp_G(B_t)$ is a compact open subgroup of G for all $t \in]0, r]$ and, moreover,

$$L(\alpha)(B_r) \subseteq V \quad \text{and} \quad \exp_G \circ L(\alpha)|_{B_r} = \alpha \circ \exp_G|_{B_r}, \tag{43}$$

whence $\alpha(B_r^{\phi}) \subseteq U$ in particular. Let $\mathfrak{g}_{<1} := \bigoplus_{\rho \in [0,1[} \mathfrak{g}_\rho$ be the indicated sum of characteristic subspaces with respect to $L(\alpha)$, and $\mathfrak{g}_{\geq 1} := \bigoplus_{\rho \geq 1} \mathfrak{g}_\rho$. Since

$$\mathfrak{g}_{<1} = \overrightarrow{\mathrm{con}}(L(\alpha)) \quad \text{and} \quad \mathfrak{g}_{\geq 1} = \overleftarrow{\mathrm{par}}(L(\alpha))$$

are Lie subalgebras of $\mathfrak{g}$ (see Theorem 3.6 (b) and Lemma 3.8) and $*$ is given by the BCH-series, we see that

$$B_r^{\mathfrak{g}_{<1}} := B_r \cap \mathfrak{g}_{<1} \quad \text{and} \quad B_r^{\mathfrak{g}_{\geq 1}} := B_r \cap \mathfrak{g}_{\geq 1}$$

are Lie subgroups of $(B_r, *)$ with Lie algebras $\mathfrak{g}_{<1}$ and $\mathfrak{g}_{\geq 1}$, respectively. After shrinking R if necessary, we may assume that

$$x * B_t^{\mathfrak{g}_{\geq 1}} = x + B_t^{\mathfrak{g}_{\geq 1}} \quad \text{for all } x \in B_R^{\mathfrak{g}_{\geq 1}} \text{ and } t \in]0, R], \tag{44}$$

see Remark 5.3 (which applies with $\mathfrak{g}_{\geq 1}$ in place of $\mathfrak{g}$ and id: $B_r^{\mathfrak{g}\geq 1} \to B_r^{\mathfrak{g}\geq 1}$ in place of ϕ). Now the mapping

$$B_r^{\mathfrak{g}\geq 1} \times B_r^{\mathfrak{g}<1} \to B_r, \quad (x, y) \mapsto x * y$$

has the derivative

$$\mathfrak{g}_{\geq 1} \times \mathfrak{g}_{<1} \to \mathfrak{g}, \quad (x, y) \mapsto x + y \tag{45}$$

at $(0, 0)$, which is an isometry if we endow $\mathfrak{g}_{<1}$ and $\mathfrak{g}_{\geq 1}$ with the norm induced by $\|.\|$ and use the maximum norm thereof on the left-hand side of (45). Hence, by the Ultrametric Inverse Function Theorem (Lemma 4.8), after shrinking r (if necessary) we may assume that

$$B_t^{\mathfrak{g}\geq 1} * B_t^{\mathfrak{g}<1} = B_t \quad \text{for all } t \in {]0, r]}. \tag{46}$$

With notation as before, we have:

Theorem 6.10 *If α is an endomorphism of a p-adic Lie group G, then*

$$s_G(\alpha) = s_{L(G)}(L(\alpha))$$

holds and B_t^{ϕ} is tidy for α, for all $t \in {]0, r]}$.

Proof Let $t \in {]0, r]}$. Applying the isomorphism $\exp_G \colon (V, *) \to U$ to both sides of (46), we see that

$$\exp_G(B_t^{\mathfrak{g}\geq 1}) \exp_G(B_t^{\mathfrak{g}<1}) = B_t^{\phi}. \tag{47}$$

In view of (43), we can apply Lemma 6.9 to G, α, $H := (\mathfrak{g}, +)$, $\beta := L(\alpha)$, $\psi := \exp_G$, $B := B_t^{\mathfrak{g}}$, and $W := B_t^{\phi}$. Hence

$$(B_t^{\phi})_+ := (B_t^{\phi})_{+,\alpha} = \exp_G((B_t^{\mathfrak{g}})_{+,\beta}) \quad \text{and} \quad (B_t^{\phi})_- := (B_t^{\phi})_{-,\alpha} = \exp_G((B_t^{\mathfrak{g}})_{-,\beta}).$$

Now

$$(B_t^{\mathfrak{g}})_{+,\beta} = B_t^{\mathfrak{g}\geq 1} \quad \text{and} \quad (B_t^{\mathfrak{g}})_{-,\beta} = B_t^{\mathfrak{g}} \cap \bigoplus_{\rho \in [0,1]} \mathfrak{g}_\rho \supseteq B_t^{\mathfrak{g}<1}$$

(cf. (17)), whence

$$(B_t^{\phi})_+ = \exp_G(B_t^{\mathfrak{g}\geq 1}) \quad \text{and} \quad (B_t^{\phi})_- \supseteq \exp_G(B_t^{\mathfrak{g}<1}).$$

Combining this with (47), we find that

$$B_t^\phi \supseteq (B_t^\phi)_+ (B_t^\phi)_- \supseteq \exp_G(B_t^{\mathfrak{g}\geq 1}) \exp_G(B_t^{\mathfrak{g}<1}) = B_t^\phi$$

and thus $B_t^\phi = (B_t^\phi)_+ (B_t^\phi)_-$, i.e., B_t^ϕ is tidy above for α and thus tidy for α, by Corollary 6.6. Note that $L(\alpha)|_{B_t^\mathfrak{g}} : (B_t^\mathfrak{g}, *) \to (V, *)$ is a group homomorphism, as $*$ is given by the BCH-series. Hence $\beta((B_t^\mathfrak{g})_+) = L(\alpha)(B_t^{\mathfrak{g}\geq 1})$ is a subgroup of the group $(V \cap \mathfrak{g}_{\geq 1}, *)$, which contains $(B_t^\phi)_{+,\beta} = B_t^{\mathfrak{g}\geq 1}$ as a subgroup. Since $\exp_G : (V, *) \to U$ is an isomorphism of groups and cosets of balls coincide in the groups $(V \cap \mathfrak{g}_{\geq 1}, +)$ and $(V \cap \mathfrak{g}_{\geq 1}, *)$ (see (44)), we obtain

$$s(L(\alpha)) = [L(\alpha)B_t^{\mathfrak{g}\geq 1} : B_t^{\mathfrak{g}\geq 1}] \quad \text{w.r.t. } +$$

$$= [L(\alpha)B_t^{\mathfrak{g}\geq 1} : B_t^{\mathfrak{g}\geq 1}] \quad \text{w.r.t. } *$$

$$= [\exp_G(L(\alpha)(B_t^{\mathfrak{g}\leq 1})) : \exp_G(B_t^{\mathfrak{g}\leq 1})]$$

$$= [\alpha(\exp_G(B_t^{\mathfrak{g}\leq 1})) : \exp_G(B_t^{\mathfrak{g}\leq 1})]$$

$$= [\alpha((B_t^\phi)_+) : (B_t^\phi)_+] = s(\alpha),$$

which completes the proof. $\square$

Remark 6.11 For *automorphisms* of p-adic Lie groups, the calculation of the scale was performed in [15].

7 Invariant Manifolds Around Fixed Points

As in the classical real case, (locally) invariant manifolds can be constructed around fixed points of time-discrete analytic dynamical systems over a totally disconnected local field (see [18] and [19]). We shall use these as a tool in our discussion of analytic endomorphisms of Lie groups over such fields. In the current section, we compile the required background.

Definition 7.1 Let E be a finite-dimensional vector space over a totally disconnected local field $\mathbb{K}$, which we endow with its natural absolute value $|.|_\mathbb{K}$. Let $\alpha : E \to E$ be lk-linear. Given $a \in \,]0, \infty]$, we call

$$E_{<a} := \bigoplus_{\rho \in [0,a[} E_\rho \quad \text{and} \quad E_{>a} := \bigoplus_{\rho \in \,]a,\infty[} E_\rho$$

the *a-stable* and *a-unstable* vector subspaces of E with respect to α, using the characteristic subspaces E_ρ with respect to α (as in 3.2). We call E_1 (i.e., E_ρ with

$\rho = 1$) the *centre subspace* of E with respect to α. A linear endomorphism α of E is called *a-hyperbolic* if $a \neq |\lambda|_{\mathbb{K}}$ for all eigenvalues λ of α in an algebraic closure $\overline{\mathbb{K}}$, i.e., if $E_a = \{0\}$ and thus

$$E = E_{<a} \oplus E_{>a}.$$

Now consider an analytic manifold M over a local field $\mathbb{K}$, an analytic mapping $f \colon M \to M$, a fixed point $p \in M$ of f and a submanifold $N \subseteq M$ such that $p \in N$. Given $a > 0$, decompose

$$T_p(M) = T_p(M)_{<a} \oplus T_p(M)_a \oplus T_p(M)_{>a}$$

with respect to the endomorphism $T_p(f)$ of $T_p(M)$, as in Definition 7.1. For our purposes, special cases of concepts in [18] and [19] are sufficient:

Definition 7.2

(a) If $a \in \,]0, 1]$ and $T_p(f)$ is a-hyperbolic, we say that the submanifold N is a *local a-stable manifold* for f around p if $T_p(N) = T_p(M)_{<a}$ and $f(N) \subseteq N$.
(b) We say that N is a *centre manifold* for f around p if $T_p(N) = T_p(M)_1$ and $f(N) = N$.
(c) If $b \geq 1$ and $T_p(f)$ is b-hyperbolic, we say that N is a *local b-unstable manifold* for f around p if $T_p(N) = T_p(M)_{>b}$ and there exists an open neighbourhood P of p in N such that $f(P) \subseteq N$.

We need a fact concerning the existence of local invariant manifolds.

Proposition 7.3 *Let M be an analytic manifold over a totally disconnected local field $\mathbb{K}$. Let $f \colon M \to M$ be an analytic mapping and $p \in M$ be a fixed point of f. Moreover, let $a \in \,]0, 1]$ and $b \in [1, \infty[$ be such that $a \neq |\lambda|_{\mathbb{K}}$ and $b \neq |\lambda|_{\mathbb{K}}$ for all eigenvalues λ of $T_p f$ in an algebraic closure $\overline{\mathbb{K}}$ of $\mathbb{K}$. Finally, let $\|.\|$ be a norm on $E := T_p(M)$ which is adapted to the endomorphism $T_p(f)$. Endow vector subspaces $F \subseteq E$ with the norm induced by $\|.\|$ and abbreviate $B_t^F := B_t^F(0)$ for $t > 0$. Then the following holds:*

(a) *There exists a local a-stable manifold W_a^s for f around p and an analytic diffeomorphism*

$$\phi_s \colon W_a^s \to B_R^{E_{<a}}$$

for some $R > 0$ such that $\phi_s(p) = 0$ holds, $W_a^s(t) := \phi_s^{-1}(B_t^{E_{<a}})$ is a local a-stable manifold for f around p for all $t \in \,]0, R]$, and $d\phi_s|_{T_p(W_a^s)} = \mathrm{id}_{E_{<a}}$.
(b) *There exists a centre manifold W^c for f around p and an analytic diffeomorphism*

$$\phi_c \colon W^c \to B_R^{E_1}$$

for some $R > 0$ *such that* $\phi_c(p) = 0$ *holds,* $W^c(t) := \phi_c^{-1}(B_t^{E_1})$ *is a centre manifold for* f *around* p *for all* $t \in \,]0, R]$, *and* $d\phi_c|_{T_p(W^c)} = \mathrm{id}_{E_1}$.

(c) *There exists a local b-unstable manifold* W_b^u *for* f *around* p *and an analytic diffeomorphism*

$$\phi_u : W_b^u \to B_R^{E_{>b}}$$

for some $R > 0$ *such that* $\phi_u(p) = 0$, $W_b^u(t) := \phi_u^{-1}(B_t^{E_{>b}})$ *is a local b-unstable manifold for* f *around* p *for all* $t \in \,]0, R]$, *and* $d\phi_u|_{T_p(W_b^u)} = \mathrm{id}_{E_{>b}}$.

Proof (a) and (c) are covered by the Local Invariant Manifold Theorem (see [19, p. 76]) and its proof. To get (b), let $\phi : U \to V$ be an analytic diffeomorphism from on open neighbourhood U of p in M onto an open 0-neighbourhood $V \subseteq E$ such that $\phi(p) = 0$ and $d\phi|_E = \mathrm{id}_E$. We can then construct centre manifolds for the analytic map

$$\phi \circ f \circ \phi^{-1} : \phi(U \cap f^{-1}(U)) \to V$$

around its fixed point 0 with [18, Proposition 4.2] and apply ϕ^{-1} to create the desired centre manifolds for f. We mention that the cited proposition only considers mappings whose derivative at the fixed point is an automorphism, but its proof never uses this hypothesis, which therefore can be omitted. □

Remark 7.4 Of course, we can use the same $R > 0$ in parts (a), (b), and (c) of Proposition 7.3 (simply take the minimum of the three numbers).

Remark 7.5 Note that, since $f(W_a^s) \subseteq W_a^s$, we have a descending sequence

$$W_a^s \supseteq f(W_a^s) \supseteq f^2(W_a^s) \supseteq \cdots$$

in Proposition 7.3 (a).

Lemma 7.6 *After shrinking R in Proposition* 7.3 (a) *if necessary, we can assume that*

$$\bigcap_{n \in \mathbb{N}_0} f^n(W_a^s) = \{p\} \quad and \quad \lim_{n \to \infty} f^n(x) = p \;\; for\;all\;\; x \in W_a^s. \tag{48}$$

Proof Abbreviate $F := E_{<a}$. The map

$$h := \phi_s \circ f|_{W_a^s} \circ \phi_s^{-1} : B_R^F \to B_R^F$$

is analytic, $h(0) = 0$, and $h'(0) = T_p(f)|_F$ has operator norm $\|h'(0)\|_{\mathrm{op}} < a$. Choose $\varepsilon > 0$ so small that

$$\theta := \|h'(0)\|_{\mathrm{op}} + \varepsilon < 1.$$

Since h is totally differentiable at 0, we find $r \in \,]0, R]$ such that

$$\|h(x) - h'(0)(x)\| \leq \varepsilon \|x\| \quad \text{for all } x \in B_r^F.$$

Then $\|h(x)\| = \|h'(0)(x) + (h(x) - h'(0)(x))\| \leq (\|h'(0)\|_{\mathrm{op}} + \varepsilon)\|x\| = \theta\|x\|$ for all $x \in B_r^F$, whence $h(B_r^F) \subseteq B_{\theta r}^F \subseteq B_r^F$ and

$$h^n(B_r^F) \subseteq B_{\theta^n r}^F \quad \text{for all } n \in \mathbb{N}_0.$$

As a consequence, $\bigcap_{n \in \mathbb{N}_0} h^n(B_r^F) = \{0\}$. Then $Q := W_a^s(r) = \phi_s^{-1}(B_r^F)$ is an open neighbourhood of p in W_a^s such that $\bigcap_{n \in \mathbb{N}_0} f^n(Q) = \{p\}$. After replacing R with r, we have (48).   □

Lemma 7.7 *After shrinking R in Proposition 7.3 (b), we may assume that the map $f|_{W^c(t)} : W^c(t) \to W^c(t)$ is an analytic diffeomorphism for each $t \in \,]0, R]$.*

Proof Abbreviate $F := T_p(M)_1 = E_1$. The mapping

$$h := \phi_c \circ f|_{W^c} \circ \phi_c^{-1} : B_R^F \to B_R^F$$

is analytic with $h(0) = 0$, and $h'(0) = T_p(f)|_F$ is an isometry. By the Ultrametric Inverse Function Theorem, after shrinking R if necessary, we can achieve that h is an analytic diffeomorphism from B_R^F onto B_R^F. Then also $f|_{W^c}$ is a diffeomorphism. Since $W^c(t)$ is a centre manifold for all $t \in \,]0, R]$, we have $f(W^c(t)) = W^c(t)$, which completes the proof.   □

Lemma 7.8 *We can always choose the open neighbourhood P around p in a local b-unstable manifold $N \subseteq M$ (as in Definition 7.2 (c)) in such a way that, for each $x \in P \setminus \{p\}$, there exists $n \in \mathbb{N}$ such that $x, f(x), \dots, f^{n-1}(x) \in P$ but $f^n(x) \in N \setminus P$.*

Proof To see this, excluding a trivial case,[12] we may assume that the b-unstable subspace $F := E_{>b} := T_p(M)_{>b}$ with respect to $T_p(f)$ is non-trivial. Let $\phi : U \to V$ be an analytic diffeomorphism from an open neighbourhood U of p in N onto an open 0-neighbourhood $V \subseteq T_p(N) = F$, such that $\phi(p) = 0$ and $d\phi|_{T_p(N)} = \mathrm{id}_F$. Then

$$h := \phi \circ f \circ \phi^{-1} : \phi(f^{-1}(U) \cap U) \to V$$

is an analytic mapping defined on an open 0-neighbourhood, such that $h'(0) = T_p(f)|_F$ is invertible and

$$\frac{1}{\|h'(0)^{-1}\|_{\mathrm{op}}} > b.$$

---

[12]Otherwise N is discrete and we can choose $P = \{p\}$.

Since h is totally differentiable at 0, there exists $r > 0$ with $B_r^F(0)$ in the domain D of h such that

$$h(B_r^F(0)) \subseteq D,$$

$$\|h(x) - h'(0)(x)\| \leq b\|x\| \tag{49}$$

for all $x \in B_r^F(0) \setminus \{0\}$, and $f(P) \subseteq U$ with $P := \phi^{-1}(B_r^F(0))$. Using (49) and (11), we deduce that

$$\|h(x)\| = \|h'(0)(x) + (h(x) - h'(0)(x))\| = \|h'(0)(x)\| > b\|x\|,$$

as $\|h'(0)(x)\| \geq \|h'(0)^{-1}\|_{\mathrm{op}}^{-1}\|x\| > b\|x\|$. So, for all $x \in B_r^F(0) \setminus \{0\}$, there is $n \in \mathbb{N}$ such that $x, h(x), \ldots, h^n(x)$ are defined and in $B_r^F(0)$, but $h^{n+1}(x) \in D \setminus B_r^F(0)$. Now P is a neighbourhood of p with the desired property. $\square$

Lemma 7.9 *Let U be an open neighbourhood of p in M and $\phi \colon U \to V$ be an analytic diffeomorphism onto an open 0-neighbourhood $V \subseteq T_p(M) =: E$ such that $\phi(P) = 0$ and $d\phi|_E = \mathrm{id}_E$. After decreasing R in Proposition 7.3 (c) if necessary, we can always assume that $B_R^E(0) \subseteq V$ and the following additional property holds for all $t \in]0, R]$:*

$W_b^u(t)$ is the set of all $x \in \phi^{-1}(B_t^E(0)) =: B_t^\phi$ for which there exists an f-regressive trajectory $(x_{-n})_{n \in \mathbb{N}_0}$ in B_t^ϕ with $x_0 = x$, such that

$$\lim_{n \to \infty} b^n \phi(x_{-n}) = 0. \tag{50}$$

Then $\lim_{n \to \infty} x_{-n} = p$ in particular, and $x_{-n} \in W_b^u(t)$ for all $n \in \mathbb{N}_0$.

Proof As before, abbreviate $B_t^E := B_t^E(0)$ and $B_t^{E>b} := B_t^{E>b}(0)$ for $t > 0$. There is $r > 0$ such that $B_r^E \subseteq V$ and $f(\phi^{-1}(B_r^E)) \subseteq U$, whence an analytic map

$$h := \phi \circ f \circ \phi^{-1}|_{B_r^E} \colon B_r^E \to E$$

can be defined with $h(0) = 0$ and $h'(0) = T_p(f)$. For $t \in]0, r]$, let Γ_t be the set of all $z \in B_t^E$ for which there exists an h-regressive trajectory $(z_{-n})_{n \in \mathbb{N}_0}$ in B_t^E with $z_0 = z$ such that

$$\lim_{n \to \infty} b^n \|z_{-n}\| = 0. \tag{51}$$

By [18, Theorem B.2] and the proof of Theorem 8.3 in [18], after shrinking r we may assume that Γ_t is a submanifold of B_t^E and

$$\phi^{-1}(\Gamma_t)$$

a local b-unstable submanifold of M for each $t \in]0, r]$; and, moreover, there is an analytic map

$$\mu: B_r^{E_{>b}} \to B_r^{E_{<b}}$$

(called ϕ there) with $\mu(0) = 0$ and

$$\mu'(0) = 0 \tag{52}$$

such that

$$\Gamma_t = \{(\mu(y), y): y \in B_t^{E_{>b}}\} \quad \text{for all } t \in]0, r],$$

identifying $E = E_{<b} \oplus E_{>b}$ with $E_{<b} \times E_{>b}$. Hence

$$v: B_r^{E_{>b}} \to \phi^{-1}(\Gamma_r), \quad y \mapsto \phi^{-1}(\mu(y), y)$$

is an analytic diffeomorphism, and thus also $v^{-1}: v(B_r^{E_{>b}}) \to B_r^{E_{>b}}$ is an analytic diffeomorphism. As a consequence of (52), we have

$$d(v^{-1})|_{E_{>b}} = \mathrm{id}_{E_{>b}}.$$

Since, like $W_b^u = \phi_u^{-1}(B_R^{E_{>b}})$, also $\phi^{-1}(\Gamma_r) = v(B_r^{E_{>b}})$ is a local b-unstable manifold for f, [18, Theorem 8.3] shows that there exists a subset $Q \subseteq W_b^u \cap v(B_r^{E_{>b}})$ which is an open neighbourhood of p in both W_b^u and $v(B_r^{E_{>b}})$. Hence, there exists $\tau > 0$ with $\tau \leq \min\{R, r\}$ such that $W_b^u(\tau) \subseteq Q$ and $v(B_\tau^{E_{>b}}) \subseteq Q$. Since Q is a submanifold of M, the manifold structures induced on Q as an open subset of W_b^u and $v(B_r^{E_{>b}})$ coincide. By Lemma 5.6, after shrinking τ if necessary we may assume that

$$W_b^u(t) = \phi_u^{-1}(B_t^{E_{>b}}) = (v^{-1})^{-1}(B_t^{E_{>b}}) = \phi^{-1}(\Gamma_t)$$

for all $t \in]0, \tau]$. Let $t \in]0, \tau]$ and $x \in B_t^\phi$.

If $x \in \phi^{-1}(\Gamma_t)$, then there exists an h-regressive trajectory $(z_{-n})_{n \in \mathbb{N}_0}$ in B_t^E with $z_0 = \phi(x)$ and $b^n \|z_{-n}\| \to 0$. Now, for $m \in \mathbb{N}_0$, the sequence $(z_{-n-m})_{n \in \mathbb{N}_0}$ is an h-regressive trajectory for z_{-m} in B_t^E such that

$$b^n \|z_{-n-m}\| = b^{-m} b^{n+m} \|z_{-n-m}\| \to 0$$

as $n \to \infty$ and thus $z_{-m} \in \Gamma_t$. As a consequence, $(\phi^{-1}(z_{-n}))_{n \in \mathbb{N}_0}$ is an f-regressive trajectory in $W_b^u(t) = \phi^{-1}(\Gamma_t)$ such that $\phi^{-1}(z_0) = x$ and $b^n \|\phi(\phi^{-1}(z_{-n}))\| = b^n \|z_{-n}\| \to 0$ as $n \to \infty$.

Conversely, assume there exists an f-regressive trajectory $(x_{-n})_{n \in \mathbb{N}_0}$ in B_t^ϕ with $x_0 = x$ and (50). Then $(\phi(x_{-n}))_{n \in \mathbb{N}_0}$ is an h-regressive trajectory in B_t^E such that (51) holds, whence $\phi(x) = \phi(x_0) \in \Gamma_t$ and thus $x \in \phi^{-1}(\Gamma_t)$.

Summing up, the conclusion of the lemma holds if we replace R with τ.   □
Let us consider a first application of invariant manifolds.

Proposition 7.10 *Let α be an analytic automorphism of a Lie group G over a totally disconnected local field $\mathbb{K}$. Let $\overline{\mathbb{K}}$ be an algebraic closure of $\mathbb{K}$. Then we have:*

(a) $\overrightarrow{\mathrm{con}}(\alpha)$ *is open in G if and only if $|\lambda|_{\mathbb{K}} < 1$ for each eigenvalue λ of $L(\alpha)$ in $\overline{\mathbb{K}}$.*
(b) α *is expansive if and only if $|\lambda|_{\mathbb{K}} \neq 1$ for each eigenvalue λ of $L(\alpha)$ in $\overline{\mathbb{K}}$.*
(c) α *is a distal automorphism if and only if $|\lambda|_{\mathbb{K}} = 1$ for each eigenvalue λ of $L(\alpha)$ in $\overline{\mathbb{K}}$.*

Proof (c) If $|\lambda|_{\mathbb{K}} < 1$ for some λ, choose $a \in \,]0, 1]$ such that $a > |\lambda|_{\mathbb{K}}$ and $L(\alpha)$ is a-hyperbolic. Then G has a local a-stable manifold $W \neq \{e\}$ for α around e, which can be chosen such that $W \subseteq \overrightarrow{\mathrm{con}}(\alpha)$ (see Lemma 7.6). Since $\alpha^n(x) \to e$ for all $x \in W \setminus \{e\}$, we see that α is not distal.

If $|\lambda|_{\mathbb{K}} > 1$ for some λ, then again we see that α is not distal, replacing α with α^{-1} and its iterates in the preceding argument.

If $|\lambda|_{\mathbb{K}} = 1$ for each λ, then $\mathfrak{g} := L(G)$ coincides with its centre subspace with respect to $L(\alpha)$, whence every centre manifold for α around e is open in G. If $x \in G \setminus \{e\}$, then Proposition 7.3 (b) provides a centre manifold W for α around e such that $x \notin W$. Since $\alpha^n(W) = W$ for all $n \in \mathbb{Z}$ and α^n is a bijection, we must have $\alpha^n(x) \notin W$ for all $n \in \mathbb{Z}$. As a consequence, the set $\{\alpha^n(x) : n \in \mathbb{Z}\}$ (and hence also its closure) is contained in the closed set $G \setminus W$. Thus $e \notin \overline{\{\alpha^n(x) : n \in \mathbb{Z}\}}$ and thus α is distal.

The proofs for (b) and the implication "⇒" in (a) are similar and again involve local invariant manifolds, see [23, Proposition 7.1] and [19, Corollary 6.1 and Proposition 3.5], respectively.

(a) To complete the proof of (a), assume that $|\lambda|_{\mathbb{K}} < 1$ for all λ. Choose $a \in \,]0, 1[$ such that $a > |\lambda|_{\mathbb{K}}$ for all λ. Then $\mathfrak{g} = \mathfrak{g}_{<a}$ with respect to $L(\alpha)$. Let W_a^s and the analytic diffeomorphism $\phi_s : W_a^s \to B_R^{\mathfrak{g}}(0)$ be as in Proposition 7.3 (a). Then W_a^s is open in G. By Lemma 7.6, after shrinking R (if necessary) we can achieve that $W_a^s \subseteq \overrightarrow{\mathrm{con}}(\alpha)$. Thus $\overrightarrow{\mathrm{con}}(\alpha)$ is an open identity neighbourhood in G and hence $\overrightarrow{\mathrm{con}}(\alpha)$ is open, being a subgroup.   □

8 Endomorphisms of Lie Groups over $\mathbb{K}$

In this section, we formulate and prove our main results concerning analytic endomorphisms of Lie groups over totally disconnected local fields.

Some Preparations

Definition 8.1 Let α be an endomorphism of a totally disconnected, locally compact group G. We say that G has *small tidy subgroups* for α if each identity neighbourhood of G contains a compact open subgroup of G which is tidy for α.

For α an automorphism, the existence of small tidy subgroups is equivalent to closedness of $\overrightarrow{\mathrm{con}}\,(\alpha)$ (see [1, Theorem 3.32] for the case of metrizable groups; the general case can be deduced with arguments from [36]). The following result concerning endomorphisms is sufficient for our Lie theoretic applications.

Lemma 8.2 *Let α be an endomorphism of a totally disconnected, locally compact group G.*

(a) *If G has small subgroups tidy for α, then $\overrightarrow{\mathrm{con}}\,(\alpha)$ is closed.*

(b) *If $\overrightarrow{\mathrm{con}}\,(\alpha)$ is closed and a compact open subgroup U of G satisfies*

$$U_- = (\overrightarrow{\mathrm{con}}\,(\alpha) \cap U_-)(U_+ \cap U_-) \tag{53}$$

and is tidy above, then U is tidy for α. Hence, if $\overrightarrow{\mathrm{con}}\,(\alpha)$ is closed and each identity neighbourhood of G contains a compact open subgroup U which satisfies (53) and is tidy above for α, then G has small tidy subgroups.

Proof

(a) Let $\mathscr{T}(\alpha)$ be the set of all tidy subgroups for α. If $\mathscr{T}(\alpha)$ is a basis of identity neighbourhoods, then

$$\overrightarrow{\mathrm{con}}\,(\alpha) = \{x \in G \colon \lim_{n \to \infty} \alpha^n(x) = e\}$$

$$= \{x \in G \colon (\forall U \in \mathscr{T}(\alpha))\,\underbrace{(\exists m)(\forall n \geq m)\ \alpha^n(x) \in U}_{\Leftrightarrow x \in U_{--}}\}$$

$$= \bigcap_{U \in \mathscr{T}(\alpha)} U_{--},$$

which is closed.

(b) Assuming that $U_- = (\overrightarrow{\mathrm{con}}\,(\alpha) \cap U_-)(U_+ \cap U_-)$, let us show that

$$U_{--} = \overrightarrow{\mathrm{con}}\,(\alpha)(U_+ \cap U_-). \tag{54}$$

If (54) holds, then U_{--} is closed (as $\overrightarrow{\mathrm{con}}\,(\alpha)$ is assumed closed and $U_+ \cap U_-$ is compact). Hence U will be tidy for α (by 2.4), and also the final assertion is then immediate.

The inclusion "$\supseteq$" in (54) is clear. To see that the converse inclusion holds, let $x \in U_{--}$. Then $\alpha^n(x) \in U_-$ for some n. As $U_- = (\overrightarrow{\mathrm{con}}\,(\alpha) \cap U_-)(U_+ \cap U_-)$ by hypothesis, we have

$$\alpha^n(x) = yz \quad \text{for some } y \in \overrightarrow{\mathrm{con}}\,(\alpha) \text{ and some } z \in U_+ \cap U_-.$$

Since $\alpha(U_+ \cap U_-) = U_+ \cap U_-$, find $w \in U_+ \cap U_-$ such that $z = \alpha^n(w)$. Then $\alpha^n(xw^{-1}) = y \in \overrightarrow{\mathrm{con}}\,(\alpha)$, entailing that $xw^{-1} \in \overrightarrow{\mathrm{con}}\,(\alpha)$ and thus $x = (xw^{-1})w \in \overrightarrow{\mathrm{con}}$ $(\alpha)(U_+ \cap U_-)$. Thus $U_{--} \subseteq \overrightarrow{\mathrm{con}}\,(\alpha)(U_+ \cap U_-)$; the proof is complete. $\qquad\square$

Remark 8.3 With much more effort, it can be shown that closedness of $\overrightarrow{\mathrm{con}}\,(\alpha)$ is always equivalent to the existence of small tidy subgroups, for every endomorphism α of a totally disconnected, locally compact group G (see [9, Theorem D]). Lemma 8.2, which is sufficient for our ends, was presented at the AMSI workshop July 25, 2016 (before the cited theorem was known).

We need a result from the structure theory of totally disconnected groups.

Lemma 8.4 *Let α be an endomorphism of a totally disconnected, locally compact group G and $V \subseteq G$ be a compact open subgroup which is tidy above for α. Then $s(\alpha)$ divides $[\alpha(V) : \alpha(V) \cap V]$.* $\qquad\square$

Proof As in [60, Definition 5], let $\mathscr{L}_V$ be the subgroup of all $x \in G$ for which there exist $y \in V_+$ and $n, m \in \mathbb{N}_0$ such that $\alpha^m(y) = x$ and $\alpha^n(y) \in V_-$. Let L_V be the closure of $\mathscr{L}_V$ in G,

$$\widetilde{V} := \{x \in V : xL_V \subseteq L_V V\}$$

(as in [60, (7)]) and $W := \widetilde{V}L_V$. Then $\widetilde{V}$ is a compact open subgroup of G which is tidy above for α and

$$[\alpha(V) : \alpha(V) \cap V] = [\alpha(\widetilde{V}) : \alpha(\widetilde{V}) \cap \widetilde{V}] \tag{55}$$

(see [60, Lemma 16]). Moreover, W is a compact open subgroup of G which is tidy for α, by the third step of the 'tidying procedure' (see [60, Step 3 following Definition 10]). Let $W_{-1} := W \cap \alpha^{-1}(W)$ and $\widetilde{V}_{-1} := \widetilde{V} \cap \alpha^{-1}(\widetilde{V})$. Then the left action

$$\widetilde{V}_+ \times Y \to Y, \quad (v, w(W_+ \cap W_{-1})) \mapsto vw(W_+ \cap W_{-1})$$

of $\widetilde{V}_+$ on $Y := W_+/(W_+ \cap W_{-1})$ is transitive (as the map ϕ defined in the proof of [60, Proposition 6 (4)] is surjective). The point $W_+ \cap W_{-1} \in Y$ has stabilizer $\widetilde{V}_+ \cap W_+ \cap W_{-1} = \widetilde{V}_+ \cap W_{-1}$. Hence

$$s(\alpha) = [\alpha(W) : \alpha(W) \cap W] = [W_+ : W_+ \cap W_{-1}] = |Y| = [\widetilde{V}_+ : \widetilde{V}_+ \cap W_{-1}], \tag{56}$$

using tidiness of W for the first equality, [60, Lemma 3 (1) and (4)] for the second and the orbit formula for the $\widetilde{V}_+$-action for the last. Since $\widetilde{V}_+ \cap W_{-1}$ contains $\widetilde{V}_+ \cap \widetilde{V}_{-1}$ as a subgroup, using [60, Lemma 3] again we deduce that

$$[\alpha(\widetilde{V}) : \alpha(\widetilde{V}) \cap \widetilde{V}] = [\widetilde{V}_+ : \widetilde{V}_+ \cap \widetilde{V}_{-1}]$$

$$= [\widetilde{V}_+ : \widetilde{V}_+ \cap W_{-1}][\widetilde{V}_+ \cap W_{-1} : \widetilde{V}_+ \cap \widetilde{V}_{-1}]. \tag{57}$$

Substituting (55) and (56) into (57), we obtain

$$[\alpha(V) : \alpha(V) \cap V] = s(\alpha)[\widetilde{V}_+ \cap W_{-1} : \widetilde{V}_+ \cap \widetilde{V}_{-1}], \tag{58}$$

which completes the proof.   □

Scale and Tidy Subgroups If $\alpha: G \to G$ is an analytic endomorphism of a Lie group G over a local field $\mathbb{K}$, we fix a norm $\|.\|$ on its Lie algebra $\mathfrak{g} := L(G) := T_e(G)$ which is adapted to the associated linear endomorphism $L(\alpha) := T_e(\alpha)$ of $\mathfrak{g}$. Let $\overline{\mathbb{K}}$ be an algebraic closure of $\mathbb{K}$. In the proof of our main result, Theorem 8.13, we want to use Lemma 5.5 to create compact open subgroups B_t^ϕ of G. To get more control over these subgroups, we now make a particular choice of ϕ.

8.5 Pick $a \in \,]0, 1]$ such that $L(\alpha)$ is a-hyperbolic and $a > |\lambda|_\mathbb{K}$ for each eigenvalue λ of $L(\alpha)$ in $\overline{\mathbb{K}}$ such that $|\lambda|_\mathbb{K} < 1$. Pick $b \in [1, \infty[$ such that $L(\alpha)$ is b-hyperbolic and $b < |\lambda|_\mathbb{K}$ for each eigenvalue λ of $L(\alpha)$ in $\overline{\mathbb{K}}$ such that $|\lambda|_\mathbb{K} > 1$. With respect to the endomorphism $L(\alpha)$, we then have

$$\mathfrak{g}_{<1} = \mathfrak{g}_{<a} \quad \text{and} \quad \mathfrak{g}_{>1} = \mathfrak{g}_{>b},$$

entailing that

$$\mathfrak{g} = \mathfrak{g}_{<a} \oplus \mathfrak{g}_1 \oplus \mathfrak{g}_{>b}. \tag{59}$$

We find it useful to identify $\mathfrak{g}$ with the direct product $\mathfrak{g}_{<a} \times \mathfrak{g}_1 \times \mathfrak{g}_{>b}$; an element (x, y, z) of the latter is identified with $x + y + z \in \mathfrak{g}$.

Let W_a^s, W^c, and W_b^u be a local a-stable manifold, centre-manifold, and local b-unstable manifold for α around $p := e$ in $M := G$, respectively, $R > 0$ and

$$\phi_s: W_a^s \to B_R^{\mathfrak{g}_{<a}}(0), \quad \phi_c: W^c \to B_R^{\mathfrak{g}_1}(0),$$

as well as $\phi_u: W_b^u \to B_R^{\mathfrak{g}_{>b}}(0)$ be analytic diffeomorphisms as described in Proposition 7.3. We abbreviate $B_t^F := B_t^F(0)$ whenever F is a vector subspace of $\mathfrak{g}$. Using the inverse maps

$$\psi_s := \phi_s^{-1}, \quad \psi_c := \phi_c^{-1}, \quad \text{and} \quad \psi_u := \phi_u^{-1},$$

we define the analytic map

$$\psi: B_R^\mathfrak{g} = B_R^{\mathfrak{g}_{<a}} \times B_R^{\mathfrak{g}_1} \times B_R^{\mathfrak{g}_{>b}} \to G, \quad (x, y, z) \mapsto \psi_s(x)\psi_c(y)\psi_u(z).$$

Then $T_0\psi = \mathrm{id}_\mathfrak{g}$ by (27) and the properties of $d\phi_s$, $d\phi_c$, and $d\phi_u$ described in Proposition 7.3 (a), (b), and (c), respectively, if we identify $T_0(\mathfrak{g}) = \{0\} \times \mathfrak{g}$ with $\mathfrak{g}$ as usual, forgetting the first component. By the Inverse Function Theorem, after shrinking R if necessary, we may assume that the image $W_a^s W^c W_b^u$ of ψ is an open identity neighbourhood in G, and that

$$\psi : B_R^{\mathfrak{g}} \to W_a^s W^c W_b^u \tag{60}$$

is an analytic diffeomorphism. We define

$$\phi := \psi^{-1}, \tag{61}$$

with domain $U := W_a^s W^c W_b^u$ and image $V := B_R^{\mathfrak{g}}$. After shrinking R further if necessary, we may assume that ϕ and R have all the properties described in 5.1 and Lemma 5.2.

8.6 In the following result and its proof, ϕ and R are as in 8.5. We let $B_t := B_t^{\mathfrak{g}} \subseteq \mathfrak{g}$ and the compact open subgroups

$$B_t^{\phi} := \phi^{-1}(B_t^{\mathfrak{g}}) = \psi(B_t^{\mathfrak{g}}) = \psi_s(B_r^{\mathfrak{g}<a})\psi_c(B_t^{\mathfrak{g}1})\psi_u(B_t^{\mathfrak{g}>b}) = W_a^s(t)W^c(t)W_b^u(t)$$

of G for $t \in {]}0, R]$ be as in Lemma 5.2 (using notation as in Proposition 7.3). The multiplication $* : B_R \times B_R \to B_R$ is as in 5.1.
We shrink R further (if necessary) to achieve the following:

Lemma 8.7 *After shrinking R, we can achieve that $W_b^u(t) = \phi_u^{-1}(B_t^{\mathfrak{g}>b})$ is a subgroup of G for all $t \in {]}0, R]$ and $W^c(t) = \phi_c^{-1}(B_t^{\mathfrak{g}1})$ normalizes $W_b^u(t)$.*

Proof Let ϕ, R and further notation be as in 8.5 and 8.6; notably, $V = B_R^{\mathfrak{g}}$. Using Lemma 7.9 with $M := G, f := \alpha$ and $p := e$, we see that, after shrinking R if necessary, we may assume the following condition $(*)$ for all $t \in {]}0, R]$:
$W_b^u(t)$ *is the set of all $x \in B_t^{\phi} = W_a^s(t)W^c(t)W_b^u(t)$ for which there exists an* α-*regressive trajectory $(x_{-n})_{n\in\mathbb{N}_0}$ in B_t^{ϕ} with $x_0 = x$ such that*

$$\lim_{n\to\infty} b^n \|\phi(x_{-n})\| = 0$$

(and then $x_{-n} \in W_b^u(t)$ for all $n \in \mathbb{N}_0$). As the analytic map

$$g : V \times V \to V \subseteq \mathfrak{g}, \quad (x, y) \mapsto x * y^{-1}$$

is totally differentiable at $(0,0)$ with $g(0,0) = 0$ and $g'(0,0)(x,y) = x - y$, after shrinking R if necessary we may assume that

$$\|x * y^{-1} - x + y\| \le \max\{\|x\|, \|y\|\}$$

for all $x, y \in B_R = V$ and thus

$$\|x * y^{-1}\| = \|x - y + (x * y^{-1} - x + y)\|$$
$$\le \max\{\|x - y\|, \|x * y^{-1} - x + y\|\} \le \max\{\|x\|, \|y\|\}, \tag{62}$$

using the ultrametric inequality. By definition, $e \in W_b^u(t)$. Hence $W_b^u(t)$ will be a subgroup of G for all $t \in]0, R]$ if we can show that $xy^{-1} \in W_b^u(t)$ for all $x, y \in W_b^u(t)$. Let $(x_{-n})_{n \in \mathbb{N}_0}$ and $(y_{-n})_{n \in \mathbb{N}_0}$ be α-regressive trajectories in B_t^ϕ such that $x_0 = x$, $y_0 = y$ and

$$b^n \|\phi(x_{-n})\|, b^n \|\phi(y_{-n})\| \to 0 \quad \text{as } n \to \infty.$$

Then $(x_{-n} y_{-n}^{-1})_{n \in \mathbb{N}_0}$ is an α-regressive trajectory in the group B_t^ϕ with $x_0 y_0^{-1} = xy^{-1}$ and

$$b^n \|\phi(x_n y_n^{-1})\| = b^n \|\phi(x_n) * \phi(y_n)^{-1}\| \le \max\{b^n \|\phi(x_n)\|, b^n \|\phi(y_n)\|\} \to 0,$$

using that $\phi \colon U \to (B_R, *)$ is a homomorphism of groups, and using the estimate (62). Thus $xy^{-1} \in W_b^u(t)$, by (∗).

We now show that, after shrinking R if necessary, $W_b^u(t)$ is normalized by $W^c(R)$ for all $t \in]0, R]$. To this end, consider the analytic map

$$h \colon V \times V \to V, \quad (x, y) \mapsto x * y * x^{-1}.$$

For $x \in V$, abbreviate $h_x := h(x, .)$. Since $h_0 = \mathrm{id}_V$, we see that $h_0'(0) = \mathrm{id}_{\mathfrak{g}}$ which is an isometry. By the Ultrametric Inverse Function Theorem with Parameters, after shrinking R we can achieve that $h_x \colon B_R^{\mathfrak{g}} \to B_R^{\mathfrak{g}}$ is an isometry for all $x \in B_R^{\mathfrak{g}}$. Hence, using that $h_x(0) = 0$,

$$\|x * y * x^{-1}\| = \|h_x(y)\| = \|y\| \quad \text{for all } x, y \in B_R^{\mathfrak{g}}. \tag{63}$$

If $x \in W^c(R)$, $t \in]0, R]$ and $y \in W_b^u(t)$, let $(y_{-n})_{n \in \mathbb{N}_0}$ be an α-regressive trajectory in B_t^ϕ such that $y_0 = y$ and $b^n \|\phi(y_{-n})\| \to 0$ as $n \to \infty$. Since $\alpha(W^c(R)) = W^c(R)$, we can find an α-regressive trajectory $(x_{-n})_{n \in \mathbb{N}_0}$ in $W^c(R)$ such that $x_0 = x$. Recall from Lemma 5.2 that B_t^ϕ is a normal subgroup of B_R^ϕ. Hence $(x_{-n} y_{-n} x_{-n}^{-1})_{n \in \mathbb{N}_0}$ is an α-regressive trajectory in B_t^ϕ such that $x_0 y_0 x_0^{-1} = xyx^{-1}$ and

$$b^n \|\phi(x_{-n} y_{-n} x_{-n}^{-1})\| = b^n \|\phi(x_{-n}) * \phi(y_{-n}) * \phi(x_{-n})^{-1}\| = b^n \|\phi(y_{-n})\| \to 0$$

as $n \to \infty$, using that ϕ is a homomorphism of groups and (63). Thus $xyx^{-1} \in W_b^u(t)$, by (∗). □

8.8 By Lemma 7.6, after shrinking R if necessary, we may assume that

$$\bigcap_{n \in \mathbb{N}_0} \alpha^n(W_a^s) = \{e\} \quad \text{and} \quad \lim_{n \to \infty} \alpha^n(x) = e \quad \text{for all } x \in W_a^s. \tag{64}$$

8.9 By Lemma 7.7, after shrinking R if necessary, we may assume that the map $\alpha|_{W^c(t)} \colon W^c(t) \to W^c(t)$ is an analytic diffeomorphism for each $t \in]0, R]$.

8.10 By Lemma 7.9, after shrinking R if necessary, we may assume that, for each $t \in]0, R]$, for each $x \in W_b^u(t)$ there exists an α-regressive trajectory $(x_{-n})_{n \in \mathbb{N}_0}$ in $W_b^u(t)$ such that $x_0 = x$ and

$$\lim_{n \to \infty} x_{-n} = e.$$

In particular, $W_b^u(t) \subseteq \alpha(W_b^u(t))$ for all $t \in]0, R]$.

8.11 By Lemma 7.8, there exists an open neighbourhood P of e in W_b^u with $\alpha(P) \subseteq W_b^u$ such that, for each $P \setminus \{e\}$, there exists $n \in \mathbb{N}_0$ such that $\alpha^n(x) \notin P$. After shrinking P, we may assume that $P = W_b^u(r)$ for some $r \in]0, R]$.
The next lemma will be applied later to $A := W^c(t)$, $B := W_b^u(t)$ and $C := \alpha(B)$.

Lemma 8.12 *Let G be a group, $B \subseteq C \subseteq G$ be subgroups and $A \subseteq G$ be a subset such that AB and AC are subgroups of G and $C \cap AB = B$. Then*

$$[AC : AB] = [C : B].$$

Proof The group C acts on $X := AC/AB$ on the left via $c'.acAB := c'acAB$ for $c, c' \in C$, $a \in A$. To see that the action is transitive, let $a \in A$ and $c \in C$. Since AC is a group, we have $(ac)^{-1} = a'c'$ for certain $a' \in A$ and $c' \in C$, entailing that $ac = (c')^{-1}(a')^{-1}$ and thus $acAB = (c')^{-1}AB = (c')^{-1}.AB$. The stabilizer of the point $AB \in X$ is $C \cap AB = B$. Now the Orbit Formula shows that the map

$$C/B \to X = AC/AB, \quad cB \mapsto cAB$$

is a well-defined bijection. The assertion follows. $\square$
Using notation as before (notably ϕ as in (61) and as in 8.11), we have:

Theorem 8.13 *Let α be an analytic endomorphism of a Lie group G over a totally disconnected local field $\mathbb{K}$. Then the scale $s(\alpha)$ divides the scale $s(L(\alpha))$ of the associated Lie algebra endomorphism $L(\alpha)$. The following conditions are equivalent:*

(a) $s_G(\alpha) = s_{L(G)}(L(\alpha))$;
(b) *There is $t_0 \in]0, r]$ such that the compact open subgroups $B_t^\phi \cong (B_t, *)$ of G are tidy for α, for all $t \in]0, t_0]$;*
(c) *G has small tidy subgroups for α;*
(d) *The contraction group $\overrightarrow{\mathrm{con}}(\alpha)$ is closed.*

Proof The implication (b)$\Rightarrow$(c) holds as the compact open subgroups B_t^ϕ for $t \in]0, t_0]$ form a basis of identity neighbourhoods in G. The implication (c)$\Rightarrow$(d) is a general fact, see Lemma 8.2(a).
   (a)$\Leftrightarrow$(b): We claim that there exists $t_0 \in]0, r]$ such that the compact open subgroups B_t^ϕ have displacement index

$$[\alpha(B_t^\phi) : \alpha(B_t^\phi) \cap B_t^\phi] = s(L(\alpha)) \quad \text{for all } t \in]0, t_0]. \tag{65}$$

If this is true, then the equivalence of (a) and (b) is clear. If α is an automorphism, then the claim holds by Lemma 5.5. For α an endomorphism, the argument is more involved. We first note that the product map

$$\pi \colon W_a^s \times W^c \times W_b^u \to W_a^s W^c W_b^u = B_R^\phi$$

is an analytic diffeomorphism as so is ψ (from (60)). Let $t \in {]}0, r]$. Since

$$\alpha(W_a^s(t)W^c(t)) \subseteq W_a^s(t)W^c(t),$$

we have $\alpha^n(W_a^s(t)W^c(t)) \subseteq W_a^s(t)W^c(t) \subseteq B_t^\phi$ for all $n \in \mathbb{N}_0$ and thus

$$W_a^s(t)W^c(t) \subseteq (B_t^\phi)_-. \tag{66}$$

Since $\alpha(W^c(t)) = W^c(t)$ and each $x \in W_b^u(t)$ has an α-regressive trajectory within $W_b^u(t)$ (see 8.10), we have

$$W^c(t)W_b^u(t) \subseteq (B_t^\phi)_+. \tag{67}$$

Thus $B_t^\phi = W_a^s(t)W^c(t)W_b^u(t) \subseteq (B_t^\phi)_-(B_t^\phi)_+ \subseteq B_t^\phi$, whence $B_t^\phi = (B_t^\phi)_-(B_t^\phi)_+$ and so $B_t^\phi = (B_t^\phi)^{-1} = (B_t^\phi)_+(B_t^\phi)_-$ is tidy above for α.

Since $B_t^\phi = W_a^s(t)W^c(t)W_b^u(t)$ is a group and $(B_t^\phi)_+$ a subgroup, (67) implies that

$$(B_t^\phi)_+ = J_t W^c(t)W_b^u(t)$$

with $J_t := (B_t^\phi)_+ \cap W_a^s(t)$. Since π is a bijection, $\alpha(J_t) \subseteq \alpha(W_a^s(t)) \subseteq W_a^s(t)$ and $\alpha(W_b^u(t)) \subseteq W_b^u$ (see 8.11)), the inclusion

$$J_t W^c(t)W_b^u(t) = (B_t^\phi)_+ \subseteq \alpha((B_t^\phi)_+) = \alpha(J_t)W^c(t)\alpha(W_b^u(t))$$

entails that $J_t \subseteq \alpha(J_t)$, whence

$$J_t \subseteq \bigcap_{n \in \mathbb{N}_0} \alpha^n(J_t) \subseteq \bigcap_{n \in \mathbb{N}_0} \alpha^n(W_a^s) = \{e\},$$

using (64). Thus $J_t = \{e\}$ and hence

$$W^c(t)W_b^u(t) = (B_t^\phi)_+, \tag{68}$$

which is a subgroup. Also

$$\alpha((B_t^\phi)_+) = W^c(t)\alpha(W_b^u(t))$$

is a subgroup, and

$$\alpha(W_b^u(t)) \cap W^c(t) W_b^u(t) = W_b^u(t)$$

since π is a bijection and $\alpha(W_b^u(t)) \subseteq W_b^u$ by 8.11. Hence, by [60, Lemma 5] and Lemma 8.12,

$$
\begin{aligned}
[\alpha(B_t^\phi) : \alpha(B_t^\phi) \cap B_t^\phi] &= [\alpha((B_t^\phi)_+) : (B_t^\phi)_+] \\
&= [W^c(t)\alpha(W_b^u(t)) : W^c(t) W_b^u(t)] \\
&= [\alpha(W_b^u(t)) : W_b^u(t)]. \quad\quad (69)
\end{aligned}
$$

Applying now Lemma 5.5 to $\alpha|_{W_b^u(r)}: W_b^u(r) \to W_b^u$ instead of $\alpha: G_0 \to G$ and $\phi_u|_{W_b^u(r)}$ instead of ϕ, we see that there is $t_0 \in \,]0, r]$ such that

$$[\alpha(W_b^u(t)) : W_b^u(t)] = s(L(\alpha)|_{\mathfrak{g}>b}) = s(L(\alpha)|_{\mathfrak{g}>1}) = s(L(\alpha)) \quad\quad (70)$$

for all $t \in \,]0, t_0]$, using Theorem 3.6 (d) for the penultimate equality. Combining (69) and (70), we get (65).

(d) $\Rightarrow$ (b): Recall that B_t^ϕ is tidy above for all $t \in \,]0, r]$; from (66) and (67), we deduce that

$$(B_t^\phi)_+ \cap (B_t^\phi)_- \supseteq W^c(t). \quad\quad (71)$$

By (64) and (66), we have

$$W_a^s(t) \subseteq \overrightarrow{\mathrm{con}}(\alpha) \cap (B_t^\phi)_-. \quad\quad (72)$$

Since $B_t^\phi = W_a^s(t) W^c(t) W_b^u(t)$ and $(B_t^\phi)_-$ is a subgroup of B_t^ϕ which contains $W_a^s(t) W^c(t)$, we have

$$(B_t^\phi)_- = W_a^s(t) W^c(t) I_t$$

with $I_t := (B_t^\phi)_- \cap W_b^u(t)$. Then $I_t = \{e\}$ as the existence of an element $x \in I_t \setminus \{e\}$ gives rise to a contradiction as follows: Since $I_t \subseteq (B_t^\phi)_-$, we must have $\alpha^n(x) \in B_t^\phi$ for all $n \in \mathbb{N}_0$. However, by 8.11, there exists $n \in \mathbb{N}$ such that $\alpha^n(x) \in W_b^u \setminus W_b^u(r)$ and thus $\alpha^n(x) \notin B_t^\phi$, which is absurd. Hence

$$(B_t^\phi)_- = W_a^s(t) W^c(t), \quad\quad (73)$$

and thus $(B_t^\phi)_- = (\overrightarrow{\mathrm{con}}(\alpha) \cap (B_t^\phi)_-)((B_t^\phi)_+ \cap (B_t^\phi)_-)$. If $\overrightarrow{\mathrm{con}}(\alpha)$ is closed, we can now use Lemma 8.2, to see that B_t^ϕ is tidy for α, for all $t \in \,]0, r]$.

Finally, as B_r^ϕ is tidy above for α, we deduce from Lemma 8.4 and (65) that $s(\alpha)$ divides $[\alpha(B_r^\phi) : \alpha(B_r^\phi) \cap B_r^\phi] = s(L(\alpha))$. $\qquad\square$

Remark 8.14 Let us collect further information on the subsets $W_a^s(t)$ and $W^c(t)$, for $t \in]0, r]$. By (68) and (73),

$$W^c(t) = W_a^s(t) W^c(t) \cap W^c(t) W_b^u(t) = (B_t^\phi)_- \cap (B_t^\phi)_+$$

is a compact *subgroup* of B_t^ϕ, for each $t \in]0, r]$. Now assume that B_t^ϕ is tidy for α. If we can show that

$$\overrightarrow{\mathrm{con}}(\alpha) \cap B_t^\phi = \overrightarrow{\mathrm{con}}(\alpha) \cap (B_t^\phi)_- = W_a^s(t), \tag{74}$$

then $W_a^s(t)$ is a compact *subgroup* of G, for all $t \in]0, r]$. Now, the first equality in (74) holds by [60, Proposition 11 (b)]. As for the second equality, the inclusion "$\supseteq$" holds by (72). Since $\overrightarrow{\mathrm{con}}(\alpha) \cap (B_t^\phi)_-$ is a subgroup of $(B_t^\phi)_- = W_a^s(t) W^c(t)$ which contains $W_a^s(t)$, it is of the form

$$\overrightarrow{\mathrm{con}}(\alpha) \cap (B_t^\phi)_- = W_a^s(t) K_t$$

with $K_t := \overrightarrow{\mathrm{con}}(\alpha) \cap (B_t^\phi)_- \cap W^c(t)$. But $K_t = \{e\}$ since $\overrightarrow{\mathrm{con}}(\alpha) \cap W^c(t) = \{e\}$; to see the latter, let $e \neq x \in W^c(t)$. There is $\tau \in]0, t]$ such that $x \notin W^c(\tau)$. Since $\alpha|_{W^c(t)}: W^c(t) \to W^c(t)$ is a bijection and $\alpha(W^c(\tau)) = W^c(\tau)$ (see 8.9), we have $\alpha^n(x) \in W^c(t) \setminus W^c(\tau)$ for all $n \in \mathbb{N}_0$ and thus $\alpha^n(x) \notin W^c(\tau)$, showing that $\alpha^n(x)$ does not converge to e in $W^c(t)$ as $n \to \infty$ (and hence neither in G).

Foliations of the 'Big Cell'

Theorem 8.15 *Let α be an analytic endomorphism of a Lie group G over a totally disconnected local field $\mathbb{K}$. If $\overleftarrow{\mathrm{con}}(\alpha)$ is closed in G, then also $\overleftarrow{\mathrm{con}}(\alpha)$ is closed and the following holds:*

(a) $\overrightarrow{\mathrm{con}}(\alpha)$, $\mathrm{lev}(\alpha)$, *and* $\overleftarrow{\mathrm{con}}(\alpha)$ *are Lie subgroups of G whose Lie algebras are* $\overrightarrow{\mathrm{con}}(L(\alpha))$, $\mathrm{lev}(L(\alpha))$, *and* $\overleftarrow{\mathrm{con}}(L(\alpha))$, *respectively;*

(b) $\Omega := \overrightarrow{\mathrm{con}}(\alpha) \mathrm{lev}(\alpha) \overleftarrow{\mathrm{con}}(\alpha)$ *is an α-invariant open identity neighbourhood in G. The product map*

$$\pi: \overrightarrow{\mathrm{con}}(\alpha) \times \mathrm{lev}(\alpha) \times \overleftarrow{\mathrm{con}}(\alpha) \to \overrightarrow{\mathrm{con}}(\alpha) \mathrm{lev}(\alpha) \overleftarrow{\mathrm{con}}(\alpha), \quad (x, y, z) \mapsto xyz$$

is an analytic diffeomorphism.

(c) $\alpha|_{\mathrm{lev}(\alpha)}$ *and* $\alpha|_{\overleftarrow{\mathrm{con}}(\alpha)}$ *are analytic automorphisms.*

Proof Openness of Ω: Note first that $\Omega := \overrightarrow{\mathrm{con}}(\alpha) \mathrm{lev}(\alpha) \overleftarrow{\mathrm{con}}(\alpha)$ contains the open identity neighbourhood $B_r^\phi = W_a^s(r) W^c(r) W_b^u(r)$ encountered in the proof of Theorem 8.13, since $W_a^s(r) \subseteq \overrightarrow{\mathrm{con}}(\alpha)$ by (64), $W_b^u(r) \subseteq \overleftarrow{\mathrm{con}}(\alpha)$ by 8.10 and $W^c(r) \subseteq \mathrm{lev}(\alpha)$ since $W^c(r)$ is compact and α-stable. As $\mathrm{lev}(\alpha)$ normalizes $\overrightarrow{\mathrm{con}}(\alpha)$

and $\overleftarrow{\mathrm{con}}\,(\alpha)$, both $P_\alpha :=\overrightarrow{\mathrm{con}}\,(\alpha)\,\mathrm{lev}(\alpha)$ and $P_\alpha^- := \mathrm{lev}(\alpha)\,\overleftarrow{\mathrm{con}}\,(\alpha)$ are subgroups of G. For each $g \in P_\alpha$, the left translation $\lambda_g\colon G \to G, x \mapsto gx$ is a homeomorphism which takes the identity neighbourhood Ω onto the g-neighbourhood

$$g\Omega = gP_\alpha \overleftarrow{\mathrm{con}}\,(\alpha) = P_\alpha \overleftarrow{\mathrm{con}}\,(\alpha) = \Omega.$$

If $h \in \Omega$, then $h = gk$ with $g \in P_\alpha$ and $k \in \overleftarrow{\mathrm{con}}\,(\alpha)$. Now the right translation $\rho_k\colon G \to G, x \mapsto xk$ is a homeomorphism which takes the g-neighbourhood Ω onto the neighbourhood

$$\Omega k = P_\alpha \overleftarrow{\mathrm{con}}\,(\alpha)k = P_\alpha \overleftarrow{\mathrm{con}}\,(\alpha) = \Omega$$

of $h = gk$. Hence Ω is a neighbourhood of each $h \in \Omega$ and thus Ω is open.

Lie Subgroups The open subset $B_r^\phi = W_a^s(r) \times W^c(r) \times W_b^u(r)$ of G has $W_a^s(r)$ as a submanifold. Since $\overrightarrow{\mathrm{con}}\,(\alpha) \cap B_r^\phi = W_a^s(r)$ (see (74)) is a submanifold, we deduce that $\overrightarrow{\mathrm{con}}\,(\alpha)$ is a Lie subgroup of G (cf. Lemma 4.6) which has $W_a^s(r)$ as an open submanifold; thus

$$L(\overrightarrow{\mathrm{con}}\,(\alpha)) = T_e(W_a^s(r)) = \mathfrak{g}_{<a} = \overrightarrow{\mathrm{con}}\,(L(\alpha)).$$

Recall from the proof of Theorem 8.13 that B_r^ϕ is tidy for α. Using the proof of [60, Proposition 19] for the first equality, we have

$$\mathrm{lev}(\alpha) \cap B_r^\phi = (B_r^\phi)_+ \cap (B_r^\phi)_- = W^c(r),$$

which is a submanifold of B_r^ϕ. Hence $\mathrm{lev}(\alpha)$ is a Lie subgroup of G which has $W^c(r)$ as an open submanifold, and thus

$$L(\mathrm{lev}(\alpha)) = \mathfrak{g}_1 = \mathrm{lev}(L(\alpha)). \tag{75}$$

Next, recall from [60, Proposition 11 (a)] that

$$\overleftarrow{\mathrm{con}}\,(\alpha) \cap B_r^\phi \subseteq (B_r^\phi)_+ = W^c(r)W_b^u(r).$$

Since $W_b^u(r) \subseteq \overleftarrow{\mathrm{con}}\,(\alpha)$, this entails that

$$\overleftarrow{\mathrm{con}}\,(\alpha) \cap B_r^\phi = N_r W_b^u(r)$$

with $N_r := \overleftarrow{\mathrm{con}}\,(\alpha) \cap W^c(r)$. Let $\mathrm{bik}(\alpha)$ be the bounded iterated kernel of α and $\mathrm{nub}(\alpha)$ be the nub subgroup (see [60] and [9]). Then $\mathrm{bik}(\alpha) \subseteq \mathrm{nub}(\alpha)$ and since α has small tidy subgroups by Theorem 8.13, we have $\mathrm{nub}(\alpha) = \{e\}$ (see [60]). If $x \in N_r$, then there exists an α-regressive trajectory $(x_{-n})_{n\in\mathbb{N}_0}$ such that $x_0 = x$ and

$$\lim_{n\to\infty} x_{-n} = e.$$

On the other hand, since $x \in W^c(r)$ which is α-stable, there exists an α-regressive trajectory $(y_{-n})_{n\in\mathbb{N}_0}$ in $W^c(r)$ with $y_0 = x$. Then $(x_{-n}y_{-n}^{-1})_{n\in\mathbb{N}_0}$ is an α-regressive trajectory such that $\{x_{-n}y_{-n}^{-1} : n \in \mathbb{N}_0\}$ is relatively compact. Thus $x_{-n}y_{-n}^{-1} \in \overleftarrow{\mathrm{par}}(\alpha)$ for each $n \in \mathbb{N}_0$ and $\alpha^n(x_{-n}y_{-n}^{-1}) = e$, whence $x_{-n}y_{-n}^{-1} \in \mathrm{bik}(\alpha) \subseteq \mathrm{nub}(\alpha) = \{e\}$. Hence $(x_{-n})_{n\in\mathbb{N}_0} = (y_{-n})_{n\in\mathbb{N}_0}$ is an $\alpha|_{W^c(r)}$-regressive trajectory which tends to e as $n \to \infty$. As $\alpha|_{W^c(r)}$ is a distal automorphism of $W^c(r)$ (cf. Proposition 7.10), the latter is only possible if $x = e$. Thus $N_r = \{e\}$ and hence

$$\overleftarrow{\mathrm{con}}(\alpha) \cap B_r^\phi = W_b^u(r),$$

which is a submanifold of Ω. Hence $\overleftarrow{\mathrm{con}}(\alpha)$ is a Lie subgroup of G with Lie algebra $\mathfrak{g}_{>b} = \overleftarrow{\mathrm{con}}(L(\alpha))$.

π *is injective.* Let $a, a' \in \overrightarrow{\mathrm{con}}(\alpha)$, $b, b' \in \mathrm{lev}(\alpha)$, and $c, c' \in \overleftarrow{\mathrm{con}}(\alpha)$ such that $abc = a'b'c'$. Then

$$x := (a')^{-1}a = b'c'c^{-1}b^{-1} \in \overrightarrow{\mathrm{con}}(\alpha) \cap \overleftarrow{\mathrm{par}}(\alpha) \subseteq \mathrm{lev}(\alpha).$$

There exists $n \in \mathbb{N}_0$ such that $\alpha^n(x) \in B_r^\phi$. Since $x \in \mathrm{lev}(\alpha)$, also $\alpha^n(x) \in \mathrm{lev}(\alpha)$ and thus

$$\alpha^n(x) \in (B_r^\phi)_+ \cap (B_r^\phi)_- = W^c(r).$$

Let $y \in W^c(r)$ such that $\alpha^n(y) = \alpha^n(x)$. Since $W^c(r) \subseteq \mathrm{lev}(\alpha)$, we then have $(a')^{-1}ay^{-1} \in \mathrm{bik}(\alpha) \subseteq \mathrm{nub}(\alpha) = \{e\}$, whence $(a')^{-1}a = y \in W^c(r) \cap \overrightarrow{\mathrm{con}}(\alpha) = \{e\}$, using that $\alpha|_{W^c(r)}$ is distal.
Thus $x = y = e$ and thus $bc = b'c'$, whence

$$(b')^{-1}b = c'c^{-1} \in \mathrm{lev}(\alpha) \cap \overrightarrow{\mathrm{con}}(\alpha).$$

Let $(z_{-n})_{n\in\mathbb{N}_0}$ be an α-regressive trajectory with $z_0 = c'c^{-1}$, such that $z_{-n} \to e$ as $n \to \infty$. For each $t \in \,]0, r]$, we have $z_{-n} \in B_t^\phi$ for some $n \in \mathbb{N}_0$, entailing that

$$z_{-n} \in B_t^\phi \cap \mathrm{lev}(\alpha) = (B_t^\phi)_+ \cap (B_t^\phi)_- = W^c(t)$$

and thus $c'c^{-1} = \alpha^n(z_{-n}) \in W^c(t)$. Therefore

$$c'c^{-1} \in \bigcap_{t\in]0,r]} B_t^\phi = \{e\},$$

whence $c' = c$ and hence also $b' = b$.

π is a diffeomorphism. Since ψ is an analytic diffeomorphism, also

$$\mu := \pi|_{W_a^s \times W^c \times W_b^u} : W_a^s \times W^c \times W_b^u \to B_R^\phi$$

is an analytic diffeomorphism. For $(a, b, c) \in \overrightarrow{\mathrm{con}}(\alpha) \times \mathrm{lev}(\alpha) \times \overleftarrow{\mathrm{con}}(\alpha) =: Y$, let us show that π is a local diffeomorphism at (a, b, c). It suffices to prove that the map

$$h: Y \to \Omega, \quad (a', b', c') \mapsto \pi(aa', bb', c'c)$$

is a local diffeomorphism at (e, e, e). Since $\mathrm{lev}(\alpha)$ normalizes $\overrightarrow{\mathrm{con}}(\alpha)$ and $\overrightarrow{\mathrm{con}}(\alpha)$ is a submanifold of G, the map

$$\beta: \overrightarrow{\mathrm{con}}(\alpha) \to \overrightarrow{\mathrm{con}}(\alpha), \quad a' \mapsto b^{-1}a'b$$

is an analytic diffeomorphism. Let $Q \subseteq \overrightarrow{\mathrm{con}}(\alpha)$ be an open identity neighbourhood such that $\beta(Q) \subseteq W_a^s$. Then the formula

$$h(a', b', c') = ab\,\mu(\beta(a'), b', c')c \quad \text{for } (a', b', c') \in Q \times W^c \times W_b^u$$

shows that h is a local diffeomorphism at (e, e, e).

To prove (c), note that $\gamma := \alpha|_{\mathrm{lev}(\alpha)}$ and $\delta := \alpha|_{\overleftarrow{\mathrm{con}}(\alpha)}$ are local diffeomorphisms at e (by the Inverse Function Theorem), since $L(\gamma) = L(\alpha)|_{\mathfrak{g}_1}$ and $L(\delta) = L(\alpha)|_{\mathfrak{g}_{>1}}$ are automorphisms of the tangent spaces $L(\mathrm{lev}(\alpha)) = \mathfrak{g}_1$ and $L(\overleftarrow{\mathrm{con}}(\alpha)) = \mathfrak{g}_{>1}$, respectively, at e. Since γ and δ are, moreover, bijective analytic endomorphisms, they are analytic automorphisms. $\square$

Remark 8.16

(a) Note that also the groups P_α and P_α^- encountered in the preceding proof are Lie subgroups since π is an analytic diffeomorphism.

(b) We mention that $P_\alpha = \overrightarrow{\mathrm{par}}(\alpha)$ and $P_\alpha^- = \overleftarrow{\mathrm{par}}(\alpha)$ (see [9, Lemma 13.1 (d) and (e)]).

(c) Since π is an analytic diffeomorphism, we see that the "big cell" Ω can be foliated into right translates of $\overrightarrow{\mathrm{con}}(\alpha)$ parametrized by $\overleftarrow{\mathrm{par}}(\alpha)$, or alternatively into right translates of $\overrightarrow{\mathrm{par}}(\alpha)$, parametrized by $\overleftarrow{\mathrm{con}}(\alpha)$. Likewise, we can foliate Ω into left translates of $\overleftarrow{\mathrm{con}}(\alpha)$ parametrized by $\overrightarrow{\mathrm{par}}(\alpha)$, or into left translates of $\overleftarrow{\mathrm{par}}(\alpha)$ parametrized by $\overrightarrow{\mathrm{con}}(\alpha)$.

8.17 Consider an analytic map $f: M \to N$ between analytic manifolds over a totally disconnected local field $\mathbb{K}$. Recall from [49, Part I, Chapter III] that f is called an *immersion* if f locally looks like a linear injection around each point, in suitable charts (or equivalently, if $T_p(f)$ is injective for all $p \in M$). If G is a Lie group over a totally disconnected local field and H a subgroup of G, endowed with an analytic manifold structure turning it into a Lie group and making the inclusion map $H \to G$ an immersion, then H is called an *immersed Lie subgroup* of G.

Remark 8.18

(a) If α is an analytic *automorphism* of a Lie group G over a totally disconnected local field $\mathbb{K}$ and $\overrightarrow{\mathrm{con}}\,(\alpha)$ is not closed, then it is still possible to turn $\overrightarrow{\mathrm{con}}\,(\alpha)$ and $\overleftarrow{\mathrm{con}}\,(\alpha)$ into *immersed* Lie subgroups of G modelled on $\overrightarrow{\mathrm{con}}\,(L(\alpha))$ and $\overleftarrow{\mathrm{con}}\,(L(\alpha))$, respectively, such that $\alpha|_{\overrightarrow{\mathrm{con}(\alpha)}}$ and $\alpha^{-1}|_{\overleftarrow{\mathrm{con}(\alpha)}}$ are contractive analytic automorphisms of these Lie groups (see [19, Proposition 6.3 (b)]).

(b) After this research was completed, it was shown in [22] that the 'big cell'

$$\Omega := \overrightarrow{\mathrm{con}}\,(\alpha)\,\mathrm{lev}(\alpha)\,\overleftarrow{\mathrm{con}}\,(\alpha)$$

is open in G for each endomorphism α of a totally disconnected locally compact group G. If G is a Lie group over a totally disconnected local field and $\alpha \colon G \to G$ an analytic endomorphism, then $\overrightarrow{\mathrm{con}}\,(\alpha)$, $\mathrm{lev}(\alpha)$ and $\overleftarrow{\mathrm{con}}\,(\alpha)$ can be turned into immersed Lie subgroups $\overrightarrow{\mathrm{con}}^*(\alpha)$, $\mathrm{lev}^*(\alpha)$ and $\overleftarrow{\mathrm{con}}^*(\alpha)$ of G modelled on $\overrightarrow{\mathrm{con}}\,(L(\alpha))$, $\mathrm{lev}(L(\alpha))$ and $\overleftarrow{\mathrm{con}}\,(L(\alpha))$, respectively, such that α induces analytic endomorphisms of the immersed Lie subgroups and the product map

$$\overrightarrow{\mathrm{con}}^*(\alpha) \times \mathrm{lev}^*(\alpha) \times \overleftarrow{\mathrm{con}}^*(\alpha) \to \Omega, \quad (a, b, c) \mapsto abc$$

is surjective and étale (i.e., a local diffeomorphism at each point), see [22].

Closedness of Contraction Groups We now mention a characterization and describe a criterion for closedness of contraction groups of endomorphisms. The next lemma is covered by [9, Theorem D and F]; for the case of automorphisms, see already [1, Theorem 3.32] (if G is metrizable).

Lemma 8.19 *Let α be an endomorphism of a totally disconnected locally compact group G. Then the contraction group $\overrightarrow{\mathrm{con}}\,(\alpha)$ is closed in G if and only if $\overrightarrow{\mathrm{con}}\,(\alpha) \cap \mathrm{lev}(\alpha) = \{e\}$.* □

Lemma 8.20 *Let $\phi \colon G \to H$ be an injective, continuous homomorphism between totally disconnected, locally compact groups and $\alpha \colon G \to G$ as well as $\beta \colon H \to H$ be endomorphisms such that $\beta \circ \phi = \phi \circ \alpha$. If $\overrightarrow{\mathrm{con}}\,(\beta)$ is closed, then also $\overrightarrow{\mathrm{con}}\,(\alpha)$ is closed.*

Proof Using Lemma 8.19, we get

$$\phi(\overrightarrow{\mathrm{con}}\,(\alpha) \cap \mathrm{lev}(\alpha)) \subseteq \overrightarrow{\mathrm{con}}\,(\beta) \cap \mathrm{lev}(\beta) = \{e\}.$$

Thus $\overrightarrow{\mathrm{con}}\,(\alpha) \cap \mathrm{lev}(\alpha) = \{e\}$, whence $\overrightarrow{\mathrm{con}}\,(\alpha)$ is closed (by Lemma 8.19). □

Proposition 8.21 *For every totally disconnected local field* $\mathbb{K}$, *every inner automorphism of a closed subgroup* $G \subseteq GL_n(\mathbb{K})$ *has a closed contraction group.*

Proof It suffices to show that each inner automorphism of $GL_n(\mathbb{K})$ has a closed contraction group. Let $e \in GL_n(\mathbb{K})$ be the identity matrix. Given $g \in GL_n(\mathbb{K})$, consider the inner automorphism

$$I_g : GL_n(\mathbb{K}) \to GL_n(\mathbb{K}), \quad h \mapsto ghg^{-1}$$

and the vector space automorphism

$$\alpha : M_n(\mathbb{K}) \to M_n(\mathbb{K}), \quad A \mapsto gAg^{-1}.$$

We know from Sect. 3 that $\overrightarrow{\mathrm{con}}(\alpha)$ is closed. Then $V := GL_n(\mathbb{K}) - e$ is an α-stable 0-neighbourhood in $M_n(\mathbb{K})$ and

$$\phi : V \to GL_n(\mathbb{K}), \quad A \mapsto A + e$$

is a homeomorphism. Now $I_g \circ \phi = \phi \circ \alpha|_V$ as

$$I_g(\phi(A)) = g(A + e)g^{-1} = gAg^{-1} + e = \phi(\alpha(A)),$$

i.e., ϕ is a topological conjugacy between the dynamical systems $(V, \alpha|_V)$ and $(GL_n(\mathbb{K}), I_g)$. Hence

$$\overrightarrow{\mathrm{con}}(I_g) = \phi(\overrightarrow{\mathrm{con}}(\alpha) \cap V),$$

which is closed in $GL_n(\mathbb{K})$.   □

Combining Lemma 8.20 and Proposition 8.21, we get:

Corollary 8.22 *If a totally disconnected, locally compact group* G *admits a faithful continuous representation* $\pi : G \to GL_n(\mathbb{K})$ *over some totally disconnected local field* $\mathbb{K}$, *then every inner automorphism of* G *has a closed contraction group.*   □

Remark 8.23 In particular, every group G of $\mathbb{K}$-rational points of a linear algebraic group over a totally disconnected local field $\mathbb{K}$ is a closed subgroup of some $GL_n(\mathbb{K})$, whence $\overrightarrow{\mathrm{con}}(\alpha)$ is closed in G by Proposition 8.21 and so, for $g \in G$

$$s(g) = s_{L(G)}(\mathrm{Ad}_g) = \prod_{|\lambda_j|_{\mathbb{K}} \geq 1} |\lambda_j|_{\mathbb{K}} \tag{76}$$

in terms of the eigenvalues $\lambda_1, \ldots, \lambda_m$ of Ad_g in an algebraic closure $\overline{\mathbb{K}}$, repeated according to their algebraic multiplicities (by Theorems 8.13 and 3.6). For Zariski-connected reductive $\mathbb{K}$-groups, this was already shown in [1, Proposition 3.23]. See also [21, Proposition 4.48 and Remark 4.49].

The following result was announced in [25] (for automorphisms).

Proposition 8.24 *Let α be an analytic endomorphism of a 1-dimensional Lie group G over a local field $\mathbb{K}$, with Lie algebra $\mathfrak{g} := L(G)$. If*

$$\overleftarrow{\mathrm{con}}\,(L(\alpha)) = \mathfrak{g},$$

assume that $\bigcup_{n \in \mathbb{N}_0} \ker(\alpha^n)$ is discrete[13]; if $\overleftarrow{\mathrm{con}}\,(L(\alpha)) \neq \mathfrak{g}$, we do not impose further hypotheses. Then $\overrightarrow{\mathrm{con}}\,(\alpha)$ is closed in G and thus $s(\alpha) = s(L(\alpha))$.

Proof Since $\mathfrak{g}$ is a 1-dimensional $\mathbb{K}$-vector space and

$$\mathfrak{g} = \overrightarrow{\mathrm{con}}\,(L(\alpha)) \oplus \mathrm{lev}(L(\alpha)) \oplus \overleftarrow{\mathrm{con}}\,(L(\alpha)),$$

we see that $\mathfrak{g}$ coincides with one of the three summands. Let a, b, R, W_a^s, W^c, and W_b^u be as in 8.5.

If $\overrightarrow{\mathrm{con}}\,(L(\alpha)) = \mathfrak{g}$, then W_a^s is a submanifold of G of full dimension 1 and hence open in G. By 8.8, we may assume that $W_a^s \subseteq \overrightarrow{\mathrm{con}}\,(\alpha)$, after shrinking R if necessary, whence the subgroup $\overrightarrow{\mathrm{con}}\,(\alpha)$ is open and hence closed in G.

If $\mathrm{lev}(L(\alpha)) = \mathfrak{g}$, then W^c is open in G and we may assume that $W^c(t)$ is a compact open subgroup of G for all $t \in\,]0, R]$, after shrinking R if necessary. The bijective analytic endomorphism $\alpha|_{W^c}\colon W^c \to W^c$ is a local analytic diffeomorphism at e (by the Inverse Function Theorem) and hence an analytic automorphism of the Lie group W^c. By Proposition 7.10(c), the automorphism $\alpha|_{W^c}$ is distal and hence

$$\overrightarrow{\mathrm{con}}\,(\alpha) \cap W^c = \overrightarrow{\mathrm{con}}\,(\alpha|_{W^c}) = \{e\}.$$

Thus $\overrightarrow{\mathrm{con}}\,(\alpha)$ is discrete, and hence closed in G.

If $\overleftarrow{\mathrm{con}}\,(L(\alpha)) = \mathfrak{g}$, then W_b^u is open and we choose an open neighbourhood P of e in W_b^u as in Lemma 7.8 (with $M := G$, $f := \alpha$, and $p := e$). Then $\overrightarrow{\mathrm{con}}\,(\alpha) = \bigcup_{n \in \mathbb{N}_0} \ker(\alpha^n)$, which is discrete (and thus closed) by hypothesis. In fact, if $x \in \overrightarrow{\mathrm{con}}\,(\alpha)$, then there exists $n_0 \in \mathbb{N}_0$ such that $\alpha^n(x) \in P$ for all $n \geq n_0$. Then $\alpha^{n_0}(x) = e$, as we chose P in such a way that the α-orbit of each $y \in P \setminus \{e\}$ leaves P.

In each case, the final assertion follows from Theorem 8.13. $\square$

---

[13]Which is, of course, automatic if α is an automorphism.

Remark 8.25 If $\mathbb{F}$ is a field of prime order p, then $G := (\mathbb{F}[[X]], +)$ is a 1-dimensional Lie group over $\mathbb{K} = \mathbb{F}((X))$ and the left shift

$$\sum_{n=0}^{\infty} a_n X^n \mapsto \sum_{n=0}^{\infty} a_{n+1} X^n$$

is an analytic endomorphism α of G, as it coincides with the linear (and hence analytic) map

$$\beta \colon \mathbb{K} \to \mathbb{K}, \quad z \mapsto X^{-1} z$$

on the open subgroup $X\mathbb{F}[[X]]$ of G. It is easy to see that $\overleftarrow{\mathrm{con}}(\alpha) = G$ and

$$\overrightarrow{\mathrm{con}}(\alpha) = \bigcup_{n \in \mathbb{N}_0} \ker(\alpha^n)$$

is the proper dense subgroup of all finitely supported sequences (see [9, Remark 10.5]). Since G is compact, $s(\alpha) = 1$ holds. As $L(\alpha) = \beta$ with scale p (by Theorem 3.6 (c)), we have $s(L(\alpha)) = p \neq s(\alpha)$.

A Non-closed Contraction Group We now describe an analytic automorphism α of a Lie group over a local field of positive characteristic such that $\overrightarrow{\mathrm{con}}(\alpha)$ is not closed. The example is taken from [21].

Example 8.26 Let $\mathbb{F}$ be a finite field, with p elements. Consider the set $G := \mathbb{F}^{\mathbb{Z}}$ of all functions $f \colon \mathbb{Z} \to \mathbb{F}$. Then G is a compact topological group under addition, with the product topology. The right shift

$$\alpha \colon G \to G, \quad \alpha(f)(n) := f(n-1)$$

is an automorphism of G. It is easy to check that $\overrightarrow{\mathrm{con}}(\alpha)$ is the set of all functions $f \in \mathbb{F}^{\mathbb{Z}}$ with support bounded below (i.e., there exists $n_0 \in \mathbb{Z}$ such that $f(n) = 0$ for all $n < n_0$). Thus $\overrightarrow{\mathrm{con}}(\alpha)$ is a dense, proper subgroup of G.

Now G can be considered as a 2-dimensional Lie group over $\mathbb{K} := \mathbb{F}((X))$, using the bijection $G \to \mathbb{F}[[X]] \times \mathbb{F}[[X]]$,

$$f \mapsto \left(\sum_{n=1}^{\infty} f(-n) X^{n-1}, \sum_{n=0}^{\infty} f(n) X^n \right)$$

as a global chart. The automorphism of $\mathbb{F}[[X]]^2$ corresponding to α coincides on the open 0-neighbourhood $X\mathbb{F}[[X]] \times \mathbb{F}[[X]]$ with the linear map

$$\beta \colon \mathbb{K}^2 \to \mathbb{K}^2, \quad \beta(v, w) = (X^{-1} v, Xw).$$

Hence α is an analytic automorphism. Since $\overrightarrow{\mathrm{con}}\,(\alpha)$ is not closed, G cannot admit a faithful continuous representation $G \to \mathrm{GL}_n(\mathbb{K})$ for any $n \in \mathbb{N}$, see Corollary 8.22.

The Scale on Closed Subgroups and Quotients For α an automorphism of a totally disconnected locally compact group G, the scale of the restriction $\alpha|_H$ to a closed α-stable subgroup $H \subseteq G$ and the scale of the induced automorphism on the quotient group G/H (for normal H) were studied in [58]; some generalizations for endomorphisms were obtained in [9] (compare also [14], if α has small tidy subgroups). The following proposition generalizes a corresponding result for inner automorphisms of p-adic Lie groups established in [15, Corollary 3.8].

Proposition 8.27 *Let G be a Lie group over a local field, α be an analytic endomorphism of G and $H \subseteq G$ be an α-invariant Lie subgroup of G. Then the following holds:*

(a) *If $\overrightarrow{\mathrm{con}}\,(\alpha)$ is closed, then $s_H(\alpha|_H)$ divides $s_G(\alpha)$.*

(b) *If H is a normal subgroup, $\overrightarrow{\mathrm{con}}\,(\alpha)$ is closed and also the induced analytic endomorphism $\overline{\alpha}$ of G/H has a closed contraction group $\overrightarrow{\mathrm{con}}\,(\overline{\alpha})$, then $s_G(\alpha) = s_H(\alpha|_H)s_{G/H}(\overline{\alpha})$.*

Proof This is immediate from Theorem 8.13 and Corollary 3.7. $\qquad\qquad\qquad\square$

When Homomorphisms Are Subimmersions Let $f: M \to N$ be an analytic mapping between analytic manifolds over a totally disconnected local field $\mathbb{K}$. Recall from [49, Part I, Chapter III] that f is called a *submersion* if f locally looks like a linear projection around each point, in suitable charts. If f locally looks like $j \circ q$ where q is a submersion and j an immersion (as in 8.17), then f is called a *subimmersion*. If char$(\mathbb{K}) = 0$, then an analytic map is a subimmersion if and only if $T_x(f)$ has constant rank for x in some neighbourhood of each point $p \in M$ (see [49, Part II, Chapter III, §10, Theorem in 4)]). As a consequence, every analytic homomorphism between Lie groups over a totally disconnected local field of characteristic 0 is a subimmersion. Analytic homomorphisms between Lie groups over local fields of positive characteristic need not be subimmersions, as the following example shows.

Example 8.28 Let $\mathbb{F}$ be a finite field with p elements and $\mathbb{K} := \mathbb{F}((X))$. Since char$(\mathbb{K}) = p$, the Frobenius homomorphism

$$\alpha: \mathbb{K} \to \mathbb{K}, z \mapsto z^p$$

is an injective endomorphism of the field $\mathbb{K}$ and an injective endomorphism of the additive topological group $(\mathbb{K}, +)$. If α was a subimmersion then α, being injective, would be an immersion which it is not as $\alpha'(z) = 0$ for all $z \in \mathbb{K}$. Thus α is not a subimmersion. Note that $\overrightarrow{\mathrm{con}}\,(\alpha)$ coincides with the subgroup $X\mathbb{F}[[X]]$, which is open; hence also $\alpha|_{\overrightarrow{\mathrm{con}}\,(\alpha)}$ is not a subimmersion.

It is not a coincidence that the endomorphism α in the preceding example is pathological also on $\overrightarrow{\mathrm{con}}(\alpha)$: If an endomorphism fails to be a subimmersion, then the trouble must be caused by its restriction to the contraction group:

Corollary 8.29 *Let α be an analytic endomorphism of a Lie group G over a local field $\mathbb{K}$ of positive characteristic, with closed contraction group $\overrightarrow{\mathrm{con}}(\alpha)$. If $\beta :=$ $\alpha|_{\overrightarrow{\mathrm{con}}(\alpha)}$ is a subimmersion, then α is a subimmersion.*

Proof If β is a subimmersion, then the restriction of α to the open set Ω from Theorem 8.15 corresponds to the self-map

$$\pi^{-1} \circ \alpha|_{\Omega} \circ \pi = \beta \times \alpha|_{\mathrm{lev}(\alpha) \times \overleftarrow{\mathrm{con}}(\alpha)}$$

of $\overrightarrow{\mathrm{con}}(\alpha) \times \mathrm{lev}(\alpha) \times \overleftarrow{\mathrm{con}}(\alpha)$ whose second factor is an analytic diffeomorphism, and thus α is a subimmersion. $\square$

9 Contractive Automorphisms

As shown in [26], p-adic Lie groups appear naturally in the classification of the simple totally disconnected contraction groups, and are among the building blocks for general contraction groups. We recall some of the results and give a new proof for the occurrence of p-adic Lie groups in the classification.

Definition 9.1 An automorphism α of a Hausdorff topological group G is called *contractive* if $\overrightarrow{\mathrm{con}}(\alpha) = G$. We then call (G, α) a *contraction group*. If, moreover, G is totally disconnected and locally compact, we say that (G, α) is a *totally disconnected contraction group*. An *isomorphism* between totally disconnected contraction groups (G, α) and (H, β) is a continuous group homomorphism $\phi: G \to H$ such that $\beta \circ \phi = \phi \circ \alpha$. A totally disconnected contraction group (G, α) is called *simple* if $G \neq \{e\}$ and G does not have closed α-stable normal subgroups other than $\{e\}$ and G.

Remark 9.2 Contraction groups $\overrightarrow{\mathrm{con}}(\alpha)$ of automorphisms arise in many contexts: In representation theory in connection with the Mautner phenomenon (see [39, Chapter II, Lemma 3.2] and (for the p-adic case) [54]); in probability theory on groups (see [27, 50, 51] and (for the p-adic case) [10]); and in the structure theory of totally disconnected, locally compact groups (see [1, 36], and [9]).

If a locally compact group G admits a contractive automorphism α, then there exists an α-stable, totally disconnected, closed normal subgroup $N \subseteq G$ such that

$$G = N \times G_e$$

internally as a topological group, where G_e is the identity component of G (see [50]). Thus (G, α) is the direct product of the totally disconnected contraction group $(N, \alpha|_N)$ and the connected contraction group $(G_e, \alpha|_{G_e})$ (which is a simply connected, nilpotent real Lie group, as shown by Siebert).

9.3 If F is a finite group and X a set, we write $F^{(X)}$ for the group of all functions $f: X \to F$ whose support $\{x \in X: f(x) \neq e\}$ is finite. We endow $F^{(X)}$ with the discrete topology.

See [26, Theorem A] for the following result.

Theorem 9.4 *If (G, α) is a simple totally disconnected contraction group, then G is either a torsion group or torsion free. We have the following classification:*

(a) *If G is a torsion group, then (G, α) is isomorphic to $F^{(-\mathbb{N})} \times F^{\mathbb{N}_0}$ with the right shift, for some finite simple group F.*

(b) *If G is torsion free, then (G, α) is isomorphic to $(\mathbb{Q}_p)^d$ with a $\mathbb{Q}_p$-linear contractive automorphism for which there are no invariant vector subspaces, for some prime number p and some $d \in \mathbb{N}$.*

Conversely, all of these are simple contraction groups.

To explain part (b) of the theorem, let us recall some concepts and facts.

9.5 If x is an element of a pro-p-group G (i.e., a projective limit of finite p-groups), then $\mathbb{Z} \to G$, $n \mapsto x^n$ is a continuous homomorphism with respect to the topology induced by $\mathbb{Z}_p$ on $\mathbb{Z}$, and hence extends to a continuous homomorphism

$$\phi: \mathbb{Z}_p \to G.$$

As usual, we write $x^z := \phi(z)$ for $z \in \mathbb{Z}_p$. If G is abelian and the group operation is written additively, we write $zx := \phi(z)$.

9.6 Let $\mathbb{T} := \mathbb{R}/\mathbb{Z}$ with the quotient topology. If G is a locally compact abelian group, we let $G^* := \mathrm{Hom}_{\mathrm{cts}}(G, \mathbb{T})$ be its *dual group*, endowed with the compact-open topology; thus, the elements of G^* are continuous homomorphisms $\xi: G \to \mathbb{T}$ (see [28, 29, 53]). We shall use the well-known fact that the dual group $\mathbb{Z}(p^\infty)^*$ of the Prüfer p-group is isomorphic to $\mathbb{Z}_p$ (compare, e.g., [53, Exercise 23.2 and Theorem 22.6]).

Some Ideas of the Proof of Theorem 9.4 As the closure C of the commutator group G' is α-stable, closed and normal in G, we must have $C = \{e\}$ (in which case G is abelian) or $C = G$, in which case G is topologically perfect. If G is abelian, then either G is torsion free, or G is a torsion group of prime exponent p: In fact, if G has a torsion element $g \neq e$, then a suitable power g^n is an element of order p for some prime number p, entailing that the p-socle $N := \{x \in G: x^p = e\}$ is a non-trivial, α-stable closed (normal) subgroup of G and thus $N = G$. As shown in [26], p-adic Lie groups occur in the case that G is abelian and torsion free, which we assume now. Like every totally disconnected contraction group, G has a compact open subgroup W such that $\alpha(W) \subseteq W$ (see [50, 3.1]). Then W can be chosen as a pro-p-group for some p. In fact, there exists a p-Sylow subgroup $P \neq \{e\}$ of W for some prime number p, which is unique as W is abelian (see [61, Proposition 2.2.2 (a) and (d)]). Since every pro-p subgroup of a pro-finite group is contained in a p-Sylow subgroup (see [61, Proposition 2.2.2 (c)]), we deduce that $\alpha(P) \subseteq P$. However, non-trivial α-invariant closed normal subgroups of the simple contraction group G must be open (see [26, Lemma 5.1]). Thus P is open. Now replace W with P if necessary.

For $0 \neq x \in W$, can define zx for $z \in \mathbb{Z}_p$ by continuity (see 9.5). Let $W(x)$ be the image of the continuous homomorphism

$$\phi: \mathbb{Z}_p^{\mathbb{N}_0} \to W, \quad (z_n)_{n \in \mathbb{N}_0} \mapsto \sum_{n=0}^{\infty} \alpha^n(z_n x).$$

Then $W(x)$ is a compact, non-trivial α-invariant subgroup of G and hence open by [26, Lemma 5.1] just mentioned. Being a torsion free abelian pro-p-group, $W(x)$ is isomorphic to $\mathbb{Z}_p^J$ for some set J. This can be shown using Pontryagin duality: Since $W(x)$ is torsion free and a projective limit of finite p-groups F, its dual group $W(x)^*$ is a divisible discrete group and a direct limit of the dual groups $F^* \cong F$, hence a p-group (see [53, Corollary 23.10] as well as [29, Proposition 7.5 (i) and (1)$\Leftrightarrow$(2) in Corollary 8.5]). By the classification of the divisible abelian groups, $W(x)^*$ is isomorphic to a direct sum $\bigoplus_{j \in J} \mathbb{Z}(p^\infty)$ of Prüfer p-groups (see [28, Theorem (A.15)], cf. also [29, Theorem A1.42]). As a consequence, $W(x) \cong W(x)^{**}$ is isomorphic to the direct product

$$\prod_{j \in J} \mathbb{Z}(p^\infty)^* \cong (\mathbb{Z}_p)^J,$$

as asserted (by [29, Theorem 7.63] and [53, Lemma 21.2 and Theorem 23.9]).

Since $pW(x) = W(px)$ is a non-trival α-invariant closed (normal) subgroup of G and hence open, $(p\mathbb{Z}_p)^J$ must be open in $\mathbb{Z}_p^J$, whence J is finite and $W(x) \cong \mathbb{Z}_p^J$ a p-adic Lie group.

Now a linearization argument[14] shows that $(G, \alpha) \cong (L(G), L(\alpha))$. $\square$

The classification implies a structure theorem for general totally disconnected contraction groups (G, α) (see [26, Theorem B]):

Theorem 9.7 *The set* tor(G) *of torsion elements and the set* div(G) *of divisible elements are fully invariant closed subgroups of G and*

$$G = \text{tor}(G) \times \text{div}(G).$$

---

[14]Let $\beta := L(\alpha)$. Using the underlying additive topological group of $L(G)$, the pair $(L(G), \beta)$ is a p-adic contraction group such that $L(\beta) = L(\alpha)$. Hence $(G, \alpha) \cong (L(G), \beta)$ by the last statement of [17, Proposition 5.1].

Moreover, tor(G) *has finite exponent and*

$$\text{div}(G) = G_{p_1} \times \cdots \times G_{p_n}$$

is a direct product of α-stable p-adic Lie groups G_p for certain primes p. □

Remark 9.8 By [54, Theorem 3.5 (iii)], each G_p is nilpotent, and it is in fact the group of $\mathbb{Q}_p$-rational points of a unipotent linear algebraic group defined over $\mathbb{Q}_p$. See [22] for algebraic properties of $\overrightarrow{\text{con}}(\alpha)$ if α is an analytic *endomorphism* of a Lie group G over a totally disconnected local field; if α is an analytic automorphism, then $\overrightarrow{\text{con}}(\alpha)$ is nilpotent (cf. Remark 8.18 and [17]).

9.9 If (G, α) is a totally disconnected contraction group with $G \neq \{e\}$, then G has a compact open subgroup U such that $\alpha(U)$ is a proper subgroup of U (cf. [50, 3.1]), whence U is a proper subgroup of $\alpha^{-1}(U)$ and thus

$$\Delta(\alpha^{-1}) = [\alpha^{-1}(U) : U] \in \{n \in \mathbb{N}: n \geq 2\}$$

(see [50, Lemma 3.2 (i)] and [26, Proposition 1.1 (e)]). If

$$\{e\} = G_0 \lhd G_1 \lhd \cdots \lhd G_n = G \tag{77}$$

is a properly ascending series of α-stable closed subgroups of G and α_j the contractive automorphism of G_j/G_{j-1} induced by α for $j \in \{1, \ldots, n\}$, then

$$\Delta(\alpha^{-1}) = \Delta(\alpha_1^{-1}) \cdots \Delta(\alpha_n^{-1}),$$

showing that n is bounded by the number of prime factors of $\Delta(\alpha^{-1})$, counted with multiplicities (see [26, Lemma 3.5]). As a consequence, we can choose a properly ascending series (77) of maximum length. Then all of the subquotients $(G_j/G_{j-1}, \alpha_j)$ are simple contraction groups. To deduce Theorem 9.7 from Theorem 9.4, one shows that the series can always be chosen in such a way that the torsion factors appear at the bottom, whence tor(α) $= G_k$ for some k. A major step then is to see that G_k is complemented in G, and that G/G_k is a product of p-adic Lie groups (see [26]).

10 Expansive Automorphisms

If α is an expansive automorphism of a totally disconnected locally compact group G, then the subset

$$\overrightarrow{\text{con}}(\alpha) \, \overleftarrow{\text{con}}(\alpha)$$

of G is an open identity neighbourhood (see [23, Lemma 1.1 (d)]). In some cases, this enables the finiteness properties of totally disconnected contraction groups (as described in 9.9) to be used with profit also for the study of expansive automorphisms. The proof of expansiveness of $\overline{\alpha}$ in part (b) of the following result from [23] is an example for this strategy.

Proposition 10.1 *Let α be an automorphism of a totally disconnected, locally compact group G.*

(a) *If α is expansive, then $\alpha|_H$ is expansive for each α-stable closed subgroup $H \subseteq G$.*

(b) *Let $N \subseteq G$ be an α-stable closed normal subgroup and $\overline{\alpha}$ be the induced automorphism of G/N which takes gN to $\alpha(g)N$. Then α is expansive if and only if $\alpha|_N$ and $\overline{\alpha}$ are expansive.* □

If a p-adic Lie group G admits an expansive automorphism α, then $L(\alpha)$ is a Lie algebra automorphism of $L(G)$ such that $|\lambda| \neq 1$ for all eigenvalues λ of $L(\alpha)$ in an algebraic closure $\overline{\mathbb{K}}$ (as recalled in Proposition 7.10), entailing that none of the λ is a root of unity. Hence $L(G)$ is nilpotent (see Exercise 21 (b) among the exercises for Part I of [7], §4, or [33, Theorem 2]). If, moreover, G is linear in the sense that it admits a faithful continuous representation $G \to \mathrm{GL}_n(\mathbb{Q}_p)$ for some $n \in \mathbb{N}$, then G has an α-stable, nilpotent open subgroup [23, Theorem D]. For closed subgroup of $\mathrm{GL}_n(\mathbb{Q}_p)$, such an open subgroup can be made explicit (see [23, Proposition 7.8]):

Proposition 10.2 *Let α be an expansive automorphism of a p-adic Lie group G. If G is isomorphic to a closed subgroup of $\mathrm{GL}_n(\mathbb{Q}_p)$ for some $n \in \mathbb{N}$, then the set $\overrightarrow{\mathrm{con}}(\alpha)\,\overleftarrow{\mathrm{con}}(\alpha)$ is a nilpotent, open subgroup of G.* □

The following example is taken from [23, Remark 7.7].

Example 10.3 Let $H = \mathbb{Q}_p^3$ be the 3-dimensional p-adic Heisenberg group with group multiplication given by

$$(x_1, y_1, z_1)(x_2, y_2, z_2) := (x_1 + x_2, y_1 + y_2, z_1 + z_2 + x_1 y_2)$$

for all $(x_1, y_1, z_1), (x_2, y_2, z_2) \in H$. Then $N = \{(0, 0, z) \in H : |z| \leq 1\}$ is a compact central subgroup of H. Identify $G = H/N$ with $\mathbb{Q}_p \times \mathbb{Q}_p \times (\mathbb{Q}_p/\mathbb{Z}_p)$ as a set. Define $\alpha : G \to G$ by

$$\alpha(x, y, z + \mathbb{Z}_p) = (px, p^{-1}y, z + \mathbb{Z}_p)$$

for all $(x, y, z + \mathbb{Z}_p) \in G$. Then α is an analytic automorphism of the p-adic Lie group G with $\mathrm{lev}(\alpha) = \{(0, 0, z + \mathbb{Z}_p) : z \in \mathbb{Q}_p\}$,

$$\overrightarrow{\mathrm{con}}(\alpha) = \{(x, 0, 0) : x \in \mathbb{Q}_p\}, \quad \text{and} \quad \overleftarrow{\mathrm{con}}(\alpha) = \{(0, y, 0) : y \in \mathbb{Q}_p\}.$$

Since $\mathrm{lev}(\alpha)$ is discrete, α is expansive (see [23, Proposition 1.3 (a)]). As

$$[\overrightarrow{\mathrm{con}}(\alpha), \overleftarrow{\mathrm{con}}(\alpha)] = \{(0,0,z+\mathbb{Z}_p): z \in \mathbb{Q}_p\}$$

and $\overrightarrow{\mathrm{con}}(\alpha)\,\overleftarrow{\mathrm{con}}(\alpha) = \{(x,y,xy+\mathbb{Z}_p): x,y \in \mathbb{Q}_p\}$, we see that the set $\overrightarrow{\mathrm{con}}(\alpha)\,\overleftarrow{\mathrm{con}}(\alpha)$ is a not a subgroup of G. Accordingly, G is not isomorphic to a closed subgroup of $\mathrm{GL}_n(\mathbb{Q}_p)$ for any $n \in \mathbb{N}$ (see Proposition 10.2).

11 Distality and Lie Groups of Type R

Following Palmer [40], a totally disconnected, locally compact group G is called *uniscalar* if $s(x) = 1$ for each $x \in G$. This holds if and only if each group element $x \in G$ normalizes some compact, open subgroup V_x of G (which may depend on x). It is natural to ask whether this condition implies that V_x can be chosen independently of x, i.e., whether G has a compact, open, *normal* subgroup. The answer is negative for a suitable p-adic Lie group which is not compactly generated (see [24, §6]). But also for some totally disconnected, locally compact groups which are compactly generated, the answer is negative (see [3] together with [37], or also [21, Proposition 11.4], where moreover all contraction groups for inner automorphisms are trivial and hence closed); the counterexamples are of the form

$$(F^{(-\mathbb{N})} \times F^{\mathbb{N}_0}) \rtimes H$$

with F a finite simple group and a suitable action of a specific finitely generated group H on $\mathbb{Z}$. Thus, to have a chance for a positive answer, one has to restrict attention to particular classes of groups (like compactly generated p-adic Lie groups). If G has the (even stronger) property that every identity neighbourhood contains an open, compact, normal subgroup of G, then G is called *pro-discrete*.[15] Finally, a Lie group G over a local field $\mathbb{K}$ is *of type R* if all eigenvalues λ of $L(\alpha)$ in an algebraic closure $\overline{\mathbb{K}}$ have absolute value $|\lambda|_{\mathbb{K}} = 1$, for each inner automorphism α (cf. [42] for $\mathbb{K} = \mathbb{Q}_p$), i.e., if each inner automorphism is distal (see Proposition 7.10).

Using the Inverse Function Theorem with Parameters and locally invariant manifolds as a tool, we can generalize results for p-adic Lie groups from [24, 42],

---

[15]Another interesting group is the semidirect product $G := F^T \rtimes T$, where F is a finite simple group and T is a Tarski monster (a certain finitely generated, infinite, simple torsion group) acting on F^T via $(x.f)(y) := f(x^{-1}y)$ for $x,y \in T$. Then, for each $x \in G$, there is a basis of identity neighbourhoods consisting of compact open subgroups of G which are normalized by x. Moreover, G has F^T as a compact open normal subgroup, but this is the only such and thus G is not pro-discrete (see [25]).

and [41] to Lie groups over local fields of arbitrary characteristic. The following result was announced in [21, Proposition 4.53].

Proposition 11.1 *Let α be an analytic automorphism of a Lie group G over a totally disconnected local field $\mathbb{K}$. Then the following properties are equivalent:*

(a) *$\overrightarrow{\mathrm{con}}(\alpha)$ is closed and $s(\alpha) = s(\alpha^{-1}) = 1$;*
(b) *All eigenvalues of $L(\alpha)$ in $\overline{\mathbb{K}}$ have absolute value 1;*
(c) *Each e-neighbourhood in G contains an α-stable compact open subgroup.*

In particular, G is of type R if and only if G is uniscalar and $\overrightarrow{\mathrm{con}}(\alpha)$ is closed for each inner automorphism α of G (in which case $\overrightarrow{\mathrm{con}}(\alpha) = \{e\}$).

Proof The implication "(a)$\Rightarrow$(b)" follows from Theorem 8.13 and Theorem 3.6 (c).

If (b) holds, then $\mathfrak{g} = \mathfrak{g}_1$ coincides with the centre subspace $\mathfrak{g}_1 = \mathrm{lev}(L(\alpha))$ with respect to $L(\alpha)$, whence $\phi_c \colon W_c \to B_R^{\mathfrak{g}_1}(0) \subseteq \mathfrak{g}_1 = \mathfrak{g}$ (as in 8.5) is a diffeomorphism with $\phi_c(e) = 0$ and $d\phi_c|_{\mathfrak{g}} = \mathrm{id}_{\mathfrak{g}}$. After shrinking R if necessary, we may assume that the sets $W^c(t) = \phi_c^{-1}(B_t^{\mathfrak{g}}(0))$, which are α-stable by 8.9, are compact open subgroups of G for all $t \in]0, R]$ (see Lemma 5.2). Thus (b) implies (c).

If (c) holds, then every identity neighbourhood of G contains a compact open subgroup V which is α-stable and hence tidy for α with

$$s(\alpha) = [\alpha(V) : \alpha(V) \cap V] = [V : V] = 1$$

and, likewise, $s(\alpha^{-1}) = 1$. Moreover, $\overrightarrow{\mathrm{con}}(\alpha)$ is closed (by Lemma 8.2 (a)), and thus (a) follows. (Since α is distal, in fact $\overrightarrow{\mathrm{con}}(\alpha) = \{e\}$.) $\square$

As shown in [41] and [24], every compactly generated, uniscalar p-adic Lie group is pro-discrete. For Lie groups over totally disconnected local fields, we have the following analogue (announced in [21, Proposition 4.54]):

Theorem 11.2 *Every compactly generated Lie group of type R over a totally disconnected local field is pro-discrete.*

Proof Let G be a Lie group over a totally disconnected local field $\mathbb{K}$, with Lie algebra $\mathfrak{g} := L(G)$, such that G is generated by a compact subset K. Then $\mathrm{Ad}(G) \subseteq GL(\mathfrak{g})$ is generated by the compact set $\mathrm{Ad}(K)$, since the adjoint representation $\mathrm{Ad} \colon G \to GL(\mathfrak{g})$ is a continuous homomorphism (as recalled in 4.7). Moreover, the subgroup generated by Ad_x is relatively compact in $GL(\mathfrak{g})$ for each $x \in G$, by Lemma 3.9. Thus $\mathrm{Ad}(G)$ is relatively compact in $GL(\mathfrak{g})$, by [41, Théorème 1]. Let $\mathbb{O}$ be the valuation ring of $\mathbb{K}$. As a consequence of Theorem 1 in Appendix 1 of [49, Chapter IV], there is a compact open $\mathbb{O}$-submodule $M \subseteq \mathfrak{g}$ with $\mathrm{Ad}_x(M) = M$ for all $x \in G$. Let

$$\|.\| \colon \mathfrak{g} \to [0, \infty[, \quad x \mapsto \inf\{|z| \colon z \in \mathbb{K} \text{ such that } x \in zM\}$$

be the Minkowski functional of M. Then $\|.\|$ is a norm on $\mathfrak{g}$ such that $\mathrm{Ad}(G) \subseteq \mathrm{Iso}(\mathfrak{g}, \|.\|)$. Using this norm, we abbreviate $B_t^{\mathfrak{g}} := B_t^{\mathfrak{g}}(0)$ for all $t > 0$. Let $\phi \colon U \to V$

be an analytic diffeomorphism from a compact open subgroup U of G onto an open 0-neighbourhood $V \subseteq \mathfrak{g}$ such that $\phi(e) = 0$ and $d\phi|_\mathfrak{g} = \mathrm{id}_\mathfrak{g}$. Let $R > 0$ and the compact open subgroups $B_t^\phi = \phi^{-1}(B_t^\mathfrak{g})$ for $t \in \,]0, R]$ be as in 5.1. Since K is compact and the mapping $G \times G \to G$, $(x, y) \mapsto xyx^{-1}$ is continuous with $xex^{-1} = e$, we find an open identity neighbourhood $W \subseteq U$ such that $xWx^{-1} \subseteq U$ for all $x \in K$. Now consider the analytic mapping

$$f: G \times \phi(W) \to V, \quad f(x, y) := \phi(x\phi^{-1}(y)x^{-1})$$

and define $f_x := f(x, .): \phi(W) \to V$ for $x \in G$. Then $f_x(0) = 0$ and $(f_x)'(0) = \mathrm{Ad}_x$ is an isometry for all $x \in G$. As a consequence of Lemma 4.9, for each $x \in K$ there exists an open neighbourhood P_x of x in G and $r_x > 0$ with $B_{r_x}^\mathfrak{g} \subseteq \phi(W)$ such that

$$f_z(B_t^\mathfrak{g}) = B_t^\mathfrak{g} \quad \text{for all } z \in P_x \text{ and } t \in \,]0, r_x].$$

We may assume that $r_x \leq R$ for all $x \in K$. There is finite subset $\Phi \subseteq K$ such that $K \subseteq \bigcup_{x \in \Phi} P_x$. Set

$$r := \min\{r_x : x \in \Phi\}.$$

Then $f_z(B_t^\mathfrak{g}) = B_t^\mathfrak{g}$ for all $z \in K$ and $t \in \,]0, r]$, entailing that

$$zB_t^\phi z^{-1} = \phi^{-1}(\phi(z\phi^{-1}(B_t^\mathfrak{g})z^{-1})) = \phi^{-1}(f_z(B_t^\mathfrak{g})) = \phi^{-1}(B_t^\mathfrak{g}) = B_t^\phi.$$

Since K generates G, we deduce that the compact open subgroup B_t^ϕ is normal in G, for each $t \in \,]0, r]$. $\qquad\qquad\square$

Related problems were also studied in [46].

Acknowledgements The author is grateful for the support provided by the University of Melbourne (Matrix Center, Creswick) and the University of Newcastle (NSW), notably George A. Willis, which enabled participation in the 'Winter of Disconnectedness.' A former unpublished manuscript concerning the scale of automorphisms dating back to 2006 was supported by DFG grant 447 AUS-113/22/0-1 and ARC grant LX 0349209.

References

1. Baumgartner, U., Willis, G.A.: Contraction groups and scales of automorphisms of totally disconnected locally compact groups. Isr. J. Math. **142**, 221–248 (2004)
2. Bertram, W., Glöckner, H., Neeb, K.-H.: Differential calculus over general base fields and rings. Expo. Math. **22**, 213–282 (2004)
3. Bhattacharjee, M., MacPherson, D.: Strange permutation representations of free groups. J. Aust. Math. Soc. **74**, 267–285 (2003)
4. Borel, A.: Linear Algebraic Groups. Springer, New York (1991)
5. Bourbaki, N.: Variétés différentielles et analytiques. Fascicule de résultats. Hermann, Paris (1967)

6. Bourbaki, N.: Topological Vector Spaces, chaps. 1–5. Springer, Berlin (1987)
7. Bourbaki, N.: Lie Groups and Lie Algebras, chaps. 1–3. Springer, Berlin (1989)
8. Bruhat, F., Tits, J.: Groupes reductifs sur un corps local. Publ. Math. Inst. Hautes Étud. Sci. **41**, 5–251 (1972)
9. Bywaters, T.P., Glöckner, H., Tornier, S., Contraction groups and passage to subgroups and quotients for endomorphisms of totally disconnected locally compact groups. Israel J. Math. (cf. arXiv:1612.06958) (to appear)
10. Dani, S.G., Shah, R.: Contraction subgroups and semistable measures on p-adic Lie groups. Math. Proc. Camb. Philos. Soc. **110**, 299–306 (1991)
11. Dani, S.G., Shah, N.A., Willis, G.A.: Locally compact groups with dense orbits under $\mathbb{Z}^d$-actions by automorphisms. Ergodic Theory Dyn. Syst. **26**, 1443–1465 (2006)
12. Dixon, J.D., du Sautoy, M.P.F., Mann, A., Segal, D.: Analytic Pro-p Groups. Cambridge University Press, Cambridge (1999)
13. du Sautoy, M., Mann, A., Segal, D.: New Horizons in Pro-p Groups. Birkhäuser, Boston (2000)
14. Giordano Bruno, A., Virili, S.: Topological entropy in totally disconnected locally compact groups. Ergod. Theory Dyn. Syst. **37**, 2163–2186 (2017)
15. Glöckner, H.: Scale functions on p-adic Lie groups. Manuscripta Math. **97**, 205–215 (1998)
16. Glöckner, H.: Implicit functions from topological vector spaces to Banach spaces. Isr. J. Math. **155**, 205–252 (2006)
17. Glöckner, H.: Contractible Lie groups over local fields. Math. Z. **260**, 889–904 (2008)
18. Glöckner, H.: Invariant manifolds for analytic dynamical systems over ultrametric fields. Expo. Math. **31**, 116–150 (2013)
19. Glöckner, H.: Invariant manifolds for finite-dimensional non-archimedean dynamical systems. In: Glöckner, H., Escassut, A., Shamseddine, K. (eds.) Advances in Non-Archimedean Analysis, pp. 73–90. Contemporary Mathematics, vol. 665. American Mathematical Society, Providence, RI (2016)
20. Glöckner, H.: Lie groups over non-discrete topological fields. Preprint (2004). arXiv:math/0408008
21. Glöckner, H.: Lectures on Lie groups over local fields. In: Caprace, P.-E., Monod, N. (eds.) New Directions in Locally Compact Groups. London Math. Soc. Lect. Notes Series 447 (to appear)
22. Glöckner, H.: Contraction groups of analytic endomorphisms and dynamics on the big cell. University of Paderborn (2017)
23. Glöckner, H., Raja, C.R.E.: Expansive automorphisms of totally disconnected, locally compact groups. J. Group Theory **20**, 589–619 (2017)
24. Glöckner, H., Willis, G.A.: Uniscalar p-adic Lie groups. Forum Math. **13**, 413–421 (2001)
25. Glöckner, H., Willis, G.A.: Directions of automorphisms of Lie groups over local fields compared to the directions of Lie algebra automorphisms. Topol. Proc. **31**, 481–501 (2007)
26. Glöckner, H., Willis, G.A.: Classification of the simple factors appearing in composition series of totally disconnected contraction groups. J. Reine Angew. Math. **634**, 141–169 (2010)
27. Hazod, W., Siebert, E.: Stable Probability Measures on Euclidean Spaces and on Locally Compact Groups. Kluwer, Dordrecht (2001)
28. Hewitt, E., Ross, K.A.: Abstract Harmonic Analysis I. Springer, Berlin (1963)
29. Hofmann, K.H., Morris, S.A.: The Structure of Compact Groups. de Gruyter, Berlin (1998)
30. Humphreys, J.E.: Linear Algebraic Groups. Springer, New York (1975)
31. Humphreys, J.E.: Introduction to Lie Algebras and Representation Theory. Springer, Berlin (1994)
32. Irwin, M.C.: On the stable manifold theorem. Bull. Lond. Math. Soc. **2**, 196–198 (1970)
33. Jacobson, N.: A note on automorphisms and derivations of Lie algebras. Proc. Am. Math. Soc. **6**, 281–283 (1955)
34. Jacobson, N.: Basic Algebra II. W. H. Freeman and Company, New York (1989)
35. Jaikin-Zapirain, A., Klopsch, B.: Analytic groups over general pro-p domains. J. Lond. Math. Soc. **76**, 365–383 (2007)

36. Jaworski, W.: On contraction groups of automorphisms of totally disconnected locally compact groups. Isr. J. Math. **172**, 1–8 (2009)
37. Kepert, A., Willis, G.A.: Scale functions and tree ends. J. Aust. Math. Soc. **70**, 273–292 (2001)
38. Lazard, M.: Groupes analytiques p-adiques. IHES Publ. Math. **26**, 389–603 (1965)
39. Margulis, G.A.: Discrete Subgroups of Semisimple Lie Groups. Springer, Berlin (1991)
40. Palmer, T.W.: Banach Algebras and the General Theory of ∗-Algebras, vol. 2. Cambridge University Press, Cambridge (2001)
41. Parreau, A.: Sous-groupes elliptiques de groupes linéaires sur un corps valué. J. Lie Theory **13**, 271–278 (2003)
42. Raja, C.R.E.: On classes of p-adic Lie groups. New York J. Math. **5**, 101–105 (1999)
43. Raja, C.R.E., Shah, R.: Some properties of distal actions on locally compact groups. Ergodic Theory Dyn. Syst., doi: 10.1017/etds.2017.58 (to appear)
44. Rathai, N.: Endomorphismen total unzusammenhängender lokal kompakter Gruppen. Bachelor's thesis, Universität Paderborn (2016)
45. Reid, C.D.: Endomorphisms of profinite groups. Groups Geom. Dyn. **8**, 553–564 (2014)
46. Reid, C.D.: Distal actions on coset spaces in totally disconnected, locally compact groups. Preprint (2016). arXiv:1610.06696
47. Schikhof, W.H.: Ultrametric Calculus. Cambridge University Press, Cambridge (1984)
48. Schneider, P.: p-Adic Lie Groups. Springer, Berlin (2011)
49. Serre, J.-P.: Lie Algebras and Lie Groups. Springer, Berlin (1992)
50. Siebert, E.: Contractive automorphisms of locally compact groups. Math. Z. **191**, 73–90 (1986)
51. Siebert, E.: Semistable convolution semigroups and the topology of contraction groups. In: Heyer, H. (ed.) Probability Measures on Groups IX. Lecture Notes in Mathematics, vol. 1379, pp. 325–343. Springer, Berlin (1989)
52. Springer, T.A.: Linear Algebraic Groups. Birkhäuser, Boston (1998)
53. Stroppel, M.: Locally Compact Groups. EMS Publishing House, Zurich (2006)
54. Wang, J.S.P.: The Mautner phenomenon for p-adic Lie groups. Math. Z. **185**, 403–412 (1984)
55. Weil, A.: Basic Number Theory. Springer, New York (1967)
56. Wells, J.C.: Invariant manifolds of non-linear operators. Pac. J. Math. **62**, 285–293 (1976)
57. Willis, G.A.: The structure of totally disconnected, locally compact groups. Math. Ann. **300**, 341–363 (1994)
58. Willis, G.A.: Further properties of the scale function on a totally disconnected group. J. Algebra **237**, 142–164 (2001)
59. Willis, G.A.: The nub of an automorphism of a totally disconnected, locally compact group. Ergod. Theory Dyn. Sys. **34**, 1365–1394 (2014)
60. Willis, G.A.: The scale and tidy subgroups for endomorphisms of totally disconnected locally compact groups. Math. Ann. **361**, 403–442 (2015)
61. Wilson, J.S.: Profinite Groups. Clarendon Press, Oxford (1998)

Compact Convex Sets with Prescribed Facial Dimensions

Vera Roshchina, Tian Sang, and David Yost

Abstract While faces of a polytope form a well structured lattice, in which faces of each possible dimension are present, this is not true for general compact convex sets. We address the question of what dimensional patterns are possible for the faces of general closed convex sets. We show that for any finite sequence of positive integers there exist compact convex sets which only have extreme points and faces with dimensions from this prescribed sequence. We also discuss another approach to dimensionality, considering the dimension of the union of all faces of the same dimension. We show that the questions arising from this approach are highly nontrivial and give examples of convex sets for which the sets of extreme points have fractal dimension.

1 Introduction

It is well known that faces of polyhedral sets have a well-defined structure (see [12, Chap. 2]). In particular, every face of a polyhedral set is a polyhedron, and there are no 'gaps' in the dimensions of their faces. On the other hand, a simple reformulation of [4, Corollary 3.7] asserts that in the compact convex set of all positive semidefinite $n \times n$ matrices with trace 1, every proper face has dimension

---

V. Roshchina (✉)
School of Science, RMIT University, Melbourne, VIC 3001, Australia

Centre for Informatics and Applied Optimisation, Federation University Australia, Ballarat, VIC 3353, Australia
e-mail: vera.roshchina@rmit.edu.au

T. Sang
School of Science, RMIT University, Melbourne, VIC 3001, Australia
e-mail: s3556268@student.rmit.edu.au

D. Yost
Centre for Informatics and Applied Optimisation, Federation University Australia, Ballarat, VIC 3353, Australia
e-mail: d.yost@federation.edu.au

© Springer International Publishing AG, part of Springer Nature 2018
D.R. Wood et al. (eds.), *2016 MATRIX Annals*, MATRIX Book Series 1,
https://doi.org/10.1007/978-3-319-72299-3_7

$\frac{k(k+1)}{2} - 1$ for some $k < n$. Thus there are naturally occurring examples with serious gaps in the dimensions of their faces. For other descriptions of this phenomenon, see Theorem 2.25 and the explanation that follows it in [10] (for the cone $\mathbb{S}_+^n$ of positive semidefinite $n \times n$ matrices), or [1, Theorem 5.36] (for the state space of a C^*-algebra). This raises the question, what are the possible patterns for the dimensions of faces of compact convex sets?

Recall that a *face* F of a closed convex set $C \subset \mathbb{R}^n$ is a closed convex subset of C such that for any point $x \in F$ and for any line segment $[a, b] \subset C$ such that $x \in (a, b)$, we have $a, b \in F$. The fact that F is a face of C is expressed as $F \lhd C$.

The difference between this definition and the definition of faces of polyhedral sets as intersections with supporting hyperplanes is due to the fact that for nonpolyhedral convex sets faces are not necessarily *exposed*: it may happen that a face cannot be represented as the intersection of a supporting hyperplane with the set. Some classic examples are shown in Figs. 1 (see [8]) and 2 (see [7]).

The dimension of a convex set is the dimension of its affine hull, same for the face. We refer the reader to the classic textbooks [5, 8]. We also would like to mention that some problems related to dimensions of convex sets were studied in the literature. For instance, [2] focusses on the dimensions of convex sets coming from optimisation problems with inequality constraints, and [3] deals with the results related to the dimensions of intersections of convex sets. However, we were unable to identify references that would address the existence of convex sets with prescribed facial dimensions.

Fig. 1 Convex hull of a torus is not facially exposed (the dashed line shows the unexposed extreme points)

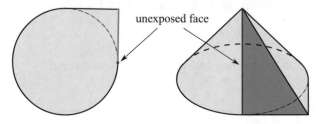

Fig. 2 An example of a two dimensional set and a three dimensional cone that have an unexposed face

The total number of possible face patterns in n dimensional space is the cardinality of the powerset of n elements. This is because every set contains zero-dimensional faces (because of the Krein-Milman theorem). We can write down face patterns either as an increasing sequence of positive numbers $(d_1, d_2, \ldots, d_k)$, which encode all possible dimensions of faces of positive dimension present in a set, or as a binary sequence $(b_1, b_2, \ldots, b_n)$, where $b_i = 1$ if a face of dimension i is present in the set, and $b_i = 0$ otherwise. For example, the dimensional pattern of a tetrahedron is either $(1, 2, 3)$ in the d-notation or $(1, 1, 1)$ in the binary notation, and the pattern of a closed Euclidean ball is either (n) or $(0, 0, \ldots, 1)$, as it does not have any faces except for zero- and n-dimensional ones. We will use the first encoding style via an increasing sequence of positive numbers in what follows.

The easiest cases to classify are the ones that we can visualise, i.e. the convex compact sets in zero- one-, two- and three-dimensional spaces. In dimension zero we have singletons $\{x\}$ for any real x with pattern $()$, in one-dimensional space there is no freedom: the only fully dimensional convex compact sets are line segments, with the only possible pattern (1). On the plane the two-dimensional possibilities are exhausted by a circle and a triangle, with patterns (2) and $(1, 2)$ respectively (see Fig. 3). Therefore for the two dimensional case we have four possibilities: $()$, (1), $(1, 2)$ and (2), which coincides with the cardinality of the powerset of two: $2^2 = 4$.

In three dimensions the possibilities for fully dimensional sets are exhausted by the unit ball (3), the tetrahedron $(1, 2, 3)$, the unit ball intersected with a closed half-space $(2, 3)$, and the convex hull of a circle in the plane and two points on opposite sides of the plane $(1, 3)$ (see Fig. 4), together with the lower dimensional examples we have in total $2^3 = 8$ possibilities.

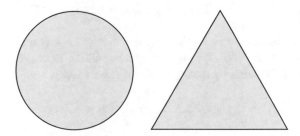

Fig. 3 All possible face patterns of fully dimensional sets in two dimensional case are given by a disk and a triangle

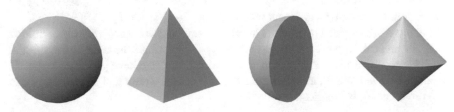

Fig. 4 All possible facial patterns for the three dimensional sets

2   Main Result

We show that all patterns of facial dimensions can be realised in a compact convex set.

Theorem 1 *For any increasing sequence of positive integers*

$$d = (d_1, d_2, \ldots, d_k)$$

there exists a compact convex set in d_k-dimensional space such that the vector d describes the pattern of facial dimensions for this set.

To prove this, we need the following technical lemma, which is surely known, but we were not able to identify it in the literature. We hence provide a short proof here as well.

Lemma 1 *Let $P, Q \subset \mathbb{R}^n$ be nonempty convex compact sets, and let $C = P + Q$. Then every face of C is the Minkowski sum of faces of P and Q. More precisely,*

$$\forall F \lhd C \quad \exists F_P \lhd P, F_Q \lhd Q \text{ such that } F = F_P + F_Q.$$

Proof Let F be a nonempty face of C. We construct two sets

$$F_P := \{x \in P \mid \exists y \in Q, x + y \in F\}, \qquad F_Q := \{y \in Q \mid \exists x \in P, x + y \in F\}.$$

Both F_P and F_Q are nonempty since F is nonempty.

First we show that $F = F_P + F_Q$. It is obvious that $F \subset F_P + F_Q$, and it remains to show the reverse inclusion. For that, pick an arbitrary $x \in F_P$, $y \in F_Q$. We will next show that $z = x + y \in F$.

By the definition of F_P and F_Q there exist $u \in P$ and $v \in Q$ such that $x + v \in F$ and $y + u \in F$. If $x = u$ or $y = v$, there is nothing to prove, as in this case $z = u + v \in F$. Otherwise, by the convexity of F we have

$$z' = \frac{x + v}{2} + \frac{y + u}{2} \in F.$$

At the same time, notice that $x + y \in P + Q \subset C$; likewise, $u + v \in P + Q \subset C$, and $z' \in (x + y, u + v)$. Since F is a face of C, this yields $z = x + y \in F$.

It remains to show that both F_P and F_Q are faces of P and Q respectively. First note that both are convex compact sets, and that $F_Q \subset Q$ and $F_P \subset P$.

Let $x \in F_P$, and pick any interval $[a, b] \subset P$ such that $x \in (a, b)$. By the definition of C, for an arbitrary $y \in F_Q$ we have $a + y, b + y \in C$. At the same time, $x + y \in F_P + F_Q = F$ and $x + y \in (a + y, b + y)$. From $F \lhd C$ we have $[a + y, b + y] \subset F$, hence, $a + y, b + y \in F$, and therefore $a, b \in F_P$. This shows that F_P is a face of P. The proof for F_Q is identical.

Fig. 5 Minkowski sum of a
line segment and a unit sphere
(on the left hand side), and of
a unit square and a sphere (on
the right hand side)

In the proof of Theorem 1 presented next we use an inductive argument to explicitly construct a compact convex set with a given facial pattern from a lower dimensional example for a truncated sequence. The key observation is that the Minkowski sum of an arbitrary compact convex set with a unit ball does not generate faces of any new dimensions (compared to the original set) other than possibly the fully dimensional face that coincides with the sum, which follows directly from Lemma 1. We sketched the Minkowski sum of two simple compact convex sets with a Euclidean ball in Fig. 5 to illustrate this argument.

Proof (Proof of Theorem 1) We use induction on d_k to demonstrate the result. Our induction base is lower dimensional examples discussed earlier. For all increasing sequences of positive numbers $(d_1, \ldots, d_k)$ with $d_k \leq 2$ we have found the relevant examples. They are realised by a point, line segment, disk and triangle.

Assume that our assertion is proven for all sequences $(d_1, \ldots, d_k)$ with $d_k \leq m$. We will show that the statement is true for $d_k = m+1$. Choose an arbitrary sequence $d = (d_1, \ldots, d_k)$, where $d_k = m + 1$. If $d = (d_k)$, the sequence is realised by the Euclidean unit ball in $\mathbb{R}^{m+1}$. If the sequence contains more than one number, consider the truncated sequence $d' = (d_1, d_2, \ldots, d_{k-1})$. Since $d_{k-1} < d_k$, we have $l := d_{k-1} \leq m$, and there exists a compact convex set $Q \subset \mathbb{R}^l$ that realises the sequence d' in $l = d_{k-1}$-dimensional space. We embed the set Q in the $m + 1$-dimensional space by letting $Q' := Q \times \{0_{m+1-l}\}$. Observe that since the definition of the face is algebraic, the facial pattern of the set Q' is identical to the one of Q. Let B be the unit ball in $\mathbb{R}^{m+1}$. We let

$$C := B + Q'$$

and claim that d is the facial pattern of C.

From Lemma 1 every face of C can be represented as the sum of faces of Q' and B. Since the only faces of B are the set itself and the singletons on the boundary, the only possible dimensions of the faces of the set C can come from the sequence $(d_1, \ldots, d_k)$. To show that no facial dimensions are lost, observe that if e denotes the unit vector $(0, 0, \ldots, 1) \in B$, then the set $\{e\} + Q'$ is a face of C (hence all its faces are also faces of C). Indeed, for the hyperplane $H = \{x \mid \langle e, x \rangle = x_{m+1} = 1\}$ supports C (notice that for every $x = q + b \in C$ with $q \in Q'$ and $b \in B$ we have

$x_{m+1} = 0 + q_{m+1} \leq 1$), moreover,

$$H \cap C = \{q + b \mid q \in Q', b \in B, q_{m+1} + b_{m+1} = 1\}$$
$$= \{q + b \mid q \in Q', b \in B, b_{m+1} = 1\}$$
$$= \{e\} + Q'.$$

It is not difficult to observe (e.g., see [8, Section 18]) that any supporting hyperplane slices off a face from a convex set, hence, $F = \{e\} + Q' \lhd C$. This face is linearly isomorphic to Q, and hence the facial structure of F coincides with the facial structure of Q, giving all possible dimensions of faces from the sequence d'. The face of the maximal dimension $m + 1$ is given by the set C itself, as it has a nonempty interior (take any point from Q' and sum it with an open ball).

3 Fractal Convex Sets

Observe that polytopes not only possess faces of all possible dimensions, but their faces are also arranged in a very regular fashion: the union of the edges of a polytope is a one-dimensional set (here we refer to a general notion of Hausdorff dimension, rather than the dimension of the affine hull that is useful for convex sets), the union of all two dimensional faces is two dimensional, and so on. More generally, the union of all faces of a polytope of a given dimension is a set of the same dimension. This is not the case for a more general setting: for instance, the dimension of the union of all extreme points of a Euclidean ball in $\mathbb{R}^n$ is $n - 1$, a stark contrast with the polyhedral case. Hence it is natural to study the dimension of the unions of equidimensional faces. The purpose of this section is to present some examples which emerged from the discussions during the MATRIX program, namely nontrivial sets with fractal facial structure and hence noninteger dimensions of the said unions; these form the foundation for our ongoing research on this topic.

Some work on fractals and convexity has been done before (see the recent work [11] and references therein), but we are not aware of any references studying the particular problems that we propose here. We focus on two examples of convex sets that are generated in a natural way by spherical fractals. The finite root system and Coxeter system are fundamental concepts in Lie algebras, which is very important in many branches of mathematics. Given a finite root system, there is a natural associated finite Coxeter group, which is the Weyl group. People in the field of geometric group theory consider finite Coxeter groups are well-studied and explained in liberature, see [6]. Therefore, we are more interested in the behaviours of infinite Coxeter groups. One such fractal comes from a recent work [9] by one of our co-authors (curiously from the study of infinite Coxeter groups), another one is constructed via projecting the Sierpinski triangle onto the unit sphere.

We first consider a fractal set on a sphere and then take its convex hull, hence generating a convex set. Our first example is constructed in a similar way to the Apollonian gasket: we take the unit sphere and construct a tetrahedron whose edges touch the sphere (see Fig. 6), then consider the intersection of the sphere with the tetrahedron. After that, we continue slicing off spherical caps in such a way that they are tangential to the existing slices (see Fig. 7). The resulting body is a spherical fractal, which is also a convex set. If we now take its convex hull, the extreme points of this convex set would be exactly the points on the fractal set, with remaining proper faces disks that result from the sliced off spherical caps. Notice that this structure is somewhat similar to the compact convex set obtained as the intersection of the cone of symmetric positive semidefinite matrices of dimension 3×3 with an affine subspace defined by matrices with a constant trace

$$C := \mathbb{S}^3_+ \cap \{M \mid \mathrm{tr}(M) = 1\}.$$

This set has dimension 5 however.

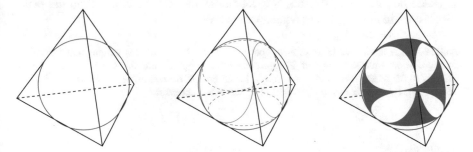

Fig. 6 Construction of the spherical gasket

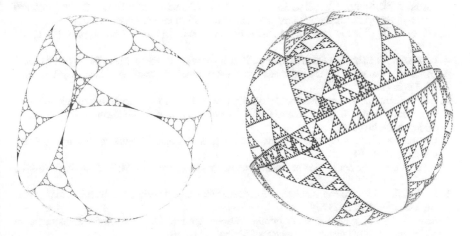

Fig. 7 Apollonian gasket on a sphere and Sierpinski triangles

Algebraically, this particular fractal set is generated by the infinite Coxeter group with following group presentation:

$$G = \langle s_1, s_2, s_3, s_4 \mid (s_i)^2 = (s_i s_j)^\infty = 1 \rangle$$

The fractal sets are generated by *limit roots*, see [9]. Limit roots exhibit peculiar geometric behaviour. Even though Coxeter groups are generated by affine reflections across hyperplanes, when we compute the roots of the group and project them down to a lower dimensional affine hyperplane, the set of limit roots behaves like a fractal set, giving self-similar patterns that cannot be obtained by reflecting across any hyperplanes.

This approach can be applied to constructing other spherical fractals. For instance, one can generalise the Sierpinski carpet by cutting out triangular pieces of the sphere in a similar fashion. The convex set obtained after taking the convex hull of this spherical fractal will have faces of all possible dimensions.

The Hausdorff dimension of the union of the extreme points is non-integer in both cases, and coincides with the dimension of the relevant two-dimensional objects. It would be interesting to study the conditions that can be imposed on the facial dimensions to define good or regular convex sets.

Acknowledgements The ideas in this paper were motivated by the discussions that took place during a recent MATRIX program in approximation and optimisation held in July 2016. We are grateful to the MATRIX team for the enjoyable and productive research stay. We would also like to thank the two referees for their insightful corrections and remarks.

References

1. Alfsen, E.M., Shultz, F.W.: State Spaces of Operator Algebras: Basic Theory, Orientations, and C^*-products. Mathematics: Theory and Applications. Birkhäuser, Boston (2001)
2. Eckhardt, U.: Theorems on the dimension of convex sets. Linear Algebra Appl. **12**(1), 63–76 (1975)
3. Grünbaum, B.: The dimension of intersections of convex sets. Pac. J. Math. **12**, 197–202 (1962)
4. Hill, R.D., Waters, S.R.: On the cone of positive semidefinite matrices. Linear Algebra Appl. **90**, 81–88 (1987)
5. Hiriart-Urruty, J.B., Lemaréchal, C.: Abridged version of convex analysis and minimization algorithms I and II. In: Fundamentals of Convex Analysis. Grundlehren Text Editions. Springer, Berlin (2001).
6. Humphreys, J.E.: Reflection Groups and Coxeter Groups. Cambridge University Press, Cambridge (1990)
7. Pataki, G.: On the connection of facially exposed and nice cones. J. Math. Anal. Appl. **400**(1), 211–221 (2013)
8. Rockafellar, R.T.: Convex Analysis. Princeton Mathematical Series, No. 28. Princeton University Press, Princeton (1970)
9. Sang, T.: Limit roots for some infinite Coxeter groups. Master's Thesis, Department of Mathematics and Statistics, University of Melbourne (2014)

10. Tunçel, L.: Polyhedral and Semidefinite Programming Methods in Combinatorial Optimization. Fields Institute Monographs, vol. 27. American Mathematical Society/Fields Institute for Research in Mathematical Sciences, Providence/Toronto (2010)
11. Vass, J.: On the Exact Convex Hull of IFS Fractals. arXiv:1502.03788v2 (2016)
12. Ziegler, G.M.: Lectures on Polytopes. Graduate Texts in Mathematics, vol. 152. Springer, New York (1995)

Chebyshev Multivariate Polynomial Approximation: Alternance Interpretation

Nadezda Sukhorukova, Julien Ugon, and David Yost

Abstract In this paper, we derive optimality conditions for Chebyshev approximation of multivariate functions. The theory of Chebyshev (uniform) approximation for univariate functions was developed in the late nineteenth and twentieth century. The optimality conditions are based on the notion of alternance (maximal deviation points with alternating deviation signs). It is not clear, however, how to extend the notion of alternance to the case of multivariate functions. There have been several attempts to extend the theory of Chebyshev approximation to the case of multivariate functions. We propose an alternative approach, which is based on the notion of convexity and nonsmooth analysis.

1 Introduction

The theory of Chebyshev approximation for univariate functions was developed in the late nineteenth (Chebyshev) and twentieth century (just to name a few [3, 6, 9]). In most cases, the authors were working on polynomial and polynomial spline approximations, however, other types of functions (for example, trigonometric polynomials) have also been used. In most cases, the optimality conditions are based on the notion of alternance: maximal deviation points with alternating deviation signs.

There have been several attempts to extend this theory to the case of multivariate functions. One of them is [7]. In this paper the author underlines the fact that the main difficulty is to extend the notion of alternance to the case of more than

N. Sukhorukova (✉)
Swinburne University of Technology, John St, Hawthorn, VIC 3122, Australia

Federation University Australia, PO Box 663, Ballarat, VIC 3353, Australia
e-mail: nsukhorukova@swin.edu.au

J. Ugon • D. Yost
Federation University Australia, PO Box 663, Ballarat, VIC 3353, Australia
e-mail: j.ugon@federation.edu.au; d.yost@federation.edu.au

© Springer International Publishing AG, part of Springer Nature 2018
D.R. Wood et al. (eds.), *2016 MATRIX Annals*, MATRIX Book Series 1,
https://doi.org/10.1007/978-3-319-72299-3_8

177

one variable, since $\mathbb{R}^d$, unlike $\mathbb{R}$, is not totally ordered. There have been also several studies in a slightly different direction, namely, in the area of multivariate interpolation [2, 4], where triangulation based approaches were used to extend the notion of polynomial splines to the case of multivariate functions.

The objective functions appearing in Chebyshev approximation optimisation problems are nonsmooth (minimisation of the maximal absolute deviation). Therefore, it is natural to use nonsmooth optimisation techniques to tackle this problem. In this paper we propose an approach, which is based on the notion of subdifferential of convex functions [8]. Subdifferentials can be considered as a generalisation of the notion of gradients for convex nondifferentiable functions.

The paper is organised as follows. In Sect. 2 we present the most relevant results from the theory of convex and nonsmooth analysis, that are essential to obtain our optimality conditions. Then, in the same section, we investigate the extremum properties of the objective function, appearing in Chebyshev approximation problems, from the points of view of convexity and nonsmooth analysis. In Sect. 2 we obtain our main results. Finally, in Sect. 3 we draw our conclusions and underline further research directions.

2 Optimality Conditions

2.1 Convexity of the Objective

Let us now define the objective function. Suppose that $Q \in \mathbb{R}^d$ is a compact set and a continuous function $f : Q \to \mathbb{R}$ is to be approximated on Q by a function

$$L(\mathbf{A}, \mathbf{x}) = a_0 + \sum_{i=1}^{n} a_i g_i(\mathbf{x}), \tag{1}$$

where g_i are the basis functions and the multipliers $\mathbf{A} = (a_1, \ldots, a_n)$ are the corresponding coefficients. At a point $\mathbf{x}$ the deviation between the function f and the approximation is defined as follows

$$p(\mathbf{A}, \mathbf{x}) = |f(\mathbf{x}) - L(\mathbf{A}, \mathbf{x})|. \tag{2}$$

Then we can define the uniform approximation error over the set Q by

$$\Psi(\mathbf{A}) = \|f - a_0 - \sum_{i=1}^{n} a_i g_i\|_\infty, \tag{3}$$

where

$$\left\| f - a_0 - \sum_{i=1}^{n} a_i g_i \right\|_\infty = \max_{\mathbf{x} \in Q} \max\{ f(\mathbf{x}) - a_0 - \sum_{i=1}^{n} a_i g_i(\mathbf{x}), a_0 + \sum_{i=1}^{n} a_i g_i(\mathbf{x}) - f(\mathbf{x}) \}.$$

The approximation problem can be formulated as follows.

$$\text{minimise } \Psi(\mathbf{A}) \text{ subject to } \mathbf{A} \in \mathbb{R}^{n+1}. \tag{4}$$

Since the function $L(\mathbf{A}, \mathbf{x})$ is linear in $\mathbf{A}$, the approximation error function $\Psi(\mathbf{A})$, as the supremum of affine functions, is convex. Convex analysis tools [8] can be applied to study this function.

Define by $E^+(\mathbf{A})$ and $E^-(\mathbf{A})$ the points of maximal positive and negative deviation:

$$E^+(\mathbf{A}) = \left\{ \mathbf{x} \in Q : L(\mathbf{A}, \mathbf{x}) - f(\mathbf{x}) = \max_{\mathbf{y} \in Q} p(\mathbf{A}, \mathbf{y}) \right\}$$

$$E^-(\mathbf{A}) = \left\{ \mathbf{x} \in Q : f(\mathbf{x}) - L(\mathbf{A}, \mathbf{x}) = \max_{\mathbf{y} \in Q} p(\mathbf{A}, \mathbf{y}) \right\}$$

and the corresponding sets $G^+(\mathbf{A})$ and $G^-(\mathbf{A})$ as

$$G^+(\mathbf{A}) = \left\{ (1, g_1(\mathbf{x}), \ldots, g_n(\mathbf{x}))^T : \mathbf{x} \in E^+(\mathbf{A}) \right\}$$

$$G^-(\mathbf{A}) = \left\{ -(1, g_1(\mathbf{x}), \ldots, g_n(\mathbf{x}))^T : \mathbf{x} \in E^-(\mathbf{A}) \right\}$$

Then the subdifferential of the approximation error function $\Psi(\mathbf{A})$ at a point $\mathbf{A}$ can be obtained using the active affine functions in the supremum [11, Theorem 2.4.18]:

$$\partial \Psi(\mathbf{A}) = \text{co} \left\{ G^+(\mathbf{A}) \cup G^-(\mathbf{A}) \right\}. \tag{5}$$

2.2 Optimality Conditions: General Case

In the case of univariate polynomial approximation, the optimality conditions are based on the notion of an alternating sequence.

Definition 1 A sequence of maximal deviation points whose deviation signs are alternating is called an alternating sequence or alternance.
The following theorem holds

Theorem 1 (Chebyshev [1]) *A degree n polynomial approximation is optimal if and only if there exist $n + 2$ alternating points.*

In the case of multivariate approximation there is no natural order and therefore the notion of alternance, as a base for optimality verification, has to be modified. The following theorem holds.

Theorem 2 *A vector* $\mathbf{A}^*$ *is an optimal solution to problem (4) if and only if the convex hulls of the vectors* $(g_1(\mathbf{x}), \ldots, g_n(\mathbf{x}))^T$, *built over corresponding positive and negative maximal deviation points, intersect, that is*

$$\operatorname{co}\{G^+(\mathbf{A})\} \cap \operatorname{co}\{-G^-(\mathbf{A})\} \neq \emptyset. \tag{6}$$

Proof The vector $\mathbf{A}^*$ is an optimal solution to the convex problem (4) if and only if

$$\mathbf{0}_{n+1} \in \partial \Psi(\mathbf{A}^*),$$

where Ψ is defined in (3). Note that due to Carathéodory's theorem, $\mathbf{0}$ can be represented as a convex combination of a finite number of points (one more than the dimension of the corresponding space). Since the dimension of the corresponding space is $n + 1$, it can be done using at most $n + 2$ points.

Assume that in this collection of $n + 2$ points k points $(h_i, \ i = 1, \ldots, k)$ are from $G^+(\mathbf{A})$ and $n + 2 - k$ $(h_i, \ i = k + 1, \ldots, n + 2)$ points are from $G^-(\mathbf{A})$. Note that $0 < k < n + 2$, since the first coordinate is either 1 or -1 and therefore $\mathbf{0}_{n+1}$ can only be formed by using both sets ($G^+(\mathbf{A})$ and $-G^-(\mathbf{A})$). Then

$$\mathbf{0}_{n+1} = \sum_{i=1}^{n+2} \alpha_i h_i, \ 0 \le \alpha \le 1.$$

Let $0 < \gamma = \sum_{i=1}^{k} \alpha_i$, then

$$\mathbf{0}_{n+1} = \sum_{i=1}^{n+2} \alpha_i h_i = \gamma \sum_{i=1}^{k} \frac{\alpha_i}{\gamma} h_i + (1 - \gamma) \sum_{i=k+1}^{n+2} \frac{\alpha_i}{1 - \gamma} h_i = \gamma h^+ + (1 - \gamma) h^-,$$

where $h^+ \in G^+$ and $h^- \in -G^-$. Therefore, it is enough to demonstrate that $\mathbf{0}_{n+1}$ is a convex combination of two vectors, one from $G^+(\mathbf{A})$ and one from $-G^-$.

By the formulation of the subdifferential of Ψ given by (5), there exists a nonnegative number $\gamma \le 1$ and two vectors

$$g^+ \in \operatorname{co}\left\{ \begin{pmatrix} 1 \\ g_1(\mathbf{x}) \\ g_2(\mathbf{x}) \\ \vdots \\ g_n(\mathbf{x}) \end{pmatrix} : \mathbf{x} \in E^+(\mathbf{A}) \right\}, \text{ and } g^- \in \operatorname{co}\left\{ \begin{pmatrix} 1 \\ g_1(\mathbf{x}) \\ g_2(\mathbf{x}) \\ \vdots \\ g_n(\mathbf{x}) \end{pmatrix} : \mathbf{x} \in E^-(\mathbf{A}) \right\}$$

such that $\mathbf{0} = \gamma g^+ - (1 - \gamma)g^-$. Noticing that the first coordinates $g_1^+ = g_1^- = 1$, we see that $\gamma = \frac{1}{2}$. This means that $g^+ - g^- = 0$. This happens if and only if

$$\text{co} \left\{ \begin{pmatrix} 1 \\ g_1(\mathbf{x}) \\ g_2(\mathbf{x}) \\ \vdots \\ g_n(\mathbf{x}) \end{pmatrix} : \mathbf{x} \in E^+(\mathbf{A}) \right\} \cap \text{co} \left\{ \begin{pmatrix} 1 \\ g_1(\mathbf{x}) \\ g_2(\mathbf{x}) \\ \vdots \\ g_n(\mathbf{x}) \end{pmatrix} : \mathbf{x} \in E^-(\mathbf{A}) \right\} \neq \emptyset. \tag{7}$$

As noted before, the first coordinates of all these vectors are the same, and therefore the theorem is true, since if γ exceeds one, the solution where all the components are divided by γ can be taken as the corresponding coefficients in the convex combination.

In the rest of this section we show how Theorem 2 can be used to formulate necessary and sufficient optimality conditions for the case of multivariate polynomial approximation. We also demonstrate how the notion of alternance can be extended to multidimensional cases. Equivalent results have been obtained in [7], however, the conditions of Theorem 2 are easier to verify. Rice's optimality verification is based on separation of positive and negative maximal deviation points by a polynomial of the same degree as the degree of the approximation m: there exists no polynomial of degree m that separates positive and negative maximal deviation points, but the removal of any maximal deviation point results in the ability to separate the remaining points by a polynomial of degree m.

2.3 Optimality Conditions for Multivariate Linear Functions

In the case of multivariate linear functions (that is $g_i(\mathbf{x}) = x_i$, $i = 1, \ldots, n$) Theorem 2 can be formulated as follows.

Theorem 3 *A multivariate linear approximation is optimal if and only if the convex hull of the maximal deviation points with positive deviation and convex hull of the maximal deviation points with negative deviation have common points.*

Theorem 3 can be considered as an alternative formulation to the necessary and sufficient optimality conditions that are based on the notion of alternance. Clearly, Theorem 3 can be used in univariate cases, since the location of the alternance points ensures the common points for the corresponding convex hulls, constructed over the maximal deviation points with positive and negative deviations respectively.

Note that in general $d \leq n$. Non-linear multivariate polynomial approximation is one of our future research priorities.

3 Conclusions and Further Research Directions

In this paper we obtained necessary and sufficient optimality conditions for best polynomial Chebyshev approximation (characterisation theorem). The main obstacle was to extend the notion of alternance to the case of multivariate polynomials. This has been done using nonsmooth calculus.

For the future we are planning to proceed in the following directions.

1. Find a necessary and sufficient optimality condition that is easy to verify in practice (currently, we only have a necessary condition, but not a sufficient one).
2. Extend these results to the case of variable polynomial degrees for each dimension.
3. Develop similar optimality conditions for multivariate trigonometric polynomials and polynomial spline Chebyshev approximations.
4. Develop an approximation algorithm to construct best multivariate approximations (similar to the famous Remez algorithm, developed for univariate polynomials [5] and extended to polynomial splines [3, 10])

Acknowledgements This paper was inspired by the discussions during a recent MATRIX program "Approximation and Optimisation" that took place in July 2016. We are thankful to the MATRIX organisers, support team and participants for a terrific research atmosphere and productive discussions.

References

1. Chebyshev, P.L.: The theory of mechanisms known as parallelograms. Selected Works, pp. 611–648. Publishing House of the USSR Academy of Sciences, Moscow (In Russian) (1955)
2. Davydov, O.V., Nurnberger, G., Zeilfelder, F.: Approximation order of bivariate spline interpolation for arbitrary smoothness. J. Comput. Appl. Math. **90**(2), 117–134 (1998)
3. Nürnberger, G.: Approximation by Spline Functions. Springer, Berlin (1989)
4. Nurnberger, G., Zeilfelder, F.: Interpolation by spline spaces on classes of triangulations. J. Comput. Appl. Math. **119**(1-2), 347–376 (2000)
5. Remez, E.Y.: General computational methods of Chebyshev approximation. At. Energy Transl. **4491** (1957)
6. Rice, J.: Characterization of Chebyshev approximation by splines. SIAM J. Numer. Anal. **4**(4), 557–567 (1967)
7. Rice, J.: Tchebycheff approximation in several variables. Trans. Am. Math. Soc. **109**, 444–466 (1963)
8. Rockafellar, R.T.: Convex Analysis. Princeton University Press, Princeton (1970)
9. Schumaker, L.: Uniform approximation by Chebyshev spline functions. II: free knots. SIAM J. Numer. Anal. **5**, 647–656 (1968)
10. Sukhorukova, N.: Vallée Poussin theorem and Remez algorithm in the case of generalised degree polynomial spline approximation. Pac. J. Optim. **6**(1), 103–114 (2010)
11. Zalinescu, C.: Convex Analysis in General Vector Spaces. World Scientific, Singapore (2002)

Spectral Triples on O_N

Magnus Goffeng and Bram Mesland

Abstract We give a construction of an odd spectral triple on the Cuntz algebra O_N, whose K-homology class generates the odd K-homology group $K^1(O_N)$. Using a metric measure space structure on the Cuntz-Renault groupoid, we introduce a singular integral operator which is the formal analogue of the logarithm of the Laplacian on a Riemannian manifold. Assembling this operator with the infinitesimal generator of the gauge action on O_N yields a θ-summable spectral triple whose phase is finitely summable. The relation to previous constructions of Fredholm modules and spectral triples on O_N is discussed.

1 Introduction

We give a geometrically inspired construction of spectral triples on the Cuntz algebra O_N with non-trivial K-homological content. One reason such spectral triples have been elusive is Connes' construction of traces from finitely summable spectral triples [1]. Purely infinite C^*-algebras such as O_N are traceless and should thus be viewed as infinite dimensional objects, at best carrying θ-summable spectral triples. Another difficulty is presented by the fact that the K-homology of O_N is torsion, so the index pairing cannot be used to detect K-homology classes.

In the literature, several approaches to noncommutative geometry on Cuntz-Krieger algebras have been explored. The crossed product C^*-algebra associated to the action of a free group on its Gromov boundary gives rise to a Cuntz-Krieger algebra, and in [5] that geometric picture is used to establish the existence of θ-summable spectral triples on such C^*-algebras. On the other hand, twisted

M. Goffeng (✉)
Department of Mathematical Sciences, Chalmers University of Technology and the University of Gothenburg, SE-412 96 Gothenburg, Sweden
e-mail: goffeng@chalmers.se

B. Mesland
Institut für Analysis, Leibniz Universität Hannover, Welfengarten 1, 30167 Hannover, Germany
e-mail: mesland@math.uni-hannover.de

© Springer International Publishing AG, part of Springer Nature 2018   183
D.R. Wood et al. (eds.), *2016 MATRIX Annals*, MATRIX Book Series 1,
https://doi.org/10.1007/978-3-319-72299-3_9

noncommutative geometries [3] circumvent the obstruction to finite summability whereas semifinite noncommutative geometries [19] allow for the extraction of index-theoretic invariants.

Recent years have seen explicit constructions of spectral triples on Cuntz-Krieger algebras [12] and more generally on Cuntz-Pimsner algebras [13], originating in the dynamics of subshifts of finite type. Their classes in K-homology were computed in [12] using Poincaré duality and extension theory (see [14, 15]), thus bypassing the difficulties discussed above. These spectral triples have the remarkable feature that they are θ-summable, but their bounded transforms $\chi(D)$, using a suitably chosen function $\chi \in C_b(\mathbb{R})$ such that $\lim_{t\to\pm\infty} \chi(t) = \pm 1$, are finitely summable. Providing a geometric context and understanding the distinct dimensional behaviours of bounded and unbounded Fredholm modules over O_N is the main problem motivating this paper.

Using the metric and Patterson-Sullivan measure on the full N-shift, we equip the Cuntz-Renault groupoid with the structure of a metric measure space. Then we consider a singular integral kernel formally similar to that of the logarithm of the Laplacian on a closed Riemannian manifold. We explicitly relate the associated integral operator to the depth-kore operator from [13], yielding a geometric construction of a K-homologically non-trivial noncommutative geometry on O_N.

2   Statement of Results on O_N

Before stating our results, we recall several notions from noncommutative geometry. The reader familiar with summability properties in noncommutative geometry and the groupoid model of O_N can proceed to page 187 for the main results.

Let A be a unital C^*-algebra. A spectral triple is a triple $(A, \mathcal{H}, D)$ where A acts unitally on the Hilbert space $\mathcal{H}$ and D is a self-adjoint operator with compact resolvent on $\mathcal{H}$ such that the $*$-subalgebra

$$\mathrm{Lip}_D(A) := \{a \in A : a\mathrm{Dom}(D) \subseteq \mathrm{Dom}(D) \text{ and } [D, a] \text{ is bounded}\} \subseteq A \,,$$

is norm dense in A. A spectral triple is sometimes called an unbounded Fredholm module. A bounded Fredholm module is a triple $(A, \mathcal{H}, F)$ as above safe the fact that F is a bounded operator assumed to satisfy that $F^2 - 1, F - F^*, [F, a] \in \mathbb{K}(\mathcal{H})$ for any $a \in A$.

Dimensional properties of (un)bounded Fredholm modules are described in terms of operator ideals. For a compact operator T on a Hilbert space $\mathcal{H}$, we denote by $\mu_k(T)$ its sequence of singular values. Given $p \in (0, \infty)$, let $\mathscr{L}^p(\mathcal{H})$, denote the pth Schatten ideal and $\mathrm{Li}^{1/p}(\mathcal{H}) \subset \mathbb{K}(\mathcal{H})$ the weak principal ideal defined by

$$\mathrm{Li}^{1/p}(\mathcal{H}) := \{T \in \mathbb{K}(\mathcal{H}) : \mu_k(T) = O((\log k)^{-1/p})\}.$$

An unbounded Fredholm module $(A, \mathcal{H}, D)$ is said to be p-summable if $(D \pm i)^{-1} \in \mathcal{L}^p(\mathcal{H})$, weakly θ-summable if $(D \pm i)^{-1} \in \mathrm{Li}^{1/2}(\mathcal{H})$ and θ-summable if $(D \pm i)^{-1}$ belongs to the closure of the finite rank operators in $\mathrm{Li}^{1/2}(\mathcal{H})$. Note that θ-summability is equivalent to requiring that $e^{-tD^2} \in \mathcal{L}^1(\mathcal{H})$ for all $t > 0$. A bounded Fredholm module $(A, \mathcal{H}, F)$ is said to be p-summable if

$$F^2 - 1, F - F^* \in \mathcal{L}^{p/2}(\mathcal{H}), \quad \text{and} \quad [F, a] \in \mathcal{L}^p(\mathcal{H}),$$

and θ-summable if

$$F^2 - 1, F - F^* \in \mathrm{Li}(\mathcal{H}), \quad \text{and} \quad [F, a] \in \mathrm{Li}^{1/2}(\mathcal{H}),$$

for all a in a dense subalgebra of A.

We emphasize the difference between the two definitions. Summability of an unbounded Fredholm module is a property of the operator D, whereas summability of a bounded Fredholm module is a property of the operator F and of its commutators with the algebra A. The two notions are related as follows. If $(A, \mathcal{H}, D)$ is a p-summable (resp. θ-summable) unbounded Fredholm module, then $(A, \mathcal{H}, \chi(D))$ is a p-summable (resp. θ-summable) bounded Fredholm module if $\chi \in C_b(\mathbb{R})$ is a function satisfying $\chi^2 = 1 + O(|x|^{-2})$ as $|x| \to \infty$. Conversely, a θ-summable bounded Fredholm module can be lifted to a θ-summable unbounded Fredholm module, see [2, Chapter IV.8, Theorem 4]. This result fails for finite summability, as is shown in particular by the examples in this paper.

Any K-homology class on a Cuntz-Krieger algebra is represented by a finitely summable bounded Fredholm module [12]. In general, Cuntz-Krieger algebras admit no finitely summable spectral triples, as discussed above. This phenomenon is widespread and, for instance, occurs for boundary crossed product algebras of hyperbolic groups [7]. The action of a free group on its Gromov boundary falls into the class of examples considered in both [7] and [12]. To our knowledge, obstructions to finite summability at the bounded level have not been studied. At present, the example [12, Lemma 6, page 95] of a K-homology class not admitting finitely summable bounded representatives is the only one known to the authors.

Before stating our main results, we recall some facts about O_N that we review in more detail in Sect. 3. For $N > 1$, the Cuntz algebra O_N [6] is defined as the universal C^*-algebra generated by N isometries with orthogonal ranges. As the C^*-algebra O_N is simple, it can be constructed in any of its Hilbert space realizations. That is, for any operators $S_1, \ldots, S_N$ such that $S_j^* S_k = \delta_{jk}$ and $1 = \sum_{j=1}^N S_j S_j^*$, O_N is canonically isomorphic to the C^*-algebra generated by $S_1, \ldots, S_N$.

An important realization of O_N is as the groupoid C^*-algebra of the Cuntz-Renault groupoid $\mathcal{G}_N$ introduced in [20, Section III.2]. The unit space of $\mathcal{G}_N$ is the full one-sided sequence space $\Omega_N := \{1, \ldots, N\}^{\mathbb{N}}$. We equip Ω_N with the product topology in which it is compact and totally disconnected. Elements $x \in \Omega_N$ are written $x = x_1 x_2 \cdots$ where $x_j \in \{1, \ldots, N\}$. The shift $\sigma : \Omega_N \to \Omega_N$ is defined by $\sigma(x_1 x_2 x_3 \cdots) = x_2 x_3 \cdots$ and is a surjective local homeomorphism. For a finite word

$\mu = \mu_1 \mu_2 \cdots \mu_k \in \{1, \ldots, N\}^k$ we define the cylinder set

$$C_\mu := \{x \in \Omega_N : x = \mu x' \text{ for some } x' \in \Omega_N\}.$$

We call $|\mu| := k$ the length of μ. As a set, the Cuntz-Renault groupoid is given by

$$\mathcal{G}_N := \{(x, n, y) \in \Omega_N \times \mathbb{Z} \times \Omega_N : \exists k \ \sigma^{n+k}(x) = \sigma^k(y)\} \rightrightarrows \Omega_N, \tag{1}$$

with domain map $d_{\mathcal{G}} : \mathcal{G} \to \Omega_N$, range map $r_{\mathcal{G}} : \mathcal{G} \to \Omega_N$ and product $\cdot$ defined by

$$d_{\mathcal{G}}(x, n, y) := y, \quad r_{\mathcal{G}}(x, n, y) := x, \quad (x, n, y) \cdot (y, m, z) = (x, n + m, z).$$

The space $\mathcal{G}_N$ admits an extended metric $\rho_{\mathcal{G}}$ defined below in Definition 2. The étale topology described in [21] coincides with the metric topology on $\mathcal{G}_N$ induced by $\rho_{\mathcal{G}}$ (see Sect. 3, Proposition 1).There is an isomorphism $O_N \cong C^*(\mathcal{G}_N)$ (see [20, 21]) and an expectation $\Phi : C^*(\mathcal{G}_N) \to C(\Omega_N)$ induced by the clopen inclusion

$$\Omega_N \subset \mathcal{G}_N, \quad x \mapsto (x, 0, x). \tag{2}$$

The algebra O_N admits a unique KMS-state ϕ (see Sect. 3.2), and we write $L^2(O_N) := L^2(O_N, \phi)$ for its GNS-representation (see below in Sect. 3.2). Under the isomorphism $O_N \cong C^*(\mathcal{G}_N)$ we have $L^2(O_N) = L^2(\mathcal{G}_N, m_{\mathcal{G}})$ for the measure $m_{\mathcal{G}} := d_{\mathcal{G}}^* m_\Omega$ induced by the Patterson-Sullivan measure m_Ω on Ω_N, characterized by $m_\Omega(C_\mu) := N^{-|\mu|}$. We often write $g = (x, n, y)$ for an element of $\mathcal{G}_N$. Note that the Hausdorff dimension of Ω_N, and hence of $\mathcal{G}_N$, equals $\log N$.

Definition 1 We define the densely defined operators c, T and $P_{\mathcal{F}}$ on $L^2(O_N)$ as follows.

1. Define c_0 by $\text{Dom}(c_0) = C_c(\mathcal{G}_N)$ and $c_0 f(x, n, y) := nf(x, n, y)$ and let c denote the closure of c_0.
2. Define T_0 by letting $\text{Dom}(T_0)$ be the compactly supported locally constant functions and

$$T_0 f(g) := \frac{1}{(1 - N^{-1})} \int_{\mathcal{G}_N} \frac{f(g) - f(h)}{\rho_{\mathcal{G}}(g, h)^{\log(N)}} \, dm_{\mathcal{G}}(h),$$

and let T denote the closure of T_0. The extended metric $\rho_{\mathcal{G}}$ is defined below in Definition 2.
3. Define the set

$$X_{\mathcal{F}} := \{(x, n, y) \in \Omega_N \times \mathbb{N} \times \Omega_N : \exists \mu \in \{1, \ldots, N\}^n \text{ s.t. } x \in C_\mu \text{ and } \sigma^n(x) = y\}. \tag{3}$$

Let $P_{\mathcal{F}}$ denote the integral operator on $L^2(\mathcal{G}_N)$ with integral kernel $\chi_{X_{\mathcal{F}}}$ (the characteristic function of $X_{\mathcal{F}}$).

There is an isomorphism $K^1(O_N) \cong \mathbb{Z}/(N-1)\mathbb{Z}$ defined from Poincaré duality for Cuntz-Krieger algebras [14] and the isomorphism $K_0(O_N) \cong \mathbb{Z}/(N-1)\mathbb{Z}$. We denote by $\widehat{[1]} \in K^1(O_N)$ the class Poincaré dual to $[1] \in K_0(O_N)$ and sometimes refer to this class as *the generator* of $K^1(O_N)$. The generator of $K^1(O_N)$ is represented by the extension considered in [8]. In the sequel we will use the operator T from Definition 1 to construct spectral triples representing the K-homology class $\widehat{[1]}$.

In the statement of our main result we will make use of the so called dispersion operator $B : L^2(O_N) \to L^2(O_N)$ which is a bounded operator defined below in Lemma 1 (see page 194). The dispersion operator measures how non-diagonal the operator T is in a particular ON-basis of $L^2(O_N)$. We also make use of a certain projection Q defined just before Theorem 4.

Theorem 1 *The operators c, T and $P_{\mathscr{F}}$ from Definition 1 are well defined self-adjoint operators. In fact $P_{\mathscr{F}}$ is an orthogonal projection, T is positive and $D := (2P_{\mathscr{F}} - 1)|c| - T$ is a self-adjoint operator with compact resolvent. Moreover,*

1. *$(O_N, L^2(O_N), D)$ is a spectral triple whose class coincides with $\widehat{[1]} \in K^1(O_N)$ and e^{-tD^2} is of trace class for all $t > 0$, i.e. D is θ-summable.*
2. *Up to finite rank operators, $P_{\mathscr{F}} = \chi_{[0,\infty)}(D)$ and for any $p > 0$,*

$$(O_N, L^2(O_N), 2P_{\mathscr{F}} - 1)$$

is a p-summable Fredholm module whose class is $\widehat{[1]} \in K^1(O_N)$.
3. *The operator $\tilde{D} := D - B + (N-1)^{-1}Q$ also defines a spectral triple on O_N, where B is the dispersion operator (see Lemma 1) and Q is a projection (see before Theorem 4). For any extended limit $\omega \in L^\infty[0,1]^*$ at 0 there is a probability measure $\tilde{m}_\omega$ on Ω_N such that*

$$\tilde{\phi}_\omega(a) := \omega \left(\frac{\mathrm{Tr}(ae^{-t\tilde{D}^2})}{\mathrm{Tr}(e^{-t\tilde{D}^2})} \right), \quad a \in O_N,$$

is computed from $\tilde{\phi}_\omega(a) = \int_{\Omega_N} \Phi(a) d\tilde{m}_\omega$.

Remark 1 The importance of part 3 of Theorem 1 is in the context of the states constructed from θ-summable spectral triples in [10]. The assumption [10, Assumption 5.4] requires the associated states to be tracial. This condition clearly fails in the purely infinite case.

A key ingredient in the proof of the theorem is the notion of the depth-kore operator from [13]. The depth-kore operator κ is a self-adjoint operator on $L^2(O_N)$ which together with c facilitates a decomposition $L^2(O_N) = \bigoplus_{n,k} \mathscr{H}_{n,k}$ into finite-dimensional subspaces with an explicit ON-basis. As we will see below in Proposition 5 of Sect. 3.2, $P_{\mathscr{F}}$ is the orthogonal projection onto the free Fock space $\mathscr{F} := \bigoplus_{n=0}^\infty \mathscr{H}_{n,0} \cong \ell^2(\mathscr{V}_N)$ where $\mathscr{V}_N = \cup_{k=0}^\infty \{1, \ldots, N\}^k$.

The structure of the paper is as follows. In Sect. 3 we describe the geometry of the Cuntz-Renault groupoid $\mathscr{G}_N$ and the GNS representation of the KMS state of the

Cuntz algebra in terms of the Cuntz-Renault groupoid. We compare the κ-function on $\mathcal{G}_N$ (cf. [12, Section 5]) to the κ-operator on its L^2-space (cf. [13, Lemma 2.13]) in Sect. 4. The integral operator T is computed in Sect. 5 and we assemble these ingredients to spectral triples in Sect. 6. The proof of Theorem 1 is found in Sects. 6 and 7.

3   Metric Measure Theory on O_N

In this section we will set the scene for the paper and describe the relevant objects. Most of this material reviews previously published results. The context we present, which to our knowledge is novel, sheds a new light on them.

3.1   The Groupoid $\mathcal{G}_N$ as a Metric Measure Space

The groupoid $\mathcal{G}_N$ was defined as a set with algebraic structure in Eq. (1) and we now describe its topology in more detail. Define the functions $\kappa_{\mathcal{G}} : \mathcal{G}_N \to \mathbb{N}$ and $c : \mathcal{G}_N \to \mathbb{Z}$ by

$$\kappa_{\mathcal{G}} : (x, n, y) \mapsto \min\{k \geq \max\{0, -n\} : \sigma^{n+k}(x) = \sigma^k(y)\}, \quad c : (x, n, y) \mapsto n.$$

For $g \in \mathcal{G}_N$ and composable $g_1, g_2 \in \mathcal{G}_N$ it holds that

$$c(g_1 \cdot g_2) = c(g_1) + c(g_2), \quad \kappa_{\mathcal{G}}(g_1 \cdot g_2) \leq \kappa(g_1) + \kappa(g_2), \quad c(g) + \kappa_{\mathcal{G}}(g) \geq 0.$$

In summary, c is a cocycle, $\kappa_{\mathcal{G}}$ is submultiplicative and their sum is a positive function.

We equip $\mathcal{G}_N$ with the smallest topology making c, $\kappa_{\mathcal{G}}$, $r_{\mathcal{G}}$ and $d_{\mathcal{G}}$ continuous. It is readily verified that a basis for the topology on $\mathcal{G}_N$ is given by the sets

$$X_{\mu,\nu} := \{(x, |\mu| - |\nu|, y) \in \mathcal{G}_N : x \in C_\mu, \ y \in C_\nu, \ \sigma^{|\mu|}(x) = \sigma^{|\nu|}(y)\}, \quad \text{for } \mu, \nu \in \mathcal{V}_N.$$

The groupoid $\mathcal{G}_N$ is étale in this topology. An étale groupoid over a totally disconnected space is again totally disconnected, so the space of compactly supported locally constant functions is dense in $C_c(\mathcal{G}_N)$. For $\mu, \nu \in \mathcal{V}_N$ we use the notation $S_\mu := S_{\mu_1} \cdots S_{\mu_{|\mu|}}, S_\mu^* := (S_\mu)^*$ and $\chi_{\mu,\nu}$ for the characteristic function of $X_{\mu,\nu}$. The following result is proven in [20, 21].

Theorem 2 *The C^*-algebras O_N and $C^*(\mathcal{G}_N)$ are isomorphic via a $*$-homomorphism $O_N \to C^*(\mathcal{G}_N)$ that maps $S_\mu S_\nu^*$ to the compactly supported locally constant function $\chi_{\mu,\nu} \in C_c(\mathcal{G}_N)$.*

Notation 1 For $g = (x, n, y) \in \mathcal{G}_N$ we have $\sigma^{n+\kappa_\mathcal{G}(g)}(x) = \sigma^{\kappa_\mathcal{G}(g)}(y)$. We will use the notation $z(g) := \sigma^{\kappa_\mathcal{G}(g)}(y)$, $\mu_\mathcal{G}(g)$ will denote the word of length $n + \kappa_\mathcal{G}(g)$ such that $x = \mu_\mathcal{G}(g)z(g)$ and $v_\mathcal{G}(g)$ will denote the word of length $\kappa_\mathcal{G}(g)$ such that $y = v_\mathcal{G}(g)z(g)$. In particular, we have

$$g = (\mu_\mathcal{G}(g)z(g), c(g), v_\mathcal{G}(g)z(g)), \quad \forall g \in \mathcal{G}_N.$$

Clearly, $z : \mathcal{G}_N \to \Omega_N$ and $\mu_\mathcal{G}, v_\mathcal{G} : \mathcal{G}_N \to \mathcal{V}_N$ are continuous. When there is no risk of confusion with fixed finite words, we write simply $\mu(g)$ and $v(g)$. We also write $y(g) := y$.

The compact space Ω_N is metrized by the metric ρ_Ω defined by

$$\rho_\Omega(x_1 x_2 \cdots, y_1 y_2 \cdots) := \inf\{e^{-l} : x_1 x_2 \cdots x_l = y_1 y_2 \cdots y_l\},$$

with the convention that $\rho_\Omega(i x_2 \cdots, j y_2 \cdots) = 1$ if $i \neq j$.

Definition 2 We define $\rho_\mathcal{G} : \mathcal{G}_N \times \mathcal{G}_N \to [0, \infty]$ by

$$\rho_\mathcal{G}(g_1, g_2) := \begin{cases} \infty, & \text{if } \kappa_\mathcal{G}(g_1) \neq \kappa_\mathcal{G}(g_2) \text{ or } \mu(g_1) \neq \mu(g_2), \\ \rho_\Omega(y(g_1), y(g_2)), & \text{if } \kappa_\mathcal{G}(g_1) = \kappa_\mathcal{G}(g_2) \text{ and } \mu(g_1) = \mu(g_2). \end{cases}$$

For $\mu \in \mathcal{V}_N$ and $k \in \mathbb{N}$, we define the set

$$\mathscr{C}_{\mu,k} := \{g \in \mathcal{G}_N : \mu_\mathcal{G}(g) = \mu, \kappa_\mathcal{G}(g) = k\}. \tag{4}$$

The set $\mathscr{C}_{\mu,k}$ is homeomorphic to a clopen subset of Ω_N via the domain mapping $d_\mathcal{G}$. We can clearly partition

$$\mathcal{G}_N = \dot{\bigcup}_{\mu,k}\mathscr{C}_{\mu,k}.$$

Moreover, for a fixed $g_1 \in \mathcal{G}_N$ we have

$$\{g_2 \in \mathcal{G}_N : \rho_\mathcal{G}(g_1, g_2) < \infty\} = C_{\mu(g_1),\kappa(g_1)}.$$

Proposition 1 *The function $\rho_\mathcal{G}$ is an extended metric on $\mathcal{G}_N$ and the topology induced by $\rho_\mathcal{G}$ coincides with the étale topology on $\mathcal{G}_N$. The functions c, $\kappa_\mathcal{G}$, $r_\mathcal{G}$ and $d_\mathcal{G}$ as well as any compactly supported locally constant function are uniformly Lipschitz continuous with respect to $\rho_\mathcal{G}$.*

Proof We start by giving the argument for why $\rho_\mathcal{G}$ is an extended metric. If $\rho_\mathcal{G}(g_1, g_2) = 0$ then $y(g_1) = y(g_2)$, $\kappa_\mathcal{G}(g_1) = \kappa_\mathcal{G}(g_2)$ and $\mu_\mathcal{G}(g_1) = \mu_\mathcal{G}(g_2)$ so $c(g_1) = c(g_2)$ and we conclude that $g_1 = g_2$. The function $\rho_\mathcal{G}$ is clearly non-negative and symmetric. The triangle inequality follows from the fact that given $\mu \in \mathcal{V}_N$ and $k \in \mathbb{N}$, the set $\mathscr{C}_{\mu,k}$ is bi-Lipschitz homeomorphic to a clopen subset of Ω_N via the domain mapping $d_{\mathscr{G}}$. The remainder of the proposition are direct consequences of the construction of the extended metric.

Remark 2 We note that $r_{\mathscr{G}}$ and $d_{\mathscr{G}}$ are locally bi-Lipschitz homeomorphisms between $\mathscr{G}_N$ and Ω_N so any local metric invariant, e.g. Hausdorff dimension, remains the same for the two spaces.

It is often fruitful to think of Ω_N as the Gromov boundary of the discrete hyperbolic space $\mathscr{V}_N$. Here we think of $\mathscr{V}_N$ as a rooted tree, with root $\emptyset \in \{1,\ldots,N\}^0 = \{\emptyset\}$ and given a directed graph structure by declaring an edge from μ to μj for any $\mu \in \mathscr{V}_N$ and $j \in \{1,\ldots,N\}$. We write $\overline{\mathscr{V}}_N := \mathscr{V}_N \cup \Omega_N$ for the corresponding compactification of $\mathscr{V}_N$; we topologize $\overline{\mathscr{V}}_N$ in such a way that $\mathscr{V}_N \subseteq \overline{\mathscr{V}}_N$ is a discrete subspace and for any $\mu \in \mathscr{V}_N$, the set $\{\nu \in \mathscr{V}_N : \nu = \mu\nu_0$ for some $\nu_0 \in \mathscr{V}_N\} \cup C_\mu$ is open. Let δ_μ denote the Dirac measure at $\mu \in \mathscr{V}_N$ and for $s > \log(N)$ define probability measures on $\overline{\mathscr{V}}_N$ via

$$m_s := \frac{\sum_{\mu \in \mathscr{V}_N} e^{-s|\mu|}\delta_\mu}{\sum_{\mu \in \mathscr{V}_N} e^{-s|\mu|}}.$$

The measures m_s are supported in $\mathscr{V}_N$. The following construction of the Patterson-Sullivan measures on Ω_N is well-known, see for instance [4].

Proposition 2 *The net of measures* $(m_s)_{s>\log(N)}$ *has a* w^*-*limit* m_Ω *as* $s \to \log(N)$. *The measure* m_Ω *is supported on* $\Omega_N \subset \overline{\mathscr{V}}_N$ *and coincides with* $\log(N)$-*dimensional Hausdorff measure. It satisfies* $m_\Omega(C_\nu) = N^{-|\nu|}$ *for any* $\nu \in \mathscr{V}_N$.

3.2 The Representation Associated with the KMS State on O_N

We will now approach O_N from an operator theoretic viewpoint. The cocycle c gives rise to a $U(1)$-action on O_N by $(z \cdot f)(g) = z^{c(g)}f(g)$ for $f \in C_c(\mathscr{G}_N)$. Under the isomorphism of Theorem 2, this action is given on the generators of O_N by $z \cdot S_i = zS_i$. The functional

$$\phi(f) := \int_{\Omega_N} f(x, 0, x)\mathrm{d}m_\Omega,$$

extends to a state on O_N. Indeed, $\phi(S_\mu S_\nu^*) = \delta_{\mu,\nu}N^{-|\nu|}$. The state ϕ is the unique KMS state on O_N (equipped with the action defined above) and its inverse temperature is $\log(N)$, see [18].

Proposition 3 *We consider the measure* $m_{\mathscr{G}} := d_{\mathscr{G}}^* m_\Omega$ *on* $\mathscr{G}_N$. *The isomorphism of Theorem 2 uniquely determines a unitary isomorphism* $L^2(O_N, \phi) \to L^2(\mathscr{G}_N, m_{\mathscr{G}})$ *compatible with the left* O_N-*action.*

The proposition follows using the fact that $1 \in O_N$, which corresponds to $\chi_\Omega \in C_c(\mathscr{G}_N) \subseteq C^*(\mathscr{G}_N)$, satisfies $\phi(f) = \langle 1, f * 1 \rangle_{L^2(\mathscr{G}_N, m_{\mathscr{G}})}$. Motivated by this result, we identify $L^2(O_N, \phi)$ with $L^2(\mathscr{G}_N, m_{\mathscr{G}})$ and write simply $L^2(O_N)$.

Definition 3 For a finite word $\mu \in \mathscr{V}_N$ we write $t(\mu) := \mu_{|\mu|}$ for μ non-empty and $t(\emptyset) = \emptyset$. We define $(e_{\mu,\nu})_{\mu,\nu \in \mathscr{V}_N} \subseteq L^2(O_N)$ by $e_{\emptyset,\emptyset} = \chi_\Omega$ and

$$
e_{\mu,\nu} := \begin{cases} N^{|\nu|/2} S_\mu S_\nu^*, & t(\mu) \neq t(\nu), \\ \\ N^{|\nu|/2} \sqrt{\frac{N}{N-1}} \left(S_\mu S_\nu^* - N^{-1} S_{\underline{\mu}} S_{\underline{\nu}}^* \right), & t(\mu) = t(\nu) \neq \emptyset. \end{cases}
$$

Here we have written $\mu = \underline{\mu} t(\mu)$ and $\nu = \underline{\nu} t(\nu)$.

Proposition 4 (Lemma 2.13 of [13]) *The collection* $(e_{\mu,\nu})_{\mu,\nu \in \mathscr{V}_N} \subseteq L^2(O_N)$ *is an ON-basis.*

Definition 4 Following [13], we define the depth-kore operator κ on $L^2(O_N)$ as the densely defined self-adjoint operator such that

$$
\kappa e_{\mu,\nu} = |\nu| e_{\mu,\nu}.
$$

We define the operator c on $L^2(O_N)$ as the densely defined self-adjoint operator such that

$$
c e_{\mu,\nu} = (|\mu| - |\nu|) e_{\mu,\nu}.
$$

We note that by construction, c commutes with κ on a common core and $c + \kappa$ is positive. We define the N^{n+2k}-dimensional space

$$
\mathscr{H}_{n,k} := \ker(c - n) \cap \ker(\kappa - k) = l.s.\{e_{\mu,\nu} : |\mu| = n + k, |\nu| = k\}. \tag{5}
$$

Proposition 5 *Let* $X_{\mathscr{F}}$ *denote the set from Eq. (3) and* $P_{\mathscr{F}}$ *the integral operator with kernel* $\chi_{X_{\mathscr{F}}}$. *The operator* $P_{\mathscr{F}}$ *is the orthogonal projection onto the Fock space* $\mathscr{F} := \ker \kappa = \bigoplus_{n=0}^{\infty} \mathscr{H}_{n,0}$. *Moreover,* $P_{\mathscr{F}}$ *preserves the domain of* c *and* κ *and commutes with* c *and* κ *on their respective domains.*

Proof The integral kernel of the orthogonal projection onto the Fock space $\mathscr{F}$ is given by the function

$$
\sum_{n=0}^{\infty} \sum_{|\mu|=n} e_{\mu,\emptyset}(g_1) e_{\mu,\emptyset}(g_2) = \sum_{n=0}^{\infty} \sum_{|\mu|=n} \chi_{X_{\mu,\emptyset} \times X_{\mu,\emptyset}}(g_1, g_2) = \chi_{\cup_\mu X_{\mu,\emptyset} \times X_{\mu,\emptyset}}(g_1, g_2).
$$

The proposition follows from the fact that $X_{\mathscr{F}} = \bigcup_\mu \left(X_{\mu,\emptyset} \times X_{\mu,\emptyset} \right)$.

Remark 3 The isometry $v : \ell^2(\mathcal{V}_N) \to L^2(O_N)$, $\delta_\mu \mapsto e_{\mu,\emptyset}$ surjects onto the Fock space $\mathcal{F} = \ker(\kappa)$ (compare [12, Remark 2.2.4]). We also note that there is an isometry $L^2(\Omega_N, m_\Omega) \to L^2(O_N)$ mapping surjectively onto the "anti-Fock space" $\mathcal{F}^{an} := \ker(c + \kappa)$ via $\chi_{C_\mu} \mapsto S_\mu S_\mu^*$. The "anti-Fock space" is often a source of trouble, see [13, Proof of Theorem 2.19]. The basis $(e_{\mu,\mu})_{\mu \in \mathcal{V}_N}$ for $L^2(\Omega_N, m_\Omega) \subseteq L^2(O_N)$ is related to the wavelet basis studied in [9].

In [13] the operators c, κ and $P_{\mathcal{F}}$ were assembled into a spectral triple. We define D_κ as the closure of $(2P_{\mathcal{F}} - 1)|c| - \kappa$. It was proven in [13] that $(O_N, L^2(O_N), D_\kappa)$ is a spectral triple whose class coincides with $[\widehat{1}] \in K^1(O_N)$. The explicit construction is motivated by the K-homological information carried by the projection $P_{\mathcal{F}} \equiv \chi_{[0,\infty)}(D_\kappa)$. The aim of this paper is to give a more geometric construction of a spectral triple on O_N, with the same K-homological content.

4   The κ-Function and κ-Operator on O_N

An important aspect in the noncommutative geometry of the Cuntz algebra O_N is the distinction between the depth-kore function $\kappa_{\mathcal{G}}$ in the groupoid $\mathcal{G}_N$ and the depth-kore operator κ. Both are invariants of Cuntz-Pimsner constructions of O_N: the depth-kore function from O_N as a Cuntz-Pimsner algebra with coefficients $C(\Omega_N)$ and the depth-kore operator from O_N as a Cuntz-Pimsner algebra with coefficients $\mathbb{C}$. These two models are discussed in [13], notably in [13, Section 2.5.3].

Let us go into the details of the other approach using $C(\Omega_N)$ as coefficients. The details can be found in [12, 13]. Let Ξ_N denote the $C(\Omega_N)$-Hilbert C^*-module completion of $C_c(\mathcal{G}_N)$ in the $C(\Omega_N)$-valued inner product $\langle f_1, f_2 \rangle_{C(\Omega)} := \Phi(f_1^* * f_2)$ where Φ denotes the conditional expectation $C_c(\mathcal{G}_N) \to C(\Omega_N)$ onto the unit space obtained from the inclusion (2). Multiplication by the functions c and $\kappa_{\mathcal{G}}$ define self-adjoint regular operators on Ξ_N. The Fock module $\mathcal{F}_\Omega := \ker \kappa_{\mathcal{G}} \subseteq \Xi_N$ is complemented and the adjointable projection

$$P_\Omega = \chi_{\{0\}}(\kappa_{\mathcal{G}}) : \Xi_N \to \Xi_N,$$

satisfies $\mathcal{F}_\Omega = P_\Omega \Xi_N$. Following the recipe above, we define a self-adjoint regular operator D_Ω on Ξ_N as the closure of $(2P_\Omega - 1)|c| - \kappa_{\mathcal{G}}$. Then $P_\Omega = \chi_{[0,\infty)}(D_\Omega)$ and this operator projects onto the $C(\Omega)$-Hilbert C^*-submodule spanned by $\{S_\mu : \mu \in \mathcal{V}_N\}$. The triple (O_N, Ξ_N, D_Ω) defines an unbounded $(O_N, C(\Omega_N))$-Kasparov module. In this instance, (O_N, Ξ_N, D_Ω) can be thought of as a bundle of spectral triples over Ω_N. We recall the following result from [12, Theorem 5.2.3].

Theorem 3 *Let $w \in \Omega_N$ and denote the discrete $d_{\mathcal{G}}$-fiber by $\mathcal{V}_w := d_{\mathcal{G}}^{-1}(w) \subseteq \mathcal{G}_N$. The C^*-algebra O_N acts on $\ell^2(\mathcal{V}_w)$ via the groupoid structure. Define D_w as a self-adjoint operator on $\ell^2(\mathcal{V}_w)$ by*

$$D_w f(x, n, w) := |n|(2P_w - 1)f(x, n, w) - \kappa_{\mathcal{G}}(x, n, w)f(x, n, w),$$

where P_w denotes the projection onto the closed linear span of the orthogonal set $\{\chi_{X_{\mu,\emptyset}}|_{\mathcal{V}_w} : \mu \in \mathcal{V}_N\} \subseteq \ell^2(\mathcal{V}_N)$. Then $(O_N, \ell^2(\mathcal{V}_N), D_w)$ is a θ-summable spectral triple whose phase $D_w|D_w|^{-1}$ defines a finitely summable Fredholm module representing the class $\widehat{[1]} \in K^1(O_N)$.

A key step in proving that (O_N, Ξ_N, D_{Ω}) is a Kasparov module is the study of the submodules $\Xi_{n,k} := \ker(c - n) \cap \ker(\kappa_{\mathscr{G}} - k) \subseteq \Xi_N$. The modules do in this instance carry geometric content as

$$\Xi_{n,k} = C(\mathscr{G}_{n,k}), \quad \text{where} \quad \mathscr{G}_{n,k} = c^{-1}(\{n\}) \cap \kappa_{\mathscr{G}}^{-1}(\{k\}) \subseteq \mathscr{G}_N.$$

The set $\mathscr{G}_{n,k}$ is compact and $C(\mathscr{G}_{n,k})$ is a finitely generated projective $C(\Omega_N)$-module. Later in the paper, we will need to make use of the interaction between the depth-kore function $\kappa_{\mathscr{G}}$ and the depth-kore operator κ.

Definition 5 For two finite words $\mu, \nu \in \mathcal{V}_N$ we write $\mu \wedge \nu$ for the longest word such that $\mu = \mu_0(\mu \wedge \nu)$ and $\nu = \nu_0(\mu \wedge \nu)$ for some words μ_0 and ν_0. We define

$$\kappa_{\mathscr{V}}(\mu, \nu) := |\nu| - |\mu \wedge \nu|.$$

Proposition 6 For $\mu, \nu \in \mathcal{V}_N$, $\kappa_{\mathscr{G}} e_{\mu,\nu} = \kappa_{\mathscr{V}}(\mu, \nu) e_{\mu,\nu}$. In particular, $e_{\mu,\nu} \in \Xi_{n,k}$ if and only if $n = |\mu| - |\nu|$ and $k = \kappa_{\mathscr{V}}(\mu, \nu)$.

The proof consists of a long inspection to verify that $\mathrm{supp}(e_{\mu,\nu}) \cap \mathscr{G}_{n,k} \neq \emptyset$ if and only if $n = |\mu| - |\nu|$ and $k = \kappa_{\mathscr{V}}(\mu, \nu)$. A key point in the proof, putting the two cases in Definition 3 on equal footing, is the identity:

$$\kappa_{\mathscr{V}}(\mu i, \nu j) = \delta_{i,j} \kappa_{\mathscr{V}}(\mu, \nu) + (1 - \delta_{i,j})(|\nu| + 1).$$

Using the fact that $0 \leq \kappa_{\mathscr{V}}(\mu, \nu) \leq |\nu|$ we deduce the next Corollary from Proposition 6.

Corollary 1 As self-adjoint operators on $L^2(O_N)$, the operator $\kappa_{\mathscr{G}}$ is relatively bounded by κ with relative norm bound 1. Moreover, $\kappa_{\mathscr{G}}$ and κ commute on a common core.

5 An Integral Operator on O_N

In this section, we define the singular integral operator that is used to construct spectral triples on O_N. The singular integral operator will at large behave like the depth-kore operator κ.

Definition 6 Define $C_c^{\infty}(\mathscr{G}_N) \subseteq C_c(\mathscr{G}_N)$ as the subspace of all compactly supported locally constant functions. Define the operator $T_0 : C_c^{\infty}(\mathscr{G}_N) \to L^2(\mathscr{G}_N)$ by

$$T_0 f(g_1) := \frac{1}{1 - N^{-1}} \int_{\mathscr{G}_N} \frac{f(g_1) - f(g_2)}{\rho_{\mathscr{G}}(g_1, g_2)^{\log(N)}} \, dm_{\mathscr{G}}(g_2).$$

In the integrand, we apply the convention that $\frac{c}{\infty} = 0$ for any finite number c.

We will compute T_0 in the basis $e_{\mu,\nu}$ of $L^2(O_N)$ and since $C_c^\infty(\mathscr{G}_N) = \text{span}\{e_{\mu,\nu}\}$, the computation shows that T_0 is well-defined and maps $C_c^\infty(\mathscr{G}_N)$ into $L^2(O_N)$. More precisely, the computation shows that T_0 is up to a bounded operator diagonal in the basis $e_{\mu,\nu}$, and as such we can extend T_0 to a densely defined self-adjoint operator

$$T : \text{Dom } T \subset L^2(O_N) \to L^2(O_N),$$

with $C_c^\infty(\mathscr{G}_N) \subseteq \text{Dom } T$.

First we define the so called dispersion operator. For two words $\mu, \nu \in \mathscr{V}_N$ we write $\mu \vee \nu$ for the finite word of maximal length such that $\mu = (\mu \vee \nu)\mu_0$ and $\nu = (\mu \vee \nu)\nu_0$ for some finite words $\mu_0, \nu_0 \in \mathscr{V}_N$.

Lemma 1 *Define the dispersion operator B on $L^2(O_N)$ by the formula*

$$Be_{\mu,\nu} = (1 - N^{-1})^{-1}(1 - \delta_{t(\mu),t(\nu)}) \sum_{m \neq t(\mu)} \sum_{\ell=0}^{|\nu|-1} \sum_{|\gamma|=|\nu|-1, |\gamma\vee\nu|=\ell} N^{\ell-|\nu|} e_{\mu,\gamma m}.$$

Then the operator B is a well defined bounded self-adjoint operator commuting with κ, c and $\kappa_\mathscr{G}$ on a common core. In fact $\kappa|_{(\ker B)^\perp} = \kappa_\mathscr{G}|_{(\ker B)^\perp}$.

Proof The operator B is defined on an ON-basis and it is clear from the expression that B is self-adjoint if B is bounded. The only non-trivial fact to prove is therefore that B is bounded. We compute that

$$\|Be_{\mu,\nu}\|^2 = (1 - N^{-1})^{-2}(1 - \delta_{t(\mu),t(\nu)}) \sum_{m \neq t(\mu)} \sum_{\ell=0}^{|\nu|-1} \sum_{|\gamma|=|\nu|-1, |\gamma\vee\nu|=\ell} N^{2\ell-2|\nu|}$$

$$= (1 - \delta_{t(\mu),t(\nu)})\frac{N}{N-1} \sum_{\ell=0}^{|\nu|-1} N^{\ell-|\nu|} = (1 - \delta_{t(\mu),t(\nu)})\frac{N}{(N-1)^2}(1 - N^{-|\nu|}).$$

It follows that B is bounded. Since $(\ker B)^\perp$ is spanned by basis vectors $e_{\mu,\nu}$ where $t(\mu) \neq t(\nu)$ and $\kappa_\mathscr{V}(\mu,\nu) = |\nu|$ if $t(\mu) \neq t(\nu)$, Proposition 6 implies $\kappa|_{(\ker B)^\perp} = \kappa_\mathscr{G}|_{(\ker B)^\perp}$.

We define Q as the orthogonal projection onto the closed linear span of the set $\{e_{\mu,\nu} : 0 \leq \kappa_\mathscr{V}(\mu,\nu) < |\nu|\}$. That is, $e_{\mu,\nu} \in QL^2(O_N)$ if and only if $t(\mu) = t(\nu) \neq \emptyset$.

Theorem 4 *As an operator on $C_c^\infty(\mathscr{G}_N)$, we have that*

$$T_0 = \kappa - \frac{1}{N}\kappa_\mathscr{G} + (N-1)^{-1}Q - B.$$

In particular, T_0 is a well-defined operator with dense domain and extends to a self-adjoint operator T on $L^2(O_N)$ with discrete spectrum and $T-\kappa$ is relatively bounded by κ and T with relative norm bound $1/N$.

The proof of this theorem will occupy the rest of this section. To prove the theorem, it suffices to prove that it holds when acting on basis elements $e_{\mu,\nu}$: they span the compactly supported locally constant functions. For $g_1 \in \mathcal{G}_N$ and $\ell \in \mathbb{N}$, we introduce the notation

$$X^\ell(g_1) := \{g_2 \in \mathcal{G}_N : \rho_\mathcal{G}(g_1, g_2) = e^{-\ell}\}.$$

Note that $(X^\ell(g_1))_{\ell \in \mathbb{N}}$ is a clopen partition of $\mathscr{C}_{\mu(g_1),\kappa_\mathcal{G}(g_1)}$. In fact, for any $g_1 \in \mathcal{G}_N$, we can make a disjoint clopen partition $\mathcal{G}_N = \{g_2 : \rho_\mathcal{G}(g_1, g_2) = \infty\} \cup (\cup_{\ell=0}^\infty X^\ell(g_1))$. From this discussion, it follows that

$$T = \frac{1}{1-N^{-1}} \sum_{\ell=0}^\infty N^\ell T_\ell, \quad \text{where } T_\ell f(g_1) := \int_{X^\ell(g_1)} (f(g_1) - f(g_2)) dm_\mathcal{G}(g_2).$$

$$(6)$$

To compute T in the ON-basis, we first compute T_ℓ on the characteristic functions $\chi_{X_{\mu,\nu}}$. To ease notation, we write $\chi_{\mu,\nu} = \chi_{X_{\mu,\nu}} \in C_c(\mathcal{G}_N)$ and $\chi_\nu = \chi_{C_\nu} \in C(\Omega_N)$. We have that

$$T_\ell \chi_{\mu,\nu}(g_1) = m_\mathcal{G}(X^\ell(g_1) \cap X_{\mu,\nu})(\chi_{\mu,\nu}(g_1) - 1)$$
$$+ m_\mathcal{G}(X^\ell(g_1) \setminus (X^\ell(g_1) \cap X_{\mu,\nu}))\chi_{\mu,\nu}(g_1).$$

We now proceed to compute the relevant volumes appearing in this expression.

Lemma 2 *For $g_1 \notin X_{\mu,\nu}$,*

$$m_\mathcal{G}(X^\ell(g_1) \cap X_{\mu,\nu}) = N^{-|\nu|} \sum_{|\gamma|=\kappa_\gamma(\mu,\nu)=\kappa_\gamma(\mu,\gamma)} \chi_{\underline{\mu}_{n+|\gamma|},\gamma}(g_1)(\chi_{\underline{\nu}_\ell}(y_1) - \chi_{\underline{\nu}_{\ell+1}}(y_1)),$$

where $\underline{\nu}_\ell$ and $\underline{\nu}_{\ell+1}$ denotes the first ℓ and $\ell+1$ letters of ν and $\underline{\mu}_{n+|\gamma|}$ the first $n+|\gamma| = n + \kappa_\gamma(\mu, \nu)$ letters of μ. We interpret $\underline{\nu}_\ell = \nu$ if $\ell \geq |\nu|$.

Proof Take $g_1 \notin X_{\mu,\nu}$. Firstly, suppose that $g_1 \in \cup_{|\gamma|=\kappa_\gamma(\mu,\nu)=\kappa_\gamma(\mu,\gamma)} X_{\underline{\mu}_{n+|\gamma|},\gamma}$ and that $y_1 \in C_{\underline{\nu}_\ell} \setminus C_{\underline{\nu}_{\ell+1}}$. This means precisely that the domain mapping defines a measure preserving bi-Lipschitz homeomorphism $X^\ell(g_1) \cap X_{\mu,\nu} \to C_\nu$. Conversely, if $g_1 \notin \cup_{|\gamma|=\kappa_\gamma(\mu,\nu)=\kappa_\gamma(\mu,\gamma)} X_{\underline{\mu}_{n+|\gamma|},\gamma}$ or $y_1 \notin C_{\underline{\nu}_\ell} \setminus C_{\underline{\nu}_{\ell+1}}$ then $X^\ell(g_1) \cap X_{\mu,\nu}$ is empty, so $X^\ell(g_1) \cap X_{\mu,\nu}$ has measure zero.

Lemma 3 *For $g_1 \in X_{\mu,\nu}$,*

$$m_\mathcal{G}(X^\ell(g_1) \setminus (X^\ell(g_1) \cap X_{\mu,\nu})) = \begin{cases} 0, & \ell \geq |\nu|, \\ N^{-\ell} - N^{-\ell-1}, & \kappa_\mathcal{G}(g_1) \leq \ell < |\nu|, \\ N^{-\ell-1}(N-1)^2, & 0 \leq \ell < \kappa_\mathcal{G}(g_1). \end{cases}$$

Proof Take $g_1 \notin X_{\mu,\nu}$. A computation with cylinder sets shows that

$$m_{\mathscr{G}}(X^\ell(g_1) \cap X_{\mu,\nu}) = \begin{cases} 0, & \ell < |\nu|, \\ N^{-\ell} - N^{-\ell-1}, & \ell \geq |\nu|, \end{cases}$$

$$\text{and} \quad m_{\mathscr{G}}(X^\ell(g_1)) = \begin{cases} N^{-\ell} - N^{-\ell-1}, & \kappa_{\mathscr{G}}(g_1) \leq \ell, \\ N^{-\ell-1}(N-1)^2, & 0 \leq \ell < \kappa_{\mathscr{G}}(g_1). \end{cases}$$

The result follows by subtracting the two expressions.

From these computations, we deduce a simple special case of Theorem 4. A short computation shows that $T_\ell \chi_{\mu,\emptyset} = 0$ for any finite word μ and $\ell \in \mathbb{N}$. Therefore

$$T e_{\mu,\emptyset} = 0.$$

We now turn to the general case.

Proof (Proof of Theorem 4) Using the decomposition (6), Lemmas 2 and 3 we write

$$(1 - N^{-1})T\chi_{\mu,\nu} = \sum_{\ell=0}^{|\nu|-1} \sum_{|\gamma|=\kappa_{\mathscr{V}}(\mu,\nu)=\kappa_{\mathscr{V}}(\mu,\gamma)} N^{\ell-|\nu|} \chi_{\underline{\mu}_{n+|\gamma|},\gamma}(g_1)(\chi_{\underline{\nu}_{\ell+1}}(y_1) - \chi_{\underline{\nu}_\ell}(y_1))$$

$$+ \left(\sum_{\ell=0}^{\kappa_{\mathscr{G}}(g_1)-1} N^{-2}(N-1)^2 + \sum_{\ell=\kappa_{\mathscr{G}}(g_1)}^{|\nu|-1} (1-N^{-1}) \right) \chi_{\mu,\nu}(g_1)$$

$$= \sum_{\ell=0}^{|\nu|-1} \sum_{|\gamma|=\kappa_{\mathscr{V}}(\mu,\nu)=\kappa_{\mathscr{V}}(\mu,\gamma)} N^{\ell-|\nu|} \chi_{\underline{\mu}_{n+|\gamma|},\gamma}(g_1)(\chi_{\underline{\nu}_{\ell+1}}(y_1) - \chi_{\underline{\nu}_\ell}(y_1))$$

$$+ (1 - N^{-1})\left(-\frac{\kappa_{\mathscr{G}}(g_1)}{N} + |\nu| \right) \chi_{\mu,\nu}(g_1). \tag{7}$$

Take two words $\mu, \nu \in \mathscr{V}_N$ with $t(\mu) = t(\nu) \neq \emptyset$. Up to normalization, $e_{\mu,\nu}$ coincides with $\chi_{\mu,\nu} - N^{-1}\chi_{\underline{\mu},\underline{\nu}}$. Here we are using the notation of Definition 3, and hence $|\underline{\nu}| = |\nu| - 1$ whenever $|\nu| > 0$. We compute that

$$(1 - N^{-1})T(\chi_{\mu,\nu} - N^{-1}\chi_{\underline{\mu},\underline{\nu}})$$

$$= (1 - N^{-1})\left(-\frac{\kappa_{\mathscr{G}}(g_1)}{N} + |\nu| \right) \chi_{\mu,\nu}(g_1)$$

$$- (1 - N^{-1})\left(-\frac{\kappa_{\mathscr{G}}(g_1)}{N} + |\nu| - 1 \right) N^{-1} \chi_{\underline{\mu},\underline{\nu}}(g_1)$$

$$+ N^{-1} \sum_{|\gamma|=\kappa_{\mathscr{V}}(\mu,\nu)=\kappa_{\mathscr{V}}(\mu,\gamma)} \chi_{\underline{\mu}_{n+|\gamma|},\gamma}(g_1)(\chi_\nu(y_1) - \chi_{\underline{\nu}_{|\nu|-1}}(y_1))$$

$$+ \sum_{\ell=0}^{|v|-2} \sum_{|\gamma|=\kappa_{\mathcal{Y}}(\mu,\underline{v})=\kappa_{\mathcal{Y}}(\mu,\gamma)} N^{\ell-|v|} \chi_{\underline{\mu}_{n+|\gamma|},\gamma}(g_1)(\chi_{\underline{v}_{\ell+1}}(y_1) - \chi_{\underline{v}_\ell}(y_1))$$

$$- N^{-1} \sum_{\ell=0}^{|v|-2} \sum_{|\gamma|=\kappa_{\mathcal{Y}}(\mu,\underline{v})=\kappa_{\mathcal{Y}}(\mu,\gamma)} N^{\ell-|v|+1} \chi_{\underline{\mu}_{n+|\gamma|},\gamma}(g_1)(\chi_{\underline{v}_{\ell+1}}(y_1) - \chi_{\underline{v}_\ell}(y_1)).$$

In the last line we are using that if $|\gamma| = \kappa_{\mathcal{Y}}(\mu, \underline{v})$ then $|\gamma| < |\mu|$ and $\kappa_{\mathcal{Y}}(\mu, \gamma) = \kappa_{\mathcal{Y}}(\mu, \gamma)$. If $t(\mu) = t(v)$ then $\kappa_{\mathcal{Y}}(\mu, v) = \kappa_{\mathcal{Y}}(\mu, \underline{v})$ and the last two terms cancel each other. We proceed with the remaining sums:

$$=(1 - N^{-1})\left(-\frac{\kappa_{\mathcal{G}}(g_1)}{N} + |v|\right)\chi_{\mu,v}(g_1)$$

$$- (1 - N^{-1})\left(-\frac{\kappa_{\mathcal{G}}(g_1)}{N} + |v| - 1\right)N^{-1}\chi_{\mu,\underline{v}}(g_1)$$

$$+ N^{-1} \sum_{|\gamma|=\kappa_{\mathcal{Y}}(\mu,\underline{v})=\kappa_{\mathcal{Y}}(\mu,\gamma)} \chi_{\underline{\mu}_{n+|\gamma|},\gamma}(g_1)(\chi_v(y_1) - \chi_{\underline{v}_{|v|-1}}(y_1))$$

$$=(1 - N^{-1})\left(-\frac{\kappa_{\mathcal{G}}(g_1)}{N} + |v| - 1 + \frac{N}{N-1}\right)(\chi_{\mu,v} - N^{-1}\chi_{\mu,\underline{v}})$$

$$- N^{-1}\chi_{\mu,v} + N^{-1} \sum_{|\gamma|=\kappa_{\mathcal{Y}}(\mu,v)=\kappa_{\mathcal{Y}}(\mu,\gamma)} \chi_{\underline{\mu}_{n+|\gamma|},\gamma}(g_1)\chi_v(y_1)$$

$$+ N^{-1}\chi_{\mu,\underline{v}} - N^{-1} \sum_{|\gamma|=\kappa_{\mathcal{Y}}(\mu,v)=\kappa_{\mathcal{Y}}(\mu,\gamma)} \chi_{\underline{\mu}_{n+|\gamma|},\gamma}(g_1)\chi_{\underline{v}}(y_1)$$

$$= (1 - N^{-1})\left(-\frac{\kappa_{\mathcal{G}}(g_1)}{N} + |v| + \frac{1}{N-1}\right)(\chi_{\mu,v} - N^{-1}\chi_{\mu,\underline{v}}).$$

Since $Be_{\mu,v} = 0$ if $t(\mu) = t(v)$ the theorem follows in this case. We now consider the case of two words $\mu, v \in \mathcal{V}_N$ with $t(\mu) \neq t(v)$. In this case, we simply note that Eq. (7) implies that

$$(1 - N^{-1})T\chi_{\mu,v} = \sum_{\ell=0}^{|v|-1} \sum_{|\gamma|=\kappa_{\mathcal{Y}}(\mu,\gamma)=|v|} N^{\ell-|v|} \chi_{\underline{\mu}_{n+|\gamma|},\gamma}(g_1)(\chi_{\underline{v}_{\ell+1}}(y_1) - \chi_{\underline{v}_\ell}(y_1))$$

$$+ (1 - N^{-1})\left(-\frac{\kappa_{\mathcal{G}}(g_1)}{N} + |v|\right)\chi_{\mu,v}(g_1)$$

$$= - \sum_{m \neq t(\mu)} \sum_{\ell=0}^{|v|-1} \sum_{|\gamma|=|v|-1, |\gamma \vee v|=\ell} N^{\ell-|v|} \chi_{\mu,\gamma m}(g_1)$$

$$+ (1 - N^{-1}) \left(-\frac{\kappa_\mathscr{G}(g_1)}{N} + |v| \right) \chi_{\mu,v}(g_1)$$

$$= (1 - N^{-1}) \left(-\frac{\kappa_\mathscr{G}(g_1)}{N} + |v| - B \right) \chi_{\mu,v}(g_1)$$

The only case left to consider is $\mu = v = \emptyset$ which holds trivially. This proves the theorem.

6   A Spectral Triple on O_N

We are now ready to assemble our operators into spectral triples on O_N. In [13] a spectral triple was constructed by defining the operator $D_\kappa = (2P_\mathscr{F} - 1)|c| - \kappa$. We proceed similarly and define D as the closure in $L^2(O_N)$ of the operator $(2P_\mathscr{F} - 1)|c| - T$ with initial domain $C_c^\infty(\mathscr{G}_N)$. In the same way, the operator $\tilde{D}$ is defined as the closure of

$$(2P_\mathscr{F} - 1)|c| - T - B + (N - 1)^{-1}Q : C_c^\infty(\mathscr{G}_N) \to L^2(O_N).$$

We note that $\tilde{D} = (2P_\mathscr{F} - 1)|c| - \tilde{T}$ where

$$\tilde{T}e_{\mu,v} = \left(-\frac{\kappa_\gamma(\mu, v)}{N} + |v| \right) e_{\mu,v}. \tag{8}$$

The following two propositions prove part 1 and 2 of Theorem 1.

Proposition 7 *The triples* $(O_N, L^2(O_N), \tilde{D})$ *and* $(O_N, L^2(O_N), D)$ *are spectral triples representing the class* $[\hat{1}] \in K^1(O_N)$.

Proof Since $D - \tilde{D}$ is a bounded self-adjoint operator, $\mathrm{Dom}(D) = \mathrm{Dom}(\tilde{D})$ and D defines a spectral triple if and only if $\tilde{D}$ does. It is easily verified from Corollary 1 and Eq. (8) that $(i \pm \tilde{D})^{-1}$ is a compact operator. Moreover, the generators $S_i \in O_N$ preserve $\mathrm{Dom}(\tilde{D})$. The operator $\tilde{D}$ has bounded commutators with the generators S_i, which is seen from combining [13, Theorem 3.19] in the model over $\mathbb{C}$ and in the model over $C(\Omega_N)$; the latter gives bounded commutators with $\kappa_\mathscr{G}$ and the former bounded commutators with $D_\kappa = \tilde{D} + \frac{\kappa_\mathscr{G}}{N}$. Thus D and $\tilde{D}$ define K-cycles for O_N. To identify their class in $K^1(O_N)$, observe that the operator inequalities

$$0 \le (1 - N^{-1})\kappa \le \tilde{T} \le \kappa, \quad 0 \le |c| + (1 - N^{-1})\kappa \le |c| + \tilde{T} \le |c| + \kappa, \tag{9}$$

hold true on the core $C_c^\infty(\mathscr{G}_N)$ and hence on all of $\mathrm{Dom}\,\tilde{T} = \mathrm{Dom}\,\kappa$ as well as on $\mathrm{Dom}\,(|c| + \tilde{T}) = \mathrm{Dom}\,(|c| + \kappa)$. By positivity, we have $\ker \kappa = \ker \tilde{T} = \mathrm{im}P_\mathscr{F}$ and thus $\tilde{T}P_\mathscr{F} = P_\mathscr{F}\tilde{T} = 0$ as well as

$$\ker(|c| + \tilde{T}) = \ker |c| \cap \ker \tilde{T} = \ker |c| \cap \ker \kappa = \ker(|c| + \kappa).$$

We can thus write the operator $\tilde{D}$ and its phase $\tilde{D}|\tilde{D}|^{-1}$ as

$$\tilde{D} = (2P_{\mathscr{F}} - 1)|c| - \tilde{T} = (2P_{\mathscr{F}} - 1)(|c| + \tilde{T}), \quad \tilde{D}|\tilde{D}|^{-1} = D_\kappa|D_\kappa|^{-1} = 2P_{\mathscr{F}} - 1,$$

which shows that $[D] = [\tilde{D}] = [D_\kappa] = [\hat{1}] \in K^1(O_N)$, by [8, 14] and [13, Theorem 3.19].

Proposition 8 *The spectral triples $(O_N, L^2(O_N), \tilde{D})$ and $(O_N, L^2(O_N), D)$ are θ-summable, $P_{\mathscr{F}} = \chi_{[0,\infty)}(\tilde{D}) = \chi_{[0,\infty)}(D_\kappa)$ and the difference $P_{\mathscr{F}} - \chi_{[0,\infty]}(D)$ is a finite rank operator. Moreover, for any $p > 0$, the set*

$$\text{Sum}^p(O_N, P_{\mathscr{F}}) := \{a \in O_N : [P_{\mathscr{F}}, a] \in \mathscr{L}^p(L^2(O_N))\},$$

is a dense $*$-subalgebra of O_N. In particular, $(O_N, L^2(O_N), 2P_{\mathscr{F}} - 1)$ is a p-summable generator of $K^1(O_N)$ for any $p > 0$.

Proof It suffices to prove θ-summability for $\tilde{D}$. This follows since $\mathscr{H}_{n,k}$ is N^{n+2k}-dimensional and $-2(|n| + k) \leq \tilde{D}|_{\mathscr{H}_{n,k}} \leq 2(|n| + k)$. The identity $P_{\mathscr{F}} = \chi_{[0,\infty)}(D_\kappa)$ follows from the construction using $P_{\mathscr{F}}L^2(O_N) = \ker(\kappa)$. Since $\tilde{D}|\tilde{D}|^{-1} = D_\kappa|D_\kappa|^{-1}$, we have $\chi_{[0,\infty)}(\tilde{D}) = \chi_{[0,\infty)}(D_\kappa)$. Moreover, for $C = \|B\|$, we have the operator inequalities

$$\tilde{D}|_{\mathscr{H}_{n,k}} - C \leq D|_{\mathscr{H}_{n,k}} \leq \tilde{D}|_{\mathscr{H}_{n,k}} + C, \quad \forall n, k.$$

Thus, by compactness of resolvents and the fact that the sign of $\tilde{D}|_{\mathscr{H}_{n,k}}$ is determined purely by n and k, it follows that $\chi_{[0,\infty)}(D) - \chi_{[0,\infty)}(\tilde{D})$ is a finite rank operator. The remaining statements follow from [12, Proposition 2.2.5], which uses methods from [11, 14]. $\blacksquare$

7 The Fröhlich Functionals on O_N

In [10], Fröhlich et al. associated a state with θ-summable spectral triples. If $(A, \mathscr{H}, D)$ is a θ-summable spectral triple, one defines the state on $a \in A$ as

$$\phi_t^D(a) := \frac{\text{tr}(ae^{-tD^2})}{\text{tr}(e^{-tD^2})}, \quad t > 0.$$

The assumption that an extended limit of ϕ_t^D as $t \to 0$ is tracial is used in [10]. This is clearly an unrealistic assumption if A admits no traces. In [17, Corollary 12.3.5] it is shown that under certain finite dimensionality conditions on $(A, \mathscr{H}, D)$, which are slightly stronger than finite summability, extended limits of ϕ_t^D are tracial. This provides interesting connections between the tracial property of the states introduced by Fröhlich et al. and Connes' tracial obstructions to finite summability.

Since D, D_0 and D_κ are θ-summable, they allow for the definition of Fröhlich functionals on O_N. For $t > 0$ we define the following states on $\mathbb{B}(L^2(O_N))$:

$$\phi_t(T) := \frac{\mathrm{tr}(Te^{-tD^2})}{\mathrm{tr}(e^{-tD^2})}, \quad \tilde{\phi}_t(T) := \frac{\mathrm{tr}(Te^{-t\tilde{D}^2})}{\mathrm{tr}(e^{-t\tilde{D}^2})} \quad \text{and} \quad \phi_t^\kappa(T) := \frac{\mathrm{tr}(Te^{-tD_\kappa^2})}{\mathrm{tr}(e^{-tD_\kappa^2})}.$$

A state $\omega \in L^\infty[0, 1]^*$ is said to be an extended limit at 0 if $\omega(f) = 0$ whenever $f = 0$ near 0. For an extended limit ω at 0, we define $\phi_\omega := \omega \circ \phi_t$, $\tilde{\phi}_\omega := \omega \circ \tilde{\phi}_t$ and $\phi_\omega^\kappa := \omega \circ \phi_t^\kappa$. The next result proves part 3 of Theorem 1.

Proposition 9 *For any extended limit ω, there exists probability measures $\tilde{m}_\omega$ and m_ω^κ on Ω_N such that for $a \in O_N$*

$$\tilde{\phi}_\omega(a) = \int_{\Omega_N} \Phi(a)\mathrm{d}\tilde{m}_\omega \quad \text{and} \quad \phi_\omega^\kappa(a) = \int_{\Omega_N} \Phi(a)\mathrm{d}m_\omega^\kappa.$$

Using the fact that $\Phi(S_\mu S_\nu^*) = \delta_{\mu,\nu} S_\mu S_\mu^*$ the proposition is immediate from the next lemma which in turn is a computational exercise.

Lemma 4 *For any finite words $\mu, \nu, \sigma, \rho \in \mathscr{V}_N$,*

$$\langle e_{\mu,\nu}, S_\rho S_\sigma^* e_{\mu,\nu}\rangle_{L^2(O_N)}$$

$$= \begin{cases} \delta_{\rho,\sigma}\delta_{|\sigma\vee\mu|,\min\{|\mu|,|\sigma|\}}N^{\min\{0,|\mu|-|\rho|\}}, & t(\mu) \neq t(\nu), \\[2ex] (N-1)^{-1}\delta_{\rho,\sigma}\Big(\delta_{|\sigma\vee\mu|,\min\{|\mu|,|\sigma|\}}(N-2)N^{\min\{0,|\mu|-|\rho|\}} + & t(\mu) = t(\nu) \\[1ex] \qquad\qquad \delta_{|\sigma\vee\underline{\mu}|,\min\{|\mu|-1,|\sigma|\}}N^{\min\{0,|\mu|-|\rho|-1\}}\Big). \end{cases}$$

Here μ starts with $\underline{\mu}$ and $|\underline{\mu}| = |\mu| - 1$.

Remark 4 It is reasonable to expect that ϕ_ω, $\tilde{\phi}_\omega$ and ϕ_ω^κ are in fact related to the KMS state ϕ. We have not been able to prove this. A simple induction procedure, or using uniqueness of KMS states on O_N combined with [16], shows that it suffices to prove that $\phi_\omega(S_\nu S_\nu^*) = N\phi_\omega(S_{\nu j} S_{\nu j}^*)$ for any finite word $\nu \in \mathscr{V}_N$ and $j = 1, \ldots, N$.

Acknowledgements We thank the MATRIX for the program *Refining C^*-algebraic invariants for dynamics using KK-theory* in Creswick, Australia (2016) where this work came into being. We are grateful to the support from Leibniz University Hannover where this work was initiated. We also thank Francesca Arici, Robin Deeley, Adam Rennie and Alexander Usachev for fruitful discussions and helpful comments, and the anonymous referee for a careful reading of the manuscript. The first author was supported by the Swedish Research Council Grant 2015-00137 and Marie Sklodowska Curie Actions, Cofund, Project INCA 600398.

References

1. Connes, A.: Compact metric spaces, Fredholm modules and hyperfiniteness. Ergodic Theory Dyn. Syst. **9**, 207–230 (1989)
2. Connes, A.: Noncommutative Geometry. Academic Press, London (1994)
3. Connes, A., Moscovici, H.: Type III and spectral triples. In: Traces in Number Theory, Geometry and Quantum Fields. Aspects of Mathematics, vol. E38, pp. 57–71. Friedrich Vieweg, Wiesbaden (2008)
4. Coornaert, M.: Mesures de Patterson-Sullivan sur le bord d'un espace hyperbolique au sens de Gromov. Pac. J. Math. **159**(2), 241–270 (1993)
5. Cornelissen, G., Marcolli, M., Reihani, K., Vdovina, A.: Noncommutative geometry on trees and buildings. In: Traces in Geometry, Number Theory, and Quantum Fields, pp. 73–98. Vieweg, Wiesbaden (2007)
6. Cuntz, J.: Simple C^*-algebras generated by isometries. Commun. Math. Phys. **57**(2), 173–185 (1977)
7. Emerson, H., Nica, B.: K-homological finiteness and hyperbolic groups. J. Reine Angew. Math. (to appear). https://doi.org/10.1515/crelle-2015-0115
8. Evans, D.E.: On O_n. Publ. Res. Inst. Math. Sci. **16**(3), 915–927 (1980)
9. Farsi, C., Gillaspy, E., Julien, A., Kang, S., Packer, J.: Wavelets and spectral triples for fractal representations of Cuntz algebras. arXiv:1603.06979
10. Fröhlich, J., Grandjean, O., Recknagel, A.: Supersymmetric quantum theory and differential geometry. Commun. Math. Phys. **193**(3), 527–594 (1998)
11. Goffeng, M.: Equivariant extensions of ∗-algebras. N. Y. J. Math. **16**, 369–385 (2010)
12. Goffeng, M., Mesland, B.: Spectral triples and finite summability on Cuntz-Krieger algebras. Doc. Math. **20**, 89–170 (2015)
13. Goffeng, M., Mesland, B., Rennie, A.: Shift tail equivalence and an unbounded representative of the Cuntz-Pimsner extension. Ergodic Theory Dyn. Syst. (to appear). https://doi.org/10.1017/etds.2016.75
14. Kaminker, J., Putnam, I.: K-theoretic duality of shifts of finite type. Commun. Math. Phys. **187**(3), 509–522 (1997)
15. Kasparov, G.G.: The operator K-functor and extensions of C^*-algebras. Izv. Akad. Nauk SSSR Ser. Mat. **44**(3), 571–636, 719 (1980)
16. Laca, M., Neshveyev, S.: KMS states of quasi-free dynamics on Pimsner algebras. J. Funct. Anal. **211**, 457–482 (2004)
17. Lord, S., Sukochev, F., Zanin, D.: Singular Traces. Theory and Applications. De Gruyter Studies in Mathematics, vol. 46. De Gruyter, Berlin (2013)
18. Olesen, D., Pedersen, G.K.: Some C^*-dynamical systems with a single KMS state. Math. Scand. **42**, 111–118 (1978)
19. Pask, D., Rennie, A.: The noncommutative geometry of graph C^*-algebras. I. The index theorem. J. Funct. Anal. **233**(1), 92–134 (2006)
20. Renault, J.: A Groupoid Approach to C^*-Algebras. Lecture Notes in Mathematics, vol. 793. Springer, Berlin (1980)
21. Renault, J.: Cuntz-like algebras. In: Operator Theoretical Methods (Timisoara, 1998), pp. 371–386. Theta Foundation, Bucharest (2000)

Application of Semifinite Index Theory to Weak Topological Phases

Chris Bourne and Hermann Schulz-Baldes

Abstract Recent work by Prodan and the second author showed that weak invariants of topological insulators can be described using Kasparov's KK-theory. In this note, a complementary description using semifinite index theory is given. This provides an alternative proof of the index formulae for weak complex topological phases using the semifinite local index formula. Real invariants and the bulk-boundary correspondence are also briefly considered.

1 Introduction

The application of techniques from the index theory of operator algebras to systems in condensed matter physics has given fruitful results, the quantum Hall effect being a key early example [3]. More recently, C^*-algebras and their K-theory (and K-homology) have been applied to topological insulator systems, see for example [6, 14, 19, 23, 30, 36].

The framework of C^*-algebras is able to encode disordered systems with arbitrary (possibly irrational) magnetic field strength, something that standard methods in solid state physics are unable to do. Furthermore, by considering the geometry of a dense subalgebra of the weak closure of the observable algebra, one can derive index formulae that relate physical phenomena, such as the Hall conductivity, to an index of a Fredholm operator.

C. Bourne (✉)
Department Mathematik, Friedrich-Alexander-Universität Erlangen-Nürnberg, Cauerstr. 11, 91058 Erlangen, Germany

Advanced Institute for Materials Research, Tohoku University, 2-1-1 Katahira, Aoba-ku, Sendai 980-8577, Japan
e-mail: christopher.j.bourne@gmail.com

H. Schulz-Baldes
Department Mathematik, Friedrich-Alexander-Universität Erlangen-Nürnberg, Cauerstr. 11, 91058 Erlangen, Germany
e-mail: schuba@mi.uni-erlangen.de

© Springer International Publishing AG, part of Springer Nature 2018
D.R. Wood et al. (eds.), *2016 MATRIX Annals*, MATRIX Book Series 1,
https://doi.org/10.1007/978-3-319-72299-3_10

Topological insulators are special materials which behave as an insulator in the interior (bulk) of the system, but have conducting modes at the edges of the system going along with non-trivial topological invariants in the bulk [33]. Influential work by Kitaev suggested that these properties are related to the K-theory of the momentum space of a free-fermionic system [22].

Recent work by Prodan and the second author considered so-called 'weak' topological phases of topological insulators [31]. In the picture without disorder or magnetic flux, a topological phase is classified by the real or complex K-theory of the torus $\mathbb{T}^d$ of dimension d. Relating Atiyah's KR-theory [1] to the K-theory of C^*-algebras and then using the Pimsner–Voiculescu sequence with trivial action allows us to compute the relevant K-groups explicitly,

$$KR^{-n}(\mathbb{T}^d, \zeta) \cong KO_n(C(i\mathbb{T}^d)) \cong KO_n(C^*(\mathbb{Z}^d)) \cong \bigoplus_{j=0}^{d} \binom{d}{j} KO_{n-j}(\mathbb{R}) . \tag{1}$$

Here n labels the universality class as described in [6, 22] and $C(i\mathbb{T}^d)$ is the real C^*-algebra $\{f \in C(\mathbb{T}^d, \mathbb{C}) : \overline{f(x)} = f(-x)\}$, which naturally encodes the involution ζ on $\mathbb{T}^d$. The 'top degree' term $KO_{n-d}(\mathbb{R})$ is said to represent the strong invariants of the topological insulator and all lower-order terms are called weak invariants.

Bounded and complex Kasparov modules were used to provide a framework to compute weak invariants in the case of magnetic field and (weak) disorder in [31]. A geometric identity is used there to derive a local formula for the weak invariants. The purpose of this paper is to provide an alternative proof of this result using semifinite spectral triples and, in particular, the semifinite local index formula in [8, 9]. This shows the flexibility of the operator algebraic approach and complements the work in [31].

The framework employed here largely follows from previous work, namely [7], where a Kasparov module and semifinite spectral triple were constructed for a unital C^*-algebra B with a twisted $\mathbb{Z}^k$-action and invariant trace. Therefore the main task here is the computation of the resolvent cocycle that represents the (semifinite) Chern character and its application to weak invariants. Furthermore, the bulk-boundary correspondence proved in [7, 30] also carries over, which allows us to relate topological pairings of the system without edge to pairings concentrated on the boundary of the sample.

2   Review: Twisted Crossed Products and Semifinite Index Theory

2.1   Preliminaries

Let us briefly recall the basics of Kasparov theory that are needed for this paper; a more comprehensive treatment can be found in [5, 31]. Due to the anti-linear symmetries that exist in topological phases, both complex and real spaces and algebras are considered.

Given a real or complex right-B C^*-module E_B, we will denote by $(\cdot \mid \cdot)_B$ the B-valued inner-product and by $\text{End}_B(E)$ the adjointable endomorphisms on E with respect to this inner product. The rank-1 operators $\Theta_{e,f}$, $e,f \in E_B$, are defined such that

$$\Theta_{e,f}(g) = e \cdot (f \mid g)_B \,, \qquad e, f, g \in E_B \,.$$

Then $\text{End}_B^{00}(E)$ denotes the span of such rank-1 operators. The compact operators on the module, $\text{End}_B^0(E)$, is the norm closure of $\text{End}_B^{00}(E)$. We will often work with $\mathbb{Z}_2$-graded algebras and spaces and denote by $\hat{\otimes}$ the graded tensor product (see [16, Section 2] and [5]). Also see [25, Chapter 9] for the basic theory of unbounded operators on C^*-modules.

Definition 1 Let A and B be $\mathbb{Z}_2$-graded real (resp. complex) C^*-algebras. A real (complex) unbounded Kasparov module $(\mathcal{A}, {}_\pi E_B, D)$ is a $\mathbb{Z}_2$-graded real (complex) C^*-module E_B, a graded homomorphism $\pi : A \to \text{End}_B(E)$, and an unbounded self-adjoint, regular and odd operator D such that for all $a \in \mathcal{A} \subset A$, a dense $*$-subalgebra,

$$[D, \pi(a)]_\pm \in \text{End}_B(E) \,, \qquad \pi(a)(1 + D^2)^{-1/2} \in \text{End}_B^0(E) \,.$$

For complex algebras and spaces, one can also remove the gradings, in which case the Kasparov module is called odd (otherwise even).

We will often omit the representation π when the left-action is unambiguous. Unbounded Kasparov modules represent classes in the KK-group $KK(A,B)$ or $KKO(A,B)$ [2].

Closely related to unbounded Kasparov modules are semifinite spectral triples. Let τ be a fixed faithful, normal, semifinite trace on a von Neumann algebra $\mathcal{N}$. Graded von Neumann algebras can be considered in an analogous way to graded C^*-algebras, though the only graded von Neumann algebras we will consider are of the form $\mathcal{N}_0 \hat{\otimes} \text{End}(\mathcal{V})$, with $\mathcal{N}_0$ trivially graded and $\text{End}(\mathcal{V})$ the graded operators on a finite dimensional and $\mathbb{Z}_2$-graded Hilbert space $\mathcal{V}$. We denote by $\mathcal{K}_\mathcal{N}$ the τ-compact operators in $\mathcal{N}$, that is, the norm closed ideal generated by the projections $P \in \mathcal{N}$ with $\tau(P) < \infty$. For graded von Neumann algebras, non-trivial projections $P \in \mathcal{N}$ are even, though the grading Ad_{σ_3} on $M_2(\mathcal{N})$ gives a grading on $M_n(\mathcal{K}_\mathcal{N})$.

Definition 2 Let $\mathcal{N}$ be a graded semifinite von Neumann algebra with trace τ. A semifinite spectral triple $(\mathcal{A}, \mathcal{H}, D)$ is given by a $\mathbb{Z}_2$-graded Hilbert space $\mathcal{H}$, a graded $*$-algebra $\mathcal{A} \subset \mathcal{N}$ with C^*-closure A and a graded representation on $\mathcal{H}$, together with a densely defined odd unbounded self-adjoint operator D affiliated to $\mathcal{N}$ such that

1. $[D, a]_\pm$ is well-defined on $\text{Dom}(D)$ and extends to a bounded operator on $\mathcal{H}$ for all $a \in \mathcal{A}$,
2. $a(1 + D^2)^{-1/2} \in \mathcal{K}_\mathcal{N}$ for all $a \in A$.

For $\mathcal{N} = \mathcal{B}(\mathcal{H})$ and $\tau = \text{Tr}$, one recovers the usual definition of a spectral triple.

If $(\mathcal{A}, E_B, D)$ is an unbounded Kasparov module and the right-hand algebra B has a faithful, semifinite and norm lower semicontinuous trace τ_B, then one can often construct a semifinite spectral triple using results from [24]. We follow this route in Sect. 2.2 below. The converse is always true, namely a semifinite spectral triple gives rise to a class in $KK(A, C)$ with C a subalgebra of $\mathcal{K}_{\mathcal{N}}$ [15, Theorem 4.1]. If A is separable, this algebra C can be chosen to be separable as well [15, Theorem 5.3], but in a largely ad-hoc fashion. Because we first construct a Kasparov module and subsequently build a semifinite spectral triple, one obtains more explicit control on the image of the semifinite index pairing defined next (see Lemma 1 below). Therefore the algebra C is not required here (as in [10, Proposition 2.13]) to assure that the range of the semifinite index pairing is countably generated, i.e. a discrete subset of $\mathbb{R}$.

Complex semifinite spectral triples $(\mathcal{A}, \mathcal{H}, D)$ with $\mathcal{A}$ trivially graded can be paired with K-theory classes in $K_*(\mathcal{A})$ via the semifinite Fredholm index. If $\mathcal{A}$ is Fréchet and stable under the holomorphic functional calculus, then $K_*(\mathcal{A}) \cong K_*(A)$ and the pairings extend to the C^*-closure. Recall that an operator $T \in \mathcal{N}$ that is invertible modulo $\mathcal{K}_{\mathcal{N}}$ has semifinite Fredholm index

$$\mathrm{Index}_\tau(T) = \tau(P_{\mathrm{Ker}(T)}) - \tau(P_{\mathrm{Ker}(T^*)}),$$

with $P_{\mathrm{Ker}(T)}$ the projection onto $\mathrm{Ker}(T) \subset \mathcal{H}$.

Definition 3 Let $(\mathcal{A}, \mathcal{H}, D)$ be a unital complex semifinite spectral triple relative to $(\mathcal{N}, \tau)$ with $\mathcal{A}$ trivially graded and D invertible. Let p be a projector in $M_n(\mathcal{A})$, which represents $[p] \in K_0(\mathcal{A})$ and u a unitary in $M_n(\mathcal{A})$ representing $[u] \in K_1(\mathcal{A})$. In the even case, define $T_\pm = \frac{1}{2}(1 \mp \gamma)T\frac{1}{2}(1 \pm \gamma)$ with Ad_γ the grading on $\mathcal{H}$. Then with $F = D|D|^{-1}$ and $\Pi = (1 + F)/2$, the semifinite index pairing is represented by

$$\langle [p], (\mathcal{A}, \mathcal{H}, D) \rangle = \mathrm{Index}_{\tau \otimes \mathrm{Tr}_{\mathbb{C}^n}}(p(F \otimes 1_n)_+ p), \qquad \text{even case},$$

$$\langle [u], (\mathcal{A}, \mathcal{H}, D) \rangle = \mathrm{Index}_{\tau \otimes \mathrm{Tr}_{\mathbb{C}^n}}((\Pi \otimes 1_n)u(\Pi \otimes 1_n)), \qquad \text{odd case}.$$

If D is not invertible, we define the double spectral triple $(\mathcal{A}, \mathcal{H} \oplus \mathcal{H}, D_M)$ for $M > 0$ and relative to $(M_2(\mathcal{N}), \tau \otimes \mathrm{Tr}_{\mathbb{C}^2})$, where the operator D_M and the action of $\mathcal{A}$ is given by

$$D_M = \begin{pmatrix} D & M \\ M & -D \end{pmatrix}, \qquad a \mapsto \begin{pmatrix} a & 0 \\ 0 & 0 \end{pmatrix},$$

for all $a \in \mathcal{A}$. If $(\mathcal{A}, \mathcal{H}, D)$ is graded by γ, then the double is graded by $\hat{\gamma} = \gamma \oplus (-\gamma)$. Doubling the spectral triple does not change the K-homology class and ensures that the unbounded operator D_M is invertible [11].

A unital semifinite spectral triple $(\mathcal{A}, \mathcal{H}, D)$ relative to $(\mathcal{N}, \tau)$ is called p-summable if $(1 + D^2)^{-s/2}$ is τ-trace-class for all $s > p$, and smooth or QC^∞ (for quantum C^∞) if for all $a \in \mathcal{A}$

$$a, [D, a] \in \bigcap_{n \geq 0} \text{Dom}(\delta^n), \qquad \delta(T) = [(1 + D^2)^{1/2}, T].$$

If $(\mathcal{A}, \mathcal{H}, D)$ is complex, p-summable and QC^∞, we can apply the semifinite local index formula [8, 9] to compute the semifinite index pairing of $[x] \in K_*(A)$ with $(\mathcal{A}, \mathcal{H}, D)$ in terms of the resolvent cocycle. Because the resolvent cocycle is a local expression involving traces and derivations, it is usually easier to compute than the semifinite Fredholm index.

2.2 Crossed Products and Kasparov Theory

2.2.1 The Algebra and Representation

Let us consider a d-dimensional lattice, so the Hilbert space $\mathcal{H} = \ell^2(\mathbb{Z}^d) \otimes \mathbb{C}^n$, and a disordered family $\{H_\omega\}_{\omega \in \Omega}$ of Hamiltonians acting on $\mathcal{H}$ indexed by disorder configurations ω drawn from a compact space Ω equipped with a $\mathbb{Z}^d$-action (possibly with twist ϕ). One can then construct the algebra of observables $M_n(C(\Omega) \rtimes_\phi \mathbb{Z}^d)$. The family of Hamiltonians $\{H_\omega\}_{\omega \in \Omega}$ are associated to a self-adjoint element $H \in M_n(C(\Omega) \rtimes_\phi \mathbb{Z}^d)$, and we always assume that H has a spectral gap at the Fermi energy. The Hilbert space fibres $\mathbb{C}^n$ and the matrices $M_n(\mathbb{C})$ are often used to implement the symmetry operators that determine the symmetry-type of the Hamiltonian. However the matrices do not play an important role in the construction of the Kasparov modules and semifinite spectral triples we consider. Hence we will work with $C(\Omega) \rtimes_\phi \mathbb{Z}^d$, under the knowledge that this algebra can be tensored with the matrices (or compact operators) without issue. The space $C(\Omega)$ can also encode a quasicrystal structure and depends on the example under consideration.

The twist ϕ is in general a twisting cocycle $\phi : \mathbb{Z}^d \times \mathbb{Z}^d \to \mathcal{U}(C(\Omega))$ such that for all $x, y, z \in \mathbb{Z}^d$,

$$\phi(x, y)\phi(x + y, z) = \alpha_x(\phi(y, z))\phi(x, y + z), \qquad \phi(x, 0) = \phi(0, x) = 1,$$

see [28]. We also assume that $\phi(x, -x) = 1$ for all $x \in \mathbb{Z}^d$ as in [20] or [30], which still encompasses most examples of physical interest.

Remark 1 (Anti-Linear Symmetries, Real Algebras and Twists) Our model always begins with a complex algebra acting on a complex Hilbert space. If the Hamiltonian satisfies anti-linear symmetries, then we restrict to a real subalgebra of the complex algebra $C(\Omega) \rtimes_\phi \mathbb{Z}^d$ that is invariant under the induced real structure by complex conjugation. This procedure is direct for time-reversal symmetry, though modifications are needed for particle-hole symmetry [14, 19, 36]. Such a restriction puts stringent constraints on the twisting cocycle ϕ and will often force the twist to be zero (e.g. if ϕ arises from an external magnetic field). For this reason, in the real case, we will only consider untwisted crossed products $C(\Omega) \rtimes \mathbb{Z}^d$. We note that this may not encompass every example of interest, but we leave the more general setting to another place.   ◇

Our focus is on weak topological invariants which have the interpretation of lower-dimensional invariants extracted from a higher-dimensional system. Using the assumption $\phi(x, -x) = 1$, one can rewrite $C(\Omega) \rtimes_\phi \mathbb{Z}^d \cong \left(C(\Omega) \rtimes_\phi \mathbb{Z}^{d-k} \right) \rtimes_\theta \mathbb{Z}^k$ with a new twist θ [20, 28]. Hence for d large enough and $1 \leq k \leq d$ one can study the lower-dimensional dynamics and topological invariants of the $\mathbb{Z}^k$-action.

With the setup in place, let B be a unital separable C^*-algebra, real or complex, and consider the (twisted) crossed product $B \rtimes_\theta \mathbb{Z}^k$ with respect to a $\mathbb{Z}^k$-action α. This algebra is generated by the elements $b \in B$ and unitary operators $\{S_j\}_{j=1}^k$ such that $S^n = S_1^{n_1} \cdots S_k^{n_k}$ for $n = (n_1, \ldots, n_k) \in \mathbb{Z}^k$ satisfy

$$S^n b = \alpha_n(b) S^n , \qquad S^m S^n = \theta(n, m) S^{m+n}$$

for multi-indices $n, m \in \mathbb{Z}^k$ and $\theta : \mathbb{Z}^k \times \mathbb{Z}^k \to \mathcal{U}(B)$ the twisting cocycle. Let $\mathscr{A}$ denote the algebra of elements $\sum_{n \in \mathbb{Z}^k} S^n b_n$, where $(\|b_n\|)_{n \in \mathbb{Z}^k}$ is in the discrete Schwartz-space $\mathscr{S}(\ell^2(\mathbb{Z}^k))$. The full crossed product completion $B \rtimes_\theta \mathbb{Z}^k$ is denoted by A. Following [7, 31] one can build an unbounded Kasparov module encoding this action. First let us take the standard C^*-module $\ell^2(\mathbb{Z}^k) \otimes B = \ell^2(\mathbb{Z}^k, B)$ with right-action given by right-multiplication and B-valued inner product

$$(\psi_1 \otimes b_1 \mid \psi_2 \otimes b_2)_B = \langle \psi_1, \psi_2 \rangle_{\ell^2(\mathbb{Z}^k)} b_1^* b_2 .$$

The module $\ell^2(\mathbb{Z}^k, B)$ has the frame $\{\delta_m \otimes 1_B\}_{m \in \mathbb{Z}^k}$ where $\{\delta_m\}_{m \in \mathbb{Z}^d}$ is the canonical basis on $\ell^2(\mathbb{Z}^k)$. Then an action on generators is defined by

$$b_1 \cdot (\delta_m \otimes b_2) = \delta_m \otimes \alpha_{-m}(b_1) b_2 ,$$

$$S^n \cdot (\delta_m \otimes b) = \theta(n, m) \cdot \delta_{m+n} \otimes b = \delta_{m+n} \otimes \alpha_{-m-n}(\theta(n, m)) b .$$

It is shown in [7, 31] that this left-action extends to an adjointable action of the crossed product on $\ell^2(\mathbb{Z}^k, B)$.

2.2.2 The Spin and Oriented Dirac Operators

Using the position operators $X_j(\delta_m \otimes b) = m_j \delta_m \otimes b$ one can now build an unbounded Kasparov module. To put things together, the real Clifford algebras $Cl_{r,s}$ are used. They are generated by r self-adjoint elements $\{\gamma^j\}_{j=1}^r$ with $(\gamma^j)^2 = 1$ and

s skew-adjoint elements $\{\rho^i\}_{i=1}^s$ with $(\rho^i)^2 = -1$. Taking the complexification we have $C\ell_{r,s} \otimes \mathbb{C} = \mathbb{C}\ell_{r+s}$.

In the complex case and k even, we may use the irreducible Clifford representation of $\mathbb{C}\ell_k = \operatorname{span}_\mathbb{C}\{\Gamma^j\}_{j=1}^k$ on the (trivial) spinor bundle $\mathfrak{S}$ over $\mathbb{T}^k$ to construct the unbounded operator $\sum_{j=1}^k X_j \hat{\otimes} \Gamma^j$ on $\ell^2(\mathbb{Z}^k, B) \hat{\otimes} \mathfrak{S}$. After Fourier transform, this is the standard Dirac operator on the spinor bundle over the torus. More concretely, $\mathfrak{S} \cong \mathbb{C}^{2^{k/2}}$ with $\{\Gamma^j\}_{j=1}^k$ self-adjoint matrices satisfying $\Gamma^i \Gamma^j + \Gamma^j \Gamma^i = 2\delta_{i,j}$. For odd k, one proceeds similarly, but there are two irreducible representations of $\mathbb{C}\ell_k$ on $\mathfrak{S} \cong \mathbb{C}^{2^{(k-1)/2}}$.

Proposition 1 *Consider a twisted $\mathbb{Z}^k$-action α, θ on a complex C^*-algebra B. Let A be the associated crossed product with dense subalgebra $\mathscr{A}$ of $\sum_{n \in \mathbb{Z}^k} S^n b_n$ with $(b_n)_{n \in \mathbb{Z}^k}$ Schwartz-class coefficients. For $v = 2^{\lfloor \frac{k}{2} \rfloor}$, the triple*

$$\lambda_k^{\mathfrak{S}} = \left(\mathscr{A}, \ell^2(\mathbb{Z}^k, B)_B \hat{\otimes} \mathbb{C}^v, \sum_{j=1}^k X_j \hat{\otimes} \Gamma^j \right)$$

is an unbounded Kasparov module that is even if k is even with grading $\operatorname{Ad}_{\Gamma_0}$ for $\Gamma_0 = (-i)^{k/2} \Gamma^1 \cdots \Gamma^k$, specifying an element of $KK(A, B)$. The triple $\lambda_k^{\mathfrak{S}}$ is odd (ungraded) if k is odd, representing a class in $KK^1(A, B) = KK(A \hat{\otimes} \mathbb{C}\ell_1, B)$ which can be specified by a graded Kasparov module

$$\left(\mathscr{A} \hat{\otimes} \mathbb{C}\ell_1, \ell^2(\mathbb{Z}^k, B) \otimes \mathbb{C}^{2^{(k-1)/2}} \hat{\otimes} \mathbb{C}^2, \begin{pmatrix} 0 & -i\sum_{j=1}^k X_j \hat{\otimes} \Gamma^k \\ i\sum_{j=1}^k X_j \hat{\otimes} \Gamma^k & 0 \end{pmatrix}\right), \quad (2)$$

where the grading is given by conjugating with $\begin{pmatrix} 1 & 0 \\ 0 & -1 \end{pmatrix}$, and $\sigma_1 = \begin{pmatrix} 0 & 1 \\ 1 & 0 \end{pmatrix}$ generates the left $\mathbb{C}\ell_1$-action.

Proof The algebra $\mathscr{A}$ is trivially graded and one computes that

$$[X_j, \sum_{m \in \mathbb{Z}^k} S^m b_m] = \sum_{m \in \mathbb{Z}^k} m_j S^m b_m,$$

which is adjointable for $(\|b_m\|)_{m \in \mathbb{Z}^k}$ in the Schwartz space over $\mathbb{Z}^k$. Therefore the commutator $[\sum_{j=1}^k X_j \hat{\otimes} \Gamma^j, a \hat{\otimes} 1_{\mathbb{C}^v}]$ is adjointable for $a \in \mathscr{A}$. The operator $(1 + |X|^2)^{-s/2}$ acts diagonally with respect to the frame $\{\delta_m \otimes 1_B\}_{m \in \mathbb{Z}^k}$ on $\ell^2(\mathbb{Z}^k, B)$. In particular,

$$(1 + |X|^2)^{-1/2} = \sum_{m \in \mathbb{Z}^k} (1 + |m|^2)^{-1/2} \Theta_{\delta_m \otimes 1_B, \delta_m \otimes 1_B},$$

which is a norm convergent sum of finite-rank operators and so it is compact on $\ell^2(\mathbb{Z}^k, B)$. In particular, $(1 + |X|^2)^{-1/2} \hat{\otimes} 1_{\mathbb{C}^v}$ is compact on $\ell^2(\mathbb{Z}^k, B) \hat{\otimes} \mathbb{C}^v$. $\square$

The triple $\lambda_k^{\mathbb{S}}$ is the unbounded representative of the bounded Kasparov module constructed in [31]. The (trivial) spin structure on the torus is used to construct the Kasparov module $\lambda_k^{\mathbb{S}}$ from Proposition 1. One can also use the torus' oriented structure. Following [16, §2], we consider $\bigwedge^* \mathbb{R}^k$ (or complex), which is a graded Hilbert space such that $\mathrm{End}_{\mathbb{R}}(\bigwedge^* \mathbb{R}^k) \cong C\ell_{0,k} \hat{\otimes} C\ell_{k,0}$, where the action of $C\ell_{0,k}$ and $C\ell_{k,0}$ is generated by the operators

$$\rho^j(w) = e_j \wedge w - \iota(e_j)w, \qquad \gamma^j(w) = e_j \wedge w + \iota(e_j)w,$$

where $\{e_j\}_{j=1}^k$ denotes the standard basis of $\mathbb{R}^k$, $w \in \bigwedge^* \mathbb{R}^k$ and $\iota(v)w$ the contraction of w along v (using the inner-product on $\mathbb{R}^k$). A careful check also shows that γ^j and ρ^k graded-commute. The grading of $\bigwedge^* \mathbb{R}^k$ can be expressed in terms of the grading operator

$$\gamma_{\bigwedge^* \mathbb{R}^k} = (-1)^k \rho^1 \cdots \rho^k \hat{\otimes} \gamma^k \cdots \gamma^1.$$

Kasparov also constructs a diagonal action of $\mathrm{Spin}_{0,k}$ (and $\mathrm{Spin}_{k,0}$) on $\mathrm{End}_{\mathbb{R}}(\bigwedge^* \mathbb{R}^k)$ [16, §2.18], though this will not be needed here.

Proposition 2 ([7, Proposition 3.2]) *Consider a $\mathbb{Z}^k$-action α on a real or complex C^*-algebra B, possibly twisted by θ. Let A be the associated crossed product with dense subalgebra $\mathscr{A}$ of elements $\sum_n S^n b_n$ with Schwartz-class coefficients. The data*

$$\lambda_k = \left(\mathscr{A} \hat{\otimes} C\ell_{0,k}, \ \ell^2(\mathbb{Z}^k, B)_B \hat{\otimes} \bigwedge^* \mathbb{R}^k, \ \sum_{j=1}^k X_j \hat{\otimes} \gamma^j \right) \tag{3}$$

defines an unbounded $A \hat{\otimes} C\ell_{0,k}$-$B$ Kasparov module and class in $KKO(A \hat{\otimes} C\ell_{0,k}, B)$ which is also denoted $KKO^k(A, B)$. The $C\ell_{0,k}$-action is generated by the operators ρ^j. In the complex case, one has to replace $\mathbb{C}\ell_k$ and $\bigwedge^* \mathbb{C}^k$ in the above formula.

For complex algebras and spaces, we have constructed two (complementary) Kasparov modules, $\lambda_k^{\mathbb{S}}$ and λ_k. We have done this to better align our results with existing literature on the topic, in particular [30, 31]. In the case $k = 1$, these Kasparov modules directly coincide.

For higher k, we can explicitly connect $\lambda_k^{\mathbb{S}}$ and λ_k by a Morita equivalence bimodule [27, 29]. For k even, there is an isomorphism $\mathbb{C}\ell_k \to \mathrm{End}(\mathbb{C}^{2^{k/2}})$ by Clifford multiplication. This observation implies that $\mathbb{C}^{2^{k/2}}$ is a $\mathbb{Z}_2$-graded Morita equivalence bimodule between $\mathbb{C}\ell_k$ and $\mathbb{C}$, where we equip $\mathbb{C}^{2^{k/2}}$ with a left $\mathbb{C}\ell_k$-valued inner-product $_{\mathbb{C}\ell_k}(\cdot \mid \cdot)$ such that $_{\mathbb{C}\ell_k}(w_1 \mid w_2) \cdot w_3 = w_1 \langle w_2, w_3 \rangle_{\mathbb{C}^\nu}$. This bimodule gives an invertible class $[(\mathbb{C}\ell_k, \mathbb{C}^{2^{k/2}}_{\mathbb{C}}, 0)] \in KK(\mathbb{C}\ell_k, \mathbb{C})$. One can take the external product of $\lambda_k^{\mathbb{S}}$ with this class on the right to obtain (complex) λ_k. That is,

$$[\lambda_k^{\mathbb{S}}] \hat{\otimes}_{\mathbb{C}} [(\mathbb{C}\ell_k, \mathbb{C}^{2^{k/2}}, 0)] = [\lambda_k] \in KK(A \hat{\otimes} \mathbb{C}\ell_k, B).$$

Similarly $[\lambda_k^{\mathfrak{S}}] = [\lambda_k]\hat{\otimes}[(\mathbb{C}, (\mathbb{C}^{2^{k/2}})^*_{\mathbb{C}\ell_k}, 0)]$ with $(\mathbb{C}^{2^{k/2}})^*_{\mathbb{C}\ell_k}$ the conjugate module providing the inverse to $[(\mathbb{C}\ell_k, \mathbb{C}^{2^{k/2}}_{\mathbb{C}}, 0)]$, see [32] for more details on Morita equivalence bimodules.

For k odd we use the graded Kasparov module (2) instead of $\lambda_k^{\mathfrak{S}}$. We can again compose this graded Kasparov module with the KK-class from the Morita equivalence bimodule $(\mathbb{C}\ell_{k-1}, \mathbb{C}^{2^{(k-1)/2}}_{\mathbb{C}}, 0)$. The external product gives $[\lambda_k] \in KK^k(A, B)$. Hence from an index-theoretic perspective, the Kasparov modules $\lambda_k^{\mathfrak{S}}$ and λ_k are equivalent up to a normalisation coming from the spinor dimension.

In the case of real spaces and algebras, a similar (but more involved) equivalence also holds for real spinor representations. Namely, for $\mathbb{K} = \mathbb{R}, \mathbb{C}$ or $\mathbb{H}$, there is a unique irreducible representation $C\ell_{r,s} \to \mathrm{End}_{\mathbb{K}}(\mathfrak{S}_{\mathbb{K}})$ if $s-r+1$ is not a multiple of 4, otherwise there are 2 irreducible representations [26, Chapter 1, Theorem 5.7]. To relate these modules to $\bigwedge^* \mathbb{R}^k$, one also uses that $\mathbb{C} \cong \mathbb{R}^2$ and $\mathbb{H} \cong \mathbb{R}^4$. Obviously there are more cases to check in the real setting, but because we do not use the spin Kasparov module in the real case, the full details are beyond the scope of this paper.

In order to consider weak invariants in the real case, we will often go beyond the limits of semifinite index theory and will need to work with the Kasparov modules and KK-classes directly. In such a setting, we prefer to work with the 'oriented' Kasparov module λ_k for several reasons:

1. The oriented structure, $\bigwedge^* \mathbb{R}^k$, and its corresponding Clifford representations is at the heart of Kasparov theory and, for example, plays a key role in the proof of Bott periodicity [16, §5] and Poincaré duality [17, §4]. This is also evidenced in Theorem 3 below (also compare with [13], where to achieve factorisation of equivariant (spin) spectral triples, a 'middle module' is required that plays of the role of the complex Morita equivalence linking $\lambda_k^{\mathfrak{S}}$ and λ_k for complex algebras).
2. The Clifford actions of $C\ell_{0,k}$ and $C\ell_{k,0}$ on $\bigwedge^* \mathbb{R}^k$ are explicit. This makes the Clifford representations more amenable to the Kasparov product as well as the Clifford index used to define real weak invariants (see Sect. 4).

2.2.3 Kasparov Module to Semifinite Spectral Triple

Returning to the example $B = C(\Omega) \rtimes_\phi \mathbb{Z}^{d-k}$, it will be assumed that Ω possesses a probability measure $\mathbf{P}$ that is invariant under the $\mathbb{Z}^d$-action and $\mathrm{supp}(\mathbf{P}) = \Omega$. Hence $\mathbf{P}$ induces a faithful trace on $C(\Omega)$ and $C(\Omega) \rtimes_\phi \mathbb{Z}^{d-k}$ by the formula

$$\tau\Big(\sum_{m\in\mathbb{Z}^{d-k}} S^m g_m \Big) = \int_\Omega g_0(\omega)\, d\mathbf{P}(\omega) \,.$$

Thus, we will assume from now on that our generic algebra B has a faithful and norm lower semicontinuous trace, τ_B, that is invariant under the $\mathbb{Z}^k$-action. This trace now allows to construct a semifinite spectral triple from the above Kasparov module. We first construct the GNS space $L^2(B, \tau_B)$ and consider the new Hilbert space $\ell^2(\mathbb{Z}^k) \otimes$

$L^2(B, \tau_B)$. Let us note that $\ell^2(\mathbb{Z}^k) \otimes L^2(B, \tau_B) \cong \ell^2(\mathbb{Z}^k, B) \otimes_B L^2(B, \tau_B)$ so the adjointable action of $A = B \rtimes_\theta \mathbb{Z}^k$ on $\ell^2(\mathbb{Z}^k, B)$ extends to a representation of A on $\ell^2(\mathbb{Z}^k) \otimes L^2(B, \tau_B)$.

Proposition 3 ([24, Theorem 1.1]) *Given $T \in \mathrm{End}_B(\ell^2(\mathbb{Z}^k, B))$ with $T \geq 0$, define*

$$\mathrm{Tr}_\tau(T) = \sup_I \sum_{\xi \in I} \tau_B[(\xi \mid T\xi)_B] \,,$$

where the supremum is taken over all finite subsets $I \subset \ell^2(\mathbb{Z}^k, B)$ with $\sum_{\xi \in I} \Theta_{\xi, \xi} \leq 1$.

1. *Then Tr_τ is a semifinite norm lower semicontinuous trace on the compact endomorphisms $\mathrm{End}_B^0(\ell^2(\mathbb{Z}^k, B))$ with the property $\mathrm{Tr}_\tau(\Theta_{\xi_1, \xi_2}) = \tau_B[(\xi_2 \mid \xi_1)_B]$.*
2. *Let $\mathscr{N}$ be the von Neumann algebra $\mathrm{End}_B^{00}(\ell^2(\mathbb{Z}^k, B))'' \subset \mathscr{B}[\ell^2(\mathbb{Z}^k) \otimes L^2(B, \tau_B)]$. Then the trace Tr_τ extends to a faithful semifinite trace on the positive cone $\mathscr{N}_+$.*

Recall that the operator $(1 + |X|^2)$ acts diagonally on the frame $\{\delta_m \otimes 1_B\}_{m \in \mathbb{Z}^k}$, so

$$(1 + |X|^2)^{-s/2} = \sum_{m \in \mathbb{Z}^k} (1 + |m|^2)^{-s/2} \Theta_{\delta_m \otimes 1_B, \delta_m \otimes 1_B}.$$

Using the properties Tr_τ, one can compute that

$$\mathrm{Tr}_\tau\left((1 + |X|^2)^{-s/2}\right) = \sum_{m \in \mathbb{Z}^k} (1 + |m|^2)^{-s/2} \tau_B((\delta_m \otimes 1_B \mid \delta_m \otimes 1_B)_B)$$

$$= \sum_{m \in \mathbb{Z}^k} (1 + |m|^2)^{-s/2} \tau_B(1_B).$$

This observation and a little more work gives the following result.

Proposition 4 ([7, Proposition 5.8]) *For $\mathscr{A} \subset B \rtimes_\theta \mathbb{Z}^k$ the algebra of operators $\sum_{n \in \mathbb{Z}^k} S^n b_n$ with Schwartz-class coefficients, the tuple*

$$\left(\mathscr{A} \hat{\otimes} C\ell_{0,k}, \, \ell^2(\mathbb{Z}^k) \otimes L^2(B, \tau_B) \hat{\otimes} \bigwedge\nolimits^* \mathbb{R}^k, \, \sum_{j=1}^k X_j \otimes 1 \hat{\otimes} \gamma^j\right)$$

is a QC^∞ and k-summable semifinite spectral triple relative to $\mathscr{N} \hat{\otimes} \mathrm{End}(\bigwedge^* \mathbb{R}^k)$ with trace $\mathrm{Tr}_\tau \hat{\otimes} \mathrm{Tr}_{\bigwedge^* \mathbb{R}^k}$.

We have the analogous result for the spin Dirac operator.

Proposition 5 *The tuple*

$$\left(\mathscr{A}, \, \ell^2(\mathbb{Z}^k) \otimes L^2(B, \tau_B) \hat{\otimes} \mathbb{C}^v, \, \sum_{j=1}^k X_j \otimes 1 \hat{\otimes} \Gamma^j\right)$$

is a QC^∞ and k-summable complex semifinite spectral triple relative to $\mathcal{N}\hat{\otimes}\mathrm{End}(\mathbb{C}^\nu)$ with trace $\mathrm{Tr}_\tau\hat{\otimes}\mathrm{Tr}_{\mathbb{C}^\nu}$. The spectral triple is even if k is even with grading operator $\Gamma_0 = (-i)^{k/2}\Gamma^1\cdots\Gamma^k$. The spectral triple is odd if k is odd.

Therefore all hypotheses required to apply the semifinite local index formula are satisfied. Furthermore, the algebra $\mathscr{A}$ is Fréchet and stable under the holomorphic functional calculus. Therefore all pairings of $K_k(\mathscr{A})$ extend to pairings with $K_k(B \rtimes_\theta \mathbb{Z}^k)$.

3 Complex Pairings and the Local Index Formula

Let us now restrict to a complex algebra $A = B \rtimes_\theta \mathbb{Z}^k$, where B is separable, unital and possesses a faithful, semifinite and norm lower semicontinuous trace τ_B that is invariant under the $\mathbb{Z}^k$-action. First, the semifinite index pairing is related to the 'base algebra' B and the dynamics of the $\mathbb{Z}^k$-action.

Lemma 1 *The semifinite index pairing of a class $[x] \in K_k(B \rtimes_\theta \mathbb{Z}^k)$ with the spin semifinite spectral triple from Proposition 5 can be computed by the K-theoretic composition*

$$K_k(B \rtimes_\theta \mathbb{Z}^k) \times KK^k(B \rtimes_\theta \mathbb{Z}^k, B) \;\to\; K_0(B) \xrightarrow{(\tau_B)_*} \mathbb{R}\,, \tag{4}$$

with the class in $KK^k(B \rtimes_\theta \mathbb{Z}^k, B)$ represented by $\lambda_k^\mathfrak{S}$ from Proposition 1.

Proof We start with the even pairing, with $p \in M_q(B \rtimes_\theta \mathbb{Z}^k)$ representing $[p] \in K_0(B \rtimes_\theta \mathbb{Z}^k)$. Taking the double $X = X_M$ if necessary, the semifinite index pairing is given by the semifinite index

$$\langle[p], [(\mathscr{A}, \mathscr{H}, X)]\rangle \;=\; (\mathrm{Tr}_\tau\otimes\mathrm{Tr}_{\mathbb{C}^l})(P_{\mathrm{Ker}(p(X\otimes 1_q)_+p)}) - (\mathrm{Tr}_\tau\otimes\mathrm{Tr}_{\mathbb{C}^l})(P_{\mathrm{Ker}(p(X\otimes 1_q)^*_+p)})\,,$$

with $P_{\mathrm{Ker}(T)}$ the projection onto the kernel of T, $\mathrm{Tr}_{\mathbb{C}^l}$ the finite trace from the spin structure and the operator X_+ comes from the decomposition $X = \begin{pmatrix} 0 & X_- \\ X_+ & 0 \end{pmatrix}$ due to the grading in even dimension. Next we compute the Kasparov product in Eq. (4) following, for example, [31, Section 4.3.1]. The product $[p]\hat{\otimes}_A[\lambda_k] \in KK(\mathbb{C}, B)$ is represented by the class of the Kasparov module

$$\left(\mathbb{C}, p\big(\ell^2(\mathbb{Z}^k, B)^{\oplus q}\big) \otimes \mathbb{C}^{2l}, \begin{pmatrix} 0 & p(X\otimes 1_q)-p \\ p(X\otimes 1_q)+p & 0 \end{pmatrix}\right), \quad \gamma = \mathrm{Ad}\begin{pmatrix} 1 & 0 \\ 0 & -1 \end{pmatrix}.$$

After regularising if necessary, $\mathrm{Ker}(p(X \otimes 1_q)_+p)$ is a finitely generated and projective submodule of $p\big(\ell^2(\mathbb{Z}^k, B)^{\oplus q}\big)\otimes\mathbb{C}^l$ and the projection onto this submodule is compact (and therefore finite-rank). We can associate a K-theory class to this

Kasparov module by noting that $\text{End}_B^0\big(p(\ell^2(\mathbb{Z}^k, B))^{\oplus q} \otimes \mathbb{C}^l\big) \cong B \otimes \mathcal{K}$ and taking the difference

$$[P_{\text{Ker}(p(X \otimes 1_q) + p)}] - [P_{\text{Ker}(p(X \otimes 1_q)^*_+ p)}] \in K_0(B)$$

Because $\text{Ker}(p(X \otimes 1_q)_+ p)$ is finitely generated, there exists a *finite* frame $\{e_j\}_{j=1}^n$ in $p(\ell^2(\mathbb{Z}^k, B)^{\oplus q}) \otimes \mathbb{C}^l$ such that $\sum_{j=1}^n \Theta_{e_j, e_j} = \text{Id}_{\text{Ker}(p(X \otimes 1_q) + p)}$. Taking the induced trace $(\tau_B)_* : K_0(B) \to \mathbb{R}$, one can use the properties of the dual trace Tr_τ to note that

$$\tau_B\big(P_{\text{Ker}(p(X \otimes 1_q) + p)}\big) = \sum_{j=1}^n \tau_B((e_j \mid e_j)_B) = \sum_{j=1}^n \text{Tr}_\tau(\Theta_{e_j, e_j}).$$

The right hand side is now a trace defined over $\text{End}_B^{00}\big(p(\ell^2(\mathbb{Z}^k, B)^{\oplus q}) \otimes \mathbb{C}^l\big) \subset \mathcal{N} \otimes \text{End}(\mathbb{C}^l)$ and by construction it is the same as $(\text{Tr}_\tau \otimes \text{Tr}_{\mathbb{C}^l})(P_{\text{Ker}(p(X \otimes 1_q) + p)})$. An analogous result holds for $\text{Ker}(p(X \otimes 1_q)^*_+ p)$, so $(\tau_B)_*([p] \hat{\otimes}_{B \rtimes_\theta \mathbb{Z}^k} [\lambda_k])$ is represented by

$$(\text{Tr}_\tau \otimes \text{Tr}_{\mathbb{C}^l})(P_{\text{Ker}(p(X \otimes 1_q) + p)}) - (\text{Tr}_\tau \otimes \text{Tr}_{\mathbb{C}^l})(P_{\text{Ker}(p(X \otimes 1_q)^*_+ p)}),$$

and thus the pairings coincide.

For the odd pairing, the same argument applies for $\text{Index}_{\text{Tr}_\tau}(\Pi u \Pi)$ with Π the positive spectral projection of X and $[u] \in K_1(B \rtimes_\theta \mathbb{Z}^k)$. For this, one has to appeal to the appendix of [15] or [31, Section 4.3.2]. □

Lemma 1 means that the semifinite pairing considered here has a concrete K-theoretic interpretation. In particular, we know that $\langle[x], [(\mathscr{A}, \mathscr{H}, X)]\rangle \subset \tau_B(K_0(B))$, which is countably generated for separable B. This is one of the reasons we build a Kasparov module first and then construct a semifinite spectral triple via the dual trace Tr_τ.

Remark 2 We may also pair K-theory classes with the Kasparov module λ_k from Proposition 2 by the composition

$$K_k(B \rtimes_\theta \mathbb{Z}^k) \times KK^k(B \rtimes_\theta \mathbb{Z}^k, B) \to KK(\mathbb{C}\ell_{2k}, B) \xrightarrow{\cong} K_0(B) \xrightarrow{(\tau_B)_*} \mathbb{R} \quad (5)$$

where $KK(\mathbb{C}\ell_{2k}, B) \cong K_0(B)$ by stability and [16, §6, Theorem 3]. We can think of Eq. (5) as the definition of the complex semifinite index pairing of K-theory with the semifinite spectral triple from Proposition 4 over the graded algebra $B \rtimes_\theta \mathbb{Z}^k \hat{\otimes} \mathbb{C}\ell_k$. Indeed, in more general circumstances, the K-theoretic composition is how the semifinite pairing is defined, where in general one pairs with the class in $KK^k(A, C)$ with C a subalgebra of $\mathscr{K}_\mathcal{N}$ [10, Section 2.3].

Equation (5) also has a natural analogue in the real case, namely

$$KO_k(B \rtimes \mathbb{Z}^k) \times KKO^k(B \rtimes \mathbb{Z}^k, B) \to KKO(C\ell_{k,0} \hat{\otimes} C\ell_{0,k}, B) \xrightarrow{\cong} KO_0(B) \xrightarrow{(\tau_B)_*} \mathbb{R}$$

as $C\ell_{k,0}\hat{\otimes}C\ell_{0,k} \cong M_l(\mathbb{R})$ which is Morita equivalent to $\mathbb{R}$. Of course, we also want to pair our Kasparov module with elements in $KO_j(B \rtimes \mathbb{Z}^k)$ for $j \neq k$, and in this situation we use the general Kasparov product (see Sect. 4). $\diamond$

To compute the local index formula, we first note some preliminary results.

Lemma 2 *The function*

$$\zeta(s) = \mathrm{Tr}_\tau\big(S^n b(1 + |X|^2)^{-s/2}\big) , \qquad s > k ,$$

has a meromorphic extension to the complex plane with

$$\mathop{\mathrm{res}}_{s=k} \mathrm{Tr}_\tau\big(S^n b(1 + |X|^2)^{-s/2}\big) = \delta_{n,0} \mathrm{Vol}_{k-1}(S^{k-1})\tau_B(b) .$$

Proof We use the frame $\{\delta_m \otimes 1_B\}_{m\in\mathbb{Z}^k}$ for $\ell^2(\mathbb{Z}^k, B)$ and note that $S^n b \cdot (\delta_m \otimes 1_B) = \delta_{m+n} \otimes \alpha_{-m-n}(\theta(n, m))\alpha_{-m}(b)$. Computing, for $s > k$,

$$\mathrm{Tr}_\tau\big(S^n b(1 + |X|^2)^{-s/2}\big) = \mathrm{Tr}_\tau\Big(S^n b \sum_{m\in\mathbb{Z}^k}(1 + |m|^2)^{-s/2}\Theta_{\delta_m\otimes 1,\delta_m\otimes 1}\Big)$$

$$= \sum_{m\in\mathbb{Z}^k}(1 + |m|^2)^{-s/2}\mathrm{Tr}_\tau\big(\Theta_{\delta_{m+n}\otimes\alpha_{-m-n}(\theta(n,m))\alpha_{-m}(b),\delta_m\otimes 1}\big)$$

$$= \sum_{m\in\mathbb{Z}^k}(1 + |m|^2)^{-s/2}\tau_B\big(\langle\delta_m, \delta_{n+m}\rangle_{\ell^2(\mathbb{Z}^k)}\alpha_{-m-n}(\theta(n, m))\alpha_{-m}(b)\big)$$

$$= \delta_{n,0} \sum_{m\in\mathbb{Z}^k}(1 + |m|^2)^{-s/2}\tau_B\big(\theta(0, m)b\big)$$

$$= \delta_{n,0} \tau_B(b) \sum_{m\in\mathbb{Z}^k}(1 + |m|^2)^{-s/2}$$

$$= \delta_{n,0} \tau_B(b) \mathrm{Vol}_{k-1}(S^{k-1}) \frac{\Gamma\big(\tfrac{k}{2}\big)\Gamma\big(\tfrac{s-k}{2}\big)}{2\Gamma\big(\tfrac{k}{2}\big)} ,$$

where the invariance of the α-action in the trace was used. By the functional equation for the Γ-function, $\zeta(s)$ has a meromorphic extension to the complex plane and is holomorphic for $\Re(s) > k$. Computing the residue obtains the result. $\square$

Next let us note that any trace on B can be extended to $\mathscr{A}$ by defining

$$\mathscr{T}\Big(\sum_n S^n b_n\Big) = \tau_B(b_0) ,$$

where $\mathscr{T}$ is faithful and norm lower semicontinuous if τ_B is faithful and norm lower semicontinuous. A direct extension of Lemma 2 then gives that

$$\mathop{\mathrm{res}}_{s=k} \mathrm{Tr}_\tau\big(a(1 + |X|^2)^{-s/2}\big) = \mathrm{Vol}_{k-1}(S^{k-1})\mathscr{T}(a) , \qquad a \in \mathscr{A} . \tag{6}$$

3.1   Odd Formula

We will compute the semifinite pairing with the spectral triple constructed from
Proposition 5, which aligns our results with [31]. The equivalence between spin
and oriented semifinite spectral triples means that we also obtain formulas for the
pairing with the semifinite spectral triple from Proposition 4, where the result would
be the same up to a normalisation.

Except for certain cases where specific results on the spinor trace of the gamma
matrices are needed, we will write the trace $\mathrm{Tr}_\tau \hat{\otimes} \mathrm{Tr}_{\mathbb{C}^\nu}$ on the von Neumann algebra
$\mathcal{N} \hat{\otimes} \mathrm{End}(\mathbb{C}^\nu)$ as just Tr_τ.

Theorem 1 (Odd Index Formula) *Let u be a complex unitary in $M_q(\mathcal{A})$ and
X_{odd} the complex semifinite spectral triple from Proposition 5 with k odd. Then the
semifinite index pairing is given by the formula*

$$\langle [u], [X_{\mathrm{odd}}] \rangle \;=\; C_k \sum_{\sigma \in S_k} (-1)^\sigma \, (\mathrm{Tr}_{\mathbb{C}^q} \otimes \mathscr{T}) \left(\prod_{i=1}^{k} u^* \partial_{\sigma(i)} u \right),$$

*where $C_{2n+1} = \frac{-2(2\pi)^n n!}{i^{n+1}(2n+1)!}$, $\mathrm{Tr}_{\mathbb{C}^q}$ is the matrix trace on $\mathbb{C}^q$, S_k is the permutation
group on $\{1, \ldots, k\}$ and $\partial_j a = -i[X_j, a]$ for any $a \in \mathcal{A}$ and $j \in \{1, \ldots, k\}$.*

Let us focus on the case $q = 1$ and then extend to matrices by taking $(D \otimes 1_q)$
with $D = \sum_{j=1}^{k} X_j \otimes \Gamma^j$. Because the semifinite spectral triple of Proposition 5 is
smooth and with spectral dimension k, the odd local index formula from [8] gives

$$\langle [u], [X_{\mathrm{odd}}] \rangle \;=\; \frac{-1}{\sqrt{2\pi i}} \mathop{\mathrm{res}}_{r=(1-k)/2} \sum_{\substack{m=1,\mathrm{odd}}}^{2N-1} \phi_m^r(\mathrm{Ch}^m(u)),$$

where u is a unitary in $\mathcal{A}$, $N = \lfloor k/2 \rfloor + 1$ and

$$\mathrm{Ch}^{2n+1}(u) \;=\; (-1)^n n! \, u^* \otimes u \otimes u^* \otimes \cdots \otimes u, \qquad (2n+2 \text{ entries}).$$

The functional ϕ_m^r is the resolvent cocycle from [8]. To compute the index pairing
we recall the following important observation.

Lemma 3 ([4, Section 11.1]) *The only term in the sum $\sum_{\substack{m=1,\mathrm{odd}}}^{2N-1} \phi_m^r(\mathrm{Ch}^m(u))$ that
contributes to the index pairing is the term with $m = k$.*

Proof We first note that the spinor trace on the Clifford generators is given by

$$\mathrm{Tr}_{\mathbb{C}^\nu}(i^k \Gamma^1 \cdots \Gamma^k) \;=\; (-i)^{\lfloor (k+1)/2 \rfloor} 2^{\lfloor (k-1)/2 \rfloor}, \tag{7}$$

and will vanish on any product of j Clifford generators with $0 < j < k$. The resolvent cocycle involves the spinor trace of terms

$$a_0 R_s(\lambda)[D, a_1] R_s(\lambda) \cdots [D, a_m] R_s(\lambda), \qquad R_s(\lambda) = (\lambda - (1 + s^2 + D^2))^{-1},$$

for $a_0, \ldots, a_m \in \mathscr{A}$. Noting that $[D, a_l] = i \sum_{j=1}^{k} \partial_j a_l \otimes \Gamma^j$ and $R_s(\lambda)$ is diagonal in the spinor representation, it follows that the product $a_0 R_s(\lambda)[D, a_1] \cdots [D, a_m] R_s(\lambda)$ will be in the span of m Clifford generators acting on $\ell^2(\mathbb{Z}^k) \otimes L^2(B, \tau_B) \hat{\otimes} \mathbb{C}^\nu$. Furthermore, the trace estimates ensure that each spinor component of ϕ_m^r

$$\int_\ell \lambda^{-k/2-r} a_0 (\lambda - (1 + s^2 + |X|^2))^{-1} \partial_{j_1} a_1 \cdots \partial_{j_m} a_m (\lambda - (1 + s^2 + |X|^2))^{-1} \, d\lambda$$

is trace-class for $a_0, \ldots, a_m \in \mathscr{A}$ and real part $\Re(r)$ sufficiently large. Hence for $0 < m < k$, the spinor trace will vanish for $\Re(r)$ large and $\phi_m^r(\mathrm{Ch}^m(u))$ analytically extends as a function holomorphic in a neighbourhood of $r = (1-k)/2$ for $0 < m < k$. Thus $\phi_m^r(\mathrm{Ch}^m(u))$ does not contribute to the index pairing for $0 < m < k$. $\quad\square$

Proof (Proof of Theorem 1) Lemma 3 simplifies the semifinite index substantially, namely it is given by the expression

$$\langle [u], [X_{\mathrm{odd}}] \rangle = \frac{-1}{\sqrt{2\pi i}} \mathop{\mathrm{res}}_{r=(1-k)/2} \phi_k^r(\mathrm{Ch}^k(u)).$$

Therefore one needs to compute the residue at $r = (k-1)/2$ of

$$\mathscr{C}_k \int_0^\infty s^k \, \mathrm{Tr}_\tau \left(\int_\ell \lambda^{-k/2-r} u^* R_s(\lambda)[D, u] R_s(\lambda)[D, u^*] \cdots [D, u] R_s(\lambda) \, d\lambda \right) ds,$$

where $k = 2n + 1$ and the constant

$$\mathscr{C}_k = -\frac{(-1)^{n+1} n!}{(2\pi i)^{3/2}} \frac{\sqrt{2i} \, 2^{d+1} \Gamma(d/2 + 1)}{\Gamma(d + 1)}$$

comes from the definition of the resolvent cocycle, see [10, Section 3.2], and $\mathrm{Ch}^k(u)$. To compute this residue we move all terms $R_s(\lambda)$ to the right, which can be done up to a function holomorphic at $r = (1-k)/2$. This allows us to take the Cauchy integral. We then observe that $\underbrace{[D, u][D, u^*] \cdots [D, u]}_{k \text{ terms}} \in \mathscr{A} \otimes 1_{\mathbb{C}^\nu}$, so Lemma 2 implies that the zeta function

$$\mathrm{Tr}_\tau \left(u^*[D, u][D, u^*] \cdots [D, u](1 + D^2)^{-z/2} \right)$$

has at worst a simple pole at $\Re(z) = k$. Therefore we can explicitly compute

$$\frac{-1}{\sqrt{2\pi i}} \operatorname*{res}_{r=(1-k)/2} \phi_k^r(\mathrm{Ch}^k(u))$$

$$= (-1)^{n+1} n! \frac{1}{k!} \tilde{\sigma}_{n,0} \operatorname*{res}_{z=k} \mathrm{Tr}_\tau\left(u^*[D,u][D,u^*]\cdots[D,u](1+D^2)^{-z/2}\right),$$

where the numbers $\tilde{\sigma}_{n,j}$ are defined by the formula

$$\prod_{j=0}^{n-1}(z+j+1/2) = \sum_{j=0}^n z^j \tilde{\sigma}_{n,j} .$$

Hence the number $\tilde{\sigma}_{n,0}$ is the coefficient of 1 in the product $\prod_{l=0}^{n-1}(z+l+1/2)$. This is the product of all the non-z terms, which can be written as

$$(1/2)(3/2)\cdots(n-1/2) = \frac{1}{\sqrt{\pi}} \, \Gamma(k/2) .$$

Putting this back together, our index pairing can be written as

$$\langle[u],[X_{\mathrm{odd}}]\rangle = (-1)^{n+1} \frac{n!\Gamma(k/2)}{k!\sqrt{\pi}} \operatorname*{res}_{z=k} \mathrm{Tr}_\tau\left(u^*[D,u][D,u^*]\cdots[D,u](1+D^2)^{-z/2}\right).$$

We make use of the identity $[D,u^*] = -u^*[D,u]u^*$, which allows us to rewrite

$$u^* \underbrace{[D,u][D,u^*]\cdots[D,u]}_{k=2n+1 \text{ terms}} = (-1)^n u^*[D,u]u^*[D,u]u^*\cdots u^*[D,u]$$

$$= (-1)^n \left(u^*[D,u]\right)^k .$$

Recall that $[D,u] = \sum_{j=1}^k [X_j,u]\hat{\otimes}\Gamma^j = i\sum_{j=1}^k \partial_j(u)\hat{\otimes}\Gamma^j$ so the relation $u^*[D,u] = i\sum_{j=1}^k u^*\partial_j(u)\hat{\otimes}\Gamma^j$ follows. Taking the kth power

$$\left(u^*[D,u]\right)^k = i^k \sum_{J=(j_1,\ldots,j_k)} u^*(\partial_{j_1}u)\cdots u^*(\partial_{j_k}u)\hat{\otimes}\Gamma^{j_1}\cdots\Gamma^{j_k}$$

where the sum is extended over all multi-indices J. Note that every term in the sum is a multiple of the identity of $\mathbb{C}^\nu$ and so has a non-zero spinor trace. Writing this product in terms of permutations,

$$(-1)^n \left(u^*[D,u]\right)^k = (-1)^n i^k \sum_{\sigma\in S_k}(-1)^\sigma \prod_{j=1}^k u^*(\partial_{\sigma(j)}u)\hat{\otimes}\Gamma^j ,$$

with S_k is the permutation group of k letters. Let's put all this back together.

$$\langle [u], [X_{odd}] \rangle = (-1)^{n+1} \frac{n! \Gamma(k/2)}{k! \sqrt{\pi}} \operatorname*{res}_{z=k} \operatorname{Tr}_\tau \left(u^*[D, u][D, u^*] \cdots [D, u](1 + D^2)^{-z/2} \right)$$

$$= -\frac{n! \Gamma(k/2)}{k! \sqrt{\pi}} \operatorname*{res}_{z=k} \operatorname{Tr}_\tau \left[i^k \left(\sum_{\sigma \in S_k} (-1)^\sigma \prod_{j=1}^{k} u^*(\partial_{\sigma(j)} u) \hat{\otimes} \Gamma^j \right) (1 + D^2)^{-z/2} \right]$$

$$= -\frac{n! \Gamma(k/2) 2^{\lfloor (k-1)/2 \rfloor}}{i^{\lfloor (k+1)/2 \rfloor} k! \sqrt{\pi}} \operatorname*{res}_{z=k} \operatorname{Tr}_\tau \left(\sum_{\sigma \in S_k} (-1)^\sigma \prod_{j=1}^{k} u^*(\partial_{\sigma(j)} u)(1 + |X|^2)^{-z/2} \right),$$

where we have used Eq. (7) and that $(1 + D^2) = (1 + |X|^2) \otimes 1_{\mathbb{C}^\nu}$. We can apply Eq. (6) to reduce the formula to

$$\langle [u], [X_{odd}] \rangle = -\frac{n! \Gamma(k/2) \operatorname{Vol}_{k-1}(S^{k-1}) 2^{\lfloor (k-1)/2 \rfloor}}{i^{\lfloor (k+1)/2 \rfloor} k! \sqrt{\pi}} \sum_{\sigma \in S_k} (-1)^\sigma \mathcal{T} \left(\prod_{i=1}^{k} u^*(\partial_{\sigma(i)} u) \right).$$

Now the identity $\operatorname{Vol}_{k-1}(S^{k-1}) = \frac{k \pi^{k/2}}{\Gamma(k/2+1)}$ allows to simplify

$$\frac{n! \Gamma(k/2) \operatorname{Vol}_{k-1}(S^{k-1}) 2^{\lfloor (k-1)/2 \rfloor}}{i^{\lfloor (k+1)/2 \rfloor} k! \sqrt{\pi}} = \frac{2(2\pi)^n n!}{i^{n+1}(2n+1)!},$$

for $k = 2n + 1$, and therefore

$$\langle [u], [X_{odd}] \rangle = C_k \sum_{\sigma \in S_k} (-1)^\sigma \mathcal{T} \left(\prod_{i=1}^{k} u^*(\partial_{\sigma(i)} u) \right), \qquad C_{2n+1} = \frac{-2(2\pi)^n n!}{i^{n+1}(2n+1)!},$$

which concludes the argument. $\qquad \square$

3.2 Even Formula

Theorem 2 (Even Index Formula) *Let p be a complex projection in $M_q(\mathscr{A})$ and X_{even} the complex semifinite spectral triple from Proposition 5 with k even. Then the semifinite index pairing can be expressed by the formula*

$$\langle [p], [X_{even}] \rangle = C_k \sum_{\sigma \in S_k} (-1)^\sigma (\operatorname{Tr}_{\mathbb{C}^q} \otimes \mathcal{T}) \left(p \prod_{i=1}^{k} \partial_{\sigma(i)} p \right),$$

where $C_k = \frac{(2\pi i)^{k/2}}{(k/2)!}$ and S_k is the permutation group of $\{1, \ldots, d\}$.

Like the setting with k odd, the computation can be substantially simplified with some preliminary results. Let us again focus on the case $q = 1$ and first recall the even local index formula [9]:

$$\langle [p], [X_{\text{even}}] \rangle \ = \ \operatorname*{res}_{r=(1-k)/2} \ \sum_{m=0,\text{even}}^{k} \phi_m^r(\text{Ch}^m(p)) \,,$$

where ϕ_m^r is the resolvent cocycle and

$$\text{Ch}^{2n}(p) \ = \ (-1)^n \frac{(2n)!}{2(n!)} \, (2p-1) \otimes p^{\otimes 2n} \,, \qquad \text{Ch}^0(p) \ = \ p \,.$$

Proof (Proof of Theorem 2) The proof of Lemma 3 also holds here to show that $\phi_m^r(\text{Ch}^m(p))$ does not contribute to the index pairing for $0 < m < k$. Therefore the index computation is reduced to

$$\langle [p], [X_{\text{even}}] \rangle \ = \ \operatorname*{res}_{r=(1-k)/2} \ \phi_k^r(\text{Ch}^k(p)) \,,$$

which is a residue at $r = (1-k)/2$ of the term

$$\mathscr{C}_k \int_0^\infty s^k \operatorname{Tr}_\tau \Big(\Gamma_0 \int_\ell \lambda^{-k/2-r}(2p-1)R_s(\lambda)[D,p]R_s(\lambda) \cdots [D,p]R_s(\lambda) \, d\lambda \Big) ds \,,$$

where $\Gamma_0 = (-i)^{k/2} \Gamma^1 \Gamma^2 \cdots \Gamma^k$ is the grading operator of $\mathbb{C}^\nu$ and

$$\mathscr{C}_k \ = \ \frac{(-1)^{k/2} k! \, 2^k \, \Gamma(k/2+1)}{i\pi(k/2)! \, \Gamma(k+1)}$$

comes from the resolvent cocycle and the normalisation of $\text{Ch}^k(p)$. Like the case of k odd, one can move the resolvent terms to the right up to a holomorphic error in order to take the Cauchy integral. Lemma 2 implies that the complex function $\operatorname{Tr}_\tau \big(\Gamma_0(2p-1)([D,p])^k(1+D^2)^{-z/2} \big)$ has at worst a simple pole at $\Re(z) = k$. Computing the residue explicitly,

$$\operatorname*{res}_{r=(1-k)/2} \phi_k^r(\text{Ch}^k(p)) \ = \ \frac{(-1)^{k/2}}{2((k/2)!)} \sigma_{k/2,1} \operatorname*{res}_{z=k} \operatorname{Tr}_\tau \big(\Gamma_0(2p-1)([D,p])^k(1+D^2)^{-z/2} \big) \,,$$

where $\sigma_{k/2,1}$ is the coefficient of z in $\prod_{j=0}^{k/2-1}(z+j)$ and is given by the number $\sigma_{k/2,1} = ((k/2)-1)!$. Putting these results back together,

$$\langle [p], [X_{\text{even}}] \rangle \ = \ (-1)^{k/2} \frac{1}{k} \operatorname*{res}_{z=k} \operatorname{Tr}_\tau \big(\Gamma_0(2p-1)([D,p])^k(1+D^2)^{-z/2} \big) \,.$$

Next we claim that $\mathrm{Tr}_\tau\big(\Gamma_0([D,p])^k(1+D^2)^{-z/2}\big)=0$ for $\Re(z)>k$. To see this, let us compute for $\Gamma_0=(-i)^{k/2}\Gamma^1\cdots\Gamma^k$,

$$[D,p]^k = \sum_{\sigma\in S_k}(-1)^\sigma\prod_{i=1}^k[X_{\sigma(i)},p]\hat\otimes\Gamma^i = i^{k/2}\Gamma_0\sum_{\sigma\in S_k}(-1)^\sigma\prod_{j=1}^k[X_{\sigma(j)},p]\hat\otimes 1_{\mathbb{C}^\nu}.$$

Because $\sum_\sigma(-1)^\sigma\prod_{j=1}^k[X_{\sigma(j)},p]$ is symmetric with respect to the ±1 eigenspaces of Γ_0, the spinor trace $\mathrm{Tr}_\tau(\Gamma_0[D,p]^k(1+D^2)^{-z/2})$ will vanish for $\Re(z)>k$. Therefore the zeta function $\mathrm{Tr}_\tau(\Gamma_0[D,p]^k(1+D^2)^{-z/2})$ analytically continues as a function holomorphic in a neighbourhood of $z=k$ and its residue does not contribute to the index.

We know that $[D,p]=\sum_{j=1}^k[X_j,p]\hat\otimes\Gamma^j = i\sum_{j=1}^k\partial_j p\hat\otimes\Gamma^j$ and so

$$p([D,p])^k = (-1)^{k/2}p\sum_{\sigma\in S_k}(-1)^\sigma\prod_{j=1}^k\partial_{\sigma(j)}p\hat\otimes\Gamma^j.$$

Therefore, recalling the spinor degrees of freedom and using Eq. (6),

$$\langle[p],[X_{\text{even}}]\rangle = (-1)^{k/2}\frac{1}{k}\,\mathop{\mathrm{res}}_{z=k}\mathrm{Tr}_\tau\big(\Gamma_0\,2p([D,p])^k(1+D^2)^{-z/2}\big)$$

$$= (-1)^{k/2}(-1)^{k/2}\frac{i^{k/2}2^{k/2}}{k}\,\mathop{\mathrm{res}}_{z=k}\mathrm{Tr}_\tau\Big(p\sum_{\sigma\in S_k}(-1)^\sigma\prod_{j=1}^k\partial_{\sigma(j)}p(1+|X|^2)^{-z/2}\Big)$$

$$= \frac{(2i)^{k/2}\mathrm{Vol}_{k-1}(S^{k-1})}{k}\,\mathscr{T}\Big(p\sum_{\sigma\in S_k}(-1)^\sigma\prod_{j=1}^k\partial_{\sigma(j)}p\Big).$$

Lastly, we use that $\mathrm{Vol}_{k-1}(S^{k-1})=\frac{k\pi^{k/2}}{(k/2)!}$ for k even to simplify

$$\langle[p],[X_{\text{even}}]\rangle = \frac{(2\pi i)^{k/2}}{(k/2)!}\sum_{\sigma\in S_k}(-1)^\sigma\,\mathscr{T}\Big(p\prod_{i=1}^k\partial_{\sigma(i)}p\Big),$$

and this concludes the proof. $\qquad\square$

The even and odd index formulas recover the generalised Connes–Chern characters for crossed products studied in [31, Section 6]. We emphasise that while we can construct both complex and real Kasparov modules and semifinite spectral triples, the local index formula only applies to complex algebras and invariants.

3.3   Application to Topological Phases

Here we return to the case of $A = \left(C(\Omega) \rtimes_\phi \mathbb{Z}^{d-k}\right) \rtimes_\theta \mathbb{Z}^k$ with $B = C(\Omega) \rtimes_\phi \mathbb{Z}^{d-k}$. If the algebra is complex and the system has no chiral symmetry, then the K-theory class of interest is the Fermi projection $P_F = \chi_{(-\infty,\mu]}(H)$, which is in A under the gap assumption. If there is a chiral symmetry present, then H can be expressed as $\begin{pmatrix} 0 & Q^* \\ Q & 0 \end{pmatrix}$ with Q invertible (assuming the Fermi energy at 0). Therefore one can take the so-called Fermi unitary $U_F = Q|Q|^{-1}$ and obtain a class in $K_1(A)$. Of course, this unitary is relative to the diagonal chiral symmetry operator $R_{ch} = \begin{pmatrix} 1 & 0 \\ 0 & -1 \end{pmatrix}$ and so the invariants are with reference to this choice, see [12, 35] for more information on this issue. Provided H is a matrix of elements in $\mathscr{A}$ (which is physically reasonable), then the above local formulas for the weak invariants will be valid.

Firstly, if $k = d$ then the index formulae are the Chern numbers for the strong invariants studied in [30]. If the measure $\mathbf{P}$ on Ω is ergodic under the $\mathbb{Z}^d$-action, then $\mathscr{T}(a) = \mathrm{Tr}_{\mathrm{Vol}}(\pi_\omega(a))$ for almost all ω, where $\mathrm{Tr}_{\mathrm{Vol}}$ is the trace per unit volume on $\ell^2(\mathbb{Z}^d)$ and $\{\pi_\omega\}_{\omega \in \Omega}$ is a family representations $C(\Omega) \rtimes_\phi \mathbb{Z}^d \to \mathscr{B}(\ell^2(\mathbb{Z}^d))$ linked by a covariance relation [30]. Under the ergodicity hypothesis, the tracial formulae become

$$\langle [U_F], [X_{\mathrm{odd}}] \rangle = C_k \sum_{\sigma \in S_k} (-1)^\sigma \, (\mathrm{Tr}_{\mathbb{C}^q} \otimes \mathrm{Tr}_{\mathrm{Vol}}) \left(\prod_{i=1}^k \pi_\omega(U_F)^*(-i)[X_{\sigma(i)}, \pi_\omega(U_F)] \right),$$

$$\langle [P_F], [X_{\mathrm{even}}] \rangle = C_k \sum_{\sigma \in S_k} (-1)^\sigma \, (\mathrm{Tr}_{\mathbb{C}^q} \otimes \mathrm{Tr}_{\mathrm{Vol}}) \left(\pi_\omega(P_F) \prod_{i=1}^k (-i)[X_{\sigma(i)}, \pi_\omega(P_F)] \right),$$

for almost all $\omega \in \Omega$. As the left hand side of the equations are independent of the disorder parameter ω, the weak invariants are stable almost surely under the disorder. Recall that we require the Hamiltonian H_ω to have a spectral gap for all $\omega \in \Omega$, so our results do not apply to the regime of strong disorder where the Fermi projection lies in a mobility gap.

The physical interpretation of our semifinite pairings has been discussed in [30]. For k even, the pairing $\langle [P_F], [X_{\mathrm{even}}] \rangle$ can be linked to the linear and non-linear transport coefficients of the conductivity tensor of the physical system. For k odd, the pairing $\langle [U_F], [X_{\mathrm{odd}}] \rangle$ is related to the chiral electrical polarisation and its derivates (with respect to the magnetic field). See [30] for more details. All algebras are separable, which implies that the semifinite pairing takes values in a discrete subset of $\mathbb{R}$. Hence we have proved that the physical quantities related to the semifinite pairings are quantised and topologically stable.

4 Real Pairings and Torsion Invariants

The local index formula is currently only valid for complex algebras and spaces. Furthermore, the semifinite index pairing involves taking a trace and thus it will vanish on torsion representatives, which are more common in the real setting. Because of the anti-linear symmetries that are of interest in topological insulator systems, we would also like a recipe to compute the pairings of interest in the case of real spaces and algebras.

Given a disordered Hamiltonian $H \in M_n(C(\Omega) \rtimes \mathbb{Z}^d)$ (considered now as a real subalgebra of a complex algebra) satisfying time-reversal or particle-hole symmetry (or both) and thus determining the symmetry class index n, one can associate a class $[H] \in KO_n(C(\Omega) \rtimes \mathbb{Z}^d)$ (see [6, 19, 23, 36]). The class can then be paired with the unbounded Kasparov module λ_k from Proposition 2. As outlined in Sect. 2.2.2, we prefer to work with the Kasparov module λ_k coming from the oriented structure $\ell^2(\mathbb{Z}^k, B) \hat{\otimes} \bigwedge^* \mathbb{R}^k$ as the Clifford actions are explicit and easier to work with. In the case of a unital algebra B and $A = B \rtimes \mathbb{Z}^k$, there is a well-defined map

$$KO_n(B \rtimes \mathbb{Z}^k) \times KKO^k(B \rtimes \mathbb{Z}^k, B) \to KKO(C\ell_{n,k}, B) .$$

The class in $KKO(C\ell_{n,k}, B)$ can be represented by a Kasparov module $(C\ell_{n,k}, E_B, \hat{X})$ which can be bounded or unbounded. Up to a finite-dimensional adjustment (see [6, Appendix B]), the topological information of interest of this Kasparov module is contained in the kernel, $\mathrm{Ker}(\hat{X})$, which is a finitely generated and projective C^*-submodule of E_B with a graded left-action of $C\ell_{n,k}$. If B is ungraded, an Atiyah–Bott–Shapiro like map then gives an isomorphism $KKO(C\ell_{n,k}, B) \to KO_{n-k}(B)$ via Clifford modules, see [34, Section 2.2].

Considering the example of $B = C(\Omega) \rtimes \mathbb{Z}^{d-k}$, then one has the Clifford module valued index

$$KO_n(C(\Omega) \rtimes \mathbb{Z}^d) \times KKO^k\big(C(\Omega) \rtimes \mathbb{Z}^d, C(\Omega) \rtimes \mathbb{Z}^{d-k}\big) \to KO_{n-k}(C(\Omega) \rtimes \mathbb{Z}^{d-k}) .$$

If $k = d$, then the pairing takes values in $KO_{n-d}(C(\Omega))$ and constitute 'strong invariants'. Furthermore, fixing a disorder configuration $\omega \in \Omega$ provides a map $KO_{n-d}(C(\Omega)) \to KO_{n-d}(\mathbb{R})$ and then a corresponding analytic index formula can be obtained as in [14] (note, however, that [14] also covers the case of a mobility gap which does not require a spectral gap).

To compute range of the weak K-theoretic pairing, let us first consider the case of Ω contractible. Then one can compute directly

$$KO_{n-k}(C(\Omega) \rtimes \mathbb{Z}^{d-k}) \cong KO_{n-k}(C^*(\mathbb{Z}^{d-k})) \cong \bigoplus_{j=0}^{d-k} \binom{d-k}{j} KO_{n-k-j}(\mathbb{R}) ,$$

which for the varying values of $k \in \{1, \ldots, d - 1\}$ recovers the weak phases described for systems without disorder in Eq. (1). Computing the range of the pairing for non-contractible Ω is much harder, see [18, Section 6] for the computation of $KO_j(C(\Omega) \rtimes \mathbb{Z}^2)$ for low j. Note also that a different action α' on Ω or a different disorder configuration space Ω' could potentially lead to different invariants.

If the K-theory class $[x] \in KO_0(B)$ is not torsion-valued and B contains a trace, then one may take the induced trace $[\tau_B(x)]$ and obtain a real-valued invariant. For $B = C(\Omega) \rtimes \mathbb{Z}^{d-k}$, the induced trace plays the role of averaging over the disorder and $(d - k)$ spatial directions. For non-torsion elements in $KO_j(B)$ with $j \neq 0$, we can apply the induced trace by rewriting $KO_j(B) \cong KO_0(C_0(\mathbb{R}^j) \otimes B) \cong KKO(\mathbb{R}, B \hat{\otimes} C\ell_{0,j})$. This equivalence comes with the limitation that one either has to work with traces on suspensions or graded traces on Clifford algebras. Of course, if $[x]$ is a torsion element the discussion does not apply as $[\tau(x)] = 0$. See [18] for recent work that aims to circumvent some of these problems.

5   The Bulk-Boundary Correspondence

We consider the (real or complex) algebra $B \rtimes_\theta \mathbb{Z}^k$ with $k \geq 2$ and the twist θ such that $\theta(m, -m) = 1$ for all $m \in \mathbb{Z}^k$ [21, 30]. Then one can decompose $B \rtimes_\theta \mathbb{Z}^k \cong (B \rtimes_\theta \mathbb{Z}^{k-1}) \rtimes \mathbb{Z}$, which gives us a short exact sequence of C^*-algebras

$$0 \to (B \rtimes_\theta \mathbb{Z}^{k-1}) \otimes \mathcal{K}(\ell^2(\mathbb{N})) \to \mathcal{T}_\mathbb{Z} \to B \rtimes_\theta \mathbb{Z}^k \to 0 . \tag{8}$$

The Toeplitz algebra $\mathcal{T}_\mathbb{Z}$ for the crossed product is described in [7, 21, 30]. In particular, the algebra $\mathcal{T}_\mathbb{Z}$ acts on the C^*-module $\ell^2(\mathbb{Z}^{k-1} \times \mathbb{N}, B)$, thought of as a space with boundary and the ideal $(B \rtimes_\theta \mathbb{Z}^{k-1}) \otimes \mathcal{K}(\ell^2(\mathbb{N}))$ can be thought of as observables concentrated at the boundary $\ell^2(\mathbb{Z}^{k-1} \times \{0\}, B)$.

Let $A_e = B \rtimes_\theta \mathbb{Z}^{k-1}$ be the edge algebra with bulk algebra $B \rtimes_\theta \mathbb{Z}^k = A_e \rtimes \mathbb{Z}$. Associated to Eq. (8) is a class in $\text{Ext}^{-1}(A_e \rtimes \mathbb{Z}, A_e) \cong KKO^1(A_e \rtimes \mathbb{Z}, A_e)$ by [16, §7].

Proposition 6 ([7, Proposition 3.3]) *The Kasparov module λ_1 from Proposition 2 with $k = 1$ and representing $[\lambda_1] \in KKO^1(A_e \rtimes \mathbb{Z}, A_e)$ or $KK^1(A_e \rtimes \mathbb{Z}, A_e)$ also represents the extension class of Eq. (8).*

Similarly, one can use Proposition 2 to build an edge Kasparov module λ_{k-1} representing a class in $KKO^{k-1}(B \rtimes_\theta \mathbb{Z}^{k-1}, B)$ or $KK^{k-1}(B \rtimes_\theta, B)$. Hence we have a map

$$KKO^1(B \rtimes \mathbb{Z}^k, B \rtimes \mathbb{Z}^{k-1}) \times KKO^{k-1}(B \rtimes \mathbb{Z}^{k-1}, B) \to KKO^k(B \rtimes \mathbb{Z}^k, B)$$

given by the Kasparov product $[\lambda_1] \hat{\otimes}_{A_e} [\lambda_{k-1}]$ at the level of classes.

Theorem 3 ([7, Theorem 3.4]) *The product $[\lambda_1]\hat{\otimes}_{A_e}[\lambda_{k-1}]$ has the unbounded representative*

$$\left(\mathscr{A}\hat{\otimes}C\ell_{0,k},\ \ell^2(\mathbb{Z}^k, B)_B\hat{\otimes}\bigwedge\nolimits^*\mathbb{R}^k,\ X_k\hat{\otimes}\gamma^1 + \sum_{j=1}^{k-1}X_j\hat{\otimes}\gamma^{j+1}\right)$$

and at the bounded level $[\lambda_1]\hat{\otimes}_{A_e}[\lambda_{k-1}] = (-1)^{k-1}[\lambda_k]$, where $-[x]$ represents the inverse of $[x]$ in the KK-group.

Recall that the weak invariants arise from the pairing of λ_k with a class $[H] \in KO_n(B \rtimes \mathbb{Z}^k)$ (or complex). Theorem 3 implies that

$$[H]\hat{\otimes}_A[\lambda_k] = [H]\hat{\otimes}_A\left([\lambda_1]\hat{\otimes}_{A_e}[\lambda_{k-1}]\right) = (-1)^{k-1}\left([H]\hat{\otimes}_A[\lambda_1]\right)\hat{\otimes}_{A_e}[\lambda_{k-1}]\,,$$

by the associativity of the Kasparov product. On the other hand, let us note that $[H]\hat{\otimes}_A[\lambda_1] = \partial[H] \in KO_{n-1}(A_e)$ as the product with $[\lambda_1]$ represents the boundary map in KO-theory associated to the short exact sequence of Eq. (8). Hence the weak pairing, up to a possible sign, is the same as a pairing over the edge algebra $A_e = B \rtimes_\theta \mathbb{Z}^{k-1}$.

Corollary 1 (Bulk-Boundary Correspondence of Weak Pairings) *The weak pairing $[H]\hat{\otimes}_A[\lambda_k]$ is non-trivial if and only if the edge pairing $\partial[H]\hat{\otimes}_{A_e}[\lambda_{k-1}]$ is non-trivial.*

In the real case we achieve a bulk-boundary correspondence of the K-theoretic pairings representing the weak invariants. The Morita equivalence between spin and oriented structures means that Theorem 3 also applies to the spin Kasparov module $\lambda_k^\mathfrak{S}$. In particular, the bulk-boundary correspondence extends to the semifinite pairing, allowing us to recover the following result from [30].

Corollary 2 (Bulk-Boundary Correspondence of Weak Chern Numbers) *The cyclic expressions for the complex semifinite index pairing are the same (up to sign) for the bulk and edge algebras. Namely for $k \geq 2$ and $p, u \in M_q(\mathscr{A})$,*

$$\langle[u], [X_{\mathrm{odd}}]\rangle = \langle\partial[u], [X_{\mathrm{even}}]\rangle\,, \qquad \langle[p], [X_{\mathrm{even}}]\rangle = -\langle\partial[p], [X_{\mathrm{odd}}]\rangle\,.$$

Proof Because the factorisation of pairings occurs at the level of the Kasparov modules $\lambda_k^\mathfrak{S}$, the result immediately follows when taking the trace. $\qquad\square$

Recall that for $B = C(\Omega) \rtimes_\phi \mathbb{Z}^k$, the complex K-theory classes of interest were the Fermi projection P_F or the Fermi unitary coming from $\mathrm{sgn}(H) = \begin{pmatrix} 0 & U_F^* \\ U_F & 0 \end{pmatrix}$ if H is chiral symmetric. We take the edge algebra, $A_e = \left(C(\Omega) \rtimes_\phi \mathbb{Z}^{d-k}\right) \rtimes_\theta \mathbb{Z}^{k-1} \cong C(\Omega) \rtimes_\phi \mathbb{Z}^{d-1}$, which is an algebra associated to a system of 1 dimension lower. The boundary maps in K-theory $\partial[P_F]$ and $\partial[U_F]$ can be written in terms of the Hamiltonian $\widehat{H} \in \mathscr{T}_\mathbb{Z}$ associated to the system with boundary. Furthermore, the pairings $\langle\partial[P_F], [X_{\mathrm{odd}}]\rangle$ and $\langle\partial[U_F], [X_{\mathrm{even}}]\rangle$ can be related to edge behaviour of the sample with boundary, e.g. edge conductance, see [21, 30]. Hence in the better-understood complex setting, the bulk-boundary correspondence has both physical and mathematical meaning.

Acknowledgements We thank our collaborators, Alan Carey, Johannes Kellendonk, Emil Prodan and Adam Rennie, whose work this builds from. We also thank the anonymous referee, whose careful reading and suggestions have improved the manuscript. We are partially supported by the DFG grant SCHU-1358/6 and C. B. is also supported by an Australian Mathematical Society Lift-Off Fellowship and a Japan Society for the Promotion of Science Postdoctoral Fellowship for Overseas Researchers (no. P16728).

References

1. Atiyah, M.F.: K-theory and reality. Q. J. Math. **17**, 367–386 (1966)
2. Baaj, S., Julg, P.: Théorie bivariante de Kasparov et opérateurs non bornés dans les C^*-modules hilbertiens. C. R. Acad. Sci. Paris Sér. I Math. **296**(21), 875–878 (1983)
3. Bellissard, J., van Elst, A., Schulz-Baldes, H.: The noncommutative geometry of the quantum Hall effect. J. Math. Phys. **35**(10), 5373–5451 (1994)
4. Benameur, M., Carey, A.L, Phillips, J., Rennie, A., Sukochev, F.A., Wojciechowski, K.P.: An analytic approach to spectral flow in von Neumann algebras. In: Booß-Bavnbek, B., Klimek, S., Lesch, M., Zhang, W. (eds.) Analysis, Geometry and Topology of Elliptic Operators, pp. 297–352. World Scientific Publishing, Singapore (2006)
5. Blackadar, B.: K-Theory for Operator Algebras. Mathematical Sciences Research Institute Publications, vol. 5. Cambridge University Press, Cambridge (1998)
6. Bourne, C., Carey, A.L., Rennie, A.: A noncommutative framework for topological insulators. Rev. Math. Phys. **28**, 1650004 (2016)
7. Bourne, C., Kellendonk, J., Rennie A.: The K-theoretic bulk-edge correspondence for topological insulators. Ann. Henri Poincaré **18**(5), 1833–1866 (2017)
8. Carey, A.L., Phillips, J., Rennie, A., Sukochev, F.A.: The local index formula in semifinite von Neumann algebras I: spectral flow. Adv. Math. **202**(2), 451–516 (2006)
9. Carey, A.L., Phillips, J., Rennie, A., Sukochev, F.A.: The local index formula in semifinite von Neumann algebras II: the even case. Adv. Math. **202**(2), 517–554 (2006)
10. Carey, A.L., Gayral, V., Rennie, A., Sukochev, F.A.: Index theory for locally compact noncommutative geometries. In: Memoirs of the American Mathematical Society, vol. 231, No. 2. American Mathematical Society, Providence (2014)
11. Connes, A.: Non-commutative differential geometry. Inst. Hautes Études Sci. Publ. Math. **62**, 41–144 (1985)
12. De Nittis, G., Gomi, K.: Chiral vector bundles: a geometric model for class AIII topological quantum systems. arXiv:1504.04863 (2015)
13. Forsyth, F., Rennie, A.: Factorisation of equivariant spectral triples in unbounded KK-theory. arXiv:1505.02863 (2015)
14. Großmann, J., Schulz-Baldes, H.: Index pairings in presence of symmetries with applications to topological insulators. Commun. Math. Phys. **343**(2), 477–513 (2016)
15. Kaad, J., Nest, R., Rennie, A.: KK-theory and spectral flow in von Neumann algebras. J. K-theory **10**(2), 241–277 (2012)
16. Kasparov, G.G.: The operator K-functor and extensions of C^*-algebras. Math. USSR Izv. **16**, 513–572 (1981)
17. Kasparov, G.G.: Equivariant KK-theory and the Novikov conjecture. Invent. Math. **91**(1), 147–201 (1988)

18. Kellendonk, J.: Cyclic cohomology for graded $C^{*,r}$-algebras and its pairings with van Daele K-theory. arXiv:1607.08465 (2016)
19. Kellendonk, J.: On the C^*-algebraic approach to topological phases for insulators. Ann. Henri Poincaré **18**(7), 2251–2300 (2017)
20. Kellendonk, J., Richard, S.: Topological boundary maps in physics. In: Boca, F., Purice, R., Strătilă, Ş. (eds.) Perspectives in Operator Algebras and Mathematical Physics. Theta Series in Advanced Mathematics, vol. 8, pp. 105–121. Theta, Bucharest (2008)
21. Kellendonk, J., Richter, T., Schulz-Baldes, H.: Edge current channels and Chern numbers in the integer quantum Hall effect. Rev. Math. Phys. **14**, 87–119 (2002)
22. Kitaev, A: Periodic table for topological insulators and superconductors. In: Lebedev, V., Feigel'Man, M. (eds.) American Institute of Physics Conference Series. American Institute of Physics Conference Series, vol. 1134, pp. 22–30 (2009)
23. Kubota, Y.: Controlled topological phases and bulk-edge correspondence. Commun. Math. Phys. **349**(2), 493–525 (2017)
24. Laca, M., Neshveyev, S.: KMS states of quasi-free dynamics on Pimsner algebras. J. Funct. Anal. **211**(2), 457–482 (2004)
25. Lance, E.C.: Hilbert C^*-Modules: A Toolkit for Operator Algebraists. London Mathematical Society Lecture Note Series, vol. 210. Cambridge University Press, Cambridge (1995)
26. Lawson, H.B., Michelsohn, M.L.: Spin Geometry. Princeton Mathematical Series, Princeton University Press, Princeton (1989)
27. Lord, S., Rennie, A., Várilly, J.C.: Riemannian manifolds in noncommutative geometry. J. Geom. Phys. **62**(2), 1611–1638 (2012)
28. Packer, J.A., Raeburn, I.: Twisted crossed products of C^*-algebras. Math. Proc. Camb. Philos. Soc. **106**, 293–311 (1989)
29. Plymen, R.J.: Strong Morita equivalence, spinors and symplectic spinors. J. Oper. Theory **16**, 305–324 (1986)
30. Prodan, E., Schulz-Baldes, H.: Bulk and Boundary Invariants for Complex Topological Insulators: From K-Theory to Physics. Springer, Cham (2016)
31. Prodan, E., Schulz-Baldes, H.: Generalized Connes-Chern characters in KK-theory with an application to weak invariants of topological insulators. Rev. Math. Phys. **28**, 1650024 (2016)
32. Raeburn, I., Williams, D.P.: Morita Equivalence and Continuous-Trace C^*-Algebras. American Mathematical Society, Providence, RI (1998)
33. Schnyder, A.P., Ryu, S., Furusaki, A., Ludwig, A.W.W.: Classification of topological insulators and superconductors in three spatial dimensions. Phys. Rev. B **78**, 195125 (2008)
34. Schröder, H.: K-Theory for Real C^*-Algebras and Applications. Taylor & Francis, New York (1993)
35. Thiang, G.C.: Topological phases: isomorphism, homotopy and K-theory. Int. J. Geom. Methods Mod. Phys. **12**, 1550098 (2015)
36. Thiang, G.C.: On the K-theoretic classification of topological phases of matter. Ann. Henri Poincaré **17**(4), 757–794 (2016)

Filtered K-Theory for Graph Algebras

Søren Eilers, Gunnar Restorff, Efren Ruiz, and Adam P.W. Sørensen

Abstract We introduce filtered algebraic K-theory of a ring R relative to a sublattice of ideals. This is done in such a way that filtered algebraic K-theory of a Leavitt path algebra relative to the graded ideals is parallel to the gauge invariant filtered K-theory for graph algebras. We apply this to verify the Abrams-Tomforde conjecture for a large class of finite graphs.

1 Introduction

Since the inception of Leavitt path algebras in [1, 4] it has been known that there is a strong connection between Leavitt path algebras and graph C^*-algebras. In particular many results for both graph C^*-algebras and Leavitt path algebras have the same hypotheses when framed in terms of the underlying graph and the conclusions about the structure of the algebras are analogous. For instance, by [13, Theorem 4.1] and [5, Theorem 4.5] the following are equivalent for a graph E.

S. Eilers (✉)
Department of Mathematical Sciences, University of Copenhagen, Universitetsparken 5, DK-2100 Copenhagen, Denmark
e-mail: eilers@math.ku.dk

G. Restorff
Department of Science and Technology, University of the Faroe Islands, Nóatún 3, FO-100 Tórshavn, Faroe Islands
e-mail: gunnarr@setur.fo

E. Ruiz
Department of Mathematics, University of Hawaii, Hilo, 200 W. Kawili St., Hilo, HI 96720-4091, USA
e-mail: ruize@hawaii.edu

A.P.W. Sørensen
Department of Mathematics, University of Oslo, PO Box 1053 Blindern, N-0316 Oslo, Norway
e-mail: apws@math.uio.no

© Springer International Publishing AG, part of Springer Nature 2018
D.R. Wood et al. (eds.), *2016 MATRIX Annals*, MATRIX Book Series 1,
https://doi.org/10.1007/978-3-319-72299-3_11

1. E satisfies Condition (K) (no vertex is the base point of exactly one return path).
2. $C^*(E)$ has real rank 0.
3. $L_{\mathbb{C}}(E)$ is an exchange ring.

That real rank 0 is the analytic analogue of the algebraic property of being an exchange ring is justified in [3, Theorem 7.2].

One of the most direct connections we could possibly have between Leavitt path algebras and graph C^*-algebras would be: If E, F are graphs then

$$L_{\mathbb{C}}(E) \cong L_{\mathbb{C}}(F) \iff C^*(E) \cong C^*(F).$$

This is called the *isomorphism question* and it is unknown if it is true. As currently stated the question is very imprecise, while it is clear what is meant by isomorphism of C^*-algebras, we could consider isomorphisms of Leavitt path algebras both as rings, algebras, and *-algebras. In the last case the forward implication of the isomorphism question holds. In [2] Abrams and Tomforde take a systematic look at the isomorphism question and many related questions, for instance whether or not the above holds with Morita equivalence in place of isomorphism. They provide evidence in favor of a positive answer to the Morita equivalence question and elevate one direction to a conjecture.

Conjecture 1 (The Abrams-Tomforde Conjecture) Let E and F be graphs. If $L_{\mathbb{C}}(E)$ is Morita equivalent to $L_{\mathbb{C}}(F)$, then $C^*(E)$ is (strongly) Morita equivalent to $C^*(F)$.

In [17] the third named author and Tomforde use ideal related algebraic K-theory to verify the Abrams-Tomforde conjecture of large classes of graphs. They introduce ideal related algebraic K-theory as a Leavitt path algebra analogue for filtered K-theory for graph C^*-algebras. This then allows them to prove the Abrams-Tomforde conjecture for all classes of graphs where the associated C^*-algebras are classified by filtered K-theory.

The authors have shown in [10] that when classifying graph C^*-algebras that do not have real rank 0, it can be useful to replace the full filtered K-theory with a version that only looks at gauge invariant ideals. Motivated by this, we develop a version of ideal related algebraic K-theory relative to a sublattice of ideals. Our goal is to get an ideal related K-theory for Leavitt path algebras that only considers graded ideals, but we try to state our result in greater generality. We look at a sublattice $\mathscr{S}$ of ideals in some ring R and consider the spectrum of these ideals, that is the set of $\mathscr{S}$-prime ideals. This set is equipped with the Jacobson (or hull-kernel) topology. In nice cases there exists a lattice isomorphism from the open sets in the spectrum to the ideals in $\mathscr{S}$. Specializing to the case of a Leavitt path algebra $L_k(E)$, we show that the spectrum associated to the graded ideals is homeomorphic to the spectrum of gauge invariant ideals in $C^*(E)$. Using this we define filtered algebraic K-theory of $L_k(E)$ relative to the graded ideals in complete analogy to the C^*-algebra definition. We then follow the work of [17] and establish the Abrams-Tomforde conjecture for all graphs where the C^*-algebras are classified by filtered K-theory of gauge invariant ideals. By [10] this includes a large class of finite graphs.

2   Preliminaries

In this section we set up the notation we will use throughout the paper and we recall the needed definitions. We begin with the definitions of graphs, graph C^*-algebras and Leavitt path algebras.

Definition 1 A graph E is a quadruple $E = (E^0, E^1, r, s)$ where E^0 is the set of vertices, E^1 is the set of edges, and r and s are maps from E^1 to E^0 giving the range and source of an edge.

> **Standing Assumption.** Unless explicitly stated otherwise, all graphs are assumed to be countable, i.e., the set of vertices and the set of edges are countable sets.

We follow the notation and definition for graph C^*-algebras in [11] and warn the reader that this is not the convention used in the monograph by Raeburn [15].

Definition 2 Let $E = (E^0, E^1, r, s)$ be a graph. The *graph C^*-algebra* $C^*(E)$ is the universal C^*-algebra generated by mutually orthogonal projections $\{p_v : v \in E^0\}$ and partial isometries $\{s_e : e \in E^1\}$ satisfying the relations

- $s_e^* s_f = 0$ if $e, f \in E^1$ and $e \neq f$,
- $s_e^* s_e = p_{r(e)}$ for all $e \in E^1$,
- $s_e s_e^* \leq p_{s(e)}$ for all $e \in E^1$, and,
- $p_v = \sum_{e \in s^{-1}(v)} s_e s_e^*$ for all $v \in E^0$ with $0 < |s^{-1}(v)| < \infty$.

We get our definition of Leavitt path algebras from [1, 4].

Definition 3 Let k be a field and let E be a graph. The *Leavitt path algebra* $L_k(E)$ is the universal k-algebra generated by pairwise orthogonal idempotents $\{v \mid v \in E^0\}$ and elements $\{e, e^* \mid e \in E^1\}$ satisfying

- $e^* f = 0$, if $e \neq f$,
- $e^* e = r(e)$,
- $s(e)e = e = er(e)$,
- $e^* s(e) = e^* = r(e)e^*$, and,
- $v = \sum_{e \in s^{-1}(v)} ee^*$, if $s^{-1}(v)$ is finite and nonempty.

Recall that graph C^*-algebras come with a natural gauge action and that Leavitt path algebras come with a natural grading. We now turn to the ideal structure of Leavitt path algebras and graph C^*-algebras, where we are particularly interested in graded ideals and gauge invariant ideals.

Standing Assumption. Unless explicitly stated otherwise, all ideals in rings are two-sided ideals and all ideals in a C^*-algebra are closed two-sided ideals.

Definition 4 For any ring R we denote by $\mathbb{I}(R)$ the lattice of ideals in R.

As per usual we write $v \geq w$ if there is a path from the vertex v to the vertex w. We call a subset $H \subseteq E^0$ *hereditary* if $v \in H$ and $v \geq w$ imply that $w \in H$, and we say that H is *saturated* if for every $v \in E^0$ with $0 < |s^{-1}(v)| < \infty$ and $r(s^{-1}(v)) \subseteq H$ we have $v \in H$. If H is saturated and hereditary we define

$$B_H = \left\{ v \in E^0 \setminus H \ : \ |s^{-1}(v)| = \infty \text{ and } 0 < |s^{-1}(v) \cap r^{-1}(E^0 \setminus H)| < \infty \right\}.$$

In other words, B_H consists of infinite emitters that are not in H and emit a non-zero finite number of edges to vertices not in H. We say that those vertices are *breaking* for H.

Definition 5 ([19, Definition 5.4]) An admissible pair (H, S) consists of a saturated hereditary subset H and a subset S of B_H. We put an order on the set of admissible pairs by letting $(H, S) \leq (H', S')$ if and only if $H \subseteq H'$ and $S \subseteq H' \cup S'$. This is in fact a lattice order.

Theorem 1 ([6, Theorem 3.6] and [19, Theorem 5.7]) *Let E be a graph and let* k *be a field.*

- *There is a canonical lattice isomorphism from the set of admissible pairs to the set of gauge invariant ideals of $C^*(E)$. We write $I_{(H,S)}^{\text{top}}$ for the image of an admissible pair.*
- *There is a canonical lattice isomorphism from the set of admissible pairs to the set of graded ideals of $L_k(E)$. We write $I_{(H,S)}^{\text{alg}}$ for the image of an admissible pair.*

One of the main reasons the sublattice of graded ideals can be used to study the Morita equivalence classes of Leavitt path algebras is that the graded ideals are preserved by (not necessarily graded) ring isomorphisms.

Lemma 1 *Let E be a graph and let* k *be a field. Suppose I is an ideal in $L_k(E)$. Then I is graded if and only if I is generated by idempotents.*

Proof Suppose I is graded. Then $I = I_{(H,S)}^{\text{alg}}$ for some admissible pair (H, S). By definition (see for instance [19, Definition 5.5]) $I_{(H,S)}^{\text{alg}}$ is generated by $\{v : v \in H\}$ and

$$\left\{ v - \sum_{\substack{s(e)=v \\ r(e) \notin H}} ee^* : v \in S \right\}.$$

Hence I is generated by idempotents.

Suppose instead I is generated by idempotents. Let $e \in I$ be an idempotent in the generating set S of idempotents for I. By [12, Theorem 3.4], e is equivalent in $M_\infty(L_k(G))$ to a finite sum of the idempotents of the form $v \in E^0$ and $w - \sum_{i=1}^n e_i e_i^*$ where $s(e) = w \in E^0$, $|s^{-1}(w)| = \infty$, and each e_i is an element of $s^{-1}(w)$. Then S_e where e is replaced by these new idempotents in the generating set S will generate the ideal I. Thus, I is generated by idempotents in the vertex set and idempotents of the form $v - \sum_{i=1}^n e_i e_i^*$, where $s(e) = v \in E^0$, $|s^{-1}(v)| = \infty$, and each e_i is an element of $s^{-1}(v)$. Therefore, I is a graded ideal.

Finally we briefly recall from [10, Section 3] the definition of $\mathrm{Prime}_\gamma(C^*(E))$ and $\mathsf{FK}^{\mathrm{top},+}(\mathrm{Prime}_\gamma(C^*(E)); C^*(E))$.

Definition 6 Let $E = (E^0, E^1, r, s)$ be a graph. Let $\mathrm{Prime}_\gamma(C^*(E))$ denote the set of all proper ideals that are prime within the set of proper gauge invariant ideals.

We give $\mathrm{Prime}_\gamma(C^*(E))$ the Jacobson topology and can then show that $C^*(E)$ has a canonical structure as a $\mathrm{Prime}_\gamma(C^*(E))$-algebra. So when E has finitely many vertices—or, more generally, $\mathrm{Prime}_\gamma(C^*(E))$ is finite—we can consider the reduced filtered ordered K-theory of $C^*(E)$: $\mathsf{FK}^{\mathrm{top},+}(\mathrm{Prime}_\gamma(C^*(E)); C^*(E))$. Loosely speaking this is the collection of the K-groups associated to certain subquotients I/J of gauge invariant ideals I, J in $C^*(E)$ together with certain maps of the associated six-term exact sequences.

3 $\mathscr{S}$-Prime Spectrum for a Ring

We will now introduce the Prime-spectrum of a ring relative to a sublattice of ideals. Our primary motivation is to look at prime graded ideals in Leavitt path algebras.

Definition 7 Let R be a ring and let $\mathscr{S}$ be a sublattice of $\mathbb{I}(R)$ containing the trivial ideals $\{0\}$ and R. An ideal $P \in \mathscr{S}$ is called $\mathscr{S}$-prime if $P \neq R$ and for any ideals $I, J \in \mathscr{S}$,

$$IJ \subseteq P \implies I \subseteq P \text{ or } J \subseteq P.$$

We denote by $\mathrm{Spec}_{\mathscr{S}}(R)$ the set of all $\mathscr{S}$-prime ideals of R.

We note that if P is $\mathscr{S}$-prime and I, J are in $\mathscr{S}$ then $IJ \subseteq I \cap J$ so we have

$$I \cap J \subseteq P \implies I \subseteq P \text{ or } J \subseteq P.$$

We will equip $\mathrm{Spec}_{\mathscr{S}}(R)$ with the Jacobson (or hull-kernel) topology. For each subset $T \subseteq \mathrm{Spec}_{\mathscr{S}}(R)$ we define the kernel of T as

$$\ker(T) = \bigcap_{\mathfrak{p} \in T} \mathfrak{p}$$

and the closure of T as

$$\overline{T} = \{\mathfrak{p} \in \operatorname{Spec}_{\mathscr{S}}(R) \, : \, \mathfrak{p} \supseteq \ker(T)\}. \tag{1}$$

Note that if R is a commutative ring and $\mathscr{S} = \mathbb{I}(R)$, then $\operatorname{Spec}_{\mathscr{S}}(R)$ is the spectrum of R with the Zariski topology.

Lemma 2 *Let R be a ring and let $\mathscr{S}$ be a sublattice of $\mathbb{I}(R)$ closed under arbitrary intersections and containing the trivial ideals $\{0\}$ and R. The closure operation defined in (1) satisfies the Kuratowski closure axioms, that is*

1. $\overline{\emptyset} = \emptyset$,
2. $T \subseteq \overline{T}$, *for all* $T \subseteq \operatorname{Spec}_{\mathscr{S}}(R)$,
3. $\overline{\overline{T}} = \overline{T}$, *for all* $T \subseteq \operatorname{Spec}_{\mathscr{S}}(R)$, *and,*
4. $\overline{T_1 \cup T_2} = \overline{T_1} \cup \overline{T_2}$, *for all* $T_1, T_2 \subseteq \operatorname{Spec}_{\mathscr{S}}(R)$.

Proof Once we recall that by definition $\ker(\emptyset) = R$ it is clear that 1. holds and since we have $\mathfrak{p} \supseteq \ker(T)$ for all $\mathfrak{p} \in T$, 2. also holds. For 3. we observe that $\ker(\overline{T}) = \ker(T)$, and then clearly $\overline{\overline{T}} = \overline{T}$.

Finally suppose that $T_1, T_2 \subseteq \operatorname{Spec}_{\mathscr{S}}(R)$. Since $\ker(T_1 \cup T_2) = \ker(T_1) \cap \ker(T_2)$ we have that

$$\begin{aligned}
\overline{T_1 \cup T_2} &= \{\mathfrak{p} \in \operatorname{Spec}_{\mathscr{S}}(R) \, : \, \mathfrak{p} \supseteq \ker(T_1 \cup T_2)\} \\
&= \{\mathfrak{p} \in \operatorname{Spec}_{\mathscr{S}}(R) \, : \, \mathfrak{p} \supseteq \ker(T_1) \cap \ker(T_2)\} \\
&= \{\mathfrak{p} \in \operatorname{Spec}_{\mathscr{S}}(R) \, : \, \mathfrak{p} \supseteq \ker(T_1) \text{ or } \mathfrak{p} \supseteq \ker(T_2)\} \\
&= \overline{T_1} \cup \overline{T_2}.
\end{aligned}$$

So 4. holds.

We now describe the open sets in the Jacobson topology. To this end we define for each $I \in \mathscr{S}$ the set

$$W(I) = \{\mathfrak{p} \in \operatorname{Spec}_{\mathscr{S}}(R) \, : \, \mathfrak{p} \not\supseteq I\}.$$

Lemma 3 *Let R be a ring and let $\mathscr{S}$ be a sublattice of $\mathbb{I}(R)$ closed under arbitrary intersections and containing the trivial ideals $\{0\}$ and R. Then for all $U \subseteq \operatorname{Spec}_{\mathscr{S}}(R)$, U is open if and only if*

$$U = W(\ker(U^c)).$$

Furthermore, if $I \in \mathscr{S}$ is such that

$$I = \ker(\{\mathfrak{p} \in \operatorname{Spec}_{\mathscr{S}}(R) \, : \, \mathfrak{p} \supseteq I\}),$$

then $W(I)$ is open.

Proof Let U be a subset of $\operatorname{Spec}_{\mathscr{A}}(R)$. Then U is open if and only if $U^c = \overline{U^c}$ if and only if

$$U^c = \{\mathfrak{p} \in \operatorname{Spec}_{\mathscr{A}}(R) \ : \ \mathfrak{p} \supseteq \ker(U^c)\}$$

if and only if

$$U = \{\mathfrak{p} \in \operatorname{Spec}_{\mathscr{A}}(R) \ : \ \mathfrak{p} \not\supseteq \ker(U^c)\} = W(\ker(U^c)).$$

Let now $I \in \mathscr{S}$ be such that

$$I = \ker(\{\mathfrak{p} \in \operatorname{Spec}_{\mathscr{A}}(R) \ : \ \mathfrak{p} \supseteq I\}).$$

To ease notation we let $H = \{\mathfrak{p} \in \operatorname{Spec}_{\mathscr{A}}(R) \ : \ \mathfrak{p} \supseteq I\}$, so that $I = \ker(H)$. Then

$$W(I)^c = \{\mathfrak{p} \in \operatorname{Spec}_{\mathscr{A}}(R) \ : \ \mathfrak{p} \not\supseteq I\}^c = \{\mathfrak{p} \in \operatorname{Spec}_{\mathscr{A}}(R) \ : \ \mathfrak{p} \not\supseteq \ker(H)\}^c$$
$$= \{\mathfrak{p} \in \operatorname{Spec}_{\mathscr{A}}(R) \ : \ \mathfrak{p} \supseteq \ker(H)\} = \overline{H}.$$

Hence $W(I)$ is open.

We now define a lattice isomorphism between the open sets of $\operatorname{Spec}_{\mathscr{A}}(R)$ and the elements of $\mathscr{S}$.

Theorem 2 *Let R be a ring and let $\mathscr{S}$ be a sublattice of $\mathbb{I}(R)$ closed under arbitrary intersections and containing the trivial ideals $\{0\}$ and R. Suppose that for each $I \in \mathscr{S}$ we have that*

$$I = \ker(\{\mathfrak{p} \in \operatorname{Spec}_{\mathscr{A}}(R) \ : \ \mathfrak{p} \supseteq I\}).$$

Define $\phi : \mathbb{O}(\operatorname{Spec}_{\mathscr{A}}(R)) \to \mathscr{S}$ by

$$\phi(U) = \ker(U^c).$$

Then ϕ is a lattice isomorphism.

Proof To show that ϕ is bijective we define $\gamma : \mathscr{S} \to \mathbb{O}(\operatorname{Spec}_{\mathscr{A}}(R))$ by $\gamma(I) = W(I)$ and check that it is an inverse. Note that by Lemma 3 the set $W(I)$ is in fact open. For each $I \in \mathscr{S}$ we have

$$\phi(\gamma(I)) = \phi(W(I)) = \ker(W(I)^c) = \ker(\{\mathfrak{p} \in \operatorname{Spec}_{\mathscr{A}}(R) \ : \ \mathfrak{p} \supseteq I\}) = I,$$

by the assumption on I. On the other hand, if $U \subseteq \operatorname{Spec}_{\mathscr{A}}(R)$ is open we can use Lemma 3 to get

$$\gamma(\phi(U)) = \gamma(\ker(U^c)) = W(\ker(U^c)) = U.$$

Hence ϕ is bijective. To show that ϕ is a lattice isomorphism it only remains to verify that both ϕ and γ preserves order. Let U, V be open subsets of $\mathrm{Spec}_{\mathscr{A}}(R)$ with $U \subseteq V$. Then $V^c \subseteq U^c$ so

$$\phi(U) = \ker(U^c) \subseteq \ker(V^c) = \phi(V),$$

and hence ϕ is order preserving. Let now $I, J \in \mathscr{S}$ be such that $I \subseteq J$. Then

$$W(I)^c = \{\mathfrak{p} \in \mathrm{Spec}_{\mathscr{A}}(R) \ : \ \mathfrak{p} \supseteq I\} \supseteq \{\mathfrak{p} \in \mathrm{Spec}_{\mathscr{A}}(R) \ : \ \mathfrak{p} \supseteq J\} = W(J)^c,$$

which implies that $\gamma(I) = W(I) \subseteq W(J) = \gamma(J)$, i.e., γ is order preserving.

In keeping with the notation from C^*-algebras we define

$$R[U] = \phi(U)$$

for every $U \in \mathbb{O}(\mathrm{Spec}_{\mathscr{A}}(R))$. Whenever we have open sets $V \subseteq U$ we can form the quotient $R[U]/R[V]$. The next lemma shows that the quotient $R[U]/R[V]$ only depends on the set difference $U \setminus V$ up to canonical isomorphism.

Lemma 4 *Let R be a ring and let $\mathscr{S}$ be a sublattice of $\mathbb{I}(R)$ closed under arbitrary intersections and containing the trivial ideals $\{0\}$ and R. Suppose that for each $I \in \mathscr{S}$ we have that*

$$I = \ker(\{\mathfrak{p} \in \mathrm{Spec}_{\mathscr{A}}(R) \ : \ \mathfrak{p} \supseteq I\}).$$

Then for all $U, V \in \mathbb{O}(\mathrm{Spec}_{\mathscr{A}}(R))$ we have

$$R[U \cup V] = R[U] + R[V] \quad and \quad R[U \cap V] = R[U] \cap R[V].$$

Consequently, if $V_1, V_2, U_1, U_2 \in \mathbb{O}(\mathrm{Spec}_{\mathscr{A}}(R))$ are such that $V_1 \subseteq U_1$, $V_2 \subseteq U_2$, and $U_1 \setminus V_1 = U_2 \setminus V_2$, then there exits an isomorphism from $R[U_1]/R[V_1]$ to $R[U_2]/R[V_2]$ and this isomorphism is natural, i.e., if also $V_3, U_3 \in \mathbb{O}(\mathrm{Spec}_{\mathscr{A}}(R))$ with $V_3 \subseteq U_3$ and $U_3 \setminus V_3 = U_1 \setminus V_1$, then the composition of the isomorphisms from $R[U_1]/R[V_1]$ to $R[U_2]/R[V_2]$ and from $R[U_2]/R[V_2]$ to $R[U_3]/R[V_3]$ is equal to the isomorphism from $R[U_1]/R[V_1]$ to $R[U_3]/R[V_3]$.

Proof The first part of the theorem follows from the fact that ϕ is a lattice isomorphism (Theorem 2) and that $\mathscr{S}$ is a sublattice.

Suppose now $V_1, V_2, U_1, U_2 \in \mathbb{O}(X)$ are as in the statement of the Lemma. Then $V_1 \cup U_2 = U_1 \cup U_2 = U_1 \cup V_2$ and therefore

$$R[U_2] + R[V_1] = R[V_1 \cup U_2] = R[U_1 \cup V_2] = R[U_1] + R[V_2].$$

Since $U_2 \cap (V_1 \cup V_2) = V_2$ we get

$$(R[U_2] + R[V_1])/(R[V_1] + R[V_2]) \cong R[U_2]/(R[U_2] \cap R[V_1 \cup V_2])$$
$$= R[U_2]/R[U_2 \cap (V_1 \cup V_2)]$$
$$= R[U_2]/R[V_2].$$

Similarly

$$(R[U_1] + R[V_2])/(R[V_1] + R[V_2]) \cong R[U_1]/R[V_1].$$

Hence

$$R[U_1]/R[V_1] \cong (R[U_1] + R[V_2])/(R[V_1] + R[V_2])$$
$$= (R[U_2] + R[V_1])/(R[V_1] + R[V_2])$$
$$\cong R[U_2]/R[V_2].$$

Suppose that we also have $V_3, U_3 \in \mathbb{O}(\operatorname{Spec}_{\mathscr{S}}(R))$ with $V_3 \subseteq U_3$ and $U_3 \setminus V_3 = U_1 \setminus V_1$. Then

$$V_1 \cup U_2 = U_1 \cup U_2 = U_1 \cup V_2,$$
$$V_2 \cup U_3 = U_2 \cup U_3 = U_2 \cup V_3,$$
$$V_1 \cup U_3 = U_1 \cup U_3 = U_1 \cup V_3,$$

$$V_1 = U_1 \cap (V_1 \cup V_2) = U_1 \cap (V_1 \cup V_3) = U_1 \cap (V_1 \cup V_2 \cup V_3),$$
$$V_2 = U_2 \cap (V_1 \cup V_2) = U_2 \cap (V_2 \cup V_3) = U_2 \cap (V_1 \cup V_2 \cup V_3), \text{ and}$$
$$V_3 = U_3 \cap (V_1 \cup V_3) = U_3 \cap (V_2 \cup V_3) = U_3 \cap (V_1 \cup V_2 \cup V_3).$$

Now, by considering the isomorphism constructed above, one then gets that the isomorphism is natural from Noether's isomorphism theorem.

Definition 8 Let X be a topological space and let Y be a subset of X. We call Y *locally closed* if $Y = U \setminus V$ where $U, V \in \mathbb{O}(X)$ with $V \subseteq U$. We let $\mathbb{LC}(X)$ be the set of locally closed subsets of X.

Definition 9 Let R be a ring and let $\mathscr{S}$ be a sublattice of $\mathbb{I}(R)$ closed under arbitrary intersections and containing the trivial ideals $\{0\}$ and R. Suppose that for each $I \in \mathscr{S}$ we have that

$$I = \ker(\{\mathfrak{p} \in \operatorname{Spec}_{\mathscr{S}}(R) : \mathfrak{p} \supseteq I\}).$$

For $Y = U \setminus V \in \mathbb{LC}(\mathrm{Spec}_{\mathscr{A}}(R))$, define

$$R[Y] := R[U]/R[V].$$

By Lemma 4, $R[Y]$ does not depend on U and V up to a canonical choice of isomorphism.

4 $\mathrm{Spec}_{\gamma}(L_{\mathsf{k}}(E))$ and $\mathrm{Prime}_{\gamma}(C^*(E))$

Having set up our notion of prime ideal spectrum relative to a sublattice, we will now apply it to the graded ideals of Leavitt path algebras.

Definition 10 Let E be a graph and let k be a field. We denote by $\mathbb{I}_{\gamma}(L_{\mathsf{k}}(E))$ the sublattice of $\mathbb{I}(L_{\mathsf{k}}(E))$ consisting of all graded ideals of $L_{\mathsf{k}}(E)$ and for brevity we let $\mathrm{Spec}_{\gamma}(L_{\mathsf{k}}(E)) = \mathrm{Spec}_{\mathbb{I}_{\gamma}(L_{\mathsf{k}}(E))}(L_{\mathsf{k}}(E))$.

Similarly we let $\mathbb{I}_{\gamma}(C^*(E))$ be the sublattice of $\mathbb{I}(C^*(E))$ consisting of all gauge invariant ideals of $C^*(E)$.

Recall from [10, Section 3] that $\mathrm{Prime}_{\gamma}(C^*(E))$ denotes the collection of prime gauge invariant ideals of $C^*(E)$. We first prove that the lattice of graded ideals and the lattice of gauge invariant ideals are isomorphic in a canonical way.

Lemma 5 *Let E be a graph. The map $\beta : \mathbb{I}_{\gamma}(L_{\mathsf{k}}(E)) \rightarrow \mathbb{I}_{\gamma}(C^*(E))$ that is given by $\beta(I_{(H,S)}^{\mathrm{alg}}) = I_{(H,S)}^{\mathrm{top}}$ is a lattice isomorphism. Furthermore β maps $\mathrm{Spec}_{\gamma}(L_{\mathsf{k}}(E))$ bijectively onto $\mathrm{Prime}_{\gamma}(C^*(E))$.*

Proof By Theorem 1 there is a lattice isomorphism β_{alg} from the set of admissible pairs to $\mathbb{I}_{\gamma}(L_{\mathsf{k}}(E))$ given by $\beta_{\mathrm{alg}}((H,S)) = I_{(H,S)}^{\mathrm{alg}}$, and a lattice isomorphism β_{top} from the set of admissible pairs to $\mathbb{I}_{\gamma}(C^*(E))$ given by $\beta_{\mathrm{top}}((H,S)) = I_{(H,S)}^{\mathrm{top}}$. Consequently, $\beta = \beta_{\mathrm{top}} \circ \beta_{\mathrm{alg}}^{-1}$ is a lattice isomorphism.

Let $\mathscr{S} = \mathbb{I}_{\gamma}(L_{\mathsf{k}}(E))$. It follows from [14, Proposition II.1.4] that a graded ideal I of $L_{\mathsf{k}}(E)$ is $\mathscr{S}$-prime if and only if I is a prime ideal of $L_{\mathsf{k}}(E)$. Thus, by [16, Theorem 3.12], every $\mathscr{S}$-prime ideal I of $L_{\mathsf{k}}(E)$ is of the form

- $I = I_{(H,S)}^{\mathrm{alg}}$, where $E^0 \setminus H$ is a maximal tail and $S = B_H$, or
- $I = I_{(H,S)}^{\mathrm{alg}}$ where $E^0 \setminus H = M(u)$ and $S = B_H \setminus \{u\}$ for some breaking vertex,

and that these ideals are distinct. In [10, Section 3] it is shown that every ideal $\mathfrak{I}$ in $\mathrm{Prime}_{\gamma}(C^*(E))$ is of the form

- $\mathfrak{I} = I_{(H,S)}^{\mathrm{top}}$, where $E^0 \setminus H$ is a maximal tail and $S = B_H$, or
- $\mathfrak{I} = I_{(H,S)}^{\mathrm{top}}$ where $E^0 \setminus H = M(u)$ and $S = B_H \setminus \{u\}$ for some breaking vertex,

and that these ideals are distinct. Hence $I_{(H,S)}^{\text{top}}$ is in $\text{Prime}_\gamma(C^*(E))$ if and only if $I_{(H,S)}^{\text{alg}}$ is in $\text{Spec}_\gamma(L_k(E))$. In other words β maps $\text{Spec}_\gamma(L_k(E))$ bijectively onto $\text{Prime}_\gamma(C^*(E))$.

We can now prove that the collection of graded ideals satisfies the kernel assumption we used in Sect. 3.

Proposition 1 *Let E be a graph. If I is a proper graded ideal of $L_k(E)$, then*

$$I = \ker\left(\{\mathfrak{p} \in \text{Spec}_\gamma(L_k(E)) \; : \; \mathfrak{p} \supseteq I\}\right).$$

Proof Let β be the lattice isomorphism from Lemma 5 and let $I \in \mathbb{I}_\gamma(L_k(E))$ be a proper ideal.

By [10, Lemma 3.5] we have that

$$\beta(I) = \bigcap_{\substack{q \in \text{Prime}_\gamma(C^*(E)) \\ q \supseteq \beta(I)}} q.$$

Since I is a graded ideal $I = I_{(H,S)}^{\text{alg}}$ for some admissible pair (H,S). As the intersection of graded ideals is again graded we also have

$$\bigcap_{\substack{\mathfrak{p} \in \text{Spec}_\gamma(L_k(E)) \\ \mathfrak{p} \supseteq I}} \mathfrak{p} = I_{(H',S')}^{\text{alg}},$$

for some admissible pair (H',S'). We will now show that $I_{(H,S)}^{\text{top}} = I_{(H',S')}^{\text{top}}$.

Since $I_{(H',S')}^{\text{alg}}$ is an intersection of ideals that all contain $I_{(H,S)}^{\text{alg}}$, $I_{(H,S)}^{\text{alg}} \subseteq I_{(H',S')}^{\text{alg}}$ which implies that $I_{(H,S)}^{\text{top}} \subseteq I_{(H',S')}^{\text{top}}$ as β is order preserving. If $q \in \text{Prime}_\gamma(C^*(E))$ is such that $I_{(H,S)}^{\text{top}} \subseteq q$, then $I_{(H,S)}^{\text{alg}} \subseteq \beta^{-1}(q)$. Therefore $\beta^{-1}(q)$ is one of the ideals whose intersection define $I_{(H',S')}^{\text{alg}}$ so

$$I_{(H',S')}^{\text{top}} = \beta(I_{(H',S')}^{\text{alg}}) \subseteq \beta(\beta^{-1}(q)) = q.$$

We now have the following inclusions

$$I_{(H,S)}^{\text{top}} \subseteq I_{(H',S')}^{\text{top}} \subseteq \bigcap_{\substack{q \in \text{Prime}_\gamma(C^*(E)) \\ q \supseteq I_{(H,S)}^{\text{top}}}} q = \beta(I) = I_{(H,S)}^{\text{top}}.$$

Therefore, $I_{(H,S)}^{\text{top}} = I_{(H',S')}^{\text{top}}$. Hence $(H,S) = (H',S')$ so

$$I = I_{(H,S)}^{\text{alg}} = I_{(H',S')}^{\text{alg}} = \bigcap_{\substack{\mathfrak{p} \in \text{Spec}_\gamma(L_k(E)) \\ \mathfrak{p} \supseteq I}} \mathfrak{p} = \ker\left(\{\mathfrak{p} \in \text{Spec}_\gamma(L_k(E)) \; : \; \mathfrak{p} \supseteq I\}\right).$$

Corollary 1 *The map*

$$U \mapsto \bigcap_{\mathfrak{p} \in \mathrm{Spec}_\gamma(L_k(E)) \setminus U} \mathfrak{p}$$

is a lattice isomorphism from $\mathbb{O}(\mathrm{Spec}_\gamma(L_k(E)))$ *to* $\mathbb{I}_\gamma(L_k(E))$.

Proof This follows from Theorem 2 which is applicable by Proposition 1 and the fact that the intersection of graded ideals is again a graded ideal.

As the final result in this section we prove that β restricts to a homeomorphism between the graded prime ideals and the gauge prime ideals.

Theorem 3 *Let E be a graph. Then* $\phi = \beta|_{\mathrm{Spec}_\gamma(L_k(E))}$ *is a homeomorphism from* $\mathrm{Spec}_\gamma(L_k(E))$ *to* $\mathrm{Prime}_\gamma(C^*(E))$, *where* β *is the lattice isomorphism from Lemma 5.*

Proof We first observe that Lemma 3 and Proposition 1 combine to show that the open sets of $\mathrm{Spec}_\gamma(L_k(E))$ are precisely the sets of the form $W(I)$ for some proper ideal $I \in \mathbb{I}_\gamma(L_k(E))$.

Let a proper ideal $I \in \mathbb{I}_\gamma(L_k(E))$ be given. Then

$$\beta(W(I)) = \beta \left(\{ \mathfrak{p} \in \mathrm{Spec}_\gamma(L_k(E)) : \mathfrak{p} \not\supseteq I \} \right)$$

$$= \{ \beta(\mathfrak{p}) : \mathfrak{p} \in \mathrm{Spec}_\gamma(L_k(E)) \text{ and } \mathfrak{p} \not\supseteq I \}$$

$$= \{ \beta(\mathfrak{p}) : \mathfrak{p} \in \mathrm{Spec}_\gamma(L_k(E)) \text{ and } \beta(\mathfrak{p}) \not\supseteq \beta(I) \}$$

$$= \{ \mathfrak{q} \in \mathrm{Prime}_\gamma(C^*(E)) : \mathfrak{q} \not\supseteq \beta(I) \} .$$

By [10, Lemma 3.6] the last set is open, and hence ϕ^{-1} is continuous.

The above computation used that β was a lattice isomorphism and that we had complete, and similar looking, descriptions of the open sets in $\mathrm{Spec}_\gamma(L_k(E))$ and $\mathrm{Prime}_\gamma(C^*(E))$. Hence a completely parallel computation will show that ϕ is also continuous. Therefore ϕ is a homeomorphism.

5 Filtered Algebraic K-Theory

In this section we define filtered algebraic K-theory for rings and show that if two Leavitt path algebras over $\mathbb{C}$ have isomorphic filtered algebraic K-theory then the associated graph C^*-algebras have isomorphic filtered K-theory. We then use this result to answer the Abrams-Tomforde conjecture for a large class of finite graphs.

Let R be a unital ring and let $BGL(R)^+$ be Quillen's +-construction (see [20, Chapter IV, Definition 1.1]). Consider $K_0(R)$ as a topological space with the discrete topology. Let $K(R) = K_0(R) \times BGL(R)^+$ with the product topology. Define $K_n^{\text{alg}}(R)$ to be

$$K_n^{\text{alg}}(R) = \begin{cases} \pi_n(K(R)) & \text{if } n \geq 0 \\ K_0(\sum^{-n} R) & \text{if } n < 0, \end{cases}$$

where $\sum A$ denotes the suspension of a ring A. For a non-unital ring R, define $K_n^{\text{alg}}(R)$ to be

$$K_n^{\text{alg}}(R) = \begin{cases} \pi_n(\text{fiber}(K(R^+) \to K(\mathbb{Z}))) & \text{if } n \geq 0 \\ \ker(K_n(R^+) \to K_n(\mathbb{Z})) & \text{if } n < 0 \end{cases}$$

where R^+ is the ring obtained from R by adjoining a unit. Therefore, $K_0^{\text{alg}}(R)$ agrees with the usual definition of $K_0(R)$ using idempotents and $K_1^{\text{alg}}(R)$ agrees with the usual definition of $K_1(R)$ using invertible matrices.

Suppose R is a ring and $\mathscr{S}$ is a sublattice of ideals. Moreover, assume that every $I \in \mathscr{S}$ has a countable approximate unit consisting of idempotents, i.e., for every $I \in \mathscr{S}$, there exists a sequence $\{e_n\}_{n=1}^{\infty}$ in I such that

- e_n is an idempotent for all $n \in \mathbb{N}$,
- $e_n e_{n+1} = e_n$ for all $n \in \mathbb{N}$, and
- for all $r \in I$, there exists $n \in \mathbb{N}$ such that $re_n = e_n r = r$.

Then for any locally closed subset $Y = U \setminus V$ of $\text{Spec}_{\mathscr{S}}(R)$, we have a collection of abelian groups $\{K_n^{\text{alg}}(R[Y])\}_{n \in \mathbb{Z}}$. Moreover, for all $U_1, U_2, U_3 \in \mathbb{O}(\text{Spec}_{\mathscr{S}}(R))$ with $U_1 \subseteq U_2 \subseteq U_3$, by [17, Lemma 3.10], we have a long exact sequence in algebraic K-theory

$$K_n^{\text{alg}}(R[U_2 \setminus U_1]) \xrightarrow{\iota_*} K_n^{\text{alg}}(R[U_3 \setminus U_1]) \xrightarrow{\pi_*} K_n^{\text{alg}}(R[U_3 \setminus U_2]) \xrightarrow{\partial_*} K_{n-1}^{\text{alg}}(R[U_2 \setminus U_1]).$$

Definition 11 Let R be a ring and let $\mathscr{S}$ be a sublattice of $\mathbb{I}(R)$ closed under arbitrary intersections and containing the trivial ideals $\{0\}$ and R. Suppose that for each $I \in \mathscr{S}$ we have that

$$I = \ker(\{\mathfrak{p} \in \text{Spec}_{\mathscr{S}}(R) : \mathfrak{p} \supseteq I\}).$$

Moreover, assume that every $I \in \mathscr{S}$ has a countable approximate unit consisting of idempotents.

1. For $k, m \in \mathbb{Z} \cup \{\pm\infty\}$ with $k \leq m$, we define $\text{FK}_{k,m}^{\text{alg}}(\text{Spec}_{\mathscr{S}}(R); R)$ to be the collection

$$\{K_n^{\text{alg}}(R[Y])\}_{k \leq n \leq m, Y \in \mathbb{LC}(\text{Spec}_{\mathscr{S}}(R))},$$

equipped with the natural transformations $\{\iota_*, \pi_*, \partial_*\}$.

2. For $k, m \in \mathbb{Z} \cup \{\pm\infty\}$ with $k \leq 0 \leq m$, we define $\mathsf{FK}_{k,m}^{\mathrm{alg},+}(\mathrm{Spec}_{\mathscr{A}}(R), R)$ to be the collection $\mathsf{FK}_{k,m}^{\mathrm{alg}}(\mathrm{Spec}_{\mathscr{A}}(R); R)$ together with the positive cone of $K_0^{\mathrm{alg}}(R[Y])$ for all $Y \in \mathbb{LC}(\mathrm{Spec}_{\mathscr{A}}(R))$.

Set

$$\mathsf{FK}^{\mathrm{alg}}(\mathrm{Spec}_{\mathscr{A}}(R); R) = \mathsf{FK}_{-\infty,\infty}^{\mathrm{alg}}(\mathrm{Spec}_{\mathscr{A}}(R); R) \quad \text{and}$$

$$\mathsf{FK}^{\mathrm{alg},+}(\mathrm{Spec}_{\mathscr{A}}(R); R) = \mathsf{FK}_{-\infty,\infty}^{\mathrm{alg},+}(\mathrm{Spec}_{\mathscr{A}}(R); R).$$

Definition 12 Let R, R' be rings, let $\mathscr{S}$ be a sublattice of $\mathbb{I}(R)$ closed under arbitrary intersections and containing the trivial ideals $\{0\}$ and R, and let $\mathscr{S}$ be a sublattice of $\mathbb{I}(R')$ closed under arbitrary intersections and containing the trivial ideals $\{0\}$ and R'. Suppose that for each $I \in \mathscr{S}$ we have that

$$I = \ker(\{\mathfrak{p} \in \mathrm{Spec}_{\mathscr{A}}(R) \; : \; \mathfrak{p} \supseteq I\}),$$

and that for each $I' \in \mathscr{S}$ we have that

$$I' = \ker(\{\mathfrak{p} \in \mathrm{Spec}_{\mathscr{A}}(R') \; : \; \mathfrak{p} \supseteq I'\}).$$

Moreover, assume that every $I \in \mathscr{S}$ and every $I' \in \mathscr{S}$ have a countable approximate unit consisting of idempotents.

For all $k, m \in \mathbb{Z} \cup \{\pm\infty\}$ with $k \leq m$, an isomorphism from $\mathsf{FK}_{k,m}^{\mathrm{alg}}(\mathrm{Spec}_{\mathscr{A}}(R); R)$ to $\mathsf{FK}_{k,m}^{\mathrm{alg}}(\mathrm{Spec}_{\mathscr{A}}(R'); R')$ consists of a homeomorphism $\phi \colon \mathrm{Spec}_{\mathscr{A}}(R) \to \mathrm{Spec}_{\mathscr{A}}(R')$ and an isomorphism $\alpha_{Y,n}$ from $K_n(R[Y])$ to $K_n(R'[\phi(Y)])$ for each n with $k \leq n \leq m$ and for each $Y \in \mathbb{LC}(\mathrm{Spec}_{\mathscr{A}}(R))$ such that the diagrams involving the natural transformations commute.

Let $k, m \in \mathbb{Z} \cup \{\pm\infty\}$ with $k \leq 0 \leq m$. If the isomorphism from $\mathsf{FK}_{k,m}^{\mathrm{alg}}(\mathrm{Spec}_{\mathscr{A}}(R); R)$ to $\mathsf{FK}_{k,m}^{\mathrm{alg}}(\mathrm{Spec}_{\mathscr{A}}(R'); R')$ restricts to an order isomorphism on $K_0(R[Y])$ for all $Y \in \mathbb{LC}(\mathrm{Spec}_{\mathscr{A}}(R))$, we write

$$\mathsf{FK}_{k,m}^{\mathrm{alg},+}(\mathrm{Spec}_{\mathscr{A}}(R); R) \cong \mathsf{FK}_{k,m}^{\mathrm{alg},+}(\mathrm{Spec}_{\mathscr{A}}(R'); R').$$

Lemma 6 *Let E be a graph and let k be a field. Then every graded-ideal of $L_{\mathsf{k}}(E)$ has a countable approximate unit consisting of idempotents. Consequently, for all $k, m \in \mathbb{Z} \cup \{\pm\infty\}$ with $k \leq 0 \leq m$, $\mathsf{FK}_{k,m}^{\mathrm{alg},+}(\mathrm{Spec}_{\gamma}(L_{\mathsf{k}}(E)); L_{\mathsf{k}}(E))$ is defined.*

Proof Let F be a graph and set $F^0 = \{v_1, v_2, \ldots\}$. Then $\{\sum_{k=1}^{n} v_k\}_{n=1}^{\infty}$ is a countable approximate unit consisting of idempotents for $L_{\mathsf{k}}(F)$. Thus, every Leavitt path algebra has a countable approximate unit consisting of idempotents. The lemma now follows since by [18, Corollary 6.2] every graded-ideal of $L_{\mathsf{k}}(E)$ is isomorphic to a Leavitt path algebra.

Lemma 7 *Let E be a directed graph and let $\phi : \mathrm{Spec}_\gamma(L_{\mathbb{C}}(E)) \to \mathrm{Prime}_\gamma(C^*(E))$ be the homeomorphism given in Theorem 3. Then for all $U \in \mathbb{O}(\mathrm{Spec}_\gamma(L_{\mathbb{C}}(E)))$, there exists an admissible pair (H, S) such that $L_{\mathbb{C}}(E)[U] = I_{(H,S)}^{\mathrm{alg}}$ and $C^*(E)[\phi(U)] = I_{(H,S)}^{\mathrm{top}}$, where $L_{\mathbb{C}}(E)[U]$ is the graded ideal corresponding to the open set U under the lattice isomorphism from $\mathbb{O}(\mathrm{Spec}_\gamma(L_{\mathbb{C}}(E)))$ to $\mathbb{I}_\gamma(L_{\mathbb{C}}(E))$ given in Theorem 2.*

Proof This follows from the construction of ϕ in Theorem 3 as the restriction of the lattice isomorphism β that sends $I_{(H,S)}^{\mathrm{alg}}$ to $I_{(H,S)}^{\mathrm{top}}$.

Let $\mathfrak{A}$ be a C^*-algebra and let A be a $*$-algebra. Suppose $\iota_{\mathfrak{A}}$ is a $*$-homomorphism from A to $\mathfrak{A}$. Denote the composition

$$K_n^{\mathrm{alg}}(A) \xrightarrow{K_n(\iota_{\mathfrak{A}})} K_n^{\mathrm{alg}}(\mathfrak{A}) \longrightarrow K_n^{\mathrm{top}}(\mathfrak{A})$$

by $\gamma_{n,\mathfrak{A}}$, where $K_n^{\mathrm{top}}(\mathfrak{A})$ is the (usual) topological K-theory of the C^*-algebra $\mathfrak{A}$.

Theorem 4 *Let E be a directed graph and let*

$$\phi : \mathrm{Spec}_\gamma(L_{\mathbb{C}}(E)) \to \mathrm{Prime}_\gamma(C^*(E))$$

be the homeomorphism given in Theorem 3. For all $U_1, U_2, U_3 \in \mathbb{O}(\mathrm{Spec}_\gamma(L_{\mathbb{C}}(E)))$ with $U_1 \subseteq U_2 \subseteq U_3$, the diagrams

$$\begin{array}{ccccccc}
K_1^{\mathrm{alg}}(L_{\mathbb{C}}(E)[U_2 \setminus U_1]) & \longrightarrow & K_1^{\mathrm{alg}}(L_{\mathbb{C}}(E)[U_3 \setminus U_1]) & \longrightarrow & K_1^{\mathrm{alg}}(L_{\mathbb{C}}(E)[U_3 \setminus U_2]) & \longrightarrow & K_0^{\mathrm{alg}}(L_{\mathbb{C}}(E)[U_2 \setminus U_1]) \\
{\scriptstyle \gamma_{1,C^*(E)[V_2 \setminus V_1]}} \downarrow & & {\scriptstyle \gamma_{1,C^*(E)[V_3 \setminus V_1]}} \downarrow & & {\scriptstyle \gamma_{1,C^*(E)[V_3 \setminus V_2]}} \downarrow & & {\scriptstyle \gamma_{0,C^*(E)[V_2 \setminus V_1]}} \downarrow \\
K_1^{\mathrm{top}}(C^*(E)[V_2 \setminus V_1]) & \longrightarrow & K_1^{\mathrm{top}}(C^*(E)[V_3 \setminus V_1]) & \longrightarrow & K_1^{\mathrm{top}}(C^*(E)[V_3 \setminus V_2]) & \longrightarrow & K_0^{\mathrm{top}}(C^*(E)[V_2 \setminus V_1])
\end{array}$$

and

$$\begin{array}{ccccc}
K_0^{\mathrm{alg}}(L_{\mathbb{C}}(E)[U_2 \setminus U_1]) & \longrightarrow & K_0^{\mathrm{alg}}(L_{\mathbb{C}}(E)[U_3 \setminus U_1]) & \longrightarrow & K_0^{\mathrm{alg}}(L_{\mathbb{C}}(E)[U_3 \setminus U_2]) \\
{\scriptstyle \gamma_{0,C^*(E)[V_2 \setminus V_1]}} \downarrow & & {\scriptstyle \gamma_{0,C^*(E)[V_3 \setminus V_1]}} \downarrow & & {\scriptstyle \gamma_{0,C^*(E)[V_3 \setminus V_2]}} \downarrow \\
K_0^{\mathrm{top}}(C^*(E)[V_2 \setminus V_1]) & \longrightarrow & K_0^{\mathrm{top}}(C^*(E)[V_3 \setminus V_1]) & \longrightarrow & K_0^{\mathrm{top}}(C^*(E)[V_3 \setminus V_2])
\end{array}$$

are commutative, where $V_i = \phi(U_i)$.

Proof This follows Lemma 7 and from [7, Theorems 2.4.1 and 3.1.9] .

Lemma 8 *Let E be a graph. Then for all $(H_1, S_1), (H_2, S_2)$ admissible pairs with $(H_1, S_1) \leq (H_2, S_2)$, we have that*

$$\gamma_{0,I_{(H_2,S_2)}^{\mathrm{top}}/I_{(H_1,S_1)}^{\mathrm{top}}} : K_0^{\mathrm{alg}}(I_{(H_2,S_2)}^{\mathrm{alg}}/I_{(H_1,S_1)}^{\mathrm{alg}}) \to K_0^{\mathrm{top}}(I_{(H_2,S_2)}^{\mathrm{top}}/I_{(H_1,S_1)}^{\mathrm{top}})$$

is an order isomorphism and

$$\gamma_{1,I^{top}_{(H_2,S_2)}/I^{top}_{(H_1,S_1)}} : K_1^{alg}(I^{alg}_{(H_2,S_2)}/I^{alg}_{(H_1,S_1)}) \to K_1^{top}(\mathfrak{I}_{(H_2,S_2)}/\mathfrak{I}_{(H_1,S_1)})$$

is surjective with kernel a divisible group.

   Suppose F is a graph and suppose there exists an order isomorphism

$$\alpha_0 : K_0^{alg}(I^{alg}_{(H_2,S_2)}/I^{alg}_{(H_1,S_1)}) \to K_0^{alg}(I^{alg}_{(H'_2,S'_2)}/I^{alg}_{(H'_1,S'_1)})$$

and there exists an isomorphism

$$\alpha_1 : K_1^{alg}(I^{alg}_{(H_2,S_2)}/I^{alg}_{(H_1,S_1)}) \to K_1^{alg}(I^{alg}_{(H'_2,S'_2)}/I^{alg}_{(H'_1,S'_1)}),$$

where (H_i, S_i) is an admissible pair of E for $i = 1, 2$ and (H'_i, S'_i) is an admissible pair of F for $i = 1, 2$ with $(H_1, S_1) \le (H_2, S_2)$ and $(H'_1, S'_1) \le (H'_2, S'_2)$. Then α_0 and α_1 induce isomorphisms

$$\widetilde{\alpha}_0 : K_0^{top}(I^{top}_{(H_2,S_2)}/I^{top}_{(H_1,S_1)}) \to K_0^{top}(I^{top}_{(H'_2,S'_2)}/I^{top}_{(H'_1,S'_1)})$$

and

$$\widetilde{\alpha}_1 : K_1^{top}(I^{top}_{(H_2,S_2)}/I^{top}_{(H_1,S_1)}) \to K_1^{top}(I^{top}_{(H'_2,S'_2)}/I^{top}_{(H'_1,S'_1)})$$

such that $\widetilde{\alpha}_0$ is an order isomorphism and

$$\gamma_{i,I^{top}_{(H'_2,S'_2)}/I^{top}_{(H'_1,S'_1)}} \circ \alpha_i = \widetilde{\alpha}_i \circ \gamma_{i,I^{top}_{(H_2,S_2)}/I^{top}_{(H_1,S_1)}}.$$

Proof Let $\iota_E : L_\mathbb{C}(E) \to C^*(E)$ be the $*$-homomorphism sending v to p_v and e to s_e. Note that for all admissible pairs (H, S), $\iota_E(I^{alg}_{(H,S)}) \subseteq I^{top}_{(H,S)}$. Therefore, for all admissible pairs $(H_1, S_1), (H_2, S_2)$ with $(H_1, S_1) \le (H_2, S_2)$, ι_E induces a $*$-homomorphism from $I^{alg}_{(H_2,S_2)}/I^{alg}_{(H_1,S_1)}$ to $I^{top}_{(H_2,S_2)}/I^{top}_{(H_1,S_1)}$. We denote this map by $\iota_{E,I^{top}_{(H_2,S_2)}/I^{top}_{(H_1,S_1)}}$. Thus, the composition of this induced map in K-theory with the homomorphism from $K_n^{alg}(I^{top}_{(H_2,S_2)}/I^{top}_{(H_1,S_1)})$ to $K_n^{top}(I^{top}_{(H_2,S_2)}/I^{top}_{(H_1,S_1)})$ is $\gamma_{n,I^{top}_{(H_2,S_2)}/I^{top}_{(H_1,S_1)}}$.

   We will show that it is enough to prove the first part of the lemma for the case $(H_2, S_2) = (\emptyset, \emptyset)$ and $(H_1, S_1) = (E^0, \emptyset)$. Let (H, S) be an admissible pair. Let $\overline{E}_{(H,S)}$ be the graph given in [18, Definition 4.1]. By the proofs of [18, Theorems 5.1 and 6.1], there exist $*$-isomorphisms

$$\beta_{(H,S)} : L_\mathbb{C}(\overline{E}_{(H,S)}) \to I^{alg}_{(H,S)} \quad \text{and} \quad \lambda_{(H,S)} : C^*(\overline{E}_{(H,S)}) \to I^{top}_{(H,S)}$$

given by

$$\beta_{(H,S)}(v) := \begin{cases} v & \text{if } v \in H \\ v^H & \text{if } v \in S \\ \alpha\alpha^* & \text{if } v = \alpha \in F_1(H,S) \\ \alpha r(\alpha)^H \alpha^* & \text{if } v = \alpha \in F_2(H,S) \end{cases}$$

$$\beta_{(H,S)}(e) := \begin{cases} e & \text{if } e \in E^1 \\ \alpha & \text{if } e = \overline{\alpha} \in \overline{F}_1(H,S) \\ \alpha r(\alpha)^H & \text{if } e = \overline{\alpha} \in \overline{F}_2(H,S) \end{cases}$$

$$\beta_{(H,S)}(e^*) := \begin{cases} e^* & \text{if } e \in E^1 \\ \alpha^* & \text{if } e = \overline{\alpha} \in \overline{F}_1(H,S) \\ r(\alpha)^H \alpha^* & \text{if } e = \overline{\alpha} \in \overline{F}_2(H,S) \end{cases}$$

and

$$\lambda_{(H,S)}(q_v) := \begin{cases} p_v & \text{if } v \in H \\ p_v^H & \text{if } v \in S \\ s_\alpha s_\alpha^* & \text{if } v = \alpha \in F_1(H,S) \\ s_\alpha p_{r(\alpha)}^H s_\alpha^* & \text{if } v = \alpha \in F_2(H,S) \end{cases}$$

$$\lambda_{(H,S)}(t_e) := \begin{cases} s_e & \text{if } e \in E^1 \\ s_\alpha & \text{if } e = \overline{\alpha} \in \overline{F}_1(H,S) \\ s_\alpha p_{r(\alpha)}^H & \text{if } e = \overline{\alpha} \in \overline{F}_2(H,S). \end{cases}$$

Note that the diagram

$$
\begin{array}{ccc}
L_{\mathbb{C}}(\overline{E}_{(H,S)}) & \xrightarrow{\;\;{}^l\overline{E}_{(H,S)}\;\;} & C^*(\overline{E}_{(H,S)}) \\
\beta_{(H,S)} \downarrow & & \downarrow \lambda_{(H,S)} \\
I_{(H,S)}^{\text{alg}} & \xrightarrow[\;\;{}^l{}_{E,I_{(H,S)}^{\text{top}}}/0\;\;]{} & I_{(H,S)}^{\text{top}}
\end{array}
$$

commutes. Therefore, for admissible pairs $(H_1, S_1), (H_2, S_2)$ with $(H_1, S_1) \leq (H_2, S_2)$, the diagram

$$
\begin{CD}
L_{\mathbb{C}}(\overline{E}_{(H_2,S_2)})/\beta_{(H_2,S_2)}^{-1}(I_{(H_1,S_1)}^{\text{alg}}) @>{\overline{\iota}_{E_{(H_2,S_2)}}}>> C^*(\overline{E}_{(H_2,S_2)})/\lambda_{(H_2,S_2)}^{-1}(I_{(H_1,S_1)}^{\text{top}}) \\
@V{\overline{\beta}_{(H_2,S_2)}}VV @VV{\overline{\lambda}_{(H_2,S_2)}}V \\
I_{(H_2,S_2)}^{\text{alg}}/I_{(H_1,S_1)}^{\text{alg}} @>>{\iota_{E,I_{(H_2,S_2)}^{\text{top}}/I_{(H_1,S_1)}^{\text{top}}}}> I_{(H_2,S_2)}^{\text{top}}/I_{(H_1,S_1)}^{\text{top}}
\end{CD}
$$

where $\overline{\beta}_{(H_2,S_2)}$ and $\overline{\lambda}_{(H_2,S_2)}$ are the induced $*$-isomorphisms on the quotient, commutes. Therefore, it is enough to prove the lemma for the graph $\overline{E}_{(H_2,S_2)}$. Hence, we may assume that $(H_2, S_2) = (E^0, \emptyset)$.

Set $(H_1, S_1) = (H, S)$ to simplify the notation. Let $E \setminus (H, S)$ be the graph defined in [19, Theorem 5.7(2)]. Then by the proof of [19, Theorem 5.7(2)] and the discussion before [6, Corollary 5.7], there are $*$-isomorphisms

$$
\delta_{(H,S)} \colon L_{\mathbb{C}}(E \setminus (H, S)) \to L_{\mathbb{C}}(E)/I_{(H,S)}
$$

and

$$
\eta_{(H,S)} \colon C^*(E \setminus (H, S)) \to C^*(E)/\mathfrak{I}_{(H,S)}
$$

such that the diagram

$$
\begin{CD}
L_{\mathbb{C}}(E \setminus (H, S)) @>{\delta_{(H,S)}}>> L_{\mathbb{C}}(E)/I_{(H,S)} \\
@V{\iota_{E \setminus (H,S)}}VV @VV{\iota_{E,C^*(E)/\mathfrak{I}_{(H,S)}}}V \\
C^*(E \setminus (H, S)) @>>{\eta_{(H,S)}}> C^*(E)/\mathfrak{I}_{(H,S)}
\end{CD}
$$

commutes. Hence, it is enough to prove the lemma for the graph $E \setminus (H, S)$. Hence, we may assume that $(H, S) = (\emptyset, \emptyset)$. Thus, proving the claim.

The fact that $\gamma_{0,C^*(E)/0}$ is an isomorphism follows from [12, Corollary 3.5]. To prove that $\gamma_{1,C^*(E)/0}$ is surjective and its kernel is a divisible group we reduce to the case that E is row-finite. Let F be a Drinen-Tomforde desingularization of E defined in [8]. Then there are embeddings $\omega \colon L_{\mathbb{C}}(E) \to L_{\mathbb{C}}(F)$ and $\rho \colon C^*(E) \to C^*(F)$ such that the diagram

$$
\begin{array}{ccc}
L_{\mathbb{C}}(E) & \xrightarrow{\ \omega\ } & L_{\mathbb{C}}(F) \\
{\scriptstyle \iota_E}\downarrow & & \downarrow{\scriptstyle \iota_F} \\
C^*(E) & \xrightarrow[\ \rho\ ]{} & C^*(F)
\end{array}
$$

commutes, $\omega(L_{\mathbb{C}}(E))$ is a full corner of $L_{\mathbb{C}}(F)$, and $\rho(C^*(E))$ is a full corner of $C^*(F)$. Hence, ω and ρ induce isomorphisms in K-theory. Therefore, it is enough to prove $\gamma_{1,C^*(E),0}$ is surjective with kernel a divisible group for the case that E is row-finite. The row-finite case follows from [17, Lemma 4.7]. The first part of the lemma now follows.

For the last part of the lemma, since $K_0(\iota_{E,I^{\mathrm{top}}_{(H_2,S_2)}/I^{\mathrm{top}}_{(H_1,S_1)}})$ is an order isomorphism, it is clear that α_0 induces an order isomorphism $\widetilde{\alpha}_0$ such that

$$
\gamma_{0,I^{\mathrm{top}}_{(H_2',S_2')}/I^{\mathrm{top}}_{(H_1',S_1')}} \circ \alpha_0 = \widetilde{\alpha}_0 \circ \gamma_{0,I^{\mathrm{top}}_{(H_2,S_2)}/I^{\mathrm{top}}_{(H_1,S_1)}}
$$

The fact that α_1 induces an isomorphism $\widetilde{\alpha}_1$ such that $\gamma_{1,I^{\mathrm{top}}_{(H_2',S_2')}/I^{\mathrm{top}}_{(H_1',S_1')}} \circ \alpha_1 = \widetilde{\alpha}_1 \circ$ $\gamma_{1,I^{\mathrm{top}}_{(H_2,S_2)}/I^{\mathrm{top}}_{(H_1,S_1)}}$ is the result of the kernel of $\gamma_{1,I^{\mathrm{top}}_{(H_2,S_2)}/I^{\mathrm{top}}_{(H_1,S_1)}}$ being a divisible group and $K_1(I^{\mathrm{top}}_{(H_2',S_2')}/I^{\mathrm{top}}_{(H_1',S_1')})$ being torsion free, thus [17, Lemma 4.8] applies.

Theorem 5 *Let E and F be graphs.*

1. *Suppose* $\mathsf{FK}^{\mathrm{alg},+}_{0,1}(\mathrm{Spec}_\gamma(L_{\mathbb{C}}(E)); L_{\mathbb{C}}(E)) \cong \mathsf{FK}^{\mathrm{alg},+}_{0,1}(\mathrm{Spec}_\gamma(L_{\mathbb{C}}(F)); L_{\mathbb{C}}(F))$.
   Then $\mathsf{FK}^{\mathrm{top},+}(\mathrm{Prime}_\gamma(C^*(E)); C^*(E)) \cong \mathsf{FK}^{\mathrm{top},+}(\mathrm{Prime}_\gamma(C^*(F)); C^*(F))$.
2. *Suppose* $|E^0|, |F^0| < \infty$. *If*

$$
\theta: \mathsf{FK}^{\mathrm{alg},+}_{0,1}(\mathrm{Spec}_\gamma(L_{\mathbb{C}}(E)); L_{\mathbb{C}}(E)) \to \mathsf{FK}^{\mathrm{alg},+}_{0,1}(\mathrm{Spec}_\gamma(L_{\mathbb{C}}(F)); L_{\mathbb{C}}(F))
$$

is an isomorphism such that θ_0 sends $[1_{L_{\mathbb{C}}(E)}]_0 \in K_0^{\mathrm{alg}}(L_{\mathbb{C}}(E))$ to $[1_{L_{\mathbb{C}}(F)}]_0 \in K_0^{\mathrm{alg}}(L_{\mathbb{C}}(F))$, then there exists an isomorphism

$$
\Theta: \mathsf{FK}^{\mathrm{top},+}(\mathrm{Prime}_\gamma(C^*(E)); C^*(E)) \to \mathsf{FK}^{\mathrm{top},+}(\mathrm{Prime}_\gamma(C^*(F)); C^*(F))
$$

such that Θ_0 sends $[1_{C^*(E)}]_0 \in K_0^{\mathrm{top}}(C^*(E))$ to $[1_{C^*(F)}]_0 \in K_0^{\mathrm{top}}(C^*(F))$.

Proof The theorem follows from Lemmas 7 and 8, and Theorem 4.

Corollary 2 *Let E and F be graphs.*

1. *If $L_{\mathbb{C}}(E)$ and $L_{\mathbb{C}}(F)$ are isomorphic as rings, then*

$$
\mathsf{FK}^{\mathrm{top},+}(\mathrm{Prime}_\gamma(C^*(E)); C^*(E)) \cong \mathsf{FK}^{\mathrm{top},+}(\mathrm{Prime}_\gamma(C^*(F)); C^*(F)).
$$

If, in addition, $|E^0|, |F^0| < \infty$, then there exists an isomorphism

$$\Theta: \mathsf{FK}^{\mathrm{top},+}(\mathrm{Prime}_\gamma(C^*(E)); C^*(E)) \to \mathsf{FK}^{\mathrm{top},+}(\mathrm{Prime}_\gamma(C^*(F)); C^*(F))$$

such that Θ_0 sends $[1_{C^*(E)}]_0 \in K_0^{\mathrm{top}}(C^*(E))$ to $[1_{C^*(F)}]_0 \in K_0^{\mathrm{top}}(C^*(F))$.
2. *If $L_\mathbb{C}(E)$ and $L_\mathbb{C}(F)$ are Morita equivalent, then*

$$\mathsf{FK}^{\mathrm{top},+}(\mathrm{Prime}_\gamma(C^*(E)); C^*(E)) \cong \mathsf{FK}^{\mathrm{top},+}(\mathrm{Prime}_\gamma(C^*(F)); C^*(F)).$$

Proof 1. Follows from Lemma 1 and Theorem 5.

Suppose $L_\mathbb{C}(E)$ and $L_\mathbb{C}(F)$ are Morita equivalent. Then by [2, Corollary 9.11], $M_\infty(L_\mathbb{C}(E)) \cong M_\infty(L_\mathbb{C}(F))$ as rings. By [2, Proposition 9.8(2)], $M_\infty(L_\mathbb{C}(E)) \cong L_\mathbb{C}(SE)$ and $M_\infty(L_\mathbb{C}(F)) \cong L_\mathbb{C}(SF)$ as $\mathbb{C}$-algebras, where SE and SF are the stabilized graphs of E and F respectively (see [2, Definition 9.4]). Note that every graded ideal $L_\mathbb{C}(SE)$ is of the from $M_\infty(I)$ for a unique graded ideal of I of $L_\mathbb{C}(E)$ and every graded ideal of $L_\mathbb{C}(SF)$ is of the from $M_\infty(J)$ for a unique graded ideal J of $L_\mathbb{C}(F)$. We also have that

$$\mathsf{FK}_{0,1}^{\mathrm{alg},+}(\mathrm{Spec}_\gamma(L_\mathbb{C}(E)); L_\mathbb{C}(E)) \cong \mathsf{FK}_{0,1}^{\mathrm{alg},+}(\mathrm{Spec}_\gamma(L_\mathbb{C}(SE)); L_\mathbb{C}(SE))$$

$$\cong \mathsf{FK}_{0,1}^{\mathrm{alg},+}(\mathrm{Spec}_\gamma(L_\mathbb{C}(SF)); L_\mathbb{C}(SF))$$

$$\cong \mathsf{FK}_{0,1}^{\mathrm{alg},+}(\mathrm{Spec}_\gamma(L_\mathbb{C}(F)); L_\mathbb{C}(F)).$$

Therefore, by Theorem 5,

$$\mathsf{FK}^{\mathrm{top},+}(\mathrm{Prime}_\gamma(C^*(E)); C^*(E)) \cong \mathsf{FK}^{\mathrm{top},+}(\mathrm{Prime}_\gamma(C^*(F)); C^*(F)).$$

Corollary 3 *The Abrams-Tomforde conjecture holds for the class of finite graphs that satisfy Condition (H) of [10, Definition 4.19]. In particular the Abrams-Tomforde conjecture holds for the class of finite graphs that satisfy Condition (K).*

Proof The first part is just a combination of Corollary 2 and [10, Theorem 6.1]. Finally, all graphs that satisfy Condition (K) satisfy Condition (H).

Remark 1 Corollary 2 will be used in [9] to show that the Abrams-Tomforde conjecture holds for the class of graphs with finitely many vertices.

Acknowledgements This work was partially supported by the Danish National Research Foundation through the Centre for Symmetry and Deformation (DNRF92), by the VILLUM FONDEN through the network for Experimental Mathematics in Number Theory, Operator Algebras, and Topology, by a grant from the Simons Foundation (# 279369 to Efren Ruiz), and by the Danish Council for Independent Research—Natural Sciences.

References

1. Abrams, G., Aranda Pino, G.: The Leavitt path algebra of a graph. J. Algebra **293**(2), 319–334 (2005). http://dx.doi.org/10.1016/j.jalgebra.2005.07.028
2. Abrams, G., Tomforde, M.: Isomorphism and Morita equivalence of graph algebras. Trans. Amer. Math. Soc. **363**(7), 3733–3767 (2011). http://dx.doi.org/10.1090/S0002-9947-2011-05264-5
3. Ara, P., Goodearl, K.R., O'Meara, K.C., Pardo, E.: Separative cancellation for projective modules over exchange rings. Israel J. Math. **105**, 105–137 (1998). http://dx.doi.org/10.1007/BF02780325
4. Ara, P., Moreno, M.A., Pardo, E.: Nonstable K-theory for graph algebras. Algebr. Represent. Theory **10**(2), 157–178 (2007). http://dx.doi.org/10.1007/s10468-006-9044-z
5. Aranda Pino, G., Pardo, E., Siles Molina, M.: Exchange Leavitt path algebras and stable rank. J. Algebra **305**(2), 912–936 (2006). http://dx.doi.org/10.1016/j.jalgebra.2005.12.009
6. Bates, T., Hong, J.H., Raeburn, I., Szymański, W.: The ideal structure of the C^*-algebras of infinite graphs. Illinois J. Math. **46**(4), 1159–1176 (2002). http://projecteuclid.org/euclid.ijm/1258138472
7. Cortiñas, G.: Algebraic vs. topological K-theory: a friendly match. In: Topics in Algebraic and Topological K-Theory. Lecture Notes in Mathematics, vol. 2008, pp. 103–165. Springer, Berlin (2011). http://dx.doi.org/10.1007/978-3-642-15708-0_3
8. Drinen, D., Tomforde, M.: The C^*-algebras of arbitrary graphs. Rocky Mountain J. Math. **35**(1), 105–135 (2005). http://dx.doi.org/10.1216/rmjm/1181069770
9. Eilers, S., Restorff, G., Ruiz, E., Sørensen, A.P.W.: The complete classification of unital graph C^*-algebras: geometric and strong. arXiv e-prints (2016, submitted)
10. Eilers, S., Restorff, G., Ruiz, E., Sørensen, A.P.W.: Geometric classification of graph C^*-algebras over finite graphs. arXiv e-prints Canad. J. Math. (2017). https://cms.math.ca/10.4153/CJM-2017-016-7
11. Fowler, N.J., Laca, M., Raeburn, I.: The C^*-algebras of infinite graphs. Proc. Amer. Math. Soc. **128**(8), 2319–2327 (2000). http://dx.doi.org/10.1090/S0002-9939-99-05378-2
12. Hay, D., Loving, M., Montgomery, M., Ruiz, E., Todd, K.: Non-stable K-theory for Leavitt path algebras. Rocky Mountain J. Math. **44**(6), 1817–1850 (2014). http://dx.doi.org/10.1216/RMJ-2014-44-6-1817
13. Jeong, J.A., Park, G.H.: Graph C^*-algebras with real rank zero. J. Funct. Anal. **188**(1), 216–226 (2002). http://dx.doi.org/10.1006/jfan.2001.3830
14. Năstăsescu, C., van Oystaeyen, F.: Graded Ring Theory. North-Holland Mathematical Library, vol. 28. North-Holland Publishing Co., Amsterdam/New York (1982)
15. Raeburn, I.: Graph Algebras. CBMS Regional Conference Series in Mathematics, vol. 103. Published for the Conference Board of the Mathematical Sciences, Washington, DC; by the American Mathematical Society, Providence, RI (2005)
16. Rangaswamy, K.M.: The theory of prime ideals of Leavitt path algebras over arbitrary graphs. J. Algebra **375**, 73–96 (2013). http://dx.doi.org/10.1016/j.jalgebra.2012.11.004
17. Ruiz, E., Tomforde, M.: Ideal-related K-theory for Leavitt path algebras and graph C^*-algebras. Indiana Univ. Math. J. **62**(5), 1587–1620 (2013). http://dx.doi.org/10.1512/iumj.2013.62.5123
18. Ruiz, E., Tomforde, M.: Ideals in graph algebras. Algebr. Represent. Theory **17**(3), 849–861 (2014). http://dx.doi.org/10.1007/s10468-013-9421-3
19. Tomforde, M.: Uniqueness theorems and ideal structure for Leavitt path algebras. J. Algebra **318**(1), 270–299 (2007). http://dx.doi.org/10.1016/j.jalgebra.2007.01.031
20. Weibel, C.A.: The K-Book. Graduate Studies in Mathematics, vol. 145. American Mathematical Society, Providence, RI (2013). An introduction to algebraic K-theory

Gysin Exact Sequences for Quantum Weighted Lens Spaces

Francesca Arici

Abstract We describe quantum weighted lens spaces as total spaces of quantum principal circle bundles, using a Cuntz-Pimsner model. The corresponding Pimsner exact sequence is interpreted as a noncommutative analogue of the Gysin exact sequence. We use the sequence to compute the K-theory and K-homology groups of quantum weighted lens spaces, extending previous results and computations due to the author and collaborators.

1 Introduction

Quantum lens spaces, both weighted and unweighted, have been the subject of increasing interest in the last years. They are Cuntz-Krieger algebras of a directed graph [16] and have played an important role in the classification program of C*-algebras [12]. Using graph algebra techniques their K-theory groups have been computed recently in [8] under very general assumptions on the weight. From a more geometric point of view, they have a natural structure of noncommutative principal circle bundles over quantum weighted projective spaces [2, 4, 6, 7, 22] and can thus be interpreted as a deformation of their classical counterparts. In this paper we focus on the noncommutative topology of quantum weighted lens spaces, realising them as Cuntz-Pimsner algebras of self-Morita equivalence bimodules. This allows us to compute their K-theory and K-homology groups, using different techniques than those in [8].

Being graph algebras, quantum weighted lens spaces admit a Cuntz-Pimsner model where the coefficient algebra is the algebra of functions on the vertex space. This picture is very well suited to encode the dynamical information contained in the graph, but has the disadvantage that the fixed point algebra for the natural gauge action does not agree with the coefficient algebra. In the Cuntz-Pimsner model

F. Arici (✉)
Faculty of Science, Institute for Mathematics, Astrophysics and Particle Physics, Radboud University Nijmegen, Heyendaalseweg 135, 6525AJ Nijmegen, The Netherlands
e-mail: f.arici@math.ru.nl

© Springer International Publishing AG, part of Springer Nature 2018
D.R. Wood et al. (eds.), *2016 MATRIX Annals*, MATRIX Book Series 1,
https://doi.org/10.1007/978-3-319-72299-3_12

we employ here, which comes from the geometric analogy described above, the coefficient algebra is the algebra of functions on a quantum weighted projective space and the resulting Cuntz-Pimsner algebra can be thought of as the total space of a noncommutative circle bundle.

The associated six term exact sequences can then be interpreted as the operator theoretic counterpart of the classical Gysin exact sequence for circle bundles (cf. [18, IV.1.13]). Under some mild assumptions on the weight, we will describe the K-theory and K-homology of quantum weighted lens spaces of any dimension, thus extending the results of [2] and [4].

2 Quantum Weighted Projective and Lens Spaces

In this section we describe the coordinate algebras of weighted projective and lens spaces, as described in [3, 7, 11] and their C^*-completions, which were extensively studied in [8].

Classically, weighted projective and lens spaces are quotients of odd-dimensional spheres by actions of the circle and of a finite cyclic group, respectively. The same is true upon replacing the sphere by a *quantum* sphere.

Let $q \in (0, 1)$. We recall from [23] that the coordinate algebra of the quantum odd-dimensional sphere $\mathcal{O}(S_q^{2n+1})$ is the universal $*$-algebra with generators the $n + 1$ elements $\{z_i\}_{i=0,\ldots,n}$ and relations:

$$z_i z_j = q^{-1} z_j z_i \qquad\qquad 0 \le i < j \le n \,,$$

$$z_i^* z_j = q z_j z_i^* \qquad\qquad i \ne j \,,$$

$$[z_n^*, z_n] = 0, \qquad [z_i^*, z_i] = (1 - q^2) \sum_{j=i+1}^{n} z_j z_j^* \qquad i = 0, \ldots, n - 1 \,,$$

and a sphere relation:

$$z_0 z_0^* + z_1 z_1^* + \ldots + z_n z_n^* = 1 \,.$$

The notation of [23] is obtained by setting $q = e^{h/2}$.

A *weight vector* $\mathsf{m} = (m_0, \ldots, m_n)$ is a finite sequence of positive integers, called *weights*. A weight vector is said to be *coprime* if g.c.d.$(m_0, \ldots, m_n) = 1$; and it is *pairwise coprime* if g.c.d.$(m_i, m_j) = 1$, for all $i \ne j$.

For any weight vector $\mathsf{m} = (m_0, \ldots, m_n)$, we define a weighted circle action $\{\sigma_\xi^{\mathsf{m}}\}_{\xi \in \mathbb{T}^1}$ on the quantum sphere, given on generators by

$$\sigma_\xi^{\mathsf{m}}(z_i) = \xi^{m_i} z_i \qquad \xi \in \mathbb{T}^1, \tag{1}$$

The $\mathbb{Z}$-grading induced by this action is equivalent to that obtained by declaring each z_i to be of degree m_i and z_i^* of degree $-m_i$.

The degree zero part or, equivalently, the fixed point algebra for the action, is the coordinate algebra of the *quantum n-dimensional weighted projective space* associated with the weight vector m, and it is denoted by $\mathcal{O}(\mathbb{WP}_q^n(\mathsf{m}))$.

By D'Andrea and Landi [11, Lemma 3.2] a set of generators for the algebra $\mathcal{O}(\mathbb{WP}_q^n(\mathsf{m}))$ is given by the elements

$$z_i z_i^*, \quad \text{and} \quad z^{\mathbf{k}} := z_0^{k_0} \cdots z_n^{k_n},$$

for $i = 0, \ldots, n$ and $\mathbf{k} \in \mathbb{Z}^{n+1}$ with $\mathbf{k} \cdot \mathsf{m} := k_0 m_0 + \cdots + k_n m_n = 0$. Note that such a set of generators is in general not minimal.

For some particular classes, one gets a complete characterisation of generators of the algebra $\mathcal{O}(\mathbb{WP}_q^n(\mathsf{m}))$. Indeed, we have two classes of weighted projective spaces, in some sense orthogonal to each other, for which it is possible to describe the generators and the representation theory.

The first class consists of those weighted projective spaces for which the weight m is of the form $\mathsf{m} = \mathsf{p}^\sharp$ for p pairwise coprime, where $\mathsf{p}^\sharp$ is defined as the weight vector whose i-th component is equal to $\prod_{j \neq i} p_j$. Classically those are the weighted projective spaces that are isomorphic, as projective varieties, to the unweighted projective space $\mathbb{CP}^n$. By D'Andrea and Landi [11, Theorem 3.8], having such a weight is a necessary and sufficient condition for the algebra $\mathcal{O}(\mathbb{WP}_q^n(\mathsf{m}))$ to be generated by the elements

$$a_{i,j} := (z_i^*)^{m_{j:i}} z_j^{m_{i:j}}, \quad m_{i:j} = m_i / \text{g.c.d.}(m_i, m_j) \quad \forall i, j = 0, \ldots, n.$$

The second class of examples, that goes in another direction with respect to the class we just described, is that of the multidimensional teardrops [7], that are obtained for the weight vector $\mathsf{m} = (1, \ldots, 1, m)$ having all but the last entry equal to 1. As described in [7, Lemma 6.1] the algebra $\mathcal{O}(\mathbb{WP}_q^n(1, \ldots, m))$ is generated, as a *-algebra, by the elements

$$b_{i,j} := (z_i^*)z_j \quad \text{and} \quad c_1 := z_0^{l_0} \cdots z_{n-1}^{l_{n-1}} z_n^*,$$

for $0 \leq i \leq j \leq n-1$ and $\mathbf{l} \in \mathbb{N}^n$ such that $\sum_{i=0}^{n-1} l_i = m$.

As a particular case of both constructions, for $\mathsf{m} = (1, \ldots, 1)$ one gets the coordinate algebra $\mathcal{O}(\mathbb{CP}_q^n)$ of the quantum projective space $\mathbb{CP}_q^n$. This is the *-subalgebra of $\mathcal{O}(S_q^{2n+1})$ generated by the elements $p_{ij} := z_i^* z_j$ for $i, j = 0, 1, \ldots, n$.

Let now $N \geq 2$ be fixed. By restriction $\mathcal{O}(S_q^{2n+1})$ admits an action of the cyclic group $\mathbb{Z}_N$ given by

$$\sigma_\zeta^{(1/N, \mathsf{m})} z_i = \zeta^{m_i} z_i,$$

where $\zeta = e^{2\pi i / N} \in \mathbb{T}$ is the generator of $\mathbb{Z}_N$.

The coordinate algebra of the quantum lens space $\mathcal{O}(L_q^{2n+1}(N; \mathsf{m}))$ is defined as the fixed point algebra for this action:

$$\mathcal{O}(L_q^{2n+1}(N; \mathsf{m})) := \mathcal{O}(S_q^{2n+1})^{\mathbb{Z}_N}. \tag{2}$$

2.1 Principal Bundle Structures

In noncommutative geometry the notion of a free action of a quantum group H on $\mathscr{A}$ is translated into that of a principal coaction on $\mathscr{A}$, which in algebraic terms amounts to having a Hopf-Galois extension. In the case of a classical Abelian group, principality is equivalent to the notion of a *strong* grading.

Given a group G, a G-graded algebra $\mathscr{A}$ is an algebra that decomposes as a direct sum $\mathscr{A} = \bigoplus_{g \in G} \mathscr{A}_g$, with $\mathscr{A}_g \mathscr{A}_h \subseteq \mathscr{A}_{gh}$. Whenever $\mathscr{A}_g \mathscr{A}_h = \mathscr{A}_{gh}$ for all $g, h \in G$, one says that the grading is strong. Note that it is enough to check this condition on a set of generators of the group.

As described in [19, Theorem 8.1.7], having a strongly graded algebra over a group G is equivalent to having a Hopf-Galois extension over the group algebra $\mathbb{C}G$. In this work we will only focus on classical Abelian groups; as said, in that case principality is equivalent to the induced grading over the Pontryagin dual being *strong*.

By Brzeziński and Fairfax [7, Lemma 2.1] strong gradings are preserved under extensions of Abelian groups, i.e. given an exact sequence

$$0 \longrightarrow K \overset{\varphi}{\longrightarrow} G \overset{\pi}{\longrightarrow} H \longrightarrow 0 \ ,$$

a G-graded algebra $\mathscr{A}$ is strongly graded if the induced H-grading on $\mathscr{A}$ and the induced K-grading on $\mathscr{A}^K := \bigoplus_{k \in K} \mathscr{A}_{\varphi(k)}$ are strong.

In our case we will be dealing with the group $\mathbb{Z} = \widehat{\mathbb{T}}$ and the finite group cyclic group $\mathbb{Z}_N = \widehat{\mathbb{Z}_N}$, so we will be interested in the short exact sequence

$$0 \longrightarrow \mathbb{Z} \overset{N \cdot}{\longrightarrow} \mathbb{Z} \overset{\pi}{\longrightarrow} \mathbb{Z}_N \longrightarrow 0.$$

As described in [7, Proposition 4.1], the $\mathbb{Z}_N$-action on $\mathcal{O}(S_q^{2n+1})$ induces a $\mathbb{Z}_N$-grading which is strong, and $\mathcal{O}(L_q^{2n+1}(N; \mathsf{m}))$ is the degree-zero subalgebra with respect to that grading. The lens space $\mathcal{O}(L_q^{2n+1}(N; \mathsf{m}))$ is a $\mathbb{Z}$-graded algebra with respect to the grading induced by that of $\mathcal{O}(S_q^{2n+1})$, by saying that $x \in \mathcal{O}(L_q^{2n+1}(N; \mathsf{m}))$ has degree n if and only if it has degree nN in $\mathcal{O}(S_q^{2n+1})$. The degree-zero part is given by the coordinate algebra of the quantum projective space $\mathcal{O}(\mathbb{WP}_q^n(\mathsf{m}))$. The induced grading is not always strong. However, by Brzeziński and Fairfax [7, Proposition 4.2], for $N_\mathsf{m} := \prod_{i=1}^{n} m_i$, the algebra $\mathcal{O}(L_q^{2n+1}(N_\mathsf{m}; \mathsf{m}))$

is strongly $\mathbb{Z}$-graded. As a consequence, the coordinate algebra of the quantum weighted lens space $\mathcal{O}(L_q^{2n+1}(N_{\mathsf{m}};\mathsf{m}))$ has the structure of a quantum principal circle bundle over the n-dimensional weighted projective space $\mathcal{O}(\mathbb{WP}_q^n(\mathsf{m}))$.

Let $d \geq 1$. By Proposition 4.6 in [4] the coordinate algebra of the quantum weighted lens space $\mathcal{O}(L_q^{2n+1}(d \cdot N_{\mathsf{m}};\mathsf{m}))$ also has the structure of a quantum principal circle bundle over the n-dimensional weighted projective space $\mathcal{O}(\mathbb{WP}_q^n(\mathsf{m}))$. This can be seen as a consequence of the aforementioned Lemma 2.1 of [7].

2.2 C^*-Completions

The C^*-algebra $C(S_q^{2n+1})$ of the odd-dimensional quantum sphere is the completion of the $*$-algebra $\mathcal{O}(S_q^{2n+1})$ in the universal C^*-norm. This C^*-algebra can be realised as a graph C^*-algebra.

The C^*-algebra $C(\mathbb{WP}^n(\mathsf{m}))$ of the quantum weighted projective space is defined as the fixed point algebra for the circle action on $C(S_q^{2n+1})$ obtained by extending σ. A complete characterisation of those C^*-algebras is not available at the moment; partial results were obtained in [8] for a large class of weighted lens spaces, those with weight vector m satisfying g.c.d.$(m_j, m_n) = 1$ for at least one $j < n$. By Brzeziński and Szymański [8, Proposition 3.2] there exists an exact sequence of C^*-algebras

$$0 \longrightarrow \mathcal{K}^{\oplus m_n} \longrightarrow C(\mathbb{WP}^n(\mathsf{m})) \longrightarrow C(\mathbb{WP}_q^{n-1}(\mathsf{m}_n)) \longrightarrow 0, \qquad (3)$$

where m_n denotes the weight vector $(m_0, \ldots, m_{n-1})$.

The K-theory groups of the C^*-algebraic weighted projective spaces can be computed by iterative use of the extension (3) under suitable assumptions on the weight vector m.

Proposition 2.1 ([8, Corollary 3.3]) *Let m be a weight vector with the property that for each $j \geq 1$ there exists $i < j$ such that* g.c.d.$(m_i, m_j) = 1$. *Then the K-theory groups of the quantum weighted projective spaces are given by*

$$K_0(C(\mathbb{WP}_q^n(\mathsf{m}))) = \mathbb{Z}^{1+\sum_{i=1}^n m_i}, \qquad K_1(C(\mathbb{WP}_q^n(\mathsf{m}))) = 0.$$

The C^*-algebraic quantum lens space is defined as the fixed point algebra for the action of $\mathbb{Z}_N$ on $C(S_q^{2n+1})$. By constructing a conditional expectation for the $\mathbb{Z}_N$-action, one can show that it agrees with the closure of the algebraic quantum lens space $\mathcal{O}(L_q^{2n+1}(N;\mathsf{m}))$ with respect to the universal C^*-norm on $C(S_q^{2n+1})$. It is isomorphic to the C^*-algebra of a directed graph.

3 A Cuntz-Pimsner Model for Quantum Lens Spaces

Cuntz-Pimsner algebras [20] are universal C*-algebras constructed out of a C*-correspondence E over a C*-algebra B. They encompass a large class of examples, like crossed product by the integers, Cuntz and Cuntz-Krieger algebras [9, 10], graph algebras and C*-algebras associated to a partial automorphism [13]. We now give a simple description of Pimsner's construction for the case of interest for this work.

Under the assumptions that B is unital and that E is a self-Morita equivalence bimodule, i.e. we have left action implemented by an isomorphism $\phi : B \to \text{End}_B(E)$, the Cuntz-Pimsner algebra O_E admits a description in terms of generators and commutation relations. This construction, which can be found for instance in [17, Section 2], works for any finitely generated projective module over a unital C*-algebra and relies on the existence of a finite frame for the module E, i.e. a finite set of elements $\{\xi_i\}_{i=1}^n$ of E satisfying

$$\xi = \sum_{j=1}^{n} \eta_j \langle \eta_j, \xi \rangle_B.$$

for any $\xi \in E$.

The algebra O_E is realised as the universal C*-algebra generated by B together with n operators $S_1, \ldots, S_n$, satisfying

$$S_i^* S_j = \langle \eta_i, \eta_j \rangle_B, \quad \sum_j S_j S_j^* = 1, \quad \text{and} \quad b S_j = \sum_i S_i \langle \eta_i, \phi(b) \eta_j \rangle_B, \tag{4}$$

for $b \in B$, and $j = 1, \ldots, n$.

Example 3.1 The module $\Gamma(\mathscr{E})$ of sections of the tautological line bundle $\mathscr{E}$ over the quantum projective line is a self-Morita equivalence bimodule over the algebra $C(\mathbb{CP}^1)$. The corresponding Cuntz-Pimsner algebra $O_{\Gamma(\mathscr{E})}$ is isomorphic to the algebra of continuous functions on the three sphere $C(S^3)$.

More generally, the Cuntz-Pimsner algebra O_E of a self-Morita equivalence bimodule can be thought of as the algebra of continuous functions on the total space of a quantum principal circle bundle. While the commutative version of this analogy was spelled out in [14], the more general case of quantum principal circle bundles was described in [4]. We also refer to the review article [3] for more details and recall the salient points here.

Given a C*-algebra A together with a strongly continuous circle action $\sigma := \{\sigma_\xi\}_{\xi \in \mathbb{T}^1}$, we define the n-th spectral subspace as

$$A_{(n)} := \{a \in A \mid \sigma_\xi(a) = \xi^n a \quad \forall \xi \in \mathbb{T}^1\}.$$

Then the invariant subspace $A_{(0)} \subseteq A$ is a C*-subalgebra and each $A_{(n)}$ is a Hilbert C*-bimodule over $A_{(0)}$. If the module $A_{(1)}$ is a self-Morita equivalence bimodule, which is equivalent to the $\mathbb{Z}$-grading given by the spectral subspaces being strong, then the action σ is said to be saturated. Then by Prop. 3.5 in [4] the Cuntz-Pimsner algebra $O_{A_{(1)}}$ of the first spectral subspace $A_{(1)}$ is isomorphic to the algebra A.

Let see what this means in our examples: if we denote by E the first spectral subspace of $C(S_q^{2n+1})$ for the weighted action of $\mathbb{T}^1$, then we get that the C^*-algebra $C(L_q^{2n+1}(N_{\mathsf{m}};\mathsf{m}))$ is isomorphic to the Pimsner algebra O_E over $C(\mathbb{WP}_q^n(\mathsf{m}))$ associated to E.

More generally, by Thm. 3.9 in [4] for any $d \geq 1$, the C*-algebraic lens space $C(L_q^{2n+1}(d \cdot N_{\mathsf{m}};\mathsf{m}))$ is isomorphic to the Pimsner algebra O_E associated to the module $E^{\otimes d}$ over $C(\mathbb{WP}_q^n(\mathsf{m}))$.

As particular cases, $C(S_q^{2n+1})$ is a Pimsner algebra over $C(\mathbb{CP}_q^n)$, and more generally the *unweighted* lens space $C(L_q^{2n+1}(d;\underline{1}))$, for the weight vector with entries identically one, is a Pimsner algebra over $C(\mathbb{CP}_q^n)$ for any $d \geq 1$. Those algebras and their K-theory group were the subject of [2].

3.1   Six-Term Exact Sequences

For a Pismner algebra one has natural exact sequences in bivariant K-theory, relating the KK-groups of the Pimsner algebra O_E with those of the base space algebra B. Those sequences were constructed by Pimsner, see [20, Theorem 4.8] and arise as six-term exact sequences associated to a semisplit extension of C*-algebras in which the Pimsner algebra is the quotient, the ideal is Morita equivalent to the base and the middle algebra is KK-equivalent to the base. Using those identifications, the resulting exact sequences in bivariant K-theory read:

$$
\begin{array}{ccccc}
KK_0(C,B) & \xrightarrow{\otimes_B(1-[E])} & KK_0(C,B) & \xrightarrow{i_*} & KK_0(C,O_E) \\
{\scriptstyle\partial}\Big\uparrow & & & & \Big\downarrow{\scriptstyle\partial} \\
KK_1(C,O_E) & \xleftarrow{i_*} & KK_1(C,B) & \xleftarrow{\otimes_B(1-[E])} & KK_1(C,B)
\end{array}
$$

and

$$
\begin{array}{ccccc}
KK_0(B,C) & \xleftarrow{(1-[E])\otimes_B} & KK_0(B,C) & \xleftarrow{i^*} & KK_0(O_E,C)\;. \\
{\scriptstyle\partial}\Big\downarrow & & & & \Big\uparrow{\scriptstyle\partial} \\
KK_1(O_E,C) & \xrightarrow{i^*} & KK_1(B,C) & \xrightarrow{(1-[E])\otimes_B} & KK_1(B,C)
\end{array}
$$

A crucial role is played by the Kasparov product with the class of the identity $1 \in KK(B,B)$ minus the class $[E] \in KK(B,B)$ of the bimodule E.

The connecting homomorphism ∂ is implemented by taking the Kasparov product with the class of the defining extension. An unbounded representative for this extension class was constructed in [21] and later generalised in [15]. A treatment of the non-unital case is in [1].

In the case of self-Morita equivalence bimodules these could be considered as a generalization of the classical *Gysin sequence* in K-theory for the *noncommutative* line bundle E over B, the Kasparov product with $1 - [E]$ plays the role of the cup product with the *Euler class* of the corresponding line bundle.

Examples of Gysin sequences in K-theory were given in [2] for line bundles over quantum projective spaces leading to a class of quantum lens spaces. These examples were generalized later in [4] for a class of quantum lens spaces as circle bundles over quantum weighted projective spaces with arbitrary weights.

To ease our notation, we let $C(L_q(d)) := C(L_q^{2n+1}(d \cdot N_{\mathsf{m}}, \mathsf{m}))$. Also, E will denote the Hilbert C*-module given by the first spectral subspace for the weighted circle action on $C(S_q^{2n+1})$.

Then, given any separable C*-algebra C, we obtain the following two six term exact sequences in KK-theory:

$$
\begin{array}{ccccc}
KK_0(C, C(\mathbb{WP}_q^n(\mathsf{m}))) & \xrightarrow{\otimes(1-[E^{\otimes d}])} & KK_0(C, C(\mathbb{WP}_q^n(\mathsf{m}))) & \xrightarrow{\ i_* \ } & KK_0\big(C, C(L_q(d))\big) \\
\Big\uparrow{\scriptstyle\partial} & & & & \Big\downarrow{\scriptstyle\partial} \\
KK_1(C, C(L_q(d))) & \xleftarrow{\ i_* \ } & KK_1(C, C(\mathbb{WP}_q^n(\mathsf{m}))) & \xleftarrow{\otimes(1-[E^{\otimes d}])} & KK_1(C, C(\mathbb{WP}_q^n(\mathsf{m}))
\end{array}
\tag{5}
$$

and

$$
\begin{array}{ccccc}
KK_0(C(\mathbb{WP}_q^n(\mathsf{m})), C) & \xleftarrow{(1-[E^{\otimes d}])\otimes} & KK_0(C(\mathbb{WP}_q^n(\mathsf{m})), C) & \xleftarrow{\ i^* \ } & KK_0\big(C(L_q(d)), C\big) \\
\Big\downarrow{\scriptstyle\partial} & & & & \Big\uparrow{\scriptstyle\partial} \\
KK_1\big(C(L_q(d)), C\big) & \xrightarrow{\ i^* \ } & KK_1(C(\mathbb{WP}_q^n(\mathsf{m})), C) & \xrightarrow{(1-[E^{\otimes d}])\otimes} & KK_1(C(\mathbb{WP}_q^n(\mathsf{m})), C)
\end{array}
\tag{6}
$$

where i_* and i^* are the maps in KK-theory induced by the inclusion of the coefficient algebra into the Pimsner algebra $i : C(\mathbb{WP}_q^n) \hookrightarrow O_{E^{\otimes d}} \simeq C(L_q(d))$.

We will refer to these two sequences as the *Gysin sequences* (in KK-theory) for the C*-algebraic quantum lens space $C(L_q^{2n+1}(d \cdot N_{\mathsf{m}}; \mathsf{m}))$.

3.2 Computing the K-Theory and K-Homology of Quantum Lens Spaces

We finish by describing how the exact sequences (5) and (6) can be used to obtain information about the KK-theory groups of quantum weighted lens spaces.

Even though those sequences exist for every choice of weight m, we will now restrict our attention to the case of weight vectors satisfying the assumptions of Proposition 2.1. This will allow us to use the computations of the K-theory groups of the weighted projective spaces in our computations.

We will now state an easy corollary of the results contained in [8].

Proposition 3.2 *Let m be a weight vector satisfying the assumptions of Proposition 2.1. Then the C^*-algebra $C(\mathbb{WP}^n(m))$ is KK-equivalent to $\mathbb{C}^{1+m_1+\cdots+m_n}$.*

Proof As a first step we use the fact that the UCT class is closed under extensions and contains the algebra of compact operators. Whenever the weight vector m satisfies the assumptions of Proposition 2.1, by iterated use of the exact sequence (3) and the fact that $C(\mathbb{WP}^0(m)) = \mathbb{C}$ we obtain that the C^*-algebraic weighted projective space is also in the UCT class. By Proposition 2.1 its K-theory groups are isomorphic to those of $\mathbb{C}^{1+m_1+\cdots+m_n}$. The claim follows from the fact that in the UCT class two C^*-algebras are KK-equivalent if and only if they have isomorphic K-theory groups (cf. [5, Corollary 23.10.2]).

Note that for $n = 1$ the extension (3) admits a completely positive splitting, hence KK-equivalence follows from the fact that the algebra $C(\mathbb{WP}_q^1(m))$ is isomorphic to the unitalisation $\mathcal{K}^{m_1} \oplus \mathbb{C}$. Explicit representatives for the two KK-equivalences were constructed in [4].

For ease of notation we will denote by $M := m_1 + \cdots + m_n$. We let $[I] \in KK(\mathbb{C}^{M+1}, C(\mathbb{WP}_q^n(m)))$ and $[\Pi] \in KK(C(\mathbb{WP}_q^n(m)), \mathbb{C}^{M+1})$ be the two classes that implement the KK-equivalence between $\mathbb{C}^{M+1}$ and $C(\mathbb{WP}_q^n(m))$, i.e., satisfy

$$[I] \otimes_{C(\mathbb{WP}_q^n(m))} [\Pi] = 1_{KK(\mathbb{C}^{M+1}, \mathbb{C}^{M+1})}, \quad [\Pi] \otimes_{\mathbb{C}^{M+1}} [I] = 1_{KK(C(\mathbb{WP}_q^n(m)), C(\mathbb{WP}_q^n(m)))}. \tag{7}$$

These KK-equivalences can be used to simplify the exact sequences (5) and (6). Indeed, one can use them to replace the KK-groups of $C(\mathbb{WP}_q^n(m))$ with those of the vector space $\mathbb{C}^{M+1}$ and then use the natural isomorphisms

$$KK_i(C, \mathbb{C}^{M+1}) \simeq \bigoplus_{k=1}^{M+1} K^i(C) \quad \text{and} \quad KK_i(\mathbb{C}^{M+1}, C) \simeq \bigoplus_{k=1}^{M+1} K_i(C), \quad i = 0, 1.$$

Tensoring the class of the Hilbert C^*-module E with the KK-equivalences $[I]$ and $[\Pi]$ one gets a class

$$[I] \otimes_{C(\mathbb{WP}_q^n(m))} [E] \otimes_{C(\mathbb{WP}_q^n(m))} [\Pi] \in KK(\mathbb{C}^{M+1}, \mathbb{C}^{M+1}). \tag{8}$$

Since the ring $KK(\mathbb{C}^{M+1}, \mathbb{C}^{M+1})$ is isomorphic to $(M+1) \times (M+1)$ matrices with entries in $\mathbb{Z}$, we can look at the corresponding matrix implementing the map (8), that we denote by A. The six term exact sequence in (5) becomes

$$
\begin{array}{ccc}
\oplus_{i=1}^{M+1} K^0(C) \xrightarrow{1-A^d} \oplus_{i=1}^{M+1} K^0(C) \longrightarrow KK_0\big(C, C(L_q(d))\big) & , \\
\uparrow & & \downarrow \\
KK_1(C, C(L_q(d))) \longleftarrow \oplus_{i=1}^{M+1} K^1(C) \underset{1-A^d}{\longleftarrow} \oplus_{i=1}^{M+1} K^1(C)
\end{array}
$$

while, denoting the transpose of A by A^t, the six term exact sequence in (6) becomes

$$
\begin{array}{ccc}
\oplus_{i=1}^{M+1} K_0(C) \underset{1-(A^t)^d}{\longleftarrow} \oplus_{r=1}^{M+1} K_0(C) \longleftarrow KK_0\big(C(L_q(d)), C\big) & . \\
\downarrow & & \uparrow \\
KK_1\big(C(L_q(d)), C\big) \longrightarrow \oplus_{i=1}^{M+1} K_1(C) \xrightarrow{1-(A^t)^d} \oplus_{i=1}^{M+1} K_1(C)
\end{array}
$$

For $C = \mathbb{C}$, using the fact that $K_1(\mathbb{C}) = K^1(\mathbb{C}) = 0$, the corresponding Gysin exact sequences in K-theory and K-homology become of the form

$$
0 \longrightarrow K_0\big(C(L_q(d)) \longrightarrow \mathbb{Z}^{M+1} \xrightarrow{1-A^d} \mathbb{Z}^{M+1} \longrightarrow K_0\big(C(L_q(d)) \longrightarrow 0 \quad ,
$$

and

$$
0 \longleftarrow K^0\big(C(L_q(d))\big) \longleftarrow \mathbb{Z}^{M+1} \underset{1-(A^t)^d}{\longleftarrow} \mathbb{Z}^{M+1} \longleftarrow K^0\big(C(L_q(d))\big) \longleftarrow 0 \quad .
$$

thus allowing for the computation of the K-theory and K-homology groups of the quantum lens spaces as kernels and cokernels of a suitable integer matrix.

Theorem 3.3 *Let* m *be a weight vector satisfying the assumptions of Proposition 2.1. Then for the matrix A constructed using (8) and any* $d \in \mathbb{N}$ *we have that*

$$
K_0\big(C(L_q(d))\big) \simeq \mathrm{Coker}(1 - A^d), \qquad K_1\big(C(L_q(d))\big) \simeq \mathrm{Ker}(1 - A^d)
$$

and

$$
K^0\big(C(L_q(dlk; k, l))\big) \simeq \mathrm{Ker}(1 - (A^t)^d), \qquad K^1\big(C(L_q(d))\big) \simeq \mathrm{Coker}(1 - (A^t)^d) .
$$

It remains an open problem to describe the precise relationship of our matrix A with the matrix used in [8] to compute the K-theory of quantum lens spaces.

4 Final Remarks

An interesting class of lens spaces that lies in the intersection of those studied in [8] and [11] is that for which the weight vector m satisfies $m_0 = 1$ and $m_i = m$ for all $i = 1, \ldots, n$. The associated coprime weight vector p for which $m = p^\sharp$ is then $p = (m, 1, \ldots, 1)$. It is straightforward to check that the number of independent Fredholm modules constructed in [11], given by the formula

$$1 + \sum_{k=1}^{n} p_0 p_1 \ldots p_{k-1},$$

equals in that case $1 + M$, the dimension of the K-homology group $K^0(C(\mathbb{WP}_q^n(m)))$.

We are also able to give, at least for this special class of examples, a positive answer to the question left open at the end of [11, Section 9], where the authors asked whether the Fredholm module they constructed actually built a complete set of generators for the K-homology group $K^0(C(\mathbb{WP}_q^n(m)))$.

Moreover these Fredholm modules can be used to give an explicit expression for the KK-equivalences $[\Pi]$ in (7), using a construction similar to the one of [4, Section 7.4], thus allowing one to write the matrix A in the form of a matrix of pairings.

Those computations go beyond the scope of the present paper and we postpone them to a later, more detailed work, where we plan to also address the problem of finding explicit representatives of K-theory and K-homology classes under less restrictive conditions on the set of weights.

Acknowledgements We thank the mathematical research institute MATRIX in Australia and the organisers of the workshop "Refining C*-algebraic invariants using KK-theory", where part of this research was performed. This work was motivated by discussions with Efren Ruiz about the structure of graph algebras. We thank Adam Rennie for helpful discussion and for his hospitality at the University of Wollongong, where part of this work was carried out. Finally, the author would like to thank Francesco D'Andrea, Giovanni Landi, Bram Mesland and Walter van Suijlekom for helpful comments on an early version of this work. This research was partially supported by NWO under the VIDI-grant 016.133.326.

References

1. Arici, F., Rennie, A.: Explicit isomorphism of mapping cone and Cuntz-Pimsner exact sequences (2016). arXiv:1605.08593
2. Arici, F., Brain, S., Landi, G.: The Gysin sequence for quantum lens spaces. J. Noncommut. Geom. **9**, 1077–1111 (2015)
3. Arici, F., D'Andrea, F., Landi, G.: Pimsner algebras and circle bundles. In: Noncommutative Analysis, Operator Theory and Applications, vol. 252, pp. 1–25. Birkhäuser, Cham (2016)
4. Arici, F., Kaad, J., Landi, G.: Pimsner algebras and Gysin sequences from principal circle actions. J. Noncommut. Geom. **10**, 29–64 (2016)

5. Blackadar, B.: *K*-Theory for Operator Algebras, 2nd edn. Cambridge University Press, Cambridge (1998)
6. Brzeziński, T., Fairfax, S.A.: Quantum teardrops. Commun. Math. Phys. **316**, 151–170 (2012)
7. Brzeziński, T., Fairfax, S.A.: Notes on quantum weighted projective spaces and multidimensional teardrops. J. Geom. Phys. **93**, 1–10 (2015)
8. Brzeziński, T., Szymański, W.: The C*-algebras of quantum lens and weighted projective spaces (2016). arXiv:1603.04678
9. Cuntz, J.: Simple C*- algebras generated by isometries. Commun. Math. Phys. **57**, 173–185 (1977)
10. Cuntz, J., Krieger, W.: A class of C*-algebras and topological Markov chains. Invent. Math. **56**, 251–268 (1980)
11. D'Andrea, F., Landi, G.: Quantum weighted projective and lens spaces. Commun. Math. Phys. **340**, 325–353 (2015)
12. Eilers, S., Restorff, G., Ruiz, E., Sørensen, A.P.W.: Geometric classification of graph C*-algebras over finite graphs (2016). arXiv:1604.05439
13. Exel, R.: Circle actions on C*-algebras, partial automorphisms, and a generalized Pimsner-Voiculescu exact sequence. J. Funct. Anal. **122**, 361–401 (1994)
14. Gabriel, O., Grensing, M.: Spectral triples and generalized crossed products (2013). arXiv:1310.5993
15. Goffeng, M., Mesland, B., Rennie, A.: Shift-tail equivalence and an unbounded representative of the Cuntz-Pimsner extension. Ergod. Theory Dyn. Syst. (2015, to appear). arXiv:1512.03455
16. Hong, J.H., Szymański, W.: Quantum lens spaces and graph algebras. Pac. J. Math. **211**, 249–263 (2003)
17. Kajiwara, T., Pinzari, C., Watatani, Y.: Ideal structure and simplicity of the C*–algebras generated by Hilbert bimodules. J. Funct. Anal. **159**, 295–322 (1998)
18. Karoubi, M.: K-Theory: An Introduction. Grundlehren der Mathematischen Wissenschaften, vol. 226. Springer, Berlin (1978)
19. Montgomery, S.: Hopf Algebras and Their Actions on Rings. Regional Conference Series in Mathematics, vol. 82. American Mathematical Society, Providence, RI (1993)
20. Pimsner, M.: A class of C*-algebras generalising both Cuntz-Krieger algebras and crossed products by $\mathbb{Z}$. In: Free Probability Theory. Fields Institute Communications, vol. 12, pp. 189–212. American Mathematical Society, Providence, RI (1997)
21. Rennie, A., Robertson, D., Sims, A.: The extension class and KMS states for Cuntz-Pimsner algebras of some bi-Hilbertian bimodules. J. Topol. Anal. **09**(02), 297 (2015). https://doi.org/10.1142/S1793525317500108
22. Sitarz, A., Venselaar, J.J.: The Geometry of quantum lens spaces: real spectral triples and bundle structure. Math. Phys. Anal. Geom. **18**, 1–19 (2015)
23. Vaksman, L., Soibelman, Ya.: The algebra of functions on the quantum group SU($n + 1$) and odd-dimensional quantum spheres. Leningr. Math. J. **2**, 1023–1042 (1991)

A Signed Version of Putnam's Homology Theory: Lefschetz and Zeta Functions

Robin J. Deeley

Abstract A signed version of Putnam homology for Smale spaces is introduced. Its definition, basic properties and associated Lefschetz theorem are outlined. In particular, zeta functions associated to an Axiom A diffeomorphism are compared.

1 Introduction

Let (M, f) be an Axiom A diffeomorphism [9]. Then there are two natural zeta functions associated to (M, f), the dynamical zeta function and the homological zeta function, see [9, Section I.4]. The former is defined as follows:

$$\zeta_{\mathrm{dym}}(s) := \exp\left(\sum_{n \geq 1} \frac{N_n}{n} t^n\right)$$

where N_n is the cardinality of the set of points with period n. The definition of latter is

$$\zeta_{\mathrm{hom}}(s) := \exp\left(\sum_{n \geq 1} \frac{\tilde{N}_n}{n} t^n\right)$$

where $\tilde{N}_n$ is obtained by counting the points of period n with "sign" (see Example 8 or [9, Section I.4] for further details).

Both these functions extend meromorphically to rational functions. For the former, this is an important theorem of Manning [6]. For the latter, it is a corollary of the Lefschetz fixed point theorem. Based on Manning's result, Bowen asked whether

R.J. Deeley (✉)
Department of Mathematics, University of Colorado, Boulder Campus Box 395, Boulder, CO 80309, USA
e-mail: robin.deeley@gmail.com

© Springer International Publishing AG, part of Springer Nature 2018
D.R. Wood et al. (eds.), *2016 MATRIX Annals*, MATRIX Book Series 1,
https://doi.org/10.1007/978-3-319-72299-3_13

there exists a homology theory for basic sets of an Axiom A diffeomorphism along with an associated Lefschetz theorem that implies that the dynamical zeta function is a rational function in the same way the classical Lefschetz theorem implies that the homological zeta function is a rational function. Recently, Ian Putnam constructed such a homology theory and proved the relevant Lefschetz theorem [7] (in particular see [7, Section 6]). For more on the relationship between zeta functions, homology, and Lefschetz theorems see [9, Section I.4] or [7, Section 6.1] for brief introductions or [5] and references therein for more details.

Putnam's homology theory is defined using the framework of Smale spaces. Smale spaces were introduced by Ruelle [8]. The reader who is unfamiliar with them can assume that any Smale space in the present paper is either the nonwandering set or a basic set of an Axiom A diffeomorphism. The precise definition of a Smale space is given in Sect. 2.

It is important to note that the dynamical zeta function of an Axiom A diffeomorphism depends only on its restriction to the nonwandering set, but this is not the case for the homological zeta function. In particular, the two zeta functions defined above are not in general equal and as such the classical homology of M and Putnam's homology of the nonwandering set of M are (again in general) not isomorphic.

The modest goal of the present paper is to outline the construction of a homology theory defined in the same spirit as Putnam's homology, but whose associated Lefschetz theorem is more closely related to the classical Lefschetz theorem; it counts periodic points with "sign", see Theorem 5 for the precise statement. This goal is achieved by considering signed Smale spaces. By definition, a signed Smale space is a Smale space along with a continuous map to $\{-1, 1\}$, which is called a sign function. Then, by following Putnam's constructions in [7] quite closely but with this additional sign function, one obtains a new "signed version" of Putnam's homology. In the case of an Axiom A diffeomorphism, the signed homology theory of the nonwandering set with a particular sign function is more closely related (in particular through the associated Lefschetz theorem) to the standard homology of the manifold, at least in particular situations, see Theorem 6 and Example 10. The notion of signed Smale space is based on work of Bowen, see in particular [1, Theorem 2].

If the Smale space is connected the only possible sign functions are constant and the signed homology is essentially the same as Putnam's homology. However, typically the nonwandering set of an Axiom A diffeomorphism is not connected and this can also occur for basic sets (e.g., shifts of finite type).

I have assumed the reader is familiar with Putnam's monograph [7] and Bowen's paper [1]. In particular, see [1] for more on filtrations and the no-cycle condition. Also, the reader should be warned that there are many definitions and a number proofs are omitted. Most notable among these are Theorems 4 and 5. Although, proofs of these theorems are long, the reader familiar with the proofs in [7] will likely see how they are proved. In particular, for Theorem 5, one follows almost verbatim the construction (which is based on Manning's proof in [6]) in [7, Section 6]. Detailed proofs of these theorems will appear elsewhere.

2 Main Results

Definition 1 A Smale space (X, φ) consists of a compact metric space (X, d) and a homeomorphism $\varphi : X \to X$ such that there exist constants $\epsilon_X > 0, 0 < \lambda < 1$ and a continuous partially defined map:

$$\{(x, y) \in X \times X \mid d(x, y) \leq \epsilon_X\} \mapsto [x, y] \in X$$

satisfying the following axioms:

B1 $[x, x] = x$,
B2 $[x, [y, z]] = [x, z]$,
B3 $[[x, y], z] = [x, z]$, and
B4 $\varphi[x, y] = [\varphi(x), \varphi(y)]$;

where in these axioms, x, y, and z are in X and in each axiom both sides are assumed to be well-defined. In addition, (X, φ) is required to satisfy

C1 For $x, y \in X$ such that $[x, y] = y$, we have $d(\varphi(x), \varphi(y)) \leq \lambda d(x, y)$ and
C2 For $x, y \in X$ such that $[x, y] = x$, we have $d(\varphi^{-1}(x), \varphi^{-1}(y)) \leq \lambda d(x, y)$.

The map $[\cdot, \cdot]$ in the definition of a Smale space is called the bracket map; it is unique (provided it exists).

Example 1 If (M, f) is an Axiom A diffeomorphism, then the restriction of f to the nonwandering set is a Smale space and likewise the restriction of f to a basic set is also a Smale space. The bracket map in the definition of a Smale space is, in these cases, given by the canonical coordinates.

An important class of Smale spaces are the shifts of finite type. They can be defined as follows. Let $G = (G^0, G^1, i, t)$ be a directed graph; that is, G^0 and G^1 are finite sets called the set of vertices and the set of edges and each edge $e \in G^1$ is given by a directed edge from $i(e) \in G^0$ to $t(e) \in G^0$, see [7, Definition 2.2.1] for further details.

From G a dynamical system is constructed by taking

$$\Sigma_G := \{(g_j)_{j \in \mathbb{Z}} \mid g_j \in G^1 \text{ and } t(g_j) = i(g_{j+1}) \text{ for each } j \in \mathbb{Z}\}$$

with the homeomorphism, $\sigma : \Sigma_G \to \Sigma_G$ given by left sided shift. Then, see for example [7], (Σ_G, σ) is a Smale space and one can define a shift of finite type to be any dynamical system that is conjugate to (Σ_G, σ) for some graph G. Often we will drop the G from the notation and denote a shift of finite type by (Σ, σ).

From G and $k \geq 2$, one can obtain a higher block presentation by constructing another graph G^k whose edges are paths in G of length k and whose vertices are paths in G of length $k - 1$; for the precise details see [7, Definition 2.2.2].

2.1 Signed Smale Spaces

Definition 2 A *signed Smale space* is a Smale space (X, φ) along with a continuous map $\Delta_X : X \to \{-1, 1\}$. Furthermore, for $n \geq 1$, we define

$$\Delta_X^{(n)}(x) = \prod_{i=0}^{n-1} \Delta_X(\varphi^i(x)).$$

A signed Smale space is denoted by (X, φ, Δ_X) and Δ_X is called the sign function; it is often denoted simply by Δ.

Example 2 Let $(\Omega, f|_\Omega)$ be a basic set of an Axiom A diffeomorphism, (M, f). We assume that the bundle $E^u|_\Omega$ can be oriented and then define $\Delta : \Omega \to \{-1, 1\}$ as follows:

$$\Delta(x) = \begin{cases} 1 : D_x(f) : (E^u|_\Omega)|_x \to (E^u|_\Omega)|_{f(x)} \text{ preserves the orientation} \\ -1 : D_x(f) : (E^u|_\Omega)|_x \to (E^u|_\Omega)|_{f(x)} \text{ reverses the orientation.} \end{cases}$$

The fact that Ω is hyperbolic implies that Δ is continuous; hence $(\Omega, f|_\Omega, \Delta)$ is a signed Smale space.

A special case of Example 2 occurs in both the statement and proof of [1, Theorem 2]. Another class of examples are hyperbolic toral automorphisms:

Example 3 Let $M = \mathbb{R}^2/\mathbb{Z}^2$ and $f = A$ where $A \in M_2(\mathbb{Z})$ with $\det(A) = \pm 1$ and eigenvalues λ_1 and λ_2 such that $0 < |\lambda_2| < 1 < |\lambda_1|$. This diffeomorphism is globally hyperbolic and the nonwandering set is the entire manifold; that is, $\Omega = M$.

The bundle E^u is isomorphic to the trivial rank one bundle. Its fiber, for example at the origin, is the eigenspace associated to λ_1. One can then show that for any $x \in M$, $\Delta(x) = \text{sign}(\lambda_1)$.

Definition 3 Let (Σ, σ) be a shift of finite type and $\Delta_\Sigma : \Sigma \to \{-1, 1\}$ be a continuous function. Then $(\Sigma, \sigma, \Delta_\Sigma)$ is called a signed shift of finite type.

2.2 The Signed Dimension Group

Proposition 1 *Let $(\Sigma, \sigma, \Delta_\Sigma)$ be a signed shift of finite type. Then, there exists a graph G such that*

1. *there is conjugacy $h : (\Sigma_G, \sigma) \to (\Sigma, \sigma)$;*
2. *for any $(g_j)_{j \in \mathbb{Z}} \in \Sigma_G$, $(\Delta_\Sigma \circ h)((g_j)_{j \in \mathbb{Z}})$ depends only on g_0.*

Proof The first item is a possible definition of a shift of finite type. Using the fact that Δ_Σ is continuous, one can obtain the second item by taking a higher block presentation.

Definition 4 Let $(\Sigma, \sigma, \Delta_\Sigma)$ be a signed shift of finite type and G is a graph which satisfies the conclusions of the previous theorem. Then, G (and the conjugacy h : $(\Sigma_G, \sigma) \to (\Sigma, \sigma)$) is called a *signed presentation* of $(\Sigma, \sigma, \Delta_\Sigma)$. We denote $\Delta_\Sigma \circ h$ by Δ_{Σ_G}.

By assumption, for $(g_j)_{j \in \mathbb{Z}} \in \Sigma_G$, $\Delta_G((g_j)_{j \in \mathbb{Z}})$ depends only on g_0. As such, the function $\Delta_G : G^1 \to \{-1, 1\}$ defined via $\Delta_G(g) := \Delta_{\Sigma_G}((g_j)_{j \in \mathbb{Z}})$ (where $g_0 = g$) is well-defined. Moreover, for a path $g_0 \cdots g_m$ in G and $n \le m$, we define

$$\Delta_{G,n}(g_0 \cdots g_m) = \prod_{j=m-n}^{m} \Delta_G(g_j).$$

Finally, given Δ_G as above, we define $\Delta_{G^k} : G^k \to \{-1, 1\}$ via

$$\Delta_{G^k}(g_0 g_1 \cdots g_{k-1}) = \Delta_G(g_{k-1})$$

where $g_0 g_1 \cdots g_{k-1}$ is an element in G^k (i.e., a path of length k in G). This choice is based on [7, Theorem 3.2.3 Part 1]. We use $(\Sigma_G, \sigma, \Delta_G)$ to denote a signed shift of finite type with a fixed signed presentation. It is important to note that Δ_{Σ_G} and Δ_G are related, but not the same; their domains are different.

Definition 5 Suppose $(\Sigma, \sigma, \Delta_G)$ is a signed shift of finite type with a fixed signed presentation. Define $\gamma^s_{G, \Delta_G} : \mathbb{Z}G^0 \to \mathbb{Z}G^0$ as follows: for each $v \in G^0$, we let

$$v \mapsto \sum_{e \in G^1, t(e) = v} i(e) \cdot \Delta_G(e).$$

Furthermore, define $D^s_{\Delta_G}(G)$ to be the inductive limit group: $\lim_{\to} (\mathbb{Z}G^0, \gamma^s_{G, \Delta_G})$.

Example 4 Let G be the graph with one vertex and two edges labelled by 0 and 1. Then the associated shift of finite type is the full two shift, (Σ_G, σ). Furthermore, let $\Delta_G : G \to \{-1, 1\}$ be the continuous map

$$\Delta_G(g) = \begin{cases} 1 & g = 1 \\ -1 & g = 0. \end{cases}$$

Then, in this case, $D^s_{\Delta_G}(\Sigma_G) \cong \{0\}$.

Theorem 1 (Reformulation of [1, Theorem 2]) *Suppose (M, f) is an Axiom A diffeomorphism satisfying the no-cycle condition, $\dim(\Omega_s) = 0$, and $q :=$ rank$(E^u|_{\Omega_s})$. Then there exists signed shift of finite type $(\Sigma_G, \sigma, \Delta_G)$ such that*

1. *(Σ_G, σ) is conjugate to $(\Omega_s, f|_{\Omega_s})$;*
2. *the map $\gamma^s_{G, \Delta_G} : \mathbb{Z}G^0 \to \mathbb{Z}G^0$ has the same nonzero eigenvalues as the map on homology: $f|_{M_s} : H_q(M_s, M_{s-1}) \to H_q(M_s, M_{s-1})$;*

3. the map $\gamma^s_{G,\Delta_G} : D^s_{\Delta_G}(G) \to D^s_{\Delta_G}(G)$ has the same nonzero eigenvalues as the
   map on homology: $f|_{M_s} : H_q(M_s, M_{s-1}) \to H_q(M_s, M_{s-1})$;

where $(M_s)^m_{s=1}$ is a fixed filtration associated to the basic sets, $(\Omega_s)^m_{s=1}$, of (M, f);
we assume it satisfies the assumptions in [1].

Proof Theorem 2 of [1] implies the existence of the signed shift of finite type
satisfying items (1) and (2) in the statement. Basic properties of inductive limits
of abelian groups imply that for any signed shift of finite type $\gamma^s_{G,\Delta_G} : D^s_{\Delta_G}(G) \to$
$D^s_{\Delta_G}(G)$ and $\gamma^s_{G,\Delta_G} : \mathbb{Z}G^0 \to \mathbb{Z}G^0$ have the same nonzero eigenvalues; item (3)
follows from this observation.

2.3 Signed Homology

Definition 6 (See [7, Definition 2.5.5]) Suppose (X, φ) and (Y, ψ) are Smale
spaces and $\pi : (X, \varphi) \to (Y, \psi)$ is a factor map. Then π is s-bijective (resp. u-
bijective) if, for each $x \in X$, $\pi|_{X^s(x)}$ (resp. $\pi|_{X^u(x)}$) is a bijection to $X^s(\pi(x))$ (resp.
$X^u(\pi(x))$).

Definition 7 (Compare with [7, Definition 2.6.2]) Suppose (X, φ, Δ_X) is a signed
Smale space. Then a signed s/u-bijective pair is the following data:

1. signed Smale spaces (Y, ψ, Δ_Y) and (Z, ζ, Δ_Z) such that $Y^s(y)$ and $Z^u(z)$ are
   totally disconnected for each $y \in Y$ and $z \in Z$;
2. s-bijective map $\pi_s : (Y, \psi) \to (X, \varphi)$;
3. u-bijective map $\pi_u : (Z, \zeta) \to (X, \varphi)$;

such that $\Delta_Y = \Delta_X \circ \pi_s$ and $\Delta_Z = \Delta_X \circ \pi_u$.

Proposition 2 (Compare with [7, Theorem 2.6.3]) *If (X, φ, Δ_X) is a nonwandering signed Smale space, then it has a signed s/u-bijective.*

Proof By Putnam [7, Theorem 2.6.3], (X, φ) has an s/u-bijective pair: $(Y, \psi, \pi_s, Z,$
$\zeta, \pi_u)$. Taking

$$\Delta_Y := \Delta_X \circ \pi_s \text{ and } \Delta_Z := \Delta_X \circ \pi_u$$

leads to a signed s/u-bijective pair.
For $L \geq 0$, $M \geq 0$, consider the Smale space (obtained via an iterated fiber product
construction):

$$\Sigma_{L,M}(\pi) := \{(y_0, \ldots, y_L, z_0, \ldots, z_M) | \pi_s(y_i) = \pi_u(z_j) \text{ for each } i, j\}$$

with σ defined to be $\psi \times \cdots \times \psi \times \zeta \times \cdots \times \zeta$. As the notation suggests $(\Sigma_{L,M}(\pi), \sigma)$
is a shift of finite type.

Moreover, again for each $L \geq 0$, $M \geq 0$, $\Delta_{\Sigma_{L,M}(\pi)} : \Sigma_{L,M}(\pi) \to \{-1, 1\}$ defined via

$$\Delta_{\Sigma_{L,M}(\pi)}(y_0, \ldots, y_L, z_0, \ldots, z_M) = \Delta_Y(y_0)$$

is a continuous map. We note that $\Delta_{\Sigma_{L,M}(\pi)}(y_0, \ldots, y_L, z_0, \ldots, z_M)$ is to equal $\Delta_Y(y_i)$ for any $0 \leq i \leq L$ and is also equal to $\Delta_Z(z_j)$ for any $0 \leq j \leq M$. In particular, $\Delta_{\Sigma_{L,M}(\pi)}$ is constant on orbits of the natural action of $S_{L+1} \times S_{M+1}$. For more details on the action (which is the natural one) see [7, Section 5.1].

Definition 8 Suppose that $\pi = (Y, \psi, \pi_s, Z, \zeta, \pi_u)$ a signed s/u-bijective pair for a signed Smale space, (X, φ, Δ). Then a graph G is a signed presentation of π if G is a presentation of π, in the sense of Definition 2.6.8 of [7], and G is also a signed presentation, in the sense of Definition 4, of $(\Sigma_{0,0}, \sigma, \Delta_{0,0})$.

Proposition 3 (Compare with [7, Theorem 2.6.9]) *If (X, φ, Δ) is a signed Smale space and $\pi = (Y, \psi, \pi_s, Z, \zeta, \pi_u)$ is a signed s/u-bijective pair for (X, φ), then there exists a presentation of π. Moreover, if G is a signed presentation of π, then, for each $L \geq 0$ and $M \geq 0$, $G_{L,M}$ is a signed presentation of $(\Sigma_{L,M}(\pi), \sigma)$.*

Proof Work of Putnam (see [7, Theorem 2.6.9]) implies that π has a presentation in the sense of [7, Definition 2.6.8]. That is, there is a graph G and conjugacy $e : \Sigma_{0,0}(\pi) \to \Sigma_G$ satisfying the conditions in [7, Definition 2.6.8]. Moreover, since $\Delta_{\Sigma_{0,0}(\pi)}$ is continuous, by possibly taking a higher block presentation of G we can ensure that this presentation leads to a signed presentation of $(\Sigma_{0,0}(\pi), \sigma, \Delta_{0,0})$. The second statement in the proposition follows as in the proof of [7, Theorem 2.6.9] and is omitted.

Definition 9 (Compare with Definition 5.2.1 of [7]) Suppose (X, φ, Δ) is a signed Smale space, $\pi = (Y, \psi, \pi_s, Z, \zeta, \pi_u)$ a signed s/u-bijective pair for (X, φ), and G is a presentation of π. Fix $k \geq 0$, $L \geq 0$, and $M \geq 0$ and let

1. $\mathscr{B}(G_{L,M}^k, S_L \times 1)$ be the subgroup of $\mathbb{Z}G_{L,M}^k$ which is generated by elements of the following forms:

   a. $p \in G_{L,M}^k$ with the property that $p \cdot ((\alpha, 1) = p$ for some non-trivial transposition, $\alpha \in S_{L+1}$;
   b. $p' = q \cdot (\alpha, 1) - \text{sign}(\alpha)q$ for some $q \in G_{L,M}^k$ and $\alpha \in S_{L+1}$;

2. $\mathscr{Q}(G_{L,M}^k, S_L \times 1)$ be the quotient of $\mathbb{Z}G_{L,M}^k$ by $\mathscr{B}(G_{L,M}^k)$; we denote the quotient map by Q;
3. $\mathscr{A}(G_{L,M}^k, 1 \times S_{M+1})$ be $\{a \in \mathbb{Z}G_{L,M}^k \mid a \cdot (1, \beta) = \text{sign}(\beta) \cdot a \text{ for all } \beta \in S_{M+1}\}$; it is a subgroup of $\mathbb{Z}G_{L,M}^k$.

Proposition 4 (See the Remark Between Definitions 5.2.1 and 5.2.2 in [7]) *Suppose $\pi = (Y, \psi, \pi_s, Z, \zeta, \pi_u)$ a signed s/u-bijective pair for a signed Smale space, (X, φ, Δ) and G is a signed presentation of π. Then, for each $k \geq 0$, $L \geq 0$,* *and* $M \geq 0$,

$$\gamma^s_{G^k_{L,M}, \Delta_{G^k_{L,M}}} (\mathscr{B}(G^k_{L,M}, S_L \times 1)) \subseteq \mathscr{B}(G^k_{L,M}, S_L \times 1)$$

$$\gamma^s_{G^k_{L,M}, \Delta_{G^k_{L,M}}} (\mathscr{A}(G^k_{L,M}, S_L \times 1)) \subseteq \mathscr{A}(G^k_{L,M}, S_L \times 1)$$

where $\gamma^s_{G^k_{L,M}, \Delta_{G^k_{L,M}}}$ *is defined in Definition 5.*

Definition 10 (Compare with [7, Definition 5.2.2]) Suppose $\pi = (Y, \psi, \pi_s, Z, \zeta, \pi_u)$ a signed s/u-bijective pair for a signed Smale space, (X, φ, Δ) and G is a signed presentation of π. Using the previous proposition, we define

$$D^s_{Q,A,G^k,\Delta_{G^k}}(G^k_{L,M}) = \lim_{\rightarrow} \left(Q(A(G^k_{L,M}, 1 \times S_{M+1})), \gamma^s_{G^k_{L,M}, \Delta_{G^k_{L,M}}} \right)$$

For each $0 \leq i \leq L$, there is a map defined at the level of graphs, $\delta^s_{i,} : G^k_{L,M} \to G^k_{L-1,M}$ obtained by removing the ith entry. Likewise, for $0 \leq j \leq M$, one has a map $\delta^s_j : G^k_{L,M} \to G^k_{L,M-1}$ that is defined by removing the $L + j$-entry. As in [7], these induce maps at the level of the abelian groups introduced in the previous definition:

Proposition 5 (Compare with [7, Lemma 5.2.4]) *Suppose* $\pi = (Y, \psi, \pi_s, Z, \zeta, \pi_u)$ *a signed s/u-bijective pair for a signed Smale space,* (X, φ, Δ) *and* G *is a signed presentation of* π. *Then, there exists* $k \in \mathbb{N}$ *such that* $\delta_{i,}$ *and* δ_j *induced group homomorphisms:*

$$\delta^s_{i,} : D^s_{Q,A,G^k,\Delta_{G^k}}(G^k_{L,M}) \to D^s_{Q,A,G^k,\Delta_{G^k}}(G^k_{L-1,M})$$

and

$$\delta^{s*}_j : D^s_{Q,A,G^k,\Delta_{G^k}}(G^k_{L,M}) \to D^s_{Q,A,G^k,\Delta_{G^k}}(G^k_{L,M+1})$$

respectively.

Definition 11 (Compare with [7, Definition 5.1.7] and [7, Sections 5.2 and 5.3]) Suppose $\pi = (Y, \psi, \pi_s, Z, \zeta, \pi_u)$ is a s/u-bijective pair for a signed Smale space, (X, φ, Δ), G is a signed presentation of π, and k is as in the statement of previous proposition. Then, we let

$$d^s_{Q,A,G^k,\Delta_{G^k}}(\pi)_{L,M} : D^s_{Q,A,\Delta_{G^k}}(G^k_{L,M}) \to D^s_{Q,A,\Delta_{G^k}}(G^k_{L-1,M}) \oplus D^s_{Q,A,\Delta_{G^k}}(G^k_{L,M+1})$$

be the map

$$\sum_{i=0}^{L}(-1)^i \delta^s_{i,} + (-1)^L \sum_{j=0}^{M}(-1)^j \delta^{s*}_j.$$

Finally, for each $N \in \mathbb{Z}$, we let $d^s_{Q,A,G^k,\Delta_{G^k}}(\pi)_N = \bigoplus_{L-M=N} d^s_{Q,A,G^k,\Delta_{G^k}}(\pi)_{L,M}$.

Theorem 2 (See [7, Sections 5.1 and 5.2]) *Assuming the setup of the previous definition,*

$$\left(\bigoplus_{L-M=N} D^s_{Q,A,\Delta_{G^k_{L,M}}}(G_{L,M}), \bigoplus_{L-M=N} d^s_{Q,A,G^k,\Delta_{G^k}}(\pi)_{L,M} \right)_{N\in\mathbb{Z}}$$

is a complex.

Definition 12 (Compare with [7, Definition 5.1.11]) Suppose $\pi = (Y, \psi, \pi_s, Z, \zeta, \pi_u)$ is a s/u-bijective pair for a signed Smale space, (X, φ, Δ) and G is a signed presentation of π. We define $H^s_*(X, \varphi, \Delta, \pi, G^k)$ to be the homology of the complex

$$\left(\bigoplus_{L-M=N} D^s_{Q,A,\Delta_{G^k_{L,M}}}(G_{L,M}), \bigoplus_{L-M=N} d^s_{Q,A,G^k,\Delta_{G^k}}(\pi)_{L,M} \right)_{N\in\mathbb{Z}}$$

from the previous theorem. We call this the signed homology and denote it by $H^s_*(X, \varphi, \Delta, \pi, G^k)$; it is a $\mathbb{Z}$-graded abelian group.

Theorem 3 (Compare with [7, Theorem 5.1.12]) *The signed homology groups have finite rank and vanish for all but finitely many $N \in \mathbb{Z}$.*

Proof Basic properties of inductive limits imply that the signed dimension groups have finite rank. Hence the homology is finite rank (see for example page 131 of [7] for further details). That the homology vanishes for all but finitely many N also follows as in [7, pages 131–132].

Theorem 4 (Compare with [7, Theorem 5.5.1]) *The signed homology is independent of the choice of signed presentation, and the choice of s/u-bijective pair.*

Definition 13 Suppose (X, φ, Δ) is a signed Smale space. Based on the previous theorem, for any choice of signed s/u-bijective pair, π, signed presentation G, and k large enough, we can define $H^s_N(X, \varphi, \Delta) := H^s_N(X, \varphi, \Delta, \pi, G^k)$.

Proposition 6 (Compare with a Special Case of [7, Theorem 5.4.1]) *Suppose (X, φ, Δ) is a signed Smale space. The homeomorphism φ and its inverse induces graded group homomorphism at the level of the signed homology groups. We denote the induced maps by φ^s and $(\varphi^{-1})^s$ respectively.*

Remark 1 General functorial properties Putnam's homology theory are nontrivial, see [3, 4, 7]. The functorial properties of the signed version are further complicated by the requirement that the map at the level of Smale space must respect the signed structure. The full details of these properties are not discussed here as they are not needed for the signed Lefschetz theorem.

Example 5 Suppose (X, φ) is a Smale space and we take Δ_X to be the constant function one. Then, it follows from the definitions involved that $H^s(X, \varphi, \Delta)$ is Putnam's stable homology theory.

Example 6 Suppose $(\Sigma_G, \sigma, \Delta_G)$ is a signed shift of finite type. The signed homology, $H^s_N(\Sigma_G, \sigma, \Delta_G)$ is the signed dimension group when $N = 0$ and is the trivial group when $N \neq 0$.

2.4 Lefschetz and Zeta Functions

Definition 14 Suppose (X, φ) is a Smale space. Then, for each $n \in \mathbb{N}$,

$$\mathrm{Per}(X, \varphi, n) := \{x \in X \mid \varphi^n(x) = x\}.$$

Definition 15 Suppose (X, φ, Δ) is a signed Smale space. Then, the signed dynamical zeta function is

$$\zeta_{(X,\Delta)}(z) = \exp\left(\sum_{n=1}^{\infty} \frac{N_n(X, \varphi, \Delta)}{n} z^n\right)$$

where $N_n(X, \varphi, \Delta) = \sum_{x \in \mathrm{Per}(X,\varphi,n)} \Delta^{(n)}(x)$.

Example 7 If (X, φ, Δ) is a signed Smale space with $\Delta \equiv 1$, then the signed dynamical zeta function is the dynamical zeta function (see the Introduction):

$$\zeta_{\mathrm{dyn}}(z) = \exp\left(\sum_{n=1}^{\infty} \frac{|\mathrm{Per}(X, \varphi, n)|}{n} z^n\right).$$

For more details on this case, see [9, Section I.4] (and also [7, Chapter 6] and references therein).

Example 8 Suppose (M, f) is an Axiom A diffeomorphism, $(\Omega, f|_\Omega)$ be the restriction of f to the nonwandering set, and for each $m \in \mathrm{Per}(\Omega, f|_\Omega, 1)$,

$$L(m) := \mathrm{sign}(\det(I - Df(m) : T_m(M) \to T_m(M))).$$

The Lefschetz fixed point formula implies that

$$\sum_{m \in \mathrm{Per}(\Omega, f|_\Omega, 1)} L(m) = \sum_{i=0}^{\dim(M)} (-1)^i \mathrm{Tr}(f_* : H_i(M; \mathbb{R}) \to H_i(M; \mathbb{R})).$$

Moreover, by for example [5, Proposition 5.7] or [9, Section I.4], $L(m) = (-1)^q \Delta(m)$ where q is the rank of E^u at the point m and Δ is as in Example 2. From this one obtains the homological zeta function discussed in the Introduction.

Theorem 5 (Compare with [7, Theorem 6.1.1]) *For each $k \in \mathbb{N}$,*

$$\sum_{N \in \mathbb{Z}} (-1)^N \mathrm{Tr}\left(((\varphi^{-1})^s_N \otimes id_{\mathbb{Q}})^n\right) = \sum_{x \in \mathrm{Per}(X,\varphi,n)} \Delta^{(n)}(x)$$

where $(\varphi^{-1})^s_N \otimes id_{\mathbb{Q}} : H^s_N(X, \varphi, \Delta) \otimes \mathbb{Q} \to H^s_N(X, \varphi, \Delta) \otimes \mathbb{Q}$ is the map on rationalized homology induced from φ^{-1}.

Definition 16 Suppose (M, f) is an Axiom A diffeomorphism satisfying the no-cycle condition, $(\Omega_s)^m_{s=1}$ are the basic sets of (M, f), and $(M_s)^m_{s=1}$ is a filtration associated to the basic sets that satisfies the assumptions in [1]. Then we let f_{even} and f_{odd} denotes the map induced by f on

$$\bigoplus_{n \text{ even}} H_n(M_s, M_{s-1}) \text{ and } \bigoplus_{n \text{ odd}} H_n(M_s, M_{s-1})$$

respectively.

Likewise if (X, φ) is a Smale space, we let φ^{-1}_{even} and φ^{-1}_{odd} denote the map induced by φ^{-1} on

$$\bigoplus_{n \text{ even}} H^s_n(X, \varphi, \Delta) \otimes \mathbb{Q} \text{ and } \bigoplus_{n \text{ :odd}} H^s_n(X, \varphi, \Delta) \otimes \mathbb{Q}$$

respectively.

Theorem 6 *Suppose (M, f) is an Axiom A diffeomorphism satisfying the no-cycle condition, and $E^u|_{\Omega_s}$ is orientable. Then (using the notation introduced in the paragraph preceding this theorem) there exists a signed Smale space, (X, φ, Δ), such that*

(1) (X, φ) is conjugate to $(\Omega_s, f|_{\Omega_s})$;
(2) (even case) if q is even, the maps $\varphi^{-1}_{even} \oplus f_{odd}$ and $\varphi^{-1}_{odd} \oplus f_{even}$ have the same nonzero eigenvalues or
(3) (odd case) if q is odd, the maps $\varphi^{-1}_{even} \oplus f_{even}$ and $\varphi^{-1}_{odd} \oplus f_{odd}$ have the same nonzero eigenvalues.

Corollary 1 *Suppose (M, f) is an Axiom A diffeomorphism, $(\Omega, f|_\Omega)$ is the non-wandering set of (M, f), $E^u|_\Omega$ is orientable, and $\Delta : \Omega \to \{-1, 1\}$ is defined as in Example 2. Let $q : \Omega \to \{0, 1\}$ be the function defined by $q(x) = \mathrm{rank}(E^u_x) \mod 2$. Then*

1. if $q \equiv 0$, then $\zeta_{\mathrm{hom}}(z) = \zeta_{(\Omega, \Delta)}(z)$;
2. if $q \equiv 1$, then $\zeta_{\mathrm{hom}}(z) = 1/\zeta_{(\Omega, \Delta)}(z)$.

Proof By definition, the signed zeta function is given by

$$\zeta_\varphi(z) = \exp\left(\sum_{n=1}^\infty \frac{N_n(X,\varphi,\Delta)}{n} z^n\right)$$

where $N_n(X,\varphi,\Delta) = \sum_{x \in \mathrm{Per}(X,\varphi,n)} \Delta^{(n)}(x)$. If $q \equiv 0$, then

$$\sum_{x \in \mathrm{Per}(\Omega,f|_\Omega,n)} L(x) = \sum_{x \in \mathrm{Per}(\Omega,f|_\Omega,n)} \Delta^{(n)}(x)$$

while if $q \equiv 1$, then

$$\sum_{x \in \mathrm{Per}(\Omega,f|_\Omega,n)} L(x) = (-1) \sum_{x \in \mathrm{Per}(\Omega,f|_\Omega,n)} \Delta^{(n)}(x).$$

The result then follows.

3 Examples

To conclude the paper, two examples are discussed. These examples point to the possibility of a stronger relationship between the signed version of Putnam's homology and the standard homology of the manifold associated with the Axiom A diffeomorphism. However, such a relationship is (at this point) highly speculative.

Example 9 (Shifts of Finite Type) In [2], Bowen and Franks prove the following results:

Theorem 7 (Reformulation of [2, Theorem 3.2]) *Suppose (M,f) is an Axiom A diffeomorphism satisfying the no-cycle condition, $\dim(\Omega_s) = 0$, and $q :=$ $\mathrm{rank}(E^u|_{\Omega_s})$. Then there exists signed shift of finite type $(\Sigma_G, \sigma, \Delta_G)$ such that*

1. *(Σ_G, σ) is conjugate to $(\Omega_s, f|_{\Omega_s})$;*
2. *the maps $\gamma^s_{G,\Delta_G} : \mathbb{Z}G^0 \to \mathbb{Z}G^0$ and $f|_{M_s} : H_q(M_s, M_{s-1}) \to H_q(M_s, M_{s-1})$ are shift equivalent;*
3. *the maps $\gamma^s_{G,\Delta_G} : D^s_{\Delta_G}(G) \to D^s_{\Delta_G}(G)$ and $f|_{M_s} : H_q(M_s, M_{s-1}) \to H_q(M_s, M_{s-1})$ are shift equivalent*

where $(M_s)_{s=1}^m$ is a fixed filtration associated to the basic sets, $(\Omega_s)_{s=1}^m$, of (M,f). The reader might notice that Bowen's and Franks' result implies [1, Theorem 2] (stated as Theorem 1 above). However, the proof in [2] uses [1, Theorem 2].

Example 10 (Two Dimensional Hyperbolic Toral Automorphisms) We give an example in which one can compute the standard homology, Putnam's homology and the relevant actions explicitly. Let

$$\varphi = A = \begin{pmatrix} 1 & 1 \\ 1 & 0 \end{pmatrix}$$

and consider the induced action on the two-torus, $\mathbb{R}^2/\mathbb{Z}^2$. In this example, Ω is the entire manifold and Δ is the constant function one and $q = 1$.

In regards to the standard homology, we have the following

$$H_N(\mathbb{R}^2/\mathbb{Z}^2; \mathbb{R}) \cong \begin{cases} \mathbb{R} & : N = 0, 2 \\ \mathbb{R}^2 & : N = 1 \\ 0 & : \text{otherwise} \end{cases}$$

and the action is given by the identity on $H_0(\mathbb{R}^2/\mathbb{Z}^2; \mathbb{R})$, A on $H_1(\mathbb{R}^2/\mathbb{Z}^2; \mathbb{R})$, and minus the identity on $H_2(\mathbb{R}^2/\mathbb{Z}^2; \mathbb{R})$.

In regards to Putnam's homology (based on [7, Example 7.4]) we have the following

$$H_N^s(\mathbb{R}^2/\mathbb{Z}^2, A) \otimes \mathbb{R} \cong \begin{cases} \mathbb{R} & : N = -1, 1 \\ \mathbb{R}^2 & : N = 0 \\ 0 & : \text{otherwise} \end{cases}$$

and the action of $((\varphi^{-1})^s) \otimes Id_{\mathbb{R}}$ is given by the identity on $H_{-1}^s(\mathbb{R}^2/\mathbb{Z}^2, A) \otimes \mathbb{R}$, A on $H_0(\mathbb{R}^2/\mathbb{Z}^2, A) \otimes \mathbb{R}$, and minus the identity on $H_1^s((\mathbb{R}^2/\mathbb{Z}^2, A) \otimes \mathbb{R}$.

Thus, in this very special case, there is an even stronger than predicted by Theorem 6 relationship between the homology of torus and Putnam's homology of the Smale space $(\mathbb{R}^2/\mathbb{Z}^2, A)$. Namely, they are the same with dimension shift of one (this is exactly the rank of bundle E^u in this case). Moreover, the actions induced by f and φ^{-1} are also the same (again with dimension shift).

Acknowledgements I thank Magnus Goffeng, Ian Putnam and Robert Yuncken for discussions. In addition, I thank Magnus for encouraging me to publish these results. I also thank the referee for a number of useful suggestions.

References

1. Bowen, R.: Entropy versus homology for certain diffeomorphism. In: Topology, vol. 13, pp. 61–67. Pergamon, Oxford (1974)
2. Bowen, R., Franks, J.: Homology for zero-dimensional nonwandering sets. Ann. Math. (2) **106**(1), 73–92 (1977)
3. Deeley, R.J., Killough, D.B., Whittaker, M.F.: Dynamical correspondences for Smale spaces. N. Y. J. Math. **22**, 943–988 (2016)
4. Deeley, R.J., Killough, D.B., Whittaker, M.F.: Functorial properties of Putnam's homology theory for Smale spaces. Ergod. Theory Dyn. Syst. **36**(5), 1411–1440 (2016)

5. Franks, J.: Homology Theory and Dynamical Systems. CBMS Regional Conference Series in Mathematics, vol. 49, viii+120 pp. American Mathematical Society, Providence, RI (1982)
6. Manning, A.: Axiom A diffeomorphisms have rational zeta functions. Bull. Lond. Math. Soc. **3**, 215–220 (1971)
7. Putnam, I.F.: A homology theory for Smale spaces. Mem. Am. Math. Soc. **232**(1094), viii+122 pp. (2014)
8. Ruelle, D.: Thermodynamic Formalism. Encyclopedia of Mathematics and Its Applications, vol. 5, xix+183 pp. Addison-Wesley, Reading, MA (1978)
9. Smale, S.: Differentiable dynamical systems. Bull. Am. Math. Soc. **73**, 747–817 (1967)

A Simple Model of 4d-TQFT

Rinat Kashaev

Abstract We show that, associated with any complex root of unity ω, there exists a particularly simple 4d-TQFT model defined on the cobordism category of ordered triangulations of oriented 4-manifolds.

1 Introduction

Pachner or bistellar moves are known to form a finite set of operations on triangulations such that arbitrary triangulations of a piecewise linear (PL) manifold can be related by a finite sequence of Pachner moves [13, 15]. As a result, the combinatorial framework of triangulated PL manifolds combined with algebraic realizations of Pachner moves can be useful for constructing combinatorial 4-dimensional topological quantum field theories (TQFT) [1, 20]. Realization of this scheme in three dimensions has been initiated in the Regge–Ponzano model [16], where the Pachner moves are realized algebraically in terms of the angular momentum $6j$-symbols satisfying the five term Biedenharn–Elliott identity [3, 7], which has eventually led to the Turaev–Viro TQFT model [18] and subsequent generalizations based on the theory of linear monoidal categories [17]. The same scheme in four dimensions is more difficult to realize, mainly because of the complicated nature of algebraic constructions generalizing those of the linear monoidal categories though some realizations are known [4–6, 11, 12]. In this paper, to any complex root of unity ω, we associate a rather simple model W_ω of 4d-TQFT defined on the cobordism category of ordered triangulations of oriented 4-manifolds. The definition is as follows.

A simplicial complex is called *ordered* if the underlying set is linearly ordered. We denote by $N := \mathrm{ord}(\omega)$ the order of ω, and we recall that in any ordered triangulation of an oriented d-manifold, each d-simplex S comes equipped with a

R. Kashaev (✉)

Section de Mathématiques, Université de Genève, 2-4 rue du Lièvre, 1211 Genève 4, Switzerland

e-mail: Rinat.Kashaev@unige.ch

© Springer International Publishing AG, part of Springer Nature 2018

D.R. Wood et al. (eds.), *2016 MATRIX Annals*, MATRIX Book Series 1,

https://doi.org/10.1007/978-3-319-72299-3_14

sign $\epsilon(S)$ taking the positive value 1 if the orientation induced by the linear order on the vertices of S agrees with the orientation of the manifold. We specify W_ω by associating the vector space $\mathbb{C}^N$ to each positive tetrahedron and the dual vector space $(\mathbb{C}^N)^*$ to each negative tetrahedron. For a pentachoron (4-simplex) P realizing an oriented 4-ball, we associate the vector

$$W_\omega(P) \in W_\omega(\partial P) = \otimes_{i=0}^4 W_\omega(\partial_i P) \tag{1}$$

defined by the formula

$$W_\omega(P) = \begin{cases} Q \text{ if } \epsilon(P) = 1; \\ \bar{Q} \text{ otherwise.} \end{cases} \tag{2}$$

where

$$Q := N^{-1/4} \sum_{k,l,m \in \mathbb{Z}/N\mathbb{Z}} \omega^{km} e_k \otimes \bar{e}_{k+l} \otimes e_l \otimes \bar{e}_{l+m} \otimes e_m, \tag{3}$$

$$\bar{Q} := N^{-1/4} \sum_{k,l,m \in \mathbb{Z}/N\mathbb{Z}} \omega^{-km} \bar{e}_k \otimes e_{k+l} \otimes \bar{e}_l \otimes e_{l+m} \otimes \bar{e}_m \tag{4}$$

with $\{e_k\}_{k \in \mathbb{Z}/N\mathbb{Z}}$ and $\{\bar{e}_k\}_{k \in \mathbb{Z}/N\mathbb{Z}}$ being the canonical dual bases of $\mathbb{C}^N$ and $(\mathbb{C}^N)^*$ respectively.

Let X be an ordered triangulation of an oriented 4-manifold. We define

$$W_\omega(X) = N^{(|X_0^{\text{int}}| - |X_1^{\text{int}}|)/2} \text{Ev}(\otimes_{P \in X} W_\omega(P)) \tag{5}$$

where the tensor product is taken over all pentachora of X, Ev is the operation of contracting along all the internal tetrahedra of X, and $|X_i^{\text{int}}|$ is the number of i-dimensional simplices in the interior of X. Our main result is the following theorem.

Theorem 1 W_ω *is a well defined 4d-TQFT.*
This TQFT is unitary in the sense that

$$W_\omega(X^*) = W_\omega(X)^* \tag{6}$$

where X^* is X with opposite orientation, while $W_\omega(X)^*$ is the Hermitian conjugate of $W_\omega(X)$ with respect to the standard Hilbert structure of the space $\mathbb{C}^N$ where the canonical basis is orthonormal. We collect a few results of calculation into Table 1 where $\chi(X)$ is the Euler characteristic.

Remark 1 Strictly speaking, the term TQFT (Topological Quantum Field Theory) here is used in an extended sense of TQFT with corners [14, 19]. In particular, for an ordered triangulation X of an oriented compact closed 3-manifold, the cylinder $X \times [0, 1]$ admits an ordered triangulation that extends that of X, see e.g. [8], and the partition function $W_\omega(X \times [0, 1])$, interpreted as an element of $\text{End}(W_\omega(X))$, is

Table 1 The values of the invariant W_ω and the Euler characteristic χ in the case of few oriented compact closed 4-manifolds

X	$\chi(X)$	$W_\omega(X)$
S^4	2	1
$S^2 \times S^2$	4	$(3 + (-1)^N)/2$
$\mathbb{C}P^2$	3	$N^{-1/2} \sum_{k=1}^{N} \omega^{k^2}$
$S^3 \times S^1$	0	1
$S^2 \times S^1 \times S^1$	0	$(3 + (-1)^N)/2$

not the identity map, as it would be if W_ω was an ordinary TQFT in the sense of Atiyah [1], but only a projection operator to a vector subspace $\tilde{W}_\omega(X) \subset W_\omega(X)$. It is this system of subspaces that can be given an interpretation of a TQFT in the sense of Atiyah. One can show that dim $\tilde{W}_\omega(S^3) = 1$, and this fact implies that the invariant is multiplicative under the connected sum.

Conjecture 1 For a given compact oriented closed 4-manifold X, the quantum invariant $W_\omega(X)$, considered as a function on the set of all complex roots of unity, takes only finitely many different values.

The first preprint version of this paper is available as [9], where a different normalization of pentachoral weight functions is used and the corresponding TQFT is denoted M_ω. In the case of closed 4-manifolds, the two TQFT's are related by the formula

$$W_\omega(X) = N^{3\chi(X)/2} M_\omega(X). \tag{7}$$

In the next two sections we prove Theorem 1 by identifying the transformation properties of W_ω under order changes and its invariance under the Pachner moves.

2   Behavior Under Order Changes

Proposition 1 *For two ordered triangulations X and Y of a compact oriented 4-manifold related by a change of ordering, one has the equality*

$$W_\omega(Y) = b(W_\omega(X)) \tag{8}$$

where

$$b: W_\omega(\partial X) \to W_\omega(\partial Y). \tag{9}$$

is an isomorphism of vector spaces.

Let us fix a square root $\sqrt{\omega}$. Following [2], we define a function

$$\Phi: \mathbb{Z}/N\mathbb{Z} \to \mathbb{C}, \quad \Phi(k) = \left(\sqrt{\omega}\right)^{k(k+N)}, \tag{10}$$

which has the properties

$$\Phi(k)^2 = \omega^{k^2}, \quad \Phi(-k) = \Phi(k), \quad \Phi(k+l) = \Phi(k)\Phi(l)\omega^{kl}. \tag{11}$$

We also denote

$$\bar{\Phi}(k) := \frac{1}{\Phi(k)}. \tag{12}$$

Next, we define two vector space isomorphisms

$$S, T: \left(\mathbb{C}^N\right)^* \to \mathbb{C}^N, \tag{13}$$

by the formulae

$$S\bar{e}_k = N^{-1/2} \sum_{l \in \mathbb{Z}/N\mathbb{Z}} \Phi(k-l)e_l, \quad T\bar{e}_k = \Phi(k)e_{-k}. \tag{14}$$

Notice that their inverses are given by the Hermitian conjugate maps :

$$S^{-1}e_k = \bar{S}e_k = N^{-1/2} \sum_{l \in \mathbb{Z}/N\mathbb{Z}} \bar{\Phi}(k-l)\bar{e}_l, \quad T^{-1}e_k = \bar{T}e_k = \bar{\Phi}(k)\bar{e}_{-k}. \tag{15}$$

We also define the permutation maps

$$P: \left(\mathbb{C}^N\right)^* \otimes \mathbb{C}^N \to \mathbb{C}^N \otimes \left(\mathbb{C}^N\right)^*, \quad \bar{P} = P^{-1}: \mathbb{C}^N \otimes \left(\mathbb{C}^N\right)^* \to \left(\mathbb{C}^N\right)^* \otimes \mathbb{C}^N. \tag{16}$$

The proof of Proposition 1 is based on the following lemma.

Lemma 1 ([10]) *One has the equalities*

$$Q = (P \otimes T \otimes \bar{T} \otimes T)\bar{Q} = (T \otimes \bar{P} \otimes \bar{S} \otimes S)\bar{Q}$$

$$= (S \otimes \bar{S} \otimes P \otimes T)\bar{Q} = (T \otimes \bar{T} \otimes T \otimes \bar{P})\bar{Q} \tag{17}$$

where the vectors Q and $\bar{Q}$ are defined in (3) and (4).

Proof Let us prove the first equality:

$$N^{1/4}(P \otimes T \otimes \bar{T} \otimes T)\bar{Q} = \sum_{k,l,m \in \mathbb{Z}/N\mathbb{Z}} \omega^{-km} e_{k+l} \otimes \bar{e}_k \otimes T\bar{e}_l \otimes \bar{T}e_{l+m} \otimes Te_m$$

$$= \sum_{k,l,m \in \mathbb{Z}/N\mathbb{Z}} \omega^{-km} \Phi(l)\bar{\Phi}(l+m)\Phi(m)e_{k+l} \otimes \bar{e}_k \otimes e_{-l} \otimes \bar{e}_{-l-m} \otimes e_{-m}$$

$$= \sum_{k,l,m\in\mathbb{Z}/N\mathbb{Z}} \omega^{-km-lm} e_{k+l} \otimes \bar{e}_k \otimes e_{-l} \otimes \bar{e}_{-l-m} \otimes e_{-m}$$

$$= \sum_{k,l,m\in\mathbb{Z}/N\mathbb{Z}} \omega^{-km} e_k \otimes \bar{e}_{k-l} \otimes e_{-l} \otimes \bar{e}_{-l-m} \otimes e_{-m}$$

$$= \sum_{k,l,m\in\mathbb{Z}/N\mathbb{Z}} \omega^{km} e_k \otimes \bar{e}_{k+l} \otimes e_l \otimes \bar{e}_{l+m} \otimes e_m = N^{1/4} Q \qquad (18)$$

where, in the third equality, we used the last relation in (11), in the forth equality we shifted the summation variable $k \to k - l$, and in the fifth equality we negated the summation variables l and m. The other relations are proved in a similar manner, see [10] for details.

Proof (of Proposition 1) For a triangle f of an ordered triangulation, we let $C(f)$ denote the set of all tetrahedra containing f. Let X and Y be two ordered triangulations differing in the orientation of only one edge e. The change of the orientation of e results in changing the sign of each pentachoron of X containing e. By applying the appropriate equality of Lemma 1 to each such pentachoron in $W_\omega(X)$ we observe that for each triangle f containing e, there is a cancellation of an inverse pair of S or T operators for each internal tetrahedron of $C(f)$. In this way, we immediately obtain the equality $W_\omega(X) = b(W_\omega(Y))$ where b is given by the product of non-canceled S or T operators acting on the boundary tetrahedra. We finish the proof by remarking that any ordering change can be obtained as a finite sequence of single edge orientation changes.

3 Invariance Under the Pachner Moves

A Pachner move in dimension 4 is associated with a splitting of the boundary of a 5-simplex into two non-empty disjoint sets of 4-simplices (pentachora). A Pachner move is called of the *type* (k, l) with $k + l = 6$, if the two disjoint subsets of pentachora consist of k and l elements respectively. Thus, altogether, we have Pachner moves of three possible types (3,3), (2,4) and (1,5). Let us discuss in more detail their algebraic realizations in terms of polynomial identities for the matrix coefficients of the vectors (3) and (4) defined by the formulae:

$$Q_{l,m}^{i,j,k} \equiv \langle \bar{e}_i \otimes e_l \otimes \bar{e}_j \otimes e_m \otimes \bar{e}_k, Q \rangle = N^{-1/4} \omega^{ik} \delta_{l,i+j} \delta_{m,j+k} \qquad (19)$$

and

$$\bar{Q}_{i,j,k}^{l,m} \equiv \langle e_i \otimes \bar{e}_l \otimes e_j \otimes \bar{e}_m \otimes e_k, \bar{Q} \rangle = N^{-1/4} \omega^{-ik} \delta_{l,i+j} \delta_{m,j+k} \qquad (20)$$

3.1 The Type (3,3)

This is the most fundamental Pachner move as it is the only one which can be written in the form involving only the pentachora of one and the same sign and, in a sense, it implies all other types.

Consider a 5-simplex with linearly ordered vertices $A = \{v_0, v_1, \ldots, v_5\}$. Its boundary is composed of six pentachora $\partial_i A = A \setminus \{v_i\}$ of which three are positive corresponding to even i's and three are negative corresponding to odd i's. All even (respectively odd) pentachora compose a 4-ball, to be called *even* (respectively *odd*) 4-ball, so that the boundary of both balls are naturally identified as simplicial complexes. Both of these balls, when considered separately, are composed only in terms of positive pentahora, and the corresponding algebraic condition on the vector Q takes the form

$$\sum_{s,t,u} Q^{i,l,m}_{s,t} Q^{s,j,n}_{p,u} Q^{t,u,k}_{q,r} = \sum_{s,t,u} Q^{m,n,k}_{s,t} Q^{l,j,t}_{u,r} Q^{i,u,s}_{p,q} \tag{21}$$

where the left hand side corresponds to the even 4-ball and the right hand side to the odd one, while the summations in both sides correspond to their own interior tetrahedra. Namely, denoting the tetrahedron $A \setminus \{v_i, v_j\}$ by A_{ij}, the indices s, t, u correspond to the tetrahedra A_{02}, A_{04} and A_{24} in the even 4-ball, and to the tetrahedra A_{15}, A_{35} and A_{13} in the odd 4-ball, while the exterior indices $i, j, k, l, m, n, p, q, r$ on both sides correspond to the boundary tetrahedra $A_{01}, A_{23}, A_{45}, A_{03}, A_{05}, A_{25}, A_{12}, A_{14}, A_{34}$ respectively. All other forms of the Pachner relation of the type (3,3) can be obtained from (21) by applying the symmetry relations (17).

Lemma 2 *The Pachner relation* (21) *holds true for the weights* (19).

Proof By substituting one after another the explicit forms from (19), we have

$$N^{3/4}(\text{l.h.s. of (21)}) = \sum_u \omega^{im} Q^{i+l,j,n}_{p,u} Q^{l+m,u,k}_{q,r}$$

$$= \omega^{im+(i+l)n} \delta_{p,i+l+j} Q^{l+m,j+n,k}_{q,r}$$

$$= \omega^{im+(i+l)n+(l+m)k} \delta_{p,i+l+j} \delta_{q,l+m+j+n} \delta_{r,j+n+k},$$

and, similarly,

$$N^{3/4}(\text{r.h.s. of (21)}) = \sum_u \omega^{mk} Q^{l,j,n+k}_{u,r} Q^{i,u,m+n}_{p,q}$$

$$= \omega^{mk+l(n+k)} \delta_{r,j+n+k} Q^{i,l+j,m+n}_{p,q}$$

$$= \omega^{mk+l(n+k)+i(m+n)} \delta_{r,j+n+k} \delta_{p,i+l+j} \delta_{q,l+j+m+n}.$$

Comparing the obtained expressions, we see that they are the same.

Remark 2 It is interesting to note that by defining three families of linear maps

$$L^i, M^j, R^k \colon \mathbb{C}^N \otimes \mathbb{C}^N \to \mathbb{C}^N \otimes \mathbb{C}^N,$$

$$Q^{i,j,k}_{l,m} = \langle \bar{e}_j \otimes \bar{e}_k, L^i(e_l \otimes e_m) \rangle = \langle \bar{e}_i \otimes \bar{e}_k, M^j(e_l \otimes e_m) \rangle$$

$$= \langle \bar{e}_i \otimes \bar{e}_j, R^k(e_l \otimes e_m) \rangle, \qquad (22)$$

we can rewrite the system (21) as a 3-index family of matrix Yang–Baxter relations in $\mathbb{C}^N \otimes \mathbb{C}^N \otimes \mathbb{C}^N$:

$$L^i_{12} M^j_{13} R^k_{23} = R^k_{23} M^j_{13} L^i_{12} \qquad (23)$$

with the standard meaning of the subscripts, for example, $L^i_{12} := L^i \otimes \mathrm{id}_{\mathbb{C}^N}$, etc. It would be interesting to understand the significance of this fact in relationships of 4d-TQFT with lattice integrable models of statistical mechanics.

Remark 3 Another equivalent form of the system (21) is given by a 3-index family of "twisted" pentagon relations either for the R^i-matrices

$$R^m_{12} R^n_{13} R^k_{23} = \sum_{s,t} Q^{m,n,k}_{s,t} R^t_{23} R^s_{12} = N^{-1/4} \omega^{mk} R^{n+k}_{23} R^{m+n}_{12}, \qquad (24)$$

or for the L^i-matrices

$$L^m_{23} L^l_{13} L^i_{12} = \sum_{s,t} Q^{i,l,m}_{s,t} L^s_{12} L^t_{23} = N^{-1/4} \omega^{im} L^{i+l}_{12} L^{l+m}_{23}, \qquad (25)$$

where we use the matrices defined in (22).

3.2   The Type (2,4)

We split the pentachora of the 5-simplex $A = \{v_0, v_1, \ldots, v_5\}$ into a subset of two pentachora $\partial_1 A$ and $\partial_3 A$ and the complementary subset of other four pentachora. The corresponding algebraic relation takes the form

$$N^{-1/2} \sum_{k,m,n,u,v,w} Q^{i,l,m}_{v,w} Q^{v,j,n}_{p,u} Q^{w,u,k}_{q,r} \bar{Q}^{s,t}_{m,n,k} = \sum_u Q^{l,j,t}_{u,r} Q^{i,u,s}_{p,q}, \qquad (26)$$

where the factor $N^{-1/2}$ in the left hand side corresponds to the internal edge $v_1 v_3$, according to our TQFT rules. All other forms of the Pachner move of the type (2,4) can be obtained from (26) combined with the symmetry relations (17).

Lemma 3 *The relation* (26) *holds true for the weights* (19) *and* (20).

Proof We rewrite (26) in the equivalent matrix form

$$\sum_{k,m,n} R_{12}^m R_{13}^n R_{23}^k \bar{Q}_{m,n,k}^{s,t} = N^{1/2} R_{23}^t R_{12}^s \tag{27}$$

and easily prove it by using (24):

$$\sum_{k,m,n} R_{12}^m R_{13}^n R_{23}^k \bar{Q}_{m,n,k}^{s,t} = N^{-1/4} \sum_n \omega^{-(s-n)(t-n)} R_{12}^{s-n} R_{13}^n R_{23}^{t-n}$$

$$= N^{-1/2} \sum_n R_{23}^t R_{12}^s = N^{1/2} R_{23}^t R_{12}^s. \tag{28}$$

Remark 4 As the proof of Lemma 3 shows, the Pachner relation of the type (2,4) given by Eq. (26) is clearly weaker than the Pachner relation of the type (3,3) given by Eq. (21). Namely, we cannot revert the argument of the proof to obtain an equivalence between the two relations.

3.3 The Type (1,5)

We split the pentachora of the 5-simplex $A = \{v_0, v_1, \ldots, v_5\}$ into the set composed of only one pentachoron $\partial_1 A$ and the complementary set of other five pentachora. The corresponding algebraic relation takes the form

$$N^{-2} \sum_{j,k,l,m,n,r,t,v,w,x} Q_{v,w}^{i,l,m} Q_{p,x}^{v,j,n} Q_{q,r}^{w,x,k} \bar{Q}_{m,n,k}^{s,t} \bar{Q}_{l,j,t}^{u,r} = Q_{p,q}^{i,u,s} \tag{29}$$

where the factor N^{-2} in the left hand side corresponds to one internal vertex v_1 and five internal edges which connect it to other five vertices, so that $N^{(1-5)/2} = N^{-2}$. As before, all other forms of the Pachner relations of the type (1,5) can be obtained from (29) by using the symmetry relations (17).

Lemma 4 *The relation* (29) *holds true for the weights* (19) *and* (20).

Proof By using (26), we write

$$N^{-2} \sum_{j,k,l,m,n,r,t,v,w,x} Q_{v,w}^{i,l,m} Q_{p,x}^{v,j,n} Q_{q,r}^{w,x,k} \bar{Q}_{m,n,k}^{s,t} \bar{Q}_{l,j,t}^{u,r}$$

$$= N^{-3/2} \sum_{j,l,r,t,x} Q_{x,r}^{l,j,t} Q_{p,q}^{i,x,s} \bar{Q}_{l,j,t}^{u,r} = N^{-2} \sum_{j,l,r,t,x} \delta_{x,u} \delta_{x,l+j} \delta_{r,j+t} Q_{p,q}^{i,x,s}$$

$$= Q_{p,q}^{i,u,s} N^{-2} \sum_{j,l,r,t} \delta_{u,l+j} \delta_{r,j+t} = Q_{p,q}^{i,u,s} N^{-2} \sum_{j,l,t} \delta_{u,l+j}$$

$$= Q_{p,q}^{i,u,s} N^{-2} \sum_{l,t} 1 = Q_{p,q}^{i,u,s}. \qquad (30)$$

Acknowledgements This work is supported in part by the Swiss National Science Foundation.

References

1. Atiyah, M.: Topological quantum field theories. Inst. Hautes Études Sci. Publ. Math. **68**, 175–186 (1988/1989). http://www.numdam.org/item?id=PMIHES_1988__68__175_0
2. Bazhanov, V.V., Baxter, R.J.: New solvable lattice models in three dimensions. J. Stat. Phys. **69**(3–4), 453–485 (1992). http://dx.doi.org/10.1007/BF01050423
3. Biedenharn, L.C.: An identity by the Racah coefficients. J. Math. Phys. **31**, 287–293 (1953)
4. Carter, J.S., Kauffman, L.H., Saito, M.: Structures and diagrammatics of four-dimensional topological lattice field theories. Adv. Math. **146**(1), 39–100 (1999). http://dx.doi.org/10.1006/aima.1998.1822
5. Crane, L., Frenkel, I.B.: Four-dimensional topological quantum field theory, Hopf categories, and the canonical bases. J. Math. Phys. **35**(10), 5136–5154 (1994). http://dx.doi.org/10.1063/1.530746. Topology and physics
6. Crane, L., Yetter, D.: A categorical construction of 4D topological quantum field theories. In: Quantum Topology. Knots Everything, vol. 3, pp. 120–130. World Scientific Publications, River Edge, NJ (1993). http://dx.doi.org/10.1142/9789812796387_0005
7. Elliott, J.P.: Theoretical studies in nuclear structure. V. The matrix elements of non-central forces with an application to the 2p-shell. Proc. R. Soc. Lond. Ser. A **218**, 345–370 (1953). http://dx.doi.org/10.1098/rspa.1953.0109
8. Hatcher, A.: Algebraic Topology. Cambridge University Press, Cambridge (2002)
9. Kashaev, R.: A simple model of 4d-TQFT (2014). arXiv:1405.5763
10. Kashaev, R.M.: On realizations of Pachner moves in 4d. J. Knot Theory Ramif. **24**(13), 1541002, 13 (2015). http://dx.doi.org/10.1142/S0218216515410023
11. Korepanov, I.G.: Euclidean 4-simplices and invariants of four-dimensional manifolds. I. Surgeries 3 → 3. Teor. Mat. Fiz. **131**(3), 377–388 (2002). http://dx.doi.org/10.1023/A:1015971322591
12. Korepanov, I.G., Sadykov, N.M.: Parameterizing the simplest Grassmann-Gaussian relations for Pachner move 3–3. Symmetry Integr. Geom. Methods Appl. **9**, Paper 053, 19 (2013)
13. Lickorish, W.B.R.: Simplicial moves on complexes and manifolds. In: Proceedings of the Kirbyfest (Berkeley, CA, 1998). Geometry & Topology Monographs, vol. 2, pp. 299–320. Geometry & Topology Publications, Coventry (1999) (electronic). http://dx.doi.org/10.2140/gtm.1999.2.299
14. Oeckl, R.: Discrete Gauge Theory. Imperial College Press, London (2005). http://dx.doi.org/10.1142/9781860947377. From lattices to TQFT
15. Pachner, U.: PL homeomorphic manifolds are equivalent by elementary shellings. Eur. J. Comb. **12**(2), 129–145 (1991). http://dx.doi.org/10.1016/S0195-6698(13)80080-7
16. Ponzano, G., Regge, T.: Semiclassical limit of Racah coefficients. In: Spectroscopic and group theoretical methods in physics, pp. 1–58. North-Holland, Amsterdam (1968)
17. Turaev, V.G.: Quantum Invariants of Knots and 3-Manifolds. de Gruyter Studies in Mathematics, vol. 18. Walter de Gruyter, Berlin (1994)

18. Turaev, V.G., Viro, O.Y.: State sum invariants of 3-manifolds and quantum 6j-symbols. Topology **31**(4), 865–902 (1992). http://dx.doi.org/10.1016/0040-9383(92)90015-A
19. Walker, K.: On Witten's 3-manifold invariants (1991). Preprint
20. Witten, E.: Topological quantum field theory. Commun. Math. Phys. **117**(3), 353–386 (1988). http://projecteuclid.org/getRecord?id=euclid.cmp/1104161738

Morse Structures on Partial Open Books with Extendable Monodromy

Joan E. Licata and Daniel V. Mathews

Abstract The first author in recent work with D. Gay developed the notion of a *Morse structure* on an open book as a tool for studying closed contact 3-manifolds. We extend the notion of Morse structure to *extendable partial* open books in order to study contact 3-manifolds with convex boundary.

1 Introduction

In [3], the first author and David Gay developed the notion of a *Morse structure* on a closed 3-manifold with an open book decomposition. Informally, a Morse structure is a nice family of functions and vector fields on the pages of the open book: the functions are Morse functions on the pages, and the vector fields are gradient-like and Liouville in an appropriate sense. In [3] it was shown that every open book admits a Morse structure.

The same paper [3] also developed the notion of a *Morse diagram*. This is a diagram consisting of some tori, one for each binding component, with some curves and decorations drawn on them. A Morse structure on an open book has a Morse diagram, and [3] (Prop. 3.7) showed that every abstract Morse diagram arises as the Morse diagram of an open book. This gives a graphical description, encoded by a finite amount of combinatorial data, of an open book and hence of a contact structure.

Morse structures and diagrams give a useful way to study *Legendrian knots and links* in a closed contact 3-manifold. A Legendrian knot or link in the standard contact $\mathbb{R}^3$ can be studied via its *front projection*, which projects the knot into a plane, and whose distance from the plane at any point is determined by the slope

J.E. Licata (✉)
Mathematical Sciences Institute, The Australian National University, Canberra, ACT, Australia
e-mail: joan.licata@anu.edu.au

D.V. Mathews
School of Mathematical Sciences, Monash University, Clayton, VIC, Australia
e-mail: Daniel.Mathews@monash.edu

© Springer International Publishing AG, part of Springer Nature 2018
D.R. Wood et al. (eds.), *2016 MATRIX Annals*, MATRIX Book Series 1,
https://doi.org/10.1007/978-3-319-72299-3_15

287

of the projection. In an analogous way, a Morse structure allows one to define a front projection for (almost) any Legendrian knot or link in any contact manifold. By flowing the link to a neighbourhood of the binding, one obtains a *front* for the link on the associated Morse diagram, and the slope of the diagram at any point determines the "distance" of the link from the binding. Fronts were defined in [3], along with a set of "Reidemeister moves": two Legendrian links represented by fronts are Legendrian isotopic if and only if their fronts are related by such moves.

The purpose of this short article is to explore a simple idea: what happens if we look at *partial* open books defined by *restrictions* of the monodromies in the closed case? We examine the consequences of [3] in this context, and extend the results to a large family of contact 3-manifolds with convex boundary. We generalise [3] to partial open books whose monodromy is *extendable* to the monodromy of an open book in the usual (non-relative) sense.

Partial open books were introduced by Honda–Kazez–Matić in [7]. They are related to open books in the same way that contact 3-manifolds with convex boundary are related to closed contact 3-manifolds. In [7] Honda–Kazez–Matić stated a relative version of the Giroux Correspondence between contact manifolds and open books [6], which was also expounded by Etgü–Ozbagci in [1].

Following [3], define a contact manifold W with a contact form α by

$$W = (0, \infty) \times S^1 \times S^1, \quad \alpha = dz + x\, dy,$$

where x, y, z are coordinates on the three factors of W. We prove the following.

Theorem 1 *Let (M, Γ, ξ) be a contact 3-manifold with convex boundary, presented by the partial open book (S, P, h), with binding B. There is a 2-complex Skel $\subset$ Int M with the property that, after modifying ξ by an isotopy through contact structures presented by (S, P, h), the interior of each connected component of $(M \setminus (Skel \cup B), \xi)$ is contactomorphic to a contact submanifold of W.*

Once sufficient notation has been established, in Sect. 5 we give a more precise description of these submanifolds in terms of the defining data (S, P, h) of an abstract open book defining (M, Γ, ξ).

In Sect. 4.2 we define a Morse structure for an extendable partial open book (S, P, h). A Morse structure consists of a function F and a vector field V, and this data can be used to define a *Morse diagram*, which is a decorated surface consisting of tori, punctured tori and annuli. A Morse diagram can be viewed as gluing instructions for assembling Skel and submanifolds of W into the original manifold M. The components of the Morse diagram are properly embedded in M and transverse to the vector field V along the pages of the partial open book. The flow of V assigns to points in the complement of Skel and the binding a well-defined image on the Morse diagram, which we call a *front*.

Theorem 2 *If Λ is a properly embedded Legendrian tangle in (M, ξ) disjoint from the binding B and transverse to Skel, then the front associated to $\Lambda \setminus Skel$ completely determines Λ. Consequently, any two Legendrian tangles with the same front are equal.*

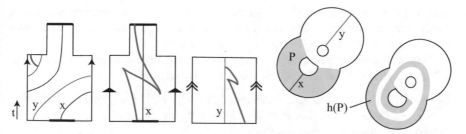

Fig. 1 Morse diagram for the extendable open book (S, P, h), shown with a front for a Legendrian tangle (bold). The bold segments at the top and bottom are identified, as are vertical edges as indicated by arrows

Fronts can effectively distinguish Legendrian tangles up to Legendrian isotopy.

Theorem 3 *The set of moves shown in Fig. 4 has the property that two Legendrian tangles in (M, Γ, ξ) are Legendrian isotopic if and only if their fronts are related by a sequence of moves and by isotopy preserving sufficiently negative slope.*

We illustrate the ideas with an example adapted from [1]; see Fig. 1. The right hand figures show $P \subset S$ and $h(P) \subset S$. The gluing map h extends to a homeomorphism of S which is a given by a Dehn twist around a curve parallel to the exterior boundary component. The three boundary components of S each correspond to a component of the Morse diagram shown on the left, and the thin curves encode the extended monodromy. The bold curve on the Morse diagram is a front projection of a Legendrian tangle with one closed component and one properly embedded interval component.

We conclude this section with a brief remark about gluing. Contact manifolds may be glued along compatible convex boundaries, and the simplest case of this is gluing contact manifolds which are products. This gluing can be represented on the Morse diagram level by stacking Morse diagrams. Front projection of Legendrian tangles also behaves nicely under this operation. In the special case of tangles braided with respect to the product structure, front projection offers a new tool for studying Legendrian braids in product manifolds.

2 Partial Open Books

We follow the definition of partial open books in [1]. All handles will be assumed two-dimensional, so a *0-handle* is a closed disc D^2 and a *1-handle* is a closed oriented 2-disc of the form $P_0 = [-1, 1] \times [-1, 1]$. To add a 1-handle to an oriented surface S, select an embedded 0-sphere $\{p, q\} \in \partial S$ called the *attaching sphere* and identify a regular neighbourhood of p, q with $[-1, 1] \times \{-1, 1\} \subset P_0$ in an orientation-reversing fashion. Any connected oriented surface with nonempty boundary can be constructed by successively attaching 1-handles to 0-handles. The *core* of a handle is $\{0\} \times [-1, 1]$ and the *co-core* is $[-1, 1] \times \{0\}$. We note that a handle attachment may be undone by cutting an attached handle through its co-core and deformation retracting it onto its attaching intervals.

Throughout this paper, (S, P) denotes a pair of compact oriented surfaces, with $P \subseteq S$, S connected and $\partial S \neq \emptyset$. We allow $P = \emptyset$ and $P = S$.

Definition 1 A *handle structure* compatible with (S, P) is a sequence of 1-handles $P_1, P_2, \ldots, P_r$ in S such that $P = P_1 \cup \cdots \cup P_r$ and S is obtained from $\overline{S \backslash P}$ by successively attaching 1-handles $P_1, \ldots, P_r$.
When we have such a handle structure, for convenience we write $R = \overline{S \backslash P}$. Thus S is obtained form R by attaching the 1-handles of P. Note then that each component of ∂P is either a component of ∂S or a concatenation of arcs alternating between $\partial P \cap \partial S$ and $\partial P \backslash \partial S$. We will denote $A = \partial P \cap \partial S$.

Definition 2 An *abstract partial open book* is a triple (S, P, h) where (S, P) admits a compatible handle structure and $h : P \to S$ is a homeomorphism onto its image such that h is the identity on A.
The function h is called the *monodromy*. Note when $P = \emptyset$, h is the null function. When $P = S$, h is a homeomorphism of S to itself fixing the boundary, and we obtain an (abstract) open book in the usual sense.

This definition of abstract partial open book differs slightly from Honda–Kazez–Matić in [7], who consider pairs (S, P) where P is a subsurface of S such that each component of ∂P is either contained in ∂S or is polygonal with every second side in ∂S. As noted above, any (S, P) admitting a compatible handle structure has this form, but the [7] definition also allows bigon components of P with one side in A. Such a *boundary-parallel bigon* deformation retracts into A and one can show that the resulting contact manifold is contactomorphic to the original one. In effect, then, the definitions are equivalent.

Clearly the existence of a compatible handle structure on (S, P) restricts the topology of S and P. For the reasons discussed above, no component of ∂P can lie in Int$S$, and no component of P is a boundary-parallel bigon.

Following [1], from a partial open book decomposition (S, P, h) we construct a sutured 3-manifold as follows. We define two handlebodies by thickening S and P and collapsing portions of their boundaries:

$$H = \frac{S \times [-1, 0]}{(x, t) \sim (x, t') \text{ for } x \in \partial S \text{ and } t \in [-1, 0]}$$

$$N = \frac{P \times [0, 1]}{(x, t) \sim (x, t') \text{ for } x \in A \text{ and } t \in [0, 1]}.$$

(Note we only collapse the part of the boundary along $A = \partial P \cap \partial S$, leaving $(\partial P \backslash \partial S) \times [0, 1]$ unscathed.) Now glue these two handlebodies together, along both the common $P \times \{0\} \subseteq S \times \{0\}$ and also by identifying points $(x, 1) \sim (h(x), -1)$ for $x \in P$.

The resulting manifold is denoted $M(S, P, h)$. It has boundary given by

$$R \times \{0\} \cup \overline{(-S \backslash h(P))} \times \{-1\} \cup (\partial P \backslash \partial S) \times [0, 1]$$

and *binding given by* $B = \partial S \times \{0\}$, modulo the identifications above, and thus has a sutured structure, with sutures Γ and complementary regions $R_\pm$ given by

$$\Gamma = \overline{\partial P \backslash \partial S} \times \{1/2\} \cup \overline{\partial S \backslash \partial P} \times \{-1/2\},$$

$$R_+ = R \times \{0\} = \overline{S \backslash P} \times \{0\}, \quad R_- = \overline{-S \backslash h(P)} \times \{-1\}.$$

Since h is a homeomorphism onto its image, $\chi(R_+) = \chi(R_-)$, so $M(S, P, h)$ is a *balanced* sutured manifold in the sense of [8]. The sutured structure on the boundary of (M, Γ) is equivalent to the structure of a dividing set for a convex surface in a contact manifold [4].

Indeed, to a partial open book (S, P, h) we associate a contact manifold with convex boundary (up to contactomorphism), given by $M(S, P, h)$, with the unique (isotopy class of) contact structure whose restrictions to H and N are both tight, with dividing sets $\partial S \times \{-1/2\}$ and $\partial P \times \{1/2\}$ respectively [1, 11]. Thus we regard $M(S, P, h)$ as a contact manifold.

Following [1], two partial open books (S, P, h) and $(\overline{S}, \overline{P}, \overline{h})$ are said to be *isomorphic* if there is a diffeomorphism $g : S \to \overline{S}$ such that $g(P) = \overline{P}$ and $\overline{h} = g \circ h \circ (g^{-1})|_{\overline{P}}$. The *relative Giroux Correspondence* establishes a bijection between isomorphism classes of partial open book decompositions, up to positive stabilisation, and compact contact 3-manifolds with convex boundary, up to contactomorphism [1, 5, 7].

In order to generalise the notion of a Morse structure from a closed contact manifold to one with convex boundary, it is helpful to discuss particular manifolds rather than isomorphism classes, so we make the following definitions.

Definition 3 A closed contact manifold (M, ξ) is *presented by the open book* (S, h) if it is contactomorphic to $M(S, h)$. A contact manifold with convex boundary (M, Γ, ξ) is *presented by the partial open book* (S, P, h) if it is contactomorphic to $M(S, P, h)$.

In the remainder of this paper we will consider manifolds of the form $M(S, h)$ or $M(S, P, h)$ so all results are up to diffeomorphism. In the case that the initial object is a manifold with an honest—as opposed to abstract—open book, the identifying diffeomorphism may be used to transfer structures from $M(S, h)$ or $M(S, P, h)$ to the given contact manifold.

3 Slices

Up to isotopy, the pair (S, P) may be encoded via a simple combinatorial diagram generated by the handle structure, which we call a *slice* and define presently.

The first step in defining a slice is to extend the core and co-core of each handle to a 1-complex. Consider a compact connected oriented surface S constructed from a finite collection of 0-handles by successively attaching 1-handles $P_1, P_2, \ldots, P_r$. Since we only consider handle structures up to isotopy, we are free to assume that the attaching spheres are disjoint from the corners where two handles meet and from the endpoints of any co-core. When a point p of the attaching sphere lies on the boundary of a 0-handle, extend the core of P_i through p via a ray to the centre of the 0-handle. Now assume that the cores of previous handles have already been extended. When p lies on the boundary of a 1-handle, there is a unique (up to isotopy) way to extend the core of P_i through 1-handles until it reaches a point on the boundary of a 0 handle and satisfies the condition that the co-core of P_j intersects the core of P_k in δ_{jk} points for all $j, k \leq i$. Then one may extend radially, as above. We call the union of the co-cores and the extended cores the *core complex* associated to the handle structure. Note that S deformation retracts onto its core complex. If, at each stage, we allow attaching points to slide along the boundary, by isotopy in the complement of the co-cores, this core complex is still determined up to isotopy.

Now consider a pair (S, P) with a compatible handle structure as in Definition 1. Then S can be constructed from 0-handles $D_1, \ldots, D_d$ by first adding 1-handles $R_1, \ldots, R_r$ to form R and then adding further 1-handles $P_1, \ldots, P_p$ to form S. That is,

$$R = D_1 \cup \cdots \cup D_d \cup R_1 \cup \cdots \cup R_r, \quad P = P_1 \cup \cdots \cup P_p, \quad S = R \cup P.$$

In the corresponding core complex, each core and co-core arises from an R_i or P_j.

The boundary ∂S consists of finitely many circles, each of which inherits a boundary orientation from S. These circles contain the endpoints of all co-cores, which form $r + p$ pairs of points. Each circle either lies in $\partial S \setminus \partial P$, or in A, or decomposes into arcs alternately in A and $\partial S \setminus \partial P$. We represent the arcs of $\partial S \setminus \partial P$ by an additional decoration—a marker denoted by an X.

Definition 4 Let $r, p, q \geq 0$ be integers. A *slice* $\mathscr{SL}$ is a collection of oriented circles, together with a set of decorations at $2(r+p)+q$ distinct points as follows:

1. r pairs of points called *antecedent pairs*
2. p pairs of points called *primary pairs*
3. q further points called *markers*.

The *slice* of a handle structure $R_1, \ldots, R_r, P_1, \ldots, P_p$ on (S, P) consists of ∂S, together with antecedent pairs given by endpoints of co-cores of the R_i, primary pairs given by endpoints of co-cores of the P_j, and a marker in each arc of $\partial S \setminus \partial P$.

Figure 2 shows two examples of pairs (S, P) with handle structures, together with their core complexes and slices.

Fig. 2 Two pairs (S, P), together with core complexes and slices. In both figures, P is white and R is shaded. In both figures, S is an annulus, and R is a disc. However on the left P is an annulus, while on the right P consists of two discs. The two handle structures are related by an isotopy of attaching points which passes through arcs of $\partial S \backslash \partial P$, resulting in distinct slices

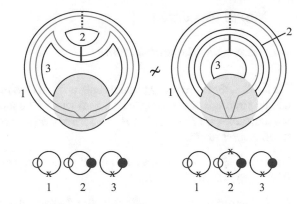

The oriented circles and pairs of points (antecedent and primary taken together) of a slice are sufficient to recover S, up to homeomorphism. To recover the pair (S, P), however, we need the distinction between antecedent and primary pairs as well as the markers.

Remark 1 Slices bear a resemblance to the *arc diagrams* of bordered Floer theory [9], especially in the *bordered sutured* case of [12] or in the context of the *quadrangulated surfaces* studied by the second author in [10]. This is not surprising, since both are essentially boundary data of handle decompositions of a surface, though slices have slightly more decoration.

Lemma 1 *If two pairs* (S, P), (S', P') *have isomorphic slices, then there is a homeomorphism of pairs* $(S, P) \cong (S', P')$.

The proof explicitly reconstructs a surface pair from a slice.

Proof First consider the slice $\mathscr{SL}$ of a pair (S, P). Surgery on $\mathscr{SL}$ at each pair of marked points (antecedent and primary) yields a 1-manifold which is the boundary of the surface formed by cutting all the 1-handles along their co-cores. This surgered surface is homeomorphic to the 0-handles, hence the number of components of the 1-manifold obtained by surgery on $\mathscr{SL}$ is equal to the number of 0-handles. In fact, the boundary of this surface naturally contains the markers, as well as the attaching spheres needed to recover R and S in turn. We note that after reattaching the antecedent handles, the boundary contains primary pairs of points and markers, and each successive primary handle is attached at points on the boundary of R or on already-attached primary handles. Up to homeomorphism preserving R and P at each stage, there is no choice where to attach handles, so it follows that the slice determines the pair (S, P).

Remark 2 The handle structures which appear in [3] were required to have a unique 0-handle, but we note that this was a choice of convenience rather than necessity. In particular, Lemma 4.5—the key technical lemma in the proof of the existence of Morse structures—explicitly covers the case of multiple index 0 critical points.

4   Morse Structures

4.1   Extendable Monodromy

For fixed S, there are many possible subsurfaces P so that (S, P) admits a compatible handle structure, and some such subsurfaces will contain others. If $P \subseteq P'$ and the monodromies $h : P \to S$, $h' : P' \to S$ satisfy $h'|_P = h$, then we say h' *extends* h or that h *extends* to P'.

Lemma 2 *If $h' : P' \to S$ extends $h : P \to S$, then there is a contact embedding of $M(S, P, h)$ into $M(S, P', h')$.*

Proof Consider the construction of the contact manifolds via handlebodies H, N and H', N', respectively. The construction of H is independent of h and P, so H, H' are contactomorphic. The construction of N, N' shows that N contact embeds in N'. Now the gluing of H and N into $M(S, P, h)$, and the gluing of H' and N' into $M(S, P', h')$, respect this contact embedding.

Definition 5 A monodromy map $h : P \to S$ is *extendable* if it extends to S, i.e., if there exists a homeomorphism $\tilde{h} : S \to S$ such that $\tilde{h}|_P = h$.

Thus, when h is extendable, $M(S, P, h)$ contact embeds into $M(S, S, \tilde{h}) = M(S, \tilde{h})$, a closed manifold. This fact will allow us to use the results of [3] in the context of partial open books.

In general, a monodromy map for a partial open book is not extendable. For instance, if h is extendable then $S \setminus P \cong S \setminus h(P)$, a condition which often fails; see, for example, Example 5. However, certain conditions guarantee that h is extendable.

Proposition 1 *If $S \setminus P$ and $S \setminus h(P)$ are both connected, then h extends to a homeomorphism of S.*

Proof Boundary components of $S \setminus P$ and $S \setminus h(P)$ are in bijective correspondence, as $\partial S \setminus \partial P$ is preserved and arcs of $\partial P \cap \mathrm{Int}\, S$ map to arcs connecting the same pairs of points on $\partial S \setminus \partial P = \partial S \setminus \partial h(P)$. Since the Euler characteristic and number of boundary components of these surfaces agree, they are homeomorphic. A homeomorphism between connected surfaces may be chosen to induce any permutation of the boundary components; this is easily seen by viewing the boundary components as marked points on a closed surface and braiding them. Thus the map fixing points of $\partial S \setminus \partial P$ may be extended to a homeomorphism of S which sends P to $h(P)$, as desired.

Figure 1 provides an example of an extendable monodromy.

4.2   Morse Diagrams for Extendable Partial Open Books

Section 3 introduced a slice as a combinatorial encoding of the pair (S, P). In order to completely encode a partial open book via slices, it remains to encode the map $h : P \to S$.

We begin by building up Morse functions on $S \times [-1, 1]$.

Definition 6   Given a homeomorphism $\widetilde{h} : S \to S$ which restricts to the identity on ∂S, a smooth function $F : S \times [-1, 1] \to (-\infty, 0]$ is a *Morse structure function* for $\widetilde{h}$ if the following properties are satisfied:

- $F^{-1}(0) = \partial S \times [-1, 1]$;
- for all values of $t \in [-1, 1]$, on the interior of the page $S \times \{t\}$, F restricts to a Morse function f_t with finitely many index 0 critical points and no index 2 critical points;
- f_t is Morse-Smale except at isolated t values, called handleslide t-values;
- $f_{-1} \circ h = f_1$, where we regard h as a function $S \times \{1\} \to S \times \{-1\}$

A Morse structure function $F : S \times [-1, 1] \to (-\infty, 0]$ descends to $M(S, \widetilde{h})$ and then restricts to a function $M(S, P, h) \to (-\infty, 0]$, also denoted F. We call a function of this form a *Morse structure function* for the partial open book.

Definition 7   A *Morse structure* on $M(S, P, h)$ is a Morse structure function F together with a vector field V such that the following conditions are satisfied:

1. the handle structures induced by f_t are isotopic for all $t \in [0, 1]$;
2. V is tangent to each page;
3. the restriction of V to the page $S \times \{t\}$ is gradient-like for f_t
4. near each component of the binding, there is a neighbourhood parameterised by (ρ, μ, λ) such that $B = \{\rho = 0\}$, $\mu = t$, $F = -\rho^2$, and $V = -(\frac{\rho}{2})\partial_\rho$.

Strictly speaking, f_0 and f_1 are defined on $S \times \{0, 1\}$, while f_t is defined only on $P \times \{t\}$ for $t \in (0, 1)$. Condition 1 above refers to $f_t|_P$ for $t \in \{0, 1\}$.

Proposition 2   *Every partial open book with extendable monodromy admits a Morse structure.*

Proof   This is immediate from Proposition 3.3 of [3]; this is a result about a (non-partial) monodromy map for a standard (non-partial) open book. It is implicit in the proof there that handleslides can happen at chosen values of t; we choose them not to happen for $t \in (0, 1)$.

A Morse structure induces a handle structure on $S \times \{t\}$. In particular, on each page the flowlines between index 0 and index 1 critical points, together with the flowlines from the index 1 critical points to $\partial S \times \{t\}$, form a core complex on $S \times \{t\}$. Thus $\widetilde{h}$ yields a slice $\mathscr{SL}_t$ on S for each value of t.

Lemma 3   *The slices on $S \times \{-1\}$ and $S \times \{0\}$ determine the mapping class of $\widetilde{h}$.*

Proof According to Proposition 2.8 in [2], there is a unique mapping class which renders the core complex of $S \times \{-1\}$ isotopic to that of $S \times \{0\}$. The lemma then follows from the observation that a slice determines these decorations up to isotopy. As the handle structures are isotopic for $t \in (0, 1)$, it is sufficient to look at t from -1 to 0.

We now consider the slices derived from the *partial* monodromy h, taking a Morse structure (F, V) as above. We restrict the slices from $\tilde{h}$ on $S \times [-1, 1]$ to $S \times [-1, 0] \cup P \times [0, 1]$. As P is a collection of handles added to R, for each $t \in [0, 1]$ we obtain a "slice" on $P \times \{t\}$, again denoted $\mathscr{SL}_t$, consisting of the oriented arcs and circles of $A = \partial P \cap \partial S$, together with pairs of points from co-cores of primary handles. (There are now no antecedent pairs, nor markers, since these arise from R, rather than P.)

Let us now consider all the slices simultaneously. For each $t \in [-1, 0]$, we have a slice $\mathscr{SL}_t$ consisting of the oriented ∂S with pairs of antecedent points, primary points, and markers. For each $t \in [0, 1]$, we have a slice $\mathscr{SL}_t$ consisting of $A \subseteq \partial S$ with pairs of primary points only. For any value of t, the associated slice embeds as a collection of curves in the corresponding page, and we may assemble these into a surface embedded in $M(S, P, h)$.

Definition 8 Given an extendable partial open book (S, P, h) and a Morse structure (F, V), the associated *Morse diagram* is the surface formed from the union of slices

$$\bigcup_{t \in [-1,1]} \mathscr{SL}_t \times \{t\}.$$

Thus, the Morse diagram consists of

$$\partial S \times [-1, 0] \cup A \times [0, 1]$$

with the identification $(x, 1) \sim (x, -1)$ for all $x \in A$, together with some decorations. (Note the gluing is straightforward since the restriction of $\tilde{h}$ to ∂S is the identity.) The decorations consist of curves, assembled from the points on each slice. Thus if a slice with $t \in [-1, 0]$ has r antecedent pairs, p primary pairs, and q markers, the Morse diagram contains r pairs of *antecedent curves*, p pairs of *primary curves*, and q *marker curves*. However, the marker curves need not be drawn, as their location is seen automatically: markers correspond to arcs of $\partial S \setminus \partial P$, which arise as segments of the boundary of the Morse diagram. Note that these curves cannot be assumed to be either connected or disjoint from each other; a handle slide of one co-core over another leads creates a *teleport* of the curve associated to the sliding co-core over the curve associated to the stationary co-core; a handleslide on the page $S \times \{t_0\}$ corresponds to a pair of trivalent vertices on the Morse diagram at height t_0. See Fig. 3.

Lemma 3 and the discussion above establish the following result:

Proposition 3 *A Morse diagram determines a partial open book (S, P, h) up to isotopy of the pair (S, P) and the mapping class of an extension $\tilde{h} \subset MCG(S)$.*

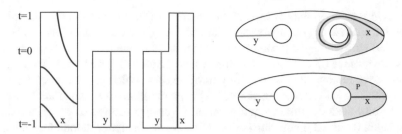

Fig. 3 Left: A Morse diagram for a partial open book. Right: The monodromy h is defined by its effect on P shown Note that h is extendable to $\widetilde{h}$ which is a single left handed Dehn twist

Remark 3 The Morse diagram of a partial open book will clearly depend on the choice of extension $\widetilde{h}$, but this mirrors the closed case which also makes no claims of uniqueness.

5 Front Projections of Legendrian Tangles

If the only goal is constructing a Morse diagram, there is a great deal of flexibility in the choice of V. However, strengthening the conditions on V allows us to prove Theorem 1 and promotes the Morse diagram to a tool for studying Legendrian tangles in $M(S, P, h)$.

Proof of Theorem 1 The main result of [3] is that for each component of the binding B of $M(S, \widetilde{h})$, the preimage of the flow of V is contactomorphic to $(0, \infty) \times S^1 \times S^1$ with coordinates $x \in (0, \infty)$, $y, z \in S^1$ and with contact structure $\xi_W = \ker(dz + x\,dy)$.

We briefly summarise the idea of the proof and refer the reader to [3] for details. The key technical ingredient is a proof that there exists a contact form α and a Morse structure (F, V) with the additional property that V is Liouville for $d(\alpha|_{\mathrm{Int}S \times \{t\}})$. By choosing α to have a specified form near the binding, we may define an explicit map which sends (ρ, μ, λ) to $\left(\frac{1}{\rho^2}, \lambda, \mu\right)$, where the latter represent (x, y, z) coordinates on W. This map identifies V near the binding with the vector field $x\partial_x$ on W and this identification extends the map to the rest of $M \setminus (\mathrm{Skel} \cup B)$.

Given this, we consider any extension $\widetilde{h}$ for h and prove Theorem 1 by considering the contact submanifold $M(S, P, h)$ inside $M(S, \widetilde{h})$. In the case of closed components of the binding of $M(S, P, h)$, the corresponding component of $M \setminus (\mathrm{Skel} \cup B)$ is contactomorphic to W itself, just as in the case of a closed contact manifold.

For binding components coming from A, we begin with a copy of W and remove points which lie in $M(S, \widetilde{h})$ but not $M(S, P, h)$. The contactomorphism described above takes pages of the open book to planes corresponding to fixed z value. For simplicity, then, we may assume that z takes values in the circle formed by identifying the endpoints of $[-1, 1]$. For each $z \in [-1, 0]$, and annulus $(0, \infty) \times S^1$ is left untouched. On the other hand, for $z \in (0, 1)$, the circle parameterised by y is identified with a boundary component of S; thus when we restrict to the partial open book, we remove $(0, \infty) \times I$ for the image of each interval I in $\partial S \backslash A$. In the language of flows, we remove the image of any flowline of V which terminates on a point of $\partial S \backslash A$, deleting $|\partial S \backslash A|$ rectangles $J \times (0, 1)$ from the Morse diagram. Finally, we note that the complete flowline from a point on A (away from the co-cores) terminates at an index 0 critical point. Since R contains an open neighborhood of each index 0 critical point, the flowline exits P after some finite amount of time. Thus for each y-interval K which remains, we also remove an open set $\big(0, g(y, z)\big) \times K \times (0, 1)$ from W; here g is a continuous function.

Having established (via appeal to the closed case) that one may always find a Morse structure which is compatible with the contact structure as described in the proof of Theorem 1, we henceforth assume all Morse structures are of this form. Suppose now that Λ is a Legendrian curve in $M(S, P, h)$ which is disjoint from the binding and meets the core complex $\mathscr{C}$ transversely. Viewing the Morse diagram as a properly embedded subsurface of the manifold, we may flow $\Lambda \backslash (\Lambda \cap \mathscr{C})$ by $\pm V$ to the Morse diagram to get a *front* $\mathscr{F}(\Lambda)$ which is sufficient to recover the original curve.

Proof of Theorem 2 In order to see that the front projection of a Legendrian tangle determines the tangle itself, it is useful to note that W is a quotient of the $x > 0$ half-space of $(\mathbb{R}^3, \xi_{\text{std}})$. Front projection for Legendrian knots is classically defined in $\mathbb{R}^3$, with the key characteristic that the slope of the tangent in the projection recovers the x coordinate of the Legendrian curve. Alternatively, one may take the perspective that front projection to the $x = c$ plane in $\mathbb{R}^3$ is the image under the flow of the vector field $x \partial_x$; this vector field is Liouville for the area form induced by $\alpha = dz + x\, dy$ on each plane $z = c$. The contactomorphism described above takes the Liouville vector field on each page $x \partial_x$ and identifies the image of an $x = c$ plane with the Morse diagram. The property that a classical front completely determines a Legendrian curve then implies the analogous statement in the context of open books.

The relationship between fronts in open books and fronts in $\mathbb{R}^3$ yields the familiar properties:

1. $\mathscr{F}(\Lambda)$ determines Λ, as the slope of the tangent to $\mathscr{F}(\Lambda)$ records the flow parameter;
2. $\mathscr{F}(\Lambda)$ is smooth away from finitely many semicubical cusps;

   On the other hand, fronts in partial open books have some new features:

1. the slope of $\mathscr{F}(\Lambda)$ is negative except where it has an endpoint on the image of $\mathscr{C}$; this follows from the description of W as a quotient of the $\{x < 0\}$ half space in $\mathbb{R}^3$.

2. for $t \in (0, 1)$, the slope of $\mathscr{F}(\Lambda)$ is bounded from above by $-\epsilon < 0$, as a slope limiting to 0 corresponds to a Legendrian curve approaching the index 0 critical point and R has an open neighbourhood around each index 0 critical point;
3. if Λ intersects a core circle C on the t_0 page, then $\mathscr{F}(\Lambda)$ will have a pair of *teleporting endpoints* at height t_0: $\mathscr{F}(\Lambda)$ will approach a curve on the Morse diagram corresponding to C from the left and the other curve corresponding to C from the right.

5.1 Reidemeister Moves

The Reidemeister moves established for Legendrian links in closed contact manifolds extend to a family of moves for fronts of properly embedded Legendrian tangles.

Proof of Theorem 3 A complete collection of Legendrian Reidemeister moves for front projections of Legnedrian knots in open books is given in [3] and shown in Fig. 4 (S, H, K moves). Since we now consider contact manifolds with convex boundary, we may extend this analysis to properly embedded Legendrian tangles. The interior of $M(S, P, h)$ is indistinguishable from the interior of a closed contact manifold, so the only new behaviour on fronts occurs at the boundary of the Morse diagram. Whether one considers these to be new moves is a question of taste; each of the moves listed below is simply the restriction to a Morse diagram for a partial open book of a planar isotopy on a Morse diagram for an ordinary open book.

The boundary of the Morse diagram has three distinct pieces: the *floor*, which is the image under the flow by V of $S \times \{-1\} \setminus h(P)$; the *ceiling*, which is the image of $S \times \{0\} \setminus P$; and the *walls*, which are the image of $\partial P \setminus (\partial S \cap \partial S) \times [0, 1]$. In addition to moves on the interior of the diagram which alter the combinatorics of the curves and projection, we see the following moves near the boundary of the Morse diagram:

Move N0: The endpoint of a curve on the front may slide freely along a component of the floor, the ceiling, or a wall, either crossing or teleporting at any trace curve encountered on a floor or ceiling.

Move N1: The endpoint of a curve on the front may slide right from the ceiling onto a wall and vice versa or left from the floor onto a wall and vice versa.

Move N2: : Given two curves whose endpoints are near each other on the boundary of the Morse diagram, one may isotope the endpoints past each other, introducing a crossing in the curves.

Move N2 move is reversible, and we note that it allows n parallel strands with adjacent endpoints may be replaced by the front projection of an arbitrary positive braid. If performing this isotopy in real time, the slopes at the endpoints must be distinct at the moment of superposition to ensure that the endpoints of the Legendrian curves remain disjoint.

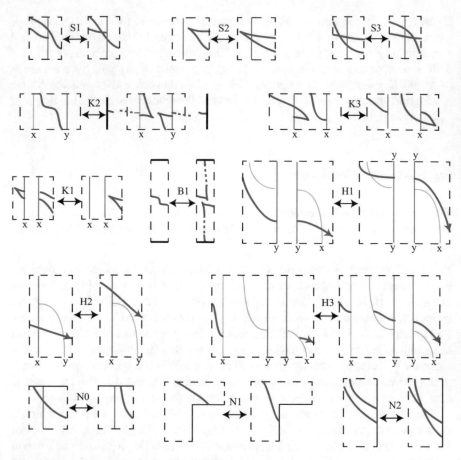

Fig. 4 Moves for Legendrian links and tangles

6 Examples

We consider some examples of simple extendable partial open books, Morse structures and front projections.

Example 1 (Empty Monodromy) Suppose we have a partial open book (S, P, h) where P is empty. Then h is trivially extendable. It is not difficult to see then that $M(S, P, h)$ is just $S \times [-1, 1]/ \sim$, with dividing set $\partial S \times \{-1/2\}$. Legendrian fronts exist for any Legendrian knots avoiding the skeleton, and as P is empty there is no issue with maximum slope.

Example 2 (Tight Ball) This example also appears in [1]. Let S be an annulus and P a thickened properly embedded arc. Let h be a positive Dehn twist, as shown in Fig. 5. A Morse diagram is shown in Fig. 6, together with initial and final pages.

Fig. 5 The tight ball of
Example 2

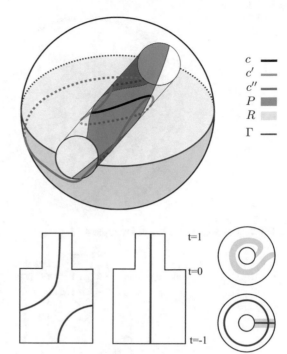

c ——
c' ——
c'' ——
P ▬
R ▨
Γ ——

Fig. 6 A Morse diagram
(left) and pages showing
initial and final slices (right).
Note that the shaded region in
the top right actually the
image of P under h^{-1}, as
required by the identification
conventions for the mapping
torus

To see why we obtain a tight 3-ball, consider a standard tight contact 3-ball B
with connected boundary dividing set Γ, and positive region R_+ a disc. Take a
Legendrian arc γ properly embedded in B, with endpoints on Γ. Drill out a small
tubular neighbourhood T of γ. Then the dividing set on the resulting surface is
shown in Fig. 5. The tube has boundary a cylinder, which is cut into two rectangles
by the dividing set. One of these rectangles is P. The tube can be regarded as $P \times$
$[0, 1]$, and its complement can be regarded as $S \times [-1, 0]$ where S is an annulus,
consisting of P together with $R = R_+$. A co-core arc c as shown, when pushed
across the tube to c', is isotopic in the complement of T to the arc c'' on S. Then the
monodromy takes c'' to c.

Example 3 ($S^2 \times I$) Let S be a disc, P a thickened properly embedded arc. Then h
must be isotopic to the identity. So $M(S, P, h)$ consists of a ball $D^2 \times [-1, 0]$, with a
disc $P \times [0, 1]$, glued to a closed curve on its boundary, forming an $S^2 \times I$. This in
fact extends to the identity $\tilde{h} : S \to S$, which produces the tight S^3, and hence the
contact structure here is the unique tight one.

Example 4 (Overtwisted Ball) Let S again be an annulus and P a thickened properly
embedded arc, as in lower right picture in Fig. 6, but now let h be a negative Dehn
twist. A Morse diagram is shown in Fig. 8, together with a $tb = 0$ unknot bounding
an overtwisted disc.

The manifold $M(S, P, h)$ is shown in Fig. 7. As in Example 2, we drill a tube
T out of a ball. However now the dividing set on the tube twists in the opposite

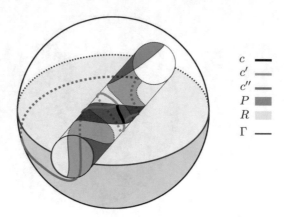

Fig. 7 The overtwisted ball of Example 4

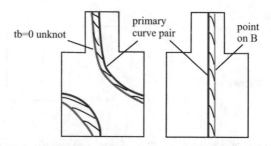

Fig. 8 An overtwisted disc in Example 4. The boundary of the disc is parallel to one of the primary curves, and the thinner lines indicate a foliation of the disc by Legendrian curves that teleport across the primary curve and meet at a point on the binding represented by a vertical line. (Example due to Dave Gay)

direction (the "wrong way") around the tube. Thus the ball is overtwisted: even if both the tube and its complement are tight, one can find an attaching arc on the tube containing bypasses on both sides. One can again take a co-core curve c, trace it through T to c' and through the complement of T to $c'' \subset S$ to show that the monodromy is the restriction of a left-handed Dehn twist.

Indeed, a Legendrian unknot of Thurston-Bennequin number zero can be seen explicitly from its front projection. The leftwards direction of the Dehn twist means that we can draw the front shown in Fig. 8. This unknot avoids all curves of the Morse diagram and bounds an overtwisted disc that lies in a subset of W. This disc can be seen explicitly on the Morse diagram, as an overtwisted disc admits a radial foliation by Legendrian curves, each of which can also be projected to the diagram. These curves terminate on a vertical line which represents a single point on B.

Example 5 We conclude with an example which breaks several of the conventions already established, but nevertheless illustrates an interesting phenomenon. The right hand pictures in Fig. 9 show initial and final pages specifying a monodromy h which is not extendable. By way of proof, consider an arc in $S \setminus P$ connecting two

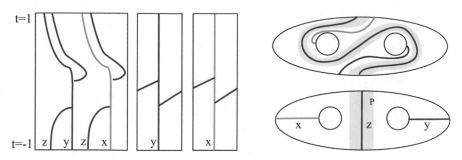

Fig. 9 Example 5

distinct boundary components; no arc with the same endpoints exists in $S \setminus h(P)$. On the other hand, this monodromy nonetheless appears to have a perfectly valid Morse diagram, in the sense that the left hand figure defines a sequence of handle slides and isotopies taking the initial core complex to the terminal one. Examples such as these may be interesting for further study.

Acknowledgements The authors would like to acknowledge the support and hospitality of MATRIX during the workshop *Quantum Invariants and Low-Dimensional Topology*. The second author is supported by Australian Research Council grant DP160103085.

References

1. Etgü, T., Ozbagci, B.: On the relative Giroux correspondence. In: Low-Dimensional and Symplectic Topology. Proceedings of Symposia in Pure Mathematics, vol. 82, pp. 65–78. American Mathematical Society, Providence, RI (2011). http://dx.doi.org/10.1090/pspum/082/2768654
2. Farb, B., Margalit, D.: A Primer on Mapping Class Groups. Princeton Mathematical Series, vol. 49. Princeton University Press, Princeton, NJ (2012)
3. Gay, D.T., Licata, J.E.: Morse structures on open books (2015). http://arxiv.org/abs/1508.05307
4. Giroux, E.: Convexité en topologie de contact. Comment. Math. Helv. **66**(4), 637–677 (1991)
5. Giroux, E.: Structures de contact en dimension trois et bifurcations des feuilletages de surfaces. Invent. Math. **141**(3), 615–689 (2000)
6. Giroux, E.: Géométrie de contact: de la dimension trois vers les dimensions supérieures. In: Proceedings of the International Congress of Mathematicians, (Beijing, 2002), vol. II, pp. 405–414. Higher Education Press, Beijing (2002)
7. Honda, K., Kazez, W.H., Matić, G.: The contact invariant in sutured Floer homology. Invent. Math. **176**(3), 637–676 (2009). http://dx.doi.org/10.1007/s00222-008-0173-3
8. Juhász, A.: Holomorphic discs and sutured manifolds. Algebr. Geom. Topol. **6**, 1429–1457 (2006) (electronic)
9. Lipshitz, R., Ozsvath, P., Thurston, D.: Bordered Heegaard Floer homology: invariance and pairing (2008). http://arxiv.org/abs/0810.0687
10. Mathews, D.V.: Strand algebras and contact categories (2016). http://arxiv.org/abs/1608.02710
11. Torisu, I.: Convex contact structures and fibered links in 3-manifolds. Int. Math. Res. Not. (9), 441–454 (2000). http://dx.doi.org/10.1155/S1073792800000246
12. Zarev, R.: Bordered Floer homology for sutured manifolds (2009). http://arxiv.org/abs/0908.1106

Counting Belyi Pairs over Finite Fields

George Shabat

Abstract Alexander Grothendieck's theory of *dessins d'enfants* relates *Belyi pairs* over $\overline{\mathbb{Q}}$ with certain graphs on compact oriented surfaces; the present paper is aimed at the extension of this correspondence. We introduce two closely related categories of Belyi pairs over arbitrary algebraically closed fields, in particular over the algebraic closures $\overline{\mathbb{F}}_p$ of finite fields. The lack of the analogs of graphs on surfaces over $\overline{\mathbb{F}}_p$ promotes the development of other tools that are introduced and discussed. The problem of counting Belyi pairs of bounded complexity is posed and illustrated by some examples; the application of powerful methods of counting dessins d'enfants together with the concept of *bad primes* is emphasized. The relations with geometry of the moduli spaces of curves is briefly mentioned.

1 Introduction

The hidden relations between seemingly different objects cause the increasing interest of mathematicians, especially since the middle of twentieth century, when it became possible to understand these relations in categorical terms. The recent explosions of activity in topological recursion, *Langlands program* (e.g., [9]), *monstrous moonshine* (e.g., [10]) provide some obvious examples.

Alexander Grothendieck's theory of *dessins d'enfants* (see [13, 24] and [17]) demonstrates yet another mixture of combinatorial topology, arithmetic geometry and group theory. In its original form it relates *Belyi pairs* (to be defined soon) over $\overline{\mathbb{Q}}$ with certain graphs on compact oriented surfaces. The concept of Belyi pair is automatically extended to the case of arbitrary algebraically closed fields, in particular they can be defined over the fields $\overline{\mathbb{F}}_p$, the algebraic closures of finite fields; the lack of the analogs of graphs on surfaces over $\overline{\mathbb{F}}_p$ promotes the development of other tools that will be introduced and discussed in the present paper.

G. Shabat (✉)

Russian State University for the Humanities, Miusskaya sq. 6, GSP-3, Moscow 125993, Russia

e mail: george.shabat@gmail.com

© Springer International Publishing AG, part of Springer Nature 2018

D.R. Wood et al. (eds.), *2016 MATRIX Annals*, MATRIX Book Series 1,

https://doi.org/10.1007/978-3-319-72299-3_16

The objects of the categories that we are going to consider are definable by finite amounts of information; hence the task of counting objects of bounded complexity arises naturally. The theory is in its infancy and therefore the consideration of some simple examples will prevail over the general theorems.

The paper is based on the author's talk at the Creswick conference in the December 2016; the author is indebted to the organizers of this conference for the stimulating atmosphere in this wonderful place. The special thanks go to P. Norbury for clarifying the matters that we are going to discuss in the last section.

2 Belyi Pairs

We shall work over ground fields $\Bbbk$, assuming forever that they are algebraically closed,

$$\overline{\Bbbk} = \Bbbk.$$

The smaller fields $\Bbbk_0$ will be considered as well, such that $\Bbbk = \overline{\Bbbk_0}$; the typical cases are $\Bbbk_0 = \mathbb{Q}$ and $\Bbbk_0 = \mathbb{F}_p$ for a prime p. The intermediate fields $\mathbb{K}$,

$$\Bbbk_0 \subset \mathbb{K} \subset \Bbbk$$

with $[\mathbb{K} : \Bbbk_0] < \infty$, will also be in the game; typically, these $\mathbb{K}$'s will be fields of *algebraic numbers* and *finite fields* $\mathbb{F}_q = \mathbb{F}_{p^r}$.

By a *curve* we always mean a *complete curve* over $\Bbbk$; it would be nice to assume that our curves are *irreducible* and *smooth* as well. However, in the cases of *bad reduction* (at least one of) these properties is lost.

For a smooth irreducible curve $\mathbf{X}$ we identify a rational function $f \in \Bbbk(\mathbf{X})$ with a *regular* map $f : \mathbf{X} \to \mathbf{P}_1(\Bbbk)$ to the projective line.

For the rest of the paper we assume that these maps are *separable*, or, equivalently, that the field extensions $\Bbbk(\mathbf{X}) \supset \Bbbk(f)$ are separable, not like $\mathbb{F}_p(\sqrt[p]{x}) \supset \mathbb{F}_p(x)$.

A non-constant $f \in \Bbbk(\mathbf{X}) \setminus \Bbbk$ defines a *surjective* map $f : \mathbf{X} \to \mathbf{P}_1(\Bbbk)$, and for *almost* all $c \in \mathbf{P}_1(\Bbbk)$—that is, except finitely many c's—the cardinality of preimages $\#f^{-1^\circ}(c)$ is the same. It is called the *degree* of f

$$\#\{c \in \mathbf{P}_1(\Bbbk) \mid \#f^{-1^\circ}(c) \neq \deg f\} < \infty.$$

Since $\mathbf{P}_1(\Bbbk)$ is infinite, the degree $\deg f$ is well-defined by this statement.

An equivalent definition is

$$\deg f := [\Bbbk(\mathbf{X}) : \Bbbk(f)].$$

The sufficient condition for f to be separable is that either $\operatorname{char}(\Bbbk) = 0$ or $\operatorname{char}(\Bbbk) \nmid \deg f$.

The points $c \in \mathbf{P}_1(\Bbbk)$ for which the number of c-preimages is non-standard, are called the *critical values* of f; the set of such points is denoted by

$$\operatorname{CritVal}(f) := \{c \in \mathbf{P}_1(\Bbbk) \mid \#f^{-1^\circ}(c) \neq \deg f\}.$$

An alternative way of expressing the inclusion $c \in \operatorname{CritVal}(f)$ is saying that f *branches* over c.

2.1 Definition

A *Belyi pair* is a pair $(\mathbf{X}, \beta)$, where $\mathbf{X}$ is a smooth irreducible curve over $\Bbbk$ and $\beta \in \Bbbk(\mathbf{X}) \setminus \Bbbk$ with $\operatorname{CritVal}(\beta) \subseteq \{\infty, 0, 1\}$. If $(\mathbf{X}, \beta)$ is a Belyi pair, then β is a *Belyi function* on $\mathbf{X}$.

In the picture below we are just trying to fix the *set-theoretical* behavior of a Belyi function—in particular, stressing the lack of ramification over a *generic* point in $\mathbf{P}_1(\Bbbk) \setminus \{0, 1, \infty\} \subseteq \mathbf{P}_1(\Bbbk) \setminus \operatorname{CritVal}(\beta)$.

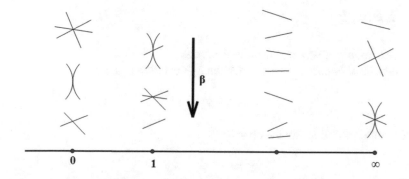

According to the Belyi theorem [1, 2], over $\Bbbk = \mathbb{C}$ a curve $\mathbf{X}$ admits a Belyi function if and only if $\mathbf{X}$ is a *complexification*, i.e. obtained via a base change, of a curve $\mathbf{X}_0$, defined over $\overline{\mathbb{Q}}$. However, finding a Belyi function on an arbitrary curve over $\overline{\mathbb{Q}}$ is a very difficult task, and the minimal possible degree of such a function can be tremendous, see [14]. The only thing we can estimate, as it will be reminded in the next section, is the *total number of Belyi pairs of bounded degree*.

2.2 Cleanness

A Belyi pair $(\mathbf{X}, \beta)$ is called *clean*, if all the branchings over $1 \in \mathbf{P}_1(\Bbbk)$ are twofold:

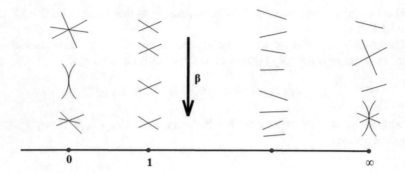

The formal definition uses the standard concepts: for a point $P \in \mathbf{X}$ denote its *local ring* $\mathcal{O}_P := \{f \in \Bbbk(\mathbf{X}) \mid f(P) \neq \infty\}$ with the maximal ideal $\mathfrak{m}_P := \{f \in \Bbbk(\mathbf{X}) \mid f(P) = 0\}$. The cleanness of β means

$$\beta - 1 \in \mathfrak{m}_P^2 \setminus \mathfrak{m}_P^3$$

at all points P with $\beta(P) = 1$. Imposing the cleanness condition is not a severe one—see below.

2.3 Examples

We give a couple of series of the simplest ones.

Generalized Fermat curves are defined by the affine equation

$$x^m + y^n = 1.$$

Under some restrictions on the char($\Bbbk$)

$$\beta := x^m = 1 - y^n$$

is a Belyi function on a generalized Fermat curve, usually not a clean one.

The concept of a curve *with many automorphisms* has two versions: in the zero and the positive characteristic of the ground field—in the latter case the cardinality of the automorphism group is *quartic* in genus (unlike the former case where the Hurwitz bound $\#\mathrm{Aut}\mathbf{X} \leq 84(g_\mathbf{X} - 1)$ holds). In many cases the factorization map

$$\beta : \mathbf{X} \longrightarrow \frac{\mathbf{X}}{\mathrm{Aut}\mathbf{X}}$$

is a Belyi function. One should be careful in the case of positive characteristic, since the factorization map is often non-separable.

A detailed treatment of the *Klein quartic* can be found in [7], and that of the Bring curve—in [28].

2.4 Two Categories of Belyi Pairs

The objects of the category $\mathcal{BELP}(\Bbbk)$ are *Belyi pairs* $(\mathbf{X}, \beta)$ over $\Bbbk$ as defined above. A morphism in $\mathcal{BELP}(\Bbbk)$ from $(\mathbf{X}, \beta)$ to $(\mathbf{X}', \beta')$ is defined as such a morphism $f : \mathbf{X} \to \mathbf{X}'$ of curves that the diagram commutes.

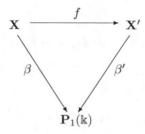

The category $\mathcal{BELP}_2(\Bbbk)$ is a full subcategory of $\mathcal{BELP}(\Bbbk)$ consisting of the *clean* Belyi pairs.

2.5 Cleaning Functor

Suppose that char($\Bbbk$) $\neq 2$. Then the introduced categories are close enough: it is easy to check that the functor

$$BP(\Bbbk) \to BP_2(\Bbbk) : (\mathbf{X}, \beta) \mapsto (\mathbf{X}, 4\beta(1 - \beta))$$

is well-defined. Thus the problems of counting the objects of bounded complexity in both categories are basically equivalent.

2.6 Fields of Definition and Galois Orbits

In the above-mentioned case $\Bbbk = \overline{\Bbbk_0}$ denote

$$\Gamma := \mathrm{Gal}(\Bbbk/\Bbbk_0)$$

the corresponding Galois group. Then the action

$$\Gamma : BP(\Bbbk)$$

is defined: take any standard model of $\Bbbk(\mathbf{X})$—planar with the simplest singularities, or tri-canonical, or whatsoever—and apply the elements of Γ coefficientwise to the equations of the curve and to the Belyi function on it. Since a Belyi pair is defined by the finite set of elements, algebraic over $\Bbbk_0$, all the Γ-orbits thus defined are finite. Therefore for each $(\mathbf{X}, \beta) \in \mathcal{BP}(\Bbbk)$ we have a stationary subgroup of Γ of finite index

$$(\mathbf{X}, \beta) \leftrightarrow \Gamma_{(\mathbf{X}, \beta)}$$

and by Galois theory

$$(\mathbf{X}, \beta) \leftrightarrow \Gamma_{(\mathbf{X}, \beta)} \leftrightarrow \mathbb{F}_{(\mathbf{X}, \beta)},$$

where the last field satisfies $\Bbbk_0 \subseteq \mathbb{F}_{(\mathbf{X}, \beta)} \subset \Bbbk$. We call this field the *field of definition*[1] of a Belyi pair $(\mathbf{X}, \beta)$. Tautologically

$$\#(\Gamma \cdot (\mathbf{X}, \beta)) = (\Gamma : \Gamma_{(\mathbf{X}, \beta)}) = (\mathbb{F}_{(\mathbf{X}, \beta)} : \Bbbk_0).$$

If a Belyi pair $(\mathbf{X}, \beta)$ can be defined over some finite extension $\mathbb{K} \supseteq \Bbbk_0$ (i.e., there exists a model of $\mathbf{X}$ over $\mathbb{K}$ with the coefficients of β belonging to $\mathbb{K}$), then it is obviously true that $\mathbb{F}_{(\mathbf{X}, \beta)} \subseteq \mathbb{K}$. However, it can happen that a Belyi pair $(\mathbf{X}, \beta)$ can <u>not</u> be defined over its field of definition; the obstruction lies in some non-commutative Galois cohomology set, see [4] or [8] for the case $\Bbbk_0 = \mathbb{Q}$. The author is unaware of similar examples over $\Bbbk_0 = \mathbb{F}_p$.

2.7 Passports and Their Realizations

The main invariant of a Belyi pair is the set of multiplicities:

$$\mathrm{div}(\beta) = a_1 A_1 + \cdots + a_\alpha A_\alpha - c_1 C_1 - \cdots - c_N C_N,$$

$$\mathrm{div}(\beta - 1) = b_1 B_1 + \cdots + b_n B_n - c_1 C_1 - \cdots - c_N C_N.$$

We collect them in a table called the *passport* of a Belyi pair:

$$\mathrm{pass}(\mathbf{X}, \beta) := \begin{pmatrix} a_1 & b_1 & c_1 \\ \cdots & \cdots & \cdots \\ \cdots & \cdots & \cdots \\ a_\alpha & b_n & c_N \end{pmatrix}$$

Lemma 1 *For any passport of a Belyi pair* $(\mathbf{X}, \beta)$

$$a_1 + \ldots + a_\alpha = b_1 + \ldots + b_n = c_1 + \ldots + c_N =: d. \tag{1a}$$

---

[1]It is often called the *field of moduli*, but we are going to use the word *moduli* in its traditional algebro-geometrical sense.

The genus g of **X** *can be defined by the equality*

$$\alpha + n + N =: d + 2 - 2g. \tag{1b}$$

Proof For (1a) we define $d := \deg \beta$ and use the well-known property of the degree of a branched covering. To establish (1b) note that, according to the definition of Belyi function

$$\mathrm{div}(d\beta) = \sum_{i=1}^{\alpha}(a_i - 1)A_i + \sum_{j=1}^{n}(b_j - 1)B_j - \sum_{k=1}^{N}(c_k + 1)C_k$$

for some points $A_1, \ldots, C_N \in \mathbf{X}$, and use $\deg(d\beta) = 2g - 2$. $\qquad \square$

Denote the *set of realizations of a passport* Π, *satisfying the above conditions* (1a) *and* (1b),

$$\mathcal{R}_\Pi(\Bbbk) := \frac{\{(\mathbf{X}, \beta) \in \mathcal{BP}(\Bbbk) \mid \mathrm{pass}(\mathbf{X}, \beta) = \Pi\}}{\text{isomorphism}}.$$

This definition makes sense since the categories $\mathcal{BP}(\Bbbk)$ are equivalent to the small ones.

Theorem 1 *For any algebraically closed field* $\Bbbk$ *and any passport* Π, *satisfying the above conditions* (1a) *and* (1b), *the set* $\mathcal{R}_\Pi(\Bbbk)$ *is finite.*

Idea of the proof The set $\mathcal{R}_\Pi(\Bbbk)$ is in a natural bijective correspondence with the corresponding 0-dimensional subscheme of the moduli space $\mathcal{M}_g(\Bbbk)$, where g is defined by (1b). The detailed proof will appear elsewhere. $\qquad \square$

So the basic counting question is to study the cardinalities of these sets:

$$\boxed{\#\mathcal{R}_\Pi(\Bbbk) = ???}$$

We don't have a complete answer even in the case $\Bbbk = \mathbb{C}$; however, see the discussion below.

As the following simple example shows, this cardinality can depend on the field:

$$\mathcal{R}_{(333)}(\overline{\mathbb{Q}}) = (y^2 = 1 - x^3, \beta = \frac{y+1}{2}),$$

while

$$\mathcal{R}_{(333)}(\overline{\mathbb{F}_3}) = \varnothing.$$

Finally, the general behavior of $\#\mathcal{R}_\Pi(\Bbbk)$'s can be studied in the Galois-theoretic terms.

Lemma 2 *For any Belyi pair* $(\mathbf{X}, \beta)$

$$\Gamma \cdot (\mathbf{X}, \beta) \subseteq \mathcal{R}_{pass(\mathbf{X}, \beta)}$$

Proof Indeed, the entries of the passports are Galois-invariant since they consist of the multiplicities of c-points of Belyi functions for the Galois-invariant points of $\mathbf{P}_1(\Bbbk)$.   □

This obvious fact should be taken into account together with the following

Observation *"Generically"* $\Gamma \cdot (\mathbf{X}, \beta) = \mathcal{R}_{\mathrm{pass}(\mathbf{X}, \beta)}$

Of course, this equality holds only in the absence of more subtle Galois-invariants—non-trivial automorphisms and others.

3 Dessins

This section is devoted to the objects whose relation with the objects studied in the previous one are far from obvious. This relation has been basically discovered by Alexander Grothendieck, see [13], and many papers and several books s were devoted to it. The books [17] and [11] are addressed to the beginners; however, we are going to use the different basic concepts, and the reason for it will be explained soon.

3.1 The Category of Grothendieck Dessins

The objects of the category $\mathcal{DESS}$ are *dessins d'enfant* in the sense of [13], i.e. such triples of topological spaces

$$X_0 \subset X_1 \subset X_2,$$

that X_0 is a non-empty finite set, whose elements are called *vertices*, X_2 is a compact connected orient*ed* surface and X_1 is an *embedded graph*, which means that the complement $X_1 \setminus X_0$ is homeomorphic to a disjoint union of real intervals, called *edges*. We demand as well that the complement $X_2 \setminus X_1$ is homeomorphic to a disjoint union of open discs, called *faces*. The difference between dessins and *two-dimensional cell complexes* lies in the concepts of *morphisms*.

In order to give a short definition of morphisms in $\mathcal{DESS}$, we add $X_{-1} = \varnothing$ to each triple as above and call a continuous mapping of surfaces *admissible*, if it respects the orientation, is *open*[2] and respects the differences, i.e. such a mapping of triples $f : (X_2, X_1, X_0) \to (Y_2, Y_1, Y_0)$ should satisfy

$$f(X_i \setminus X_j) \subseteq Y_i \setminus Y_j$$

---

[2]According to the somewhat forgotten theory, developed by S. Stoilow, any open mapping of Riemann surfaces is locally topologically conjugated to a holomorphic one, see [26].

for $-1 \leq j < i \leq 3$. The two admissible mappings are called *admissibly equivalent*, if they are homotopic in the class of admissible mappings, and the morphisms in $\mathcal{DESS}$ are defined as classes of admissible equivalence of admissible mappings.

3.2 The Category of Colored Triangulations

The objects of the category $\mathcal{DESS}_3$ are the *tricolored* dessins, i.e. the dessins $X_0 \subset X_1 \subset X_2$ endowed with a *coloring mapping*

$$\mathrm{col}_3 : X_1 \longrightarrow \{blue, green, red\},$$

constant on the edges. It is demanded that

(0) any vertex is incident to edges of only <u>two</u> colors;
(1) any edge has <u>two</u> vertices in its closure;
(2) any face has <u>three</u> edges in its closure, colored pairwise differently.

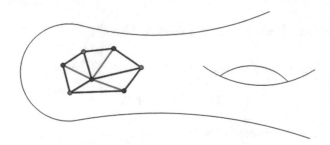

Taking into account the assumption **(0)**, we color every vertex by the (only remaining) color, that is different from the colors of incident edges. Due to the assumption **(2)** the connected components of $X_2 \setminus X_1$ will be called (topological) *triangles*. It can be deduced from the *orientability* of X_2 that these triangles can also be colored, now in *black* and *white*, in such a way that the *neighboring* triangles—i.e., having a common edge—will be colored differently.

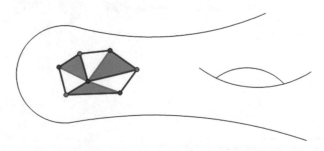

So the coloring mapping col_3 can be extended to

$$col_5 : X_2 \longrightarrow \{black, blue, green, red, white\},$$

with exactly two choices of black/white coloring, corresponding to the orientations of X_2. We agree that the positive-counter-clockwise orientation of the white triangles corresponds to the *blue-green-red-blue* cyclic order of the colors of edges in its closure; this choice will be motivated below.

The objects of $\mathcal{DESS}_3$ will be called *colored triangulations*; we note, however, that there is precisely one object of this category, that is <u>not</u> a triangulation of a surface in the usual sense; this object is formed by a pair of black and white triangles with colored edges after identifying edges with the same color.

The morphisms in $\mathcal{DESS}_3$ are defined in the same way as in $\mathcal{DESS}$ with the additional assumption of *color-respecting*.[3]

The theory is fundamentally symmetric with respect to the three colors involved; this is the reason why the traditional approach, developed in [17] and [11], where two of them are distinguished, does not satisfy us completely.

3.3   Relations Between Two Types of Dessins

There is an obvious color-forgetting functor

$$\mathcal{DESS}_3 \longrightarrow \mathcal{DESS}.$$

In the other direction there is a non-trivial one

$$\mathcal{DESS} \hookrightarrow \mathcal{DESS}_3,$$

which we introduce by the picture:

---

[3]The "same" category was considered in [15] under the name *oriented hypermaps*; our vertexes of three colors were called *hypervertices, hyperedges* and *hyperfaces*.

3.4 Counting Dessins

For several decades the powerful "physical" methods are used in the study of the quantities of dessins of bounded complexity; the corresponding key words are *matrix integrals* and *map enumeration*. The progress is still impressive. E.g., recently the generating function for the weighted[4] quantities of dessins with the prescribed set of degrees of 2-valencies has been (in a certain sense) written down— see [16].

However, the quantities of dessins with prescribed sets of both 0- and 2- valencies are still out of reach. As it will follow from the results of the next section, this problem is equivalent to counting Belyi pairs over $\mathbb{C}$ with a prescribed passport.

4 Correspondence Between Belyi Pairs and Dessins

In this section we work over $\mathbb{k} = \mathbb{C}$.

4.1 The Functor "draw"

We define the functor

$$\mathbf{draw} : \mathcal{BELP}_2(\mathbb{C}) \longrightarrow \mathcal{DESS}.$$

To a clean Belyi pair $(\mathbf{X}, \beta) \in \mathcal{BELP}_2(\mathbb{C})$ a dessin d'enfant with

$$X_2 := \mathbf{top}(\mathbf{X})$$

is assigned; here **top** means the forgetful functor that assigns to a complex algebraic curve (= Riemann surface) the underlying topological oriented surface.

Next define

$$X_1 := \beta^{-1\circ}([0, 1]) \text{ and } X_0 := \beta^{-1\circ}(\{0\}).$$

The branching condition imposed on β over 1 implies that while $P \in X_2$ moves along some edge (a connected component of $X_1 \setminus X_0$) from one vertex (an element of X_0) to another, the point $\beta(P)$ moves from 0 to 1 and back, the edge being *folded* in the point of $\beta^{-1\circ}(1)$; a local coordinate z centered at this point can be chosen so that $\beta = 1 + z^2$ in its domain.

---

[4]A dessin D is counted with the weight $\frac{1}{\#\mathrm{Aut}D}$.

A morphism of Belyi pairs obviously defines the corresponding morphism of dessins.

Theorem 2 *The functor* **draw** *defines the equivalence of the categories* $\mathcal{BELP}_2(\mathbb{C})$ *and* $\mathcal{DESS}$.

A detailed proof can be found in [22].

4.2 The Functor "paint"

In order to define the functor

$$\textbf{paint} : \mathcal{BELP}(\mathbb{C}) \longrightarrow \mathcal{DESS}_3;$$

we introduce the *Belyi sphere* $\mathbf{P}_1(\mathbb{C})^{\text{Bel}}$ which is the *colored Riemann sphere* $\mathbf{P}_1(\mathbb{C})$. Decomposing $\mathbf{P}_1(\mathbb{C}) = \mathbb{C} \coprod \{\infty\}$, we define this coloring as

$$\text{col}_5^{\text{Bel}} : \mathbf{P}_1(\mathbb{C}) \longrightarrow \{black, blue, green, red, white\} :$$

$$z \mapsto \begin{cases} black \text{ if } z \in \mathbb{C} \setminus \mathbb{R} \text{ and } \Im z < 0, \\ white \text{ if } z \in \mathbb{C} \setminus \mathbb{R} \text{ and } \Im z > 0, \\ blue \text{ if } z \in \mathbb{R}_{<0} \text{ or } z = 1, \\ green \text{ if } z \in (0, 1) \text{ or } z = \infty, \\ red \text{ if } z \in \mathbb{R}_{>1} \text{ or } z = 0. \end{cases}$$

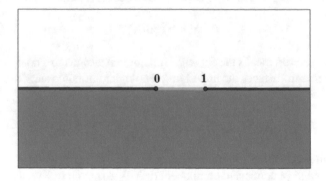

The choice of the colors is motivated as follows. The *black* and *white* for the lower and the upper parts is quite traditional (hell and heaven...), while the real line is colored in such a way that *blue* (symbolizing *cold*) corresponds to negative numbers, while *red* (symbolizing *hot*) corresponds to positive ones. The *green* is just in between and is assigned no meaningful association. The vertices of the colored topological "triangle" $\mathbf{P}_1(\mathbb{R})$ have the same color as the opposite side.

Furthermore, the colors of the pieces of the real line occur in the *alphabetical* order. The motivation of the choice of "colored" orientation can be given now: the traditional counter-clockwise detour around the <u>white</u> triangle correspond to moving along the real line form $-\infty$ to ∞.

Now we can finalize the definition of the functor **paint**: for a Belyi pair $(\mathbf{X}, \beta)$ the surface $X_2 := \mathbf{top}(\mathbf{X})$ is colored by $\mathrm{col}_5 := \beta^* \mathrm{col}_5^{\mathsf{Bel}}$, i.e. the points of the surface are colored according to the colors of their images under the Belyi mapping: for any $P \in X_2$

$$\mathrm{col}_5(P) := \mathrm{col}_5^{\mathsf{Bel}}(\beta(P)).$$

Obviously, the set X_1 turns out to be the closure of the union of the green edges and X_0 the set of isolated red points.

Theorem 3 *The functor **paint** defines the equivalence of the categories $\mathcal{BELP}(\mathbb{C})$ and $\mathcal{DESS}_3$.*

A detailed proof can be found in [22].

4.3 Implications of Belyi Theorem

According to the above-quoted theorem, the category inclusion

$$\mathcal{BP}_2(\overline{\mathbb{Q}}) \xrightarrow{\simeq} \mathcal{BP}_2(\mathbb{C})$$

is *a* category equivalence. We emphasize that it is <u>not canonical</u>: introduce the *absolute Galois group*

$$\Gamma := \mathrm{Aut}(\overline{\mathbb{Q}})$$

and note that the inclusion $\overline{\mathbb{Q}} \hookrightarrow \mathbb{C}$ is defined only up to the Γ-action.

According to the previously formulated results, we have the Γ-set of category equivalences

$$\boxed{\mathcal{BP}_2(\overline{\mathbb{Q}}) \xleftrightarrow{\simeq} \mathcal{DESS}}$$

and, as we have just seen, it has some invariant meaning only being considered together with the enigmatic action of Γ on $\mathcal{DESS}$.

So the true arithmetic meaning can be given not to individual dessins, but only to their Γ-orbits.

5   Belyi Pairs over Finite Fields

The theory is in its infancy. However, it is inevitable, and we start this last section
with the demonstration of the occurrence of Belyi pairs over $\mathbb{F}_p$'s in the course of
the constructive realization of the equivalence $\mathcal{BP}_2(\overline{\mathbb{Q}}) \xleftrightarrow{\simeq} \mathcal{DESS}$.

5.1   Example

The Belyi pairs, corresponding to the clean unicellular 4-edged toric dessins, were
calculated in [23]. In the course of calculations it was impossible to ignore the flows
of powers of small primes in the denominators. It turned out that in all the cases these
primes have the invariant meaning: they are the *bad primes* of the corresponding
elliptic curves, see [25]. The results are summarized in the following table.

Dessins	Bad primes
	2,3
	2
	2,3
	3,7
	2,5,7
	3
	2,3,7

Here all the toric dessins are drawn either in the square or in the hexagon; it is
meant that the opposite sides are identified. They are grouped in the raws according
to the sets of valencies; these raws constitute the Galois orbits, except the two cases
(in the second and the penultimate raws) where the Galois orbits are split due to the
obvious symmetries.

In most cases the bad primes have an obvious combinatorial meaning; they divide one of the valencies. However, the occurrence of 7 can not be explained this way. Instead we see the *sum of valencies* phenomenon: the badness of 7 is explained by $7 = 2 + 5$ and $7 = 3 + 4$. The similar phenomenon in the case of plane trees was explained by the author's students [27] and [20].

5.2   Good and Bad Primes

This subsection is written in a somewhat informal style, since some details of the corresponding concepts have not yet been written up (however, see [12]).

If for a Belyi pair $(\mathbf{X}, \beta)$ over $\overline{\mathbb{Q}}$ both the equations of $\mathbf{X}$ and the coefficients of β can be chosen in a finite extension $\mathbb{K} \supseteq \mathbb{Q}$, such a field $\mathbb{K}$ is called a *field of realization* of $(\mathbf{X}, \beta)$. Let $\mathcal{O}$ be the *ring of integers* of $\mathbb{K}$; it is clear that $(\mathbf{X}, \beta)$ then can be *realized* over $\mathcal{O}$ (nobody claims any kind of *uniqueness* of such a realization).

Given a nonzero prime ideal $\mathfrak{p} \lhd \mathcal{O}$, we can construct the pair $(\mathbf{X}, \beta) \bmod \mathfrak{p}$ over the algebraic closure of the finite field $\frac{\mathcal{O}}{\mathfrak{p}}$. If the curve $\mathbf{X} \bmod \mathfrak{p}$ is smooth (or, equivalently, has the same genus as $\mathbf{X}$) and $\deg(\beta \bmod \mathfrak{p}) = \deg(\beta)$ then $\mathbb{K}$, a model and $\mathfrak{p}$ are called *good* for $(\mathbf{X}, \beta)$. A prime p is called *good* for $(\mathbf{X}, \beta)$, if such a good choice exists with $\mathrm{char}\left(\frac{\mathcal{O}}{\mathfrak{p}}\right) = p$. Otherwise p is *bad* for $(\mathbf{X}, \beta)$.

For a dessin D denote $(\mathbf{X}_D, \beta_D)$ the corresponding Belyi pair over $\overline{\mathbb{Q}}$ and introduce the *set of primes of bad reduction*

$$\mathrm{bad}_D := \{p \in \{2, 3, 5, \dots\} \mid p \text{ is bad for } (\mathbf{X}_D, \beta_D).$$

As for many other objects of arithmetic geometry, all the sets bad_D are *finite*.

In the previous subsection the sets of bad primes were presented without any attempts of precise definitions; the reason is that in the case of genus 1 the prime is bad if and only if it divides the *discriminant* of the curve.

It is an outstanding problem

to define the sets bad_D *in terms of the combinatorics of* D.

5.3   Counting

The author is currently unaware of the passports that are realizable over $\overline{\mathbb{F}_p}$ but not realizable over $\mathbb{C}$. Hence typically

$$\#\mathcal{R}_\Pi(\mathbb{F}_p) \leq \#\mathcal{R}_\Pi(\mathbb{C}),$$

and the inequality often becomes strict due to the bad reduction.

The numbers of bad reductions often behave systematically in *families* of dessins. Unfortunately, a mathematical definition of a family of dessins (similar, say, to the definition of a family of algebraic varieties) hardly exists, so we just consider an example.

The passports $\begin{pmatrix} n & n & 3 \\ & & 1 \\ & & \cdots \\ & & 1 \end{pmatrix}$ with natural $n \geq 3$ correspond to the unicellular toric

dessins

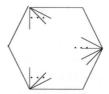

(the opposite sides identified). In terms of [5] these are the dessins whose *pruning* is a *toric hexagon*, defined by the passport (333).

Now, using the notation $\#\#\mathcal{Z} := \sum_{z \in \mathcal{Z}} \frac{1}{\#\mathrm{Aut}_z}$ for the weighted sum, introduce for a field $\Bbbk$

$$\mathrm{Hex}_n(\Bbbk) := \#\#\mathcal{R}_{\begin{pmatrix} n & n & 3 \\ & & 1 \\ & & \cdots \\ & & 1 \end{pmatrix}}(\Bbbk),$$

and give the promised example:

$$\mathrm{Hex}_n(\mathbb{C}) - \mathrm{Hex}_n(\overline{\mathbb{F}_\mathrm{i}}) = \sum_{0 < k < \frac{n}{p}} (n - kp).$$

The proof can be found in [21]. The summing of the arithmetic progression in the right-hand side has not been performed in order to emphasize the nature of the bad reduction which is explained in terms of geometry of the modular curves.

5.4   On the Cohomology of Moduli Spaces

The geometry of moduli spaces $\mathcal{M}_{g,N}$ of N-pointed curves of genus g is related to dessins in more than one way. The famous decomposition (constructed by Mumford, Harer, Penner, Witten and others)

$$\mathcal{M}_{g,N}(\mathbb{C}) \times \mathbb{R}^N_{>0} \simeq \coprod_{D \in \mathbf{DESS}_{g,N}} \mathbb{R}^{E(D)}_{>0},$$

where $\mathbf{DESS}_{g,N}$ stands for the set of isomorphism classes of N-cellular dessins of genus g with all the 0-valencies ≥ 3 and $E(D)$ is the set of edges of a dessin D, provides a direct way to the singular cohomology of $\mathcal{M}_{g,N}(\mathbb{C})$. In the case $(g, N) = (2, 1)$ this approach (modified a bit for a level-3 smooth cover) was realized in [6].

In [18] it was shown that replacing $\mathbb{R}_{>0}$ be $\mathbb{N}$ (i.e. considering ribbon graphs with only integer edge lengths) results in replacing $\mathbb{C}$ by $\overline{\mathbb{Q}}$, so more "arithmetic" cohomology theories become available. The Witten-Kontsevich integrals then are replaced by counting the integral points in the polytopes, the perfect techniques for which was developed in [19].

The methods of calculating cohomology of moduli spaces by *counting curves* over finite fields, i.e. determining $\#\mathcal{M}_{g,N}(\mathbb{F}_{p^r})$, are based on the (now proved) Weil conjectures. The applications of these methods can be found for example in [3].

Since counting Belyi pairs is closely related to counting curves and counting dessins (together with the principles of bad reduction, the first steps of understanding which were mentioned above), there is a fundamental hope of blending all these approaches.

Acknowledgement The paper is supported in part by the Simons foundation.

References

1. Belyi, G.V.: Galois extensions of a maximal cyclotomic fields. Math. USSR Izv. **14**(2), 247–256 (1980)
2. Belyi, G.V.: A new proof of the three-point theorem. Mat. Sb. **193**(3), 21–24 (2002)
3. Bergstrom J., Tommasi O.: The rational cohomology of $\mathcal{M}_4$. Math. Ann. **338**(1), 207–239 (2007)
4. Couveignes, J.-M.: Calcul et rationalité de fonctions de Belyi en genre 0. Annales de l'inst. Fourier **44**(1), 1–38 (1994)
5. Do, N., Norbury, P.: Pruned Hurwitz numbers (2013). arXiv:1312.7516 [math.GT]
6. Dunin-Barkowski, P., Popolitov, A., Shabat, G., Sleptsov, A.: On the homology of certain smooth covers of moduli spaces of algebraic curves. Differ. Geom. Appl. **40**, 86–102 (2015)
7. Elkies, N.D.: The Klein quartic in number theory. In: The Eightfold Way: The Beauty of Klein's Quartic Curve, pp. 51–102. Cambridge University Press, Cambridge (1999)
8. Filimonenkov, V.O., Shabat, G.B.: Fields of definition of rational functions of one variable with three critical values. Fundam. Prikl. Mat. **1**(3), 781–799 (1995)
9. Frenkel, E.: Lectures on the Langlands Program and Conformal Field Theory. In: Cartier, P., Moussa, P., Julia, B., Vanhove, P. (eds.) Frontiers in Number Theory, Physics, and Geometry II. Springer, Berlin, Heidelberg (2007)
10. Gannon, T.: Moonshine Beyond the Monster: The Bridge Connecting Algebra, Modular Forms and Physics. Cambridge University Press, Cambridge (2006)
11. Girondo, E., Gonzalez-Diez, G.: Introduction to Compact Riemann Surfaces and Dessins d'Enfants. London Mathematical Society Student Texts. Cambridge University Press, Cambridge (2012)
12. Goldring, W.: Unifying themes suggested by Belyi's theorem. In: Number Theory, Analysis and Geometry (Serge Lang Memorial Volume), pp. 181–214. Springer, Boston, MA (2011)
13. Grothendieck, A.: Esquisse d'un Programme. Unpublished manuscript (1984). English translation by Lochak, P., Schneps, L. in Geometric Galois actions, vol. 1. London Mathematical Society Lecture Note Series, vol. 242, pp. 5–48. Cambridge University Press, Cambridge (1997)

14. Javanpeykar, A., Bruin, P.: Polynomial bounds for Arakelov invariants of Belyi curves. Algebr. Number Theory **8**(1), 89–140 (2014)
15. Jones, G., Singerman D.: Maps, hypermaps and triangle groups. In: The Grothendieck Theory of Dessins d'Enfant. London Mathematical Society Lecture Note Series, vol. 200, pp.115–146. Cambridge University Press, Cambridge (1994)
16. Kazarian, M., Zograf, P.: Virasoro constraints and topological recursion for Grothendieck's dessin counting. Lett. Math. Phys. **105**(8), 1057–1084 (2015)
17. Lando, S., Zvonkin, A.: Graphs on Surfaces and Their Applications. Springer, Berlin, Heidelberg (2004)
18. Mulase, M., Penkava, M.: Ribbon graphs, quadratic differentials on Riemann surfaces, and algebraic curves defined over $\overline{\mathbb{Q}}$. Asian J. Math. **2**(4), 875–920 (1998)
19. Norbury, P.: Counting lattice points in the moduli space of curves. Math. Res. Lett. **17**, 467–481 (2010)
20. Oganesyan, D.: Zolotarev polynomials and reduction of Shabat polynomials into a positive characteristic. Mosc. Univ. Math. Bull. **71**(6), 248–252 (2016)
21. Oganesyan, D.: Abel pairs and modular curves. Zap. Nauchn. Sem. POMI **446**, 165–181 (2016)
22. Shabat, G.: Combinatorial-topological methods in the theory of algebraic curves. Theses, Lomonosov Moscow State University (1998)
23. Shabat, G.B.: Unicellular four-edged toric dessins. Fundamentalnaya i prikladnaya matematika **18**(6), 209–222 (2013)
24. Shabat, G.B., Voevodsky, V.A.: Drawing curves over number fields. In: Cartier, P., Illusie, L., Katz, N., Laumon, G., Manin, Y., Ribet, K. (eds.) The Grothendieck Festschrift, vol. 3, 5th edn., pp. 199–227. Birkhauser, Basel (1990)
25. Silverman, J.H.: The Arithmetic of Elliptic Curves. Graduate Texts in Mathematics, vol. 106, 2nd edn. Springer, New York (2009)
26. Stoilow, S.: Leçons sur les Principes Topologiques de la Théorie des Fonctions Analytiques. Gauthier-Villars, Paris (1956)
27. Vashevnik, A.M.: Prime numbers of bad reduction for dessins of genus 0. J. Math. Sci. **142**(2), 1883–1894 (2007)
28. Weber, M.: Kepler's small stellated dodecahedron as a Riemann surface. Pac. J. Math. **220**, 167–182 (2005)

Unravelling the Dodecahedral Spaces

Jonathan Spreer and Stephan Tillmann

Abstract The hyperbolic dodecahedral space of Weber and Seifert has a natural non-positively curved cubulation obtained by subdividing the dodecahedron into cubes. We show that the hyperbolic dodecahedral space has a 6-sheeted irregular cover with the property that the canonical hypersurfaces made up of the mid-cubes give a very short hierarchy. Moreover, we describe a 60-sheeted cover in which the associated cubulation is special. We also describe the natural cubulation and covers of the spherical dodecahedral space (aka Poincaré homology sphere).

1 Introduction

A *cubing* of a 3-manifold M is a decomposition of M into Euclidean cubes identified along their faces by Euclidean isometries. This gives M a singular Euclidean metric, with the singular set contained in the union of all edges. The cubing is *non-positively curved* if the dihedral angle along each edge in M is at least 2π and each vertex satisfies Gromov's *link condition*: The link of each vertex is a triangulated sphere in which each 1-cycle consists of at least three edges, and if a 1-cycle consists of exactly three edges, then it bounds a unique triangle. In this case, we say that M has an *NPC cubing*.

The universal cover of an NPC cubed 3-manifold is CAT(0). Aitchison et al. [5] showed by a direct construction that if each edge in an NPC cubed 3-manifold has even degree, then the manifold is virtually Haken. Moreover, Aitchison and

J. Spreer (✉)
Discrete Geometry Group, Mathematical Institute, Freie Universität Berlin, Arnimallee 2,
14195 Berlin, Germany
e-mail: jonathan.spreer@fu-berlin.de

S. Tillmann
School of Mathematics and Statistics F07, The University of Sydney, Camperdown,
NSW 2006, Australia
e-mail: stephan.tillmann@sydney.edu.au

Rubinstein [3] showed that if each edge degree in such a cubing is a multiple of four, then the manifold is virtually fibred.

A cube contains three canonical squares (or 2-dimensional cubes), each of which is parallel to two sides of the cube and cuts the cube into equal parts. These are called *mid-cubes*. The collection of all mid-cubes gives an immersed surface in the cubed 3-manifold M, called the *canonical (immersed) surface*. If the cubing is NPC, then each connected component of this immersed surface is π_1-injective. If one could show that one of these surface subgroups is separable in $\pi_1(M)$, then a well-known argument due to Scott [14] shows that there is a finite cover of M containing an embedded π_1-injective surface, and hence M is virtually Haken. In the case where the cube complex is *special* (see Sect. 2), a *canonical completion and retraction construction* due to Haglund and Wise [11] shows that these surface subgroups are indeed separable because the surfaces are convex. Whence a 3-manifold with a special NPC cubing is virtually Haken. The missing piece is thus to show that an NPC cubed 3-manifold has a finite cover such that the lifted cubing is special. This is achieved in the case where the fundamental group of the 3-manifold is hyperbolic by the following cornerstone in Agol's proof of Waldhausen's Virtual Haken Conjecture from 1968:

Theorem 1 (Virtual Special; Agol [2], Thm 1.1) *Let G be a hyperbolic group which acts properly and cocompactly on a CAT(0) cube complex X. Then G has a finite index subgroup F so that X/F is a special cube complex.*

In general, it is known through work of Bergeron and Wise [6] that if M is a closed hyperbolic 3-manifold, then $\pi_1(M)$ is isomorphic to the fundamental group of an NPC cube complex. However, the dimension of this cube complex may be arbitrarily large and it may not be a manifold. Agol's theorem provides a finite cover that is a special cube complex, and the π_1-injective surfaces of Kahn and Markovic [13] are quasi-convex and hence have separable fundamental group. Thus, the above outline completes a sketch of the proof that M is virtually Haken. An embedding theorem of Haglund and Wise [11] and Agol's virtual fibring criterion [1] then imply that M is also virtually fibred.

Weber and Seifert [16] described two closed 3-manifolds that are obtained by taking a regular dodecahedron in a space of constant curvature and identifying opposite sides by isometries. One is hyperbolic and known as the *Weber-Seifert dodecahedral space* and the other is spherical and known as the *Poincaré homology sphere*. Moreover, antipodal identification on the boundary of the dodecahedron yields a third closed 3-manifold which naturally fits into this family: the real projective space.

The dodecahedron has a natural decomposition into 20 cubes, which is a NPC cubing in the case of the Weber-Seifert dodecahedral space. The main result of this note can be stated as follows.

Theorem 2 *The hyperbolic dodecahedral space WS of Weber and Seifert admits a cover of degree 60 in which the lifted natural cubulation of WS is special.*

In addition, we exhibit a 6-sheeted cover of WS in which the canonical immersed surface consists of six embedded surface components and thus gives a very short hierarchy of $\widetilde{\text{WS}}$. The special cover from Theorem 2 is the smallest regular cover of WS that is also a cover of this 6-sheeted cover. Moreover, it is the smallest regular cover of WS that is also a cover of the 5-sheeted cover with positive first Betti number described by Hempel [12].

We conclude this introduction by giving an outline of this note. The dodecahedral spaces are described in Sect. 3. Covers of the hyperbolic dodecahedral space are described in Sect. 4, and all covers of the spherical dodecahedral space and the real projective space in Sect. 5.

2 Cube Complexes, Injective Surfaces and Hierarchies

A *cube complex* is a space obtained by gluing Euclidean cubes of edge length one along subcubes. A cube complex is $CAT(0)$ if it is $CAT(0)$ as a metric space, and it is *non-positively curved (NPC)* if its universal cover is $CAT(0)$. Gromov observed that a cube complex is NPC if and only if the link of each vertex is a flag complex.

We identify each n-cube as a copy of $[-\frac{1}{2}, \frac{1}{2}]^n$. A mid-cube in $[-\frac{1}{2}, \frac{1}{2}]^n$ is the intersection with a coordinate plane $x_k = 0$. If X is a cube complex, then a new cube complex Y is formed by taking one $(n-1)$-cube for each midcube of X and identifying these $(n-1)$-cubes along faces according to the intersections of faces of the corresponding n-cubes. The connected components of Y are the *hyperplanes* of X, and each hyperplane H comes with a canonical immersion $H \to X$. The image of the immersion is termed an *immersed hyperplane* in X. If X is $CAT(0)$, then each hyperplane is totally geodesic and hence embedded.

The NPC cube complex X is *special* if

1. Each immersed hyperplane embeds in X (and hence the term "immersed" will henceforth be omitted).
2. Each hyperplane is 2-sided.
3. No hyperplane self-osculates.
4. No two hyperplanes inter-osculate.

The prohibited pathologies are shown in Fig. 1 and are explained now. An edge in X is dual to a mid-cube if it intersects the midcube. We say that the edge of X is dual to the hyperplane H if it intersects its image in X. The hyperplane dual to edge a is unique and denoted $H(a)$. Suppose the immersed hyperplane is embedded. It

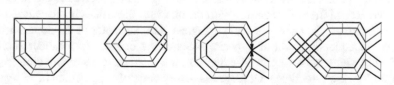

Fig. 1 Not embedded; 1-sided; self-osculating; inter-osculating (compare to [17, Figure 4.2])

is 2-sided if one can consistently orient all dual edges so that all edges on opposite sides of a square have the same direction. Using this direction on the edges, H *self-osculates* if it is dual to two distinct edges with the same initial or terminal vertex. Hyperplanes H_1 and H_2 *inter-osculate* if they cross and they have dual edges that share a vertex but do not lie in a common square.

The situation is particularly nice in the case where the NPC cube complex X is homeomorphic to a 3-manifold. Work of Aitchison and Rubinstein (see §3 in [4]) shows that each immersed hyperplane is mapped π_1-injectively into X. Hence if one hyperplane is embedded and 2-sided, then X is a Haken 3-manifold. Moreover, if each hyperplane embeds and is 2-sided, then one obtains a hierarchy for X. This is well-known and implicit in [4]. One may first cut along a maximal union of pairwise disjoint hypersurfaces to obtain a manifold X_1 (possibly disconnected) with incompressible boundary. Then each of the remaining hypersurfaces gives a properly embedded surface in X_1 that is incompressible and boundary incompressible. This process iterates until one has cut open X along all the mid-cubes, and hence it terminates with a collection of balls. In particular, if Y consists of three pairwise disjoint (not necessarily connected) surfaces, each of which is embedded and 2-sided, then one has a very short hierarchy.

3   The Dodecahedral Spaces

The main topic of this paper is a study of low-degree covers of the hyperbolic dodecahedral space. However, we also take the opportunity to extend this study to the spherical dodecahedral space in the hope that this will be a useful reference. When the sides are viewed combinatorially, there is a third dodecahedral space which naturally fits into this family and again gives a spherical space form: the real projective space. The combinatorics of these spaces is described in this section.

3.1   The Weber-Seifert Dodecahedral Space

The Weber-Seifert Dodecahedral space WS is obtained by gluing the opposite faces of a dodecahedron with a $3\pi/5$-twist. This yields a decomposition $\mathscr{D}_{WS}$ of the space into one vertex, six edges, six pentagons, and one cell (see Fig. 2 on the left). The dodecahedron can be decomposed into 20 cubes by (a) placing a vertex at the centre of each edge, face, and the dodecahedron, and (b) placing each cube around one of the 20 vertices of the dodecahedron with the other seven vertices in the centres of the three adjacent edges, three adjacent pentagons, and the center of the dodecahedron. Observe that identification of opposite faces of the original dodecahedron with a $3\pi/5$-twist yields a 14-vertex, 54-edge, 60 square, 20-cube decomposition $\hat{\mathscr{D}}_{WS}$ of WS (see Fig. 2 on the right). Observe that every edge of $\hat{\mathscr{D}}_{WS}$ occurs in ≥ 4 cubes, and each vertex satisfies the link condition. We therefore have an NPC cubing.

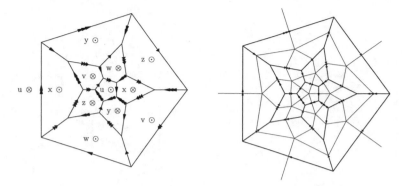

Fig. 2 Left: face and edge-identifications on the dodecahedron yielding the Weber-Seifert dodecahedron space. Right: decomposition of the Weber-Seifert dodecahedral space into 20 cubes

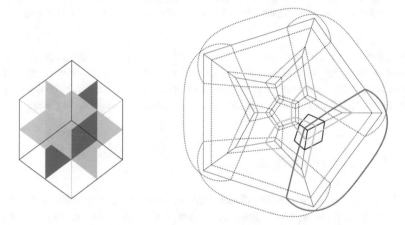

Fig. 3 Left: Immersed canonical surface in one cube. Right: intersection pattern of one cube, the
immersed canonical surface, and the boundary of the dodecahedron in $\mathscr{D}_{\mathrm{WS}}$

The mid-cubes form pentagons parallel to the faces of the dodecahedron, and
under the face pairings glue up to give a 2-sided immersed surface of genus four.
The immersion of this canonical surface into $\mathscr{D}_{\mathrm{WS}}$ in the neighbourhood of one
cube is shown in Fig. 3. We wish to construct a cover in which the canonical surface
splits into embedded components—which neither self-osculate with themselves, nor
inter-osculate with other surface components.

3.2   The Poincaré Homology Sphere

The Poincaré homology sphere Σ^3 is obtained from the dodecahedron by gluing
opposite faces by a $\pi/5$-twist. This results in a decomposition $\mathscr{D}_{\Sigma^3}$ of Σ^3 into one

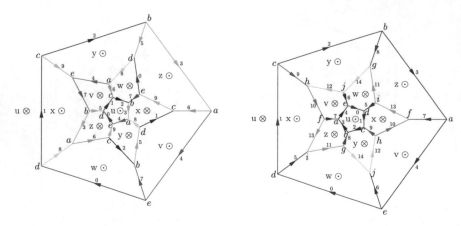

Fig. 4 Left: face and edge-identifications on the dodecahedron yielding the Poincaré homology sphere. Right: face and edge-identifications on the dodecahedron yielding the real projective space

vertex, ten edges, six pentagons, and one cell (see Fig. 4 on the left). Again, we can decompose $\mathscr{D}_{\Sigma^3}$ into 20 cubes. Note, however, that in this case some of the cube-edges only have degree three (the ones coming from the edges of the original dodecahedron). This is to be expected since Σ^3 supports a spherical geometry.

3.3 Real Projective Space

Identifying opposite faces of the dodecahedron by a twist of π results in identifying antipodal points of a 3-ball (see Fig. 4 on the right). Hence, the result is a decomposition $\mathscr{D}_{\mathbb{R}P^3}$ of $\mathbb{R}P^3$ into ten vertices, 15 edges, six faces, and one cell. As in the above cases, this decomposition can be decomposed into 20 cubes, with some of the cube-edges being of degree two.

4   Covers of the Weber-Seifert Space

In order to obtain a complete list of all small covers of the Weber-Seifert space WS, we need a list of all low index subgroups of $\pi_1(WS)$ in a presentation compatible with $\mathscr{D}_{WS}$ and its cube decomposition $\hat{\mathscr{D}}_{WS}$.

The complex $\mathscr{D}_{WS}$ has six pentagons u, v, w, x, y, and z. These correspond to antipodal pairs of pentagons in the original dodecahedron, see Fig. 2 on the left. Passing to the dual decomposition, these six pentagons corresponds to loops which naturally generate $\pi_1(WS)$. The six edges of $\mathscr{D}_{WS}$ ⵏ, ⵏ, ⵏ, ⵏ, ⵏ, and ⵏ each give rise to a relator in this presentation of the fundamental group of WS in the following way: fix edge ⵏ and start at a pentagon containing ⵏ, say u. We start at the pentagon labelled u with a back of an arrow $\otimes$—the outside in Fig. 2 on the left. We traverse the dodecahedron, resurface on the other pentagon labelled u with an arrowhead $\odot$ (the innermost pentagon in Fig. 2). We then continue with the unique pentagon adjacent to the center pentagon along edge $\dagger$. In this case v labelled with the tail of an arrow, we traverse the dodecahedron, resurface at $(v, \odot)$, and continue with $(w, \odot)$ which we follow through the dodecahedron in reverse direction, and so on. After five such traversals we end up at the outer face where we started. The relator is now given by the labels of the pentagons we encountered, taking into account their orientation (arrowhead or tail). In this case the relator is $r(\dagger) = uvw^{-1}y^{-1}z$.

Altogether we are left with

$$\pi_1(\text{WS}) = \langle\, u, v, w, x, y, z \mid uxy^{-1}v^{-1}w, \, uyz^{-1}w^{-1}x, \, uzv^{-1}x^{-1}y, \\ uvw^{-1}y^{-1}z, \, uwx^{-1}z^{-1}v, \, vxzwy \,\rangle.$$

Using this particular representation of the fundamental group of the Weber-Seifert dodecahedral space we compute subgroups of $\pi_1(\text{WS})$ of index k $(k < 10)$ via GAP function `LowIndexSubgroupsFpGroup` [10], and Magma function `LowIndexSubgroups` [7] and use their structure to obtain explicit descriptions of their coset actions (using GAP function `FactorCosetAction` [10]) which, in turn, can be transformed into a gluing table of k copies of the dodecahedron (or $20k$ copies of the cube). Given such a particular decomposition, we can track how the canonical surface evolves and whether it splits into embedded components.

We provide a *GAP* script for download from [15]. The script takes a list of subgroups as input (presented each by a list of generators from $\pi_1(\text{WS})$) and computes an array of data associated to the corresponding covers of $\mathcal{D}_{\text{WS}}$. The script comes with a sample input file containing all subgroups of $\pi_1(\text{WS})$ of index less than ten. The subgroups are presented in a form compatible with the definition of $\pi(\text{WS})$ discussed above.

4.1 Covers of Degree Up to Five

A computer search reveals that there are no covers of degrees 2, 3, and 4, and 38 covers of degree 5. Their homology groups are listed in Table 1. For none of them, the canonical surface splits into embedded components. Moreover, in all but one case it does not even split into multiple immersed components, with the exception being the 5-sheeted cover with positive first Betti number described by Hempel [12], where it splits into five immersed components.

4.2 Covers of Degree Six

There are 61 covers of degree six, for 60 of which the canonical surface does not split into multiple connected components (see Table 1 below for their first homology

Table 1 First homology groups of all 490 covers of degree up to nine

Degree	$H_1(X)$	$\beta_1(X)$	# surf. comp.	$\exists$ emb. surf. comp.	# of covers
1	$\mathbb{Z}_5^3$	0	1	No	1
					$\Sigma = 1$
5	$\mathbb{Z}_5^2 \oplus \mathbb{Z}_{25}^2$	0	1	No	25
	$\mathbb{Z}_3 \oplus \mathbb{Z}_5 \oplus \mathbb{Z}_{25}^3$	0	1	No	6
	$\mathbb{Z}_5^6 \oplus \mathbb{Z}_{25}$	0	1	No	6
	$\mathbb{Z}^4 \oplus \mathbb{Z}_3^2$	4	5	No	1
					$\Sigma = 38$
6	$\mathbb{Z}_4 \oplus \mathbb{Z}_5^3$	0	1	No	6
	$\mathbb{Z}_3^2 \oplus \mathbb{Z}_4 \oplus \mathbb{Z}_5^3$	0	1	No	15
	$\mathbb{Z}_3 \oplus \mathbb{Z}_5^3 \oplus \mathbb{Z}_{11}^2$	0	1	No	24
	$\mathbb{Z}_3^2 \oplus \mathbb{Z}_4 \oplus \mathbb{Z}_5^3 \oplus \mathbb{Z}_{16}^2$	0	1	No	15
	$\mathbb{Z}^5 \oplus \mathbb{Z}_2^2 \oplus \mathbb{Z}_5^3$	5	6	Yes (all)	1
					$\Sigma = 61$
7	$\mathbb{Z}_2^3 \oplus \mathbb{Z}_5^3$	0	1	No	20
	$\mathbb{Z}_3^2 \oplus \mathbb{Z}_5^3 \oplus \mathbb{Z}_7 \oplus \mathbb{Z}_9 \oplus \mathbb{Z}_{11}$	0	1	No	30
					$\Sigma = 50$
8	$\mathbb{Z}_2^3 \oplus \mathbb{Z}_5^3$	0	1	No	40
	$\mathbb{Z}_2^3 \oplus \mathbb{Z}_3 \oplus \mathbb{Z}_5^3 \oplus \mathbb{Z}_9$	0	1	No	20
	$\mathbb{Z}_2 \oplus \mathbb{Z}_5^3 \oplus \mathbb{Z}_7^3$	0	1	No	40
	$\mathbb{Z}_3 \oplus \mathbb{Z}_4 \oplus \mathbb{Z}_5^3 \oplus \mathbb{Z}_{19}^2$	0	1	No	15
	$\mathbb{Z}_2^5 \oplus \mathbb{Z}_3 \oplus \mathbb{Z}_5^3 \oplus \mathbb{Z}_7^2$	0	1	No	10
	$\mathbb{Z} \oplus \mathbb{Z}_2^3 \oplus \mathbb{Z}_3^2 \oplus \mathbb{Z}_5^3$	1	2	No	20
	$\mathbb{Z} \oplus \mathbb{Z}_2^3 \oplus \mathbb{Z}_3 \oplus \mathbb{Z}_5^3 \oplus \mathbb{Z}_{13}^2$	1	1	No	40
					$\Sigma = 185$
9	$\mathbb{Z}_2 \oplus \mathbb{Z}_3 \oplus \mathbb{Z}_4^2 \oplus \mathbb{Z}_5^3$	0	1	No	60
	$\mathbb{Z}_2 \oplus \mathbb{Z}_3 \oplus \mathbb{Z}_5^3 \oplus \mathbb{Z}_8^2$	0	1	No	40
	$\mathbb{Z}_2^4 \oplus \mathbb{Z}_5^3 \oplus \mathbb{Z}_9 \oplus \mathbb{Z}_{89}$	0	1	No	15
	$\mathbb{Z} \oplus \mathbb{Z}_3 \oplus \mathbb{Z}_4 \oplus \mathbb{Z}_5^3 \oplus \mathbb{Z}_7$	1	2	Yes (one)	10
	$\mathbb{Z} \oplus \mathbb{Z}_3 \oplus \mathbb{Z}_4^2 \oplus \mathbb{Z}_5^3 \oplus \mathbb{Z}_{19}$	1	1	No	30
					$\Sigma = 155$

groups, obtained using GAP function AbelianInvariants [10]). However, the single remaining example leads to an irregular cover $\mathscr{C}$ with deck transformation group isomorphic to A_5, for which the canonical surface splits into six embedded components. The cover is thus a Haken cover (although this fact also follows from the first integral homology group of $\mathscr{C}$ which is isomorphic to $\mathbb{Z}^5 \oplus \mathbb{Z}_2^2 \oplus \mathbb{Z}_5^3$, see also Table 1), and the canonical surface defines a very short hierarchy.

The subgroup is generated by

$$u, \quad v^{-1}w^{-1}, \; w^{-1}x^{-1}, \; x^{-1}y^{-1}, \; y^{-1}z^{-1},$$
$$z^{-1}v^{-1}, \; vuy^{-1}, \; v^2z^{-1}, \; vwy^{-1}, \; vxv^{-1}$$

face	orbit
u	$(2,5,3,6,4)$
v	$(1,2,6,4,3)$
w	$(1,3,2,5,4)$
x	$(1,4,3,6,5)$
y	$(1,5,4,2,6)$
z	$(1,6,5,3,2)$

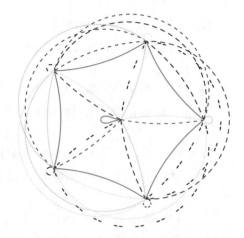

Fig. 5 Left: gluing orbits of face classes from $\mathcal{D}_{WS}$ in 6-sheeted Haken cover $\mathcal{C}$. Right: face pairing graph of $\mathcal{C}$. Colours encode face classes in the base $\mathcal{D}_{WS}$. Note that each dodecahedron has one self-identification and, in particular, that the cover is not cyclic

and the complex is given by gluing six copies $1, 2, \ldots, 6$ of the dodecahedron with the orbits for the six faces as shown in Fig. 5 on the left (the orientation of the orbit is given as in Fig. 2 on the left). The dual graph of $\mathcal{C}$ (with one vertex for each dodecahedron, one edge for each gluing along a pentagon, and one colour per face class in the base $\mathcal{D}_{WS}$) is given in Fig. 5 on the right.

The six surfaces consist of 60 mid-cubes each. All surfaces can be decomposed into 12 "pentagonal disks" of five quadrilaterals each, which are parallel to one of the pentagonal faces of the complex, but slightly pushed into one of the adjacent dodecahedra. The six surfaces are given by their pentagonal disks and listed below. Since all of their vertices (which are intersections of the edges of the dodecahedra) must have degree 5, each surface must have 12 such vertices, 12 pentagonal disks, and 30 edges of pentagonal disks, and thus is of Euler characteristic -6. Moreover, since the Weber-Seifert space is orientable and the surface is 2-sided, it must be orientable of genus 4.

Every pentagonal disk is denoted by the corresponding pentagonal face it is parallel to, and the index of the dodecahedron it is contained in. The labelling follows Fig. 2.

$$S_1 = \Big\langle \, (z, \odot)_1, \ (y, \otimes)_1, \ (v, \odot)_2, \ (y, \odot)_2, \ (w, \odot)_3, \ (w, \otimes)_3,$$
$$(x, \otimes)_4, \ (z, \otimes)_4, \ (v, \otimes)_5, \ (u, \otimes)_5, \ (u, \odot)_6, \ (x, \odot)_6 \, \Big\rangle$$

$$S_2 = \Big\langle \, (w, \otimes)_1, \ (x, \odot)_1, \ (u, \otimes)_2, \ (y, \otimes)_2, \ (u, \odot)_3, \ (v, \odot)_3,$$
$$(w, \odot)_4, \ (y, \odot)_4, \ (z, \otimes)_5, \ (z, \odot)_5, \ (x, \otimes)_6, \ (v, \otimes)_6 \, \Big\rangle$$

$$S_3 = \Big\langle (w, \odot)_1, (v, \otimes)_1, (w, \otimes)_2, (z, \otimes)_2, (x, \otimes)_3, (u, \otimes)_3,$$
$$(z, \odot)_4, (u, \odot)_4, (x, \odot)_5, (v, \odot)_5, (y, \otimes)_6, (y, \odot)_6 \Big\rangle$$

$$S_4 = \Big\langle (x, \otimes)_1, (y, \odot)_1, (w, \odot)_2, (u, \odot)_2, (x, \odot)_3, (z, \odot)_3,$$
$$(v, \odot)_4, (v, \otimes)_4, (w, \otimes)_5, (y, \otimes)_5, (z, \otimes)_6, (u, \otimes)_6 \Big\rangle$$

$$S_5 = \Big\langle (u, \otimes)_1, (u, \odot)_1, (v, \otimes)_2, (z, \odot)_2, (z, \otimes)_3, (y, \odot)_3,$$
$$(x, \odot)_4, (y, \otimes)_4, (x, \otimes)_5, (w, \odot)]_5, (w, \otimes)_6, (v, \odot)_6 \Big\rangle$$

$$S_6 = \Big\langle (v, \odot)_1, (z, \otimes)_1, (x, \otimes)_2, (x, \odot)_2, (v, \otimes)_3, (y, \otimes)_3,$$
$$(u, \otimes)_4, (w, \otimes)_4, (y, \odot)_5, (u, \odot)_5, (w, \odot)_6, (z, \odot)_6 \Big\rangle$$

Note that the 12 pentagonal disks of every surface component intersect each dodecahedron exactly twice (A priori, given a sixfold cover of $\mathscr{D}_{WS}$ with six embedded surface components, such an even distribution is not clear: an embedded surface can intersect a dodecahedron in up to three pentagonal disks.). Moreover, every surface component can be endowed with an orientation, such that all of its dual edges point towards the centre of a dodecahedron. Hence, all surface components must be self-osculating through the centre points of some dodecahedron.

Remark 1 The fact that there must be some self-osculating surface components in $\mathscr{C}$ can also be deduced from the fact that the cover features self-identifications (i.e., loops in the face pairing graph). To see this, assume w.l.o.g. that the top and the bottom of a dodecahedron are identified. Then, for instance, pentagonal disk P_1 (which must be part of some surface component) intersecting the innermost pentagon in edge $\langle v_1, v_2 \rangle$ must also intersect the dodecahedron in pentagon P_2, and the corresponding surface component must self-osculate (see Fig. 6).

4.3   A Special Cover of Degree 60

The (non-trivial) normal cores of the subgroups of index up to 6 are all of index either 60 or 360 in $\pi_1(WS)$. For the computation of normal cores we use the GAP function Core [10]. One of the index 60 subgroups is the index 12 normal core of Hempel's cover mentioned in Sect. 4.1. This is equal to the index 10 normal core of $\pi_1(\mathscr{C})$ from Sect. 4.2, and we now show that it produces a special cover $\mathscr{S}$ of WS of degree 60. The deck transformation group is the alternating group A_5 and the abelian invariant of the cover is $\mathbb{Z}^{41} \oplus \mathbb{Z}_2^{12}$.

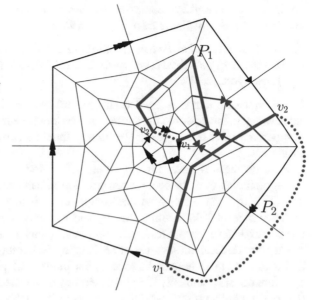

Fig. 6 Self-identifications (as indicated by the arrows) always result in a self-osculating component of the canonical surface—as indicated by the two pentagonal disks P_1 and P_2 of the canonical surface, glued along the dotted edge $\langle v_1, v_2 \rangle$ as a result of the self-identification

The generators of $\pi_1(\mathscr{S})$ are

$$uv^{-1}w^{-1}, \quad uw^{-1}x^{-1}, \quad ux^{-1}y^{-1}, \quad uy^{-1}z^{-1}, \quad uz^{-1}v^{-1}, \quad u^{-1}vz$$
$$u^{-1}wv, \quad u^{-1}xw, \quad u^{-1}yx, \quad u^{-1}zy, \quad vuy^{-1}u^{-1}, \quad vu^{-1}w$$
$$vwy^{-1}, \quad vxv^{-1}u^{-1}, \quad vx^{-1}z, \quad vy^{-1}x^{-1}, \quad v^{-1}uz^{-1}, \quad v^{-1}u^{-1}xu$$
$$v^{-1}xy, \quad v^{-1}y^{-1}vu, \quad wuz^{-1}u^{-1}, \quad wu^{-1}x, \quad wv^{-1}xv, \quad wxz^{-1}$$
$$wyw^{-1}u^{-1}, \quad wzw^{-1}v, \quad wz^{-1}y^{-1}, \quad w^{-1}u^{-1}yu, \quad w^{-1}vzv^{-1}, \quad w^{-1}xwv^{-1}$$
$$w^{-1}z^{-1}wu, \quad xuv^{-1}u^{-1}, \quad xvwv^{-1}, \quad xw^{-1}yw, \quad xzx^{-1}u^{-1}, \quad x^{-1}u^{-1}zu$$
$$x^{-1}v^{-1}xu, \quad x^{-1}wxv, \quad x^{-1}yxw^{-1}, \quad yuw^{-1}u^{-1}, \quad yvy^{-1}u^{-1}, \quad ywxw^{-1}$$
$$yx^{-1}zx, \quad y^{-1}u^{-1}vu, \quad y^{-1}v^{-1}z^{-1}v, \quad y^{-1}w^{-1}yu, \quad y^{-1}xyw, \quad zux^{-1}u^{-1}$$
$$z^{-1}u^{-1}wu, \quad u^5, \quad u^2v^{-1}w^{-1}u^{-1}, \quad uvuy^{-1}u^{-2}, \quad uvw^{-1}x^{-2}, \quad uvxv^{-1}u^{-2}$$
$$uvyv^{-2}, \quad uwuz^{-1}u^{-2}, \quad uwx^{-1}y^{-2}, \quad uwzw^{-2}, \quad uxuv^{-1}u^{-2}, \quad uxy^{-1}z^{-2}$$
$$uyuw^{-1}u^{-2}, u^{-2}vzu, \quad u^{-2}v^{-1}w^{-1}u^{-2}, u^{-1}v^{-1}u^{-1}y^{-1}u^{-2}, u^{-1}v^{-1}x^{-1}v^2.$$

In order to see that $\mathscr{S}$ is in fact a special cover, we must establish a number of observations on embedded surface components in covers of $\hat{\mathscr{D}}_{WS}$. In the following paragraphs we always assume that we are given a finite cover $\mathscr{B}$ of $\mathscr{D}_{WS}$ together with its canonical immersed surface defined by the lift of $\hat{\mathscr{D}}_{WS}$ in $\mathscr{B}$. Whenever we refer to faces of the decomposition of $\mathscr{B}$ into dodecahedra, we explicitly say so. Otherwise we refer to the faces of the lift of the natural cubulation in $\mathscr{B}$. We start with a simple definition.

Definition 1 A dodecahedral vertex is said to be *near* a component S of the canonical immersed surface of $\mathscr{B}$ if it is the endpoint of an edge of the cubulation dual to S.

Lemma 1 *An embedded component S of the canonical immersed surface of $\mathscr{B}$ self-osculates if and only if at least one of the following two situations occurs.*

a) *There exists a dodecahedron containing more than one pentagonal disk of S.*
b) *The number of dodecahedral vertices near S is strictly smaller than its number of pentagonal disks.*

Proof First note that S is 2-sided and can be transversely oriented such that one side always points towards the centres of the dodecahedra it intersects. From this it is apparent that if one of (a) or (b) occurs, then the surface component must self-osculate.

Assume that (a) does not hold; that is, all dodecahedra contain at most one pentagonal disk of S. Hence, no self-osculation can occur through the centre of a dodecahedron. Since every surface component S is made out of pentagonal disks, with five of such disks meeting in every vertex, S has as many pentagonal disks as it has pentagonal vertices. Moreover, every such pentagonal vertex of S must be near exactly one dodecahedral vertex of $\mathscr{B}$. Hence, the number of dodecahedral vertices that S is near to is bounded above by its number of pentagonal disks. Equality therefore occurs if and only if S is not near any dodecahedral vertex twice. Hence, if (b) does not hold, no self-osculation can occur through a vertex of a dodecahedron.

It remains to prove that if S self-osculates, then it must self-osculate through a centre point of a dodecahedron or through a vertex of a dodecahedron. The only other possibilities are that it self-osculates through either the midpoint of a dodecahedral edge or through the centre point of a dodecahedral face.

First assume that the surface self-osculates through the midpoint of a dodecahedral edge e. Then either the surface has two disjoint pentagonal disks both parallel to e and hence also self-osculates through the two dodecahedral endpoints of e; or the surface has two disjoint pentagonal disks both intersecting e, in which case there exists a pair of pentagonal disks in the same dodecahedron—and the surface self-osculates through the centre of that dodecahedron.

Next assume the surface self-osculates through the centre point of a dodecahedral face f. Then either the surface has two disjoint pentagonal disks both parallel to f and hence also self-osculates through the five dodecahedral vertices of f; or the surface has two disjoint pentagonal disks both intersecting f, in which case there exists a pair of pentagonal disks in the same dodecahedron and the surface self-osculates through the centre of that dodecahedron.

Lemma 2 *A pair of intersecting, embedded, and non-self-osculating components S and T of the canonical immersed surface of $\mathscr{B}$ inter-osculates if and only if at least one of the following two situations occurs.*

a) *Some dodecahedron contains pentagonal disks of both S and T which are disjoint.*
b) *The number of all dodecahedral vertices near S or T minus the number of all pairs of intersecting pentagonal disks is strictly smaller than the number of all pentagonal disks in S or T.*

Proof We first need to establish the following three claims.

Claim 1: If S and T inter-osculate, then they inter-osculate through the centre of a dodecahedron or a vertex of a dodecahedron.

This follows from the arguments presented in the second part of the proof of Lemma 1 since inter-osculation locally behaves exactly like self-osculation.

Claim 2: Every pentagonal disk of S intersects T in at most one pentagonal disk and vice versa.

A pentagonal disk can intersect another pentagonal disk in five different ways. Every form of multiple intersection causes either S or T to self-osculate or even self-intersect.

Claim 3: A dodecahedral vertex near an intersection of S and T cannot be near any other pentagonal disk of S or T, other than the ones close to the intersection.

Assume otherwise, then this causes either S or T to self-osculate or even self-intersect.

We now return to the proof of the main statement. If (a) is satisfied, then the surface pair inter-osculates through the centre of the dodecahedron (see also the proof of Lemma 1). If (b) is satisfied, then by Claim 2 and Claim 3, both S and T must be near a dodecahedral vertex away from their intersections and thus S and T inter-osculate.

For the converse assume that neither (a) nor (b) holds. By Claim 1, it suffices to show that S and T do not inter-osculate through the centre of a dodecahedron or a vertex of a dodecahedron.

We first show that S and T do not inter-osculate through the centre of a dodecahedron. If at most one of S or T meets a dodecahedron, then this is true for its centre. Hence assume that both S and T meet a dodecahedron in pentagonal discs. By Claim 2 the dodecahedron contains exactly one pentagonal disc from each surface. These intesect since (a) is assumed false. The only dual edges to S (resp. T) with a vertex at the centre of the cube run from the centre of the pentagonal face of the dodecahedron dual to S (resp. T) to the centre of the dodecahedron. But these two edges lie in the boundary of a square in the dodecahedron since the pentagonal discs intersect and hence the pentagonal faces are adjacent. Hence S and T do not inter-osculate through the centre of a dodecahedron.

We next show that S and T do not inter-osculate through the vertex of a dodecahedron. The negation of (b) is that the number of all dodecahedral vertices near S or T minus the number of all pairs of intersecting pentagonal disks equals the number of all pentagonal disks of S and T. Suppose a dodecahedral vertex is the endpoint of dual edges to squares in S and T. If the dual edges are contained in the same dodecahedron then they are in the boundary of a common square. Hence assume they are contained in different dodecahedra. Then the equality forces at least one of the dual edges to be in the boundary of a cube intersected by both S and T. But then at least one of the surfaces self-osculates.

Due to Lemmata 1 and 2, checking for self-osculating embedded surface components is a straightforward task. Furthermore, as long as surface components are embedded and non-self-osculating, checking for inter-osculation of a surface pair is simple as well.

In the cover $\mathscr{S}$ we have:

(a) the canonical immersed surface splits into 60 embedded components,
(b) every surface component of $\mathscr{S}$ is made up of 12 pentagonal disks (and thus is orientable of genus 4, see the description of the canonical surface components of $\mathscr{C}$ in Sect. 4.2 for details),
(c) every surface component distributes its 12 pentagonal disks over 12 distinct dodecahedra,
(d) every surface component is near 12 dodecahedral vertices, and
(e) every pair of intersecting surface components intersects in exactly three pentagonal disks (and hence in exactly three dodecahedra), and for each such pair both surface components combined are near exactly 21 dodecahedral vertices.

These properties of $\mathscr{S}$ can be checked using the GAP script available from [15]. From them, and from Lemmata 1 and 2 it follows that $\mathscr{S}$ is a special cover. The gluing orbits for $\mathscr{S}$ of the face classes from $\mathscr{D}_{WS}$, as well as all 60 surface components are listed in Sect. 6.

4.4 Covers of Higher Degree

An exhaustive enumeration of all subgroups up to index 9 reveals a total of 490 covers, but no further examples of covers where the canonical surface splits into embedded components (and in particular no further special covers). There are, however, 20 examples of degree 8 covers where the canonical surface splits into two immersed connected components (all with first homology group $\mathbb{Z} \oplus \mathbb{Z}_2^3 \oplus \mathbb{Z}_3^2 \oplus \mathbb{Z}_5^3$). Moreover, there are 10 examples of degree 9 covers, where the canonical surface splits into two components, one of which is embedded (all with first homology group $\mathbb{Z} \oplus \mathbb{Z}_3 \oplus \mathbb{Z}_4 \oplus \mathbb{Z}_5^3 \oplus \mathbb{Z}_7$). All of them are Haken, as can be seen by their first integral homology groups.

In an attempt to obtain further special covers we execute a non-exhaustive, heuristic search for higher degree covers. This is necessary since complete enumeration of subgroups quickly becomes infeasible for subgroups of index larger than 9. This more targeted search is done in essentially two distinct ways.

In the first approach we compute normal cores of all irregular covers of degrees 7, 8, and 9 from the enumeration of subgroups of $\pi_1(\text{WS})$ of index at most 9 described above. This is motivated by the fact that the index 60 normal core of $\pi_1(\mathscr{C})$ yields a special cover. The normal cores have indices 168, 504, 1344, 2520, 20,160, and 181,440. Of the ones with index at most 2520, we construct the corresponding cover. Very often, the covers associated to these normal cores exhibit a single (immersed) surface component. However, the normal cores of the 10 subgroups corresponding to the covers of degree 9 with two surface components yield (regular) covers where the canonical immersed surface splits into nine embedded components. All of these covers are of degree 504 with deck transformation group PSL(2, 8). Each of the surface components has 672 pentagons. Accordingly, each of them must be (orientable) of genus 169. All nine surface components necessarily self-osculate (they are embedded and contain more pentagonal disks than there are dodecahedra in the cover). The first homology group of all of these covers is given by

$$\mathbb{Z}^8 \oplus \mathbb{Z}_2^{10} \oplus \mathbb{Z}_3 \oplus \mathbb{Z}_4^9 \oplus \mathbb{Z}_5^{17} \oplus \mathbb{Z}_7 \oplus \mathbb{Z}_8^6 \oplus \mathbb{Z}_9^7 \oplus \mathbb{Z}_{17}^{28} \oplus \mathbb{Z}_{27}^7 \oplus \mathbb{Z}_{29}^9 \oplus \mathbb{Z}_{83}^{18}.$$

In addition, there are 120 subgroups with a core of order 1344, and factor group isomorphic to a semi-direct product of $\mathbb{Z}_2^3$ and PSL(3, 2). For 40 of them the corresponding (regular) cover splits into 8 immersed components. These include the covers of degree 8 where the canonical immersed surface splits into two immersed components.

In the second approach we analyse low degree covers of $\mathscr{C}$ from Sect. 4.2. This is motivated by the fact that, in such covers, the canonical surface necessarily consists of embedded components.

There are 127 twofold covers of $\mathscr{C}$, 64 of which are fix-point free (i.e., they do not identify two pentagons of the same dodecahedron—a necessary condition for a cover to be special, see the end of Sect. 4.2). For 40 of them the canonical surface still only splits into six embedded components. For the remaining 24, the surface splits into 7 components. For more details, see Table 2.

The 127 twofold covers of $\mathscr{C}$ altogether have 43,905 twofold covers. Amongst these 24-fold covers of $\mathscr{D}_{WS}$, 16,192 are fix-point free. They admit 6–14 surface components with a single exception where the surface splits into 24 components. This cover is denoted by $\mathscr{E}$. Details on the number of covers and surface components can be found in Table 3.

We have for the generators of the subgroup corresponding to the cover $\mathscr{E}$

$$
\begin{array}{lllllll}
u^{-2}, & uvz, & uv^{-1}w^{-1}, & uwv, & uw^{-1}x^{-1}, & uxw, & ux^{-1}y^{-1}, \\
uyx, & uy^{-1}z^{-1}, & uzy, & uz^{-1}v^{-1}, & vux, & vu^{-1}w, & vwy^{-1}, \\
vxz, & vx^{-1}z, & vy^{-1}x^{-1}, & z^{-1}uy^{-1}, & z^{-1}u^{-1}x^{-1}, & z^{-1}vy^{-1}v^{-1}, & z^{-1}wx, \\
z^{-1}w^{-1}zv^{-1}, & z^{-1}yv^{-2}, & z^{-1}y^{-1}w, & z^{-2}wv^{-1}, & wuy, & wv^{-1}xv.
\end{array}
$$

Table 2 Summary of all 64 fix-point free double covers of $\mathscr{C}$

$H_1(X)$	# surf. comp.	# covers
$\mathbb{Z}^5 \oplus \mathbb{Z}_4^2 \oplus \mathbb{Z}_5^3$	6	20
$\mathbb{Z}^5 \oplus \mathbb{Z}_2^2 \oplus \mathbb{Z}_4 \oplus \mathbb{Z}_5^3$	6	20
$\mathbb{Z}^6 \oplus \mathbb{Z}_2^2 \oplus \mathbb{Z}_5^3$	7	12
$\mathbb{Z}^7 \oplus \mathbb{Z}_5^3$	7	12
		$\Sigma = 64$

By construction, all surface components are embedded

Table 3 Summary of all 16,192 fix-point free double covers of the 127 double covers of $\mathscr{C}$

# surf. comp.	6	7	7	8	8	8	9	9	12	14	14	24	
# covers	8960	3240	2160	180	720	540	240	24	85	24	18	1	$\Sigma = 16,192$

By construction, all surface components are embedded

Surface components in $\mathcal{E}$ are small (12 pentagonal disks per surface, as also observed in the degree 60 special cover $\mathcal{S}$, see Sect. 4.3). This motivates an extended search for a degree 48 special cover by looking at degree 2 covers of $\mathcal{E}$. However, amongst the 131,071 fix-point free covers of degree 2, no special cover exists. More precisely, there are 120,205 covers with 24 surface components, 10,200 with 25 surface components, 240 with 26 and 27 surface components each, 162 with 28, and 24 with 33 surface components. For most of them, most surface components self-osculate.

5 Poincaré Homology Sphere and Projective Space

The Poincaré homology sphere has as fundamental group the binary icosahedral group of order 120, which is isomorphic to $SL(2, 5)$. From its subdivision given by the dodecahedron, we can deduce a presentation with six generators dual to the six pentagons of the subdivision, and one relator dual to each of the ten edges:

$$\pi_1(\mathcal{D}_{\Sigma^3}) = \langle\ u,\ v,\ w,\ x,\ y,\ z\ |\ uxz,\quad uyv,\quad uzw,\quad uvx,\quad uwy,$$
$$xy^{-1}z,\ yz^{-1}v,\ zv^{-1}w,\ vw^{-1}x,\ wx^{-1}y\ \rangle.$$

$SL(2, 5)$ has 76 subgroups falling into 12 conjugacy classes forming the subgroup lattice shown in Fig. 7 on the left hand side. For the corresponding hierarchy of covers together with the topological types of the covering 3-manifolds see Fig. 7 on the right hand side.

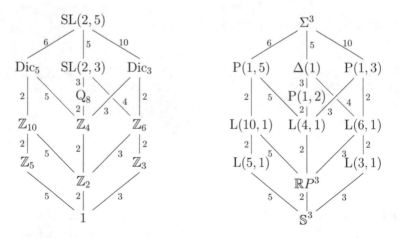

Fig. 7 Left: subgroup lattice of $SL(2, 5)$ with indices. Right: covers of Σ^3 with degrees. Here $P(n, m)$ denotes the prism space with parameters n and m, and $\Delta(n)$ denotes the tetrahedral space with parameter n

By construction, the universal cover of $\mathscr{D}_{\Sigma^3}$ is the 120-cell, which is dual to the simplicial 600-cell. In particular, the dual cell decomposition of any of the 12 covers is a (semi-simplicial) triangulation. The dual of $\mathscr{D}_{\Sigma^3}$ itself is isomorphic to the minimal five-tetrahedron triangulation of the Poincaré homology sphere.

Most of the topological types are determined by the isomorphism type of the subgroups. The only two non-trivial cases are the lens spaces $L(5, 1)$ and $L(10, 1)$. For the former, we passed to the dual triangulation of the cover of degree 24 using the *GAP*-package simpcomp [9], and then fed the result to the 3-manifold software *Regina* [8] to determine the topological type of the cover to be $L(5, 1)$. The latter is then determined by the observation that there is no 2-to-1-cover of $L(10, 3)$ to $L(5, 1)$.

Regarding the canonical immersed surface, the situation is quite straightforward. Since all edges of $\mathscr{D}_{\Sigma^3}$, or of any of its covers, are of degree three, the canonical surface is a surface decomposed into pentagonal disks with three such disks meeting in each vertex. Consequently, all surface components must be 2-spheres isomorphic to the dodecahedron, thus have 12 pentagons, and the number of connected components of the canonical surface must coincide with the degree of the cover. Moreover, each surface component runs parallel to the 2-skeleton of a single dodecahedron, and the surface components are embedded if and only if there are no self-intersections of dodecahedra.

In more detail the relevant properties of all covers are listed in Table 4.

The case of the projective space is rather simple. The only proper cover (of degree > 1) is the universal cover of degree 2. Since the edges of $\mathscr{D}_{\mathbb{R}P^3}$ are all of degree two, the canonical surface of $\mathscr{D}_{\mathbb{R}P^3}$ has six embedded sphere components, each consisting of two pentagons glued along their boundary, surrounding one of the six

Table 4 Covers of Σ^3

Deg.	Top. type.	Subgroup	f-vec.	Embedded	# surf.	Regular	Deck trafo grp.
1	Σ^3	$SL(2, 5)$	$(5, 10, 6, 1)$	No	1	Yes	1
5	$\Delta(1)$	$SL(2, 3)$	$(25, 50, 30, 5)$	Yes	5	No	A_5
6	$P(1, 5)$	$Dic(5)$	$(30, 60, 36, 6)$	No	6	No	A_5
10	$P(1, 3)$	$Dic(3)$	$(50, 100, 60, 10)$	Yes	10	No	A_5
12	$L(10, 1)$	$\mathbb{Z}_{10}$	$(60, 120, 72, 12)$	No	12	No	A_5
15	$P(1, 2)$	Q_8	$(75, 150, 90, 15)$	Yes	15	No	A_5
20	$L(6, 1)$	$\mathbb{Z}_6$	$(100, 200, 120, 20)$	Yes	20	No	A_5
24	$L(5, 1)$	$\mathbb{Z}_5$	$(120, 240, 144, 24)$	Yes	24	No	$SL(2, 5)$
30	$L(4, 1)$	$\mathbb{Z}_4$	$(150, 300, 180, 30)$	Yes	30	No	A_5
40	$L(3, 1)$	$\mathbb{Z}_3$	$(200, 400, 240, 40)$	Yes	40	No	$SL(2, 5)$
60	$\mathbb{R}P^3$	$\mathbb{Z}_2$	$(300, 600, 360, 60)$	Yes	60	Yes	A_5
120	$\mathbb{S}^3$	1	$(600, 1200, 720, 120)$	Yes	120	Yes	$SL(2, 5)$

$P(n, m)$ denotes the prism space with parameters n and m, and $\Delta(n)$ denotes the tetrahedral space with parameter n. "f-vec." denotes the f-vector of the cover as a decomposition into dodecahedra. I.e., $(25, 50, 30, 5)$ means that the corresponding cover contains 25 vertices, 50 edges, 30 pentagons, and 5 dodecahedra. "deck trafo grp." denotes the deck transformation group of the cover

pentagonal faces each. Consequently, the universal cover is a 3-sphere decomposed into two balls along a dodecahedron with the canonical surface splitting into 12 sphere components.

6 The Special Cover $\mathscr{S}$

The special cover $\mathscr{S}$ from Sect. 4.3 is of degree 60 with deck transformation group A_5 and abelian invariant $\mathbb{Z}^{41} \oplus \mathbb{Z}_2^{12}$. The subgroup is generated by

$$
\begin{array}{llllll}
uv^{-1}w^{-1}, & uw^{-1}x^{-1}, & ux^{-1}y^{-1}, & uy^{-1}z^{-1}, & uz^{-1}v^{-1}, & u^{-1}vz \\
u^{-1}wv, & u^{-1}xw, & u^{-1}yx, & u^{-1}zy, & vuy^{-1}u^{-1}, & vu^{-1}w \\
vwy^{-1}, & vxv^{-1}u^{-1}, & vx^{-1}z, & vy^{-1}x^{-1}, & v^{-1}uz^{-1}, & v^{-1}u^{-1}xu \\
v^{-1}xy, & v^{-1}y^{-1}vu, & wuz^{-1}u^{-1}, & wu^{-1}x, & wv^{-1}xv, & wxz^{-1} \\
wyw^{-1}u^{-1}, & wzw^{-1}v, & wz^{-1}y^{-1}, & w^{-1}u^{-1}yu, & w^{-1}vzv^{-1}, & w^{-1}xwv^{-1} \\
w^{-1}z^{-1}wu, & xuv^{-1}u^{-1}, & xvwv^{-1}, & xw^{-1}yw, & xzx^{-1}u^{-1}, & x^{-1}u^{-1}zu \\
x^{-1}v^{-1}xu, & x^{-1}wxv, & x^{-1}yxw^{-1}, & yuw^{-1}u^{-1}, & yvy^{-1}u^{-1}, & ywxw^{-1} \\
yx^{-1}zx, & y^{-1}u^{-1}vu, & y^{-1}v^{-1}z^{-1}v, & y^{-1}w^{-1}yu, & y^{-1}xyw, & zux^{-1}u^{-1} \\
z^{-1}u^{-1}wu, & u^5, & u^2v^{-1}w^{-1}u^{-1}, & uvuy^{-1}u^{-2}, & uvw^{-1}x^{-2}, & uvxv^{-1}u^{-2} \\
uvyv^{-2}, & uwuz^{-1}u^{-2}, & uwx^{-1}y^{-2}, & uwzw^{-2}, & uxuv^{-1}u^{-2}, & uxy^{-1}z^{-2} \\
uyuw^{-1}u^{-2}, & u^{-2}vzu, & u^{-2}v^{-1}w^{-1}u^{-2}, & u^{-1}v^{-1}u^{-1}y^{-1}u^{-2}, & u^{-1}v^{-1}x^{-1}v^2. &
\end{array}
$$

The gluing orbits of face classes from $\mathscr{D}_{WS}$ are given by

Face	Orbit
u	$(1,2,14,20,3)(4,18,47,24,7)(5,12,17,46,23)(6,19,48,25,9)$
	$(8,15,49,21,11)(10,16,50,22,13)(26,54,43,37,27)(28,42,52,36,29)$
	$(30,33,39,38,51)(31,44,53,40,32)(34,55,45,41,35)(56,59,60,58,57)$
v	$(1,4,26,30,5)(2,15,51,32,6)(3,13,28,54,21)(7,29,56,33,11)$
	$(8,27,57,31,12)(9,10,18,49,23)(14,46,40,35,16)(17,38,58,34,19)$
	$(20,25,41,42,47)(22,45,59,43,24)(36,55,44,39,37)(48,53,60,52,50)$
w	$(1,6,34,36,7)(2,16,52,37,8)(3,5,31,55,22)(4,10,35,58,27)$
	$(9,32,57,29,13)(11,12,19,50,24)(14,47,43,39,17)(15,18,42,60,38)$
	$(20,21,33,44,48)(23,30,56,45,25)(26,28,41,40,51)(46,49,54,59,53)$
x	$(1,8,38,40,9)(2,17,53,41,10)(3,7,27,51,23)(4,15,46,25,13)$
	$(5,11,37,58,32)(6,12,39,60,35)(14,48,45,28,18)(16,19,44,59,42)$
	$(20,22,29,26,49)(21,24,36,57,30)(31,33,43,52,34)(47,50,55,56,54)$
y	$(1,10,42,43,11)(2,18,54,33,12)(3,9,35,52,24)(4,28,59,39,8)$
	$(5,6,16,47,21)(7,13,41,60,37)(14,49,30,31,19)(15,26,56,44,17)$
	$(20,23,32,34,50)(22,25,40,58,36)(27,29,45,53,38)(46,51,57,55,48)$
z	$(1,12,44,45,13)(2,19,55,29,4)(3,11,39,53,25)(5,33,59,41,9)$
	$(6,31,56,28,10)(7,8,17,48,22)(14,50,36,27,15)(16,34,57,26,18)$
	$(20,24,37,38,46)(21,43,60,40,23)(30,54,42,35,32)(47,52,58,51,49)$

The 60 surfaces are given by their pentagonal disks. Every pentagonal disk is denoted by the corresponding pentagonal face in the lift of $\mathscr{D}_{WS}$ it is parallel to, and the index of the dodecahedron it is contained in. The labelling follows Fig. 2.

$$S_1 = \Big\langle (u, \odot)_{60}, \ (x, \odot)_{42}, \ (w, \odot)_{39}, \ (v, \odot)_{41}, \ (z, \odot)_{43}, \ (y, \odot)_{53},$$
$$(y, \otimes)_{54}, \ (w, \otimes)_{28}, \ (x, \otimes)_{44}, \ (v, \otimes)_{33}, \ (z, \otimes)_{45}, \ (u, \otimes)_{56} \ \Big\rangle$$

$$S_2 = \Big\langle (x, \otimes)_{60}, \ (w, \odot)_{52}, \ (u, \otimes)_{41}, \ (y, \odot)_{58}, \ (v, \otimes)_{42}, \ (z, \otimes)_{40},$$
$$(z, \odot)_{16}, \ (u, \odot)_{34}, \ (w, \otimes)_{9}, \ (y, \otimes)_{10}, \ (v, \odot)_{32}, \ (x, \odot)_{6} \ \Big\rangle$$

$$S_3 = \Big\langle (y, \otimes)_{60}, \ (x, \odot)_{40}, \ (u, \otimes)_{39}, \ (z, \odot)_{58}, \ (w, \otimes)_{53}, \ (v, \otimes)_{37},$$
$$(v, \odot)_{46}, \ (u, \odot)_{51}, \ (x, \otimes)_{8}, \ (z, \otimes)_{17}, \ (w, \odot)_{27}, \ (y, \odot)_{15} \ \Big\rangle$$

$$S_4 = \Big\langle (z, \otimes)_{60}, \ (y, \odot)_{37}, \ (u, \otimes)_{42}, \ (v, \odot)_{58}, \ (x, \otimes)_{43}, \ (w, \otimes)_{35},$$
$$(w, \odot)_{24}, \ (u, \odot)_{36}, \ (v, \otimes)_{47}, \ (y, \otimes)_{16}, \ (x, \odot)_{34}, \ (z, \odot)_{50} \ \Big\rangle$$

$$S_5 = \Big\langle (v, \otimes)_{60}, \ (z, \odot)_{35}, \ (u, \otimes)_{53}, \ (w, \odot)_{58}, \ (y, \otimes)_{41}, \ (x, \otimes)_{38},$$
$$(x, \odot)_{9}, \ (u, \odot)_{32}, \ (w, \otimes)_{25}, \ (z, \otimes)_{46}, \ (y, \odot)_{51}, \ (v, \odot)_{23} \ \Big\rangle$$

$$S_6 = \Big\langle (w, \otimes)_{60}, \ (v, \odot)_{38}, \ (u, \otimes)_{43}, \ (x, \odot)_{58}, \ (z, \otimes)_{39}, \ (y, \otimes)_{52},$$
$$(y, \odot)_{8}, \ (u, \odot)_{27}, \ (v, \otimes)_{24}, \ (x, \otimes)_{11}, \ (z, \odot)_{36}, \ (w, \odot)_{7} \ \Big\rangle$$

$$S_7 = \Big\langle (v, \odot)_{60}, \ (x, \odot)_{52}, \ (y, \odot)_{39}, \ (w, \otimes)_{42}, \ (u, \odot)_{37}, \ (z, \otimes)_{59},$$
$$(z, \odot)_{24}, \ (y, \otimes)_{47}, \ (w, \odot)_{11}, \ (x, \otimes)_{33}, \ (u, \otimes)_{54}, \ (v, \otimes)_{21} \ \Big\rangle$$

$$S_8 = \Big\langle (w, \odot)_{60}, \ (y, \odot)_{40}, \ (z, \odot)_{42}, \ (x, \otimes)_{53}, \ (u, \odot)_{35}, \ (v, \otimes)_{59},$$
$$(v, \odot)_{9}, \ (z, \otimes)_{25}, \ (x, \odot)_{10}, \ (y, \otimes)_{28}, \ (u, \otimes)_{45}, \ (w, \otimes)_{13} \ \Big\rangle$$

$$S_9 = \Big\langle (x, \odot)_{60}, \ (z, \odot)_{37}, \ (v, \odot)_{53}, \ (y, \otimes)_{43}, \ (u, \odot)_{38}, \ (w, \otimes)_{59},$$
$$(w, \odot)_{8}, \ (v, \otimes)_{11}, \ (y, \odot)_{17}, \ (z, \otimes)_{44}, \ (u, \otimes)_{33}, \ (x, \otimes)_{12} \ \Big\rangle$$

$$S_{10} = \Big\langle (y, \odot)_{60}, \ (v, \odot)_{35}, \ (w, \odot)_{43}, \ (z, \otimes)_{41}, \ (u, \odot)_{52}, \ (x, \otimes)_{59},$$
$$(x, \odot)_{16}, \ (w, \otimes)_{10}, \ (z, \odot)_{47}, \ (v, \otimes)_{54}, \ (u, \otimes)_{28}, \ (y, \otimes)_{18} \ \Big\rangle$$

$$S_{11} = \Big\langle (z, \odot)_{60}, (w, \odot)_{38}, (x, \odot)_{41}, (v, \otimes)_{39}, (u, \odot)_{40}, (y, \otimes)_{59},$$
$$(y, \odot)_{46}, (x, \otimes)_{17}, (v, \odot)_{25}, (w, \otimes)_{45}, (u, \otimes)_{44}, (z, \otimes)_{48} \Big\rangle$$

$$S_{12} = \Big\langle (u, \otimes)_{60}, (v, \otimes)_{52}, (w, \otimes)_{40}, (x, \otimes)_{37}, (y, \otimes)_{35}, (z, \otimes)_{38},$$
$$(z, \odot)_{34}, (w, \odot)_{36}, (v, \odot)_{51}, (x, \odot)_{32}, (y, \odot)_{27}, (u, \odot)_{57} \Big\rangle$$

$$S_{13} = \Big\langle (u, \odot)_{59}, (x, \odot)_{54}, (w, \odot)_{44}, (v, \odot)_{28}, (z, \odot)_{33}, (y, \odot)_{45},$$
$$(y, \otimes)_{30}, (w, \otimes)_{26}, (x, \otimes)_{55}, (v, \otimes)_{31}, (z, \otimes)_{29}, (u, \otimes)_{57} \Big\rangle$$

$$S_{14} = \Big\langle (v, \odot)_{59}, (x, \odot)_{43}, (y, \odot)_{44}, (w, \otimes)_{54}, (u, \odot)_{39}, (z, \otimes)_{56},$$
$$(z, \odot)_{11}, (y, \otimes)_{21}, (w, \odot)_{12}, (x, \otimes)_{31}, (u, \otimes)_{30}, (v, \otimes)_{5} \Big\rangle$$

$$S_{15} = \Big\langle (w, \odot)_{59}, (y, \odot)_{41}, (z, \odot)_{54}, (x, \otimes)_{45}, (u, \odot)_{42}, (v, \otimes)_{56},$$
$$(v, \odot)_{10}, (z, \otimes)_{13}, (x, \odot)_{18}, (y, \otimes)_{26}, (u, \otimes)_{29}, (w, \otimes)_{4} \Big\rangle$$

$$S_{16} = \Big\langle (x, \odot)_{59}, (z, \odot)_{39}, (v, \odot)_{45}, (y, \otimes)_{33}, (u, \odot)_{53}, (w, \otimes)_{56},$$
$$(w, \odot)_{17}, (v, \otimes)_{12}, (y, \odot)_{48}, (z, \otimes)_{55}, (u, \otimes)_{31}, (x, \otimes)_{19} \Big\rangle$$

$$S_{17} = \Big\langle (y, \odot)_{59}, (v, \odot)_{42}, (w, \odot)_{33}, (z, \otimes)_{28}, (u, \odot)_{43}, (x, \otimes)_{56},$$
$$(x, \odot)_{47}, (w, \otimes)_{18}, (z, \odot)_{21}, (v, \otimes)_{30}, (u, \otimes)_{26}, (y, \otimes)_{49} \Big\rangle$$

$$S_{18} = \Big\langle (z, \odot)_{59}, (w, \odot)_{53}, (x, \odot)_{28}, (v, \otimes)_{44}, (u, \odot)_{41}, (y, \otimes)_{56},$$
$$(y, \odot)_{25}, (x, \otimes)_{48}, (v, \odot)_{13}, (w, \otimes)_{29}, (u, \otimes)_{55}, (z, \otimes)_{22} \Big\rangle$$

$$S_{19} = \Big\langle (u, \otimes)_{59}, (v, \otimes)_{43}, (w, \otimes)_{41}, (x, \otimes)_{39}, (y, \otimes)_{42}, (z, \otimes)_{53},$$
$$(z, \odot)_{52}, (w, \odot)_{37}, (v, \odot)_{40}, (x, \odot)_{35}, (y, \odot)_{38}, (u, \odot)_{58} \Big\rangle$$

$$S_{20} = \Big\langle (x, \otimes)_{58}, (w, \odot)_{34}, (u, \otimes)_{40}, (y, \odot)_{57}, (v, \otimes)_{35}, (z, \otimes)_{51},$$
$$(z, \odot)_{6}, (u, \odot)_{31}, (w, \otimes)_{23}, (y, \otimes)_{9}, (v, \odot)_{30}, (x, \odot)_{5} \Big\rangle$$

$$S_{21} = \Big\langle (y, \otimes)_{58}, (x, \odot)_{51}, (u, \otimes)_{37}, (z, \odot)_{57}, (w, \otimes)_{38}, (v, \otimes)_{36},$$
$$(v, \odot)_{15}, (u, \odot)_{26}, (x, \otimes)_{7}, (z, \otimes)_{8}, (w, \odot)_{29}, (y, \odot)_{4} \Big\rangle$$

$$S_{22} = \Big\langle (z, \otimes)_{58}, \ (y, \odot)_{36}, \ (u, \otimes)_{35}, \ (v, \odot)_{57}, \ (x, \otimes)_{52}, \ (w, \otimes)_{32},$$
$$(w, \odot)_{50}, \ (u, \odot)_{55}, \ (v, \otimes)_{16}, \ (y, \otimes)_{6}, \ (x, \odot)_{31}, \ (z, \odot)_{19} \ \Big\rangle$$

$$S_{23} = \Big\langle (v, \otimes)_{58}, \ (z, \odot)_{32}, \ (u, \otimes)_{38}, \ (w, \odot)_{57}, \ (y, \otimes)_{40}, \ (x, \otimes)_{27},$$
$$(x, \odot)_{23}, \ (u, \odot)_{30}, \ (w, \otimes)_{46}, \ (z, \otimes)_{15}, \ (y, \odot)_{26}, \ (v, \odot)_{49} \ \Big\rangle$$

$$S_{24} = \Big\langle (w, \otimes)_{58}, \ (v, \odot)_{27}, \ (u, \otimes)_{52}, \ (x, \odot)_{57}, \ (z, \otimes)_{37}, \ (y, \otimes)_{34},$$
$$(y, \odot)_{7}, \ (u, \odot)_{29}, \ (v, \otimes)_{50}, \ (x, \otimes)_{24}, \ (z, \odot)_{55}, \ (w, \odot)_{22} \ \Big\rangle$$

$$S_{25} = \Big\langle (u, \otimes)_{58}, \ (v, \otimes)_{34}, \ (w, \otimes)_{51}, \ (x, \otimes)_{36}, \ (y, \otimes)_{32}, \ (z, \otimes)_{27},$$
$$(z, \odot)_{31}, \ (w, \odot)_{55}, \ (v, \odot)_{26}, \ (x, \odot)_{30}, \ (y, \odot)_{29}, \ (u, \odot)_{56} \ \Big\rangle$$

$$S_{26} = \Big\langle (x, \otimes)_{57}, \ (w, \odot)_{31}, \ (u, \otimes)_{51}, \ (y, \odot)_{56}, \ (v, \otimes)_{32}, \ (z, \otimes)_{26},$$
$$(z, \odot)_{5}, \ (u, \odot)_{33}, \ (w, \otimes)_{49}, \ (y, \otimes)_{23}, \ (v, \odot)_{54}, \ (x, \odot)_{21} \ \Big\rangle$$

$$S_{27} = \Big\langle (y, \otimes)_{57}, \ (x, \odot)_{26}, \ (u, \otimes)_{36}, \ (z, \odot)_{56}, \ (w, \otimes)_{27}, \ (v, \otimes)_{55},$$
$$(v, \odot)_{4}, \ (u, \odot)_{28}, \ (x, \otimes)_{22}, \ (z, \otimes)_{7}, \ (w, \odot)_{45}, \ (y, \odot)_{13} \ \Big\rangle$$

$$S_{28} = \Big\langle (z, \otimes)_{57}, \ (y, \odot)_{55}, \ (u, \otimes)_{32}, \ (v, \odot)_{56}, \ (x, \otimes)_{34}, \ (w, \otimes)_{30},$$
$$(w, \odot)_{19}, \ (u, \odot)_{44}, \ (v, \otimes)_{6}, \ (y, \otimes)_{5}, \ (x, \odot)_{33}, \ (z, \odot)_{12} \ \Big\rangle$$

$$S_{29} = \Big\langle (v, \otimes)_{57}, \ (z, \odot)_{30}, \ (u, \otimes)_{27}, \ (w, \odot)_{56}, \ (y, \otimes)_{51}, \ (x, \otimes)_{29},$$
$$(x, \odot)_{49}, \ (u, \odot)_{54}, \ (w, \otimes)_{15}, \ (z, \otimes)_{4}, \ (y, \odot)_{28}, \ (v, \odot)_{18} \ \Big\rangle$$

$$S_{30} = \Big\langle (w, \otimes)_{57}, \ (v, \odot)_{29}, \ (u, \otimes)_{34}, \ (x, \odot)_{56}, \ (z, \otimes)_{36}, \ (y, \otimes)_{31},$$
$$(y, \odot)_{22}, \ (u, \odot)_{45}, \ (v, \otimes)_{19}, \ (x, \otimes)_{50}, \ (z, \odot)_{44}, \ (w, \odot)_{48} \ \Big\rangle$$

$$S_{31} = \Big\langle (y, \otimes)_{55}, \ (x, \odot)_{29}, \ (u, \otimes)_{50}, \ (z, \odot)_{45}, \ (w, \otimes)_{36}, \ (v, \otimes)_{48},$$
$$(v, \odot)_{7}, \ (u, \odot)_{13}, \ (x, \otimes)_{20}, \ (z, \otimes)_{24}, \ (w, \odot)_{25}, \ (y, \odot)_{3} \ \Big\rangle$$

$$S_{32} = \Big\langle (w, \otimes)_{55}, \ (v, \odot)_{22}, \ (u, \otimes)_{19}, \ (x, \odot)_{45}, \ (z, \otimes)_{50}, \ (y, \otimes)_{44},$$
$$(y, \odot)_{20}, \ (u, \odot)_{25}, \ (v, \otimes)_{17}, \ (x, \otimes)_{14}, \ (z, \odot)_{53}, \ (w, \odot)_{46} \ \Big\rangle$$

$$S_{33} = \Big\langle\ (v, \odot)_{55},\ (x, \odot)_{44},\ (y, \odot)_{50},\ (w, \otimes)_{31},\ (u, \odot)_{48},\ (z, \otimes)_{34},$$
$$(z, \odot)_{17},\ (y, \otimes)_{12},\ (w, \odot)_{14},\ (x, \otimes)_{16},\ (u, \otimes)_{6},\ (v, \otimes)_{2}\ \Big\rangle$$

$$S_{34} = \Big\langle\ (x, \odot)_{55},\ (z, \odot)_{48},\ (v, \odot)_{36},\ (y, \otimes)_{19},\ (u, \odot)_{22},\ (w, \otimes)_{34},$$
$$(w, \odot)_{20},\ (v, \otimes)_{14},\ (y, \odot)_{24},\ (z, \otimes)_{52},\ (u, \otimes)_{16},\ (x, \otimes)_{47}\ \Big\rangle$$

$$S_{35} = \Big\langle\ (x, \otimes)_{54},\ (w, \odot)_{21},\ (u, \otimes)_{18},\ (y, \odot)_{43},\ (v, \otimes)_{49},\ (z, \otimes)_{42},$$
$$(z, \odot)_{20},\ (u, \odot)_{24},\ (w, \otimes)_{16},\ (y, \otimes)_{14},\ (v, \odot)_{52},\ (x, \odot)_{50}\ \Big\rangle$$

$$S_{36} = \Big\langle\ (z, \otimes)_{54},\ (y, \odot)_{33},\ (u, \otimes)_{49},\ (v, \odot)_{43},\ (x, \otimes)_{30},\ (w, \otimes)_{47},$$
$$(w, \odot)_{5},\ (u, \odot)_{11},\ (v, \otimes)_{23},\ (y, \otimes)_{20},\ (x, \odot)_{24},\ (z, \odot)_{3}\ \Big\rangle$$

$$S_{37} = \Big\langle\ (w, \odot)_{54},\ (y, \odot)_{42},\ (z, \odot)_{49},\ (x, \otimes)_{28},\ (u, \odot)_{47},\ (v, \otimes)_{26},$$
$$(v, \odot)_{16},\ (z, \otimes)_{10},\ (x, \odot)_{14},\ (y, \otimes)_{15},\ (u, \otimes)_{4},\ (w, \otimes)_{2}\ \Big\rangle$$

$$S_{38} = \Big\langle\ (y, \odot)_{54},\ (v, \odot)_{47},\ (w, \odot)_{30},\ (z, \otimes)_{18},\ (u, \odot)_{21},\ (x, \otimes)_{26},$$
$$(x, \odot)_{20},\ (w, \otimes)_{14},\ (z, \odot)_{23},\ (v, \otimes)_{51},\ (u, \otimes)_{15},\ (y, \otimes)_{46}\ \Big\rangle$$

$$S_{39} = \Big\langle\ (y, \otimes)_{53},\ (x, \odot)_{25},\ (u, \otimes)_{17},\ (z, \odot)_{40},\ (w, \otimes)_{48},\ (v, \otimes)_{38},$$
$$(v, \odot)_{20},\ (u, \odot)_{23},\ (x, \otimes)_{15},\ (z, \otimes)_{14},\ (w, \odot)_{51},\ (y, \odot)_{49}\ \Big\rangle$$

$$S_{40} = \Big\langle\ (v, \otimes)_{53},\ (z, \odot)_{41},\ (u, \otimes)_{48},\ (w, \odot)_{40},\ (y, \otimes)_{45},\ (x, \otimes)_{46},$$
$$(x, \odot)_{13},\ (u, \odot)_{9},\ (w, \otimes)_{22},\ (z, \otimes)_{20},\ (y, \odot)_{23},\ (v, \odot)_{3}\ \Big\rangle$$

$$S_{41} = \Big\langle\ (x, \odot)_{53},\ (z, \odot)_{38},\ (v, \odot)_{48},\ (y, \otimes)_{39},\ (u, \odot)_{46},\ (w, \otimes)_{44},$$
$$(w, \odot)_{15},\ (v, \otimes)_{8},\ (y, \odot)_{14},\ (z, \otimes)_{19},\ (u, \otimes)_{12},\ (x, \otimes)_{2}\ \Big\rangle$$

$$S_{42} = \Big\langle\ (w, \otimes)_{52},\ (v, \odot)_{37},\ (u, \otimes)_{47},\ (x, \odot)_{36},\ (z, \otimes)_{43},\ (y, \otimes)_{50},$$
$$(y, \odot)_{11},\ (u, \odot)_{7},\ (v, \otimes)_{20},\ (x, \otimes)_{21},\ (z, \odot)_{22},\ (w, \odot)_{3}\ \Big\rangle$$

$$S_{43} = \Big\langle\ (y, \odot)_{52},\ (v, \odot)_{34},\ (w, \odot)_{47},\ (z, \otimes)_{35},\ (u, \odot)_{50},\ (x, \otimes)_{42},$$
$$(x, \odot)_{19},\ (w, \otimes)_{6},\ (z, \odot)_{14},\ (v, \otimes)_{18},\ (u, \otimes)_{10},\ (y, \otimes)_{2}\ \Big\rangle$$

$$S_{44} = \Big\langle\ (x, \otimes)_{51},\ (w, \odot)_{32},\ (u, \otimes)_{46},\ (y, \odot)_{30},\ (v, \otimes)_{40},\ (z, \otimes)_{49},$$
$$(z, \odot)_9,\ \ (u, \odot)_5,\ \ (w, \otimes)_{20},\ (y, \otimes)_{25},\ (v, \odot)_{21},\ (x, \odot)_3\ \Big\rangle$$

$$S_{45} = \Big\langle\ (z, \odot)_{51},\ (w, \odot)_{26},\ (x, \odot)_{46},\ (v, \otimes)_{27},\ (u, \odot)_{49},\ (y, \otimes)_{38},$$
$$(y, \odot)_{18},\ (x, \otimes)_4,\ \ (v, \odot)_{14},\ (w, \otimes)_{17},\ (u, \otimes)_8,\ \ (z, \otimes)_2\ \Big\rangle$$

$$S_{46} = \Big\langle\ (w, \otimes)_{50},\ (v, \odot)_{24},\ (u, \otimes)_{14},\ (x, \odot)_{22},\ (z, \otimes)_{47},\ (y, \otimes)_{48},$$
$$(y, \odot)_{21},\ (u, \odot)_3,\ \ (v, \otimes)_{46},\ (x, \otimes)_{49},\ (z, \odot)_{25},\ (w, \odot)_{23}\ \Big\rangle$$

$$S_{47} = \Big\langle\ (v, \odot)_{50},\ (x, \odot)_{48},\ (y, \odot)_{47},\ (w, \otimes)_{19},\ (u, \odot)_{20},\ (z, \otimes)_{16},$$
$$(z, \odot)_{46},\ (y, \otimes)_{17},\ (w, \odot)_{49},\ (x, \otimes)_{18},\ (u, \otimes)_2,\ \ (v, \otimes)_{15}\ \Big\rangle$$

$$S_{48} = \Big\langle\ (v, \otimes)_{45},\ (z, \odot)_{28},\ (u, \otimes)_{22},\ (w, \odot)_{41},\ (y, \otimes)_{29},\ (x, \otimes)_{25},$$
$$(x, \odot)_4,\ \ (u, \odot)_{10},\ (w, \otimes)_7,\ (z, \otimes)_3,\ \ (y, \odot)_9,\ \ (v, \odot)_1\ \Big\rangle$$

$$S_{49} = \Big\langle\ (v, \odot)_{44},\ (x, \odot)_{39},\ (y, \odot)_{19},\ (w, \otimes)_{33},\ (u, \odot)_{17},\ (z, \otimes)_{31},$$
$$(z, \odot)_8,\ \ (y, \otimes)_{11},\ (w, \odot)_2,\ \ (x, \otimes)_6,\ \ (u, \otimes)_5,\ \ (v, \otimes)_1\ \Big\rangle$$

$$S_{50} = \Big\langle\ (w, \otimes)_{43},\ (v, \odot)_{39},\ (u, \otimes)_{21},\ (x, \odot)_{37},\ (z, \otimes)_{33},\ (y, \otimes)_{24},$$
$$(y, \odot)_{12},\ (u, \odot)_8,\ \ (v, \otimes)_3,\ \ (x, \otimes)_5,\ \ (z, \odot)_7,\ \ (w, \odot)_1\ \Big\rangle$$

$$S_{51} = \Big\langle\ (w, \odot)_{42},\ (y, \odot)_{35},\ (z, \odot)_{18},\ (x, \otimes)_{41},\ (u, \odot)_{16},\ (v, \otimes)_{28},$$
$$(v, \odot)_6,\ \ (z, \otimes)_9,\ \ (x, \odot)_2,\ \ (y, \otimes)_4,\ \ (u, \otimes)_{13},\ (w, \otimes)_1\ \Big\rangle$$

$$S_{52} = \Big\langle\ (v, \otimes)_{41},\ (z, \odot)_{10},\ (u, \otimes)_{25},\ (w, \odot)_{35},\ (y, \otimes)_{13},\ (x, \otimes)_{40},$$
$$(x, \odot)_1,\ \ (u, \odot)_6,\ \ (w, \otimes)_3,\ \ (z, \otimes)_{23},\ (y, \odot)_{32},\ (v, \odot)_5\ \Big\rangle$$

$$S_{53} = \Big\langle\ (w, \otimes)_{39},\ (v, \odot)_{17},\ (u, \otimes)_{11},\ (x, \odot)_{38},\ (z, \otimes)_{12},\ (y, \otimes)_{37},$$
$$(y, \odot)_2,\ \ (u, \odot)_{15},\ (v, \otimes)_7,\ \ (x, \otimes)_1,\ \ (z, \odot)_{27},\ (w, \odot)_4\ \Big\rangle$$

$$S_{54} = \Big\langle\ (w, \otimes)_{37},\ (v, \odot)_8,\ (u, \otimes)_{24},\ (x, \odot)_{27},\ (z, \otimes)_{11},\ (y, \otimes)_{36},$$
$$(y, \odot)_1,\ \ (u, \odot)_4,\ \ (v, \otimes)_{22},\ (x, \otimes)_3,\ \ (z, \odot)_{29},\ (w, \odot)_{13}\ \Big\rangle$$

$$S_{55} = \Big\langle \ (x, \otimes)_{35}, \ (w, \odot)_{16}, \ (u, \otimes)_9, \ (y, \odot)_{34}, \ (v, \otimes)_{10}, \ (z, \otimes)_{32},$$
$$(z, \odot)_2, \ (u, \odot)_{19}, \ (w, \otimes)_5, \ (y, \otimes)_1, \ (v, \odot)_{31}, \ (x, \odot)_{12} \ \Big\rangle$$

$$S_{56} = \Big\langle \ (v, \odot)_{33}, \ (x, \odot)_{11}, \ (y, \odot)_{31}, \ (w, \otimes)_{21}, \ (u, \odot)_{12}, \ (z, \otimes)_{30},$$
$$(z, \odot)_1, \ (y, \otimes)_3, \ (w, \odot)_6, \ (x, \otimes)_{32}, \ (u, \otimes)_{23}, \ (v, \otimes)_9 \ \Big\rangle$$

$$S_{57} = \Big\langle \ (v, \otimes)_{29}, \ (z, \odot)_{26}, \ (u, \otimes)_7, \ (w, \odot)_{28}, \ (y, \otimes)_{27}, \ (x, \otimes)_{13},$$
$$(x, \odot)_{15}, \ (u, \odot)_{18}, \ (w, \otimes)_8, \ (z, \otimes)_1, \ \ (y, \odot)_{10}, \ (v, \odot)_2 \ \Big\rangle$$

$$S_{58} = \Big\langle \ (v, \otimes)_{25}, \ (z, \odot)_{13}, \ (u, \otimes)_{20}, \ (w, \odot)_9, \ (y, \otimes)_{22}, \ (x, \otimes)_{23},$$
$$(x, \odot)_7, \ \ (u, \odot)_1, \ \ (w, \otimes)_{24}, \ (z, \otimes)_{21}, \ (y, \odot)_5, \ \ (v, \odot)_{11} \ \Big\rangle$$

$$S_{59} = \Big\langle \ (v, \odot)_{19}, \ (x, \odot)_{17}, \ (y, \odot)_{16}, \ (w, \otimes)_{12}, \ (u, \odot)_{14}, \ (z, \otimes)_6,$$
$$(z, \odot)_{15}, \ (y, \otimes)_8, \ (w, \odot)_{18}, \ (x, \otimes)_{10}, \ (u, \otimes)_1, \ \ (v, \otimes)_4 \ \Big\rangle$$

$$S_{60} = \Big\langle \ (v, \otimes)_{13}, \ (z, \odot)_4, \ (u, \otimes)_3, \ \ (w, \odot)_{10}, \ (y, \otimes)_7, \ (x, \otimes)_9,$$
$$(x, \odot)_8, \ \ (u, \odot)_2, \ \ (w, \otimes)_{11}, \ (z, \otimes)_5, \ \ (y, \odot)_6, \ (v, \odot)_{12} \ \Big\rangle$$

Acknowledgements Research of the first author was supported by the Einstein Foundation (project Einstein Visiting Fellow Santos). Research of the second author was supported in part under the Australian Research Council's Discovery funding scheme (project number DP160104502). The authors thank Schloss Dagstuhl Leibniz-Zentrum für Informatik and the organisers of Seminar 17072, where this work was completed.

The authors thank Daniel Groves and Alan Reid for their encouragement to write up these results, and the anonymous referee for some insightful questions and comments which triggered us to find a special cover.

References

1. Agol, I.: Criteria for virtual fibering. J. Topol. **1**(2), 269–284 (2008)
2. Agol, I.: The virtual Haken conjecture. With an appendix by Agol, Daniel Groves, and Jason Manning. Doc. Math. **18**, 1045–1087 (2013)
3. Aitchison, I.R., Rubinstein, H.: Polyhedral metrics and 3-manifolds which are virtual bundles. Bull. Lond. Math. Soc. **31**(1), 90–96 (1999)
4. Aitchison, I.R., Rubinstein, H.: Combinatorial Dehn surgery on cubed and Haken 3-manifolds. In: Proceedings of the Kirbyfest (Berkeley, CA, 1998). Geometry & Topology Monographs, vol. 2, pp. 1–21. Geometry & Topology Publications, Coventry (1999)
5. Aitchison, I.R., Matsumoto, S., Rubinstein, H.: Immersed surfaces in cubed manifolds. Asian J. Math. **1**, 85–95 (1997)

6. Bergeron, N., Wise, D.T.: A boundary criterion for cubulation. Am. J. Math. **134**(3), 843–859 (2012)
7. Bosma, W., Cannon, J., Playous, C.: The Magma algebra system. I. The user language. J. Symb. Comput. **24**, 235–265 (1997)
8. Burton, B.A., Budney, R., Pettersson, W., et al.: Regina: Software for low-dimensional topology (1999–2016). http://regina.sourceforge.net/
9. Effenberger, F., Spreer, J.: simpcomp – a GAP toolkit for simplicial complexes, version 2.1.6 (2016). https://github.com/simpcomp-team/simpcomp/
10. GAP – Groups, Algorithms, and Programming, version 4.8.7 (2017). http://www.gap-system.org/
11. Haglund, F., Wise, D.T.: Special cube complexes. Geom. Funct. Anal. **17**(5), 1551–1620 (2008)
12. Hempel, J.: Orientation reversing involutions and the first Betti number for finite coverings of 3-manifolds. Invent. Math. **67**(1), 133–142 (1982)
13. Kahn, J., Markovic, V.: Immersing almost geodesic surfaces in a closed hyperbolic three manifold. Ann. Math. (2) **175**(3), 1127–1190 (2012)
14. Scott, P.: Subgroups of surface groups are almost geometric. J. Lond. Math. Soc. (2) **17**(3), 555–565 (1978)
15. Spreer, J., Tillmann, S.: Ancillary files to *Unravelling the Dodecahedral Spaces* (2017). https://arxiv.org/src/1702.08080/anc
16. Weber, C., Seifert, H.: Die beiden Dodekaederräume. Math. Z. **37**(1), 237–253 (1933)
17. Wise, D.: From riches to raags: 3-manifolds, right-angled Artin groups, and cubical geometry. In: CBMS Regional Conference Series in Mathematics. Published for the Conference Board of the Mathematical Sciences, Washington, DC, vol. 117. American Mathematical Society, Providence, RI (2012)

Part II
Other Contributed Articles

Lecture Notes on Infinity-Properads

Philip Hackney and Marcy Robertson

Abstract These are notes for three lectures on higher properads given at a program at the mathematical institute MATRIX in Australia in June 2016. The first lecture covers the case of operads, and provides a brief introduction to the Moerdijk-Weiss theory of dendroidal sets. The second lecture extends the discussion to properads and our work with Donald Yau on graphical sets. These two lectures conclude with models for higher (pr)operads given by an inner horn filling condition. Finally, in the last lecture, we explore some properties of the graphical category and use them to propose a Segal-type model for higher properads.

1 Introduction

The main goal of this lecture series is to provide a brief introduction to the theory of higher operads and properads. As these informal lecture notes stay very close to our presentations, which occupied only three hours in total, we were necessarily extremely selective in what is included. It is important to reiterate that this is *not* a survey paper on this area, and the reader will necessarily have to use other sources to get a 'big picture' overview.

Various models of infinity-operads have been developed in work of Barwick, Cisinski, Lurie, Moerdijk, Weiss and others [1, 8–10, 18, 20, 21]. In these lectures we focus on the combinatorial models which arise when one extends the simplicial category Δ by a category of trees Ω. This 'dendroidal category' leads immediately to the category of dendroidal sets [20], namely the presheaf category $\mathrm{Set}^{\Omega^{op}}$. A dendroidal set $X \in \mathrm{Set}^{\Omega^{op}}$ which satisfies an inner horn-filling condition is called a quasi-operad (see Definition 2.14). We briefly review these objects in Sect. 2.

P. Hackney
Macquarie University, Sydney, NSW, Australia
e-mail: philip@phck.net

M. Robertson (✉)
The University of Melbourne, Parkville, VIC, Australia
e-mail: marcy.robertson@unimelb.edu.au

© Springer International Publishing AG, part of Springer Nature 2018   351
D.R. Wood et al. (eds.), *2016 MATRIX Annals*, MATRIX Book Series 1,
https://doi.org/10.1007/978-3-319-72299-3_18

Properads are a generalization of operads introduced by Vallette [23] which parametrize algebraic structures with several inputs and several outputs. These types of algebraic structures include Hopf algebras, Frobenius algebras and Lie bialgebras. In our monograph [12] with D. Yau and in subsequent papers, we work to generalize the theory of infinity-operads to the properad setting. In Sect. 3 we explain the appropriate replacement of the dendroidal category Ω the graphical category Γ and define quasi-properads as graphical sets which satisfy an inner horn-filling condition. This material (and much more) can be found in the monograph [12]. It is worth mentioning that J. Kock, while reading the manuscript of [12], realized that one can give an alternative definition of the category Γ. The interested reader can find more details of this construction in [17].

In the final section, we propose a Segal-type model for infinity properads. There are clear antecedents for models of this form in several other settings [4, 6, 9, 16]. We recall the C. Berger and I. Moerdijk theory of generalized Reedy categories from [3]. The graphical category Γ is such a category, so the category of graphical spaces sSet$^{\Gamma^{op}}$ possesses a cofibrantly generated model structure with levelwise weak equivalences and relatively few fibrant objects. Finally, we discuss the Segal condition in the context of graphical sets and spaces.

2   Colored Operads, Dendroidal Sets, and Quasi-Operads

This section is a brief overview of dendroidal sets, introduced by Moerdijk and Weiss [20], which allow us to discuss the 'quasi-operad' model for infinity categories [8, 20]. Throughout this section, we are using the formal language that we will need to extend to the more subtle case of properads. For those who are unfamiliar with dendroidal sets we recommend the original paper [20] and the lecture notes by Moerdijk [19] as references.

Definition 2.1 A *graph* is a connected, directed graph G which admits legs and does not admit directed cycles. A *leg* is an edge attached to a vertex at only one end. We also want our graphs to have an ordering given by bijections

$$\mathrm{ord}_G^{in} : \{1, \ldots, m\} \longrightarrow \mathrm{in}(G)$$

$$\mathrm{ord}_G^{out} : \{1, \ldots, n\} \longrightarrow \mathrm{out}(G)$$

as well as bijections

$$\mathrm{ord}_v^{in} : \{1, \ldots, k\} \longrightarrow \mathrm{in}(v)$$

and

$$\mathrm{ord}_v^{out} : \{1, \ldots, j\} \longrightarrow \mathrm{out}(v)$$

for each v in $\mathrm{Vt}(G)$.

If we say that G is a $\mathfrak{C}$-colored graph then we are including the extra data of an edge coloring function $\eta : \mathrm{Edge}(G) \longrightarrow \mathfrak{C}$.

When we draw pictures of graphs, we will omit the arrows, and always assume the direction in the direction of gravity.

Definition 2.2 A *tree* is a simply connected graph with a unique output (the root).

For any vertex v in a $\mathfrak{C}$-colored tree T, $\mathrm{in}(v)$ is written as a list $\underline{c} = c_1, \ldots, c_k$ of colors $c_i \in \mathfrak{C}$. A list of colors like $\underline{c}$ is called a *profile* of the vertex v. Similarly, $\mathrm{out}(v) = d$ identifies the element $d \in \mathfrak{C}$ which colors the output of the vertex v. The complete input-output data of a vertex v is given by the *biprofile* $(\underline{c}; d)$.

Example 2.3 In the following picture the tree has legs labeled $3, 4, 5, 6$ and 0. The leg 0 is the single output of this graph. Internal edges are labeled $1, 2$ and 7. The edges at each vertex all come equipped with an ordering, and if we wish to list the inputs to the vertex v we would write $\mathrm{in}(v) = (1, 2)$.

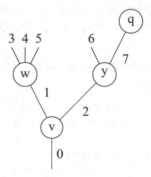

If we wanted to consider T as a $\mathfrak{C}$-colored tree, we would add the data of a coloring function $\mathrm{Edge}(T) \to \mathfrak{C}$ which would result in our picture looking like

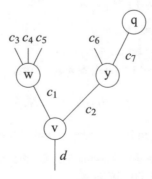

where d and each of the c_i are elements of $\mathfrak{C}$. The profile of v is $\mathrm{in}(v) = (c_1, c_2) = \underline{c}$ and the biprofile of v is written as $(\underline{c}; d)$, where the semi-colon differentiates between inputs and outputs.

2.1 Colored Operads

A *colored operad* is a generalization of a category in which we have a set of objects
(or colors) but where we allow for morphisms which have a finite list of inputs and
a single output. When we visualize these morphisms we write them as colored trees,
so that the morphism $(x_1, x_2) \xrightarrow{f} y$ looks like

Notice that in this depiction the edges of the tree are colored by the objects (hence
the name colors). A modern comprehensive treatment of colored operads appears in
the book of Yau [24].

Definition 2.4 A *colored operad P* consists of the following data:

1. A set of colors $\mathfrak{C} = \mathrm{col}(P)$;
2. for all $n \geq 0$ and all biprofiles $(\underline{c}; d) = (c_1, \ldots, c_n; d)$ in $\mathfrak{C}$, a set $P(\underline{c}; d)$;
3. for $\sigma \in \Sigma_n$, maps $\sigma^* : P(\underline{c}; d) \rightarrow P(\underline{c}\sigma; d) = P(c_{\sigma(1)}, \ldots, c_{\sigma(n)}; d)$ so that $(\sigma\tau)^* = \tau^*\sigma^*$;
4. for each $c \in \mathfrak{C}$ a unit element $\mathrm{id}_c \in P(c; c)$;
5. associative, equivariant and unital compositions

$$P(\underline{c}; d) \circ_i P(\underline{d}; c_i) \rightarrow P(c_1, \ldots, c_{i-1}, (d_1 \ldots, d_k), c_{i+1}, \ldots, c_m; d)$$

where $\underline{d} = (d_1, \ldots, d_k)$ and $1 \leq i \leq m$.

A morphism $f : P \rightarrow Q$ consists of:

1. a map of color sets $f : \mathrm{col}(P) \rightarrow \mathrm{col}(Q)$;
2. for all $n \geq 0$ and all biprofiles $(\underline{c}; d)$, a map of sets

$$f : P(\underline{c}, d) \rightarrow Q(f\underline{c}, fd)$$

which commutes with symmetric group actions, composition and units.

The category of colored operads is denoted by **Operad**.
  Examples of colored operads include:

- The 2-colored operad $\mathsf{O}^{[1]}$, whose algebras are morphisms of O-algebras for a specified uncolored operad O [2, 1.5.3]
- The $\mathbb{N}$-colored operad whose algebras are all one colored operads [2, 1.5.6], [24], [25, §14.1].

We now focus on operads which are generated by uncolored trees. Explicitly, given any uncolored tree T, one can generate a colored operad $\Omega(T)$ so that

- the set of colors of $\Omega(T)$ is taken to be the set of edges of T;
- the operations of $\Omega(T)$ are freely generated by vertices in the tree.

Example 2.5 Consider the uncolored tree T

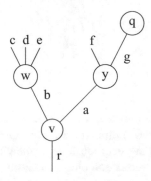

where we have labeled the edges by letters, but do not mean there is a coloring.

The associated colored operad $\Omega(T)$ will have color set

$$\mathfrak{C} = \{a, b, c, d, e, f, g, r\} = \text{Edge}(T)$$

and operations freely generated by the vertices. In this example, generating operations are $v \in \Omega(T)(a, b; r)$, $y \in \Omega(T)(f, g; a)$, $w \in \Omega(T)(c, d, e; b)$ and $q \in \Omega(T)(-; g)$. Composition of operations are given by formal graph substitutions (see Definition 2.10) into appropriate partially grafted corollas (Definition 2.8). To give a specific example, the operation $v \circ_a y \in \Omega(T)(b, f, g; r)$ is a composition of v and y which we visualize as being the result of collapsing along the edge marked a.

Definition 2.6 ([20]) The dendroidal category Ω is the full subcategory of **Operad** whose objects are colored operads of the form $\Omega(T)$. When no confusion can arise, we often write T for $\Omega(T)$.

Definition 2.7 ([20, Definition 4.1]) A *dendroidal set* is a functor $X : \Omega^{op} \to$ Set. Collectively these form a category $\text{Set}^{\Omega^{op}}$ of dendroidal sets.

An element of $x \in X_T$ is called a *dendrex* of shape T. We also have the representable functors $\Omega[T] = \Omega(-, T)$.

2.2 Coface Maps and Graph Substitution

Quasi-operads are similar in spirit to quasi-categories. In particular, they are dendroidal (rather than simplicial) sets which satisfy an inner Kan condition. This requires that we define coface and codegeneracy maps in Ω which we will make precise by using formal graph substitution.

Definition 2.8 ([12, 2.16]) A *partially grafted corolla P* is a graph with two vertices u and v in which a nonempty finite list of outputs of u are inputs of v.

Example 2.9 The following graph P is a partially grafted corolla.

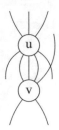

Partially grafted corollas play a key role in describing operadic and properadic composition as it arises from graph substitution. Graph substitution is a formal language for saying something very intuitive, namely that in a given graph G, you can drill a little hole at any vertex and plug in a graph H and assemble to get a new graph.

Definition 2.10 ([12, 2.4]) We can substitute a graph H into a graph G at vertex v if:

1. there is a specified bijection $\mathrm{in}(H) \xrightarrow{\cong} \mathrm{in}(v)$,
2. a specified bijection $\mathrm{out}(H) \xrightarrow{\cong} \mathrm{out}(v)$, and
3. the coloring of inputs and outputs of H matches the local coloring of G at the vertex v.

The resulting graph is denoted as $G(H_v)$ and we say that $G(H_v)$ was obtained from G via graph substitution. The subscript on H_v indicates that we substituted H into vertex v. If $S \subseteq \mathrm{Vt}(G)$, we will write $G\{H_v\}_{v\in S}$ when we perform graph substitution at several vertices simultaneously.

Graph substitution induces maps in Ω. For example consider T

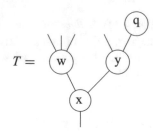

and the partially grafted corolla P

$$P = $$

Since the total number of inputs of P matches the total number of inputs of the vertex $w \in \mathrm{Vt}(G)$ and the number of outputs of P matches the number of outputs of w we can preform graph substitution.

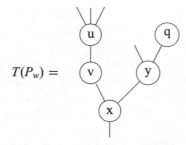

$$T(P_w) = $$

Graph substitution induces a map $T \to T(P_w)$ in Ω which sends the w to $u \circ v$, x to x, y to y, and q to q. This example generalizes, in that if we take any tree S we can expand a vertex to create an additional internal edge by substitution of the proper partially grafted corolla. The expansion of an internal edge can be written as an internal graph substitution, and we have an induced Ω-map $d^{uv} : S \to T = S(P)$ where P is the appropriate partially grafted corolla. Maps of the type d^{uv} are called *inner coface* maps [12, 6.1.1], [20, p. 6].

Let's look at another example of graph substitution. Consider the partially grafted corolla P

$$P = $$

and the tree S

$$S = $$

We can substitute S into the vertex v in the partial grafted corolla P since S has the same number of inputs and outputs as v. The resulting picture is the tree $P(S_v)$

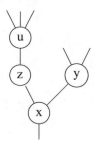

and there is a natural map $d^u : S \to P(S_v)$ which is an inclusion of S as a subtree in $P(S_v)$. For any tree T we can write all subtree inclusions by (possibly iterated) substitution of the subtree into a partially grafted corolla and maybe relabeling [12, Definition 6.32]. Maps like these which are induced by graph substitution where the partially grafted corolla is on the "outside" are called *outer coface maps* [12, 6.1.2], [20, p. 6]. The third class of maps we will concern ourselves with are called *codegeneracies* and are given by the substitution of a graph with no vertices $\downarrow$ into a bivalent vertex v, i.e. the maps $\sigma^v : H \to H(\downarrow)$ [12, 6.1.3], [20, p. 6]. The cofaces and codegeneracies satisfy identities reminiscent of the simplicial identities.

Lemma 2.11 ([20, Lemma 3.1]) *The category Ω is generated by the inner and outer coface maps, codegeneracies and isomorphisms.*

In other words, any every map in Ω can be factored as a composition of inner and outer coface maps, codegeneracies and isomorphisms. These factorizations will be more carefully discussed in Sect. 4.

2.3   Boundaries and Horns

Now that we have defined inner and outer coface maps, we can describe faces and boundaries of dendroidal sets.

Definition 2.12 ([20, pg 16]) Let $\alpha : T \to S$ be an (inner or outer) coface map in Ω. Then the α-face of $\Omega[T]$ is the image of the induced map $\alpha^* : \Omega[S] \to \Omega[T]$. We will write $\partial_\alpha[T]$ for the α-face of $\Omega[T]$.

Definition 2.13 The *boundary* of $\Omega[T]$ is the union over all the faces $\partial[T] = \bigcup_\alpha \partial_\alpha[T]$. If we omit the β-face, we have the β-horn $\Lambda^\beta[T] = \bigcup_{\alpha \neq \beta} \partial_\alpha[T]$. If, moreover, β is the image of an inner coface map then $\Lambda^\beta[T]$ is called an *inner horn*.

A quasi-operad is now defined as a dendroidal set satisfying an inner Kan lifting property.

Definition 2.14 ([21, pg 352]) A dendroidal set X is a *quasi-operad* if for every diagram given by the solid arrows admits a lift

where T ranges over all trees and β ranges over all inner coface maps.

Definition 2.15 ([8, Proposition 1.5]) A monomorphism of dendroidal sets $X \to Y$ is said to be *normal* if and only if for any tree T, the action of $\mathrm{Aut}(T)$ on $Y_T \setminus X_T$ is free.

In analogy to the Joyal model structure on sSet for quasi-categories (see [5] for references), we have the following.

Theorem 2.16 ([8, Theorem 2.4]) *There is a model category structure on* $\mathrm{Set}^{\Omega^{op}}$ *such that the quasi-operads are the fibrant objects and the normal monomorphisms are the cofibrations.*

3 Colored Properads, Graphical Sets, and Quasi-Properads

In the previous section we gave a very quick introduction to the dendroidal category using some of the formal language of graph substitution. We will now extend this language to a larger class of graphs to describe properads.

Isomorphisms between graphs preserve all the structure (including orderings) and *weak isomorphisms* between graphs preserve all the structure except the ordering. We denote the category of graphs up to strict isomorphism as **Graph**. The category **Graph**(m, n) is a subcategory of **Graph** whose objects are graphs G where $|\mathrm{in}(G)| = m$ and $|\mathrm{out}(G)| = n$. The category **Graph**$(\underline{c}, \underline{d})$ similarly consists of all $\mathfrak{C}$-colored graphs with $\mathrm{in}(G) = \underline{c} = (c_1, .., c_m)$ and $\mathrm{out}(G) = \underline{d} = (d_1, ..., d_m)$.

3.1 Properads

Like an operad, a colored properad is a generalization of a category. We have a set of objects, called colors, and now we allow our morphisms to have finite lists of inputs and finite lists of outputs. When we write down a visual representation of a morphism $(x_1, x_2) \xrightarrow{f} (y_1, y_2, y_3)$ in a properad we usually write a colored graph

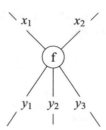

but it really could be any graph with 2 inputs and 3 outputs that is colored by the objects of the properad P. In other words, a morphism $(x_1, x_2) \xrightarrow{g} (y_1, y_2, y_3)$ in P is a graph $g \in \mathsf{Graph}(x_1, x_2; y_1, y_2, y_3)$. Composition of morphisms follows the same basic principle of operad composition. In an operad you think of the $\circ_i$ composition as plugging the root of a tree into the ith leaf of another tree. For properads we want to be able to take any sub-list of outputs of a graph and glue them to appropriately matched sub-list of inputs in another graph.

Definition 3.1 ([12, Definition 3.5]) An $\mathfrak{C}$-colored properad P consists of

- a set $\mathfrak{C} = \mathrm{col}(P)$ of colors;
- for each biprofile $(\underline{c}; \underline{d}) = (c_1, \ldots, c_m; d_1, \ldots, d_n)$, a set $P(\underline{c}; \underline{d})$;
- for $\sigma \in \Sigma_m$ and $\tau \in \Sigma_n$, maps

$$P(\underline{c}; \underline{d}) \to P(\underline{c}\sigma; \tau\underline{d}) = P(c_{\sigma(1)}, \ldots, c_{\sigma(m)}; d_{\tau^{-1}(1)}, \ldots, d_{\tau^{-1}(n)})$$

which assemble into a $\Sigma_m^{op} \times \Sigma_n$ action on the collection $\coprod_{|\underline{c}|=m, |\underline{d}|=n} P(\underline{c}; \underline{d})$;
- for all $c \in \mathfrak{C}$, a unit $\mathrm{id}_c \in P(c; c)$;
- an associative, unital and equivariant composition

$$\boxtimes_{b'}^{a'} : P(\underline{c}; \underline{d}) \otimes P(\underline{a}; \underline{b}) \to P(\underline{a} \circ_{a'} \underline{c}; \underline{b} \circ_{b'} \underline{d})$$

where $\underline{a}'$ and $\underline{b}'$ denote some non-empty finite sublist of $\underline{c}$ and $\underline{b}$, respectively. The notation $\underline{a} \circ_{a'} \underline{c}$ denotes identifying some sublist of $\underline{a}$ with the appropriate sublist of $\underline{c}$.

A map of colored properads $f : P \to Q$ consists of

- $f_0 : \mathrm{Col}(P) \to \mathrm{Col}(Q)$;
- $f_1 : P(\underline{c}; \underline{d}) \to Q(f_0\underline{c}; f_0\underline{d})$ for all biprofiles $(\underline{c}, \underline{d})$ in $\mathfrak{C}$.

We denote the category of all colored properads and properad maps between them as Properad.

Properadic composition is easiest to write down in terms of graph substitution. In the previous talk we described a formal process called *graph substitution*, which now repeat in the case of graphs.

Definition 3.2 ([12, 2.4]) Given a graph $G \in \mathsf{Graph}(\underline{c}; \underline{d})$, and a graph $H_v \in \mathsf{Graph}(\mathrm{in}(v); \mathrm{out}(v))$ so that each H_v is equipped with bijections

- in$(H_v) \to$ in(v) and
- out$(H_v) \to$ out(v)

one constructs a new graph $G(H_v) \in$ Graph$(\underline{c}; \underline{d})$ by formally identifying H_v with $v \in G$. In this case we say that $G(H_v)$ is obtained from G by substitution.

The following is an example of (uncolored) graph substitution. Let G and P be the graphs below.

Graph$(5, 6) \ni G =$ 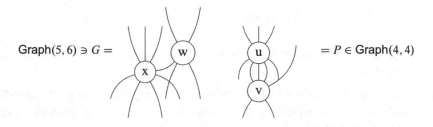 $= P \in$ Graph$(4, 4)$

The graph $G(P_x)$ is still a member in the category Graph$(5, 6)$, but now has a additional three internal edges.

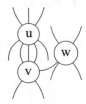

To see how this might encode composition, notice that if we squish down the three internal edges between the vertex u and v we would have something that captures our description of composition.

Following this discussion, one would say that a $\mathfrak{C}$-colored properad P is the object you get if you consider the set $\mathfrak{C}$ as objects (or colors) and morphisms between objects $P(\underline{c}; \underline{d})$ are a set of (possibly decorated) $\mathfrak{C}$-colored graphs in Graph$(\underline{c}, \underline{d})$. Composition of a G-configuration of morphisms is given by graph substitution

$$\gamma_P^G : P[G] = \prod_{\mathrm{Vt}(G)} P(\mathrm{in}(v); \mathrm{out}(v)) \to P(\mathrm{in}\, G; \mathrm{out}\, G)$$

where we are ranging over all maps that arise from graph substitution and look like $G(P_x) \to G$ in the example above. Properadic composition defined in this way is associative and unital because graph substitution is associative and unital [12, 2.2.4]. Symmetric group actions come from weak isomorphisms of graphs and properadic composition is equivariant because graph substitution is an operation which is defined up to weak isomorphism class of graphs.

Remark 3.3 Because graph substitution is associative, we observe that it is possible to define properadic composition one operation at a time. In fact, properadic composition is completely determined by the operations described by partially grafted corollas, γ_P^G, the graph with just an edge (for identities), and the one vertex graphs (for symmetric group actions).

3.2 The Graphical Category Γ

It should by now be unsurprising to hear that given an uncolored graph G we can freely generate a properad $\Gamma(G)$.

Definition 3.4 ([12, Section 5.1]) Given an uncolored graph G, the properad $\Gamma(G)$ is a colored properad which has the set Edge(G) as colors and morphisms are generated by the vertices.

More explicitly, an operation in $\Gamma(G)(\underline{c}; \underline{d})$ is a $\hat{G}$-*decorated graph*, meaning:

- a graph H in Graph($\underline{c}; \underline{d}$) whose edges are colored by edges of G;
- a function from the vertices of H to the vertices of G which is compatible with the coloring of H.

Example 3.5 ([12, Lemma 5.13]) Given the following graph G,

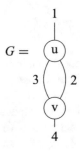

the $\hat{G}$-decorated graph H below is an example of a morphism in $\Gamma(G)(1, 1; 4, 4)$.

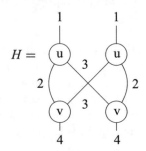

Notice that there are many, many more operations in the properads $\Gamma(G)$ than there were in the operads $\Omega(T)$ that we discussed in the first lecture. This isn't because we forgot to mention operations in $\Omega(T)$ but rather because of the following lemma.

Lemma 3.6 ([12, Lemma 5.10]) *If G is a simply connected graph, then each vertex in G can appear in a morphism in the properad $\Gamma(G)$ at most once.*

As we mentioned in Remark 3.3, properadic composition is generated by the composites of partially grafted corollas, the graph with one edge, and one vertex graphs. To see that our definition of $\Gamma(G)$ actually is a properad, it then suffices to check the following lemma.

Lemma 3.7 *All $\hat{G}$-decorated graphs can be built iteratively using partially grafted corollas.*

The naive guess, based on what we expect from understanding Δ and Ω, would be to define a category Γ which has as objects the graphical properads $\Gamma(G)$ and morphisms all properad maps between them. This is, unfortunately, not the appropriate definition of Γ as there maps between graphical properads that exhibit idiosyncratic behavior.

Definition 3.8 A properad morphism $f : \Gamma(G) \to \Gamma(H)$ consists of:

- a function $f_0 : \mathrm{Edge}(G) \to \mathrm{Edge}(H)$ together with
- a map $f_1 : \mathrm{Vt}(G) \to \{\mathrm{Vt}(H)\text{-decorated graphs}\}$ such that for every $v \in \mathrm{Vt}(G)$, $f_1(v)$ is an $\hat{H}$-decorated graph in $\mathsf{Graph}(f_0 \text{ in } v; f_0 \text{ out } v)$.

Definition 3.9 The image of $f : \Gamma(G) \to \Gamma(H)$ is $f_0 G\{f_1(v)\}_{v \in \mathrm{Vt}(G)}$ which is naturally $\mathrm{Vt}(H)$-decorated. The notation $G\{f_1(v)\}_{v \in \mathrm{Vt}(G)}$ stands for performing iterated graph substitution of $\mathrm{Vt}(H)$-decorated graphs at each vertex v in G.

Morphisms between graphical properads are very strange, so we will pause here and give an explicit description of the image of a map $f : \Gamma(G) \to \Gamma(H)$.

Example 3.10 Suppose that $G = \downarrow$ is the graph with no vertices and let Q be a $\mathfrak{C}$-colored properad. Then a properad map $f : \Gamma(\downarrow) \to Q$ is a choice of color $c \in \mathfrak{C}$.

Example 3.11 An example of a morphism of graphical properads that behaves poorly is the following. Suppose G is the graph

$$G = \quad \text{}$$

let $f : \Gamma(G) \to \Gamma(G)$ be the morphism where f_0 is the identity on edge sets and

- $f_1(v)$ is the $\hat{G}$ decorated graph

$$f_1(v) =$$

and
- $f_1(u)$ is the $\hat{G}$ decorated graph

$$f_1(u) =$$

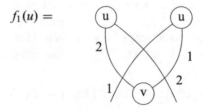

The image of f in $\Gamma(G)$ is then the $\hat{G}$-decorated graph

$$\mathrm{im}(G) =$$

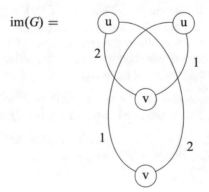

As we saw in Example 3.11, properad maps $f : \Gamma(H) \to \Gamma(G)$ need not have the property that the image of H is a subgraph of G. This kind of behavior does not show up in dendroidal sets. In fact, for maps into simply connected graphical properads behaves exactly as we would expect from the dendroidal case.

Proposition 3.12 ([12, Proposition 5.32]) *If the target of $f : \Gamma(H) \to \Gamma(G)$ is simply connected (eg any object of Ω), then f is uniquely determined by what it does on edges.*

As we will explain in Sect. 4, in order for our graphical category to have the proper sense of homotopy theory, we will want to force a property of this kind on the category Γ.

Proposition 3.13 *If the image of H under $f : \Gamma(H) \to \Gamma(G)$ is a subgraph of G, then f is uniquely determined by what it does on edges.*

Definition 3.14 The *graphical category* Γ is the category with objects graphical properads and morphisms the subset of properad maps $f : \Gamma(H) \to \Gamma(G)$ consisting of those f with the property that $\mathrm{im} f$ is a subgraph of G.

Definition 3.15 The category of *graphical sets* is the category of presheaves on Γ, that is $\mathrm{Set}^{\Gamma^{op}}$.

For every graph G an element in the set X_G is called a *graphex* with shape G. The plural form of graphex is graphices. The representable objects of shape G are $\Gamma[G] = \Gamma(-, G)$.

3.3 The Properadic Nerve

The obvious question to ask at this point is how do we know that by throwing out badly behaved properad maps that we are still looking at a reasonable definition of graphical sets? The *properadic nerve* [12, Definition 7.5] is the functor

$$N : \mathrm{Properad} \longrightarrow \mathrm{Set}^{\Gamma^{op}}$$

defined by

$$(NP)_G = \mathrm{Properad}(\Gamma(G), P)$$

for P a properad. A graphex in $(NP)_G$ is really a P-decoration of G, which consists of a coloring of the edges in G by the colors of P and a decoration of each vertex in G by an element in P with the corresponding profiles.

Proposition 3.16 ([12, Proposition 7.39]) *The properadic nerve*

$$N : \mathrm{Properad} \longrightarrow \mathrm{Set}^{\Gamma^{op}}$$

is fully faithful.

This proposition implies that while we have lost some maps in Γ we have still enough information so that the entire category Properad sits inside of $\mathrm{Set}^{\Gamma^{op}}$.

3.4 Cofaces and Codegeneracies

As in our first lecture, the coface and codegeneracy maps are given by graph substitutions of various kinds. A *codegeneracy* map $\sigma^v : H \to H(\downarrow)$ is a map induced by substitution of the graph with one edge $\downarrow$ into $(1, 1)$-vertex $v \in \mathrm{Vt}(H)$.

This has the effect of deleting a vertex. Like in Ω, an *inner coface* map will have the effect of "blowing up" the graph between two vertices by an inner substitution of a partially grafted corolla $d^{uv} : G \to G(P)$.

Example 3.17 As an example of an inner coface map consider the graph substitution we have already seen,

where the partially grafted corolla $P \in \mathsf{Graph}(4, 4)$ is pictured below.

Example 3.18 When restricted to linear graphs, an inner coface map as above is the same as an inner coface map in the simplicial category Δ [12, Example 6.4].

An *outer coface* map $d^v : G \to P(G)$ is an outer substitution of a graph G into a partially grafted corolla. In the next section, we will discuss how these maps generate the whole category Γ in the sense that all morphisms in Γ are compositions of (inner or outer) coface maps, codegeneracies and isomorphisms.

Definition 3.19 A *face* of a representable $\Gamma[H]$ is given by considering the image of an inner or outer coface map. The boundary of $\Gamma[H]$ is defined as $\partial[H] = \bigcup_\alpha \partial_\alpha[H] \subset \Gamma[H]$ where α ranges over all inner and outer coface maps. The β-*horn* is then defined as $\Lambda^\beta[H] \subset \Gamma[H] = \bigcup_{\beta \neq \alpha} \partial_\alpha[H]$ where β is a coface map.

Definition 3.20 A graphical set X is a *quasi-properad* if, for all inner coface maps α and all H in Γ, the diagram

$$
\begin{array}{ccc}
\Lambda^\alpha[H] & \longrightarrow & X \\
\downarrow & \nearrow & \\
\Gamma[H] & &
\end{array}
$$

admits a lift.

A model category structure on $\mathrm{Set}^{\Gamma^{op}}$ in which quasi-properads are the fibrant objects is work in progress between the authors and D. Yau.

4 Generalized Reedy Structures and a Segal Model

In the previous section we described the graphical category Γ and quasi-properads. For more details on why this is precisely a properad "up to homotopy" see the description in [12, 7.2]. In this section we will describe the Reedy structure of Γ and use it as a starting point to construct one model category structure for infinity properads.

4.1 Generalized Reedy Categories

Definition 4.1 ([3, Definition 1.1]) A dualizable generalized Reedy structure on a small category $\mathbb{R}$ consists of two subcategories $\mathbb{R}^+$ and $\mathbb{R}^-$ which each contain all objects of $\mathbb{R}$, together with a degree function $\mathrm{Ob}(\mathbb{R}) \to \mathbb{N}$ satisfying:

1. non-invertible morphisms in $\mathbb{R}^+$ (respectively $\mathbb{R}^-$) raise (respectively lower degree). Isomorphisms preserve degree.
2. $\mathbb{R}^+ \cap \mathbb{R}^- = \mathrm{Iso}(\mathbb{R})$
3. Every morphism f factors as $f = gh$ such that $g \in \mathbb{R}^+$ and $h \in \mathbb{R}^-$ and this factorization is unique up to isomorphism.
4. If $\theta f = f$ for $\theta \in \mathrm{Iso}(\mathbb{R})$ and $f \in \mathbb{R}^-$ then θ is an identity.
5. $f\theta = f$ for $\theta \in \mathrm{Iso}(\mathbb{R})$ and $f \in \mathbb{R}^+$ then θ is an identity.

Remark 4.2 A category $\mathbb{R}$ that satisfies axioms (1)–(4) is a generalized Reedy category. If, in addition, $\mathbb{R}$ satisfies axiom (5) then $\mathbb{R}$ is said to be dualizable, which implies that $\mathbb{R}^{op}$ is also a generalized Reedy category.

A (classical) Reedy category is a generalized Reedy category $\mathbb{R}$ in which every element of $\mathrm{Iso}(\mathbb{R})$ is an identity. Examples of classical Reedy categories include Δ and Δ^{op}. Examples of generalized Reedy categories include the dendroidal category Ω, finite sets, pointed finite sets, and the cyclic category Λ.

The main idea of Reedy categories is that we can think about lifting morphisms from $\mathbb{R}$ to $\mathcal{M}^{\mathbb{R}}$ by induction on the degree of our objects. To formalize this idea we introduce the notion of latching and matching objects.

For any $r \in \mathbb{R}$, the category $\mathbb{R}^+(r)$ is defined to be a full subcategory of $\mathbb{R}^+ \downarrow r$ consisting of those maps with target r which are not invertible. Similarly, the category $\mathbb{R}^-(r)$ is the full subcategory of $r \downarrow \mathbb{R}^-$ consisting of maps $\alpha : r \to s$ which are non-invertible. One can now define the *latching object*

$$L_r(X) = \operatorname*{colim}_{\alpha \in \mathbb{R}^+(r)} X_s,$$

for each X in $\mathcal{M}^{\mathbb{R}}$ which comes equipped with a map $L_r(X) \to X_r$. Similarly, for each $X \in \mathcal{M}^{\mathbb{R}}$ we define the *matching object*

$$\lim_{\alpha \in \mathbb{R}^-(r)} X_s = M_r(X)$$

which comes equipped with a map $X_r \to M_r(X)$.

Definition 4.3 If $\mathcal{M}$ is a cofibrantly generated model category, and $\mathbb{R}$ is generalized Reedy, we say that a morphism $f : X \to Y$ in $\mathcal{M}^{\mathbb{R}}$ is:

- a Reedy cofibration if $X_r \cup_{L_r X} L_r Y \to Y_r$ is a cofibration in $\mathcal{M}^{\mathrm{Aut}(r)}$ for all $r \in \mathbb{R}$;
- a Reedy weak equivalence if $X_r \to Y_r$ is a weak equivalence in $\mathcal{M}^{\mathrm{Aut}(r)}$ for all $r \in \mathbb{R}$;
- a Reedy fibration if $X_r \to M_r X \times_{M_r Y} Y_r$ is a fibration in $\mathcal{M}^{\mathrm{Aut}(r)}$ for all $r \in \mathbb{R}$.

Theorem 4.4 ([3, Theorem 1.6]) *If $\mathcal{M}$ is a cofibrantly generated model category and $\mathbb{R}$ is a generalized Reedy category then the diagram category $\mathcal{M}^{\mathbb{R}}$ is a model category with the Reedy fibrations, Reedy cofibrations, and Reedy weak equivalences defined above.*

4.2   The Graphical Category is Generalized Reedy

Theorem 4.5 ([12, 6.4]) *The graphical category Γ is a dualizable generalized Reedy category.*

The degree function $d : \mathrm{Ob}(\Gamma) \to \mathbb{N}$ is defined as $d(G) = |\mathrm{Vt}(G)|$. The positive maps are then those morphisms in Γ which are injective on edge sets. The negative maps are those $H \to G$ which are surjective on edge sets and which, for every vertex $v \in \mathrm{Vt}(G)$, there is a vertex $\tilde{v} \in \mathrm{Vt}(H)$ so that $f_1(\tilde{v})$ is a corolla containing v. An alternate, more illuminating, description is given by the following proposition.

Lemma 4.6 ([12, 6.65])

- *A map $f : H \to G$ is in Γ^+ if we can write it as a composition of isomorphisms and coface maps.*
- *A map $f : H \to G$ is in Γ^- if we can write it as a composition of isomorphisms and codegeneracy maps.*

The proof of this lemma isn't entirely trivial, but the general idea is that codegeneracy maps decrease degree and satisfy the extra condition; coface maps increase degree and are injective on edges.

We will not fully prove here that Γ is Reedy. However, we can show where the decompositions in the third axiom of Definition 4.1 come from.

Proposition 4.7 ([12, 6.68]) *Every map in $f \in \Gamma$ factors as $f = g \circ h$, where $h \in \Gamma^-$ and $g \in \Gamma^+$ and this factorization is unique up to isomorphism.*

Proof (Sketch of Existence) Given a morphism $f: G \to K$ in Γ we know that for all $v \in \mathrm{Vt}(G), f_1(v)$ is a subgraph of K.

Let us consider $T \subset \mathrm{Vt}(G)$, the subset of vertices of G such that $f_1(v) = \downarrow$. We can define a graph $G_1 = G\{\downarrow_w\}_{w \in \mathrm{Vt}(G)}$ which is the graph obtained by substitution of an edge into each $w \in T$ and a corolla substituted into each additional vertex. There is then a map $G \to G_1$ which is a composition of codegeneracy maps, one for each $w \in T$. Next, define a subgraph G_2 of K as $G_2 = f_0(G_1)$. In other words, G_2 is the subgraph obtained by applying f_0 to the edges of G_1, which makes sense because for each $w \in T$ the incoming edge and outgoing edge of w will have the same image under f_0. There is an isomorphism $G_1 \to G_2$ which is just the changing the names of edges via the assignment given by f_0. The vertices of G_2 are in bijection with the set $\mathrm{Vt}(G_1) \setminus T$.

It is now the case that the image of f, $\mathrm{im}(f) = G_2\{f_1(u)\}_{u \in \mathrm{Vt}(G)\setminus T}$ where each $f_1(u)$ has at least one vertex. Summarizing, (ignoring coloring) there exists a factorization:

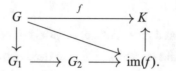

$$G \xrightarrow{f} K$$
$$\downarrow \qquad \uparrow$$
$$G_1 \longrightarrow G_2 \dashrightarrow \mathrm{im}(f).$$

This shows the existence of the decomposition.

Example 4.8 Let us turn to an example of how we generate G_1 for the example of f below.

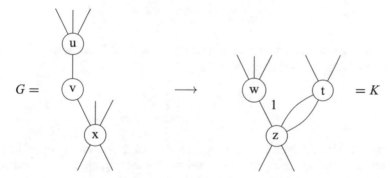

$$G = \quad \longrightarrow \quad = K$$

Notice that the vertex v is the only vertex in G which has exactly one input and one output, and is mapped by f to the edge in K we have labeled 1. It follows then that $G_1 = G(\downarrow_v)$ and looks like

$$G_1 =$$

The subgraph G_2 is now a relabeling and $\text{im}(f) = G_2(f_1(u), f_1(x))$ where $f_1(u)$ is a corolla and $f_1(x)$ is the appropriate partially grafted corolla.

4.3 A Segal Model Structure for Infinity-Properads

In this section, we attempt to describe a model structure for infinity-properads. In preparing these notes, we realized the model structure is more complicated than what we presented in the original lectures, for reasons we outline in Remarks 4.10 and 4.11.

We begin with a description of the Segal condition for a graphical set $X \in \text{Set}^{\Gamma^{op}}$. For $G \in \Gamma$, there is a natural map

$$X_G \to \prod_{v \in \text{Vt}(G)} X_{C_v} \tag{1}$$

by using all of the (iterated outer coface) maps $C_v \to G$. Of course if there is an edge e between two vertices v and w, then the two composites $\downarrow_e \overset{i_e}{\to} C_v \to G$ and $\downarrow_e \overset{i_e}{\to} C_w \to G$ are equal, so (1) factors through a subspace[1]

$$X_G^1 = \lim_{\substack{C_v \leftarrow \downarrow_e \to C_w \\ e \text{ an internal} \\ \text{edge of } G}} \left(\begin{array}{ccc} X_{C_v} & & X_{C_w} \\ & \searrow \quad \swarrow & \\ & X_\downarrow & \end{array} \right)$$

consisting of those sequences (x_v) so that $i_e^*(x_v) = i_e^*(x_w)$ whenever e is an edge between v and w. The *Segal map* is

$$X_G \overset{\chi_G}{\longrightarrow} X_G^1 \subseteq \prod_{v \in \text{Vt}(G)} X_{C_v}.$$

If $X = N(P)$ is the nerve of a properad P, then χ_G is an isomorphism [12, Lemma 7.38]. In fact, this property *characterizes* those graphical sets which are isomorphic to the nerve of a properad [12, Theorem 7.42].

If we allow ourselves to work with graphical *spaces* instead of just graphical sets, then we can replace the isomorphism condition on the Segal maps by a homotopy condition (this type of idea goes all the way back to Segal [22]).

---

[1]This is not a condition when $X_\downarrow = *$ is a one-point set; in that case, X_G^1 is just the product from (1).

Definition 4.9 A graphical space $X \in \mathrm{sSet}^{\Gamma^{op}}$ is said to satisfy the *Segal condition* if the Segal map

$$X_G \xrightarrow{\chi_G} X_G^1 \subseteq \prod_{v \in \mathrm{Vt}(G)} X_{C_v}$$

is a weak homotopy equivalence of simplicial sets between X_G and X_G^1 for each graph G.

As in the classical cases, the Segal condition is not categorically well-behaved. To study the homotopy theory of graphical spaces satisfying the Segal condition, we will build a model structure which allows us to identify such graphical spaces (or, at least those which possess an additional fibrancy condition).

Since Γ is a dualizable Reedy category [12, Theorem 6.70], we know that Γ^{op} is also generalized Reedy. Hence, by Berger and Moerdijk [3, Theorem 1.6], the diagram category $\mathrm{sSet}^{\Gamma^{op}}$ admits a generalized Reedy model structure.

Remark 4.10 During the lecture, we stated that we could modify this so that the diagram category $\mathrm{sSet}_{disc}^{\Gamma^{op}}$ admits a Reedy-type model structure, where the subscript *disc* means that $X_{\downarrow}$ is discrete as a simplicial set. Indeed, there is such a model structure: the inclusion functor $\mathrm{sSet}_{disc}^{\Gamma^{op}} \hookrightarrow \mathrm{sSet}^{\Gamma^{op}}$ admits a left adjoint given by sending X to the pushout of $\pi_0(\mathrm{sk}_0(X)) \leftarrow \mathrm{sk}_0(X) \to X$, where the skeleton is taken in the Γ direction. One can then lift the model structure from $\mathrm{sSet}^{\Gamma^{op}}$ using [15, 11.3.2]. Unfortunately, one of the generating cofibrations is not a monomorphism, hence this model structure on $\mathrm{sSet}_{disc}^{\Gamma^{op}}$ is not cellular.

The following remark is essentially adapted from the end of [4, §3.12].

Remark 4.11 There is no model structure on $\mathrm{sSet}_{disc}^{\Gamma^{op}}$ where weak equivalences are levelwise and cofibrations are monomorphisms, as one can see by attempting to factor $\Gamma[\downarrow] \amalg \Gamma[\downarrow] \to \Gamma[\downarrow]$ as a cofibration followed by an acyclic fibration:

$$\Gamma[\downarrow] \amalg \Gamma[\downarrow] \rightarrowtail X \xrightarrow{\sim} \Gamma[\downarrow].$$

Since $\Gamma[\downarrow]_{\downarrow}$ is a set of cardinality one, the object $X \in \mathrm{sSet}_{disc}^{\Gamma^{op}}$ would satisfy $2 \leq |X_{\downarrow}| = 1$.

Definition 4.12 (Segal Core Inclusions) [12, Definition 7.35] Given a graph G with at least one vertex let C_v denote the corolla at each $v \in \mathrm{Vt}(G)$ and let $\Gamma[C_v]$ denote the representable graphical set on C_v. Define the *Segal core* $\mathrm{Sc}[G]$ as the graphical subset

$$\mathrm{Sc}[G] = \bigcup_{v \in \mathrm{Vt}(G)} \mathrm{Im}\left(\Gamma[C_v] \xrightarrow{i_v} \Gamma[G] \right) \subseteq \Gamma[G]$$

where i_v is an iterated outer coface map. Denote by $\mathrm{Sc}[G] \xrightarrow{c} \Gamma[G]$ the *Segal core inclusion*.

The reader should compare this definition with [9, Definition 2.2]. Notice how suggestive this is in light of Definition 4.9: the map χ_G is exactly c^* : $\mathrm{map}(\Gamma[G], -) \to \mathrm{map}(\mathrm{Sc}[G], -)$ when X is fibrant. As we saw above, we cannot guarantee the existence of a left Bousfield localization of the non-cellular category $\mathrm{sSet}_{disc}^{\Gamma^{op}}$ at the set of Segal core inclusions. Despite that, we still expect that the following holds.

Conjecture 4.13 There is a model structure on $\mathrm{sSet}_{disc}^{\Gamma^{op}}$ analogous to those given in [9, 8.13] and [4, 5.1].

In [13], D. Yau and the authors gave a model structure on the category sProperad of simplicially-enriched properads. The properadic nerve functor that we discussed earlier extends to a functor

$$N : \mathrm{sProperad} \to \mathrm{sSet}_{disc}^{\Gamma^{op}}$$

since $N(P)_\downarrow$ is the *set* of colors of the simplicially-enriched properad P. One should compare the conjectural model structure on $\mathrm{sSet}_{disc}^{\Gamma^{op}}$ with the model structure on sProperad.

Conjecture 4.14 The properadic nerve functor from simplicial properads to graphical spaces,

$$N : \mathrm{sProperad} \to \mathrm{sSet}_{disc}^{\Gamma^{op}}$$

is the right adjoint in a Quillen equivalence.

4.4 A Diagrammatic Overview

We conclude with a diagram which was provided as a handout at our lectures. It indicates some interconnectedness of many models of categories, operads, properads, and props.

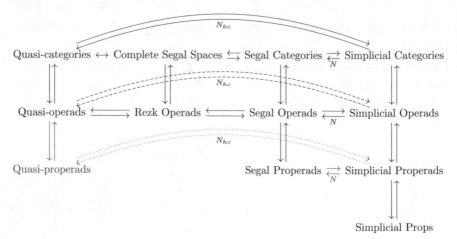

The vertical uncolored adjunctions are Quillen adjunctions. The horizontal adjunctions are Quillen equivalences;

- precise references for the top row may be found in [5], and
- the middle row is contained in [8–10].

In addition, the model structure for quasi-operads is equivalent to a model structure for Lurie's infinity operads [1, 7, 14]. The existence of the model structures in the bottom two slots on the right are [11, 13].

Acknowledgements These lectures were given in the inaugural workshop at the mathematical research institute MATRIX in Australia called "Higher Structures in Geometry and Physics" in June 2016; needless to say, these notes would not exist had MATRIX not supported us and allowed us to host the program in the first place. We would like to thank all the participants of the workshop for asking interesting questions and forcing us to refine these ideas, and also to Jon Beardsley, Julie Bergner, and Joachim Kock for offering feedback on earlier drafts of these notes. A special thank you goes to Gabriel C. Drummond-Cole who generously shared his liveTEXed notes which formed the backbone of this document. We are also grateful to the Hausdorff Research Institute for Mathematics and the Max Planck Institute for Mathematics for their hospitality while we were finishing the writing and editing of these notes.

References

1. Barwick, C.: From operator categories to topological operads (2013). https://arxiv.org/abs/1302.5756
2. Berger, C., Moerdijk, I.: Resolution of coloured operads and rectification of homotopy algebras. In: Categories in Algebra, Geometry and Mathematical Physics. Contemporary Mathematics, vol. 431, pp. 31–58. American Mathematical Society, Providence, RI (2007). http://dx.doi.org/10.1090/conm/431/08265
3. Berger, C., Moerdijk, I.: On an extension of the notion of Reedy category. Math. Z. **269**(3–4), 977–1004 (2011). http://dx.doi.org/10.1007/s00209-010-0770-x
4. Bergner, J.E.: Three models for the homotopy theory of homotopy theories. Topology **46**(4), 397–436 (2007). http://dx.doi.org/10.1016/j.top.2007.03.002
5. Bergner, J.E.: A survey of $(\infty, 1)$-categories. In: Towards Higher Categories. The IMA Volumes in Mathematics and its Applications, vol. 152, pp. 69–83. Springer, New York (2010). http://dx.doi.org/10.1007/978-1-4419-1524-5_2
6. Bergner, J.E., Hackney, P.: Group actions on Segal operads. Israel J. Math. **202**(1), 423–460 (2014). http://dx.doi.org/10.1007/s11856-014-1075-2
7. Chu, H., Haugseng, R., Heuts, G.: Two models for the homotopy theory of ∞-operads (2016). https://arxiv.org/abs/1606.03826
8. Cisinski, D.C., Moerdijk, I.: Dendroidal sets as models for homotopy operads. J. Topol. **4**(2), 257–299 (2011). http://dx.doi.org/10.1112/jtopol/jtq039
9. Cisinski, D.C., Moerdijk, I.: Dendroidal Segal spaces and ∞-operads. J. Topol. **6**(3), 675–704 (2013). http://dx.doi.org/10.1112/jtopol/jtt004
10. Cisinski, D.C., Moerdijk, I.: Dendroidal sets and simplicial operads. J. Topol. **6**(3), 705–756 (2013). http://dx.doi.org/10.1112/jtopol/jtt006
11. Hackney, P., Robertson, M.: The homotopy theory of simplicial props. Israel J. Math. **219**(2), 835–902 (2017). https://doi.org/10.1007/s11856-017-1500-4
12. Hackney, P., Robertson, M., Yau, D.: Infinity Properads and Infinity Wheeled Properads. Lecture Notes in Mathematics, vol. 2147. Springer, Berlin (2015). arXiv:1410.6716 [math.AT]. http://dx.doi.org/10.1007/978-3-319-20547-2

13. Hackney, P., Robertson, M., Yau, D.: A simplicial model for infinity properads. To appear in Higher Structures (2015). http://arxiv.org/abs/1502.06522
14. Heuts, G., Hinich, V., Moerdijk, I.: On the equivalence between Lurie's model and the dendroidal model for infinity-operads (2013). https://arxiv.org/abs/1305.3658
15. Hirschhorn, P.S.: Model Categories and Their Localizations. Mathematical Surveys and Monographs, vol. 99. American Mathematical Society, Providence, RI (2003)
16. Hirschowitz, A., Simpson, C.: Descente pour les n-champs (descent for n-stacks) (1998). https://arxiv.org/abs/math/9807049
17. Kock, J.: Graphs, hypergraphs, and properads. Collect. Math. **67**(2), 155–190 (2016). http://dx.doi.org/10.1007/s13348-015-0160-0
18. Lurie, J.: Higher Algebra (May 18, 2011, preprint). www.math.harvard.edu/~lurie/papers/higheralgebra.pdf
19. Moerdijk, I.: Lectures on dendroidal sets. In: Simplicial Methods for Operads and Algebraic Geometry. Advanced Courses in Mathematics. CRM Barcelona, pp. 1–118. Birkhäuser/Springer Basel AG, Basel (2010). http://dx.doi.org/10.1007/978-3-0348-0052-5. Notes written by Javier J. Gutiérrez
20. Moerdijk, I., Weiss, I.: Dendroidal sets. Algebra Geom. Topol. **7**, 1441–1470 (2007). http://dx.doi.org/10.2140/agt.2007.7.1441
21. Moerdijk, I., Weiss, I.: On inner Kan complexes in the category of dendroidal sets. Adv. Math. **221**(2), 343–389 (2009). http://dx.doi.org/10.1016/j.aim.2008.12.015
22. Segal, G.: Categories and cohomology theories. Topology **13**, 293–312 (1974)
23. Vallette, B.: A Koszul duality for PROPs. Trans. Am. Math. Soc. **359**(10), 4865–4943 (2007). http://dx.doi.org/10.1090/S0002-9947-07-04182-7
24. Yau, D.: Colored Operads. Graduate Studies in Mathematics, vol. 170. American Mathematical Society, Providence, RI (2016)
25. Yau, D., Johnson, M.W.: A Foundation for PROPs, Algebras, and Modules. Mathematical Surveys and Monographs, vol. 203. American Mathematical Society, Providence, RI (2015)

Lectures on Feynman Categories

Ralph M. Kaufmann

Abstract These are expanded lecture notes from lectures given at the Workshop on higher structures at MATRIX Melbourne. These notes give an introduction to Feynman categories and their applications. Feynman categories give a universal categorical way to encode operations and relations. This includes the aspects of operad-like theories such as PROPs, modular operads, twisted (modular) operads, properads, hyperoperads and their colored versions. There is more depth to the general theory as it applies as well to algebras over operads and an abundance of other related structures, such as crossed simplicial groups, the augmented simplicial category or FI-modules. Through decorations and transformations the theory is also related to the geometry of moduli spaces. Furthermore the morphisms in a Feynman category give rise to Hopf- and bi-algebras with examples coming from topology, number theory and quantum field theory. All these aspects are covered.

1 Introduction

1.1 Main Objective

The main aim is to provide a *lingua universalis* for operations and relations in order to understand their structure. The main idea is just like what Galois realized for groups. Namely, one should separate the theoretical structure from the concrete realizations and representations. What is meant by this is worked out below in the Warm Up section.

In what we are considering, we even take one more step back, namely we provide a theoretical structure for theoretical structures. Concretely the theoretical structures

Workshop on higher structures at MATRIX in Creswick, June 7–9, 2016.

R.M. Kaufmann (✉)
Department of Mathematics, Purdue University, 150 N. University St., West Lafayette, IN 47907, USA
e-mail: rkaufman@purdue.edu

© Springer International Publishing AG, part of Springer Nature 2018
D.R. Wood et al. (eds.), *2016 MATRIX Annals*, MATRIX Book Series 1,
https://doi.org/10.1007/978-3-319-72299-3_19

375

are encoded by a Feynman category and the representations are realized as functors from a given Feynman category $\mathfrak{F}$ to a target category C. It turns out, however, that to a large extent there are constructions which pass up and down the hierarchy of theoretical structure vs. representation. In concrete examples, we have a Feynman category whose representations in C are say algebras. Given a concrete algebra, then there is a new Feynman category whose functors correspond to representations of the algebra. Likewise, for operads, one obtains algebras over the operad as functors.

This illustrates the two basic strategies for acquiring new results. The first is that once we have the definition of a Feynman category, we can either analyze it further and obtain internal applications to the theory by building several constructions and getting further higher structures. The second is to apply the found results to concrete settings by choosing particular representations.

1.1.1 Internal Applications

Each of these will be discussed in the indicated section.

1. Realize universal constructions (e.g. free, push-forward, pull-back, plus construction, decorations); see Sects. 5 and 7.
2. Construct universal transforms (e.g. bar, co-bar) and model category structures; see Sect. 8.
3. Distill universal operations in order to understand their origin (e.g. Lie brackets, BV operators, Master Equations); see Sect. 7.
4. Construct secondary objects, (e.g. Lie algebras, Hopf algebras); see Sects. 7 and 10.

1.1.2 Applications

These are mentioned or discussed in the relevant sections and in Sect. 9.

1. Transfer to other areas such as algebraic geometry, algebraic topology, mathematical physics, number theory.
2. Find out information of objects with operations. E.g. Gromov-Witten invariants, String Topology, etc.
3. Find out where certain algebra structures come from naturally: pre-Lie, BV, etc.
4. Find out origin and meaning of (quantum) Master Equations.
5. Construct moduli spaces and compactifications.
6. Find background for certain types of Hopf algebras.
7. Find formulation for TFTs.

1.2 References

The lectures are based on the following references.

1. With B. Ward. *Feynman categories* [33].
2. With J. Lucas. *Decorated Feynman categories* [30].
3. With B. Ward. and J. Zuniga. *The odd origin of Gerstenhaber brackets, Batalin-Vilkovisky operators and Master Equations* [35].
4. With I. Galvez-Carrillo and A. Tonks. *Three Hopf algebras and their operadic and categorical background* [14].
5. With C. Berger. *Derived Feynman categories and modular geometry* [5].

We also give some brief information on works in progress [25] and further developments [50].

1.3 Organization of the Notes

These notes are organized as follows. We start with a warm up in Sect. 2. This explains how to understand the concepts mentioned in the introduction. That is, how to construct the theoretical structures in the basic examples of group representations and associative algebras. The section also contains a glossary of the terms used in the following. This makes the text more self-contained. We give the most important details here, but refrain from the lengthy full fledged definitions, which can be found in the standard sources.

In Sect. 3, we then give the definition of a Feynman category and provide the main structure theorems, such as the monadicity theorem and the theorem establishing push-forward and pull-back. We then further explain the concepts by expanding the notions and providing details. This is followed by a sequence of examples. We also give a preview of the examples of operad-like structures that are discussed in detail in Sect. 4. We end Sect. 3 with a discussion of the connection to physics and a preview of the various constructions for Feynman categories studied in later sections.

Section 4 starts by introducing the category of graphs of Borisov–Manin and the Feynman category 𝔊 which is a subcategory of it. We provide an analysis of this category, which is pertinent to the following sections as a blue print for generalizations and constructions. The usual zoo of operad-like structures is obtained from 𝔊 by decorations and restrictions, as we explain. We also connect the language of Feynman categories to that of operads and operad-like structures. This is done in great detail for the readers familiar with these concepts. We end with omnibus theorems for these structures, which allow us to provide all the three usual ways of introducing these structures (a) via composition along graphs, (b) as algebras over a triple and (c) by generators and relations.

Decoration is actually a technical term, which is explained in Sect. 5. This paragraph also contains a discussion of so-called non-Sigma, aka. planar versions. We also give the details on how to define the decorations of Sect. 4 as decorations in the technical sense. We then discuss how with decorations one can obtain the three formal geometries of Kontsevich and end the section with an outlook of further applications of this theory.

The details of enrichments are studied in Sect. 6. We start by motivating these concepts through the concrete consideration of algebras over operads. After this prelude, we delve into the somewhat involved definitions and constructions. The central ones are Feynman categories indexed enriched over another Feynman category, the $+$ and hyp constructions. These are tied together in the fact that enrichments indexed over $\mathfrak{F}$ are equivalent to strict symmetric monoidal functors with source $\mathfrak{F}^{hyp}$. This is the full generalization of the construction of the Feynman category for algebras over a given operad. Further constructions are the free monoidal construction $\mathfrak{F}^{\boxtimes}$ for which strict symmetric monoidal functors from $\mathfrak{F}^{\boxtimes}$ to C are equivalent to ordinary functors from $\mathfrak{F}$. And the nc-construction $\mathfrak{F}^{nc}$ for which the strict symmetric monoidal functors from $\mathfrak{F}^{nc}$ to C are equivalent to lax monoidal functors from $\mathfrak{F}$.

Universal operations, transformations and Master Equations are treated in Sect. 7. Examples of universal operations are the pre-Lie bracket for operads or the BV structure for non-connected modular operads. These are also the operations that appear in Master Equations. We explain that these Master Equations are equations which appear in the consideration of Feynman transforms. These are similar to bar-and cobar constructions that are treated as well. We explain that the fact that the universal operations appear in the Master Equation is not a coincidence, but rather is a reflection of the construction of the transforms. The definition of the transforms involves odd versions for the Feynman categories, the construction of which is also spelled out.

As for algebras, the bar-cobar or the double Feynman transformation are expected to give resolutions. In order to make these statements precise, one needs a Quillen model structure. These model structures are discussed in Sect. 8 and we give the conditions that need to be satisfied in order for the transformations above to yield a cofibrant replacement. These model structures are on categories of strict symmetric monoidal functors from the Feynman category into a target category C. The conditions for C are met for simplicial sets, dg-vector spaces in characteristic 0 and for topological spaces. The latter requires a little extra work. We also give a W-construction for the topological examples.

The geometric counterpart to some of the algebraic constructions is contained in Sect. 9. Here we show how the examples relate to various versions of moduli spaces and how Master Equations correspond to compactifications.

Finally, in Sect. 10 we expound the connection of Feynman categories to Hopf algebras. Surprisingly, the examples considered in Sect. 3 already yield Hopf algebras that are fundamental to number theory, topology and physics. These are the Hopf algebras of Goncharov, Baues and Connes–Kreimer. We give further generalizations and review the full theory.

2 Warm Up and Glossary

Here we will discuss how to think about operations and relations in terms of theoretical structures and their representations by looking at two examples.

2.1 Warm Up I: Categorical Formulation for Representations of a Group G

Let $\underline{G}$ the category with one object $*$ and morphism set G. The composition of morphisms is given by group multiplication $f \circ g := fg$. This is associative and has the group identity e as a unit $e = id_*$.

There is more structure though. Since G is a group, we have the extra structure of inverses. That is every morphism in $\underline{G}$ is invertible and hence $\underline{G}$ is a groupoid. Recall that a category in which every morphism is invertible is called a groupoid.

2.1.1 Representations as Functors

A representation (ρ, V) of the group G is equivalent to a functor ρ from $\underline{G}$ to the category of k-vector spaces $\mathcal{V}ect_k$. Giving the values of the functor on the sole object and the morphisms provides: $\rho(*) = V$, $\rho(g) := \underline{\rho}(g) \in Aut(V)$. Functoriality then says $\bar{\rho}(G) \subset Aut(V)$ is a subgroup and all the relations for a group representation hold.

2.1.2 Categorical Formulation of Induction and Restriction

Given a morphism $f : H \to G$ between two groups. There are the restriction and induction of any representation ρ: $Res_H^G \rho$ and $Ind_H^G \rho$. The morphism f induces a functor $\underline{f}$ from $\underline{H}$ to $\underline{G}$ which sends the unique object to the unique object and a morphism g to $f(g)$. In terms of functors restriction simply becomes pull-back $\underline{f}^*(\rho) := \rho \circ \underline{f}$ while induction becomes push-forward, $\underline{f}_*$, for functors. These even form an adjoint pair.

2.2 Warm Up II: Operations and Relations—Description of Associative Algebras

An associative algebra in a tensor category $(C, \otimes)$ is usually given by the following data: An object A and one operation: a multiplication $\mu : A \otimes A \to A$ which satisfies the axiom of the associativity equation:

$$(ab)c = a(bc)$$

2.2.1 Encoding

Think of μ as a 2-linear map. Let $\circ_1$ and $\circ_2$ be substitution in the 1st respectively the 2nd variable. This allows us to rewrite the associativity equation as

$$(\mu \circ_1 \mu)(a, b, c) := \mu(\mu(a, b), c) = (ab)c = a(bc) = \mu(a, \mu(b, c)) := (\mu \circ_2 \mu)(a, b, c)$$

The associativity hence becomes

$$\mu \circ_1 \mu = \mu \circ_2 \mu \tag{1}$$

as morphisms $A \otimes A \otimes A \to A$. The advantage of (1) is that it is independent of elements and of C and merely uses the fact that in multi-linear functions one can substitute. This allows the realization that associativity is an equation about iteration.

In order to formalize this, we have to allow all possible iterations. The realization this description affords is that all iterations of μ resulting in an n-linear map are equal. On elements one usually writes $a_1 \otimes \cdots \otimes a_n \to a_1 \ldots a_n$.

In short: for an associative algebra one has one basic operation and the relation is that all n-fold iterates agree.

2.2.2 Variations

If C is symmetric, one can also consider the permutation action. Using elements the permutation action gives the opposite multiplication $\tau\mu(a, b) = \mu \circ \tau(a, b) = ba$.

This give a permutation action on the iterates of μ. It is a free action and there are $n!$ n-linear morphisms generated by μ and the transposition. One can also think of commutative algebras or unital versions.

2.2.3 Categories and Functors

In order to construct the data, we need to have the object A, its tensor powers and the multiplication map. Let $\mathbf{1}$ be the category with one object $*$ and one morphism id_*. We have already seen that the functors from $\mathbf{1}$ correspond to objects of C. To get the tensor powers, we let $\underline{\mathbf{N}}$ be the category whose objects are the natural numbers including 0 with only identity morphisms. This becomes a monoidal category with the tensor product given by addition $m \otimes n = m + n$. Strict monoidal functors O from $\underline{\mathbf{N}} \to C$ are determined by their value on 1. Say $O(1) = A$ then $O(n) = A^{\otimes n}$.

To model associative algebras, we need a morphisms $\pi : 2 \to 1$. A monoidal functor O will assign a morphism $\mu := O(\pi) : A \otimes A \to A$. If we look for the "smallest monoidal category" that has the same objects as $\underline{\mathbf{N}}$ and contains π as a morphism, then this is the category $sk(\mathcal{S}urj_<)$ of order preserving surjections between the sets $\underline{n}$ in their natural order. Here we think of n as $\underline{n} = \{1, \ldots, n\}$.

Indeed any such surjection is an iteration of π. Alternatively, $sk(Surj_<)$ can be constructed from $\underline{\mathbf{N}}$ by adjoining the morphism π to the strict monoidal category and modding out by the equation analogous to (1): $\pi \circ id \otimes \pi = \pi \circ \pi \otimes id$.

It is easy to check that functors from $sk(Surj_<)$ to C correspond to associative algebras (aka. monoids) in C. From this we already gained that starting from say k-algebras, i.e. $C = Vect_k$ (the category of k vector spaces), we can go to any other monoidal category C and have algebra objects there.

2.2.4 Variations

The variation in which we consider the permutation operations is very important. In the first step, we will need to consider $\mathbb{S}$, which has the same objects as $\underline{\mathbf{N}}$, but has additional isomorphisms. Namely $Hom(n, n) = \mathbb{S}_n$ the symmetric group on n letters. The functors out of $\mathbb{S}$ one considers are strict symmetric monoidal functors O into symmetric monoidal categories C. Again, these are fixed by $O(1) =: A$, but now every $O(n) = A^{\otimes n}$ has the $\mathbb{S}_n$ action of permuting the tensor factors according to the commutativity constraints in C.

Adding the morphisms π to $\mathbb{S}$ and modding out by the commutativity equations, leaves the "smallest symmetric monoidal category" that contains the necessary structure. This is the category of all surjections $sk(Surj)$ on the sets $\underline{n}$. Functors from this category are commutative algebra objects, since $\pi \circ \tau = \pi$ if τ is the transposition.

In order to both have symmetry and not force commutativity, one formally does not mod out by the commutativity equations. The result is then equivalent to the category $sk(Surj_{ord})$ of ordered finite sets with surjections restricted to the sets $\underline{n}$. The objects of $Surj_{ord}$ are a finite set S with an order $<$. The bijections of S with itself act simply transitively on the orders by push-forward.

The second variation is to add an identity. An identity in a k-algebra A is described by an element 1_A, that is a morphism $\eta : k \to A$ with $\eta(1_k) = 1_A$. Coding this means that we will have to have one more morphism in the source category. Since $k = \mathbb{1}$ is the unit of the monoidal structure of $Vect_k$, we see that we need a morphism $u : 0 \to 1$. We then need to mod out by the appropriate equations, which are given by $\eta \circ_1 \mu = \eta \circ_2 \mu = id$ which translate to $\pi \circ u \otimes id_1 = \pi \circ id_1 \otimes u = id_1$.

2.3 Observations

There is a graphical calculus that goes along with the example above. This is summarized in Fig. 1. Adding in the orders corresponds to regarding planar corollas.

We have dealt with strict structures and actually skeletal structures in the examples. This is not preferable for a general theory. Just as it is preferable to work with all finite dimensional vector spaces in lieu of just considering the collection of k^n with matrices as morphisms.

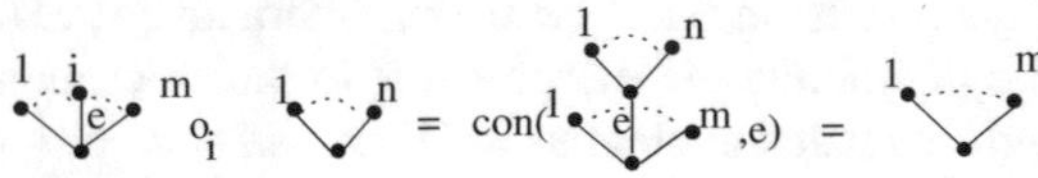

Fig. 1 Example of grafting two (planar) corollas. First graft at a leaf and then contract the edge

2.4 Glossary: Key Concepts and Notations

Here is a brief description of key concepts. For more information and full definitions see e.g. [23, 39].

Groupoid A category in which every morphism is an isomorphism.

As we have seen, every group defines a groupoid. Furthermore for any category C, the subcategory $Iso(C)$ which has the same objects as C but only includes the isomorphisms of C is a groupoid.

Monoidal Category A category C with a functor $\otimes : C \times C \to C$, associativity constraints and unit constraints. That is an operation on objects $(X, Y) \to X \otimes Y$ and on morphisms $(\phi : X \to Y, \psi : X' \to Y') \to \phi \otimes \psi : X \otimes X' \to Y \otimes Y'$. Furthermore a unit object $\mathbb{1}$ with isomorphisms $\mathbb{1} \otimes X \simeq X \simeq X \otimes \mathbb{1}$ called left and right unit constraints and associativity constraints, which are isomorphisms $a_{X,Y,Z} : X \otimes (Y \otimes Z) \to (X \otimes Y) \otimes Z$. These have to satisfy extra conditions called the pentagon axiom and the triangle equation ensuring the compatibilities. In particular, it is the content of Mac Lane's coherence Theorem that due to these axioms any two ways to iteratively rebracket and add/absorb identities to go from one expression to another are equal as morphisms.

A monoidal category is called *strict* if the associativity and unit constraints are identities. Again, due to Mac Lane, every monoidal category is monoidally equivalent to a strict monoidal category (see below).

An example is $\mathcal{Vect}_k$ the category of k-vector spaces with tensor product $\otimes$. Strictly speaking, the associativity constraint $a_{U,V,W}$ acts on elements as $a_{U,V,W}((u \otimes v) \otimes w)) = u \otimes (v \otimes w)$. The unit is k and the unit constraints are $k \otimes U \simeq U \simeq U \otimes k$.

Monoidal Functor A (lax) monoidal functor between two monoidal categories C and $\mathcal{D}$ is an ordinary functor $F : C \to \mathcal{D}$ together with a morphisms $\phi_0 : \mathbb{1}_{\mathcal{D}} \to F(\mathbb{1}_C)$ and a family of natural morphisms $\phi_2 : F(X) \otimes_{\mathcal{D}} F(Y) \to F(X \otimes_C Y)$, which satisfy compatibility with associativity and the unit. A monoidal functor is called strict if these morphisms are identities and strong if the morphism are isomorphisms. If the morphisms go the other way around, the functor is called co-monoidal.

Symmetric Monoidal Category A monoidal category C with all the structures above together with commutativity constraints which are isomorphisms $c_{X,Y} : X \otimes Y \to Y \otimes X$. These have to satisfy the axioms of the symmetric group, i.e. $c_{Y,X} \circ c_{X,Y} = id$ and the braiding for three objects. Furthermore, they are compatible with the associativity constraints, which is expressed by the so-called hexagon equation.

For $\mathcal{V}ect_k$, the symmetric structure $c_{U,V}$ is given on elements as $c_{U,V}(u \otimes v) = v \otimes u$. We can also consider $\mathbb{Z}$-graded vector spaces. In this category, the commutativity constraint on elements is given by $c_{U,V}(u \otimes v) = (-1)^{deg(u)deg(v)} v \otimes u$ where $deg(u)$ is the $\mathbb{Z}$-degree of u.

Symmetric Monoidal Functors A symmetric monoidal functor is a monoidal functor, for which the ϕ_2 commute with the commutativity constraint.

Free Monoidal Categories There are several versions of these depending on whether one is using strict or non-strict and symmetric versions or non-symmetric versions.

Let $\mathcal{V}$ be a category. A free (strict/symmetric) category on $\mathcal{V}$ is a (strict/symmetric) monoidal category $\mathcal{V}^\otimes$ and a functor $\jmath : \mathcal{V} \to \mathcal{V}^\otimes$ such that any functor $\imath : \mathcal{V} \to \mathcal{F}$ to a (strict/symmetric) category $\mathcal{F}$ factors as

$$(2)$$

where $\imath^\otimes$ is a (strict/symmetric) monoidal functor.

The free strict monoidal category is given by words in objects of $\mathcal{V}$ and words of morphisms in $\mathcal{V}$. The free monoidal category is harder to describe. Its objects are iteratively build up from $\otimes$ and the constraints, see [23], where it is also shown that:

Proposition 2.1 *There is a strict monoidal equivalence between the free monoidal category and the strict free monoidal category.*

This allows us some flexibility when we are interested in data given by a category up to equivalence.

If one includes "symmetric" into the free monoidal category, then one (iteratively) adds morphisms to the free categories that are given by the commutativity constraints. In the strict case, one gets commutative words, but extra morphisms from the commutativity constraints. As an example, regard the trivial category **1**: $\mathbf{1}^{\otimes,strict} = \mathbf{N}$ while $\mathbf{1}^{\otimes symmetric,strict} = \mathbb{S}$.

Skeleton of a Category A skeleton $sk(C)$ of a category C is a category that is equivalent to C, but only has one object in each isomorphism class.

An example is the category of ordered finite sets *FinSet* and morphisms between them with the disjoint union as a symmetric monoidal category. A skeleton for this category is given by the category whose objects are natural numbers, where each such object n is thought of as the set $\underline{n} = \{1, \ldots, n\}$ and all morphisms between them. This category is known as the (augmented) crossed simplicial group $\Delta_+ S$.

Underlying Discrete Category The underlying discrete category of a category C is the subcategory which has the same objects as C, but retains the identity maps. It will be denoted by C_0. For instance $\mathbb{S}_0 = \underline{\mathbf{N}}$.

Underlying Groupoid of a Category For a category C the underlying groupoid $Iso(C)$ is the subcategory of C which has the same objects as C buy only retains all the isomorphisms in C.

Comma Categories Recall that for two functors $\imath : \mathcal{D} \to C$ and $\jmath : \mathcal{E} \to C$, the comma category $(\jmath \downarrow \imath)$ is the category whose objects are triples (X, Y, ϕ) with $X \in \mathcal{D}, Y \in \mathcal{E}$ and $\phi \in Hom_C(\jmath(X), \imath(Y))$. A morphism between such ϕ and ψ is given by a commutative diagram.

$$
\begin{array}{ccc}
\jmath(X) & \xrightarrow{\phi} & \imath(Y) \\
{\scriptstyle \jmath(f)} \downarrow & & \downarrow {\scriptstyle \imath(g)} \\
\jmath(X') & \xrightarrow[\psi]{} & \imath(Y')
\end{array}
$$

with $f \in Hom_{\mathcal{D}}(X, X'), g \in Hom_{\mathcal{E}}(Y, Y')$. We will write $(\imath(f), \imath(g))$ for such morphisms or simply (f, g).

If a functor, say $\imath : \mathcal{V} \to \mathcal{F}$, is fixed we will just write $(\mathcal{F} \downarrow \mathcal{V})$, and given a category $\mathcal{G}$ and an object X of $\mathcal{G}$, we denote the respective comma category by $(\mathcal{G} \downarrow X)$. I.e. objects are morphisms $\phi : Y \to X$ with Y in $\mathcal{G}$ and morphisms are morphisms over X, that is morphisms $Y \to Y'$ in $\mathcal{G}$ which commute with the base maps to X. This is sometimes also called the slice category or the category of objects over X.

3   Feynman Categories

With the examples and definitions of the warm up in mind, we give the definition of Feynman categories and then discuss several basic examples. The Feynman categories will give the operations and relations part. The concrete examples of the structures thus encoded are then given via functors, just like discussed above.

3.1   Definition

3.1.1   Data for a Feynman Category

1. $\mathcal{V}$ a groupoid
2. $\mathcal{F}$ a symmetric monoidal category
3. $\imath : \mathcal{V} \to \mathcal{F}$ a functor.

Let $\mathcal{V}^{\otimes}$ be the free symmetric category on $\mathcal{V}$ and $\imath^{\otimes}$ the functor in (2).

3.2 Feynman Category

Definition 1 The data of triple $\mathfrak{F} = (\mathcal{V}, \mathcal{F}, \iota)$ as above is called a Feynman category if the following conditions hold.

i. $\iota^{\otimes}$ induces an equivalence of symmetric monoidal categories between $\mathcal{V}^{\otimes}$ and $Iso(\mathcal{F})$.
ii. ι and $\iota^{\otimes}$ induce an equivalence of symmetric monoidal categories between $(Iso(\mathcal{F} \downarrow \mathcal{V}))^{\otimes}$ and $Iso(\mathcal{F} \downarrow \mathcal{F})$.
iii. For any $* \in \mathcal{V}$, $(\mathcal{F} \downarrow *)$ is essentially small.

Condition (i) is called the *isomorphisms condition*, (ii) is called the *hereditary condition* and (iii) the *size condition*. The objects of $(\mathcal{F} \downarrow \mathcal{V})$ are called *one-comma generators*.

3.2.1 Non-symmetric Version

Now let $(\mathcal{V}, \mathcal{F}, \iota)$ be as above with the exception that $\mathcal{F}$ is only a monoidal category, $\mathcal{V}^{\otimes}$ the free monoidal category, and $\iota^{\otimes}$ is the corresponding morphism of monoidal groupoids.

Definition 3.1 A non-symmetric triple $\mathfrak{F} = (\mathcal{V}, \mathcal{F}, \iota)$ as above is called a *non-Σ* Feynman category if

i. $\iota^{\otimes}$ induces an equivalence of monoidal groupoids between $\mathcal{V}^{\otimes}$ and $Iso(\mathcal{F})$.
ii. ι and $\iota^{\otimes}$ induce an equivalence of monoidal groupoids $Iso(\mathcal{F} \downarrow \mathcal{V})^{\otimes}$ and $Iso(\mathcal{F} \downarrow \mathcal{F})$.
iii. For any object $*_v$ in $\mathcal{V}$, $(\mathcal{F} \downarrow *_v)$ is essentially small.

3.3 Ops and Mods

Definition 2 Fix a symmetric monoidal category C and $\mathfrak{F} = (\mathcal{V}, \mathcal{F}, \iota)$ a Feynman category.

• $\mathcal{F}\text{-}Ops_C := Fun_{\otimes}(\mathcal{F}, C)$ is defined to be the category of strong symmetric monoidal functors which we will call $\mathcal{F}$-ops in C. An object of the category will be referred to as an $\mathcal{F}$-*op* in C.
• $\mathcal{V}\text{-}Mods_C := Fun(\mathcal{V}, C)$, the set of (ordinary) functors will be called $\mathcal{V}$-mods in C with elements being called a $\mathcal{V}$-mod in C.

There is an obvious forgetful functor $G : Ops \to Mods$ given by restriction.

Theorem 3.2 *The forgetful functor $G : Ops \to Mods$ has a left adjoint F (free functor) and this adjunction is monadic. This means that the category of the algebras over the triple $\mathbb{T} = GF$ in C are equivalent to the category of $\mathcal{F}\text{-}Ops_C$.*

Morphisms between Feynman categories are given by strong monoidal functors that preserve the structures. Natural transformations between them give 2-morphisms. The categories $\mathcal{F}\text{-}Ops_C$ and $\mathcal{F}\text{-}Mods_C$ again are symmetric monoidal categories, where the symmetric monoidal structure is inherited from C. E.g. the tensor product is pointwise, $(O \otimes O')(X) := O(X) \otimes O'(X)$, and the unit is the functor $\mathbb{1}_{Ops} : \mathcal{F} \to C$. I.e. the functor that assigns $\mathbb{1}_C \in Obj(C)$ to any object in $\mathcal{V}$, and which sends morphisms to the identity morphism. This is a strong monoidal functor by using the unit constraints.

Theorem 3.3 *Feynman categories form a 2-category and it has push-forwards and pull-backs for Ops. That is, for a morphism of Feynman categories f, both push-forward f_* and pull-back f^* are adjoint symmetric monoidal functors $f_* :$ $\mathcal{F}\text{-}Ops_C \leftrightarrows \mathcal{F}'\text{-}Ops_C : f^*$.*

3.4 Details

3.4.1 Details on the Definition

The conditions can be expanded and explained as follows.

1. Since $\mathcal{V}$ is a groupoid, so is $\mathcal{V}^{\otimes}$. Condition (i) on the object level says, that any object X of $\mathcal{F}$ is isomorphic to a tensor product of objects coming from $\mathcal{V}$. $X \simeq \bigotimes_{v \in I} \imath(*_v)$. On the morphisms level it says that all the isomorphisms in $\mathcal{F}$ basically come from $\mathcal{V}$ via tensoring basic isomorphisms of $\mathcal{V}$, the commutativity and the associativity constraints. In particular, any two decompositions of X into $\bigotimes_{v \in I} \imath(*_v)$ and $\bigotimes_{v' \in I'} \imath(*'_v)$ there is a bijection $\Psi : I \leftrightarrow I'$ and an isomorphism $\sigma_v : \imath(*_v) \to \imath(*_{v'})$. This implies that for any X there is a unique length $|I|$, where I is any index set for a decomposition of X as above, which we denote by $|X|$. The monoidal unit $\mathbb{1}_F$ has length 0 as the tensor product over the empty index set.

2. Condition (ii) of the definition of a Feynman category is to be understood as follows: An object in $(\mathcal{F} \downarrow \mathcal{V})$ is a morphism $\phi : X \to \imath(*)$, with $*$ in $Obj(\mathcal{V})$. An object in $(\mathcal{F} \downarrow \mathcal{V})^{\otimes}$ is then a formal tensor product of such morphisms, say $\phi_v : X_v \to \imath(*_v)$, $v \in I$ for some index set I. To such a formal tensor product, the induced functor assigns $\bigotimes_{v \in V} \phi_v : \bigotimes_v X_v \to \bigotimes_v *_v$, which is a morphisms in $\mathcal{F}$ and hence an object of $(\mathcal{F} \downarrow \mathcal{F})$.

   The functor is defined in the same fashion on morphisms. Recall that an isomorphism in a comma category is given by a commutative diagram, in which the vertical arrows are isomorphisms, the horizontal arrows being source and target. In our case the equivalence of the categories on the object level says that any morphisms $\phi : X \to X'$ in $\mathcal{F}$ has a "commutative decomposition diagram" as follows

$$
\begin{array}{ccc}
X & \xrightarrow{\quad\phi\quad} & X' \\
\simeq \downarrow & & \downarrow \simeq \\
\bigotimes_{v \in I} X_v & \xrightarrow{\bigotimes_{v \in I} \phi_v} & \bigotimes_{v \in I} \imath(*_v)
\end{array}
\tag{3}
$$

which means that when $\phi : X \to X'$ and $X' \simeq \bigotimes_{v \in I} \imath(*_v)$ are fixed there are $X_v \in \mathcal{F}$, and $\phi_v \in Hom(X_v, *_v)$ s.t. the above diagram commutes.

The morphisms part of the equivalence of categories means the following:

a. For any two such decompositions $\bigotimes_{v \in I} \phi_v$ and $\bigotimes_{v' \in I'} \phi'_{v'}$ there is a bijection $\psi : I \to I'$ and isomorphisms $\sigma_v : X_v \to X'_{\psi(v)}$ s.t. $P_\psi^{-1} \circ \bigotimes_v \sigma_v \circ \phi_v = \bigotimes \phi'_{v'}$ where P_ψ is the permutation corresponding to ψ.

b. These are the only isomorphisms between morphisms.

As it is possible that $X_v = \mathbb{1}$, the axiom allows to have morphisms $\mathbb{1} \to X'$, which are decomposable as a tensor product of morphisms $\mathbb{1} \to \imath(*_v)$. On the other hand, there can be no morphisms $X \to \mathbb{1}$ for any object X with $|X| \geq 1$. If $\mathbb{1}$ is the target, the index set I is empty and hence $X \simeq \mathbb{1}$, since the tensor product over the empty set is the monoidal unit.

We set the length of a morphisms to be $|\phi| = |X| - |X'|$. This can be positive or negative in general. In many interesting examples, it is, however, either non-positive or non-negative.

3. The last condition is a size condition, which ensures that certain colimits over these comma-categories to cocomplete categories exist.

3.4.2 Details on the Adjoint Free Functor

The free functor F is defined as follows: Given a $\mathcal{V}$-module Φ, we extend Φ to all objects of $\mathcal{F}$ by picking a functor $\jmath$ which yields the equivalence of $\mathcal{V}^\otimes$ and $Iso(\mathcal{F})$. Then, if $\jmath(X) = \bigotimes_{v \in I} *_v$, we set

$$
\Phi(X) := \bigotimes_{v \in I} \Phi(*_v)
\tag{4}
$$

Now, for any $X \in \mathcal{F}$ we set

$$
F(\Phi)(X) = \operatorname{colim}_{Iso(\mathcal{F} \downarrow X)} \Phi \circ s
\tag{5}
$$

where s is the source map in $\mathcal{F}$ from $Hom_{\mathcal{F}} \to Obj_{\mathcal{F}}$ and on the right hand side or (5), we mean the underlying object. These colimits exist due to condition (iii). For a given morphism $X \to Y$ in $\mathcal{F}$, we get an induced morphism of the colimits and it is straightforward that this defines a functor. This is actually nothing but the left Kan extension along the functor $\iota^{\otimes}$ due to (i). What remains to be proven is that this functor is actually a strong symmetric monoidal functor, that is that $f_*(O) : \mathcal{F}' \to C$ is strong symmetric monoidal. This can be shown by using the hereditary condition (ii).

The fact that f_* is itself symmetric monoidal amounts to a direct check as does the fact that f^* and f_* are adjoint functors. The fact that f^* is symmetric monoidal is clear.

3.4.3   Details on Monadicity

A triple aka. monad on a category is the categorification of a unital semigroup. I.e. a triple $\mathbb{T}$ on a category C is an endofunctor $T : C \to C$ together with two natural transformations, $\eta : Id_C \to T$, where id_C is the identity functor and a multiplication natural transformation $\mu : T \circ T \to T$, which satisfy the associativity equation $\mu \circ T\mu = \mu \circ \mu T$ as natural transformations $T^3 \to T$, and the unit equation $\mu \circ T\eta = \mu \circ \eta T = id_T$, where id_T is the identity natural transformation of the functor T to itself. The notation is to be read as follows: $\mu \circ T\mu$ has the components $T(T^2(X)) \overset{T(\mu_X)}{\to} T^2X \overset{\mu_X}{\to} TX$, where $\mu_X : T^2X \to TX$ is the component of μ.

An algebra over such a triple is an object X of C and a morphism $h : TX \to X$ which satisfies the unital algebra equations. $h \circ Th = h \circ \mu_X : T^2X \to X$ and $id_X = h \circ \eta_X : X \to X$.

3.4.4   Details on Morphisms, Push-Forward and Pull-Back

A morphisms of Feynman categories $(\mathcal{V}, \mathcal{F}, \iota)$ and $(\mathcal{V}', \mathcal{F}', \iota')$ is a pair of functors (v, f) where $v \in Fun(\mathcal{V}, \mathcal{V}')$ and $f \in Fun_{\otimes}(\mathcal{F}, \mathcal{F}')$ which commute with the structural maps ι, ι' and $\iota^{\otimes}, \iota'^{\otimes}$ in the natural fashion. For simplicity, we assume that this means strict commutation. In general, these should be 2-commuting, see [33]. Given such a morphisms the functor $f^* : \mathcal{F}'$-$Ops_C \to \mathcal{F}'$-Ops_C is simply given by precomposing $O \mapsto f \circ O$.

The push-forward is defined to be the left Kan extension $Lan_f O$. It has a similar formula as (5). One could also write $f_!$ for this push-forward. Thinking geometrically f_* is more appropriate.

We will reserve $f_!$ for the right Kan extension, which need not exist and need not preserve strong symmetric monoidality. However, when it does it provides an extension by 0 and hence a triple of adjoint functors $(f_*, f^*, f_!)$. This situation is characterized in [50] which also gives a generalization of $f_!$ and its left adjoint in those cases where the right Kan extension does not preserve strong symmetry.

3.5   Examples

3.5.1   Tautological Example

$(\mathcal{V}, \mathcal{V}^{\otimes}, \jmath)$. Due to the universal property of the free symmetric monoidal category, we have $Mods_{\mathcal{C}} \simeq Ops_{\mathcal{C}}$.

Example If $\mathcal{V} = \underline{G}$, that is $\mathcal{V}$ only has one object, we recover the motivating example of group theory in the Warm Up. For a functor $f : \underline{G} \to \underline{H}$ we have the functor $f^{\otimes}$ and the pair $(\underline{f}, f^{\otimes})$ gives a morphism of Feynman categories. Pull-back becomes restriction and push-forward becomes induction under the equivalence $Mods_{\mathcal{C}} \simeq Ops_{\mathcal{C}}$.

Given any Feynman category $(\mathcal{V}, \mathcal{F}, \imath)$ there is always the morphism of Feynman categories given by $\imath$ and $\imath^{\otimes}: (\mathcal{V}, \mathcal{V}^{\otimes}, \jmath) \to (\mathcal{V}, \mathcal{F}, \imath)$ and the push-forward along it is the free functor F.

3.5.2   Finite Sets and Surjections: $\mathcal{F} = Surj, \mathcal{V} = 1$

An instructive example for the hereditary condition (ii) is the following. As above let $Surj$ the category of finite sets and surjection with disjoint union $\amalg$ as monoidal structure and let $\mathbf{1}$ the trivial category with one object $*$ and one morphism id_*.

$\mathbf{1}^{\otimes}$ is equivalent to the category $\underline{\mathbf{N}}$, where we think $\underline{n} = \{1, \ldots, n\} = \{1\} \amalg \cdots \amalg \{1\}$, $1 = \imath(*)$. This identification ensures condition (i): indeed $\mathbf{1}^{\otimes} \simeq Iso(Surj)$.

Condition (ii) is more interesting. The objects of $(\mathcal{F} \downarrow \mathcal{V})$ are the surjections $S \twoheadrightarrow \imath(*)$. Now consider an arbitrary morphism of $Surj$ that is a surjection $f : S \twoheadrightarrow T$ and pick an identification $T \simeq \{1, \ldots, n\}$, where $n = |T|$. Then we can decompose the morphism f as follows.

$$
\begin{array}{ccc}
S & \xrightarrow{\ \ f\ \ } & T \\
\downarrow{\scriptstyle \simeq} & & \downarrow{\scriptstyle \simeq} \\
\amalg_{i=1}^{|T|} f^{-1}(i) & \xrightarrow{\ \amalg f|_{f^{-1}(i)}\ } & \amalg_{i=1}^{|T|} \imath(*)
\end{array}
\tag{6}
$$

Notice that both conditions (a) and (b) of Sect. 3.4.1 hold for these diagrams. This is because the fibers of the morphisms are well defined. Condition (iii) is immediate. So indeed $Surj = (Surj, \mathbf{1}, \imath)$ is a Feynman category.

$\mathbf{1}$-$Mods_{\mathcal{C}}$ is just $Obj(\mathcal{C})$ and $Surj$-Ops are commutative and associative algebra objects or monoids in $\mathcal{C}$ as discussed in the Warm Up. The commutativity follows from the fact that if π is the surjection $\underline{2} \to \underline{1}$, as above, and τ_{12} is the permutation of 1 and 2 in $\underline{2} = \{1, 2\}$, which is also the commutativity constraint, then $\pi \circ \tau_{12} = \pi$.

The functor G forgets the algebra structure and the functor F associates to every object X in $\mathcal{C}$ the symmetric tensor algebra of X in $\mathcal{C}$. In general, the commutativity constraints define what "symmetric tensors" means.

The monadicity can be read as in the Warm Up. Being an algebra over GF means that there is one morphism for each symmetric tensor power $A^{\odot n} \to A$, that on elements is given by $a_1 \odot \cdots \odot a_n \to a_1 \ldots a_n$. This is equivalent to defining a commutative algebra structure.

The length of the morphisms is always non-negative and only isomorphisms have length 0.

3.5.3 Similar Examples

There are more examples in which $\mathcal{V}$ is trivial and $V^{\otimes} \simeq \mathbb{S}$.

Let $\mathcal{F} = Inj$ the category of finite sets and injections. This is a Feynman category in which all the morphisms have non-positive length, with the isomorphisms being the only morphisms of length 0. If we regard $(\mathcal{F} \downarrow \mathcal{V})$, we see that the injection $i : \emptyset \to \imath(*)$ is a non-isomorphism, where $\emptyset = \mathbb{1}$ is the monoidal unit with respect to $\amalg$. By basic set theory, any other injection can be written as $id \amalg \cdots \amalg id \amalg i \cdots \amalg i$ followed by a permutation. This gives the decomposition for axiom (ii). The other two axioms are straightforward.

Using both injections and surjections, that is $\mathcal{F} = FinSet$, the category of finite sets and all set maps, we get the Feynman category $\mathcal{F}inSet = (\mathbf{1}, FinSet, \imath)$.

3.5.4 Skeletal Versions: Biased vs. Unbiased

Notice that the skeletal versions of Feynman categories do give different *ops*, although the categories *Ops* are equivalent. This is sometimes distinguished by calling the skeletal definition biased vs. the general set definition which is called unbiased. This terminology is prevalent in the graph based examples, see Sects. 3.7 and 4.

3.5.5 FI-modules and Crossed Simplicial Groups, and Free Monoidal Feynman Category

We can regard the skeletal versions of the $\mathcal{F}$ above. For $sk(Inj)$ the ordinary functors $Fun(sk(Inj), C)$ are exactly the FI-modules of [9]. Similarly, for $\Delta_+ S$ the augmented crossed simplicial group, $Fun(\Delta_+ S, C)$ are augmented symmetric simplicial sets in C.

In order to pass to symmetric monoidal functors, that is *Ops*, one can use a free monoidal construction $\mathcal{F}^{\boxtimes}$. This associates to any Feynman category $\mathcal{F}$ a new Feynman category $\mathcal{F}^{\boxtimes}$ for which $\mathcal{F}^{\boxtimes}\text{-}Ops_C$ is equivalent to the category of functors (not necessarily monoidal) $Fun(\mathcal{F}, C)$, see Sect. 6.4.

3.5.6   Ordered Examples

As in the warm up, we can consider $\mathcal{V} = \mathbf{1}$, but look at ordered finite sets $\mathcal{F}inSet_{ord}$ with morphisms being surjections/injections/all set morphisms. In this case the automorphisms of a set act transitively on all orders. For surjections we obtain not necessarily commutative algebras in C as *ops*.

3.6   Units

Adding units corresponds to adding a morphisms $u : \emptyset \to \imath(*)$ and the modding out by the unit constraint $\pi \circ id_1 \otimes u = id_1$. An *op* O will take u to $\eta = O(u) : \mathbf{1} \to A = O(1)$.

3.7   Graph Examples

3.7.1   *Ops*

There are many examples based on graphs, which are explained in detail in the next Sect. 4. Here the graphs we are talking about are not objects of $\mathcal{F}$, but are part of the underlying structure of the morphisms, which is why they are called ghost graphs. The maps themselves are morphisms between aggregates (collections) of corollas. Recall that a corolla is a graph with one vertex and no edges, only tails. These morphisms come from an ambient category of graphs and morphisms of graphs. In this way, we obtain several Feynman categories by restricting the morphisms to those morphisms whose underlying graphs satisfy certain (hereditary) conditions. The *Ops* will then yield types of operads or operad like objects. As a preview:

Ops	Graph, i.e. underlying ghost graphs are of the form
Operads	Rooted trees
Cyclic operads	Trees
Modular operads	Connected graphs (add genus marking)
PROPs	Directed graphs (and input output marking)
NC modular operad	Graphs (and genus marking)
Broadhurst-Connes -Kreimer	1-PI graphs
...	...

Here the last entry is a new class. There are further decorations, which yield the Hopf algebras appearing in [7], see [30].

3.7.2 Non-Σ Feynman Categories: The Augmented Simplicial Category

If we use $\mathcal{V} = \mathbf{1}$ as before, we can see that $\mathcal{F} = \Delta_+$ yields a Feynman category. Now the non-symmetric $\mathcal{V}^{\otimes} = \underline{\mathbf{N}}$ and the analog of *Surj* and *Inj* will then be order-preserving surjections and injections. These are Joyal dual to each other and play a special role in the Hopf algebra considerations.

Another non-Σ example comes from planar trees where $\mathcal{V}$ are rooted planar corollas and all morphisms preserve the orders given in the plane. The $\mathcal{F}$-Ops_C are then non-sigma operads. Notice that a skeleton of $\mathcal{V}$ is given by corollas, whose in flags are labelled $\{1, \ldots, n\}$ in their order and these have no automorphisms.

3.7.3 Dual Notions: Co-operads, etc.

In order to consider dual structure, such as co-operads, one simply considers $\mathcal{F}$-$Ops_{C^{op}}$. Of course one can equivalently turn around the variance in the source and obtain the triple: $\mathfrak{F}^{op} = (\mathcal{V}^{op}, \mathcal{F}^{op}, \iota^{op})$. Now $\mathcal{V}^{op}$ is still a groupoid and $\iota^{\cdot \otimes}$ still induces an equivalence, but $\mathcal{F}^{op}$ will satisfy the dual of (ii). At this stage, we thus choose not to consider $\mathfrak{F}^{op}$, but it does play a role in other constructions.

3.8 Physics Connection

The name Feynman category was chosen with physics in mind. $\mathcal{V}$ are the interaction vertices and the morphisms of $\mathcal{F}$ are Feynman graphs. Usually one decorates these graphs by fields.

In this setup, the categories $(\mathcal{F} \downarrow *)$ are the channels in the S matrix. The external lines are given by the target of the morphism. The comma/slice category over a given target is then a categorical version of the S-matrix.

The functors $O \in \mathcal{F}$-Ops_C are then the correlation functions. The constructions of the Hopf algebras agrees with these identifications and leads to further questions about identifications of various techniques in quantum field theory to this setup and vice-versa. What corresponds to algebras and plus construction, functors? Possible answers could be accessible via Rota–Baxter equations and primitive elements [25].

3.9 Constructions for Feynman Categories

There are several constructions which will be briefly discussed below.

1. Decoration $\mathcal{F}_{decO}$: this allows to define non-Sigma and dihedral versions. It also yields all graph decorations needed for the zoo; see Sect. 5.

2. $+$ construction and its quotient $\mathfrak{F}^{hyp}$: This is used for twisted modular operad and twisted versions of any of the previous structures; see Sect. 6.
3. The free constructions $\mathfrak{F}^{\boxtimes}$, for which $\mathfrak{F}^{\boxtimes}\text{-}Ops_C = Fun(\mathcal{F}, C)$, see Sect. 6. Used for the simplicial category, crossed simplicial groups and FI-algebras.
4. The non-connected construction $\mathfrak{F}^{nc}$, whose $\mathcal{F}^{nc}\text{-}Ops$ are equivalent to lax monoidal functors of $\mathcal{F}$, see Sect. 6.
5. The Feynman category of universal operations on $\mathfrak{F}\text{-}Ops$; see Sect. 7.
6. Cobar/bar, Feynman transforms in analogy to algebras and (modular) operads; see Sect. 7.
7. W-construction, which gives a topological cofibrant replacement; see Sect. 8.
8. Bi- and Hopf algebras from Feynman categories; see Sect. 10.

4 Graph Based Examples: Operads and All of the Zoo

In this section, we consider graph based examples of Feynman categories. These include operads, cyclic operads, modular operads, PROPs, properads, their wheeled and colored versions, operads with multiplication, operads with A_∞ multiplications, etc., see Table 1. They all come from a standard example of a Feynman category called $\mathfrak{G}$ via decorations and restrictions [30, 33]. The category $\mathfrak{G}$ is a subcategory of the category of graphs of Borisov–Manin [6] and decoration is a technical term explained in Sect. 5.4.

Caveat Although $\mathfrak{G}$ is obtained from a category whose objects are graphs, the objects of the Feynman category are rather boring graphs; they have no edges or loops. The usual graphs that one is used to in operad theory appear as underlying (or ghost) graphs of morphisms defined in [33]. These two levels should not be confused and differentiate our treatment from that of [6].

4.1 The Borisov–Manin Category of Graphs

We start out with a brief recollection of the category of graphs given in [6]

1. A graph Γ is a tuple $(F_\Gamma, V_\Gamma, \partial_\Gamma, \imath_\Gamma)$ of flags F_Γ, vertices V_Γ, an incidence relation $\partial_\Gamma : F \to V$ and an involution $\imath : F_\Gamma^{\circlearrowleft}, \imath_\Gamma^2 = id$ which exhibits that either two flags, aka. half-edges are glued to an edge in the case of an orbit of order 2, or a flag is an unpaired half-edge, aka. a tail if its orbit is of order one.
2. A graph morphism $\phi : \Gamma \to \Gamma'$ is a triple $(\phi_V, \phi^F, \imath_\phi)$, where $\phi_V : V_\Gamma \to V_{\Gamma'}$ is a surjection on vertices, $\phi^F : F_{\Gamma'} \to F_\Gamma$ is an injection and $\imath_\phi : F_\Gamma \setminus \phi^F(F_{\Gamma'})^{\circlearrowleft}$ is a self-pairing ($\imath_\phi^2 = id$ and there are no orbits of order 1). This pairs together flags that "disappeared" from $\mathcal{F}_\Gamma$ to ghost edges.

Table 1 List of Feynman categories with conditions and decorations on the graphs, yielding the zoo of examples

$\mathfrak{F}$	Feynman category for	Condition on ghost graphs $\mathbb{\Gamma}_v$ and additional decoration
$\mathfrak{O}$	(Pseudo)-operads	Rooted trees
$\mathfrak{O}_{May}$	May operads	Rooted trees with levels
$\mathfrak{O}^{-\Sigma}$	Non-Sigma operads	Planar rooted trees
$\mathfrak{O}_{mult}$	Operads with mult.	B/w rooted trees
$\mathfrak{C}$	Cyclic operads	Trees
$\mathfrak{C}^{-\Sigma}$	Non-Sigma cyclic operads	Planar trees
$\mathfrak{G}$	Unmarked nc modular operads	Graphs
$\mathfrak{G}^{ctd}$	Unmarked modular operads	Connected graphs
$\mathfrak{M}$	Modular operads	Connected + genus marking
$\mathfrak{M}^{nc,}$	nc Modular operads	Genus marking
$\mathfrak{D}$	Dioperads	Connected directed graphs w/o directed loops or parallel edges
$\mathfrak{P}$	PROPs	Directed graphs w/o directed loops
$\mathfrak{P}^{ctd}$	Properads	Connected directed graphs w/o directed loops
$\mathfrak{D}^{\circlearrowleft}$	Wheeled dioperads	Directed graphs w/o parallel edges
$\mathfrak{P}^{\circlearrowleft,ctd}$	Wheeled properads	Connected directed graphs
$\mathfrak{P}^{\circlearrowleft}$	Wheeled props	Directed graphs
$\mathfrak{F}_{1PI}$	1-PI algebras	1-PI connected graphs

3. These morphisms have to satisfy obvious compatibilities, see [6] or [33]. One of these is preservation of incidence $\phi_V \circ \partial_\Gamma \circ \phi^F(f') = \partial_{\Gamma'}(f')$ and ghost edges are indeed contracted $\phi_V(\partial_\Gamma(f)) = \phi_V\partial_\Gamma(\iota_\phi(f))$.

We will call an edge $\{f, \iota(f) \neq f\}$ with two vertices $(\partial(f) \neq \partial(\iota(f))$ a simple edge and an edge with one vertex $(\partial(f) = \partial(\iota(f)))$ a simple loop.

As objects, the corollas are of special interest. We will write $*_S = (S, \{*\}, \partial : S \twoheadrightarrow \{*\}, id)$ for the corolla with vertex $*$ and flags S. This also explains our notation for elements of $\mathcal{V}$ in general.

An essential new definition [33] is that of a ghost graph of a morphism.

Definition 4.1 The ghost graph (or underlying graph) of a morphisms $\phi = (\phi_V, \phi^F, \iota_\phi)$ is the graph $\mathbb{\Gamma}(\phi) = (V_\Gamma, F_\Gamma, \hat{\iota}_\phi)$, where $\hat{\iota}_\phi$ is the extension of ι_ϕ to all of F_Γ by the identity on $F_\Gamma \setminus \phi^F(F_{\Gamma'})$.

Example 4.2 Typical examples are isomorphisms—which only change the names of the labels—forming of new edges, contraction of edges and mergers. The latter are morphisms which identify vertices. These identifications are kept track of by ϕ_V. Composing the forming of a new edge and then subsequently contracting it, makes the two flags that form the edge "disappear" in the resulting graph. This is what ι_ϕ keeps track of. The "disappeared" flags form a ghost edge and this is the only way that flags may "disappear". The ghost graph says that the morphism factors through

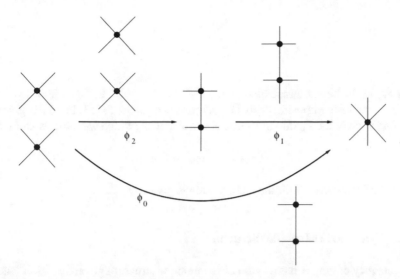

Fig. 2 A composition of morphisms and the respective ghost graphs. The first morphism glues two flags to an edge, the second contracts an edge. The result is a morphism in $\mathcal{A}gg$

a sequence of edge formations and subsequent contractions, namely those edges in the ghost graph, see Fig. 2.

Remark 4.3 As can be seen from these examples: The ghost graph does not determine the morphism. All the information about isomorphisms and almost all information about mergers is forgotten when passing from a morphism to the underlying graph.

What the ghost graph does, however, is keep track of are edge/loop contractions and this can be used to restrict morphisms. Further information is provided by the connectivity of the ghost graph, especially when mapping to a corolla. In this case, we see that mergers have non-connected ghost graphs. Likewise, if we know that there are no mergers, then each component of the ghost graph corresponds to a vertex $v \in V_{\Gamma'}$.

4.1.1 Composition of Ghost Graphs Corresponds to Insertion of Graphs into Vertices

The operation of inserting a graph Γ_v into a vertex v of a graph Γ_1, is well defined for a given identification of the tails of Γ_v with the flags F_v incident to v. The result is the graph $\Gamma_v \circ_v \Gamma_1$ whose vertex set is $V = V_{\Gamma_1} \setminus \{v\} \amalg V_{\Gamma_v}$, the flags $F = F_{\Gamma_1} \amalg F_{\Gamma_v} \setminus tails(\Gamma_v)$ with $\imath$ given by the disjoint union and ∂ given by the disjoint union and the identification of F_v with the tails of Γ_1.

Consider two composable morphisms and their composition:

$$X \xrightarrow{\phi_2} Y \xrightarrow{\phi_1} Z$$
$$\underbrace{\phantom{X \xrightarrow{\phi_2} Y}}_{\phi_0}$$

Now let Γ_i be the associated graphs of ϕ_i, $i = 0, 1, 2$. Decomposing, $Y = \amalg_{v \in V_Y} *_v$, and decomposing ϕ_2 as $\amalg_{v \in V} \phi_v$ one can calculate [33] that Γ_0 is given by inserting each of the Γ_v into the vertices v of $\Gamma_{\phi_1} = V$, which we write as $\amalg_v \Gamma_v \circ \Gamma_1$.

$$\Gamma(\phi_0) = \Gamma(\phi_2) \circ \Gamma(\phi_1) \tag{7}$$

where the identification for the composition is given by ϕ_2^F.

4.1.2 Symmetric Monoidal Structure

The category of graphs has a symmetric monoidal structure given by disjoint union. The unit is the empty graph $(\emptyset, \emptyset, id_\emptyset, id_\emptyset)$ where $id_\emptyset : \emptyset \to \emptyset$ is the unique morphism from the empty set to itself.

4.2   The Feynman Category $\mathfrak{G} = (Crl, \mathcal{A}gg, \iota)$

Let Crl be the subgroupoid of corollas with isomorphisms and $\mathcal{A}gg$. $\mathcal{A}gg$ the full subcategory whose objects are aggregates of corollas. An aggregate of corollas is a graph without any edges $\iota_\Gamma = id$. Any aggregate of corollas is a (possibly empty) disjoint union of corollas and vice-versa. Including corollas into the aggregates as one vertex aggregates gives an inclusion $\iota : Crl \to \mathcal{A}gg$.

Proposition 4.4 $\mathfrak{G} = (Crl, \mathcal{A}gg, \iota)$ *is a Feynman category.*

In this example the one-comma generators $(\mathcal{F} \downarrow \mathcal{V})$ are morphisms from an aggregate to a simple corolla $*_v$

Proof Looking at the definition of morphisms it follows that $Crl^\otimes \simeq Iso(\mathcal{A}gg)$. Condition (iii) is clear. For condition (ii) let $\phi : \Gamma \to \Gamma'$. We will write any such morphism this as a disjoint union of one-comma generators.

For $v \in V_{\Gamma'}$ define Γ_v to be the restriction of Γ to the vertices mapping to v. That is $\Gamma_v = (V_{\Gamma,v} = \phi_V^{-1}(v), F_{\Gamma,v} = \partial_\Gamma^{-1}(V_{\Gamma,v}), \partial_{\Gamma,v} = id)$. We let $\phi_v : \Gamma_v \to v_{F_v}$ be the restriction of ϕ, where v_{F_v} is the corolla with vertex v and its incident flags $F_v = \partial_{\Gamma'}^{-1}(v)$. $\Gamma = (\phi_V|_{V_{\Gamma,v}}, \phi^F|_{F_v}, \iota_\phi|_{F_{\Gamma,v} \setminus (\phi^F)^{-1}(F_v)})$. It then follows that $\Gamma = \amalg_{v \in V_\Gamma'} \Gamma_v$, $\Gamma' = \amalg_{v \in V_{\Gamma'}} v_{F_v}$ and $\phi = \amalg_{v \in V_{\Gamma'}} \phi_v$. This yields the decomposition. It is easy to check conditions (a) and (b).

Notice that forming an edge or a loop is not a morphism in $\mathcal{A}gg$. However the composition of the two morphisms, forming an edge or a loop *and then subsequently contracting it* is a morphism in $\mathcal{A}gg$, see Fig. 2. One could call this a virtual or ghost edge contractions. For simplicity we will call these simply edge or loop contractions.

4.2.1 Morphisms in $\mathcal{A}gg$

1. *Simple edge contraction.* ϕ^F is the identity and the complement of the image ϕ^F is given by two flags s, t, which form a unique ghost edge. The two flags are not adjacent to the same vertex and these two vertices are identified by ϕ_V. The ghost graph is obtained from the source aggregate by adding the edge $\{s, t\}$. We will denote this by ${}_s\circ_t$.
2. *Simple loop contraction.* As above, but the two flags of the ghost edge are adjacent to the same vertex. That is both ϕ_V and ϕ^F are identities. This is called a simple loop contraction. We will denote this by $\circ_{st}$.
3. *Simple merger.* This is a merger in which ϕ_V only identifies two vertices v and w. ϕ^F is an isomorphism. Its degree is 0 and the weight is 1. The ghost graph is simply the source graph. We will denote this by ${}_v\boxminus_w$.
4. *Isomorphism.* This is a relabelling preserving the incidence conditions. Here ϕ_V and ϕ_F are bijections. The ghost graph is the original graph.

Typical examples of such morphisms are shown in Fig. 3.

Actually any morphism is a composition of such morphisms [33]. The relations between these types of morphisms are spelled out below. In order to make things canonical, we will call a morphism pure $\phi : \Gamma \to \Gamma'$, if $\phi^F = id$ when restricted to its image, and the vertices of Γ' are the fibers of ϕ_V, that is $\phi_V(v) = \{w \in V_\Gamma | \phi_V(w) = \phi_V(v)\}$. With this terminology any morphism decomposes as

$$\phi = \sigma \circ \phi_m \circ \phi_c \tag{8}$$

were ϕ_c is a pure contraction, ϕ_m is a pure merger, and σ is an isomorphism.

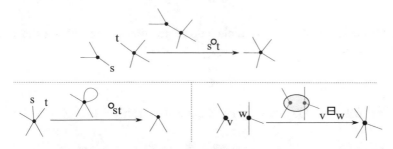

Fig. 3 The three basic morphisms in $\mathfrak{G}$: an edge contraction (top), a loop contraction (left), and a merger (right). In the morphism, we give the ghost graph and label it by the standard notation. The shaded region is for illustration only, to indicate the merger

4.2.2 Ghost Graphs for $\mathcal{A}gg$

In the case of morphism in $\mathcal{A}gg$, we can say more about the morphisms that have a fixed underlying ghost graph. First, the source of a morphism ϕ has the same vertices and flags as its ghost graph $\Gamma(\phi)$ and is hence completely determined. If the ghost graph is connected, then up to isomorphism the target is the vertex obtained from Γ by contracting all edges. If $\Gamma(\phi)$ is not connected, one needs the information of ϕ_V to obtain the target up to isomorphism. This is due to possible vertex mergers that are not recorded by the connected components of Γ. This information is encoded in a decomposition $\Gamma = \amalg_{v \in V} \Gamma_v$. The $\Gamma_v = \Gamma(\phi_v)$ are the ghost graphs of one-comma generators of the decomposition $\phi = \amalg_v \phi_v$.

Stated in another fashion: in the decomposition (8), $\Gamma(\phi)$ fixes ϕ_c, the decomposition $\Gamma(\phi) = \amalg_v \Gamma_v$ fixes ϕ_m.

4.2.3 Relations

All relations among morphisms in $\mathfrak{G}$ are homogeneous in both weight and degree. We will not go into the details here, since they follow directly from the description in the appendix of [33]. There are the following types.

1. *Isomorphisms.* Isomorphisms commute with any ϕ in the following sense. For any ϕ and any isomorphism σ there are unique ϕ' and σ' with $\Gamma(\phi \circ \sigma) = \Gamma(\phi')$ such that

$$\phi \circ \sigma = \sigma' \circ \phi' \qquad (9)$$

2. *Simple edge/loop contractions.* All edge contractions commute in the following sense: If two edges do not form a cycle, then the simple edge contractions commute on the nose

$$_s\circ_t {}_{s'}\circ_{t'} = {}_{s'}\circ_{t'} {}_s\circ_t \qquad (10)$$

The same is true if one is a simple loop contraction and the other a simple edge contraction:

$$_s\circ_t \circ_{s't'} = \circ_{s't'} {}_s\circ_t \qquad (11)$$

If there are two edges forming a cycle, this means that

$$_s\circ_t \circ_{s't'} = {}_{s'}\circ_{t'} \circ_{st} \qquad (12)$$

This is pictorially represented in Fig. 4.

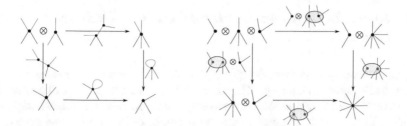

Fig. 4 Squares representing commuting edge contractions and commuting mergers. The ghost graphs are shown. The shaded region is for illustrative purposes only, to indicate the merger

3. *Simple mergers.* Mergers commute amongst themselves

$$_v\boxminus_w {_{v'}}\boxminus_{w'} = {_{v'}}\boxminus_{w'} {_v}\boxminus_w \qquad (13)$$

If $\{\partial(s), \partial(t)\} \neq \{v, w\}$ then

$$_s\circ_t {_v}\boxminus_w = {_v}\boxminus_w {_s}\circ_t, \quad \circ_{st} {_v}\boxminus_w = {_v}\boxminus_w \circ_{st} \qquad (14)$$

If $\partial(s) = v$ and $\partial(t) = w$ then for a simple edge contraction, we have the following relation

$$_s\circ_t = \circ_{st} {_v}\boxminus_w \qquad (15)$$

This is pictorially represented in Fig. 5.

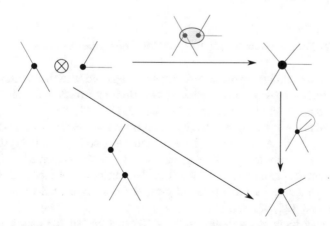

Fig. 5 A triangle representing commutation between edge contraction and a merger followed by a loop contraction. The ghost graphs are shown. The shaded region is for illustrative purposes only, to indicate the merger

4.3 Examples Based on $\mathfrak{G}$: Morphisms Have Underlying Graphs

We are now ready to present the zoo of operad-like structures in a structured way using the Feynman category $\mathfrak{G}$. The different Feynman categories will be obtained by decoration and restriction. Restriction often involves the underlying ghost graphs—to be precise, the underlying ghost graphs of the one-comma generators. What one needs to check is that any such restriction is stable under composition and the decorations compose, whence the term hereditary. For this it suffices to check compositions $X \to Y \to \iota(*)$. In other words, verify that $\amalg_v \Gamma_v \circ \Gamma$ satisfies a given restriction whenever Γ and the Γ_v are composable ghost graphs of one-comma generators satisfying this restriction. Likewise, one also has to define how the decorations compose and check that this gives an associative composition. The usual way is to induce the decoration on $\amalg_v \Gamma_v \circ \Gamma$ whenever the decorations on Γ and the Γ_v are given. This can be done in the following cases (Table 1) in a straightforward fashion, see [33] for details. For readers unfamiliar with some of these structures, the table may serve as a definition. We will discuss decorations, such as roots or directions in a more general fashion in Sect. 5. For instance all these examples have *colored* versions by decorating the flags with colors.

We will say that $\mathfrak{F}$ is a Feynman category for a structure X if $\mathcal{F}\text{-}Ops_C$ are the X-structures in C. E.g. $\mathfrak{O}$ is the Feynman category for operads means that $\mathfrak{O}\text{-}Ops_C$ is the category of operads in C.

New examples can also be constructed in this fashion. The first is the 1-PI (one particle irreducible) condition. A graph is 1PI if it is connected furthermore even after remains connected after cutting any one edge the graph. There are more new examples of this type coming from quantum field theory and number theory, like the ones used in [7], see [14].

4.3.1 Push-Forwards and Pull-Backs: Non-connected Versions

There are obvious inclusion maps and forgetful maps between these categories. E.g. $\mathfrak{C} \to \mathfrak{M}$, which assigns $g = 0$ to each vertex. Here pull-back is the restriction and push-forward is the modular envelope. Looking at $\mathfrak{O} \to \mathfrak{P}$, the root being "out", the push-forward is the PROP generated by and operad and the restriction is the operad contained in a PROP. An examples that has been described by hand [28] is the PROP obtained from a modular operad. For this there is the morphism $\mathfrak{P} \to \mathfrak{M}$, which forgets the directions and adds genus 0 to the vertices. Another is the inclusion $\mathfrak{M} \to \mathfrak{M}^{nc}$ which under push-forward gives the non-connected versions used for moduli spaces in [21, 35, 47].

Analogously there is an inclusion $\mathfrak{F} \to \mathfrak{F}^{nc}$ for any of the candidates $\mathfrak{F}$ with connected graphs, where $\mathfrak{F}^{nc}$ allows non-connected graphs of the same type. Even more generally for and $\mathfrak{F}$ there is such a non-connected version $\mathfrak{F}^{nc}$ whose category Ops is equivalent to lax monoidal functors from $\mathfrak{F}$, see Sect. 6.5.

4.4 Details

4.4.1 Operad-Lingo and Notation: Composition Along Graphs, Self Gluing, Non-self Gluing and Horizontal Composition

Let us unravel the data involved in an $O \in \mathcal{F}\text{-}Ops$. Given a one-comma generator $\phi : X = \amalg_i *_{S_i} \to *_T$ we get a morphisms $O(\phi) : O(X) = \bigotimes_i O(*_{S_i}) \to O(*_T)$. Here $X = s(\phi)$ is also the set of vertices of $\Gamma(\phi)$. If $\phi = \phi_c$ it is completely determined by its ghost graph and for pure contractions to corollas, which have connected ghost graphs, we can set $O(\Gamma(\phi)) := O(\phi)$. This yields usual operad-like notations as follows. Define $O(S) := O(*_S)$. Then one can use the abbreviated notation

$$O(\Gamma) := O(\Gamma(\phi)) : \bigotimes O(S_i) \to O(T)$$

for the composition "along any connected graph Γ".

For a simple edge contraction $_s\circ_t : *_S \otimes *_T \to *_{(S\setminus s)\amalg(T\setminus t)}$ we get the standard non-self gluing pseudo operad compositions $O(S) \otimes O(T) \to (S \setminus s) \amalg (T \setminus t)$, which is often denoted by $_s\circ_t$ as well. In a similar manner, one obtains the May operations γ for a rooted tree whose internal edges are all incident to the root. A simple loop contraction $\circ_{s,s'} : *_S \to *_{S\setminus\{s,s\}}$ becomes the self gluing operation $O(S) \to O(S \setminus \{s, s'\})$; again by abuse of notation simply denoted $\circ_{s,s'}$.

If $\boxminus : *_S \amalg *_T \to *_{S\amalg T}$ is a simple merger then in the usual PROP notation this becomes the horizontal composition $O(S) \otimes O(T) \to O(S \amalg T)$ usually also denoted by $\boxminus$.

Finally there are the isomorphisms. These are already incorporated into the $\mathcal{V}$-$Mods$ structure and not mentioned as structure operations in the operad-lingo. They are pushed into the underlying notion of $\mathbb{S}$-module, or $\mathcal{V}$-$Mods$ in general, on which operads are built. Thus by using (8) we can write any $O(\phi)$ in the usual operad-lingo. The downside is that we *have to make this decomposition first*.

4.4.2 Biased and Unbiased Versions

Sending $S \to *_S$ provides an equivalence from $\mathcal{F}inSet$ to Crl. We see that a skeleton of Crl is given by $\mathbb{S}$. Choosing $\mathcal{V} = \mathbb{S}$, the $\mathcal{V}$-$Mods$ become $\mathbb{S}$-modules. Here usually one identifies n with $\{0, 1, \ldots, n\}$ with 0 indexing the root if there is one present.

If we fix $Iso(\mathcal{F}) = \mathcal{V}^{\otimes}$ with $\mathcal{V} = \mathbb{S}$, we obtain the biased notions of operads, etc., that is objects $O(n)$ with extra operations. Using $\mathcal{V} = \mathcal{F}inSet$, we get $O(S)$ with extra operations indexed by flags.

If there is an extra decoration, then this is part of $\mathcal{V}$ and the set of vertices becomes bigger. An example is the genus marking in the modular operad case, so that we get $O(n, g)$ or $O(n, m)$ for Props, where n are the incoming flags and m are the outgoing flags in the biased version and $O(S, g)$ and $O(S, T)$ in the unbiased one.

For instance, in the directed case a typical element of $\mathcal{V}$ is $*_{S,T}$ where S are the in-flags and T are the out flags. Hence one obtains $O(S, T)$ as for PROPs. Similarly if there is a genus marking a typical element is $*_{S,g}$ and hence in operad-lingo, we get $O(S, g)$.

Variations If one is dealing with roots, often one uses the sets $n_+ = \{0, \dots, n\}$ with the 0 being the label of the root. An isomorphism must fix the roots, so that $Aut(*_{n_+}) = \mathbb{S}_n$. For operads, we then have the translation $\circ_i := {}_i\circ_0$. In cyclic and modular operads, one commonly writes $O((n))$ for $O((n-1)_+)$ when using cyclic or modular operads, but does not insist that the maps are pointed, i.e. that the label 0 is preserved, so that $Aut((n)) = \mathbb{S}_n$.

4.4.3 A Special Case: PROP(erad)s vs. Di-operads and Wheeled Versions

PROPs and properads are a special case. Here the generators are not only the single edge contraction, but all multiparallel edge contractions. In the graphs, parallel edges in the same direction are allowed. These cannot be factored into single edge constructions, so that there are generators $\circ^k_{v,w}$ which simultaneously contract k ghost edges of (necessarily) the same orientation between v and w.

Allowing only the single edge contractions, one arrives at di-operads. Allowing wheels also allows to factor a multi-edge contraction and a single edge contraction followed by single loop contractions.

4.4.4 Identities, Multiplications, etc. as Morphisms and Decorations

We will briefly describe how to incorporate these operations. Say, we want to add a "unit" as to get the Feynman category for unital operads. Recall that for and operad O a unit is an element $\eta : \mathbb{1}_C \to O(1)$ which satisfies $u \circ_1 a = a = a \circ_i u$.

Since $\mathbb{1}_C = O(\mathbb{1}_\mathcal{F})$, we adjoint a morphisms $u : \emptyset \to *_{1_+}$ to the Feynman category for operads $\mathfrak{O}$ with source the empty graph. This can be graphically noted by putting a u on a binary vertex of a ghost tree, whenever we want to use the morphism u, as illustrated in Fig. 6. This does not yet constitute putting in a unit, but rather asking for the data of an element in $O(1)$. This is actually what is needed in the case of the Hopf algebra of Connes and Kreimer [14], see also Sect. 10. In order to get a unit, we have to quotient by the relation given above. The simplest graphical

Fig. 6 Graphically adding a morphism as marked binary vertex of the ghost graph

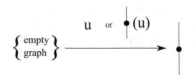

way to do this is to remove all the vertices u from the graph. Technically this is given by an equivalence relation. If one does this, one can create a new "degenerate graph" consisting of a lone flag, which represents any tree whose vertices are all marked by u. This explains the notation of e.g. [41].

In this fashion, one sees that one gets an isomorphism of Feynman categories between the Feynman category for unital May operads and that for unital operads, see [33] for details.

Similarly, for multiplications one needs an extra morphism $\mu : \mathbb{1}_C \to O(2)$. Consequently, one adjoins a morphism $\emptyset \to *_{2_+}$. In the graphical version, the (ghost) graphs will now have a possible decoration on 3-valent vertices by μ. This just gives a multiplication, one can then quotient out by the associativity equation. This amounts to graphs with black and white vertices, where black indicates an iteration of μ. Here associativity induces an equivalence relation, which allows to contract all edges of any subtree of vertices marked solely by μ. A similar procedure adds the μ_n for A_∞ multiplications as black vertices of arity n, see e.g. [26, 32, 33].

Furthermore all these kinds of extra morphisms can be collected and turned into a decoration in the technical sense. This is detailed in [33].

4.5 Omnibus Theorems

For any of these, we have a general triple of graphs $\mathbb{T} = GF$. We immediately obtain a general theorem for all of the zoo and all new species of this kind; see also, Sect. 5. These give the usual three ways of describing these objects (a) via composition along graphs, (b) as algebras over a triple or (c) via generators and relations for the morphisms.

Theorem 4.5 *The biased and unbiased Ops_C are equivalent. Moreover the $\mathcal{F}$-Ops_C are equivalent to algebras over the relevant triple of graphs.*

Notice the usual triples of graph, see e.g. [43], match up exactly with the triples above, when one considers the ghost graphs and their composition. Moreover, the whole semi-simplicial structure of iterating the endofunctors, cf. [19, 43], coincides as demonstrated in [33].

Theorem 4.6 *Generators and relations description. All the examples have a generator and relations description. The generators always contain the isomorphisms, the edge contractions ${}_s\circ_t$. If non-connected graphs are allowed, the morphisms include the mergers $\boxminus_{v,w}$ and if loops are allowed, then they contain the loop contractions. In the presence of decorations, these are restricted to respect the decorations (cf. Sect. 5). The relations are the ones given above.*

If one adds additional morphisms with relations, these are be included in the list. This can be formalized using Feynman categories indexed over another Feynman category, see [33].

Example For instance, when adding units, the morphism u is a generator and the relations with u are the unit relations. This way, one can, for example, get the Feynman category for unital cyclic operads in all three definitions.

Remark 4.7 In the PROP(erad) case, which is special, the generators are not only the simple edge contraction, but multi-edge contractions, see Sect. 4.4.3.

5 Decorating Feynman Categories $\mathfrak{F}_{decO}$

Decorations can be made into a technical definition. The details for this section are in [30]. The basic idea is that one can decorate a Feynman category by using elements of $\mathcal{F}$-*Ops*. The reason this works is that in order to define a composition, one has to give a composition for the decorations, but this is precisely the data of an $O \in \mathcal{F}$-*Ops*. These decorations actually decorate the elements of $\mathcal{V}$. In the graph example above, this means that one can decorate vertices and flags.

5.1　Main Theorems

The main constructive theorem is the following.

Theorem 5.1 *Given an $O \in \mathcal{F}$-Ops, then there is a Feynman category $\mathcal{F}_{decO}$ which is indexed over $\mathcal{F}$. It objects are pairs $(X, dec \in O(X))$ and $Hom_{\mathcal{F}_{decO}}((X, dec), (X', dec'))$ is the set of $\phi : X \to X'$, s.t. $O(\phi) : dec \to dec'$.*

Remark 5.2 This theorem also works in the enriched setting, where one considers enrichment over C, confer Sect. 6. This construction works directly for Cartesian C, and with modifications it also works for the non-Cartesian case.

Example 5.3 All planar structures: Non-sigma operads, cyclic non-Sigma operads, non-Sigma modular operads. Here O is $\mathcal{A}ssoc$, $Cyc\mathcal{A}ssoc$, $ModCyc\mathcal{A}ssoc$. These are actually all obtained by functoriality, see below. This recovers e.g. that the modular envelope of $Cyc\mathcal{A}ssoc$ factors through non-Sigma modular operads [42].

Theorem 5.4 (Functoriality in $\mathfrak{F}$ and O) *Given a morphism of Feynman categories $f : \mathfrak{F} \to \mathfrak{F}'$ and a morphisms $\sigma : O \to \mathcal{P}$. There are commutative squares which are natural in O*

$$
\begin{array}{ccc}
\mathfrak{F}_{decO} & \xrightarrow{f^O} & \mathfrak{F}'_{dec f_*(O)} \\
{\scriptstyle forget}\downarrow & & \downarrow{\scriptstyle forget'} \\
\mathfrak{F} & \xrightarrow{f} & \mathfrak{F}'
\end{array}
\qquad
\begin{array}{ccc}
\mathfrak{F}_{decO} & \xrightarrow{\sigma_{dec}} & \mathfrak{F}_{dec\mathcal{P}} \\
{\scriptstyle f^O}\downarrow & & \downarrow{\scriptstyle f^{\mathcal{P}}} \\
\mathfrak{F}'_{dec f_*(O)} & \xrightarrow{\sigma'_{dec}} & \mathfrak{F}'_{dec f_*(\mathcal{P})}
\end{array}
\tag{16}
$$

On the categories of monoidal functors to C, we get the induced diagram of adjoint functors.

$$
\begin{array}{ccc}
\mathcal{F}_{decO}\text{-}Ops & \xrightarrow{\ f^O_*\ } & \mathcal{F}'_{decf_*(O)}\text{-}Ops \\[4pt]
forget_* \Big\downarrow\Big\uparrow\, forget^* & {}^{f^O{}^*} & forget'^* \Big\downarrow\Big\uparrow\, forget'_* \\[4pt]
\mathcal{F}\text{-}Ops & \xleftarrow[f^*]{f_*} & \mathcal{F}'\text{-}Ops
\end{array}
\tag{17}
$$

5.2 Terminal Objects and Minimal Extensions

Theorem 5.5 *If $\mathcal{T}$ is a terminal object for $\mathcal{F}$-Ops and forget : $\mathcal{F}_{decO} \to \mathcal{F}$ is the forgetful functor, then $forget^*(\mathcal{T})$ is a terminal object for $\mathcal{F}_{decO}$-Ops. We have that $forget_* forget^*(\mathcal{T}) = O$.*

Definition 5.6 We call a morphism of Feynman categories $i : \mathfrak{F} \to \mathfrak{F}'$ a minimal extension over C if $\mathfrak{F}$-Ops_C has a terminal/trivial functor $\mathcal{T}$ and $i_*\mathcal{T}$ is a terminal/trivial functor in $\mathfrak{F}'$-Ops_C.

Example 5.7 There are two examples that appear naturally. The first is *CycCom* and *ModCycCom* for $\mathfrak{C} \to \mathfrak{M}$ and the second is the decorated version $forget^*(CycAssoc)$ and $i^O_*(forget^*(CycAssoc))$.

Proposition 5.8 *If $f : \mathfrak{F} \to \mathfrak{F}'$ is a minimal extension over C, then $f^O : \mathfrak{F}_{decO} \to \mathfrak{F}'_{decf_*(O)}$ is as well. This condition has more recently been further analyzed and has been identified as part of a factorization system in [4].*

5.3 Example

5.3.1 Markl's Non-Σ Modular (See Also [31])

$$
\begin{array}{ccc}
\mathcal{F}_{dec\,CycAssoc} = \mathfrak{C}^{\neg\Sigma} & \xrightarrow{\ i_{CycAssoc}\ } & \mathfrak{M}_{dec\,i_*(CycAssoc)} = \mathfrak{M}^{\neg\Sigma} \\[4pt]
forget \Big\downarrow & & \Big\downarrow forget \\[4pt]
\mathfrak{C} & \xrightarrow[\ i\ ]{} & \mathfrak{M}
\end{array}
\tag{18}
$$

1. The commutative square exists simply by Theorem 5.4.
2. On the left side, if $*_C$ is final for $\mathfrak{C}$ and hence $forget^*(*_C) = \underline{*}_C$ is final for $\mathfrak{C}^{\neg\Sigma}$. The pushforward $forget_*(\underline{*}_C) = CycAssoc$.

3. On the right side, if $*_M$ is final for $\mathfrak{M}$ and hence $forget^*(*_M) = \underline{*}_M$ is final for $\mathfrak{M}^{\neg\Sigma}$. The pushforward $forget_*(\underline{*}_M) = ModAssoc$.
4. The inclusion i is a minimal extension. This is a fact explained by basic topology. Namely gluing together polygons in their orientation by gluing edges pairwise yields all closed oriented surfaces, see e.g. [46].
5. Hence $i^{CycAssoc}$ is also a minimal extension. which explains why indeed the pushforward of the terminal op is up to that point still terminal. It also reflects the fact that not gluing all edges pairwise, but preserving orientation, does yield all surfaces with boundary.

5.4 Examples on $\mathfrak{G}$ with Extra Decorations, Non-sigma, Colored Versions, etc.

We now give the details on how to understand the decorations in Sect. 4 as decorations in the technical sense. Decoration and restriction allows to generate the whole zoo and even new species. Examples of the needed decorations are listed in Table 2.

5.4.1 Flag Labelling, Colors, Direction and Roots as a Decoration

Recall that $*_S$ is the one vertex graph with flags labelled by S and these are the objects of $\mathcal{V} = Crl$ for $\mathfrak{G}$. For any set X introduce the following $\mathfrak{G}$-op: $X(*_S) = X^S$. The compositions are simply given by restricting to the target flags.

If the decoration is by $d : F_\Gamma \to X$ then $d(f) = \overline{d(\iota_\phi(f))}$. Then a natural subcategory $\mathfrak{F}_{decX}^{dir}$ of $\mathfrak{G}_{decX}$ is given by the wide subcategory, whose morphisms

Table 2 List of decorated Feynman categories with decorating O and possible restriction

$\mathfrak{F}_{decO}$	Feynman category for	Decorating O	Restriction
$\mathfrak{F}^{dir}$	Directed version	$\mathbb{Z}/2\mathbb{Z}$ set	Edges contain one input and one output flag
$\mathfrak{F}^{rooted}$	Root	$\mathbb{Z}/2\mathbb{Z}$ set	Vertices have one output flag
$\mathfrak{F}^{genus}$	Genus marked	$\mathbb{N}$	
$\mathfrak{F}^{c-col}$	Colored version	c Set	Edges contain flags of same color
$\mathfrak{O}^{\neg\Sigma}$	Non-sigma-operads	$Assoc$	
$\mathfrak{C}^{\neg\Sigma}$	Non-Sigma-cyclic operads	$CycAssoc$	
$\mathfrak{M}^{\neg\Sigma}$	Non-sigma-modular	$ModAssoc$	
$\mathfrak{C}^{dihed}$	Dihedral	$Dihed$	
$\mathfrak{M}^{dihed}$	Dihedral modular	$ModDihed$	

$\mathfrak{F}$ stands for an example based on $\mathfrak{G}$ in the list or more generally indexed over $\mathfrak{G}$ (see [33])

additionally satisfy that only flags marked by elements x and $\bar{x}$ are glued and then contracted; viz ι_ϕ only pairs flags of marked x with edges marked by $\bar{x}$. That is the underlying ghost graph has edges whose two flags are labelled accordingly. In the notation of graphs: $X(f) = \overline{\iota_\phi(f)}$.

If X is pointed by x_0, there is the subcategory of $\mathfrak{G}_{decX}$ whose objects are those generated by $*_S$ with exactly one flag labelled by x_0 and where the restriction on graphs is that for the underlying graph additionally, each edge has one flag labelled by x_0.

Now if $X = \mathbb{Z}/2\mathbb{Z} = \{0, 1\}$ with the involution $\bar{0} = 1$, we can call 0 "out" and 1 "in". As a result, we obtain the category of directed graphs $\mathfrak{G}_{dec\mathbb{Z}/2\mathbb{Z}}$. Furthermore, if 0 is the distinguished element, we get the rooted version. This explains the relevant examples Table 2.

More generally, in quantum field theory the involution sends a field to its anti-field and this is what decorates the lines or propagators in a Feynman graph.

5.4.2 Genus Decoration

Let $\mathbb{N}$ be the $\mathfrak{G}$-op which on objects of $\mathcal{V}$ has constant value the natural numbers $\mathbb{N}(*_S) = \mathbf{N_0}$. On morphisms $\mathbb{N}$ is defined to behave like the genus marking. That is for $\phi : X \to *_S$, we define $\mathbb{N}(\phi) : \mathbb{N}(X) = \mathbf{N_0}^{|X|} \to \mathbf{N_0} = \mathbb{N}(*_S)$ as the concatenation $\mathbf{N_0}^{|X|} \overset{\Sigma}{\to} \mathbf{N_0} \overset{+\bar{\gamma}(\phi)}{\to} \mathbf{N_0}$ where $\bar{\gamma}(\phi)$ equals one minus the Euler characteristic of the graph underlying ϕ. If this graph is connected this is just first Betti number also sometimes called the genus. This coincides with the description in [33, Appendix A]. Hence, if $\mathfrak{F}$ is a subcategory of $\mathfrak{G}$, then the genus marked version is just $\mathfrak{F}_{dec\mathbb{N}}$. Examples are listed in Table 2.

5.4.3 Assoc-Decorated, aka. Non-Sigma, aka. Non-planar

Likewise, we can regard the cyclic associative operad, *CycAssoc*. The pull back of *CycAssoc* under *forget* : $\mathfrak{O} \to \mathfrak{C}$ is the associative operad *Assoc*. Now $\mathfrak{O}_{dec\,Assoc} = \mathfrak{O}^{\neg\Sigma}$ is the Feynman category for non-Sigma operads. Indeed, the elements of $Assoc(*_s)$ are the linear orders on S, which means that we are dealing with planar corollas as objects. Likewise, for the morphisms the condition that $\phi(a_X) = a_Y$ means that the trees are also planar. The story for cyclic operads is similar $\mathfrak{C}_{decCycAssoc} = \mathfrak{C}^{\neg\Sigma}$.

Things are more interesting in the modular case. In this case, we have *ModAssoc* := $i_*(CycAssoc)$ as a possible decoration and we get the decorated Feynman category $\mathfrak{M}^{\neg\Sigma} := \mathfrak{M}_{dec\,ModAssoc}$. Indeed using this decoration, we recover the definition of [42] of non-sigma modular operads, which is the special case of a brane-labelled c/o system, with trivial closed part and only one brane color [31, Appendix A.6]; see also [34], the appendix of [29] and [42] for details about the correspondence between stable or almost ribbon graphs and surfaces.

Here we can understand these constructions in a more general framework. First, the diagram considered in [42] is exactly a diagram of Theorem 5.4. Then the fact that the non-Sigma modular envelope of *CycAssoc* is terminal is obvious from Theorem 5.5 and Proposition 5.8. The key observations are that the terminal object of $\mathfrak{C}^{\neg\Sigma}$ pushed forward is indeed *CycAssoc* and that *ModAssoc* is the pushforward of the terminal object of $\mathfrak{M}^{\neg\Sigma}$. Notice *CycAssoc* is not a modular operad, so it is not a valid decoration for $\mathfrak{M}$. This is reflected in the treatments of [31, 42]. We see that we do get a planar aka. non-Sigma version by pushing forward *Assoc*.

5.5 Kontsevich's Three Geometries

In this framework, one can also understand Kontsevich's three geometries [37] as follows.

5.5.1 Com, or Trivially Decorated

The operad *CycCom*, the operad for cyclic commutative algebras, is the terminal/trivial object in $\mathfrak{C}$-*Ops*. Thus by Theorem 5.5, we have that $\mathfrak{O}_{decCom} = \mathfrak{O}$. The analogous statement holds for $\mathfrak{C}$. Indeed, there is a forgetful functor $\mathfrak{O} \to \mathfrak{C}$ and the pull-back of *CycCom* is *Com* and hence $\mathfrak{C}_{decCycCom} = \mathfrak{C}$. Finally using the inclusion $i : \mathfrak{C} \to \mathfrak{M}$ means that the modular envelope $i_*(Com)$ is a modular operad. Tracing around the trivially decorated diagram, we see that this is again a terminal/trivial operad. Indeed this is the content of Proposition 5.8.

5.5.2 Lie, etc. or Graph Complexes

For this we actually need the enriched version.

One of the most interesting generalizations is that of Lie or in general of Kontsevich graph complexes. Here notice that *Assoc*, *Com* and *Lie* are all three cyclic operads, so that they all can be used to decorate the Feynman category for cyclic operads. For *Lie* it is important that we can also work over k-Vect. Thus, answering a question of Willwacher (Private communication), indeed there is a Feynman category for the Lie case.

To go to the case of graph complexes, one needs to first shift to the odd situation and then take colimits as described in detail in [33], see especially section 6.9 of *loc. cit.*

5.6   Further Applications

Further forthcoming applications will be

1. Infinity versions of the Assoc, Com and Lie and their transformations.
2. New decorated interpretation of moduli space operations generalizing those of [27, 28].
3. The new Stolz–Teichner–Dwyer setup for twisted field theories.
4. Kontsevich's graph complexes.
5. Actions of the Grothendieck–Teichmüller group.

6   Enrichment, Algebras, Odd Versions and Further Constructions

6.1   Enriched Versions, Plus Construction, and Algebras over $\mathfrak{F}$-Ops: Overview and Examples

There are several reasons why one would like to consider enriched versions of Feynman categories. They are necessary to define the transforms and resolutions. Here it is necessary to introduce signs or anti-commuting morphisms. They are also natural from an algebra over operads point of view. We will start with this construction.

6.1.1   The Feynman Category for an Algebra over an Operad

Recall that an algebra over an operad O in C is an object A and a morphism of operads $\rho : O \rightarrow End(A)$. For this to make sense, one assumes that C is closed monoidal. Then $\mathcal{E}nd(A)(n) = \underline{Hom}(A^{\otimes n}, A)$. One can simply think of $C = \mathcal{V}ect$ or Set. Substitutions then give the operad structure.

Algebras as Natural Transformations   Generally, given a reference target $\mathfrak{F}$-op $\mathcal{E}$, then for another $O \in \mathcal{F}$-Ops_C we define an O-algebra relative to $\mathcal{E}$ as a natural transformation of functors $\rho : O \rightarrow \mathcal{E}$.

Indeed, for instance in the operad case with $\mathcal{E} = \mathcal{E}nd$, we obtain $\rho(n) : O(n) \rightarrow Hom(A^{\otimes n}, A)$ which commute with compositions.

Algebras over Operads as Functors   We will start with the operad case. Given a May operad O, we will construct a Feynman category $\mathfrak{F}_O$ whose *ops* are algebras over O. The data we have to encode are $A \in C$ and $\rho(n) : O(n) \rightarrow Hom(A^{\otimes n}, A)$. Now if we take $\mathcal{V}_O = \mathbf{1}$ and $Iso(\mathcal{F}_O) = \mathbb{S}$, then we see that a strict symmetric monoidal functor $\rho : \mathbb{S} \rightarrow C$ will send n to $A^{\otimes n}$ and the $\sigma \in Aut(n) = \mathbb{S}_n$ to the permutations of the factors of $A^{\otimes n}$.

We now add more morphisms. A morphisms from $\phi : n \to 1$ will be sent to a morphism $\rho(\phi) : Hom(A^{\otimes n}, A)$. Thus, we set the one-comma generators as $O(n) =: Hom_{\mathcal{F}_O}(n, 1)$. This fixes data of the $\rho(n)$ is and vice-versa. Notice that when adding in these morphisms, $O(n)$ is—and has to be—an $\mathbb{S}_n$-module to fix the pre-composition with the isomorphisms $Aut(n)$.

Here we assume that we can also work with enriched categories. In particular, we need to be enriched over C if O is an operad in C, see details below.

With these one-comma generators, due to condition (ii), we get that $Hom_{\mathcal{F}_O}(n, m) = \bigotimes_{(n_1,\ldots,n_m): \sum n_i = n} O(n_1) \otimes \cdots \otimes O(n_m)$. Here $\bigoplus$ is the colimit, which we assume to exist. There is more data. In order to compose $Hom_{\mathcal{F}_O}(m, 1) \otimes Hom_{\mathcal{F}_O}(n, m) \to Hom_{\mathcal{F}_O}(n, 1)$, we need morphisms

$$\gamma_{n_1,\ldots,n_k} : O(m) \otimes O(n_1) \otimes \cdots \otimes O(n_m) \to O(n) \quad n = \sum n_i \tag{19}$$

These have to be compatible with the isomorphisms. This data is the composition of a May operad and vice-versa defines a category structure on $\mathcal{F}_O$.

This category has a special structure, namely that

$$Hom_{\mathcal{F}_O}(n, m) = \bigoplus_{\phi: n \twoheadrightarrow m} O(\phi) \text{ where } O(\phi) = \bigotimes_{i \in m} O(f^{-1}(i)) \tag{20}$$

Caveats In order to obtain a Feynman category, we will need to define what an enriched Feynman category over C is. This is straightforward if C is Cartesian. In the non-Cartesian case, we have to be a bit more careful, see below. There we will see that the isomorphism condition will dictate that $O(1)$ has only $\mathbb{1}$, that is a copy of $\mathbb{1}_C$ corresponding to id as the "invertible element". Also, the relevant notion is that of a Feynman category indexed enriched over another Feynman category. In our example, we are indexed enriched over a skeleton of $Surj$.

Clearing these up leads to the theorem:

Theorem 6.1 *The category of Feynman categories enriched over $\mathcal{E}$ indexed over Surj is equivalent to the category of operads (with the only iso in $O(1)$ being the identity) in $\mathcal{E}$ with the correspondence given by $O(n) = Hom(\underline{n}, \underline{1})$. The Ops are now algebras over the underlying operad.*

Remark 6.2 We can also deal with algebras over operads which have isomorphisms in $O(1)$ by enlarging $\mathcal{V}$. For this one needs a splitting $O(1) = O(1)^{iso} \oplus \bar{O}(1)$, where no element of $\bar{O}(1)$ is invertible and $O(1)^{iso} = \bigoplus_{g \in G} \mathbb{1}_C$ for an index group G is the free algebra on G. Then we enlarge $\mathcal{V}$ by letting 1 have isomorphisms G. The construction is then analogous to the one above and that of K-algebras [19]. Another way is to use lax monoidal functors, see [33].

6.1.2 General Situation for Algebras: Plus Construction

There is a "+" construction, not unlike that for polynomial monads [2], that produces a new Feynman category out of an old one. Inverting morphisms stemming from isomorphisms one obtains $\mathfrak{F}^{hyp}$ and there is a further reduction to an equivalent category $\mathfrak{F}^{hyp,rd}$. Details will be provided below.

The main theorem is that enrichments of $\mathfrak{F}$ are in 1-1 correspondence with $\mathfrak{F}^{hyp}$-*Ops*.

Example 6.3 $\mathfrak{M}^{hyp} = \mathfrak{F}_{hyper}$, the Feynman category for hyper-operads as defined by Getzler and Kapranov [19], whence the name. $Surj^{+} = \mathfrak{F}_{Mayoperads}$, $\mathfrak{F}^{hyp,rd}_{surj} = \mathfrak{O}_0$, the category for operads whose $O(1)$, has only (multiples of) *id* as an invertible element. $\mathfrak{F}^{+}_{triv} = Surj$, $\mathfrak{F}^{hyp,rd}_{triv} = \mathfrak{F}_{triv}$.

Definition 6.4 Let $\mathfrak{F}$ be a Feynman category and $\mathfrak{F}^{hyp,rd}$ its reduced hyper category, O an $\mathfrak{F}^{hyp,rd}$-*op* and $\mathcal{D}_O$ the corresponding enrichment functor. Then we define an O-algebra to be a $\mathfrak{F}_{\mathcal{D}_O}$-*op*.

6.1.3 Odd Feynman Categories over Graphs

In the case of underlying graphs for morphisms, odd usually means that edges get degree 1, that is we use a Kozsul sign with that degree. In particular, in these discussions, one is augmented over $\mathcal{A}b$, the category of Abelian groups. Then there is an indexed enriched version of the Feynman categories. In order to write this down, one needs an ordered presentation.

For graphs this amounts to adding signs in the relations Sect. 4.2.3. In particular, the following quadratic relations become anti-commutative:

$$s^{\circ}_t \, s'^{\circ}_{t'} = -s'^{\circ}_{t'} \, s^{\circ}_t \tag{21}$$

$$s^{\circ}_t \circ_{s't'} = -\circ_{s't'} \, s^{\circ}_t \tag{22}$$

$$s^{\circ}_t \circ_{s't'} = -s'^{\circ}_{t'} \circ_{st} \tag{23}$$

Since (15) is not quadratic and hence the degree of a merger must be 0 and the relation does not get a sign

$$s^{\circ}_t = \circ_{st} \, {}_v\boxminus_w \tag{24}$$

Consequently, the following quadratic relations also remain without sign

$${}_v\boxminus_w \, {}_{v'}\boxminus_{w'} = {}_{v'}\boxminus_{w'} \, {}_v\boxminus_w \tag{25}$$

$$s^{\circ}_t \, {}_v\boxminus_w = {}_v\boxminus_w \, s^{\circ}_t \tag{26}$$

$$\circ_{st} \, {}_v\boxminus_w = {}_v\boxminus_w \circ_{st} \tag{27}$$

Isomorphisms also naturally have degree 0 and hence there is no change in the relevant relation:

$$\phi \circ \sigma = -\sigma' \circ \phi' \tag{28}$$

6.1.4 Orders and Orientations

In order to pictorially represent this, one can add decorations. This is very similar to the construction of ordered and oriented simplices, see e.g. [46]. The first step is to give an order on all the edges of the ghost graph. The second step is to define orientations as orbits under even permutations. Finally one can impose the relation that two opposite orientations differ by a sign. Algebraically, one also uses the determinant line on the edges [19]. It is only at this last step that the enrichment is needed. Furthermore one can push this last step into the functor, that is only regard functors to Abelian C that take different change of orientations to sign changes. These constructions are discussed in detail in [33].

6.1.5 Graph Examples

A list of examples is given in Table 3.

6.1.6 Suspension vs. Odd

In operad-lingo, one can suspend operads, etc. On the Feynman category side this corresponds to certain twists. I.e. there is a twist Σ and a Σ twisted Feynman category $\mathfrak{F}_\Sigma$ such that $O \in \mathcal{F}\text{-}Ops_C$ iff the suspension $\Sigma O \in \mathfrak{F}_\Sigma\text{-}Ops_C$. For general

Table 3 List of Feynman categories with conditions and decorations on the graphs

$\mathfrak{F}$	Feynman category for	Condition on graphs + additional decoration
$\mathfrak{C}^{odd}$	Odd cyclic operads	Trees + orientation of set of edges
$\mathfrak{M}^{odd}$	$\mathfrak{K}$-modular	Connected + orientation on set of edges + genus marking
$\mathfrak{M}^{nc,odd}$	nc $\mathfrak{K}$-modular	Orientation on set of edges + genus marking
$\mathfrak{D}^{\circlearrowright odd}$	Odd wheeled dioperads	Directed graphs w/o parallel edges + orientations of edges
$\mathfrak{P}^{\circlearrowright,ctd,odd}$	Odd wheeled properads	Connected directed graphs w/o parallel edges + orientation of set of edges
$\mathfrak{P}^{\circlearrowright,odd}$	Odd wheeled props	Directed graphs w/o parallel edges + orientation of set of edges

twistings of this type see Sect. 6.2.3. These are equivalent to the odd version *if* we are in the directed case and there is a bijection between vertices and out flags, see [35]. Even in the directed case, as explained in [35]. the odd versions are actually more natural and yield the correct degrees in the Hochschild complex and correct signs and Master Equations, see Sect. 7 below. A well known example for unexpected, but correct, signs is the Gerstenhaber bracket. It is *odd* Poisson.

In the same vein for the bar/cobar and Feynman transforms, it is not the suspended structures that are pertinent, but the odd structures, see Sect. 7.

6.1.7 Examples

1. Operads are very special, in the respect that their Feynman category is equivalent to the one for their odd version.
2. The odd cyclic operads are equivalent to anti-cyclic operads.
3. For modular operads the suspended version is not equivalent to the odd versions a.k.a. $\mathfrak{K}$-modular operads. The difference is given by the twist $H_1(\mathbb{T}(\phi))$.

6.2 Enriched Versions: Details

We can consider Feynman categories and target categories enriched over another monoidal category, such as $\mathcal{T}op$, $\mathcal{A}b$ or $dg\mathcal{V}ect$. Note that there are two cases. Either the enrichment is Cartesian, then we simply have to replace the free (symmetric) monoidal category by the enriched version. There is also a more categorical version of the definition with a condition going back to [16]. For that definition one simply replaces all limits by indexed limits. Or, the enrichment is not Cartesian, then we will replace the groupoid condition by an indexing just like above.

6.2.1 Cartesian Case: Categorical Version

In [33] we proved that in the non-enriched case we can equivalently replace (ii) by (ii′).

(ii′) The pull-back of presheaves $\imath^{\otimes\wedge}: [\mathcal{F}^{op}, Set] \rightarrow [\mathcal{V}^{\otimes op}, Set]$ *restricted to representable presheaves* is monoidal.

This then yields a definition in the Cartesian case if one replaces (iii) by the appropriate indexed limit condition.

6.2.2 Non-Cartesian Case Indexed Enrichment

In the non-Cartesian case, the notion of groupoid ceases to make sense. The first option is to drop the groupoid condition and simply ask that the inclusion $\iota^{\otimes}$ is essentially surjective. This is possible and called a weak Feynman category, which is very close to the notion of a pattern and explains that notion in more down to earth terms. This is, however, not adequate for the bar/cobar and Feynman transforms or the twists.

The better notion is that of a Feynman category enriched over $\mathcal{E}$, indexed over another Feynman category $\mathfrak{F}$. The idea is that the Feynman category $\mathfrak{F}_O$ for algebras over an operad O is a Feynman category enriched over C indexed over $Surj$. The precise definition goes via enrichment functors, which are 2-functors.

In general, we will call the enrichment category $\mathcal{E}$. This is a monoidal category and hence can be thought of as a 2-category with one object, which we denote by $\underline{\mathcal{E}}$. Here the 1-morphisms of $\underline{\mathcal{E}}$ are the objects of $\mathcal{E}$ with the composition being $\otimes$, the monoidal structure of $\mathcal{E}$. The 2-morphisms are then the 2-morphisms of $\mathcal{E}$, their horizontal composition being $\otimes$ and their vertical composition being $\circ$. Also, we can consider any category $\mathcal{F}$ to be a 2-category with the two morphisms generated by triangles of composable morphisms.

Definition 6.5 Let $\mathfrak{F}$ be a Feynman category. An enrichment functor is a lax 2-functor $\mathcal{D} : \mathcal{F} \to \underline{\mathcal{E}}$ with the following properties

1. $\mathcal{D}$ is strict on compositions with isomorphisms.
2. $\mathcal{D}(\sigma) = \mathbb{1}_{\mathcal{E}}$ for any isomorphism.
3. $\mathcal{D}$ is monoidal, that is $\mathcal{D}(\phi \otimes_{\mathcal{F}} \psi) = \mathcal{D}(\phi) \otimes_{\mathcal{E}} \mathcal{D}(\psi)$

Given a monoidal category $\mathcal{F}$ considered as a 2-category and lax 2-functor $\mathcal{D}$ to $\underline{\mathcal{E}}$ as above, we define an enriched monoidal category $\mathcal{F}_{\mathcal{D}}$ as follows. The objects of $\mathcal{F}_{\mathcal{D}}$ are those of $\mathcal{F}$. The morphisms are given as

$$Hom_{\mathcal{F}_{\mathcal{D}}}(X, Y) := \bigoplus_{\phi \in Hom_{\mathcal{F}}(X,Y)} \mathcal{D}(\phi) \tag{29}$$

The composition is given by

$$Hom_{\mathcal{F}_{\mathcal{D}}}(X, Y) \otimes Hom_{\mathcal{F}_{\mathcal{D}}}(Y, Z) \tag{30}$$

$$= \bigoplus_{\phi \in Hom_{\mathcal{F}}(X,Y)} \mathcal{D}(\phi) \otimes \bigoplus_{\psi \in Hom_{\mathcal{F}}(Y,Z)} \mathcal{D}(\psi) \tag{31}$$

$$\simeq \bigoplus_{(\phi,\psi) \in Hom_{\mathcal{F}}(X,Y) \times Hom_{\mathcal{F}}(Y,Z)} \mathcal{D}(\phi) \otimes \mathcal{D}(\psi) \tag{32}$$

$$\xrightarrow{\oplus \mathcal{D}(\circ)} \bigoplus_{\chi \in Hom_{\mathcal{F}}(X,Z)} \mathcal{D}(\chi) = Hom_{\mathcal{F}_{\mathcal{D}}}(X, Z) \tag{33}$$

The image lies in the components $\chi = \psi \circ \phi$. Using this construction on $\mathcal{V}$, pulling back $\mathcal{D}$ via ι, we obtain $\mathcal{V}_{\mathcal{D}} = \mathcal{V}_{\mathcal{E}}$, the freely enriched $\mathcal{V}$. The functor ι then is naturally upgraded to an enriched functor $\iota_{\mathcal{E}} : \mathcal{V}_{\mathcal{D}} \to \mathcal{F}_{\mathcal{D}}$.

Definition 6.6 Let $\mathfrak{F}$ be a Feynman category and let $\mathcal{D}$ be an enrichment functor. We call $\mathfrak{F}_{\mathcal{D}} := (\mathcal{V}_{\mathcal{E}}, \mathcal{F}_{\mathcal{D}}, \iota_{\mathcal{E}})$ a Feynman category enriched over $\mathcal{E}$ indexed by $\mathcal{D}$.

Theorem 6.7 $\mathfrak{F}_{\mathcal{D}}$ *is a weak Feynman category. The forgetful functor from* $\mathcal{F}_{\mathcal{D}}$*-Ops to* $\mathcal{V}_{\mathcal{E}}$*-Mods has a left adjoint and more generally push-forwards among indexed enriched Feynman categories exist. Finally there is an equivalence of categories between algebras over the triple (aka. monad) GF and* $\mathcal{F}_{\mathcal{D}}$*-Ops.*

Example 6.8 The freely enriched Feynman category. The functor $\mathcal{D}$ is simply the identity. This is the triple $\mathfrak{F}_{\mathcal{E}} := (\mathcal{V}_{\mathcal{E}}, \mathcal{F}_{\mathcal{E}}, \iota_{\mathcal{E}})$ where $\mathfrak{F} = (\mathcal{V}, \mathcal{F}, \iota)$ is a Feynman category and the subscript $\mathcal{E}$ means free enrichment.

Theorem 6.9 *The indexed enriched (over $\mathcal{E}$) Feynman category structures on a given FC $\mathfrak{F}$ are in 1-1 correspondence with $\mathfrak{F}^{hyp}$-Ops and these are in 1-1 correspondence with enrichment functors.*

Example 6.10 (Twisted (Modular) Operads) Looking at $\mathfrak{F} = \mathfrak{M}$, we recover the notion of twisted modular operad. There is a twist for each hyper-operad $\mathcal{D}$. We have the Feynman category $\mathfrak{M}_{\mathcal{D}}$. The triple then corresponds to $\mathbb{M}_{\mathcal{D}}$ in the notation of [19]. What we add is the descriptions (a) and (c) mentioned in paragraph 1.3, that is via compositions along graphs and generators and relations. Here the graphs are actually decorated on the set of edges according to (29). To see this one decomposes ϕ into simple edge or loop contractions as defined in Sect. 4.

Example 6.11 Algebras over operads. In this case $\mathfrak{F} = Surj$ and $\mathfrak{F}^{hyp,rd} = \mathfrak{O}_0$. An operad $O \in \mathfrak{O}_0\text{-}Ops_{\mathcal{C}}$ then gives an enrichment functor $\mathcal{D}_O$ of $Surj$. In particular $\mathcal{D}_O(n \twoheadrightarrow 1) = O(n)$ as in Sect. 6.1.1.

6.2.3 Coboundaries and $\mathcal{V}$-twists

Coboundaries in the sense of [19] are generalized to $\mathcal{V}$-twists. Let $\mathfrak{L}: \mathcal{V} \to Pic(\mathcal{E})$, that is the full subcategory of $\otimes$-invertible elements of $\mathcal{E}$. A twist of a Feynman category indexed by $\mathcal{D}$ by $\mathfrak{L}$ is given by setting the new twist-system to be $\mathcal{D}_{\mathfrak{L}}(\phi) = \mathfrak{L}(t(\phi))^{-1} \otimes \mathcal{D}(\phi) \otimes \mathfrak{L}(s(\phi))$.

The suspension functor s is such a coboundary twist, see [19, 35]. Here $\mathfrak{L} = s$ with $s(*_{(n-1)_+}) = \Sigma^{2-n} sign_n$ in dg $\mathcal{V}ect$ for cyclic operads, or $s(*_{n_+}) = \Sigma^{1-n} sign_n$ for operads, or in general $s(*_{(n-1)_+}) = \Sigma^{-2(g-1)+n} sign_n$ where Σ is the suspension and $sign_n$ is the sign representation, see [35] for a detailed explanation.

6.2.4 Odd Versions and Shifts

Given a well-behaved presentation of a Feynman category (generators+relations for the morphisms) we can define an odd version which is enriched over $\mathcal{A}b$ by giving a twist. To obtain the odd versions, we use $\mathcal{D}(\phi) = \det(Edges(\Gamma(\phi)))$. In the cyclic case, an example are anti-cyclic operads and the theory of modular operads this twist is called $\mathfrak{K}$. It is *not* a coboundary in general. Rather up to the suspension coboundary and the shift coboundary, this twist is a twist by $H_1(\Gamma)$ in the modular case, see [19, 35] for details.

6.3 Feynman Level Category $\mathfrak{F}^+$, Hyper Category $\mathfrak{F}^{hyp}$ and Its Reduction $\mathfrak{F}^{hyp,rd}$

6.3.1 Feynman Level Category $\mathfrak{F}^+$

Given a Feynman category $\mathfrak{F}$, and a choice of basis for it, we will define its Feynman level category $\mathfrak{F}^+ = (\mathcal{V}^+, \mathcal{F}^+, \iota^+)$ as follows. The underlying objects of $\mathcal{F}^+$ are the morphisms of $\mathcal{F}$. The morphisms of $\mathcal{F}^+$ are given as follows: given ϕ and ψ, consider their decompositions

$$
\begin{array}{ccc}
X \xrightarrow{\phi} Y & \qquad & X' \xrightarrow{\psi} Y' \\
\sigma \downarrow \simeq \qquad \simeq \downarrow \hat{\sigma} & & \tau \downarrow \simeq \qquad \simeq \downarrow \hat{\tau} \\
\bigotimes_{v\in I}\bigotimes_{w\in I_v} *_w \xrightarrow{\bigotimes_{v\in I}\phi_v} \bigotimes_{v\in I} *_v & & \bigotimes_{v'\in I'}\bigotimes_{w\in I'_{v'}} *_w \xrightarrow{\bigotimes_{v'\in I'}\psi_{v'}} \bigotimes_{v'\in I'} *_{v'}
\end{array}
$$

$$\tag{34}$$

where we have dropped the ι from the notation, $\sigma, \hat{\sigma}, \tau$ and $\hat{\tau}$ are given by the choice of basis and the partition I_v of the index set for X and $I'_{v'}$ for the index set of Y is given by the decomposition of the morphism.

A morphism from ϕ to ψ is a two level partition of $I : (I_{v'})_{v'\in I'}$, and partitions of $I_{v'} : (I^1_{v'}, \ldots, I^{k_{v'}}_{v'})$ such that if we set $\phi^i_{v'} := \bigotimes_{v\in I^i_{v'}} \phi_v$ then $\psi_{v'} = \phi^k_{v'} \circ \cdots \circ \phi^1_{v'}$.

To compose two morphisms $f : \phi \to \psi$ and $g : \psi \to \chi$, given by partitions of $I : (I_{v'})_{v'\in I'}$ and of the $I_{v'} : (I^1_{v'}, \ldots, I^{k_{v'}}_{v'})$ respectively of $I' : (I_{v''})_{v''\in I''}$ and the $I_{v''} : (I'^1_{v''}, \ldots, I'^{k_{v''}}_{v''})$, where I'' is the index set in the decomposition of χ, we set the compositions to be the partitions of $I : (I_{v''})_{v''\in I''}$ where $I_{v''}$ is the set partitioned by $(I_{v'})_{v'\in I'^j_{v''}, j=1,\ldots,k_{v''}}$. That is, we replace each morphism $\psi_{v'}$ by the chain $\phi^{v'}_1 \circ \cdots \circ \phi^{v'}_k$.

Morphisms alternatively correspond to rooted forests of level trees thought of as flow charts, see Fig. 7. Here the vertices are decorated by the ϕ_v and the composition along the rooted forest is ψ. There is exactly one tree $\tau_{v'}$ per $v' \in I'$ in the forest and accordingly the composition along that tree is ψ'_v.

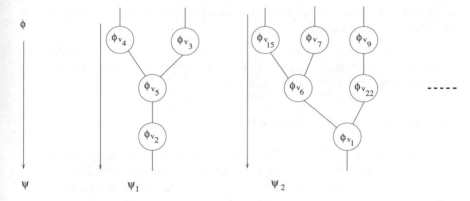

Fig. 7 The level forest picture for morphisms in $\mathfrak{F}^+$. Indicated is a morphism from $\phi \simeq \bigotimes_v \phi_v$ to $\Psi \simeq \bigotimes_i \Psi_i$

Technically, the vertices are the $v \in I$. The flags are the union $\amalg_v \amalg_{w \in I_v} *_w \amalg \amalg_{v \in I} *_v$ with the value of ∂ on $*_w$ being v if $w \in I_v$ and v on $*_v$ for $v \in I$. The orientation at each vertex is given by the target being out. The involution ι is given by matching source and target objects of the various ϕ_v. The level structure of each tree is given by the partition $I_{v'}$. The composition is the composition of rooted trees by gluing trees at all vertices—that is we blow up the vertex marked by $\psi_{v'}$ into the tree $\tau_{v'}$.

6.3.2 $\mathcal{F}^+$-*Ops*

After passing to the equivalent strict Feynman category, an element $\mathcal{D}$ in $\mathcal{F}^+$-*Ops* is a symmetric monoidal functor that has values on each morphism $\mathcal{D}(\phi) = \bigotimes \mathcal{D}(\phi_v)$ and has composition maps $\mathcal{D}(\phi_0 \otimes \phi) \to \mathcal{D}(\phi_1)$ for each decomposition $\phi_1 = \phi \circ \phi_0$. Further decomposing $\phi = \bigotimes \phi_v$ where the decomposition is according to the target of ϕ_0, we obtain morphisms

$$\mathcal{D}(\phi_0) \otimes \bigotimes_v \mathcal{D}(\phi_v) \to \mathcal{D}(\phi_1) \tag{35}$$

It is enough to specify these functors for $\phi_1 \in (\mathcal{F} \downarrow \mathcal{V})$ and then check associativity for triples.

Example 6.12 If we start from the tautological Feynman category on the trivial category $\mathfrak{F} = (1, 1^{\otimes}, \iota)$ then $\mathfrak{F}^+$ is the Feynman category $\mathfrak{Sur}$ of surjections. Indeed the possible trees are all linear, that is only have 2-valent vertices, and there is only one decoration. Such a rooted tree is specified by its total length n and the permutation which gives the bijection of its vertices with the set n_i. Looking at a forest of these trees we see that we have the natural numbers as objects with morphisms being surjections.

Example 6.13 We also have $\mathfrak{Sur}^+ = \mathfrak{O}_{May}$, which is the Feynman category for May operads. Indeed the basic maps (35) are precisely the composition maps γ. To be precise, these are May operads without units.

6.3.3 Feynman Hyper Category $\mathfrak{F}^{hyp}$

There is a "reduced" version of $\mathfrak{F}^+$ which is central to our theory of enrichment. This is the universal Feynman category through which any functor $\mathcal{D}$ factors if it satisfies the following restriction: $\mathcal{D}(\sigma) \simeq \mathbb{1}$ *for any isomorphism σ where $\mathbb{1}$ is the unit of the target category $\mathcal{C}$.*

For this, we invert the morphisms corresponding to composing with isomorphisms, see [33] for details.

6.3.4 $\mathcal{F}^{hyp}$-Ops

An element $\mathcal{D} \in \mathcal{F}^{hyp}$-*Ops* corresponds to the data of functors from $Iso(\mathcal{F} \downarrow \mathcal{F}) \rightarrow \mathcal{C}$ together with morphisms (35) which are associative and satisfy the condition that all the following diagrams commutes:

$$
\begin{array}{ccc}
\mathcal{D}(\phi) \otimes \bigotimes_v \mathcal{D}(\sigma_v) & \xrightarrow[\sim]{\mathcal{D}(\tau)} & \mathcal{D}(^\sigma\phi) \\[2mm]
{\scriptstyle\sim}\Big\uparrow & & \Big\uparrow{\scriptstyle\mathcal{D}(\sigma)} \\[2mm]
\mathcal{D}(\phi) \otimes \bigotimes_v \mathbb{1} & \xleftarrow[\sim]{id\otimes\bigotimes_v r_{\mathbb{1}}^{-1}} & \mathcal{D}(\phi)
\end{array}
\tag{36}
$$

see [33] for details.

Example 6.14 The paradigmatic examples are hyper-operads in the sense of [19]. Here $\mathfrak{F} = \mathfrak{M}$ and $\mathfrak{F}^{hyp}$ is the Feynman category for hyper-operads.

6.3.5 A Reduced Version $\mathfrak{F}^{hyp,rd}$

One may define $\mathfrak{F}^{hyp,rd}$, a Feynman subcategory of $\mathfrak{F}^{hyp}$ which is equivalent to it by letting $\mathcal{F}^{hyp,rd}$ and $\mathcal{V}^{hyp,rd}$ be the respective subcategories whose objects are morphisms that do not contains isomorphisms in their decomposition. In view of the isomorphisms $\emptyset \rightarrow \sigma$ this is clearly an equivalent subcategory. In particular, the respective categories of *Ops* and *Mods* are equivalent.

The morphisms are described by rooted forests of trees whose vertices are decorated by the ϕ_v as above—none of which is an isomorphism—, with the additional decoration of an isomorphism per edge and tail. Alternatively, one can think of the decoration as a black 2-valent vertex. Indeed, using maps from $\emptyset \rightarrow \sigma$, we can introduce as many isomorphisms as we wish. These give rise to 2-valent vertices, which we mark black. All other vertices remain labeled by ϕ_v. If there are sequences of such black vertices, the corresponding morphism is isomorphic to the morphism resulting from composing the given sequence of these isomorphisms.

Example 6.15 For $\mathfrak{F}_{surj}^{hyp,rd} = \mathfrak{O}_0$, the Feynman category whose morphisms are trees with at least trivalent vertices (or identities) and whose *Ops* are operads whose $O(1) = \mathbb{1}$. Indeed the basic non-isomorphism morphisms are the surjections $\underline{n} \to \underline{1}$, which we can think of as rooted corollas. Since for any two singleton sets there is a unique isomorphism between them, we can suppress the black vertices in the edges. The remaining information is that of the tails, which is exactly the map ϕ^F in the morphism of graphs.

Example 6.16 For the trivial Feynman category, we obtain back the trivial Feynman category as the reduced hyper category, since the trees all collapse to a tree with one black vertex.

6.4 Free Monoidal Construction $\mathcal{F}^{\boxtimes}$

Sometimes it is convenient to construct a new Feynman category from a given one whose vertices are the objects of $\mathcal{F}$. Formally, we set $\mathfrak{F}^{\boxtimes} = (\mathcal{V}^{\otimes}, \mathcal{F}^{\boxtimes}, \imath^{\otimes})$ where $\mathcal{F}^{\boxtimes}$ is the free monoidal category on $\mathcal{F}$ and we denote the "outer" free monoidal structure by $\boxtimes$. This is again a Feynman category. There is a functor $\mu : \mathcal{F}^{\boxtimes} \to \mathcal{F}$ which sends $\boxtimes_i X_i \mapsto \bigotimes_i X_i$ and by definition $Hom_{\mathcal{F}^{\boxtimes}}(\mathbf{X} = \boxtimes_i X_i, \mathbf{Y} = \boxtimes_i Y_i) = \bigotimes_i Hom_{\mathcal{F}}(X_i, Y_i)$. The only way that the index sets can differ, without the Hom-sets being empty, is if some of the factors are $\mathbb{1} \in \mathcal{F}^{\boxtimes}$. Thus the one-comma generators are simply the elements of $Hom_{\mathcal{F}}(X, Y)$. Using this identification one obtains: $Iso(\mathcal{F}^{\boxtimes}) \simeq Iso(\mathcal{F})^{\boxtimes} \simeq (\mathcal{V}^{\otimes})^{\boxtimes}$. The factorization and size axiom follow readily from this description.

Proposition 6.17 $\mathcal{F}^{\boxtimes}$-*Ops*$_C$ *is equivalent to the category of functors (not necessarily monoidal)* $Fun(\mathcal{F}, C)$.

Example 6.18 Examples are *FI* modules and (crossed) simplicial objects for the free monoidal Feynman categories for *FI* and Δ_+ where for the latter one uses the non-symmetric version.

6.5 NC-construction

For any Feynman category one can define its nc (non-connected) version. It plays a crucial role in physics and mathematics and manifests itself through the BV equation [35]. Namely, for the operator Δ in the case of modular operads to become a differential, one needs a multiplication. This, on the graph level, is given by disjoint union for the one-comma generators. This amounts to dropping the condition of connectedness. Astonishingly this works in full generality for any Feynman category.

Let $\mathfrak{F} = (\mathcal{V}, \mathcal{F}, \iota)$, then we set $\mathfrak{F}^{nc} = (\mathcal{V}^{\otimes}, \mathcal{F}^{nc}, \iota^{\otimes})$ where $\mathcal{F}^{nc}$ has objects $\mathcal{F}^{\boxtimes}$, the free monoidal product. We however add more morphisms. The one-comma generators will be $Hom_{\mathcal{F}^{nc}}(\mathbf{X}, Y) := Hom_{\mathcal{F}}(\mu(\mathbf{X}), Y)$, where for $\mathbf{X} = \boxtimes_{i \in I} X_i$, $\mu(\mathbf{X}) = \bigotimes_{i \in I} X_i$. This means that for $\mathbf{Y} = \boxtimes_{j \in J} Y_j$, $Hom_{\mathcal{F}}(\mathbf{X}, \mathbf{Y}) \subset Hom_{\mathcal{F}}(\mu(\mathbf{X}), \mu(\mathbf{Y}))$, includes only those morphisms for which there is a partition $I_j, j \in J$ of I such that the morphism factors through $\bigotimes_{j \in J} Z_j$ where $Z_j \xrightarrow{\sigma_j} \bigotimes_{k \in I_j} X_k$ is an isomorphism. That is $\psi = \bigotimes_{j \in J} \phi_j \circ \sigma_j$ with $\phi_j : Z_j \to Y_j$. Notice that there is a map of "disjoint union" or "exterior multiplication" given by $\mu : X_1 \boxtimes X_2 \to X_1 \otimes X_2$ via $id \otimes id$.

Example 6.19 The terminology "non-connected" has its origin in the graph examples. Examples can be found in [35], where also a box-picture for graphs is presented. The connection is that morphisms in $\mathcal{F}^{nc}$ have an underlying graph that is disconnected and the connected components are those of the underlying $\mathcal{F}$.

Proposition 6.20 ([33]) *There is an equivalence of categories between $\mathcal{F}^{nc}$-Ops_C and symmetric lax monoidal functors $Fun_{lax \otimes}(\mathcal{F}, C)$.*

Using lax-monoidal functors, is also a way to deal with algebras over operads whose $O(1)$ has isomorphisms.

7 Universal Operations, Transforms and Master Equations

7.1 Universal Operations

7.1.1 Universal Operations for Operads, etc.

A well known result in operad theory is that for an operad O there is an odd Lie bracket defined on $\bigoplus O(n)$ [15]. This actually descends to coinvariants $\bigoplus O(n)_{\mathbb{S}_n}$ [24]. For anti-cyclic operads there is again an odd Lie bracket on the coinvariants $\bigoplus O((n))_{\mathbb{S}_n}$ with lifts to the smaller coinvariants w.r.t. the cyclic groups C_n, namely on $\bigoplus_n O((n))_{C_n}$ [35]. Similarly there are operations Δ on $\bigoplus O((n, g))_{\mathbb{S}_n}$ for modular operads [1, 35]. Here we show that these operations can be understood purely from the Feynman category and we can explain why exactly these operations turn up in the Master Equations.

7.1.2 Cocompletion

Let $\hat{\mathcal{F}}$ be the cocompletion of $\mathcal{F}$. This is monoidal with the monoidal structure given by the Day convolution $\circledast$. If C is cocomplete then $O \in Ops$ factors:

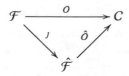

Theorem 7.1 *Let* $\mathbb{1} := \text{colim}_{\mathcal{V}} \jmath \circ \imath \in \hat{\mathcal{F}}$ *and let* $\mathcal{F}_{\mathcal{V}}$ *the symmetric monoidal subcategory generated by* $\mathbb{1}$*. Then* $\mathfrak{F}_{\mathcal{V}} := (\mathcal{F}_{\mathcal{V}}, \mathbb{1}, \imath_{\mathcal{V}})$ *is a Feynman category. (This gives an underlying operad of universal operations).*

If $\mathcal{E}$ is Abelian, we say $\mathfrak{F}_{\mathcal{V}}$ is weakly generated by morphisms $\phi \in \Phi$ if the summands of the components $[\phi_{X_j,i}]$ generate the morphisms of $\mathfrak{F}_{\mathcal{V}}$. Here different summands are indexed by different isomorphism classes of morphisms.

7.1.3 Example: Operads

$\mathfrak{O}$ the Feynman category for operads, $C = dg\mathcal{V}ect$.

Then $\hat{O}(\mathbb{1}) = \bigoplus_n O(n)_{\mathbb{S}_n}$ and the Feynman category is (weakly) generated by $\circ := [\sum \circ_i]$. (This is a two-line calculation). This gives rise to the Lie bracket by using the anti-commutator. It lifts to the non-Sigma case along the forgetful $\mathfrak{O}^{\neg \Sigma} \to \mathfrak{O}$ and gives the pre-Lie structure on $\bigoplus_n O(n)$, which goes back to [15]. In [24] it was shown that the pre-Lie structure descends to the coinvariants. In [35] it is argued that the pre-Lie structure lives naturally on the coinvariants and lifts to the invariants.

In general these kinds of lifts are possible if there is a non-Sigma version.

7.1.4 Example: Odd/Anti-cyclic Operad

The universal operations are (weakly) generated by a Lie bracket. $[,] := [\sum_{st} \circ_{st}]$, (see [35]). This actually lifts to cyclic coinvariants (non-sigma cyclic operads) that is along the map $\mathfrak{C}^{odd,pl} \to \mathfrak{C}^{odd}$. Here we also see that one cannot expect a further lift, since the planar version for $\mathfrak{C}^{odd}$ still has a non discrete $\mathcal{V}$.

7.1.5 The Three Geometries of Kontsevich

The endomorphism operad $End(V)$ for a symplectic vector space is anti-cyclic. Any tensor product: $(O \otimes P)(n) := O(n) \otimes P(n)$ with O a cyclic operad and P an anti-cyclic operad is anti-cyclic and hence has the odd Lie bracket discussed above.

Fix V^n n-dim symplectic $V^n \to V^{n+1}$. For each n get Lie algebras

1. *Comm* $\otimes End(V^n)$
2. *Lie* $\otimes End(V^n)$
3. *Assoc* $\otimes End(V^n)$

Taking the limit as $n \to \infty$ one obtains the formal geometries of [10, 37].

Table 4 Here $\mathfrak{F}_{\cdot V}$ and $\mathfrak{F}^{nt}_{\cdot V}$ are given as $\mathfrak{F}_O$ for the operad O, the composition as discussed being insertion

$\mathfrak{F}$	Feynman cat for	$\mathfrak{F}, \mathfrak{F}_{\cdot V}, \mathfrak{F}^{nt}_{\cdot V}$	Weak gen. subcat.
$\mathfrak{O}$	Operads	Rooted trees	$\mathfrak{F}_{pre\text{-}Lie}$
$\mathfrak{O}^{odd}$	Odd operads	Rooted trees + orientation of set of edges	Odd pre-Lie
$\mathfrak{O}^{\neg \Sigma}$	Non-Sigma operads	Planar rooted trees	All $\circ_i$ operations
$\mathfrak{O}_{mult}$	Operads with mult.	B/w rooted trees	Pre-Lie + mult.
$\mathfrak{C}$	Cyclic operads	Trees	Com. mult.
$\mathfrak{C}^{odd}$	Odd cyclic operads	Trees + orientation of set of edges	Odd Lie
$\mathfrak{M}^{odd}$	$\mathfrak{K}$-Modular	Connected + orientation on set of edges	Odd dg Lie
$\mathfrak{M}^{nc,odd}$	nc $\mathfrak{K}$-modular	Orientation on set of edges	BV
$\mathfrak{D}$	Dioperads	Connected directed graphs w/o directed loops or parallel edges	Lie-admissible

The former is for the type of graph with unlabelled tails and the latter for the version with no tails

Our construction is more general and works for any anti-cyclic operad. For instance another family of Lie algebras can be obtained as follows, [35]. Let V^n be a vector space with a symmetric non-degenerate form. $End(V)$ is a cyclic operad. Since the *PreLie* operad is anti-cyclic [8], for each n we get a Lie algebra *PreLie* $\otimes End(V)$. It is not known what geometry we get when we take the limit as $n \to \infty$.

7.1.6 Further Examples

For further examples, see Table 4.

7.2 Transforms and Master Equations

There are three transforms we will consider: the bar-, the cobar transform and the Feynman transform aka. dual transform.

7.2.1 Motivating Example: Algebras

If A is an associative algebra, then the bar transform is the dg-coalgebra given by the free coalgebra $BA = T\Sigma^{-1}\bar{A}$ together with co-differential from algebra structure. The usual notation for an element in BA is $a_0|a_1|\ldots|a_n$.

Likewise let C be an associative co-algebra. The co-bar transform is the dg-algebra $\Omega C := Free_{alg}(\Sigma^{-1}\bar{C})$ together with a differential coming from co-algebra structure. The bar-cobar transform ΩBA is a resolution of A.

For the Feynman transform consider A a finite-dimensional algebra or graded algebra with finite dimensional pieces and let $\check{A}$ be its dual co-algebra. Then the dual or Feynman transform of A is $FA := \Omega\check{A}$ + differential from multiplication. Now, the double Feynman transform FFA a resolution.

7.2.2 Transforms

These transforms take $O \in \mathcal{F}\text{-}Ops_C$ and transform it to an op for the odd version of the Feynman category $\mathfrak{F}^{odd}$ either in C^{op} or C. All these are free constructions, which, however, also have the extra structure of an additional (co)differential. Thus the resulting Feynman category is actually enriched over chain complexes and one can start out there as well. Furthermore, for the (co)differential to work, we have to have signs. These are exactly what is provided by the odd versions. In order to be able to define the transforms, one has to fix an odd version $\mathfrak{F}^{odd}$ of $\mathfrak{F}$, just as in Sect. 6.1.3. This is analogous to the suspension in the usual bar transforms. In fact, the following is more natural, see [33, 35]. The degree is 1 for each bar and in the graph case the edges get degree 1; see Fig. 8. We can generalize the construction of $\mathfrak{F}^{odd}$ to so-called well-presented Feynman categories, see below and [33]. In this case, we can define the transformations for elements of Ops.

The Feynman transform is of particular interest. Since the construction is free, any $\mathcal{V} \in Mods$ will yield an op. On the other hand, this need not be compatible with the dg structure. It turns out that it is, if it satisfies a Master Equation.

The transforms are of interest in themselves, but one common application is that the bar-cobar transform as well as the double Feynman transform give a "free" resolution. In general, of course, "free" means co-fibrant. For this kind of statement one needs a Quillen model structure, which is provided in Sect. 8.

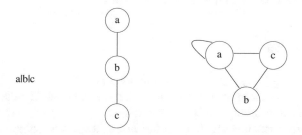

Fig. 8 The sign mnemonics for the bar construction, traditional version with the symbols | of degree 1, the equivalent linear tree with edges of degree 1, and a more general graph with edges of degree 1. Notice that in the linear case there is a natural order of edges, this ceases to be the case for more general graphs

Remark 7.2 As before one can ask the question of how much of the structure of these transforms can be pulled back to the Feynman category side. The answer is: "Pretty much all of it". We shall not discuss this here, but it can be found in [33].

7.2.3 Presentations

In order to define the transforms, we have to give what is called an ordered presentation [33]. Rather then giving the technical conditions, we will consider the graph case and show these structures in this case.

7.2.4 Basic Example 𝔊

In 𝔊 the presentation comes from the following set of morphisms Φ

1. There are four types of basic morphisms: Isomorphisms, simple edge contractions, simple loop contractions and simple mergers. Call this set Φ.
2. These morphisms generate all one-comma generators upon iteration. Furthermore, isomorphisms act transitively on the other classes. The relations on the generators are given by commutative diagrams.
3. The relations are quadratic for edge contractions as are the relations involving isomorphisms. Finally there is a non-homogenous relation coming from a simple merger and a loop contraction being equal to an edge contraction.
4. We can therefore assign degrees as 0 for isomorphisms and mergers, 1 for edge or loop contractions and split Φ as $\Phi^0 \amalg \Phi^1$. This gives a degree to any morphism.

Up to isomorphism any morphism of degree n can be written in $n!$ ways up to morphisms of degree 0. These are the enumerations of the edges of the ghost graph.

There is also a standard order in which isomorphisms come before mergers which come before edge contractions as in (8). This gives an ordered presentation.

In general, an ordered presentation is a set of generators Φ and extra data such as the subsets Φ^0 and Φ^1; we refer to [33] for details.

7.2.5 Differential

Given a $d_{\Phi^1} = \sum_{[\phi_1] \in \Phi^1 / \sim} \phi_1 \circ$ defines an endomorphism on the Abelian group generated by the isomorphism classes morphisms. The non-defined terms are set to zero. Φ^1 is called resolving if this is a differential.

In the graph case, this amounts to the fact that for any composition of edge contractions $\phi_e \circ \phi_{e'}$, there is precisely another pair of edge contractions $\phi_{e''} \circ \phi_{e'''}$ which contracts the edges in the opposite order.

This differential will induce differentials for the transforms, which we call by the same name. We again refer to [33] for details.

7.2.6 Setup

$\mathfrak{F}$ be a Feynman category enriched over $\mathcal{A}b$ and with an ordered presentation and let $\mathfrak{F}^{odd}$ be its corresponding odd version. Furthermore let Φ^1 be a resolving subset of one-comma generators and let C be an additive category, i.e. satisfying the analogous conditions above. In order to give the definition, we need a bit of preparation. Since $\mathcal{V}$ is a groupoid, we have that $\mathcal{V} \simeq \mathcal{V}^{op}$. Thus, given a functor $\Phi : \mathcal{V} \to C$, using the equivalence we get a functor from $\mathcal{V}^{op}$ to C which we denote by Φ^{op}. Since the bar/cobar/Feynman transform adds a differential, the natural target category from $\mathcal{F}$-Ops is not C, but complexes in C, which we denote by $Kom(C)$. Thus any O may have an internal differential d_O.

7.2.7 The Bar Construction

This is the functor

$$\mathsf{B} \colon \mathcal{F}\text{-}Ops_{Kom(C)} \to \mathcal{F}^{odd}\text{-}Ops_{Kom(C^{op})}$$

$$\mathsf{B}(O) := \iota_{\mathfrak{F}^{odd}\,*}(\iota_{\mathfrak{F}}^*(O))^{op}$$

together with the differential $d_{O^{op}} + d_{\Phi^1}$.

7.2.8 The Cobar Construction

This is the functor

$$\Omega \colon \mathcal{F}^{odd}\text{-}Ops_{Kom(C^{op})} \to \mathcal{F}\text{-}Ops_{Kom(C)}$$

$$\Omega(O) := \iota_{\mathfrak{F}\,*}(\iota_{\mathfrak{F}^{odd}}^*(O))^{op}$$

together with the co-differential $d_{O^{op}} + d_{\Phi^1}$.

7.2.9 Feynman Transform

Assume there is a duality equivalence $\vee \colon C \to C^{op}$. The Feynman transform is a pair of functors, both denoted FT,

$$\mathsf{FT} \colon \mathcal{F}\text{-}Ops_{Kom(C)} \leftrightarrows \mathcal{F}^{odd}\text{-}Ops_{Kom(C)} \colon \mathsf{FT}$$

defined by

$$FT(O) := \begin{cases} \vee \circ B(O) & \text{if } O \in \mathcal{F}\text{-}Ops_{Kom(C)} \\ \vee \circ \Omega(O) & \text{if } O \in \mathcal{F}^{odd}\text{-}Ops_{Kom(C)} \end{cases}$$

Proposition 7.3 *The bar and cobar construction form an adjunction.*

$$\Omega : \mathcal{F}^{odd}\text{-}Ops_{Kom(C^{op})} \rightleftarrows \mathcal{F}\text{-}Ops_{Kom(C)} : B$$

The quadratic relations in the graph examples are a feature that can be generalized to the notion of *cubical* Feynman categories. The name reflects the fact that in the graph example the $n!$ ways to decompose a morphism whose ghost graph is connected and has n edges into simple edge contractions correspond to the edge paths of I^n going from $(0, \ldots, 0)$ to $(1, \ldots, 1)$. Each edge flip in the path represent one of the quadratic relations and furthermore the $\mathbb{S}_n$ action on the coordinates is transitive on the paths, with transposition acting as edge flips.

This is a convenient generality in which to proceed.

Theorem 7.4 *Let $\mathfrak{F}$ be a cubical Feynman category and $O \in \mathcal{F}\text{-}Ops_{Kom(C)}$. Then the counit $\Omega B(O) \to O$ of the above adjunction is a levelwise quasi-isomorphism.*

Remark 7.5 In the case of $C = dgVect$, the Feynman transform can be intertwined with the aforementioned push-forward and pull-back operations to produce new operations on the categories $\mathcal{F}\text{-}Ops_C$. A lifting (up to homotopy) of these new operations to $C = Vect$ is given in [50]. In particular this result shows how the Feynman transform of a push-forward (resp. pull-back) may be calculated as the push-forward (resp. pull-back) of a Feynman Transform. One could thus assert that the study of the Feynman transform belongs to the realm of Feynman categories as a whole and not just to the representations of a particular Feynman category.

7.3 Master Equations

In [35], we identified the common background of Master Equations that had appeared throughout the literature for operad-like objects and extended them to all graphs examples. An even more extensive theorem for Feynman categories can also be given.

The Feynman transform is quasi-free. An algebra over FO is dg—if and only if it satisfies the relevant Master Equation. First, we have the tabular theorem from [35] for the usual suspects.

Theorem 7.6 ([1, 35, 44, 45]) *Let $O \in \mathcal{F}\text{-}Ops_C$ and $\mathcal{P} \in \mathcal{F}^{odd}\text{-}Ops_C$ for an $\mathcal{F}$ represented in Table 5. Then there is a bijective correspondence:*

$$Hom(FT(\mathcal{P}), O) \cong ME(lim_\vee(\mathcal{P} \otimes O))$$

Table 5 Collection of Master Equations for operad-type examples

Name of $\mathcal{F}$-Ops_C	Algebraic structure of FO	Master Equation (ME)
Operad [17]	Odd pre-Lie	$d(-) + - \circ - = 0$
Cyclic operad [18]	Odd Lie	$d(-) + \frac{1}{2}[-, -] = 0$
Modular operad [19]	Odd Lie + Δ	$d(-) + \frac{1}{2}[-, -] + \Delta(-) = 0$
Properad [49]	Odd pre-Lie	$d(-) + - \circ - = 0$
Wheeled properad [44]	Odd pre-Lie + Δ	$d(-) + - \circ - + \Delta(-) = 0$
Wheeled prop [35]	dgBV	$d(-) + \frac{1}{2}[-, -] + \Delta(-) = 0$

Here ME is the set of solutions of the appropriate Master Equation set up in each instance.

With Feynman categories this tabular theorem can be compactly written and generalized. The first step is the realization that the differential specifies a natural operation, in the above sense, for each arity n. Furthermore, in the Master Equation there is one term form each generator of Φ^1 up to isomorphism. This is immediate from comparing Table 5 with Table 4. The natural operation which lives on a space associated to an $Q \in \mathcal{F}$-Ops is denoted $\Psi_{Q,n}$ and is formally defined as follows:

Definition 7.7 For a Feynman category $\mathfrak{F}$ admitting the Feynman transform and for $Q \in \mathcal{F}$-Ops_C we define the formal Master Equation of $\mathfrak{F}$ with respect to Q to be the completed cochain $\Psi_Q := \prod \Psi_{Q,n}$. If there is an N such that $\Psi_{Q,n} = 0$ for $n > N$, then we define the Master Equation of $\mathfrak{F}$ with respect to Q to be the finite sum:

$$d_Q + \sum_n \Psi_{Q,n} = 0$$

We say $\alpha \in lim_V(Q)$ is a solution to the Master Equation if $d_Q(\alpha) + \sum_n \Psi_{Q,n}(\alpha^{\otimes n}) = 0$, and we denote the set of such solutions as $ME(lim_V(Q))$. Here the first term is the internal differential and the term for $n = 1$ is the differential corresponding to d_{Φ^1}, where Φ^1 is the subset of odd generators.

Theorem 7.8 *Let $O \in \mathcal{F}$-Ops_C and $\mathcal{P} \in \mathcal{F}^{odd}$-$Ops_C$ for an $\mathcal{F}$ admitting a Feynman transform and Master Equation. Then there is a bijective correspondence:*

$$Hom(\mathrm{FT}(\mathcal{P}), O) \cong ME(lim_V(\mathcal{P} \otimes O))$$

8 Model Structures, Resolutions and the W-constructions

In this section we discuss Quillen model structures for $\mathcal{F}$-Ops_C. It turns out that these model structures can be defined if C satisfies certain conditions and if this is the case work for all $\mathfrak{F}$, e.g. all the previous examples.

8.1 Model Structure

Theorem 8.1 *Let $\mathfrak{F}$ be a Feynman category and let C be a cofibrantly generated model category and a closed symmetric monoidal category having the following additional properties:*

1. *All objects of C are small.*
2. *C has a symmetric monoidal fibrant replacement functor.*
3. *C has $\otimes$-coherent path objects for fibrant objects.*

Then $\mathcal{F}$-Ops_C is a model category where a morphism $\phi: O \to Q$ of $\mathcal{F}$-ops is a weak equivalence (resp. fibration) if and only if $\phi: O(v) \to Q(v)$ is a weak equivalence (resp. fibration) in C for every $v \in \mathcal{V}$.

8.1.1 Examples

1. Simplicial sets. (Straight from Theorem 8.1)
2. $dgVect_k$ for $char(k) = 0$ (Straight from Theorem 8.1)
3. *Top* (More work, see below.)

8.1.2 Remark

Condition (i) is not satisfied for *Top* and so we can not directly apply the theorem. In [33] this point was first cleared up by following [13] and using the fact that all objects in *Top* are small with respect to topological inclusions.

Theorem 8.2 *Let C be the category of topological spaces with the Quillen model structure. The category $\mathcal{F}$-Ops_C has the structure of a cofibrantly generated model category in which the forgetful functor to $\mathcal{V}$-Seq_C creates fibrations and weak equivalences.*

8.2 Quillen Adjunctions from Morphisms of Feynman Categories

8.2.1 Adjunction from Morphisms

We assume C is a closed symmetric monoidal and model category satisfying the assumptions of Theorem 8.1. Let $\mathfrak{E}$ and $\mathfrak{F}$ be Feynman categories and let $\alpha: \mathfrak{E} \to \mathfrak{F}$ be a morphism between them. This morphism induces an adjunction

$$\alpha_*: \mathcal{E}\text{-}Ops_C \leftrightarrows \mathcal{F}\text{-}Ops_C : \alpha^*$$

where $\alpha^*(\mathcal{A}) := \mathcal{A} \circ \alpha$ is the right adjoint and $\alpha_*(\mathcal{B}) := Lan_\alpha(\mathcal{B})$ is the left adjoint.

Lemma 8.3 *Suppose α_R restricted to $\mathcal{V}_\mathfrak{F}\text{-}Mods_C \to \mathcal{V}_\mathfrak{E}\text{-}Mods_C$ preserves fibrations and acyclic fibrations, then the adjunction (α_L, α_R) is a Quillen adjunction.*

8.3 Example

1. Recall that $\mathfrak{C}$ and $\mathfrak{M}$ denote the Feynman categories whose *ops* are cyclic and modular operads, respectively, and that there is a morphism $i: \mathfrak{C} \to \mathfrak{M}$ by including $*_S$ as genus zero $*_{S,0}$.
2. This morphism induces an adjunction between cyclic and modular operads

$$i_*: \mathfrak{C}\text{-}Ops_C \leftrightarrows \mathfrak{M}\text{-}Ops_C: i^*$$

and the left adjoint is called the modular envelope of the cyclic operad.
3. The fact that the morphism of Feynman categories is inclusion means that i_R restricted to the underlying $\mathcal{V}$-modules is given by forgetting, and since fibrations and weak equivalences are levelwise, i_R restricted to the underlying $\mathcal{V}$-modules will preserve fibrations and weak equivalences.
4. Thus by the Lemma above this adjunction is a Quillen adjunction.

8.4 Cofibrant Replacement

Theorem 8.4 *The Feynman transform of a non-negatively graded dg $\mathcal{F}$-op is cofibrant.*

The double Feynman transform of a non-negatively graded dg $\mathcal{F}$-op in a quadratic Feynman category is a cofibrant replacement.

8.5 W-construction

8.5.1 Setup

In this section we start with a quadratic Feynman category $\mathfrak{F}$.

8.5.2 The Category $w(\mathfrak{F}, Y)$, for $Y \in \mathcal{F}$

Objects The objects are the set $\coprod_n C_n(X, Y) \times [0, 1]^n$, where $C_n(X, Y)$ are chains of morphisms from X to Y with n degree ≥ 1 maps modulo contraction of isomorphisms.

An object in $w(\mathfrak{F}, Y)$ will be represented (uniquely up to contraction of isomorphisms) by a diagram

$$X \xrightarrow[f_1]{t_1} X_1 \xrightarrow[f_2]{t_2} X_2 \to \cdots \to X_{n-1} \xrightarrow[f_n]{t_n} Y$$

where each morphism is of positive degree and where $t_1, \ldots, t_n$ represents a point in $[0, 1]^n$. These numbers will be called weights. Note that in this labeling scheme isomorphisms are always unweighted.

Morphisms

1. Levelwise commuting isomorphisms which fix Y, i.e.:

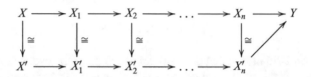

2. Simultaneous $\mathbb{S}_n$ action.
3. Truncation of 0 weights: morphisms of the form $(X_1 \xrightarrow{0} X_2 \to \cdots \to Y) \mapsto (X_2 \to \cdots \to Y)$.
4. Decomposition of identical weights: morphisms of the form $(\cdots \to X_i \xrightarrow{t} X_{i+2} \to \ldots) \mapsto (\cdots \to X_i \xrightarrow{t} X_{i+1} \xrightarrow{t} X_{i+2} \to \ldots)$ for each (composition preserving) decomposition of a morphism of degree ≥ 2 into two morphisms each of degree ≥ 1.

Definition 8.5 Let $\mathcal{P} \in \mathcal{F}\text{-}Ops_{\mathcal{T}op}$. For $Y \in ob(\mathcal{F})$ we define

$$W(\mathcal{P})(Y) := colim_{w(\mathfrak{F},Y)} \mathcal{P} \circ s(-)$$

Theorem 8.6 *Let $\mathfrak{F}$ be a simple Feynman category and let $\mathcal{P} \in \mathcal{F}\text{-}Ops_{\mathcal{T}op}$ be ρ-cofibrant. Then $W(\mathcal{P})$ is a cofibrant replacement for $\mathcal{P}$ with respect to the above model structure on $\mathcal{F}\text{-}Ops_{\mathcal{T}op}$.*

Here "simple" is a technical condition satisfied by all graph examples.

9 Geometry

9.1 Moduli Space Geometry

Although many of the examples up to now have been algebraic or combinatorial in nature, there are very important and deep links to the geometry of moduli spaces. We will discuss these briefly.

9.1.1 Modular Operads

The typical topological example for modular operads are the Deligne–Mumford compactifications $\bar{M}_{gn}$ of Riemann's moduli space of curves of genus g with n marked points.

These give rise to chain and homology operads. An important application comes from enumerative geometry. Gromov–Witten invariants make $H^*(V)$ an algebra over $H_*(\bar{M}_{g,n})$ [40].

9.1.2 Odd Modular

As explained in [35], the canonical geometry for odd modular operads is given by $\bar{M}^{KSV}$ which are real blowups of $\bar{M}_{gn}$ along the boundary divisors [36].

On the topological level one has 1-parameter gluings parameterized by S^1. Taking the full S^1 family on chains or homology gives us the structure of an odd modular operad. That is the gluing operations have degree 1 and in the dual graph, the edges have degree 1.

9.2 Master Equation and Compactifications

Going back to Sen and Zwiebach [48], a viable string field theory action S is a solution of the quantum Master Equation. Rephrasing this one can say "The Master Equation drives the compactification", which is one of the mantras of [35].

In particular, the constructions of [36] and [21] give the correct compactification.

9.3 W-construction

In [5] we will prove the fact that the derived modular envelope defined via the W-construction of the cyclic associative operads is the Kontsevich/Penner compactification $M_{g,n}^{comb}$.

We will also give an A_∞ version of this theorem and a 2-categorical realization that gives our construction of string topology and Hochschild operations from Moduli Spaces [27, 28] via the Feynman transform.

10   Bi- and Hopf Algebras

We will give a brief overview of the constructions of [14].

10.1   Overview

Consider a non-Sigma Feynman category $\mathcal{B} = Hom(Mor(\mathcal{F}), \mathbb{Z})$.

Product Assume that $\mathfrak{F}$ is strict monoidal, that is $\mathcal{F}$ is strict monoidal, then $\otimes$ is an associative unital product on $\mathcal{B}$ with unit $id_{\mathbb{1}_{\mathcal{F}}}$.

Coproduct Assume that $\mathcal{F}$ decomposition finite, i.e. that the sum below is finite. Set

$$\Delta(\phi) = \sum_{(\phi_0, \phi_1): \phi = \phi_1 \circ \phi_0} \phi_0 \otimes \phi_1 \tag{37}$$

and $\epsilon(\phi) = 1$ if $\phi = id_X$ and 0 else.

Theorem 10.1 ([14]) *$\mathcal{B}$ together with the structures above is a bi-algebra. Under certain mild assumptions, a canonical quotient is a Hopf algebra.*

Remark 10.2 Now, it is not true that any strict monoidal category with finite decomposition yields a bi-algebra. Also, if $\mathfrak{F}$ is a Feynman category, then $\mathfrak{F}^{op}$, although not necessarily a Feynman category, does yield a bi-algebra.

10.1.1   Examples

The Hopf algebras of Goncharov for multi-zeta values [20] can be obtained in this way starting with the Joyal dual of the surjections in the augmented simplicial category. In short, this Hopf algebra structures follows from the fact that simplices form an operad. In a similar fashion, but using a graded version, we recover a Hopf algebra of Baues that he defined for double loop spaces [3]. We can also recover the non-commutative Connes–Kreimer Hopf algebra of planar rooted trees, see e.g. [12] in this way.

Remark 10.3 This coproduct for any finite decomposition category appeared in [38] and was picked up later in [22]. We realized with hindsight that the co-product we first constructed on indecomposables, as suggested to us by Dirk Kreimer, is equivalent to this coproduct.

10.1.2   Symmetric Version

There is a version for symmetric Feynman categories, but the constructions are more involved. In this fashion, we can reproduce Connes–Kreimer's Hopf algebra. There is a threefold hierarchy. A bialgebra version, a commutative Hopf algebra version and an "amputated" version, which is actually the algebra considered in [11]. A similar story holds for the graph versions and in general.

10.2 Details: Non-commutative Version

We use non-symmetric Feynman categories whose underlying tensor structure is only monoidal (not symmetric). $\mathcal{V}^{\otimes}$ is the free monoidal category.

Lemma 10.4 (Key Lemma) *The bi-algebra equation holds due to the hereditary condition (ii).*

The proof is a careful check of the diagrams that appear in the bialgebra equation.

For $\Delta \circ \mu$ the sum is over diagrams of the type

$$X \otimes X' \xrightarrow{\Phi = \phi \otimes \psi} Z \otimes Z'$$

with Φ_0, Φ_1, Y (38)

where $\Phi = \Phi_1 \circ \Phi_0$.

When considering $(\mu \otimes \mu) \circ \pi_{23} \circ (\Delta \otimes \Delta)$ the diagrams are of the type

$$X \otimes X' \xrightarrow{\phi \otimes \psi} Z \otimes Z'$$

with $\phi_0 \otimes \psi_0$, $\phi_1 \otimes \psi_1$, $Y \otimes Y'$ (39)

where $\phi = \phi_1 \circ \phi_0$ and $\psi = \psi_1 \circ \psi_0$. In general, there is no reason for there to be a bijection of such diagrams, but there is for non-symmetric Feynman categories.

For simplicity, we assume that $\mathcal{F}$ is skeletal.

10.3 Hopf Quotient

Even after quotienting out by the isomorphisms, the bi-algebra is usually not connected. The main obstruction is that there are many identities and that there are still automorphisms. The main point is that in the skeletal case:

$$\Delta(id_X) = \sum_{\sigma \in Aut(X)} \sigma \otimes \sigma^{-1} \tag{40}$$

where here and in the following we assume that if σ has a one-sided inverse then it is invertible. This is the case in all examples.

10.3.1 Almost Connected Feynman Categories

In the skeletal version, consider the ideal generated by $\mathfrak{C} = |Aut(X)|[id_X] - |Aut(Y)|[id_Y] \subset B$, this is closed under Δ, but not quite a co-ideal. Rescaling ϵ by $\frac{1}{|Aut(X)|}$, $\mathcal{H} = B/\mathfrak{C}$ becomes a bi-algebra. We call $\mathfrak{F}$ almost connected if $\mathcal{H}$ is connected.

Theorem 10.5 *For the almost connected version $\mathcal{H}$ is a connected bi-algebra and hence a Hopf-algebra.*

10.4 Symmetric/Commutative Version

In the case of a symmetric Feynman category, the bi-algebra equation does not hold anymore, due to the fact that $Aut(X) \otimes Aut(Y) \subset Aut(X \otimes Y)$ may be a proper subgroup due to the commutativity constraints. The typical example is $\mathbb{S}$ where $Aut(n) \times Aut(m) = \mathbb{S}_n \times \mathbb{S}_m$ while $Aut(n + m) = \mathbb{S}_{n+m}$. In order to rectify this, one considers the co-invariants. Since commutativity constraints are isomorphisms the resulting algebra structure is commutative.

Let $\mathcal{B}^{iso}$ the quotient by the ideal defined by the equivalence relation generated by isomorphism. That is $f \sim g$ if there are isomorphisms σ, σ' such that $f = \sigma \circ g \circ \sigma'$. This ideal is again closed under co-product. As above one can modify the co-unit to obtain a bialgebra structure on $\mathcal{B}^{iso}$. Now the ideal generated by $\mathfrak{C} = \langle |Aut(X)|[id_X] - |Aut(Y)|[id_Y]$ is a co-ideal and $\mathcal{H} = B/\mathfrak{C}$ becomes a bi-algebra. We call $\mathfrak{F}$ almost connected if $\mathcal{H}$ is connected.

The main theorem is

Theorem 10.6 *If $\mathfrak{F}$ is almost connected, the coinvariants $\mathcal{B}^{iso}$ are a commutative Hopf algebra.*

This allows one to construct Hopf algebras with external legs in the graph examples. It also explains why the Connes–Kreimer examples are commutative.

10.4.1 Amputated Version

In order to forget the leg structure, aka. amputation, one needs a semi-cosimplicial structure, i.e. one must be able to forget external legs coherently. This is always possible by deleting flags in the graph cases. Then there is a colimit, in which all the external legs can be forgotten. Again, one obtains a Hopf algebra. The example *par excellence* is of course, Connes–Kreimer's Hopf algebra without external legs (e.g. the original version).

10.5 Restriction and Generalization of Special Case: Co-operad with Multiplication

In a sense the above examples were free. One can look at a more general setting where this is not the case. This is possible in the simple cases of enriched Feynman categories over $Surj$. Here the morphisms are operads, and $\mathcal{B}$ has the dual co-operad structure for the one-comma generators. The tensor product $\otimes$ makes $\mathcal{B}$ have the structure of a free algebra over the one-comma generators $O(n)$ with the co-operad structure being distributive or multiplicative over $\otimes$. Now one can generalize to a general co-operad structure with multiplication.

10.5.1 Coproduct for a Cooperad with Multiplication

Theorem 10.7 ([14]) *Let $\check{O}$ be a co-operad with compatible associative multiplication. $\mu : \check{O}(n) \otimes \check{O}(m) \to \check{O}(n + m)$ in an Abelian symmetric monoidal category with unit $\mathbb{1}$. Then $\mathcal{B} := \bigoplus_n \check{O}(n)$ is a (non-unital, non-co-unital) bialgebra, with multiplication μ and comultiplication Δ given by $(\mathbb{I} \otimes \mu)\check{\gamma}$:*

$$
\check{O}(n) \xrightarrow{\quad \check{\gamma} \quad} \bigoplus_{\substack{k \geq 1, \\ n = m_1 + \cdots + m_k}} \left(\check{O}(k) \otimes \bigotimes_{r=1}^{k} \check{O}(m_r) \right)
$$

$$
\Delta := (\mathbb{I} \otimes \mu)\check{\gamma} \searrow \qquad \Big\downarrow {\scriptstyle \mathbb{I} \otimes \mu^{k-1}}
$$

$$
\bigoplus_{k \geq 1} \check{O}(k) \otimes \check{O}(n). \tag{41}
$$

10.5.2 Free Cooperad with Multiplication on a Cooperad

The guiding example is:

$$
\check{O}^{nc}(n) = \bigoplus_{k} \bigoplus_{(n_1, \ldots, n_k) : \sum n_i = n} \check{O}(n_1) \otimes \cdots \otimes \check{O}(n_k)
$$

Multiplication is given by $\mu = \otimes$. This structure coincides with one of the constructions of a non-connected operad in [35].

The example is the one that is relevant for the three Hopf algebras of Baues, Goncharov and Connes–Kreimer. It also shows how a cooperad with multiplications generalizes an enrichment of F_{surj}.

This is most apparent in Connes–Kreimer, where the Hopf algebra is not actually on rooted trees, but rather on forests. The extension of the co-product to a forest is tacitly given by the bi-algebra equations.

In the symmetric case, one has to further induce the natural $(\mathbb{S}_{n_1} \times \cdots \times \mathbb{S}_{n_k}) \wr \mathbb{S}_k$ action to an $\mathbb{S}_n$ action for each summand. The coinvariants constituting $\mathcal{B}^{iso}$ are then the symmetric products $\breve{O}(n_1)_{\mathbb{S}_{n_1}} \odot \cdots \odot \breve{O}(n_k)_{\mathbb{S}_{n_k}}$.

The following is the list of motivating examples:

Hopf algebras	(co)operads	Feynman category
H_{Gont}	$Inj_{*,*} = Surj^*$	$\mathfrak{F}_{Surj}$
H_{CK}	Leaf labelled trees	$\mathfrak{F}_{Surj,O}$
$H_{CK,graphs}$	Graphs	$\mathfrak{F}_{graphs}$
H_{Baues}	$Inj_{*,*}^{gr}$	$\mathfrak{F}_{Surj,odd}$

10.5.3 Grading/Filtration, the q Deformation and Infinitesimal Version

We will only make very short remarks, the details are in [14].

The length of an object in the Feynman category setting is replaced by a depth filtration. The algebras are then deformations of their associated graded, see [14]. In the amputated version one has to be more careful with the grading.

Co-operad with multiplication	Operad degree − depth
Amputated version	Co-radical degree + depth

Taking a slightly different quotient, one can get a non-unital, co-unital bi-algebra and a q-filtration. Sending $q \to 1$ recovers $\mathcal{H}$.

Acknowledgements I thankfully acknowledge my co-authors with whom it has been a pleasure to work. I furthermore thank the organizers of the MATRIX workshop for providing the opportunity to give these lectures and for arranging the special issue.

The work presented here has at various stages been supported by the Humboldt Foundation, the Institute for Advanced Study, the Max–Planck Institute for Mathematics, the IHES and by the NSF. Current funding is provided by the Simons foundation.

References

1. Barannikov, S.: Modular operads and Batalin-Vilkovisky geometry. Int. Math. Res. Not. IMRN (19), Art. ID rnm075, 31 (2007)
2. Batanin, M., Berger, C.: Homotopy theory for algebras over polynomial monads (2013). http://arxiv.org/abs/1305.0086
3. Baues, H.J.: The double bar and cobar constructions. Compos. Math. **43**(3), 331–341 (1981)
4. Berger, C., Kaufmann, R.M.: Comprehensive factorisation systems. Special issue in honor of Professors Peter J. Freyd and F. William Lawvere on the occasion of their 80th birthdays. Tbillisi Math. J. **10**(3), 255–277 (2017)
5. Berger, C., Kaufmann, R.M.: Derived Feynman categories and modular geometry, (2018)
6. Borisov, D.V., Manin, Y.I.: Generalized operads and their inner cohomomorphisms. In: Geometry and Dynamics of Groups and Spaces. Progress in Mathematics, vol. 265 pp. 247–308. Birkhäuser, Basel (2008)

7. Brown, F.: Feynman amplitudes and cosmic Galois group (2015). Preprint. arxiv:1512.06409
8. Chapoton, F.: On some anticyclic operads. Algebr. Geom. Topol. **5**, 53–69 (electronic) (2005)
9. Church, T., Ellenberg, J.S., Farb, B.: FI-modules: a new approach to stability for S_n-representations (2012). arXiv:1204.4533
10. Conant, J., Vogtmann, K.: On a theorem of Kontsevich. Algebr. Geom. Topol. **3**, 1167–1224 (2003)
11. Connes, A., Kreimer, D.: Hopf algebras, renormalization and noncommutative geometry. Commun. Math. Phys. **199**(1), 203–242 (1998)
12. Foissy, L.: Les algèbres de Hopf des arbres enracinés décorés. II. Bull. Sci. Math. **126**(4), 249–288 (2002)
13. Fresse, B.: Props in model categories and homotopy invariance of structures. Georgian Math. J. **17**(1), 79–160 (2010)
14. Gálvez-Carrillo, I., Kaufmann, R.M., Tonks, A.: Hopf algebras from cooperads and Feynman categories (2016). arXiv:1607.00196
15. Gerstenhaber, M.: The cohomology structure of an associative ring. Ann. of Math. (2) **78**, 267–288 (1963)
16. Getzler, E.: Operads revisited. In: Algebra, Arithmetic, and Geometry: In Honor of Yu. I. Manin, vol. I. Progress in Mathematics, vol. 269 pp. 675–698. Birkhäuser, Boston, MA (2009)
17. Getzler, E., Jones J.D.S.: Operads, homotopy algebra and iterated integrals for double loop spaces (1994). http://arxiv.org/abs/hep-th/9403055
18. Getzler, E., Kapranov, M.M.: Cyclic operads and cyclic homology. In: Geometry, Topology, & Physics, Conference Proceedings and Lecture Notes in Geometry and Topology, vol. IV, pp. 167–201. International Press, Cambridge, MA (1995)
19. Getzler, E., Kapranov, M.M.: Modular operads. Compos. Math. **110**(1), 65–126 (1998)
20. Goncharov, A.B.: Galois symmetries of fundamental groupoids and noncommutative geometry. Duke Math. J. **128**(2), 209–284 (2005)
21. Harrelson, E., Voronov, A.A., Javier Zúñiga, J.: Open-closed moduli spaces and related algebraic structures. Lett. Math. Phys. **94**(1), 1–26 (2010)
22. Joni, S.A., Rota, G.-C.: Coalgebras and bialgebras in combinatorics. Stud. Appl. Math. **61**(2), 93–139 (1979)
23. Joyal, A., Street, R.: Braided tensor categories. Adv. Math. **102**(1), 20–78 (1993)
24. Kapranov, M., Manin, Yu.: Modules and Morita theorem for operads. Am. J. Math. **123**(5), 811–838 (2001)
25. Kaufmann, R.M.: Feynman categories in quantum field theory (in progress)
26. Kaufmann, R.M.: Operads, moduli of surfaces and quantum algebras. In: Woods Hole Mathematics. Ser. Knots Everything, vol. 34, pp. 133–224. World Scientific Publishing, Hackensack, NJ (2004)
27. Kaufmann, R.M.: Moduli space actions on the Hochschild co-chains of a Frobenius algebra. I. Cell operads. J. Noncommut. Geom. **1**(3), 333–384 (2007)
28. Kaufmann, R.M.: Moduli space actions on the Hochschild co-chains of a Frobenius algebra. II. Correlators. J. Noncommut. Geom. **2**(3), 283–332 (2008)
29. Kaufmann, R.M.: Dimension vs. genus: a surface realization of the little k-cubes and an E_∞ operad. In: Algebraic Topology—Old and New. Banach Center Publications, vol. 85, pp. 241–274. Polish Academy of Sciences, Institute of Mathematics, Warsaw (2009)
30. Kaufmann, R., Lucas, J.: Decorated Feynman categories. J. Noncommut. Geom. **11**(4), 1437–1464 (2017)
31. Kaufmann, R.M., Penner, R.C.: Closed/open string diagrammatics. Nuclear Phys. B **748**(3), 335–379 (2006)
32. Kaufmann, R.M., Schwell, R.: Associahedra, cyclohedra and a topological solution to the A_∞ Deligne conjecture. Adv. Math. **223**(6), 2166–2199 (2010)
33. Kaufmann, R.M., Ward, B.C.: Feynman categories. Astérisque **387**, x+161 pp. (2017)
34. Kaufmann, R.M., Livernet, M., Penner, R.C.: Arc operads and arc algebras. Geom. Topol. **7**, 511–568 (electronic) (2003)

35. Kaufmann, R.M., Ward, B.C., Javier Zúñiga, J.: The odd origin of Gerstenhaber brackets, Batalin-Vilkovisky operators, and master equations. J. Math. Phys. **56**(10), 103504, 40 (2015)
36. Kimura, T., Stasheff, J., Voronov, A.A.: On operad structures of moduli spaces and string theory. Commun. Math. Phys. **171**(1), 1–25 (1995)
37. Kontsevich, M.: Formal (non)commutative symplectic geometry. In: The Gel'fand Mathematical Seminars, 1990–1992, pp. 173–187. Birkhäuser, Boston, MA (1993)
38. Leroux, P.: Les catégories de Möbius. Cahiers Topologie Géom. Différentielle **16**(3), 280–282 (1975)
39. Mac Lane, S.: Categories for the Working Mathematician. Graduate Texts in Mathematics, 2nd edn., vol. 5. Springer, New York (1998)
40. Manin, Y.I.: Frobenius manifolds, quantum cohomology, and moduli spaces. American Mathematical Society Colloquium Publications, vol. 47. American Mathematical Society, Providence, RI (1999)
41. Markl, M.: Operads and PROPs. In: Handbook of Algebra, vol. 5, pp. 87–140. Elsevier/North-Holland, Amsterdam (2008)
42. Markl, M.: Modular envelopes, OSFT and nonsymmetric (non-σ) modular operads (2014). arXiv:1410.3414
43. Markl, M., Shnider, S., Stasheff, J.: Operads in Algebra, Topology and Physics. Mathematical Surveys and Monographs, vol. 96. American Mathematical Society, Providence, RI (2002)
44. Markl, M., Merkulov, S., Shadrin, S.: Wheeled PROPs, graph complexes and the master equation. J. Pure Appl. Algebra **213**(4), 496–535 (2009)
45. Merkulov, S., Vallette, B.: Deformation theory of representations of prop(erad)s. I. J. Reine Angew. Math. **634**, 51–106 (2009)
46. Munkres, J.R.: Topology: A First Course. Prentice-Hall, Englewood Cliffs, NJ (1975)
47. Schwarz, A.: Grassmannian and string theory. Commun. Math. Phys. **199**(1), 1–24 (1998)
48. Sen, A., Zwiebach, B.: Quantum background independence of closed-string field theory. Nucl. Phys. B **423**(2–3), 580–630 (1994)
49. Vallette, B.: A Koszul duality for PROPs. Trans. Am. Math. Soc. **359**(10), 4865–4943 (2007)
50. Ward, B.C.: Six operations formalism for generalized operads. Preprint. arxiv.org/abs/1701.01374

Moduli Spaces of (Bi)algebra Structures in Topology and Geometry

Sinan Yalin

Abstract After introducing some motivations for this survey, we describe a formalism to parametrize a wide class of algebraic structures occurring naturally in various problems of topology, geometry and mathematical physics. This allows us to define an "up to homotopy version" of algebraic structures which is coherent (in the sense of ∞-category theory) at a high level of generality. To understand the classification and deformation theory of these structures on a given object, a relevant idea inspired by geometry is to gather them in a moduli space with nice homotopical and geometric properties. Derived geometry provides the appropriate framework to describe moduli spaces classifying objects up to weak equivalences and encoding in a geometrically meaningful way their deformation and obstruction theory. As an instance of the power of such methods, I will describe several results of a joint work with Gregory Ginot related to longstanding conjectures in deformation theory of bialgebras, E_n-algebras and quantum group theory.

1 Introduction

To motivate a bit the study of algebraic structures and their moduli spaces in topology, we will simply start from singular cohomology. Singular cohomology provides a first approximation of the topology of a given space by its singular simplices, nicely packed in a cochain complex. Computing the cohomology of spaces already gives us a way to distinguish them and extract some further information like characteristic classes for instance. Singular cohomology has the nice property to be equipped with an explicit commutative ring structure given by the cup product. This additional structure can distinguish spaces which have the same cohomology groups, illustrating of the following idea: adding finer algebraic structures is a way to parametrize finer invariants of our spaces. In the case of

S. Yalin (✉)
Laboratoire angevin de recherche en mathématiques (LAREMA), University of Angers, 2 Boulevard Lavoisier, 49045 Angers, France
e-mail: yalinprop@gmail.com

© Springer International Publishing AG, part of Springer Nature 2018
D.R. Wood et al. (eds.), *2016 MATRIX Annals*, MATRIX Book Series 1,
https://doi.org/10.1007/978-3-319-72299-3_20

manifolds, it can also be used to get more geometric data (from characteristic classes and Poincaré duality for instance). Such an algebraic structure determined by operations with several inputs and one single output (the cup product in our example) satisfying relations (associativity, commutativity) is parametrized by an operad (here the operad *Com* of commutative associative algebras). More generally, the notion of operad has proven to be a fundamental tool to study algebras playing a key role in algebra, topology, category theory, differential and algebraic geometry, mathematical physics (like Lie algebras, Poisson algebras and their variants).

We can go one step further and relax such structures *up to homotopy* in an appropriate sense. Historical examples for this include higher Massey products, Steenrod squares and (iterated) loop spaces.

Higher Massey products organize into an A_∞-algebra structure on the cohomology of a space and give finer invariants than the cup product. For instance, the trivial link with three components has the same cohomology ring as the Borromean link (in both cases, the cup product is zero), but the triple Massey product vanishes in the second case and not in the first one, implying these links are not equivalent.

Loop spaces are another fundamental example of A_∞-algebras (in topological spaces this time). When one iterates this construction by taking the loop space of the loop space and so on, one gets an E_n-algebra (more precisely an algebra over the little n-disks operad). These algebras form a hierarchy of "more and more" commutative and homotopy associative structures, interpolating between A_∞-algebras (the E_1 case, encoding homotopy associative structures) and E_∞-algebras (the colimit of the E_n's, encoding homotopy commutative structures). Algebras governed by E_n-operads and their deformation theory play a prominent role in a variety of topics, not only the study of iterated loop spaces but also Goodwillie-Weiss calculus for embedding spaces, deformation quantization of Poisson manifolds, Lie bialgebras and shifted Poisson structures in derived geometry, and factorization homology of manifolds [4, 20, 21, 24, 28, 32, 39, 42, 45, 46, 50, 51, 55, 61, 66, 69, 73].

The cup product is already defined at a chain level but commutative only up to homotopy, meaning that there is an infinite sequence of obstructions to commutativity given by the so called higher cup products. That is, these higher cup products form an E_∞-algebra structure on the singular cochain complex. This E_∞-structure classifies the rational homotopy type of spaces (this comes from Sullivan's approach to rational homotopy theory [68]) and the integral homotopy type of finite type nilpotent spaces (as proved by Mandell [53]). Moreover, such a structure induces the Steenrod squares acting on cohomology and, for Poincaré duality spaces like compact oriented manifolds for example, the characteristic classes that represent these squares (the Wu classes). This is a first instance of how a homotopy algebraic structure can be used to build characteristic classes. Then, to study operations on generalized cohomology theories, one moves from spaces to the stable homotopy theory of spectra, the natural recipient for (generalized) cohomology theories, and focuses on theories represented by E_∞-ring spectra (more generally, highly structured ring spectra). In this setting, the moduli space approach (in a homotopy theoretic way) already proved to be useful [33] (leading to a non trivial improvement of the Hopkins-Miller theorem in the study of highly structured ring spectra).

However, algebraic structures not only with products but also with coproducts, play a crucial role in various places in topology, geometry and mathematical physics. One could mention for instance the following important examples: Hopf algebras in representation theory and mathematical physics, Frobenius algebras encompassing the Poincaré duality phenomenon in algebraic topology and deeply related to field theories, Lie bialgebras introduced by Drinfeld in quantum group theory, involutive Lie bialgebras as geometric operations on the equivariant homology of free loop spaces in string topology. A convenient way to handle such kind of structures is to use the formalism of props, a generalization of operads encoding algebraic structures based on operations with several inputs and several outputs.

A natural question is then to classify such structures (do they exist, how many equivalence classes) and to understand their deformation theory (existence of infinitesimal perturbations, formal perturbations, how to classify the possible deformations). Understanding how they are rigid or how they can be deformed provides information about the objects on which they act and new invariants for these objects. For this, a relevant idea inspired from geometry come to the mind, the notion of moduli space, a particularly famous example being the moduli spaces of algebraic curves (or Riemann surfaces). The idea is to associate, to a collection of objects we want to parametrize equipped with an equivalence relation (surfaces up to diffeomorphism, vector bundles up to isomorphism...), a space $\mathcal{M}$ whose points are these objects and whose connected components are the equivalence classes of such objects. This construction is also called a classifying space in topology, a classical example being the classifying space BG of a group G, which parametrizes isomorphism classes of principal G-bundles. If we are interested also in the deformation theory of our collection of objects (how do we allow our objects to modulate), we need an additional geometric structure which tells us how we can move infinitesimally our points (tangent spaces). To sum up, the guiding lines of the moduli space approach are the following:

- To determine the non-emptiness of $\mathcal{M}$ and to compute $\pi_* \mathcal{M}$ solve existence and unicity problems;
- The geometric structure of $\mathcal{M}$ imply the existence of tangent spaces. The tangent space over a given point x of $\mathcal{M}$ is a dg Lie algebra controlling the (derived) deformation theory of x (deformations of x form a derived moduli problem) in a sense we will precise in Sect. 4;
- One can "integrate" over $\mathcal{M}$ to produce invariants of the objects parametrized by $\mathcal{M}$. Here, the word "integrate" has to be understood in the appropriate sense depending on the context: integrating a differential form, pairing a certain class along the (virtual) fundamental class of $\mathcal{M}$, etc.

We already mentioned [29] (inspired by the method of [2]) as an application of the first item in the list above. In the second one, we mention *derived* deformation theory and *derived* moduli problems, which implicitly assume that, in some sense, our moduli space $\mathcal{M}$ lives in a (∞-)category of derived objects (e.g. derived schemes, derived stacks...) where tangent spaces are actually complexes. This is due to the fact that we want to encompass the whole deformation theory of points, and this cannot be done in the classical setting: for varieties or schemes, the tangent space is a vector space which consists just of the equivalence classes of infinitesimal deformations of the point. For stacks, the tangent space is a two-term complex whose H^0 is the set of equivalences classes of infinitesimal deformations, and H^{-1} is the Lie algebra of automorphisms of the point (infinitesimal automorphisms). But obstruction theory does not appear on the tangent structure here, because it has to live in positive degrees (we will go back to this remark in Sect. 4).

As a last remark about the third item in the list above, let us say that the idea of using moduli spaces to produce topological invariants got also a lot of inspiration from quantum field theory and string theory in the 1980s. By the Feynman path integral approach, the equations describing the evolution of a quantum system are determined by the minimas of a functional integral over all the possible paths of this system, that is, by integrating a certain functional over the space of fields. This independence from a choice of path led to the idea that one could build a topological invariant of a geometric object by computing an integral over the moduli space of all possible geometric structures of this kind, ensuring automatically the desired invariance property. This is the principle underlying two important sorts of invariants of manifolds. First, Witten's quantization of the classical Chern-Simmons invariant in the late 1980s [77], which provided topological invariants for 3-dimensional manifolds (including known invariants such as the Jones polynomial) by integrating a geometric invariant over a moduli space of connections. Second, Kontsevich's formalization of Gromov- Witten invariants in symplectic topology (counting pseudo-holomorphic curves) and algebraic geometry (counting algebraic curves), defined by a pairing along the virtual fundamental class of the moduli space (stack) of stable maps (an analogue of the fundamental class for singular objects suitably embedded in a derived setting).

Organization of the Paper The first section is devoted to the formalism of props and algebras over props, accompanied by relevant examples of topological or geometric origin in Sect. 3. The third section focuses on algebraic structures up to homotopy, defined as algebras over a cofibrant resolution of the prop, and the fundamental theorem asserting that this notion does not rely on the choice of such a resolution (up to an equivalence of ∞-categories). This lays down the coherent foundations to study homotopy bialgebras. We then provide a little introduction to derived algebraic geometry and formal moduli problems in Sect. 5, without going too far in the details (we refer the reader to [72] for a more thorough survey on this topic), before formalizing the idea of moduli spaces of algebraic structures in Sect. 6 as well as their most important properties. The way to recover geometrically deformation theory and obstruction theory for such structures is explained in Sect. 6. Section 6 describes a joint work with Gregory Ginot, merging the homotopical and geometric theory of such moduli spaces with several features of factorization homology (higher Hochschild (co)homology) to solve several open conjectures in deformation theory of E_n-algebras and bialgebras related to quantum group theory.

2 Parametrizing Algebraic Structures

To simplify our exposition, we will work in the base category $\mathscr{C}h$ of $\mathbb{Z}$-graded cochain complexes over a field $\mathbb{K}$ of characteristic zero. Before stating the general definition of a prop, let us give a few examples of algebraic structures the reader may have encountered already.

Example 2.1 Differential graded (dg for short) associative algebras are complexes A equipped with an associative product $A \otimes A \rightarrow A$. We can represent such an operation by an oriented graph with two inputs and one output $\vee$ satisfying the associativity relation

$$\bigvee\!\!\!\diagdown \;=\; \diagdown\!\!\!\bigvee$$

Common examples of such structures include algebras $\mathbb{K}[G]$ of finite groups G in representation theory, or the singular cochains $C^*(X; \mathbb{Z})$ of a topological space equipped with the cup product $\cup$ of singular simplices. In the first case we have an associative algebra in $\mathbb{K}$-modules, in the second case this is a dg associative algebra, so the cup product is a cochain morphism determined by linear maps

$$\cup : C^m(X; \mathbb{Z}) \otimes C^n(X; \mathbb{Z}) \rightarrow C^{m+n}(X; \mathbb{Z}).$$

Example 2.2 In certain cases, the product is not only associative but also commutative, and one call such algebras commutative dg algebras or cdgas. To represent graphically this symmetry condition, we index the inputs of the product $\underset{1}{\diagdown}\underset{2}{\diagup}$ and

add the symmetry condition

$$\underset{1}{\overset{1 \quad 2}{\vee}} = \underset{1}{\overset{2 \quad 1}{\vee}}.$$

A way to rephrase this symmetry is to say that Σ_2 acts trivially on $\underset{1}{\overset{1 \quad 2}{\vee}}$. In the dg setting, this symmetry has to be understood in the graded sense, that is $ab = (-1)^{deg(a)deg(b)} ba$. Commutative algebras are very common objects, for instance the singular cohomology of spaces equipped with the cup product defined previously at the chain level, or the de Rham cohomology for manifolds. Commutative rings also represent affine schemes in algebraic geometry or rings of functions on differentiable manifolds. Cdgas over $\mathbb{Q}$ also model the rational homotopy type of simply connected spaces.

Example 2.3 Another example of product is the bracket defining Lie algebras 1 2 satisfying an antisymmetry condition

$$1 \diagdown 2 = -2 \diagdown 1$$

and the Jacobi identity

$$1 \diagdown 2 \diagdown 3 + 3 \diagdown 1 \diagdown 2 + 2 \diagdown 3 \diagdown 1 = 0.$$

A way to rephrase the antisymmetry is to say that the action of Σ_2 on $1 \diagdown 2$

is given by the signature representation sgn_2. Lie algebras appear for instance as tangent spaces of Lie groups in differential geometry (Lie's third theorem gives an equivalence between the category of finite dimensional Lie algebras in vector spaces and the category of simply connected Lie groups), in Quillen's approach to rational homotopy theory and in deformation theory ("Deligne principle" relating formal moduli problems to dg Lie algebras).

In these three first examples, we see that the algebraic structure is defined only by operations with several inputs and one single output. Such structures can be encoded by a combinatorial object called an operad, and a given kind of algebra is an algebra over the associated operad. We refer the reader to [49] for more details about this formalism. However, there are more general algebraic structures involving operations with several inputs and several outputs. We give below two fundamental examples of these, before unwrapping the general definition of the combinatorial structure underlying them (props).

Example 2.4 Poisson-Lie groups are Lie groups with a compatible Poisson structure, which occur in mathematical physics as gauge groups of certain classical mechanical systems such as integrable systems. Because of the Poisson bracket, the tangent space T_eG of a Poisson-Lie group G at the neutral element e is equipped with a "Lie cobracket" compatible with its Lie algebra structure, so that T_eG forms something called a Lie bialgebra. The compatibility relation between the bracket and the cobracket is called the Drinfeld's compatibility relation or the cocycle relation. In terms of graphical presentation, we have a bracket and a cobracket

which are antisymmetric, that is, with the signature action of Σ_2. These two operations satisfy the following relations:

Jacobi

$$1 \diagdown 2 \diagup 3 + 3 \diagdown 1 \diagup 2 + 2 \diagdown 3 \diagup 1 = 0$$

co-Jacobi

$$\diagup 1 \diagdown 2 \diagdown 3 \quad + \quad \diagup 3 \diagdown 1 \diagdown 2 \quad + \quad \diagup 2 \diagdown 3 \diagdown 1 \quad = 0$$

The cocycle relation

$$1 \diagdown 2 \diagup 1 \diagdown 2 = 1 \diagdown 2 \quad + 2 \diagdown 1 \quad - 1 \diagdown 2 \quad + 2 \diagdown 1$$

The cocycle relation means that the Lie cobracket of a Lie bialgebra g is a cocycle in the Chevalley-Eilenberg complex $C^*_{CE}(g, \Lambda^2 g)$, where $\Lambda^2 g$ is equipped with the structure of g-module induced by the adjoint action. Let us note that there is an analogue of Lie's third theorem in this context, namely the category of finite dimensional Lie bialgebras in vector spaces is equivalent to the category of simply connected Poisson-Lie groups [11]. Deformation quantization of Lie bialgebras produces quantum groups, which turned out to be relevant for mathematical physics and for low-dimensional topology (quantum invariants of knots and 3-manifolds). This process also deeply involves other kind of objects such as Grothendieck-Teichmüller groups, multizeta values via the Drinfeld associators [11], or graph complexes. The problem of a universal quantization of Lie bialgebras raised by Drinfeld was solved by Etingof and Kazhdan [16, 17]. A deformation quantization of a Lie bialgebra g is a topologically free Hopf algebra H over the ring of formal power series $\mathbb{K}[[\hbar]]$ such that $H/\hbar H$ is isomorphic to $U(g)$ (the enveloping algebra of g)as a co-Poisson bialgebra. Such a Hopf algebra is called a quantum universal enveloping algebra (QUE for short). The general idea underlying this process is to tensor the $\mathbb{K}$-linear category of g-modules by formal power series, equip it with a braided monoidal structure induced by the choice of a Drinfeld associator and an r-matrix, and make the forgetful functor from $g[[\hbar]]$-modules to $\mathbb{K}[[\hbar]]$-modules braided monoidal. Applying the Tannakian formalism to this functor, the category of $g[[\hbar]]$-modules is equivalent to the category of modules over the QUE algebra of g. Deformation quantization of Lie bialgebras can be formulated in the formalism of props and their algebras, see for instance the introduction of [14] explaining quantization/de-quantization problems in terms of prop morphisms. Another point of view is the prop profile approach of [57], particularly useful to relate the results of [30] to deformation quantization of Lie bialgebras.

A variant of Lie bialgebras called involutive Lie bialgebras arose in low dimensional topology, in the work of Goldman [35] and Turaev [75]. Given a surface

S, one considers the $\mathbb{K}$-module generated by the free homotopy classes of loops on S. Let us note $L : S^1 \to S$ a free loop on the surface S (that is, a continuous map which is not pointed, contrary to based loops) and $[L]$ its free homotopy class. Up to homotopy, we can make two loops intersect transversely, so we suppose that two given loops L and K intersect only at a finite number of points, and we note $L \cap K$ this finite set. The Lie bracket of $[L]$ and $[K]$ is then defined by

$$\{[L], [K]\} = \sum_{p \in L \cap K} \epsilon_p [L \cup_p K]$$

where $L \cup_p K$ is the loop parametrized by going from p to p along L, then going again from p to p along K. The symbol ϵ_p denote a number which is -1 or 1, depending on the way L and K intersect at p with respect to a choice of orientation. The cobracket is then defined similarly, by considering this time the self-intersections of L (that we can take transverse, up to homotopy):

$$\delta([L]) = \sum_{p \in L \cap L} \epsilon_p ([L_{1,p}] \otimes [L_{2,p}] - [L_{2,p}] \otimes [L_{1,p}])$$

where L_1 and L_2 are the two loops obtained by separating L in two parts at the self-intersection point p. These two operations define a Lie bialgebra structure, satisfying moreover $(\{,\} \circ \delta)([L]) = 0$. From the graphical presentation viewpoint, this means that we add the involutivity relation

$$\diamondsuit = 0$$

Links defined in the cylinder $S \times [0; 1]$ over S can be presented by diagrams of loops on S via the canonical projection $S \times [0; 1] \to S$. Explicit quantizations of the Lie bialgebra of loops on S have been used to produce (quantum) invariants of those links [75] and the corresponding 3-dimensional TQFTs [64].

Ten years after, algebraic structures on free loop spaces for more general manifolds were introduced by Chas and Sullivan, giving birth to string topology [7], a very active field of research nowadays. In the equivariant setting, the Lie bialgebra of Goldman and Turaev has been generalized to loop spaces of smooth manifolds [6]. The string homology of a smooth manifold M is defined as the reduced equivariant homology (i.e. relative to constant loops) of the free loop space LM of M. The word equivariant refers here to the action of S^1 on loops by rotation. According to [6], the string homology of a smooth manifold forms an involutive Lie bialgebra. Let us note that for an n-dimensional manifold, the bracket and the cobracket of this structure are of degree $2 - n$. In particular, the string homology of a surface is isomorphic to Goldman-Turaev Lie bialgebra as a graded Lie bialgebra. Let us note that such a structure is also related to very active research topics in symplectic topology. Precisely, the string homology of M is isomorphic as a graded Lie bialgebra to the contact homology of its cotangent bundle (equipped with the standard symplectic form) [8]. This result is part of a larger program aimed at relating string topology and symplectic field theory.

Example 2.5 A dg Frobenius algebra is a unitary dg commutative associative algebra of finite dimension A endowed with a symmetric non-degenerate bilinear form $< .,. >: A \otimes A \rightarrow \mathbb{K}$ which is invariant with respect to the product, i.e $< xy, z >=< x, yz >$.

A dg Frobenius bialgebra of degree m is a triple (B, μ, Δ) such that:

(i) (B, μ) is a dg commutative associative algebra;
(ii) (B, Δ) is a dg cocommutative coassociative coalgebra with $deg(\Delta) = m$;
(iii) the map $\Delta : B \rightarrow B \otimes B$ is a morphism of left B-module and right B-module, i.e in Sweedler's notations we have the Frobenius relations

$$\sum_{(x.y)} (x.y)_{(1)} \otimes (x.y)_{(2)} = \sum_{(y)} x.y_{(1)} \otimes y_{(2)}$$

$$= \sum_{(x)} (1)^{m|x|} x_{(1)} \otimes x_{(2)}.y$$

The two definitions are strongly related. Indeed, if A is a Frobenius algebra, then the pairing $< .,. >$ induces an isomorphism of A-modules $A \cong A^*$, hence a map

$$\Delta : A \xrightarrow{\cong} A^* \xrightarrow{\mu^*} (A \otimes A)^* \cong A^* \otimes A^* \cong A \otimes A$$

which equips A with a structure of Frobenius bialgebra. Conversely, one can prove that every unitary counitary Frobenius bialgebra gives rise to a Frobenius algebra, so the two notions are equivalent. In terms of graphical presentation, we have a product of degree 0 and a coproduct of degree m presented by

and satisfying the following relations:

Associativity and coassociativity

Frobenius relations

In the unitary and counitary case, one adds a generator for the unit, a generator for the counit and the necessary compatibility relations with the product and the coproduct. We refer the reader to [44] for a detailed survey about the role of these operations and relations in the classification of two-dimensional topological quantum field theories. Let us note that a variant of Frobenius bialgebras called special Frobenius bialgebra is closely related to open-closed topological field theories [48] and conformal field theories [19].

A classical example of Frobenius (bi)algebra comes from Poincaré duality. Let M be an oriented connected closed manifold of dimension n. Let $[M] \in H_n(M; \mathbb{K}) \cong H^0(M; \mathbb{K}) \cong \mathbb{K}$ be the fundamental class of $[M]$. Then the cohomology ring $H^*(M; \mathbb{K})$ of M inherits a structure of commutative and cocommutative Frobenius bialgebra of degree n with the following data:

1. the product is the cup product

$$\mu : H^k M \otimes H^l M \to H^{k+l} M$$

$$x \otimes y \mapsto x \cup y$$

2. the unit $\eta : \mathbb{K} \to H^0 M \cong H_n M$ sends $1_{\mathbb{K}}$ on the fundamental class $[M]$;
3. the non-degenerate pairing is given by the Poincaré duality:

$$\beta : H^k M \otimes H^{n-k} M \to \mathbb{K}$$

$$x \otimes y \mapsto <x \cup y, [M]>$$

i.e the evaluation of the cup product on the fundamental class;
4. the coproduct $\Delta = (\mu \otimes id) \circ (id \otimes \gamma)$ where

$$\gamma : \mathbb{K} \to \bigoplus_{k+l=n} H^k M \otimes H^l M$$

is the dual copairing of β, which exists since β is non-degenerate;
5. the counit $\epsilon = <., [M]> : H^n M \to \mathbb{K}$ i.e the evaluation on the fundamental class.

A natural question after looking at all these examples is the following: can we extract a common underlying pattern, analogue to representation theory of groups or to operad theory, which says that an algebraic structure of a given kind is an algebra over a corresponding combinatorial object? A formalism that include algebras over operads as well as more general structures like Lie bialgebras and Frobenius bialgebras? We answer this question with the following definition, originally due to MacLane [52]. A Σ-biobject is a double sequence $\{M(m, n) \in \mathscr{C}h\}_{(m,n) \in \mathbb{N}^2}$ where each $M(m, n)$ is equipped with a right action of Σ_m and a left action of Σ_n commuting with each other.

Definition 2.1 A prop is a Σ-biobject endowed with associative horizontal composition products

$$\circ_h : P(m_1, n_1) \otimes P(m_2, n_2) \to P(m_1 + m_2, n_1 + n_2),$$

associative vertical composition products

$$\circ_v : P(k,n) \otimes P(m,k) \to P(m,n)$$

and maps $\mathbb{K} \to P(n,n)$ which are neutral for $\circ_v$ (representing the identity operations). These products satisfy the exchange law

$$(f_1 \circ_h f_2) \circ_v (g_1 \circ_h g_2) = (f_1 \circ_v g_1) \circ_h (f_2 \circ_v g_2)$$

and are compatible with the actions of symmetric groups.

Morphisms of props are equivariant morphisms of collections compatible with the composition products.

A fundamental example of prop is given by the following construction. To any complex X we can associate an endomorphism prop End_X defined by

$$End_X(m,n) = Hom_{\mathscr{C}h}(X^{\otimes m}, X^{\otimes n}).$$

The prop structure here is crystal clear: the actions of the symmetric groups are the permutations of the tensor powers, the vertical composition is the composition of homomorphisms and the horizontal one is the tensor product of homomorphisms.

Definition 2.2 A P-algebra on a complex X is a prop morphism $P \to End_X$. That is, a P-algebra structure on X is a collection of equivariant cochain morphisms

$$\{P(m,n) \to Hom_{\mathscr{C}h}(X^{\otimes m}, X^{\otimes n})\}_{m,n\in\mathbb{N}}$$

commuting with the vertical and horizontal composition products. Hence the formal operations of P are sent to actual operations on X, and the prop structure of P determines the relations satisfied by such operations.

Remark 2.3 MacLane's original definition is more compact: a prop P in a closed symmetric monoidal category $\mathscr{C}$ as a symmetric monoidal category enriched in $\mathscr{C}$, with the natural integers as objects and the tensor product $\otimes$ defined by $m \otimes n = m + n$. A morphism of props is then an enriched symmetric monoidal functor. An algebra over a prop is an enriched symmetric monoidal functor $P \to \mathscr{C}$, and a morphism of algebras is an enriched symmetric monoidal transformation (see also [82, Section 2.1] for the colored case).

There is an adjunction between the category of Σ-biobjects and the category of props, with the right adjoint given by the forgetful functor and the left adjoint given by a free prop functor. Briefly, given a Σ-biobject M, the free prop $\mathscr{F}(M)$ on M is defined by

$$\mathscr{F}(M)(m,n) = \bigoplus_{G\in Gr(m,n)} (\bigotimes_{v\in Vert(G)} M(|In(v)|, |Out(v)|))_{Aut(G)}$$

where

- The direct sums runs over the set $Gr(m, n)$ of directed graphs with m inputs, n outputs and no loops;
- The tensor products are indexed by the sets $Vert(G)$ of vertices of such graphs G;
- For each vertex v of G, the numbers $|In(v)|$ and $|Out(v)|$ are respectively the number of inputs and the number of outputs of v;
- These tensor products are mod out by the action of the group $Aut(G)$ of automorphisms of the graph G.

We refer the reader to [23, Appendix A] for more details about this construction. Moreover, there is an obvious notion of ideal in a prop P, defined as a Σ-biobject I such that $i \circ_v p \in I$ for $i \in I$ and $p \in P$, and $i \circ_h p \in I$ for $i \in I$ and $p \in P$. This means that each prop admits a *presentation by generators and relations*, something particularly useful to describe an algebraic structure. For instance, all the operations $A^{\otimes n} \to A$ on an associative algebra A induced by the algebra structure are entirely determined by a product $A \otimes A \to A$ and the associativity condition. Actually, the graphical presentations we gave in the examples above are exactly presentations of the corresponding props by generators and relations ! For instance, if we denote by *BiLie* the prop of Lie bialgebra, we have

$$BiLie = \mathscr{F}(M)/I$$

where $M(2, 1) = sgn_2 \otimes \mathbb{K} . 1 \bigvee 2, M(1, 2) = sgn_2 \otimes \mathbb{K} . \bigwedge_{1 \ 2}$ and $M(m, n) = 0$ for $(m, n) \notin \{(2, 1), (1, 2)\}$ (recall here that sgn_2 is the signature representation of Σ_2). The ideal I is generated by the graphs defining the relations in Example 4 (Jacobi, co-Jacobi, cocycle relation). A Lie bialgebra g is then the datum of a prop morphism

$$\{BiLie(m, n) \to Hom_{\mathscr{C}h}(g^{\otimes m}, g^{\otimes n})\}_{m, n \in \mathbb{N}}.$$

According to the presentation of *BiLie* by generators and relations, this prop morphism is completely determined by its values on the generators. That is, we send the generator $1 \bigvee 2$ to a cochain map $[,] : g \otimes g \to g$, the generator $\bigwedge_{1 \ 2}$ to a cochain map $\delta : g \to g \otimes g$, and the graphs of I to zero. This implies that $[,]$ is a Lie bracket, δ a Lie cobracket and they satisfy moreover the cocycle relation.

Actually, for a wide range of algebraic structures, a well defined grafting operation on connected graphs is sufficient to parametrize the whole structure. Such a grafting is defined by restricting the vertical composition product of props to connected graphs. The unit for this connected composition product $\boxtimes_c$ is the Σ-biobject I given by $I(1, 1) = \mathbb{K}$ and $I(m, n) = 0$ otherwise. The category of Σ-biobjects then forms a symmetric monoidal category $(Ch_{\mathbb{K}}^S, \boxtimes_c, I)$.

Definition 2.4 A dg properad (P, μ, η) is a monoid in $(Ch_{\mathbb{K}}^S, \boxtimes_c, I)$, where μ denotes the product and η the unit. It is augmented if there exists a morphism of properads $\epsilon : P \to I$. In this case, there is a canonical isomorphism $P \cong I \oplus \overline{P}$ where $\overline{P} = ker(\epsilon)$ is called the augmentation ideal of P.

Morphisms of properads are morphisms of monoids in $(Ch_{\mathbb{K}}^{\mathbb{S}}, \boxtimes_c, I)$.
Properads have also their dual notion, namely coproperads:

Definition 2.5 A dg coproperad (C, Δ, ϵ) is a comonoid in $(Ch_{\mathbb{K}}^{\mathbb{S}}, \boxtimes_c, I)$.
As in the prop case, there exists a free properad functor $\mathscr{F}$ forming an adjunction

$$\mathscr{F} : Ch_{\mathbb{K}}^{\mathbb{S}} \rightleftarrows Properad : U$$

with the forgetful functor U. There is an explicit construction of the free properad
analogous to the free prop construction, but restricted to connected directed graphs
instead of all directed graphs. Dually, there exists a cofree coproperad functor
denoted $\mathscr{F}_c(-)$ having the same underlying Σ-biobject. There is also a notion of
algebra over a properad similar to an algebra over a prop, since the endomorphism
prop restricts to an endomorphism properad. Properads are general enough to
encode a wide range of bialgebra structures such as associative and coassociative
bialgebras, Lie bialgebras, Poisson bialgebras, Frobenius bialgebras for instance.

Remark 2.6 There is a free-forgetful adjunction between properads and props [76].

3 Homotopy Theory of (Bi)algebras

We already mentioned before the natural occurrence of "relaxed" algebraic structures, like A_∞-algebras or E_∞-algebras, in various situations where a given
relation (associativity, commutativity) is satisfied only up to an infinite sequence of
obstructions vanishing at the cohomology level. More generally, one can wonder
how to set up a coherent framework to define what it means to "relax" a P-
algebra structure, encompassing in particular the previous examples. Moreover, we
will see later that deformation theory of differential graded P-algebras can not
be defined without working in the larger context of P-algebras up to homotopy
(or homotopy P-algebras). This is due to the fact that the base category $\mathscr{C}h$ itself
manifests a non trivial homotopy theory. A natural way to define homotopy P-
algebras is to resolve the prop P itself by means of homotopical algebra. For this,
we recall briefly that $\mathscr{C}h$ has all the homotopical properties needed for our purposes,
namely, it forms a cofibrantly generated symmetric monoidal model category. We
refer the reader to Hirschhorn [40] and Hovey [41] for a comprehensive treatment
of homotopical algebra and monoidal model categories. The Σ-biobjects form a
category of diagrams in $\mathscr{C}h$ and inherit thus a cofibrantly generated model structure
with pointwise weak equivalence and fibrations (the projective model structure). The
free prop functor allows to transfer the projective model structure of Σ-biobjects
along the free-forgetful adjunction:

Theorem 3.1 (cf. [23, Theorem 5.5]) *The category of dg props Prop equipped
with the classes of componentwise weak equivalences and componentwise fibrations
forms a cofibrantly generated model category.*

Remark 3.2 According to [59], the similar free-forgetful adjunction between Σ-biojects and dg properads equips dg properads with a cofibrantly generated model category structure with componentwise fibrations and weak equivalences.
Hence we can define homotopy algebras over props as follows:

Definition 3.3 A homotopy P-algebra is a P_∞-algebra, where $P_\infty \xrightarrow{\sim} P$ is a cofibrant resolution of P.

Homotopy algebra structures appear naturally in plenty of topological and geometric situations, especially for transfer and realization problems:

- **Transfer problems**: given a quasi-isomorphism $X \xrightarrow{\sim} Y$, if Y forms a P-algebra, then X can not inherit a P-algebra structure as well (since this is not a strict isomorphism) but rather a P_∞-algebra structure. In the converse way, a choice of quasi-isomorphism $X \xrightarrow{\sim} H^*X$ from a complex to its cohomology allows to transfer any P-algebra structure on X to a P_∞-algebra structure on H^*X. That is, the data of a big complex with a strict structure can transferred to a smaller complex with a bigger structure up to homotopy.
- **Realization problems**: A P-algebra structure on the cohomology H^*X is induced by a finer P_∞-algebra structure on X, which consists in a family of higher operations on cochains.

Let us name a few applications of such ideas:

- A_∞-structures (associative up to homotopy) appeared very early in the study of loop spaces and monoidal categories (Stasheff's associahedra), and the A_∞-structure induced on the singular cohomology of a topological space by the cochain-level cup product gives the higher Massey products. Such products are topological invariants, for instance the triple Massey product differentiate the Borromean rings from the trivial link, even though their respective cohomologies are isomorphic as associative algebras.
- E_∞-structures (commutative up to homotopy) on ring spectra play a key role to encode cohomological operations in stable homotopy theory. Realization problems for such structures have been the subject of a consequent work by Goerss-Hopkins [33], following the idea of [2] to study the homotopy type of the moduli space of all realizations on a given spectrum by decomposing it as the limit of a tower of fibrations, and determining the obstruction groups of the corresponding spectral sequence (which turns out to be André-Quillen cohomology groups).
- The E_∞-structure on singular cochains classifies the homotopy type of nilpotent spaces (see Sullivan over $\mathbb{Q}$, Mandell over $\mathbb{Z}$ and $\mathbb{F}_p$).
- L_∞-structures (Lie up to homotopy) encode the deformation theory of various algebraic, topological or geometric structures, a striking application being Kontsevich's deformation quantization of Poisson manifolds [46].
- In string topology, the homology of a loop space ΩM on a manifold M is equipped with a natural Batalin-Vilkovisky algebra (BV-algebra) structure [7]. On the other hand, the Hochschild cohomology of the singular cochains on M

is also a BV-algebra (extending the canonical Gerstenhaber algebra structure). In characteristic zero, when M is a simply connected closed manifold, both are known to be isomorphic as BV-algebras [18]. It turns out that this structure lifts to a BV_∞-structure on Hochschild cochains (a result called the cyclic Deligne conjecture). Homotopy BV-algebras are related not only to string topology but also to topological conformal field theories and vertex algebras [26].

- Homotopy Gerstenhaber algebras, or equivalently E_2-algebras, are the natural structures appearing on Hochschild complexes by Deligne's conjecture, which has been generalized to the existence of E_{n+1}-algebra structures on higher Hochschild complexes of E_n-algebras. These results have applications to deformation quantization but also to factorization homology of manifolds and generalizations of string topology [32]. The proof of Deligne's conjecture relies on a transfer of structures combined with an obstruction theoretic method. Let us note that a bialgebra version of this conjecture obtained recently in [30] relies in particular on this "transfer+obstruction" method in the case of E_3-algebras and has applications to open problems in quantum group theory.

Moreover, homotopy algebra structures are the structures controlled by the cohomology theories of algebras, when one works in the dg setting. For instance, the Hochschild complex of a dg associative algebra A controls (in a sense we will precise later) not the strict algebra deformations but the A_∞ deformations of A.

However, there is a quite obvious problem in the definition of homotopy algebra we gave above. Indeed, it relies a priori on the choice of a resolution. For instance, two homotopy P-algebras could be weakly equivalent for a certain choice of P_∞ but not for another choice. In order to make sense of this notion and of the various deformation theoretic, transfer and realization problems in which it naturally arises, we have to prove an invariance result for the homotopy theory of homotopy P-algebras:

Theorem 3.4 ([79, Theorem 0.1]) *A weak equivalence* $\varphi : P_\infty \overset{\sim}{\to} Q_\infty$ *between cofibrant props induces an equivalence of* $(\infty, 1)$-*categories*

$$\varphi^* : (Q_\infty - Alg, q - isos) \overset{\sim}{\to} (P_\infty - Alg, q - isos),$$

where $(P_\infty - Alg, q - isos)$ *is the* $(\infty, 1)$-*category associated to the category of dg* P_∞-*algebras with quasi-isomorphisms as weak equivalences.*

In the case of algebras over operads, this result is already known by using classical methods of homotopical algebra. A weak equivalence $\varphi : P \to Q$ of dg operads induces an adjunction

$$\varphi_! : P_\infty - Alg \rightleftarrows Q_\infty - Alg : \varphi^*,$$

where φ^* is the functor induced by precomposition $P_\infty \to Q_\infty \to End_X$ and $\varphi_!$ is a certain coequalizer. The functor φ^* is a right Quillen functor since weak equivalences and fibrations of algebras over operads are determined in complexes, so this is a Quillen adjunction. One can then prove that the unit and the counit of this adjunction are weak equivalences, hence the desired result (a Quillen equivalence induces an equivalence of the associated $(\infty, 1)$-categories. We refer the reader to [22, Chapter 16] for a detailed proof of this result. This method completely fails in the case of algebras over props for two reasons:

- Algebras over props are a priori not stable under all colimits, so the left adjoint $\varphi_!$ does not exist in general;
- There is no free P-algebra functor, hence no way to transfer a model category structure from the one of cochain complexes (and by the previous point, the first axiom of model categories already fails).

To overcome these difficulties, one has to go through a completely new method based on the construction of a functorial path object of P-algebras and a corresponding equivalence of classification spaces proved in [78], then an argument using the equivalences of several models of $(\infty, 1)$-categories [79]. The equivalence of Theorem 3.4 is stated and proved in [79] as an equivalence of hammock localizations in the sense of Dwyer-Kan [12].

Theorem 3.4 means that the notion of algebraic structure up to homotopy is coherent in a very general context, and in particular that transfer and realization problems make sense also for various kinds of bialgebras. Two motivating examples are the realizations of Poincaré duality of oriented closed manifolds as homotopy Frobenius algebra structures at the cochain level, and realizations of the Lie bialgebra structure on string homology at the chain level. Let us note that an explicit realization has been recently obtained in [9] (with interesting relationships with symplectic field theory and Lagrangian Floer theory), using a notion of homotopy involutive Lie bialgebra which actually matches with the minimal model of the associated properad obtained in [5] (see [9, Remark 2.4]). However, classification and deformation theory of such structures, as well as the potential new invariants that could follow, are still to be explored.

4 Deformation Theory and Moduli Problems in a Derived Framework

Geometric Idea A common principle in algebraic topology and algebraic geometry is the following.

- In order to study a collection of objects (or structures) equipped with an equivalence relation, one construct a space (classifying space in topology, moduli space in geometry) whose points are given by this collection of objects and connected components are their equivalence classes.
- The set of equivalence classes is not enough. Indeed, understanding the deformation theory of these objects amounts to studying the infinitesimal deformations (formal neighbourhood) of the corresponding points on the moduli space. For

this, one needs the existence of some tangent structure, thus the existence of a geometry on such a moduli space.

- The deformation theory of a given point is then described by the associated formal moduli problem, which consists, roughly speaking, of a functor from augmented Artinian cdgas to simplicial sets with nice gluing properties, so that its evaluation on an algebra R is the space of R-deformations of this point.

- One would like an algebraic description of this deformation theory in terms of deformation complexes and obstruction theory. For this, one has to move in the derived world and use Lurie's equivalence theorem between formal moduli problems and dg Lie algebras. The corresponding dg Lie algebra is called the tangent Lie algebra.

In the two sections below, we describe some key ideas to work out the construction above in a derived framework, and motivate the necessity to introduce these additional derived data.

4.1 Derived Algebraic Geometry in a Nutshell

A usual geometric approach to moduli problems is to build an algebraic variety, scheme, or stack parameterizing a given type of structures or objects (complex structures on a Riemann surface, vector bundles of fixed rank...). However, the usual stacks theory shows its limits when one wants to study families of objects related by an equivalence notion weaker than isomorphisms (for instance, complexes of vector bundles) and capture their full deformation theory on the tangent spaces. Derived algebraic geometry is a conceptual framework to solve such problems, that can be seen as a homotopical perturbation or thickening of algebraic geometry [74].

Recall that as a ringed space, a usual scheme is a couple $(X, \mathcal{O}_X)$, where X is a topological space and $\mathcal{O}_X$ a sheaf of commutative algebras over X called the structural sheaf of the scheme. That is, *schemes are structured spaces locally modelled by commutative algebras*. From the "functor of points" perspective, schemes are sheaves $Aff \to Set$ on the category Aff of affine schemes, which is the opposite category of the category $Com_{\mathbb{K}}$ of commutative algebras: they are functors $Com_{\mathbb{K}} \to Set$ satisfying a gluing condition (also called descent condition) with respect to a specified collection of families of maps in $Com_{\mathbb{K}}$ called a Grothendieck topology on $Com_{\mathbb{K}}$. The notion of Grothendieck topology can be seen as a categorical analogue of the notion of covering of a topological space, and like sheaves on a topological space, we want sheaves on a given category to satisfy a gluing condition along the "coverings" given by this Grothendieck topology. Stack theory goes one step further, replacing Set by the 2-category of groupoids $Grpd$. Stacks are then functors $Com_{\mathbb{K}} \to Grpd$ satisfying a 2-categorical descent condition (gluing on objects of the groupoids and compatible gluing on sets of isomorphisms between these objects). A motivation for such a complicated generalization of scheme theory is to handle all the interesting moduli problems that cannot be represented by a moduli space in the category of schemes, due to the fact that the families of objects parametrized by this moduli problem have *non trivial automorphisms* (consider for instance fiber bundles on a variety).

To give a geometric meaning and good properties for such moduli spaces, one has to go further and work with geometric stacks, a subcategory of stacks which can be obtained by gluing (taking quotients of) representable stacks along a specified class **P** of maps. An important example of Grothendieck topology is the étale topology. In this topology, the geometric stacks obtained by choosing for **P** the class of étale maps are the Deligne-Mumford stacks, and the geometric stacks obtained by choosing for **P** the class of smooth maps are the Artin stacks. To satisfy the corresponding conditions forces the points of such a stack to have "not too wild" automorphism groups: the points of a Deligne-Mumford stack have finite groups of automorphisms (the historical example motivating the introduction of stack theory is the moduli stack of stable algebraic curves), and Artin stacks allow more generally algebraic groups of automorphisms (for example a quotient of a scheme by the action of an algebraic group).

A derived scheme is a couple $S = (X, \mathcal{O}_X)$, where X is a topological space and $\mathcal{O}_X$ a sheaf of differential graded commutative algebras over X, such that $t_0 S = (X, H^0 \mathcal{O}_X)$ (the zero truncation of S) is a usual scheme and the $H^{-i}\mathcal{O}_X$ are quasi-coherent modules over $H^0 \mathcal{O}_X$. That is, *derived schemes are structured spaces locally modelled by cdgas*. Using the "functor of points" approach, we can present derived geometric objects in the diagram

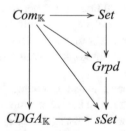

where $CDGA_\mathbb{K}$ is the ∞-category of non-positively graded commutative differential graded algebras, *Set* the category of sets, *Grpds* the (2-)category of groupoids and *sSet* the ∞-category of simplicial sets (∞-groupoids).

- Schemes are sheaves $Com_\mathbb{K} \to Set$ over the category of affine schemes (the opposite category of $Com_\mathbb{K}$) for a choice of Grothendieck topology.
- Stacks are "sheaves" $Com_\mathbb{K} \to Grpd$ for a 2-categorical descent condition, and landing in groupoids allows to represent moduli problems for which objects have **non-trivial automorphisms**.
- Higher stacks are "sheaves up to homotopy" $Com_\mathbb{K} \to sSet$, and landing in simplicial sets allows to represent moduli problems for which objects are related by **weak equivalences** instead of isomorphisms.

- Derived stacks are "sheaves up to homotopy" $CDGA_{\mathbb{K}} \to sSet$ over the ∞-category of non-positively graded cdgas (in the cohomological convention) with a choice of Grothendieck topology on the associated homotopy category. They capture the **derived data** (obstruction theory via (co)tangent complexes, non-transverse intersections, K-theoretic virtual fundamental classes [72, Section 3]) and convey richer geometric structures (shifted symplectic structures for instance [60]).

It is important to be precise that, to get derived stacks with geometric properties, we have to restrict to a sub-∞-category of these, called derived Artin stacks. Derived 1-Artin stacks are geometric realizations of smooth groupoid objects in derived affine schemes, and derived n-Artin stack are recursively defined as the geometric realization of smooth groupoid object in derived $n - 1$-Artin stacks. An alternate way is to define n-Artin stacks as smooth n-hypergroupoid objects in derived affine schemes [63]. This is the natural generalization, in the derived setting, of the geometric stacks we mentionned earlier: we obtain them by gluing representables along smooth maps, and this gluing is defined as the realization of a (higher) "groupoid-like" object. Such stacks are also said to be n-geometric. Derived Artin stacks admit cotangent complexes, an associated obstruction theory and various properties for which we refer the reader to [72]. Concerning in particular the obstruction theory, the cotangent complex of a derived n-Artin stack is cohomologically concentrated in degrees $] - \infty; n]$. If the derived Artin stack X is locally of finite presentation, then it admits a tangent complex (the dual of the cotangent complex in the ∞-category $L_{qcoh}(X)$ of quasi-coherent complexes over X) cohomologically concentrated in degrees $[-n; \infty[$. The geometric meaning of the cohomological degree is the following: at a given point x of X, the cohomology of the tangent complex in positive degrees controls the *obstruction theory* of x (extensions of infinitesimal deformations to higher order deformations), the 0th-cohomology group is the group of *equivalence classes of infinitesimal deformations*, and the cohomology of the tangent complex in negative degrees controls the *(higher) symmetries* of x (the homotopy type of its automorphisms is bounded by n). This last part generalizes to derived geometry the idea of the usual theory of algebraic stacks, that we have to control the automorphisms of the points to get a nice geometric object.

Remark 4.1 Derived Artin stacks satisfy the "geometricity" condition for a derived analogue of the class of smooth maps. Similarly, one can define derived Deligne-Mumford stacks by a geometricity condition for a derived analogue of the class of étale maps.

To illustrate this homotopical enhancement of algebraic geometry, let us give some interesting examples.

Example 4.1 Let X and Y be two subvarieties of a smooth variety V. Their intersection is said to be transverse if and only if for every point $p \in X \cap Y$, we have $T_p V = T_p X + T_p Y$ where T_p is the tangent space at p. This means that $X \cap Y$ is still a subvariety of V. Transverse intersections are very useful:

- In algebraic topology, to define the intersection product $[X].[Y] = [X \cap Y]$ on the homology H_*M of a manifold M (classes being represented by submanifolds X, Y of M).
- In algebraic geometry, classes represented by subvarieties are called algebraic classes, and the formula of the intersection product above equip algebraic classes with a ring structure. This is called the Chow ring.

It is thus natural to ask what happens when intersections are not transverse. The idea is to deform X to another subvariety X' and Y to another subvariety Y' such that X' and Y' intersect transversely, and to define $[X].[Y] = [X' \cap Y']$. The drawback is that $X \cap Y$ is not a geometric object anymore but just a homology class.

Another natural question is to count multiplicity (in some sense, the "degree of tangency") of non transverse intersections. For example, consider $X = \{y = 0\}$ a line tangent to $Y = \{y - x^2 = 0\}$ the parabola in the affine plane, and look at the intersection point $p = (0,0)$ of X and Y. If we deform this situation to a generic case by moving the line along the parabola, the line intersects the parabola at two distinct points. This means that the multiplicity of p is 2. In general, the multiplicity of the intersection of two subvarieties X and Y at a generic point p is given by Serre's intersection formula

$$I(p; X, Y) = \sum_i (-1)^i dim_{\mathcal{O}_{V,p}}(Tor_i^{\mathcal{O}_{V,p}}(\mathcal{O}_{X,p}, \mathcal{O}_{Y,p}))$$

$$= dim(\mathcal{O}_{X,p} \otimes_{\mathcal{O}_{V,p}} \mathcal{O}_{Y,p}) + \text{correction terms}$$

where $\mathcal{O}_{V,p}$ is the stalk of $\mathcal{O}_V$ at p, and $\mathcal{O}_{X,p}, \mathcal{O}_{Y,p}$ are $\mathcal{O}_{V,p}$-modules for the structures induced by the inclusions $X \hookrightarrow V, Y \hookrightarrow V$. In certain cases, the multiplicity is determined by the dimension of $\mathcal{O}_{X,p} \otimes_{\mathcal{O}_{V,p}} \mathcal{O}_{Y,p}$, but in general this is not sufficient and we have to introduce correction terms given by the derived functors Tor with no geometric meaning.

Non transverse intersections have a natural geometric construction in derived geometry. The idea is to realize $X \cap Y$ as a derived scheme by using a derived fiber product

$$X \times_V^h Y = (X \cap Y, \mathcal{O}_{X \times_V^h Y} = \mathcal{O}_X \otimes_{\mathcal{O}_V}^{\mathbb{L}} \mathcal{O}_Y)$$

where $\otimes^{\mathbb{L}}$ is the left derived tensor product of sheaves of cdgas and $\otimes_{\mathcal{O}_V}^{\mathbb{L}}$ is the left derived tensor product of dg $\mathcal{O}_V$-modules. Then

$$I(p; X, Y) = \sum_i (-1)^i dim(H^{-i} \mathcal{O}_{X \times_V^h Y}),$$

that is, the intersection number naturally and geometrically arises as the Euler characteristic of the structure sheaf of the derived intersection. In a sentence, *the transversality failure is measured by the derived part of the structure sheaf.*

Example 4.2 Another kind of application is Kontsevich's approach to Gromov-Witten theory in symplectic topology and algebraic geometry (which has also applications in string theory). On the algebraic geometry side, the problem is the following. When we want to count the intersection points of two curves in $\mathbb{P}^2$, we use intersection theory on $\mathbb{P}^2$ and Bezout theorem. More generally, one could wonder how to count rational curves of a given degree in $\mathbb{P}^N$ that intersect a given number of points $p_1, \cdots, p_n$, or replace $\mathbb{P}^N$ by a more general variety X. The idea to address this question is to define a moduli space of such curves and do intersection theory on this moduli space. But for this, one has to define a moduli space with good geometric properties, a constraint that leads to the notion of stable map. Let C be a curve of genus g and degree d with marked points $p_1, \cdots, p_n$. A stable map is a map $f : C \to X$ satisfying an additional "stability condition" we do not precise here. Counting rational curves of genus g and degree d in X passing through n fixed points $x_1, \cdots, x_n$ of X amounts to count such stable maps, and this defines the Gromov-Witten invariants of X. A classical idea is to define an invariant by integrating some function on the appropriate moduli space (via intersection theory). Here, this is the moduli space of stable maps $\overline{\mathcal{M}}_{g,n}(X, d)$. In the case $X = \mathbb{P}^n$, this is a smooth and compact Deligne-Mumford stack. In the general case of a smooth proper variety, the moduli space $\overline{\mathcal{M}}_{g,n}(X, d)$ is not smooth anymore and this is a major trouble.

Indeed, we would like to define Gromov-Witten invariants by

$$GW_d(x_1, \cdots, x_n) = \int_{\overline{\mathcal{M}}_{g,n}(X,d)} ev_1^*[x_1] \cup \cdots \cup ev_n^*[x_n]$$

$$= < [\overline{\mathcal{M}}_{g,n}(X, d)], ev_1^*[x_1] \cup \cdots \cup ev_n^*[x_n] >$$

where $ev_i : \overline{\mathcal{M}}_{g,n}(X, d) \to X, f \mapsto f(p_i)$ is the evaluation map at the ith marked point of curves, the class $[x_i]$ is the cohomology class associated to the homology class of the point x_i by Poincaré duality, and $<, >$ is the Poincaré duality pairing. Intuitively, the class $ev_1^*[x_i]$ represents curves in X whose ith marked point coincide (up to deformation of the curve) with x_i, that is, equivalences classes of curves passing through x_i. The product $ev_1^*[x_1] \cup \cdots \cup ev_n^*[x_n]$ then correspond to equivalence classes of curves passing through $x_1, \cdots, x_n$, and counting such curves amounts to pair it along the fundamental class $[\overline{\mathcal{M}}_{g,n}(X, d)]$ of $\overline{\mathcal{M}}_{g,n}(X, d)$. And this is the problem: there is no such thing as a "'fundamental class of $\overline{\mathcal{M}}_{g,n}(X, d)$", since $\overline{\mathcal{M}}_{g,n}(X, d)$ is not smooth.

Briefly, Kontsevich's idea is to see $\overline{\mathcal{M}}_{g,n}(X, d)$ as a "derived space" (i.e. equipped with a differential graded structure sheaf), that is, to make $\overline{\mathcal{M}}_{g,n}(X, d)$ formally behave like a smooth space by replacing the tangent spaces by tangent complexes. Then one associates to its dg sheaf a "virtual fundamental class" $[\overline{\mathcal{M}}_{g,n}(X, d)]^{vir}$, generalizing the fundamental class of smooth objects to singular objects (by taking the Euler characteristic of this dg sheaf in K-theory, and sending this K-theory virtual class to a class in the Chow ring of $\overline{\mathcal{M}}_{g,n}(X, d)$, thanks to the existence of a Chern character). This allows to properly define

$$GW_d(x_1, \cdots, x_n) = < [\overline{\mathcal{M}}_{g,n}(X, d)]^{vir}, ev_1^*[x_1] \cup \cdots \cup ev_n^*[x_n] > .$$

Example 4.3 Another very interesting application is the possibility to define a derived version of character varieties. Let M be a smooth manifold and G a Lie group (or an algebraic group). We know that a G-local system on M is given by a G-bundle with flat connection, and those bundles are equivalent to representations $\pi_1 M \to G$ by the Riemann-Hilbert correspondence. The variety of G-characters of M is defined by

$$Loc_G(M) = Hom(\pi_1 M, G)/G$$

where G acts by conjugation. This is the moduli space of G-local systems on M. Character varieties are of crucial importance in various topics of geometry and topology, including

- Teichmüller geometry: for a Riemann surface S, the variety $Loc_{SL_2}(S)$ contains the Teichmüller space of S as a connected component.
- Low dimensional topology: for $dim(M) = 3$, the variety $Loc_G(M)$ is related to quantum Chern-Simons invariants of M (there are various conjectures about how the properties of $Loc_G(M)$ could determine the behaviour of the 3-TQFT associated to G and M and associated invariant such as the colored Jones polynomial).

However, this is generally a highly singular object, and one would like to apply the principle shown in the previous example: treat this singular object as a smooth object in a derived framework. To formalize this idea, one defines a derived stack

$$RLoc_G(M) = Map(Betti(M), BG)$$

where $Betti(M)$ is the Betti stack of M, BG is the derived classifying stack of M and Map is the internal mapping space in the ∞-category of derived stacks [72]. This new object satisfies the following important properties:

- Its zero truncation gives the usual character variety

$$T_0 RLoc_G(M) = Loc_G(M).$$

- The tangent complex over a point computes the cohomology of M with coefficients in the associated G-local system.
- There is a nice new geometric structure appearing on such objects, which is typically of derived nature: it possesses a canonical $2 - dim(M)$-shifted symplectic structure [60]. Briefly, shifted symplectic structures are the appropriate generalization of symplectic structure from smooth manifolds to derived stacks. Here, since tangent spaces are complexes, differential forms come with a cohomological degree in addition to their weight. An n-shifted symplectic structure is a cohomology class of degree n in the de Rham complex of closed 2-forms satisfying a weak non-degeneracy condition: for every point $Spec(\mathbb{K}) \to$

X, it induces a quasi-isomorphism $\mathbb{T}_{X/\mathbb{K}} \overset{\sim}{\to} \mathbb{L}_{X/\mathbb{K}}[-n]$ between the tangent complex and the shifted cotangent complex.

If X is a smooth manifold, a 0-shifted symplectic structure on X is a usual symplectic structure. Let us note that in the case of a surface, the 0-shifted symplectic form $RLoc_G(M)$ restricts to Goldman's symplectic form on the smooth locus of $Loc_G(M)$ [34], so this is really an extension of Goldman's form to the whole moduli space.

Finally, to come back to the main topic of our survey and to motivate a bit the use of homotopy theory for moduli spaces of algebraic structures, let us see on a very simple example what happens if we build such a space with usual algebraic geometry:

Example 4.4 Let V be a vector space of dimension n, and let us consider a basis $\{e_1, \cdot, e_n\}$ of V. An associative product on V is a linear map $\mu : V \otimes V \to V$ satisfying the associativity condition, hence it is determined by its values on the basis vectors

$$\mu(e_i, e_j) = \sum_{k=1}^{n} c_{ij}^k e_k,$$

where the c_{ij}^k's satisfy moreover a certain set of relations R determined by the associativity of μ. We can build an affine scheme whose $\mathbb{K}$-points are the associative algebra structures on V: its ring of functions is simply given by $A = \mathbb{K}[c_{ij}^k]/(R)$. But we would like to classify such structures up to isomorphism, hence up to base change in V. For this, we have to mod out by the action of GL_n on V. In order to have a well defined quotient of $Spec(A)$ by GL_n, we take the quotient stack $[Spec(A)/GL_n]$ as our moduli space of associative algebra structures up to isomorphisms.

Now let R be an associative algebra with underlying vector space V, which represents a $\mathbb{K}$-point of V (given by the orbit of the action of $GL(V)$ on R). Then the truncated tangent complex $\mathbb{T}_R$ of $[Spec(A)/GL_n]$ over the orbit of R is given by a map

$$d\psi : gl(V) \to T_R Spec(A),$$

where $gl(V)$ is the Lie algebra of $GL(V)$ (the Lie algebra of matrices with coefficients in V) sitting in degree -1, and $T_R Spec(A)$ is the tangent space of $Spec(A)$ over R, sitting in degree 0. This map is the tangent map of the scheme morphism

$$\phi : GL(V) \to Spec(A)$$

which sends any $f \in GL(V)$ to $f.R$, the action of f on R, defined by transferring the algebra structure of R along f. This is what one should expect for the tangent complex: two associative algebra structures are equivalent if and only if they are related by the action of $GL(V)$ (also called action of the "gauge group"). We then get $H^{-1}\mathbb{T}_R = End_{alg}(R)$ (the Lie algebra of algebra endomorphisms of R, tangent to $Aut_{alg}(R)$) and $H^0\mathbb{T}_R = HH^2(R, R)$ the second Hochschild cohomology group of R. Let us note that this computation is a very particular case of [80, Theorem 5.6]. The group $HH^2(R, R)$ classifies equivalence classes of infinitesimal deformations of R. In particular, if $HH^2(R, R) = 0$ then the algebra R is rigid, in the sense that any infinitesimal deformation of R is equivalent to the trivial one.

The construction above has two main drawbacks. First, the tangent complex does not give us any information about the obstruction theory of R, for instance, obstruction groups for the extension of infinitesimal deformations to formal ones. Second, in the differential graded case this construction does not make sense any more, and gives no way to classify structures up to quasi-isomorphisms.

4.2   Derived Formal Moduli Problems

Formal moduli problems arise when one wants to study the infinitesimal deformation theory of a point x of a given moduli space X (variety, scheme, stack, derived stack) in a formal neighbourhood of this point (that is, the formal completion of the moduli space at this point). Deformations are parametrized by augmented Artinian rings, for example $\mathbb{K}[t]/(t^2)$ for infinitesimal deformations of order one, or $\mathbb{K}[t]/(t^n)$ for polynomial deformations of order n. The idea is to pack all the possible deformations of x in the datum of a deformation functor

$$Def_{X,x} : Art_{\mathbb{K}}^{aug} \rightarrow Set$$

from augmented Artinian algebras to sets, sending an Artinian algebra R to the set of equivalence classes of R-deformations of x, that is, equivalence classes of lifts

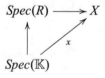

(where the morphism $Spec(\mathbb{K}) \rightarrow Spec(R)$ is induced by the augmentation $R \rightarrow \mathbb{K}$). These are nothing but the fiber of the map $X(R) \rightarrow X(\mathbb{K})$ induced by the augmentation $R \rightarrow \mathbb{K}$ and taken over the base point x. Later on, several people realized that one could use Lie theory of dg Lie algebra to describe these deformation functors. Precisely, given a dg Lie algebra g, we consider the functor

$$Def_g : Art_{\mathbb{K}}^{aug} \rightarrow Set$$

$$R \longmapsto MC(g \otimes_{\mathbb{K}} m_R)$$

where m_R is the maximal ideal of g and $MC(g \otimes_{\mathbb{K}} m_R)$ is the set of Maurer-Cartan elements of the dg Lie algebra $g \otimes_{\mathbb{K}} m_R$, that is, elements x of degree 1 satisfying the Maurer-Cartan equation $dx + \frac{1}{2}[x,x] = 0$. The functor Def_g is a formal moduli problem called the deformation functor or deformation problem associated to g. This characterization of formal moduli problems arose from unpublished work of Deligne, Drinfed and Feigin, and was developed further by Goldman-Millson, Hinich, Kontsevich, Manetti among others. Defining deformation functors via dg Lie algebras led to striking advances, for instance in the study of representations of fundamental groups of varieties [36, 67] and in deformation quantization of Poisson manifolds [46].

It turned out that all known deformation problems related to moduli spaces in geometry were of this form, which led these people to conjecture that there should be a general correspondence between formal moduli problems and dg Lie algebras. However, there was no systematic recipe to build a dg Lie algebra from a given moduli problem (the construction above is the converse direction of this hypothetical equivalence), and even worse, different dg Lie algebras could represent the same moduli problem. Moreover, the obstruction theory associated to a moduli problem, given by the positive cohomology groups of its Lie algebra, has no natural interpretation in terms of the deformation functor. Indeed, deformation theory is actually of derived nature. For instance, if we want to study the extension of polynomial deformations of order n to order $n + 1$, we have to study the properties of the natural projection $\mathbb{K}[t]/(t^{n+1}) \rightarrow \mathbb{K}[t]/(t^n)$ and under which conditions the induced map $X(\mathbb{K}[t]/(t^{n+1})) \rightarrow X(\mathbb{K}[t]/(t^n))$ is surjective, or bijective. This projection actually fits in a *homotopy pullback* (not a strict pullback) of *augmented dg Artinian algebras* (not augmented commutative algebras in $\mathbb{K}$-modules)

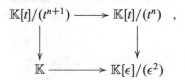

where ϵ is of cohomological degree 1 (not 0). If we could define formal moduli problems in this dg setting, we would like to apply the formal moduli problem X_x, associated to a given point x of a moduli space X, to the diagram above to get a fiber sequence

$$X_x(\mathbb{K}[t]/(t^{n+1})) \rightarrow X_x(\mathbb{K}[t]/(t^n)) \rightarrow X(\mathbb{K}[\epsilon]/(\epsilon^2))$$

and study the obstruction theory by understanding $X(\mathbb{K}[\epsilon]/(\epsilon^2))$ in an algebraic way.

These problems hint towards the necessity to introduce some homotopy theory in the study of formal moduli problems. For this, one replaces augmented Artinian algebras $Art_{\mathbb{K}}^{aug}$ by augmented dg Artinian algebras $dgArt_{\mathbb{K}}^{aug}$, and sets Set by simplicial sets $sSet$:

Definition 4.2 A derived formal moduli problem is a functor $F : dgArt^{aug}_{\mathbb{K}} \to sSet$ from augmented Artinian commutative differential graded algebras to simplicial sets, such that

1. We have an equivalence $F(\mathbb{K}) \simeq pt$.
2. The functor F sends quasi-isomorphisms of cdgas to weak equivalences of simplicial sets.
3. Let us consider a homotopy pullback of augmented dg Artinian algebras

$$
\begin{array}{ccc}
A & \longrightarrow & B \\
\downarrow & & \downarrow \\
C & \longrightarrow & D
\end{array}
$$

and suppose that the induced maps $H^0C \to H^0D$ and $H^0B \to H^0D$ are surjective.

Then F sends this homotopy pullback to a homotopy pullback of simplicial sets.
Formal moduli problems form a full sub-∞-category noted $FMP_{\mathbb{K}}$ of the ∞-category of simplicial presheaves over augmented Artinian cdgas. To make explicit the link with derived algebraic geometry, the formal neighbourhood of a point x in a derived stack X (formal completion of X at x) gives the derived formal moduli problem X_x controlling the deformation theory of x. Given an Artinian algebra R with augmentation $\epsilon : R \to \mathbb{K}$, the homotopy fiber

$$X_x(R) = hofib(X(\epsilon) : X(R) \to X(\mathbb{K}))$$

taken over the $\mathbb{K}$-point x is the space of R-deformations of X, and equivalence classes of R-deformations are determined by $\pi_0 X_x(R)$. In particular, applying X_x to the homotopy pullback

$$
\begin{array}{ccc}
\mathbb{K}[t]/(t^{n+1}) & \longrightarrow & \mathbb{K}[t]/(t^n) \\
\downarrow & & \downarrow \\
\mathbb{K} & \longrightarrow & \mathbb{K}[\epsilon]/(\epsilon^2)
\end{array}
\quad,
$$

we get a homotopy fiber sequence of spaces

$$X_x(\mathbb{K}[t]/(t^{n+1})) \to X_x(\mathbb{K}[t]/(t^n)) \to X(\mathbb{K}[\epsilon]/(\epsilon^2)),$$

hence a fiber sequence

$$\pi_0 X_x(\mathbb{K}[t]/(t^{n+1})) \to \pi_0 X_x(\mathbb{K}[t]/(t^n)) \to \pi_0 X_x(\mathbb{K}[\epsilon]/(\epsilon^2)) \cong H^1 \mathfrak{g}_{X_x},$$

where $\mathfrak{g}_{X_x}$ is the tangent Lie algebra of the formal moduli problem X_x. We can take equivalently the cohomology of the shifted tangent complex $\mathbb{T}_{X,x}[-1]$ of the stack X at x.

Remark 4.3 Actually, as proved in [37], for any derived Artin stack X locally of finite presentation (so that we can dualize the cotangent complex to define the tangent complex), there exists a quasi-coherent sheaf $\mathfrak{g}_X$ of $\mathcal{O}_X$-linear dg Lie algebras over X such that

$$\mathfrak{g}_X \simeq \mathbb{T}_{X/\mathbb{K}}[-1]$$

in the ∞-category $L_{qcoh}(X)$ of quasi-coherent complexes over X, where $\mathbb{T}_{X/\mathbb{K}}$ is the global tangent complex of X over $\mathbb{K}$. Pulling back this equivalence along a point $x : Spec(\mathbb{K}) \to X$, we get a quasi-isomorphism $\mathfrak{g}_{X_x} \simeq \mathbb{T}_{X,x}[-1]$. The sheaf $\mathfrak{g}_X$ thus encodes the family of derived formal moduli problems parametrized by X which associates to any point of X its deformation problem (the formal completion of X at this point).

The rigorous statement of an equivalence between derived formal moduli problems and dg Lie algebras was proved independently by Lurie in [50] and by Pridham in [62]:

Theorem 4.4 (Lurie, Pridham) *The ∞-category $FMP_{\mathbb{K}}$ of derived formal moduli problems over $\mathbb{K}$ is equivalent to the ∞-category $dgLie_{\mathbb{K}}$ of dg Lie $\mathbb{K}$-algebras.*

Moreover, one side of the equivalence is made explicit, and is equivalent to the nerve construction of dg Lie algebras studied thoroughly by Hinich in [38]. The homotopy invariance of the nerve relies on nilpotence conditions on the dg Lie algebra. In the case of formal moduli problems, this nilpotence condition is always satisfied because one tensors the Lie algebra with the maximal ideal of an augmented Artinian cdga. In this article, what we will call moduli problems are actually derived moduli problems.

4.2.1 Extension to L_∞-Algebras

Certain deformation complexes of interest are not strict Lie algebras but homotopy Lie algebras, that is L_∞-algebras. There is a strictification theorem for homotopy Lie algebras (more generally, for dg algebras over any operad when $\mathbb{K}$ is of characteristic zero), so any L_∞-algebra is equivalent to a dg Lie algebra, but this simplification of the algebraic structure goes with an increased size of the underlying complex, which can be very difficult to make explicit. This is why one would like the theory of derived formal moduli problems to extend to L_∞-algebras, and fortunately it does. There are two equivalent definitions of an L_∞-algebra:

Definition 4.5

(1) An L_∞-algebra is a graded vector space $g = \{g_n\}_{n \in \mathbb{Z}}$ equipped with maps $l_k : g^{\otimes k} \to g$ of degree $2 - k$, for $k \geq 1$, satisfying the following properties:

- $l_k(\ldots, x_i, x_{i+1}, \ldots) = -(-1)^{|x_i||x_{i+1}|} l_k(\ldots, x_{i+1}, x_i, \ldots)$
- for every $k \geq 1$, the generalized Jacobi identities

$$\sum_{i=1}^{k} \sum_{\sigma \in Sh(i,k-i)} (-1)^{\epsilon(i)} l_k(l_i(x_{\sigma(1)}, \ldots, x_{\sigma(i)}), x_{\sigma(i+1)}, \ldots, x_{\sigma(k)}) = 0$$

where σ ranges over the $(i, k-i)$-shuffles and

$$\epsilon(i) = i + \sum_{j_1 < j_2, \sigma(j_1) > \sigma(j_2)} (|x_{j_1}||x_{j_2}| + 1).$$

(2) An L_∞-algebra structure on a graded vector space $g = \{g_n\}_{n \in \mathbb{Z}}$ is a coderivation $Q : \hat{Sym}^{\bullet \geq 1}(g[1]) \to \hat{Sym}^{\bullet \geq 1}(g[1])$ of degree 1 of the cofree cocommutative coalgebra $\hat{Sym}^{\bullet \geq 1}(g[1])$ such that $Q^2 = 0$.

The bracket l_1 is actually the differential of g as a cochain complex. When the brackets l_k vanish for $k \geq 3$, then one gets a dg Lie algebra. The dg algebra $C^*(g)$ obtained by dualizing the dg coalgebra of (2) is called the Chevalley-Eilenberg algebra of g.

A L_∞ algebra g is filtered if it admits a decreasing filtration

$$g = F_1 g \supseteq F_2 g \supseteq \ldots \supseteq F_r g \supseteq \ldots$$

compatible with the brackets: for every $k \geq 1$,

$$l_k(F_r g, g, \ldots, g) \in F_r g.$$

We suppose moreover that for every r, there exists an integer $N(r)$ such that $l_k(g, \ldots, g) \subseteq F_r g$ for every $k > N(r)$. A filtered L_∞ algebra g is complete if the canonical map $g \to \lim_r g / F_r g$ is an isomorphism.

The completeness of a L_∞ algebra allows to define properly the notion of Maurer-Cartan element:

Definition 4.6

(1) Let g be a dg L_∞-algebra and $\tau \in g^1$, we say that τ is a Maurer-Cartan element of g if

$$\sum_{k \geq 1} \frac{1}{k!} l_k(\tau, \ldots, \tau) = 0.$$

The set of Maurer-Cartan elements of g is noted $MC(g)$.

(2) The simplicial Maurer-Cartan set is then defined by

$$MC_\bullet(g) = MC(g \hat{\otimes} \Omega_\bullet),$$

where $\Omega_\bullet$ is the Sullivan cdga of de Rham polynomial forms on the standard simplex $\Delta^\bullet$ (see [68]) and $\hat\otimes$ is the completed tensor product with respect to the filtration induced by g.

The simplicial Maurer-Cartan set is a Kan complex, functorial in g and preserves quasi-isomorphisms of complete L_∞-algebras. The Maurer-Cartan moduli set of g is $\mathscr{MC}(g) = \pi_0 MC_\bullet(g)$: it is the quotient of the set of Maurer-Cartan elements of g by the homotopy relation defined by the 1-simplices. When g is a complete dg Lie algebra, it turns out that this homotopy relation is equivalent to the action of the gauge group $exp(g^0)$ (a prounipotent algebraic group acting on Maurer-Cartan elements), so in this case this moduli set coincides with the one usually known for Lie algebras. We refer the reader to [80] for more details about all these results. The notion of Maurer-Cartan space allows to define the classical deformation functor of g given by

$$\underline{\mathscr{MC}}(g) : Art_{\mathbb{K}} \rightarrow Set$$
$$R \longmapsto \mathscr{MC}(g \otimes m_R)$$

and the derived deformation functor or derived formal moduli problem of g given by

$$\underline{MC_\bullet}(g) : dgArt_{\mathbb{K}}^{aug} \rightarrow sSet$$
$$R \longmapsto MC_\bullet(g \otimes m_R)$$

(which belongs indeed to $FMP_{\mathbb{K}}$). By [80, Corollary 2.4], the tensor product $MC_\bullet(g \otimes m_R)$ does not need to be completed because R is Artinian. To see why Theorem 3.3 extends to L_∞-algebras, let $\pi : L_\infty \overset{\sim}{\rightarrow} Lie$ be a cofibrant resolution of the operad Lie. This morphism induces a functor $p^* : dgLie \rightarrow L_\infty - Alg$ which associates to any dg Lie algebra the L_∞-algebra with the same differential, the same bracket of arity 2 and trivial higher brackets in arities greater than 2. This functor fits in a Quillen equivalence

$$p_! : L_\infty - Alg \leftrightarrows dgLie : p^*,$$

where the left adjoint is a certain coequalizer (see [22, Theorem 16.A]), and Quillen equivalences induce equivalences of the corresponding ∞-categories, so we have a commutative triangle of ∞-categories

$$
\begin{array}{ccc}
 & L_\infty - Alg & \\
p^* \uparrow & & \searrow \tilde\psi \\
dgLie & \xrightarrow[\psi]{} & FMP_{\mathbb{K}}
\end{array}
$$

where ψ and $\tilde{\psi}$ send a Lie algebra, respectively an L_∞-algebra, to its derived formal moduli problem. The maps p^* and ψ are weak equivalences of ∞-categories, so $\tilde{\psi} : L_\infty - Alg \to FMP_\mathbb{K}$ is a weak equivalence of ∞-categories as well (here, by weak equivalence we mean a weak equivalence in the chosen model category of ∞-categories, say quasi-categories for instance).

4.2.2 Twistings of L_∞-Algebras

We recall briefly the notion of twisting by a Maurer-Cartan element. The twisting of a complete L_∞ algebra g by a Maurer-Cartan element τ is the complete L_∞ algebra g^τ with the same underlying graded vector space and new brackets l_k^τ defined by

$$l_k^\tau(x_1, \ldots, x_k) = \sum_{i \geq 0} \frac{1}{i!} l_{k+i}(\underbrace{\tau, \ldots, \tau}_{i}, x_1, \ldots, x_k)$$

where the l_k are the brackets of g. The twisted L_∞-algebra g^φ is the *deformation complex of* φ, that is, the derived formal moduli problem of g^φ controls the deformation theory of φ. To see this, let us define another kind of Maurer-Cartan functor

$$\tilde{MC}_\bullet(g \otimes -) : dgArt_\mathbb{K}^{aug} \to sSet$$
$$R \longmapsto MC_\bullet(g \otimes R).$$

We replaced the maximal ideal m_R in the definition of the deformation functor by the full algebra R. That is, the functor $\tilde{MC}_\bullet(g \otimes -)$ sends R to the space of R-linear extensions of Maurer-Cartan elements of g. Then, for every augmented dg Artinian algebra R one has

$$MC_\bullet(g^\varphi \otimes m_R) = hofib(MC_\bullet(g \otimes R) \to MC_\bullet(g), \varphi)$$

where the map in the right side is induced by the augmentation $R \to \mathbb{K}$ and the homotopy fiber is taken over the base point φ. That is, the space $MC_\bullet(g^\varphi \otimes m_R)$ is the space of R-linear extensions of φ as Maurer-Cartan elements of $g \otimes R$.

5 Moduli Spaces of Algebraic Structures

5.1 First Version: A Simplicial Construction

We refer the reader to [40, Chapter 16, Chapter 17] and [24] for some prerequisites about simplicial mapping spaces in model categories. We use this notion of simplicial mapping space and the model category structure on props to define our moduli spaces. Let us define a first version of this moduli space as a simplicial set. This was originally defined in the setting of simplicial operads [65], and can be extended to algebras over differential graded props as follows (see [82]):

Definition 5.1 Let P_∞ be a cofibrant prop and X be a cochain complex. The (simplicial) moduli space of P_∞-algebra structures on X is the simplicial set $P_\infty\{X\}$ defined in each simplicial dimension k by

$$P_\infty\{X\}_k = Mor_{prop}(P_\infty, End_X \otimes \Omega_k),$$

where $(End_X \otimes \Omega_k)(m, n) = Hom(X^{\otimes m}, X^{\otimes n}) \otimes \Omega_k$.

The Sullivan algebras Ω_k gather into a simplicial commutative differential graded algebra $\Omega_\bullet$ whose faces and degeneracies induce the simplicial structure on $P_\infty\{X\}$. The functor $(-) \otimes \Omega_\bullet$ is a functorial simplicial resolution in the model category of props [81, Proposition 2.5], so this simplicial moduli space is a homotopy mapping space in this model category. In particular, this means that this simplicial set is a Kan complex whose points are the P_∞-algebra structures $P_\infty \to End_X$ and 1-simplices are the homotopies between such structures (the prop $End_X \otimes \Omega_1$ forms a path object of End_X in the model category of props). The later property implies that

$$\pi_0 P_\infty\{X\} = [P_\infty, End_X]_{Ho(Prop)}$$

is the set of homotopy classes of P_∞-algebra structures on X. So our simplicial moduli space has the two first properties one expects from a moduli space: its points are the objects we want to classify and its connected components are the equivalence classes of these objects. Moreover, the fact that this is a homotopy mapping space implies that it is homotopy invariant with respect to the choice of a cofibrant resolution for the source, that is, any weak equivalence of cofibrant props $P_\infty \xrightarrow{\sim} Q_\infty$ induces a weak equivalence of Kan complexes

$$Q_\infty\{X\} \xrightarrow{\sim} P_\infty\{X\}.$$

So this is a well defined classifying object for homotopy P-algebra structures on X.

Another interesting homotopy invariant is the classification space of P_∞-algebras, defined as the nerve $\mathcal{N}wP_\infty - Alg$ of the subcategory whose objects are P_∞-algebras and morphisms are quasi-isomorphisms of P_∞-algebras. By [12, 13], this classification space admits a decomposition

$$\mathcal{N}wP_\infty - Alg \simeq \sqcap_{[X]\in\pi_0\mathcal{N}wP_\infty-Alg}\overline{W}L^H wP_\infty - Alg(X, X).$$

Here the product ranges over weak equivalence classes of P_∞-algebras, and $\overline{W}L^H wP_\infty - Alg(X, X)$ is the classifying complex of the simplicial monoid of zigzags of weak equivalences $X \xleftarrow{\sim} \bullet \xrightarrow{\sim} X$ in the hammock localization (or equivalently in the simplicial localization) of $P_\infty - Alg$ in the sense of Dwyer-Kan, i.e. the self equivalences of X in the ∞-category of P_∞-algebras. Let us note that when P_∞

is an operad and X is a cofibrant P_∞-algebra, this space is equivalent to the usual simplicial monoid $haut_{P_\infty}(X)$ of self weak equivalences of X. This means that the classification space of P_∞-algebras encodes symmetries and higher symmetries of P_∞-algebras in their homotopy theory. Homotopy invariance of the classification space for algebras over props is a non trivial theorem:

Theorem 5.2 ([78, Theorem 0.1]) *Let $\varphi : P_\infty \xrightarrow{\sim} Q_\infty$ be a weak equivalence between two cofibrant props. The map φ gives rise to a functor*

$$\varphi^* : wQ_\infty - Alg \to wP_\infty - Alg$$

which induces a weak equivalence of simplicial sets

$$\mathcal{N}\varphi^* : \mathcal{N}wQ_\infty - Alg \xrightarrow{\sim} \mathcal{N}wP_\infty - Alg.$$

Moreover, it turns out that the simplicial moduli space defined above gives a local approximation of this classification space, precisely we have the following result:

Theorem 5.3 ([82, Theorem 0.1]) *Let P_∞ be a cofibrant dg prop and X be a cochain complex. The commutative square*

$$
\begin{array}{ccc}
P_\infty\{X\} & \longrightarrow & \mathcal{N}wP_\infty - Alg \\
\downarrow & & \downarrow \\
\{X\} & \longrightarrow & \mathcal{N}w\mathcal{C}h
\end{array}
$$

is a homotopy pullback of simplicial sets.

This homotopy fiber theorem has been applied to study the homotopy type of realization spaces in [83] in terms of derivation complexes and to count equivalence classes of realizations (of Poincaré duality for example).

The reader has probably noticed that we used the following property to define our simplicial moduli space: tensoring a prop by a cdga componentwise preserves the prop structure. This allows us to extend the definition of this moduli space and make it a simplicial presheaf of cdgas

$$\underline{Map}(P_\infty, Q) : R \in CDGA_\mathbb{K} \mapsto Map_{Prop}(P_\infty, Q \otimes A).$$

Moreover, the notion of classification space defined above in the sense of Dwyer-Kan can also be extended to a simplicial presheaf. For this, we use that for any cdga R, the category Mod_R is a (cofibrantly generated) symmetric monoidal model category tensored over chain complexes, so that one can define the category $P_\infty - Alg(Mod_R)$ of P_∞-algebras in Mod_R. The assignment

$$A \mapsto wP_\infty - Alg(Mod_R)$$

defines a weak presheaf of categories in the sense of [1, Definition I.56]. It sends a morphism $A \to B$ to the symmetric monoidal functor $- \otimes_A B$ lifted at the level of P_∞-algebras. This weak presheaf can be strictified into a presheaf of categories (see [1, Section I.2.3.1]). Applying the nerve functor then defines a simplicial presheaf of Dwyer-Kan classification spaces that we note $\mathcal{N}w P_\infty - Alg$. The simplicial presheaf $\mathcal{N}w Ch_{\mathbb{K}}$ associated to $A \mapsto Mod_A$ is the simplicial presheaf of quasi-coherent modules of [74, Definition 1.3.7.1]. The constructions above then make the following generalization of Theorem 4.3 meaningful:

Proposition 5.4 ([30, Proposition 2.13]) *Let P_∞ be a cofibrant prop and X be a chain complex. The forgetful functor $P_\infty - Alg \to Ch_{\mathbb{K}}$ induces a homotopy fiber sequence*

$$P_\infty\{X\} \to \mathcal{N}w P_\infty - Alg \to \mathcal{N}w Ch_{\mathbb{K}}$$

of simplicial presheaves over cdgas, taken over the base point X.

5.2 Second Version: A Stack Construction and the Associated Deformation Theory

If P is a properad with cofibrant resolution $(\mathscr{F}(s^{-1}C), \partial) \xrightarrow{\sim} P$ for a certain homotopy coproperad C (see [58, Section 4] for the definition of homotopy coproperads), and Q is any properad, then we consider the total complex $g_{P,Q} = Hom_\Sigma(\overline{C}, Q)$ given by homomorphisms of Σ-biobjects from the augmentation ideal of C to Q. In the case $Q = End_X$ we will note it $g_{P,X}$. By [59, Theorem 5], it is a complete dg L_∞ algebra whose Maurer-Cartan elements are prop morphisms $P_\infty \to Q$. This L_∞-structure was also independently found in [54, Section 5], where it is proved that such a structure exists when replacing our cofibrant resolution above by the minimal model of a $\mathbb{K}$-linear prop (and its completeness follows by [54, Proposition 15]). In [81], we prove a non trivial generalization of this result at the level of simplicial presheaves:

Theorem 5.5 ([81, Theorem 2.10,Corollary 4.21]) *Let P be a dg properad equipped with a minimal model $P_\infty := (\mathscr{F}(s^{-1}C), \partial) \xrightarrow{\sim} P$ and Q be a dg properad. Let us consider the simplicial presheaf*

$$Map(P_\infty, Q) : R \in CDGA_{\mathbb{K}} \mapsto Map_{Prop}(P_\infty, Q \otimes A)$$

where $CDGA_{\mathbb{K}}$ is the category of commutative differential graded $\mathbb{K}$-algebras and $Q \otimes A$ is the componentwise tensor product defined by $(Q \otimes A)(m, n) = Q(m, n) \otimes A$. This presheaf is equivalent to the simplicial presheaf

$$\widetilde{MC}_\bullet(Hom_\Sigma(\overline{C}, Q)) : A \in CDGA_{\mathbb{K}} \mapsto MC_\bullet(Hom_\Sigma(\overline{C}, Q) \otimes A)$$

associated to the complete L_∞-algebra $Hom_\Sigma(\overline{C}, Q)$.

In the case $Q = End_X$, we get the simplicial presheaf which associates to A the moduli space of P_∞-algebra structures on $X \otimes A$. Let us note that $\underline{Map}(P_\infty, Q)$ can be alternately defined by

$$A \mapsto Map_{Prop(Mod_A)}(P_\infty \otimes A, Q \otimes A),$$

where $Map_{Prop(Mod_A)}$ is the simplicial mapping space in the category of props in dg A-modules. In the case $Q = End_X$, we have $Q \otimes A \cong End_{X \otimes A}^{Mod_A}$ where $End_{X \otimes A}^{Mod_A}$ is the endormorphism prop of $X \otimes A$ taken in the category of A-modules. That is, it associates to A the simplicial moduli space of A-linear P_∞-algebra structures on $X \otimes A$ in the category of A-modules. This theorem applies to a large class of algebraic structures, including for instance Frobenius algebras, Lie bialgebras and their variants such as involutive Lie bialgebras, as well as the properad $Bialg$ encoding associative and coassociative bialgebras.

Under additional assumptions, we can equip such a presheaf with a stack structure:

Theorem 5.6 ([81, Corollary 0.8])

(1) Let $P_\infty = (\mathscr{F}(s^{-1}C), \partial) \xrightarrow{\sim} P$ be a cofibrant resolution of a dg properad P and Q be any dg properad such that each $Q(m, n)$ is a bounded complex of finite dimension in each degree. The functor

$$\underline{Map}(P_\infty, Q) : A \in CDGA_{\mathbb{K}} \mapsto Map_{Prop}(P_\infty, Q \otimes A)$$

is an affine stack in the setting of complicial algebraic geometry of [74].

(2) Let $P_\infty = (\mathscr{F}(s^{-1}C), \partial) \xrightarrow{\sim} P$ be a cofibrant resolution of a dg properad P in non positively graded cochain complexes, and Q be any properad such that each $Q(m, n)$ is a finite dimensional vector space. The functor

$$\underline{Map}(P_\infty, Q) : A \in CDGA_{\mathbb{K}} \mapsto Map_{Prop}(P_\infty, Q \otimes A)$$

is an affine stack in the setting of derived algebraic geometry of [74], that is, an affine derived scheme.

In the derived algebraic geometry context, the derived stack $\underline{Map}(P_\infty, Q)$ is not affine anymore whenever the $Q(m, n)$ are not finite dimensional vector spaces. However, we expect these stacks to be derived n-Artin ind-stacks for the $Q(m, n)$ being perfect complexes with finite amplitude n, using the characterization of derived n-Artin stacks via resolutions by Artin n-hypergroupoids given in [62].

We denote by $\mathbb{T}_{\underline{Map}(P_\infty, Q), x_\varphi}$ the tangent complex of $\underline{Map}(P_\infty, Q)$ at an A-point x_φ associated to a properad morphism $\varphi : P_\infty \to Q \otimes_e A$. As we explained before in Sect. 4, non-positive cohomology groups of the deformation complex correspond to negative groups of the tangent complex, which computes the higher automorphisms (higher symmetries) of the point, and the positive part which computes the obstruction theory. Adding some finiteness assumptions on the resolution P_∞, we can make explicit the ring of functions of this affine stack:

Theorem 5.7 ([81, Theorem 0.14]) *Let P be a dg properad equipped with a cofibrant resolution $P_\infty := \Omega(C) \xrightarrow{\sim} P$, where C admits a presentation $C = \mathcal{F}(E)/(R)$, and Q be a dg properad such that each $Q(m, n)$ is a bounded complex of finite dimension in each degree. Let us suppose that each $E(m, n)$ is of finite dimension, and that there exists an integer N such that $E(m, n) = 0$ for $m + n > N$. Then*

(1) The moduli stack $\underline{Map}(P_\infty, Q)$ is isomorphic to $\mathbb{R}Spec_{C^*(Hom_\Sigma(\overline{C}, Q))}$, where $C^*(Hom_\Sigma(\overline{C}, Q))$ is the Chevalley-Eilenberg algebra of $Hom_\Sigma(\overline{C}, Q)$.

(2) The cohomology of the tangent dg Lie algebra at a $\mathbb{K}$-point $\varphi : P_\infty \to Q$ is explicitly determined by

$$H^*(\mathbb{T}_{\underline{Map}(P_\infty, Q), x_\varphi}[-1]) \cong H^*(Hom_\Sigma(\overline{C}, Q)^\varphi).$$

This theorem applies to a wide range of structures including for instance Frobenius algebras, Lie bialgebras and their variants such as involutive Lie bialgebras, and associative-coassociative bialgebras.

5.3 Properties of the Corresponding Formal Moduli Problems and Derived Deformation Theory

Before turning to formal moduli problems, a natural question after reading the previous section is the following: how are the tangent complexes of our moduli spaces related to the usual cohomology theories of well-known sorts of algebras such as Hochschild cohomology of associative algebras, Harrison cohomology of commutative algebras, Chevalley-Eilenberg cohomology of Lie algebras, or Gerstenhaber-Schack cohomology of associative-coassociative bialgebras (introduced to study the deformation theory of quantum groups [27]). It turns out that these tangent Lie algebras do not give exactly the usual cohomology theories, but rather shifted truncations of them. For instance, let us consider the Hochschild complex $Hom(A^{\otimes > 0}, A)$ of a dg associative algebra A. This Hochschild complex is bigraded with a cohomological grading induced by the grading of A and a weight grading given by the tensor powers $A^{\otimes \bullet}$. It turns out that the part $Hom(A, A)$ of weight 1 in the Hochschild complex is the missing part in $g^\varphi_{Ass,A}$ (the L_∞-algebra of Theorem 5.7, where $\varphi : Ass \to End_A$ is the associative algebra structure of A). There is also a "full" version of the Hochschild complex defined by $Hom(A^{\otimes \geq 0}, A)$. These three variants of Hochschild complexes give a sequence of inclusions of three dg Lie algebras

$$Hom(A^{\otimes \geq 0}, A)[1] \supset Hom(A^{\otimes > 0}, A)[1] \supset Hom(A^{\otimes > 1}, A)[1].$$

All of these have been considered in various places in the literature, but without comparison of their associated moduli problems. For the full complex, it is known that it controls the linear deformation theory of Mod_A as a dg category [43, 61].

The same kind of open question arises for other cohomology theories and their variants, and one of the achievements of our work with Gregory Ginot [30] was to describe precisely the moduli problems controlled by these variants and how they are related in the general context of P_∞-algebras.

The formal moduli problem $P_\infty\{X\}^\varphi$ controlling the formal deformations of a P_∞-algebra structure $\varphi : P_\infty \to End_X$ on X is defined, on any augmented dg Artinian algebra R, by the homotopy fiber

$$\underline{P_\infty\{X\}^\varphi}(R) = hofib(\underline{P_\infty\{X\}}(R) \to \underline{P_\infty\{X\}}(\mathbb{K}))$$

taken over the base point φ, where the map is induced by the augmentation $R \to \mathbb{K}$. The twisting of the complete L_∞-algebra $Hom_\Sigma(\overline{C}, End_X)$ by a properad morphism $\varphi : P_\infty \to End_X$ is the deformation complex of φ, and we have an isomorphism

$$g_{P,X}^\varphi = Hom_\Sigma(\overline{C}, End_X)^\varphi \cong Der_\varphi(\Omega(C), End_X)$$

where the right-hand term is the complex of derivations with respect to φ [59, Theorem 12], whose L_∞-structure induced by the twisting of the left-hand side is equivalent to the one of [54, Theorem 1]. Section 4.2.2 combined with Theorem 5.5 tells us which formal moduli problem this deformation complex controls:

Proposition 5.8 ([30, Proposition 2.11]) *The tangent L_∞-algebra of the formal moduli problem $P_\infty\{X\}^\varphi$ is given by*

$$g_{P,X}^\varphi = Hom_\Sigma(\overline{C}, End_X)^\varphi.$$

In derived algebraic geometry, a Zariski open immersion of derived Artin stacks $F \hookrightarrow G$ induces a weak equivalence between the tangent complex over a given point of F and the tangent complex over its image in G [74]. It is thus natural to wonder more generally whether an "immersion" of an ∞-category $\mathscr{C}$ into another ∞-category $\mathscr{D}$ induces an equivalence between the deformation problem of an object X of $\mathscr{C}$ (which should be in some sense a tangent space of $\mathscr{C}$ at X) and the deformation problem of its image in $\mathscr{D}$, in particular an equivalence of the corresponding tangent dg Lie algebras when such a notion makes sense. Here the word "immersion" has to be understood as "fully faithful conservative ∞-functor", that is, a fully faithful ∞-functor $\mathscr{C} \to \mathscr{D}$ such that a map of $\mathscr{C}$ is an equivalence if and only if its image in $\mathscr{D}$ is a weak equivalence. In the case of ∞-categories of algebras over props, Proposition 5.4 tells us that the formal moduli problem $P_\infty\{X\}^\varphi$ is the "tangent space" over (X, φ) to the Dwyer-Kan classification space of the ∞-category of P_∞-algebras, with associated tangent L_∞-algebra $g_{P,X}^\varphi$. In this setting, we can thus transform the intuition above into a precise statement:

Theorem 5.9 ([30, Theorem 2.16]) *Let $F : P_\infty - Alg \to Q_\infty - Alg$ be a fully faithful and conservative ∞-functor inducing functorially in A, for every augmented Artinian cdga A, a fully faithful and conservative ∞-functor $F : P_\infty - Alg(Mod_A) \to Q_\infty - Alg(Mod_A)$. Then F induces an equivalence of formal moduli problems*

$$\underline{P_\infty\{X\}^\varphi} \sim \underline{Q_\infty\{F(X)\}^{F(\varphi)}},$$

where $F(\varphi)$ is the Q_∞-algebra structure on the image $F(X, \varphi)$ of X, φ under F, hence an equivalence of the associated L_∞-algebras

$$g_{P,X}^\varphi \sim g_{Q,F(X)}^{F(\varphi)}.$$

Proposition 5.4 also hints towards the fact that $g_{P,X}^\varphi$ does not control the deformation theory of homotopy automorphisms of (X, φ) in the infinitesimal neighbourhood of $id_{(X,\varphi)}$, but should be closely related to it, since classification spaces decompose into disjoint unions of homotopy automorphisms. These homotopy automorphisms form a derived algebraic group [21], and as for underived algebraic groups, one can associate Lie algebras to such objects. Indeed, as explained in [21], given a moduli functor F and a point $x \in F(\mathbb{K})$, the reduction of F at x is the functor F_x defined by the homotopy fiber

$$F_x(R) = hofib(F(R) \to F(\mathbb{K}))$$

where the map is induced by the augmentation of R and the homotopy fiber is taken over the base point x. A point x of $F(\mathbb{K})$ such that the reduction of F at x is a formal moduli problem (called an infinitesimal moduli problem in [21, Definition 4.5]) is called formally differentiable [21, Definition 4.10], so there is a tangent Lie algebra of F at x defined as the Lie algebra of the formal moduli problem F_x. In the case of derived algebraic groups, the neutral element is a formally differentiable point and the Lie algebra of a derived algebraic group is the Lie algebra of its reduction at the neutral element. This is the natural extension to a derived framework of the well known Lie algebra of a Lie group. Consequently, there should be a homotopy fiber sequence of L_∞-algebras relating $Lie(\underline{haut_{P_\infty}}(X, \varphi))$ to the tangent L_∞-algebra $g_{P,X}^\varphi$ of $\underline{P_\infty\{X\}^\varphi}$.

Let us make explicit a bit the construction of derived algebraic groups of homotopy automorphisms. Given a complex X, its homotopy automorphism group is denoted $haut(X)$. Given a P_∞-algebra (X, φ), its homotopy automorphism group in the ∞-category of P_∞-algebras is denoted $haut_{P_\infty}(X, \varphi)$. In the general case, it is defined by Dwyer-Kan's hammock localization $L^H wP_\infty((X, \varphi), (X, \varphi))$, since we do not have a model category structure on the category of P_∞-algebras. However in the particular case where P_∞ is an operad, it turns out that this construction is equivalent to the usual simplicial monoid of homotopy automorphisms of (X, φ) in the model category of P_∞-algebras (the simplicial sub-monoid of self weak equivalences in the usual homotopy mapping space $Map_{P_\infty - Alg}(X, X)$).

Remark 5.10 What we mean here by a homotopy automorphism is a self weak equivalence, not the homotopy class of a strict automorphism.

The derived algebraic group $\underline{haut}(X)$ of homotopy automorphisms of X is defined by the strictification of the weak simplicial presheaf

$$R \mapsto haut_{Mod_R}(X \otimes R),$$

where $haut_{Mod_A}$ is the simplicial monoid of homotopy automorphisms in the category of A-modules. The derived algebraic group $\underline{haut}_{P_\infty}(X, \varphi)$ of homotopy automorphisms of (X, φ) is defined by the strictification of the weak simplicial presheaf

$$R \mapsto haut_{P_\infty}(X \otimes R, \varphi \otimes R)_{Mod_R}$$

where $haut_{P_\infty}(X \otimes R, \varphi \otimes R)_{Mod_R}$ is the simplicial monoid of homotopy automorphisms of $(X \otimes R, \varphi \otimes R) \in P_\infty - Alg(Mod_R)$. The reduction of $\underline{haut}_{P_\infty}(X, \varphi)$ at $id_{(X,\varphi)}$ associates to any augmented dg Artinian algebra R the space of R-linear extensions of homotopy automorphisms living in the connected component of $id_{(X,\varphi)}$, that is, homotopy isotopies. Finally, the deformation complex of φ in the ∞-category of props and the deformation complex of homotopy isotopies of (X, φ) in the ∞-category of P_∞-algebras are related by the expected fiber sequence:

Proposition 5.11 ([30, Proposition 2.14]) *There is a homotopy fiber sequence of L_∞-algebras*

$$g_{P,X}^\varphi \to Lie(\underline{haut}_{P_\infty}(X, \varphi)) \to Lie(\underline{haut}(X)).$$

Moreover, we can make explicit $Lie(\underline{haut}_{P_\infty}(X, \varphi))$ as a slight modification $g_{P+,X}^{\varphi^+}$ of $g_{P,X}^\varphi$, which consists in adding a component $Hom(X, X)$ to $g_{P,X}^\varphi$ (we refer to [30, Section 3]):

Theorem 5.12 ([30, Theorem 3.5]) *There is a quasi-isomorphism of L_∞-algebras*

$$g_{P+,X}^{\varphi^+} \simeq Lie(\underline{haut}_{P_\infty}(X, \varphi)).$$

The conceptual explanation underlying this phenomenon is that $g_{P,X}^\varphi$ controls the deformations of the P_∞-algebra structure over a fixed complex X, whereas $g_{P+,X}^{\varphi^+}$ controls deformations of this P_∞-algebra structure plus compatible deformations of the differential of X, that is, deformations of the P_∞-algebra structure up to self quasi-isomorphisms of X. This is the role of the part $Hom(X, X)$ appearing for instance in Hochschild cohomology. For instance, given an associative dg algebra A, the complex $g_{Ass+,A}^{\varphi^+} \cong Hom(A^{\otimes >0}, A)[1]$ computes the Hochschild cohomology of A and the complex $g_{Ass,A}^\varphi \cong Hom(A^{\otimes >1}, A)[1]$ is the one controlling the formal moduli problem of deformations of A with fixed differential. The full shifted

Hochschild complex $Hom(A^{\otimes \geq 0}, A)[1]$ controls the linear deformations of the dg category Mod_A.

For an n-Poisson algebra A (Poisson algebras with a Poisson bracket of degree $1 - n$), we have the same kind of variants of L_∞-algebras: the full shifted Poisson complex $CH_{Pois_n}(A)[n]$ [3], the deformation complex $CH_{Pois_n}^{(\bullet > 0)}(A)[n]$ introduced by Tamarkin [71] which is the part of positive weight in the full Poisson complex, and the further truncation $CH_{Pois_n}^{(\bullet > 1)}(A)[n]$. In [30, Section 6], we solve the open problem to determine which deformation problems these L_∞-algebras control:

Theorem 5.13 *Let A be an n-Poisson algebra.*

(1) The truncation $CH_{Pois_n}^{(\bullet > 1)}(A)[n]$ is the deformation complex $g_{Pois_n,R}^{\varphi}$ of the formal moduli problem $\underline{Pois_{n\infty}\{A\}^\varphi}$ of homotopy n-Poisson algebra structures deforming φ.

(2) Tamarkin's deformation complex controls deformations of A into dg-$Pois_n$-algebras, that is, it is the tangent Lie algebra $g_{Pois_n^+,A}^{\varphi^+}$ of $\underline{haut_{Pois_n}(A)}$.

Remark 5.14 We conjecture that the L_∞-algebra structure of the full shifted Poisson complex $CH_{Pois_n}^*(A)[n]$ controls the deformations of Mod_A into E_{n-1}-monoidal dg categories. This should have interesting consequences for deformation quantization of n-shifted Poisson structures in derived algebraic geometry [4, 73].

6 Gerstenhaber-Schack Conjecture, Kontsevich Formality Conjecture and Deformation Quantization

6.1 E_n-Operads, Higher Hochschild Cohomology and the Deligne Conjecture

Recall that an E_n-operad is a dg operad quasi-isomorphic to the singular chains C_*D_n of the little n-disks operad. We refer the reader to [24, Volume I] for a comprehensive treatment of the construction and main properties of the little n-disks operads. These E_n-operads satisfy the following properties:

- There is an isomorphism $H_*E_1 \cong Ass$, where Ass is the operad of associative algebras.
- For $n \geq 2$, there is an isomorphism $H_*E_n \cong Pois_n$ where $Pois_n$ is the operad of n-Poisson algebras.
- For $n \geq 2$, the E_n-operads are formal, i.e. there is a quasi-isomorphism of operads

$$E_n \xrightarrow{\sim} Pois_n.$$

Modulo a technical assumption satisfied in particular by C_*D_n, this formality holds over $\mathbb{Q}$ [25].

The formality of the little n-disks operad has a long story of intermediate formality results (for $n = 2$ over $\mathbb{Q}$ in [70], for $n \geq 2$ over $\mathbb{R}$ in [45, 47], finally an intrinsic formality result over $\mathbb{Q}$ in [25]). This formality is the key point to prove Deligne conjecture, which states the existence of a homotopy Gerstenhaber structure (that is, the E_2-algebra structure) of the Hochschild complex with product given by the usual cup product. This result provided in turn an alternative method for deformation quantization of Poisson manifolds [45, 46, 69, 70].

The cohomology theory of E_n-algebras is called the higher Hochschild cohomology or E_n-Hochschild cohomology:

Definition 6.1 The (full) E_n-Hochschild complex of an E_n-algebra A is the derived hom $CH^*_{E_n}(A, A) = \mathbb{R}Hom^{E_n}_A(A, A)$ in the category of (operadic) A-modules over E_n. Given an ordinary associative (or E_1) algebra A, the category of (operadic) A-modules over E_1 is the category of A-bimodules, so one recovers the usual Hochschild cohomology. Moreover, the aforementioned Deligne conjecture generalizes to E_n-algebras:

Theorem 6.2 (see [32, Theorem 6.28] or [21, 51]) *The E_n-Hochschild complex $CH^*_{E_n}(A, A)$ of an E_n-algebra A forms an E_{n+1}-algebra.*

The endomorphisms $Hom_{biMod_A}(A, A)$ of A in the category $biMod_A$ of A-bimodules form nothing but the center $Z(A)$ of A. Deriving this hom object gives the Hochschild complex, and the Hochschild cohomology of A satisfies $HH^0(A, A) = Z(A)$. One says that the Hoschchild complex is the derived center of A, and the result above can then be reformulated as "the derived center of an E_n-algebra forms an E_{n+1}-algebra". This sentence has actually a precise meaning, because higher Hochschild cohomology can be alternately defined as a centralizer in the ∞-category of E_n-algebras. We refer the reader to [51] for more details about this construction. Associated to an E_n-algebra A, one also has its cotangent complex L_A, which classifies square-zero extensions of A [21, 51], and its dual the tangent complex $T_A := Hom^{E_n}_A(L_A, A) \cong \mathbb{R}Der(A, A)$.

Theorem 6.3 (see [21, 51]) *The shifted tangent complex $T_A[-n]$ of an E_n-algebra is an E_{n+1}-algebra, and is related to its E_n-Hochschild complex by a homotopy fiber sequence of E_{n+1}-algebras*

$$A[-n] \to T_A[-n] \to CH^*_{E_n}(A, A).$$

6.2   From Bialgebras to E_2-Algebras

A bialgebra is a complex equipped with an associative algebra structure and a coassociative coalgebra structure, such that the product is a coalgebra map and equivalently the coproduct is an algebra map. That is, bialgebras are equivalently algebras in coalgebras or coalgebras in algebras. Their cohomology theory is the Gerstenhaber-Schack cohomology [27], intertwinning Hochschild cohomology of algebras and co-Hochschild cohomology of coalgebras. Such structures naturally occur in algebraic topology (homology or cohomology of an H-space, for instance loop spaces), Lie theory (universal enveloping algebras, cohomology of Lie groups), representation theory (group rings, regular functions on algebraic groups, Tannaka-Krein duality), quantum field theory (renormalization Hopf algebras, AdS/CFT formalism)...Here we are going to focus on their prominent role in quantum group theory, [10, 11, 16, 17, 27, 56, 57]. As explained in Example 4, deformation quantization of Lie bialgebras produce quantum groups, whose categories of representations are particularly well behaved (modular tensor categories) and used to produce topological invariants via 3-TQFTs [64]. It turns out that bialgebras are deeply related to E_n-algebras, via the natural occurrence of E_n-structures in deformation quantization and representation theory of quantum groups for instance, leading people to investigate the relationship between these two kinds of structures to understand various related problems on both sides. A hope in particular was to establish some equivalence between their respective deformation theories, maybe even their homotopy theories. One of the first goals of [30] was to embody this long-standing hope in a precise mathematical incarnation. A first crucial step is to relate bialgebras to a "half-restricted" kind of E_2-coalgebras by the following equivalence of ∞-categories:

Theorem 6.4 ([30, Theorem 0.1])

(1) There exists a bar-cobar adjunction

$$\mathscr{B}_{E_1}^{enh} : E_1 - Alg^{0-con}(dgCog^{conil}) \rightleftarrows E_1 - Cog^{conil}(dgCog^{conil}) : \Omega_{E_1}^{enh}$$

inducing an equivalence of ∞-categories between nilpotent homotopy associative algebras in conilpotent dg coalgebras (0-connected conilpotent homotopy associative bialgebras) and conilpotent homotopy coassociative coalgebras in conilpotent dg coalgebras.

(2) The equivalence above induces an equivalence of $(\infty, 1)$-categories

$$E_1 - Alg^{aug,nil}(dgCog^{conil}) \rightleftarrows E_1 - Cog^{conil,pt}(dgCog^{conil})$$

between nilpotent augmented conilpotent homotopy associative bialgebras and pointed conilpotent homotopy coassociative coalgebras in conilpotent dg coalgebras.

In part (1), the notation $0 - con$ means 0-connected, that is dg bialgebras concentrated in positive degrees. In part (2), the notation aug, nil stands for augmented and nilpotent. A typical example of such a bialgebra is the total complex of the symmetric algebra over a cochain complex. The notation pt stands for pointed coalgebras, that is, a coalgebra C equipped with a counit $\epsilon : C \to \mathbb{K}$ and a coaugmentation $e : \mathbb{K} \to C$ such that $\epsilon \circ e = id_{\mathbb{K}}$. More generally, one can wonder, working in a given stable symmetric monoidal ∞-category (not necessarily cochain complexes), under which conditions a bar-cobar adjunction induces an equivalence of ∞-categories between algebras over an operad and conilpotent coalgebras over its bar construction [20]. Theorem 6.3 solves this conjecture of Francis-Gaitsgory [20] in the case where the base category is the category of conilpotent dg coalgebras, respectively the category of pointed conilpotent dg coalgebras.

Remark 6.5 This example is also interesting with respect to the conditions imposed on the ground symmetric monoidal ∞-category in [20], since the categories considered here are a priori not pronilpotent in the sense of [20, Definition 4.1.1].
Using Koszul duality of E_n-operads and an ∞-categorical version of Dunn's theorem [31, 51], we deduce from these equivalences the precise and long awaited relationship between homotopy theories of bialgebras and E_2-algebras. The correct answer to this problem needs an appropriate notion of "cobar construction for bialgebras", which intertwines a bar construction on the algebra part of the structure with a cobar construction on the resulting E_2-coalgebra:

Theorem 6.6 ([30, Corollary 0.2]) *The left adjoint of Theorem 0.1(2) induces a conservative fully faithful ∞-functor*

$$\tilde{\Omega} : E_1 - Alg^{aug,nil}(dgCog^{conil}) \hookrightarrow E_2 - Alg^{aug}$$

embedding augmented nilpotent and conilpotent homotopy associative bialgebras into augmented E_2-algebras.
By Theorem 5.12, this "immersion" of ∞-categories induces equivalences of formal moduli problems between the moduli problem of homotopy bialgebra structures on a bialgebra B and the moduli problem of E_2-algebra structures on its cobar construction $\tilde{\Omega}B$. Moreover, at the level of formal moduli problems controlling homotopy isotopies of B and homotopy isotopies of $\tilde{\Omega}B$, the tangent L_∞-algebras can be identified respectively with the shifted Gerstenhaber-Schack complex of B and the shifted (truncated) higher Hochschild complex (or E_2-tangent complex) of ΩB as L_∞-algebras:

Theorem 6.7 ([30, Theorem 0.6]) *Let B be a pointed conilpotent homotopy associative dg bialgebra. Let $\varphi : Bialg_\infty \rightarrow End_B$ be this homotopy bialgebra structure on B (where Bialg is the prop of associative-coassociative bialgebras) , and let $\tilde{\Omega}\varphi : E_2 \rightarrow End_{\tilde{\Omega}B}$ be the corresponding E_2-algebra structure on its cobar construction $\tilde{\Omega}B$.*

(1) There is a homotopy equivalence of formal moduli problems

$$\underline{Bialg_\infty\{B\}^\varphi} \simeq \underline{E_2\{\tilde{\Omega}B\}}^{\tilde{\Omega}\varphi}.$$

This homotopy equivalence induces a quasi-isomorphism of L_∞-algebras

$$g^\varphi_{Bialg,B} \xrightarrow{\sim} g^{\tilde{\Omega}\varphi}_{E_2,\tilde{\Omega}B}.$$

(2) There is a homotopy equivalence of formal moduli problems

$$\underline{Bialg_\infty^+}\{B\}^{\varphi^+} \simeq \underline{E_2^+}\{\tilde{\Omega}B\}^{\tilde{\Omega}\varphi^+}.$$

This homotopy equivalence induces a quasi-isomorphism of L_∞-algebras

$$C_{GS}^*(B,B)[2] \xrightarrow{\sim} T_{\tilde{\Omega}(B)}$$

between the shifted Gerstenhaber-Schack complex of B and the (truncated) E_2-Hochschild complex or E_2-tangent complex of $\tilde{\Omega}(B)$.

The L_∞ structure on the E_2-Hochschild complex $T_{\tilde{\Omega}(B)}$ is the one induced by the E_3 structure on $T_{\tilde{\Omega}(B)}[-2]$ (see Theorem 6.3). Proving that the higher Hochschild complex of the cobar construction of a bialgebra is a deformation complex of this bialgebra is important, since it allows to reduce questions of deformations of bialgebras to those of E_2-structures for which more tools are available.

6.3 Gerstenhaber-Schack Conjecture

At the beginning of the 1990s, Gerstenhaber and Schack enunciated (in a wrong way) a conjecture [27] characterizing the structure of the complex controlling the deformation theory of bialgebras, which remained quite mysterious for a while. It is a dg bialgebra version of the Deligne conjecture. In [27, Section 8], we extended the equivalences of Theorem 6.7 to an equivalence of homotopy fiber sequences of E_3-algebras, getting a much stronger version of the longstanding Gerstenhaber-Schack conjecture for the different versions of the Gerstenhaber-Schack and E_2-Hochschild complexes:

Theorem 6.8 (Generalized Gerstenhaber-Schack Conjecture [30, Corollary 0.7])

(1) There is an E_3-algebra structure on $C_{GS}^*(B,B)$ and a unital E_3-algebra structure on $C_{GS}^{full}(B,B)$) such that the following diagram

$$
\begin{array}{ccc}
\tilde{\Omega}B[-1] & \longrightarrow T_{\tilde{\Omega}(B)} & \longrightarrow CH_{E_2}^*(\tilde{\Omega}B, \tilde{\Omega}B) \\
\| & \simeq \uparrow & \simeq \uparrow \\
\tilde{\Omega}B[-1] & \longrightarrow C_{GS}^*(B,B) & \longrightarrow C_{GS}^{full}(B,B)
\end{array}
$$

is a commutative diagram of non-unital E_3-algebras with vertical arrows being equivalences.

(2) The E_3-algebra structure on $C_{GS}^*(B,B)$ is a refinement of its L_∞-algebra structure controlling the deformation theory of the bialgebra B.

Let us note that the upper fiber sequence of part (1) is the fiber sequence of Theorem 6.3. In particular, the E_3-algebra structure on the deformation complex of dg bialgebra B comes from the E_3-algebra structure on the E_2-Hochschild complex of $\tilde{\Omega}B$ given by the higher Deligne conjecture.

6.4 Kontsevich Formality Conjecture and Deformation Quantization of Lie Bialgebras

Let us first recall briefly how deformation quantization of Poisson manifolds works in Kontsevich's work [46]. We fix a finite dimensional Poisson manifold M, and we consider two complexes one can associate to such a manifold. First, the Hochschild complex $CH^*(\mathscr{C}^\infty(M), \mathscr{C}^\infty(M))$, second, the complex of polyvector fields $T_{poly}(M) = \left(\bigoplus_{k\geq 0} \bigwedge^k \Gamma T(M)[-k]\right)[1]$ where $\Gamma T(M)$ is the space of sections of the tangent bundle on M. The Poisson structure we fixed on M is the datum of a bivector $\Pi \in \bigwedge^2 \Gamma T(M)$ satisfying the Maurer-Cartan equation, that is, a Maurer-Cartan element of weight 2 in the Lie algebra of polyvector fields $T_{poly}(M)[1]$ (equipped with the Schouten-Nihenjuis bracket). To get the equivalent definition of Poisson manifold as a manifold whose ring of functions is a Poisson algebra, set $\{f, g\} = \Pi(df, dg)$. A well known theorem called the Hochschild-Kostant-Rosenberg theorem (HKR for short) states that the cohomology of $CH^*(\mathscr{C}^\infty(M), \mathscr{C}^\infty(M))$ is precisely $T_{poly}(M)$. In [46], Kontsevich proved that there exists a L_∞-quasi-isomorphism

$$T_{poly}(M)[1] \xrightarrow{\sim} CH^*(\mathscr{C}^\infty(M), \mathscr{C}^\infty(M))[1]$$

realizing in particular the isomorphism of the HKR theorem. We did not use the notion of L_∞-quasi-isomorphism before, let us just say briefly that it is a quasi-isomorphism of cdgas between the Chevalley-Eilenberg algebra of $T_{poly}(M)[1]$ and the Chevalley-Eilenberg algebra of $CH^*(\mathscr{C}^\infty(M), \mathscr{C}^\infty(M))[1]$. In particular, it is determined by an infinite collection of maps $T_{poly}(M)[1] \to \Lambda^k(CH^*(\mathscr{C}^\infty(M), \mathscr{C}^\infty(M))[1])$ for $k \in \mathbb{N}$, whose first map is the HKR quasi-isomorphism.

Remark 6.9 An L_∞-quasi-isomorphism of dg Lie algebras is actually equivalent to a chain of quasi-isomorphisms of dg Lie algebras.

This formality theorem then implies the deformation quantization of Poisson manifolds by the following arguments. First, noting $g[[\hbar]]_+ = \bigoplus \hbar g^n[[\hbar]]$, one proves that the Maurer-Cartan set $MC(T_{poly}(M)[[\hbar]]_+)$ is the set of Poisson algebra structures on $\mathscr{C}^\infty(M)[[\hbar]]$ and that the Maurer-Cartan set $MC(D_{poly}(M)[[\hbar]]_+)$ is the set of $*_\hbar$-products, which are associative products on $\mathscr{C}^\infty(M)[[\hbar]]$ of the form $a.b + B_1(a, b)t + \dots$ (i.e these products restrict to the usual commutative associative product on $\mathscr{C}^\infty(M)$). Second, an L_∞-quasi-isomorphism of nilpotent dg Lie algebras induces a bijection between the corresponding moduli sets of Maurer-Cartan elements, so there is a one-to-one correspondence between gauge equivalence classes of both sides. Consequently, isomorphism classes of formal Poisson structures on M are in bijection with equivalence classes of $*_\hbar$-products.

Kontsevich builds explicit formality morphisms in the affine case $M = \mathbb{R}^d$, with formulae involving integrals on compactification of configuration spaces and deeply related to the theory of multi-zeta functions. An alternative proof of the formality theorem for $M = \mathbb{R}^d$ due to Tamarkin [69], relies on the formality of E_2-operads (hence on the choice of a Drinfeld associator) and provides a formality quasi-isomorphism of homotopy Gerstenhaber algebras (that is E_2-algebras)

$$T_{poly}(\mathbb{R}^n) \xrightarrow{\sim} CH^*(\mathscr{C}^\infty(\mathbb{R}^n), \mathscr{C}^\infty(\mathbb{R}^n)).$$

His method works as follows:

- Prove the Deligne conjecture stating the existence of an E_2-algebra structure on the Hochschild complex;
- Transfer this structure along the HKR quasi-isomorphism to get an E_2-quasi-isomorphism between $CH^*(\mathscr{C}^\infty(\mathbb{R}^n), \mathscr{C}^\infty(\mathbb{R}^n))$ with its E_2-algebra structure coming from the Deligne conjecture, and $T_{poly}(\mathbb{R}^n)$.
- By the formality of E_2, this means that we have two E_2-structures on $T_{poly}(\mathbb{R}^n)$, the one coming from the Deligne conjecture and the one coming from the $Pois_2$-structure given by the wedge product and the Schouten-Nijenhuis bracket. One proves that $T_{poly}(M)$ has a unique homotopy class of E_2-algebra structures by checking that it is intrinsically formal (precisely, the $Aff(\mathbb{R}^n)$-equivariant $Pois_2$-cohomology of $T_{poly}(\mathbb{R}^n)$ is trivial).

This "local" formality for affine spaces is then globalized to the case of a general Poisson manifold by means of formal geometry [46, Section 7].

In the introduction of his celebrated work on deformation quantization of Poisson manifolds [46], Kontsevich conjectured that a similar picture should underline the deformation quantization of Lie bialgebras. Etingof-Kazhdan quantization (see [11, 16, 17]) should be the consequence of a deeper formality theorem for the deformation complex $Def(Sym(V))$ of the symmetric bialgebra $Sym(V)$ on a vector space V. This deformation complex should possess an E_3-algebra structure whose underlying L_∞-structure controls the deformations of $Sym(V)$, and should be formal as an E_3-algebra. Then this formality result should imply a one-to-one correspondence between gauge classes of Lie bialgebra structures on V and gauge classes of their quantizations. In [30], we solved this longstanding conjecture at a greater level of generality than the original statement, and deduced a generalization of Etingof-Kadhan's deformation quantization theorem. Here we consider not a vector space, but a $\mathbb{Z}$-graded cochain complex V whose cohomology is of finite dimension in each degree. By the results explained in Sect. 6.3, we know the existence of such an E_3-algebra structure (interestingly coming from the higher Hochschild complex of $\tilde{\Omega}Sym(V)$). It remains to prove the E_3-formality of $Def(Sym(V))$ by proving the homotopy equivalence between two E_3-structures on the Gerstenhaber-Schack cohomology of $Sym(V)$: the one transferred from $Def(Sym(V))$, and the canonical one coming from the action of $Pois_3$ (giving an E_3-structure via the formality $E_3 \xrightarrow{\sim} Pois_3$). Indeed, the cohomology of $Def(Sym(V))$ (which is precisely the Gerstenhaber-Schack complex of V) is explicitly computable, and given by

$$H^*_{GS}(Sym(V), Sym(V)) \cong \hat{Sym}(H^*V[-1] \oplus H^*V^\vee[-1])$$

where $\hat{Sym}$ is the completed symmetric algebra and $H^*V^\vee$ is the dual of H^*V as a graded vector space. This symmetric algebra has a canonical 3-Poisson algebra structure induced by the evaluation pairing between H^*V and $H^*V^\vee$. In the spirit of Tamarkin's method, we have to use obstruction theoretic methods to show that $Def(Sym(V))$ is rigid as an E_3-algebra. We thus get a generalization of Kontsevich's conjecture (originally formulated in the case where V is a vector space):

Theorem 6.10 *(Kontsevich formality conjecture [30, Theorem 0.8]) The deformation complex of the symmetric bialgebra $Sym(V)$ on a $\mathbb{Z}$-graded cochain complex V whose cohomology is of finite dimension in each degree is formal over $\mathbb{Q}$ as an E_3-algebra.*

We prove it by using in particular the relationship between Gerstenhaber-Schack cohomology and E_2-Hochschild cohomology and the higher HKR-theorem for the latter [3]. We then obtain a new proof of Etingof-Kazhdan quantization theorem from the underlying L_∞-formality given by our E_3-formality. Indeed, this formality induces an equivalence of the associated derived formal moduli problems, in particular we have an equivalence of Maurer-Cartan moduli sets (suitably extended over formal power series in one variable). On the right hand side, the Maurer-Cartan moduli set is identified with equivalence classes of homotopy Lie bialgebra structures on the cochain complex $V[[\hbar]]$. On the right hand side, it is identified with deformation quantization of these Lie bialgebras (formal deformations of $Sym(V)$ as a homotopy dg bialgebra). Moreover, what we get is actually a generalization of Etingof-Kazhdan quantization to homotopy dg Lie bialgebras:

Corollary 6.11 *([30, Corollary 0.9]) The L_∞-formality underlying Theorem 5.9 induces a generalization of Etingof-Kazdhan deformation quantization theorem to homotopy dg Lie bialgebras whose cohomology is of finite dimension in each degree. In the case where V is a vector space, this gives a new proof of Etingof-Kazdhan's theorem.*

This result encompasses the case of usual Lie bialgebras, because if V is concentrated in degree 0, then homotopy Lie bialgebra structures on V are exactly Lie bialgebra structures on V.

Remark 6.12 Actually, what we prove in [30] is even stronger. We get a sequence of E_3-formality morphisms for the three variants of the Gerstenhaber-Shack complex [30, Theorem 7.2], indicating that important variants of deformation quantization like [15] should also follow from such E_3-formality morphisms.

Acknowledgements The idea of writing such a survey originates in the inaugural 2-week program at the mathematical research institute MATRIX in Australia called Higher Structures in Geometry and Physics, which took place in June 2016. The author gave a talk at this program about moduli spaces of algebraic structures and their application to the recent paper [30]. The present article is somehow a (largely) extended version of his talk, which will be eventually part of a Proceedings Volume devoted to this workshop. The author would like to thank the MATRIX institute for supporting this program, the organizers of this programme for inviting him, and all the participants for their interest and for the very enjoyable atmosphere during the 2 weeks spent there. Last but not least, kangaroos are very much thanked for their natural awesomeness.

References

1. Anel, M.: Champs de modules des catégories linéaires et abéliennes, PhD thesis, Université Toulouse III - Paul Sabatier (2006)
2. Blanc, D., Dwyer, W.G., Goerss, P.G.: The realization space of a Π-algebra: a moduli problem in algebraic topology. Topology **43**, 857–892 (2004)
3. Calaque, D., Willwacher, T.: Triviality of the higher Formality Theorem. Proc. Am. Math. Soc. **143**(12), 5181–5193 (2015)
4. Calaque, D., Pantev, T., Toen, B., Vaquié, M., Vezzosi, G.: Shifted Poisson structures and deformation quantization, J. Topology **10**, 483–484 (2017)
5. Campos, R., Merkulov, S., Willwacher, T.: The Frobenius properad is Koszul. Duke Math. J. **165**(15), 2921–2989 (2016)
6. Chas, M., Sullivan, D.: Closed string operators in topology leading to Lie bialgebras and higher string algebra. In: The Legacy of Niels Henrik Abel, pp. 771–784. Springer, Berlin (2004)
7. Chas, M., Sullivan, D.: String Topology (1999). Preprint. arXiv:math/9911159
8. Cieliebak, K., Latschev, J.: The role of string topology in symplectic field theory. In: New Perspectives and Challenges in Symplectic Field Theory. CRM Proceedings of Lecture Notes, vol. 49, pp. 113–146. American Mathematical Society, Providence, RI (2009)
9. Cieliebak, K., Fukaya, K., Latschev, J.: Homological algebra related to surfaces with boundary (2015). Preprint. arXiv:1508.02741
10. Drinfeld, V.G.: Quantum groups. In: Proceedings of the International Congress of Mathematicians, vols. 1, 2, Berkeley, CA, 1986, pp. 798–820. American Mathematical Society, Providence, RI (1987)
11. Drinfeld, V.G.: On quasitriangular quasi-Hopf algebras and on a group that is closely connected with $Gal(\overline{Q}/Q)$. Algebra i Analiz **2**, 149–181 (1990). English translation in Leningrad Math. J. **2** (1991), 829–860
12. Dwyer, W.G., Kan, D.: Function complexes in homotopical algebra. Topology **19**, 427–440 (1980)
13. Dwyer, W.G., Kan, D.: A classification theorem for diagrams of simplicial sets. Topology **23**, 139–155 (1984)
14. Enriquez, B., Etingof, P.: On the invertibility of quantization functors. J. Algebra **289**(2), 321–345 (2005)
15. Enriquez, B., Halbout, G.: Quantization of quasi-Lie bialgebras. J. Am. Math. Soc. **23**, 611–653 (2010)
16. Etingof, P., Kazhdan, D.: Quantization of Lie bialgebras I. Sel. Math. (N. S.) **2**(1), 1–41 (1996)
17. Etingof, P., Kazhdan, D.: Quantization of Lie bialgebras. II, III. Sel. Math. (N.S.) **4**(2), 213–231, 233–269 (1998)
18. Félix, Y., Thomas, J.-C.: Rational BV-algebra in string topology. Bull. Soc. Math. France **136**, 311–327 (2008)

19. Fjelstad, J., Fuchs, J., Runkel, I., Schweigert, C.: Topological and conformal field theory as Frobenius algebras. In: Categories in Algebra, Geometry and Mathematical Physics. Contemporary Mathematics, vol. 431, pp. 225–247. American Mathematical Society, Providence, RI (2007)

20. Francis, J., Gaitsgory, D.: Chiral Koszul duality. Sel. Math. New Ser. **18**, 27–87 (2012)

21. Francis, J.: The tangent complex and Hochschild cohomology of E_n-rings. Compos. Math. **149**(3), 430–480 (2013)

22. Fresse, B.: Modules Over Operads and Functors. Lecture Notes in Mathematics 1967. Springer, Berlin (2009)

23. Fresse, B.: Props in model categories and homotopy invariance of structures. Georgian Math. J. **17**, 79–160 (2010)

24. Fresse, B.: Homotopy of operads and Grothendieck-Teichmuller groups I and II. Mathematical Surveys and Monographs (to appear). AMS. http://math.univ-lille1.fr/~fresse/OperadHomotopyBook/

25. Fresse, B., Willwacher, T.: The intrinsic formality of E_n-operads (2015). Preprint. arXiv:1503.08699

26. Galvez-Carillo, I., Tonks, A., Vallette, B.: Homotopy Batalin-Vilkovisky algebras. J. Noncommut. Geom. **6**, 539–602 (2012)

27. Gerstenhaber, M., Schack, S.D.: Algebras, bialgebras, quantum groups, and algebraic deformations. In: Deformation Theory and Quantum Groups with Applications to Mathematical Physics (Amherst, MA, 1990). Contemporary Mathematics, vol. 134, pp. 51–92. American Mathematical Society, Providence, RI (1992)

28. Getzler, E., Jones, J.D.S.: Operads, homotopy algebra and iterated integrals for double loop spaces (1994). Preprint. arXiv:hep-th/9403055

29. Ginot, G., Halbout, G.: A formality theorem for poisson manifolds. Lett. Math. Phys. **66**, 37–64 (2003)

30. Ginot, G., Yalin, S.: Deformation theory of bialgebras, higher Hochschild cohomology and formality (2016). Preprint. arXiv:1606.01504

31. Ginot, G., Tradler, T., Zeinalian, M.: Higher hochschild homology, topological chiral homology and factorization algebras. Commun. Math. Phys. **326**(3), 635–686 (2014)

32. Ginot, G., Tradler, T., Zeinalian, M.: Higher Hochschild cohomology of E-infinity algebras, Brane topology and centralizers of E-n algebra maps (2014). Preprint. arXiv:1205.7056

33. Goerss, P.G., Hopkins, M.J.: Moduli spaces of commutative ring spectra. In: Structured Ring Spectra. London Mathematical Society Lecture Note Series, vol. 315, pp. 151–200. Cambridge University Press, Cambridge (2004)

34. Goldman, W.M.: The symplectic nature of fundamental groups of surfaces. Adv. Math. **54**, 200–225 (1984)

35. Goldman, W.M.: Invariant functions on Lie groups and Hamiltonian flows of surface group representations. Invent. Math. **85**(2), 263–302 (1986)

36. Goldman, W.M., Millson, J.: The deformation theory of representations of fundamental groups of compact Kähler manifolds. Publ. Math. IHES **67**, 43–96 (1988)

37. Hennion, B.: Tangent Lie algebra of derived Artin stacks, to appear in J. für die reine und angewandte Math., published online (2015). DOI 10.1515/crelle-2015-0065

38. Hinich, V.: DG coalgebras as formal stacks. J. Pure Appl. Algebra **162**, 209–250 (2001)

39. Hinich, V.: Tamarkin's proof of Kontsevich formality theorem. Forum Math. **15**, 591–614 (2003)

40. Hirschhorn, P.S.: Model Categories and Their Localizations. Mathematical Surveys and Monographs, vol. 99. American Mathematical Society, Providence, RI (2003)

41. Hovey, M.: Model Categories. Mathematical Surveys and Monographs, vol. 63. American Mathematical Society, Providence, RI (1999)

42. Kapustin, A.: Topological field theory, higher categories, and their applications. In: Proceedings of the International Congress of Mathematicians, vol. III, pp. 2021–2043. Hindustan Book Agency, New Delhi (2010)

43. Keller, B., Lowen, W.: On Hochschild cohomology and Morita deformations. Int. Math. Res. Not. IMRN **2009**(17), 3221–3235 (2009)
44. Kock, J.: Frobenius Algebras and 2D Topological Quantum Field Theories. London Mathematical Society Student Texts, vol. 59. Cambridge University Press, Cambridge (2003)
45. Kontsevich, M.: Operads and motives in deformation quantization, Moshé Flato (1937–1998). Lett. Math. Phys. **48**(1), 35–72 (1999)
46. Kontsevich, M.: Deformation quantization of Poisson manifolds. Lett. Math. Phys. **66**(3), 157–216 (2003)
47. Lambrechts, P., Volic, I.: Formality of the little N-disks operad. Mem. Am. Math. Soc. vol. 230, no. 1079. American Mathematical Society, Providence, RI (2014)
48. Lauda, A.D., Pfeiffer, H.: Open-closed strings: two-dimensional extended TQFTs and Frobenius algebras. Topol. Appl. **155**, 623–666 (2008)
49. Loday, J-L., Vallette, B.: Algebraic Operads. Grundlehren der mathematischen Wissenschaften, vol. 346. Springer, Berlin (2012)
50. Lurie, J.: Derived Algebraic Geometry X (2011). http://www.math.harvard.edu/~lurie/
51. Lurie, J.: Higher Algebra (2017). http://www.math.harvard.edu/~lurie/
52. MacLane, S.: Categorical algebra. Bull. Am. Math. Soc. **71**, 40–106 (1965)
53. Mandell, M.A.: Cochains and homotopy type. Publ. Math. IHES **103**, 213–246 (2006)
54. Markl, M.: Intrinsic brackets and the L-infinity deformation theory of bialgebras. J. Homotopy Relat. Struct. **5**(1), 177–212 (2010)
55. May, J.P.: The Geometry of Iterated Loop Spaces. Springer, Berlin (1972)
56. Merkulov, S.: Prop profile of Poisson geometry. Commun. Math. Phys. **262**, 117–135 (2006)
57. Merkulov, S.: Formality theorem for quantization of Lie bialgebras. Lett. Math. Phys. **106**(2), 169–195 (2016)
58. Merkulov, S., Vallette, B.: Deformation theory of representation of prop(erad)s I. J. für die reine und angewandte Math. (Crelles Journal) **634**, 51–106 (2009)
59. Merkulov, S., Vallette, B.: Deformation theory of representation of prop(erad)s II. J. für die reine und angewandte Math. (Crelles Journal) **636**, 125–174 (2009)
60. Pantev, T., Toen, B., Vaquié, M., Vezzosi, G.: Shifted symplectic structures. Inst. Hautes Études Sci. Publ. Math. **117**, 271–328 (2013)
61. Preygel, A.: Thom-Sebastiani and Duality for Matrix Factorizations, and Results on the Higher Structures of the Hochschild Invariants. Thesis (Ph.D.), M.I.T. (2012)
62. Pridham, J.P.: Unifying derived deformation theories. Adv. Math. **224**, 772–826 (2010)
63. Pridham, J.P.: Presenting higher stacks as simplicial schemes. Adv. Math. **238**, 184–245 (2013)
64. Reshetikhin, N., Turaev, V.G.: Invariants of 3-manifolds via link polynomials and quantum groups. Invent. Math. **103**(3), 547–597 (1991)
65. Rezk, C.W.: Spaces of algebra structures and cohomology of operads, Thesis, MIT (1996)
66. Shoikhet, B.: Tetramodules over a bialgebra form a 2-fold monoidal category. Appl. Categ. Struct. **21**(3), 291–309 (2013)
67. Simpson, C.: Moduli of representations of the fundamental group of a smooth projective variety. Publ. Math. IHES **80**, 5–79 (1994)
68. Sullivan, D.: Infinitesimal computations in topology. Inst. Hautes Études Sci. Publ. Math. **47**, 269–331 (1977)
69. Tamarkin, D.: Another proof of M. Kontsevich formality theorem (1998). Preprint. arXiv:math/9803025
70. Tamarkin, D.: Formality of chain operad of little discs. Lett. Math. Phys. **66**, 65–72 (2003)
71. Tamarkin, D.: Deformation complex of a d-algebra is a (d+1)-algebra (2000). Preprint. arXiv:math/0010072
72. Toën, B.: Derived algebraic geometry. EMS Surv. Math. Sci. **1**(2), 153–240 (2014)
73. Toën, B.: Derived Algebraic Geometry and Deformation Quantization. ICM Lecture (2014). arXiv:1403.6995v4
74. Toën, B., Vezzosi, G.: Homotopical algebraic geometry II. Geometric stacks and applications. Mem. Am. Math. Soc. **193**(902), x+224 (2008)

75. Turaev, V.G.: Skein quantization of Poisson algebras of loops on surfaces. Ann. Sci. École Norm. Sup. **24**(6), 635–704 (1991)
76. Vallette, B.: A Koszul duality for props. Trans. Am. Math. Soc. **359**, 4865–4943 (2007)
77. Witten, E.: Quantum field theory and the Jones polynomial. Commun. Math. Phys. **121**, 351–399 (1989)
78. Yalin, S.: Classifying spaces of algebras over a prop. Algebr. Geom. Topol. **14**(5), 2561–2593 (2014)
79. Yalin, S.: Simplicial localization of homotopy algebras over a prop. Math. Proc. Cambridge Philos. Soc. **157**(3), 457–468 (2014)
80. Yalin, S.: Maurer-Cartan spaces of filtered L_∞ algebras. J. Homotopy Relat. Struct. **11**, 375–407 (2016)
81. Yalin, S.: Moduli stacks of algebraic structures and deformation theory. J. Noncommut. Geom. **10**, 579–661 (2016)
82. Yalin, S.: Function spaces and classifying spaces of algebras over a prop. Algebr. Geom. Topology **16**, 2715–2749 (2016)
83. Yalin, S.: Realization spaces of algebraic structures on chains. Int. Math. Res. Not. **2018**, 236–291 (2018)

Embedding Calculus and the Little Discs Operads

Victor Turchin

Abstract This note describes recent development in the study of embedding spaces from the manifold calculus viewpoint. An important progress that has been done was the discovery and application of the connection to the theory of operads. This allows one to describe embedding spaces as certain derived operadic module maps and to produce their explicit deloopings.

1 Manifold Functor Calculus, Little Discs Operads, Embedding Spaces

Manifold calculus appeared as a tool to study spaces of embeddings between manifolds [12, 26]. This is also a very nice application of the operad theory. The main operad that appears is the little disks operad. The calculus itself was invented by Goodwillie and Weiss.

Assume that we have a smooth manifold M. We can consider the category $\mathcal{O}(M)$ of open subsets of M, and then we can look at the functors $\mathcal{O}(M) \to$ Top in both the covariant and the contravariant case. The functors are supposed to be isotopy invariant, so that the functor should send isotopy equivalences to homotopy equivalences. The functor calculus provides a sequence of polynomial approximations. In the covariant case, we have a tower $T_0F \to T_1F \to T_2F \to \cdots$, all of which come with a map to F. The T_kF is the kth polynomial approximation. For the contravariant case all the arrows go in the opposite direction, $T_0F \leftarrow T_1F \leftarrow \cdots$.

There is a version of this calculus which is so-called "context-free." Consider the category Man_m of all smooth manifolds of dimension m. The morphisms are codimension 0 embeddings. Then we similarly study functors $\mathrm{Man}_m \to$ Top.

V. Turchin (✉)
Math Department, Kansas State University, Manhattan, KS 66506, USA
e-mail: turchin@ksu.edu

© Springer International Publishing AG, part of Springer Nature 2018   489
D.R. Wood et al. (eds.), *2016 MATRIX Annals*, MATRIX Book Series 1,
https://doi.org/10.1007/978-3-319-72299-3_21

Definition 1 A covariant functor $F : \mathrm{Man}_n \to \mathrm{Top}$ is *polynomial of degree k* if for any manifold M and for any collection of closed and pairwise disjoint subsets $A_0, \ldots, A_k$, the cubical diagram assigning to $S \subset \{0 \ldots k\}$ the space

$$S \mapsto F(M \setminus \bigcup_{i \in S} A_i),$$

is homotopy cocartesian.

As example, in case $k = 2$ we get the cube

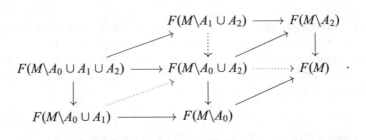

One of the main properties of being polynomial is that one can build the value of the functor on M out of its value on smaller pieces.

Here are a few examples. The functor $M \mapsto M^{\times k}$ is polynomial of degree k. If you take the functor $M \mapsto M^{\times 2}$, this is not linear. Indeed, for the diagram

$$(M \backslash A_0 \cup A_1)^2 \longrightarrow (M \backslash A_0)^2$$
$$\downarrow$$
$$(M \backslash A_1)^2$$

the colimit will be $(M \backslash A_0)^2 \cup (M \backslash A_1)^2$, but this is not M^2. But if you do this in the three dimensional cube, then $M^2 = (M \backslash A_0)^2 \cup (M \backslash A_1)^2 \cup (M \backslash A_2)^2$, which shows that the functor is indeed quadratic.

For the functor $F(M) = M^{\times k}$ we can actually describe explicitly the kth polynomial approximation. In this case $T_i F(M) = \{(x_1, \ldots, x_k) \in M^{\times k} | \#\{x_1, \ldots, x_k\} \leq i\}$, where # denotes the cardinal of a set. So this functor is not homogeneous.

As another example, can take $M \mapsto M^{\times k}/\Sigma_k$, this is polynomial of degree k. Or $\binom{M}{k}$, the unlabeled configuration spaces of k points, this is also polynomial of degree k. Or you could take the spherical tangent bundle of M, or $M \mapsto M \times A$, these functors are linear.

For the contravariant functors the definition is dual: similar cubes must be homotopy cartesian instead of cocartesian. Linear examples would be $M \mapsto \mathrm{Maps}(M, A)$ or $M \mapsto \Gamma(p)$ where p is a functorial bundle $E_M \to M$. So the first example is a trivial example of the second.

Another example would be immersions of M in some larger dimension space N, because this is equivalent to sections of a certain fiber bundle, formal immersions,

$\Gamma(p, M)$. Here to any manifold M we assign a fibration $p: E_M \to M$, where E_M is the space of triples

$$(m, n, \alpha : T_m M \to T_n N)$$

with α a monomorphism. Smale proved his famous immersion theorem, that $\mathrm{Imm}(S^2, \mathbb{R}^3)$ is connected, which follows from seeing the sphere as the union of disks and then seeing that the obtained square is homotopy cartesian.

As another example of a degree k functor, one has $M \mapsto \mathrm{Maps}(M^{\times k}, A)$ (we could also ask for Σ_k-equivariance if A is acted on by Σ_k).

The good news is that there is a theorem by Goodwillie and Klein saying that the map $\mathrm{Emb}(M, N) \to T_k \mathrm{Emb}(M, N)$ is $(1 - m + k(n - m - 2))$-connected, provided $n - m > 2$ [11]. In other words, the Taylor tower becomes closer and closer to the initial space of embeddings.

Now let us recall the operadic interpretation appearing in the context free setting [2, 23]. Consider the full subcategory $\mathrm{Disc}_{\leq k} \subset \mathrm{Man}_m$ of manifolds with objects disjoint unions of up to k disks. Then according to Weiss, the kth Taylor approximation is described as follows

$$T_k F(M) = \underset{\mathrm{Disc}_{\leq k} \downarrow M}{\mathrm{holim}} F.$$

In other words, it is the homotopy right Kan extension

$$\mathrm{Disc}_{\leq k} \xrightarrow{F \circ i} \mathrm{Top}$$
$$\downarrow i \qquad \nearrow$$
$$\mathrm{Man}_m \qquad \mathrm{hRan}$$

The category Man_m is monoidal and enriched in topological spaces. Thus one can consider the topological operad $\mathrm{End}(D^m)$ of endomorphisms of D^m. Its kth component is the space of embeddings of a disjoint union of k disks into a disk. In the little disks operad, the embeddings should be just translation and scaling. Here we allow all transformations. One can easily see that this operad is equivalent to the framed discs operad $B_m^{fr}(k)$.

Theorem 1 (Boavida de Brito–Weiss [2], T. [23])

$$T_k F(M) = \mathrm{hRmod}_{\mathrm{End}(D^m)}^{\leq k}(\mathrm{Emb}(-, M), F(-))$$

So if a functor is contravariant, the sequence $\{F(\mathbf{1}), F(X), F(X^{\otimes 2}), F(X^{\otimes 3}), \ldots\}$ becomes a right module over $\mathrm{End}(X)$, that abusing notation we denote by $F(-)$. As a particular example, $\mathrm{Emb}(-, M)$ is also a right module. In the above formula we look at the space of derived maps of truncated up to arity k right modules. For $k = \infty$ we get a formula similar to factorization homology.

Just as a remark, if we look at the initial definition, one gets [23]

$$\operatorname*{holim}_{\mathrm{Disc}_{\leq k}\downarrow M} F \cong \mathrm{hRmod}^{\leq k}_{\mathrm{End}(D^m)^\delta}(\mathrm{Emb}(-, M)^\delta, F(-)),$$

where δ means "with the discrete topology". Thus Theorem 1 was to understand the continuous version of the same result.

One can also consider functors from manifolds to chain complexes. In this case one also gets the enriched version:

Theorem 2 (Boavida de Brito–Weiss [2])

$$T_k F(M) = \mathrm{hRmod}^{\leq k}_{C_* B^{\mathrm{fr}}_m}(C_*(\mathrm{Emb}(-, M)), F(-)).$$

An interesting space of embeddings is the space $\mathrm{Emb}(S^m, S^n)$, and assuming $n-m \geq 2$ this has the same π_0 as $\mathrm{Emb}_\partial(D^m, D^n)$. So it would be interesting to study the space of embeddings of disks and the calculus of the closed disk in general.

The functor calculus in the closed case works similarly, we should just change the category Disc to $\widetilde{\mathrm{Disc}}$, whose objects are disjoint unions of discs and one *anti-disc* $S^{m-1} \times [0, 1)$. Using this idea together with Arone, we showed that the Taylor tower on a closed disc can be expressed in terms of maps of truncated infinitesimal bimodules. Notice that here we use the usual (non-framed) operad of little discs. Informally speaking we can do so because the disc is parallelizable.

Theorem 3 (Arone–T. [1])

$$T_k F(D^m) \cong \mathrm{hInfBim}^{\leq k}_{B_m}(B_m, F(-)).$$

So what are infinitesimal bimodules over an operad? We have so-called infinitesimal left action. The structure is Abelian, you can only insert in one input. The right action is just usual, since the right action is also unital, we can insert only in one of the inputs. For more details, see [1].

Given a functor F on the category $\widetilde{\mathrm{Disc}}$, we get an infinitesimal bimodule $F(-)$ whose kth component is $F\left((S^{m-1} \times [0, 1)) \sqcup \coprod_k D^m\right)$. The left action comes from the embeddings of discs in the collar component $S^{m-1} \times [0, 1)$. Now the inclusion of operads $B_m \to B_n$ induces an infinitesimal B_m-bimodule structure on the target.

As a corollary, we get the following.

Corollary 1 (Arone–T. [1])

$$T_k \overline{\mathrm{Emb}}_\partial(D^m, D^n) \cong \mathrm{hInfBim}^{\leq k}_{B_m}(B_m, B_n).$$

Here $\overline{\mathrm{Emb}}_\partial$ is the homotopy fiber of $\mathrm{Emb}_\partial(D^m, D^n) \rightarrow \mathrm{Imm}_\partial(D^m, D^n) \cong \Omega^m V_m(\mathbb{R}^n)$.

The right-hand side of the equation above is the derived mapping space between B_m and B_n in the category of infinitesimal bimodules over B_m. Now the question is, what about the derived mapping space between these objects in the category of operads? We can also look at the truncated case, where we look at the category of truncated operads with no more than k inputs. We can study this algebraic structure.

Theorem 4 (Dwyer–Hess [7], Boavida de Brito–Weiss [3], Ducoulombier–T. [5])

$$T_k \overline{\text{Emb}_\partial}(D^m, D^n) \cong \Omega^{m+1} \text{hOper}_{\leq k}(B_m, B_n)$$

The second talk/section is devoted to different proofs of this result. One should mention that only the second one (by Boavida de Brito and Weiss) appeared already as a preprint.

For Dwyer–Hess, they proved it first for $m = 1$ [6]. They don't consider the case of truncation, i.e. they only look at the case $k = \infty$. Boavida and Weiss understand the truncated case, and we (Ducoulombier and I) also do the truncated case. However, our approaches are very different. They don't use our theorem from above, but Dwyer–Hess and Ducoulombier and I, we do use it. This really becomes a theory of operads, not calculus.

The rational homology and homotopy groups can be computed for the embedding spaces. The main reason that things work nicely is the relative formality of the little disks operad.

Theorem 5 (Tamarkin [22], Kontsevich [15], Lambrecht–Volić [16], T.– Willwacher [25], Fresse–Willwacher [9]) *The map of operads $C_* B_m \to C_* B_n$ of singular chains is rationally formal if and only if $n - m \neq 1$.*

So what does the statement mean? The claim is that we can find a zigzag of equivalences of maps of operads from the morphism $C_* B_m \to C_* B_n$ to the induced map $H_* B_m \to H_* B_n$. An equivalence is a commutative square, which in every degree for both source and target, induces an isomorphism on homology.

What is the homology of the little disks operad? This is a theorem of Fred Cohen, it's either the associative operad when $m = 1$ or it's the Poisson operad (with bracket of degree $m - 1$) for $m \geq 2$. What is $B_m(2)$? It's a configuration space of two disks and is homotopy equivalent to an $(m-1)$-sphere. The degree 0 class gives the product and the degree $m - 1$ class gives the bracket of the Poisson structure, which disappears when you map to $B_n(2) \cong S^{n-1}$, $n > m$.

The formality theorem together with the operadic approach to the manifold calculus outlined above allows one to compute the rational homology and homotopy groups of embedding spaces. Recall the categories Fin of finite sets and Fin_* of pointed finite sets. It is easy to see that a contravariant functor from Fin is the same thing as the right module over the commutative operad Com; and a contravariant functor from Fin_* is the same thing as an infinitesimal bimodule over Com. Thus in particular for $n \geq 2$, $H_* B_n$ is a right and infinitesimal bimodule over $H_0 B_n = \text{Com}$ and can be viewed as both Fin and Fin_* module.

Given a topological space (respectively, pointed space) X and a cofunctor L from Fin (respectively Fin$_*$) to chain complexes, Pirashvili defines the higher order Hochschild homology $HH^X(L)$ [18]. In the operadic language $HH^X(L) = H(\mathrm{hRmod}_{\mathrm{Com}}(C_*(X^{\times \bullet}), L))$ (respectively $HH^X(L) = H(\mathrm{hInfBim}_{\mathrm{Com}}(C_*(X^{\times \bullet}), L)))$. For a smooth m-manifold M let $\overline{\mathrm{Emb}}(M, \mathbb{R}^n)$ denote similarly the homotopy fiber of $\mathrm{Emb}(M, \mathbb{R}^n) \hookrightarrow \mathrm{Imm}(M, \mathbb{R}^n)$.

Theorem 6 (Arone-T. [1]) *Let $n \geq 2m + 2$ and let M be a smooth m-manifold. Then*

$$H_*(\overline{\mathrm{Emb}}(M, \mathbb{R}^n), \mathbb{Q}) \simeq HH^M(H_* B_n),$$

(this is the non-pointed version of higher Hochschild homology);

$$H_*(\overline{\mathrm{Emb}}_\partial(D^m, D^n), \mathbb{Q}) = HH^{S^m}(H_* B_n)$$

(here and below is the pointed version) and

$$\pi_*(\overline{\mathrm{Emb}}_\partial(D^m, D^n)) \otimes \mathbb{Q} = HH^{S^m}(\pi_* B_n \otimes \mathbb{Q}).$$

Together with G. Arone we describe $HH^{S^m}(\pi_* B_n \otimes \mathbb{Q})$ as the homology of a graph-complex obtained as the invariant space of the modular closure of the L_∞ operad.

In the recent work of Fresse–T.–Willwacher [10] using the delooping result Theorem 4, we improve the last statement of the theorem above to the range $n - m > 2$, i.e. the whole range in which the manifold calculus works. Another crucial point that we use is the strong Hopf statement of the relative little discs formality: the map of operads $C_* B_m \to C_* B_n$ is formal in the category of Hopf operads—operads in coalgebras (over $\mathbb{Q}$). In particular, in our graph-complex we can see a cycle which corresponds to the Haefliger trefoil [13, 14] appearing when $m = 4k-1, n = 6k$. This knot is the only one in codimension > 2, which is trivial as immersion and has infinite order. So it's known that $\pi_0(\mathrm{Emb}(S^m, S^n))$ is an Abelian group for $n - m > 2$ of rank at most one. This is a generator which is not torsion.

The result that we obtained in [10] is in fact deeper than mere computations of the rational homotopy groups. We showed the theorem

Theorem 7 (Fresse–T.–Willwacher [10]) *For $n - m \geq 3$ (respectively $n - m \geq 2$), hOper(B_m, B_n) (respectively hOper$_{\leq k}(B_m, B_n)$) is $n - m - 1$-connected and its rational homotopy type is described by the L_∞ algebra of homotopy biderivations of the map $H_*(B_m) \to H_*(B_n)$ (respectively, truncated to $\leq k$).*

Essentially all the rational information is encoded by this homology map $H_* B_m \to H_* B_n$ of Hopf operads. These are maps of (truncated) Hopf operads, so we need cofibrant and fibrant replacements for these guys. Hopf cofibrant essentially means cofibrant in chain complexes; Hopf fibrant means all components of the operad are fibrant coalgebras. Then we look at maps which are derivations of both structures: operadic composition and levelwise for the coalgebra structure. At the limit when $k \to \infty$, we need codimension three. The problem is that the maps between stages in the tower don't become higher and higher connected when codimension is 2, but the projective limit of groups doesn't commute with tensoring with rational numbers.

2 Delooping Results

The goal of this section/talk is to give insight into different proofs of Theorem 4. Let me reiterate that only the proof of Boavida de Brito–Weiss [3] already appeared.[1] Their approach will be explained at the very end. Both Dwyer-Hess [7] and Ducoulombier-T. [5] use Corollary 1 in their proof. In fact we prove a purely operadic statement Theorem 9 below, which together with Corollary 1 implies Theorem 4.

Before going any further let us consider the special case $m = 1$, for which the result described by Corollary 1 is really due to Dev Sinha [21]. Indeed, B_1 is naturally equivalent to the associative operad Ass. In fact, B_n is equivalent to a certain operad K_n, called *Kontsevich operad*,[2] and we have a zigzag of equivalences of operad maps

$$
\begin{array}{ccccc}
B_1 & \xleftarrow{\ \cong\ } & W_1 & \xrightarrow{\ \cong\ } & \text{Ass} \\
\downarrow & & \downarrow & & \downarrow \\
B_n & \xleftarrow{\ \cong\ } & W_n & \xrightarrow{\ \cong\ } & K_n.
\end{array}
$$

An infinitesimal bimodule over Ass is a cosimplicial object, and $\text{hInfBim}_{\text{Ass}}(\text{Ass}, K_n) = \text{hTot}\, K_n(\bullet)$. Thus we recover Sinha's theorem:

$$
T_k\overline{\text{Emb}}_\partial(D^1, D^n) \cong \text{hTot}_k\, K_n(\bullet), \quad k \le \infty.
$$

For $m = 1$, Theorem 4 was first proved by Dwyer and Hess [6] and then I gave a different proof [24]. It was obtained as a combination of Sinha's theorem and the following result. Given a map of operads Ass $\to \mathcal{O}$, the sequence $\mathcal{O}(\bullet)$ becomes a cosimplicial object.[3] Moreover, provided $\mathcal{O}(0) \cong \mathcal{O}(1) \cong *$,

$$
\text{hTot}\, \mathcal{O}(\bullet) \cong \Omega^2\, \text{hOper}(\text{Ass}, \mathcal{O}). \tag{1}
$$

All this business is actually related to Deligne's Hochschild cohomology conjecture (now theorem), that on the Hochschild complex of an associative algebra one gets an action of the operad of chains on little squares. Afterward, McClure and Smith generalized this to the topological setting showing that for any multiplicative operad

---

[1] By the time of the latest revision, the proof of Ducoulombier and Turchin appeared in [5].

[2] This operad was invented by D. Sinha.

[3] In this case $\mathcal{O}$ is called a multiplicative operad.

$\mathscr{O}$, the space hTot $\mathscr{O}(\bullet)$ admits an explicit B_2-action [17]. The result (1) is an explicit delooping (conjecturally to this action).

Let me give a brief sketch of ideas of Dwyer and Hess' proof for the case $m = 1$. They prove a theorem

Theorem 8 (Dwyer–Hess [6]) *Let $M_1 \to M_2$ be a morphism of monoids in a monoidal model category with unit $\mathbf{1}$ and satisfying natural axioms. (Thus M_2 gets an induced structure of an M_1-bimodule.) Then, provided the mapping space from $\mathbf{1}$ to M_2 is contractible, we get an equivalence of spaces*

$$\mathrm{hBim}_{M_1}(M_1, M_2) \cong \Omega\,\mathrm{hMon}(M_1, M_2).$$

The two spaces above are derived mapping spaces respectively of bimodules and monoids.

So how does this help to prove (1)? We can consider a map of operads $P \to Q$. Operads are monoids in the category of symmetric sequences with respect to the $\circ$ product. (Dwyer and Hess considered non-symmetric operads, then we have just sequences of spaces.) So Q becomes a bimodule over P, and then $\mathrm{hBim}_P(P, Q) \cong \Omega\,\mathrm{hOper}(P, Q)$, provided $Q(1) \cong *$.[4] We take $P = \mathrm{Ass}$, and we obtain that

$$\mathrm{hBim}_{\mathrm{Ass}}(\mathrm{Ass}, \mathscr{O}) \cong \Omega\,\mathrm{hOper}(\mathrm{Ass}, \mathscr{O}). \tag{2}$$

Now we need a second delooping, which is the following statement:

$$\mathrm{hTot}\,\mathscr{O}(\bullet) \cong \Omega\,\mathrm{hBim}_{\mathrm{Ass}}(\mathrm{Ass}, \mathscr{O}) \tag{3}$$

provided that $\mathscr{O}(0) \cong *$. This delooping takes place always when $\mathscr{O}$ is an Ass-bimodule endowed with a map Ass $\to \mathscr{O}$. To prove this delooping from Theorem 8, we consider the following monoidal model category: right modules over Ass with tensor product $(P \boxtimes Q)(n) = \bigsqcup_{i+j=n} P(i) \times Q(j)$. Then monoids with respect to this structure are Ass-bimodules, and bimodules over the monoid Ass in this category are cosimplicial objects.

My proof of the case $m = 1$ also proceeds in the same two steps (2), (3) by providing explicit cofibrant replacement for Ass, see [24].

Now for high dimensions.

Theorem 9 (Dwyer–Hess [7], Ducoulombier–T. [5])

1. If $B_m \to \mathscr{O}$ is an operad map and $\mathscr{O}(0) \cong \mathscr{O}(1) \cong *$, then

$$\mathrm{hInfBim}_{B_m}^{\leq k}(B_m, \mathscr{O}) \cong \Omega^{m+1}\,\mathrm{hOper}_{\leq k}(B_m, \mathscr{O}).$$

---

[4]This statement is also true in the setting of coloured operads [20].

2. If $B_m \to M$ is a B_m-bimodule map and $M(0) \cong *$, then

$$\text{hInfBim}_{B_m}^{\leq k}(B_m, M) \cong \Omega^m \, \text{hBim}_{B_m}^{\leq k}(B_m, \mathscr{O}).$$

So the second one implies the first one by the Dwyer-Hess-Robertson theorem. This has more implications than to the study of embeddings. Let me give some motivation for this result and then the ideas of the proofs.

We can consider any space of maps $\text{Maps}_{\partial}^{\mathbb{S}}(D^m, D^n)$, where these maps avoid certain multisingularity $\mathbb{S}$, for example triple intersections or something like that. For these spaces, it's a difficult question whether the Goodwillie tower converges. Still we can apply the theorem, and get the delooping of the corresponding Taylor towers.

Consider the sequence $\{\text{Maps}^{\mathbb{S}}(\sqcup_k D^m, D^n), \ k \geq 0\}$. This is a B_m-bimodule under B_m. Therefore the tower $T_{\bullet}$ for the corresponding space $\text{Maps}_{\partial}^{\mathbb{S}}(D^m, D^n)$ can also be delooped in this way. As a more concrete example, one could look at $\text{Imm}_{\partial}^{(\ell)}(D^m, D^n)$—the space of immersions which avoid ℓ-self intersections. One has an obvious inclusion $\text{Imm}_{\partial}^{(\ell)}(D^m, D^n) \hookrightarrow \text{Imm}_{\partial}(D^m, D^n)$. We denote its homotopy fiber space by $\overline{\text{Imm}}_{\partial}^{(\ell)}(D^m, D^n)$. Let $B_n^{(\ell)}(k)$ be the space of collections of k labeled open disks which can overlap but no ℓ of them have a common point. The collection $B_n^{(\ell)}(\bullet)$ is a bimodule over B_n. Then the tower $T_{\bullet}$ of the space $\overline{\text{Imm}}_{\partial}^{(\ell)}(D^m, D^n)$ is described as follows

$$T_k \overline{\text{Imm}}_{\partial}^{(\ell)}(D^m, D^n) \cong \text{hInfBim}_{B_m}^{\leq k}(B_m, B_n^{(\ell)}) \cong \Omega^m \, \text{hBim}_{B_m}^{\leq k}(B_m, B_n^{(\ell)}).$$

Note that in these examples the spaces $\text{Maps}_{\partial}^{\mathbb{S}}(D^m, D^n)$, $\text{Imm}_{\partial}^{(\ell)}(D^m, D^m)$, $\overline{\text{Imm}}_{\partial}^{(\ell)}(D^m, D^n)$ are naturally acted on by B_m. We conjecture that this action is compatible with the delooping of their towers. One should also mention that for embedding spaces $\text{Emb}_{\partial}^{fr}(D^m, D^n)$, $\overline{\text{Emb}}_{\partial}(D^m, D^n)$ we have not just an action of B_m but also of B_{m+1}. Where does this come from? Morally speaking, it comes from the fact that we can make knots small and pull ones through the others. This action was rigorously defined by Budney [4].

The approach of Dwyer–Hess to this theorem, they are using the fact that $B_m \cong \underbrace{\text{Ass} \otimes \cdots \otimes \text{Ass}}_{m}$, the Boardman–Vogt tensor product [8]. They use this decomposition and apply iteratively Theorem 8. How exactly it works I don't know. It is probably technical, that's why they are slow in writing it down.

Our approach is more direct, and the proof is very similar to my proof of the second delooping (3) in the case $m = 1$, with an explicit cofibrant replacement. For any operad $\mathscr{P}$ (doubly reduced $\mathscr{P}(0) \cong \mathscr{P}(1) \cong *$), and any $\mathscr{P}$-bimodule map $P \to M$, we construct a natural map

$$\text{Maps}_*(\Sigma \mathscr{P}(2), \text{hBim}_{\mathscr{P}}(\mathscr{P}, M)) \to \text{hInfBim}_{\mathscr{P}}(\mathscr{P}, M)$$

and we write down when this is an equivalence. Then we check that for the little disks this condition is satisfied.

Our approach works for the truncated case as well. In Dwyer–Hess, it is more difficult. One has to look at the tensor product of truncated operads and then it is not clear how well it works.

Now let us discuss the approach of Boavida de Brito and Weiss. How do they prove that $\overline{\mathrm{Emb}}(D^m, D^n) \cong \Omega^{m+1} \mathrm{hOper}(B_m, B_n)$.

Their result is weaker and stronger. Their approach can not be applied to other spaces like non-(ℓ)-equal immersions or spaces avoiding a given multisingularity, but it's stronger because their deloopings respect the action of the little disks. We have $\mathrm{Emb}_\partial(D^m, D^n)$, which is mapped to $\Omega^m V_m(\mathbb{R}^n)$, the m-loop space on the Stiefel manifold $V_m(\mathbb{R}^n)$. By the Smale-Hirsch principle, $\mathrm{Imm}_\partial(D^m, D^n) \cong \Omega^m V_m(\mathbb{R}^n)$, which is also equivalent to the linear approximation $T_1 \mathrm{Emb}_\partial(D^m, D^n)$. Thus we have a map $T_k \mathrm{Emb}_\partial(D^m, D^n) \to \Omega^m V_m(\mathbb{R}^n)$. There is also a natural map $V_m(\mathbb{R}^n) \to \mathrm{hOper}(B_m, B_n)$. The theorem of Boavida de Brito and Weiss is:

Theorem 10 *The sequence*

$$T_k \mathrm{Emb}_\partial(D^m, D^n) \to \Omega^m V_m(\mathbb{R}^n) \to \Omega^m \mathrm{hOper}_{\leq k}(B_m, B_n)$$

is a fiber sequence.
In particular for $n - m > 2$, they get

$$\mathrm{Emb}_\partial(D^m, D^n) \cong \Omega^m \mathrm{hofib}(V_m(\mathbb{R}^n) \to \mathrm{hOper}(B_m, B_n)). \tag{4}$$

Theorem 4 is an obvious consequence of the theorem above, when we take the homotopy fiber of the first map, we get Ω^{m+1}, as stated. Notice that in the fiber sequence

$$\mathrm{Emb}_\partial(D^m, D^n) \to \Omega^m V_m(\mathbb{R}^n) \to \Omega^m \mathrm{hOper}(B_m, B_n),$$

both maps respect the B_m-action. Therefore, the delooping (4) is compatible with the B_m-action.[5]

To give an idea of the techniques that Boavida de Brito-Weiss are using, the crucial things are configuration categories. They don't need M to be smooth, and they define $\mathrm{Con}(M)$, as a topological category. The objects of the category are the disjoint union of embeddings of k labeled points to M, $\mathrm{Emb}(\underline{k}, M)$. If $x \in \mathrm{Emb}(\underline{k}, M)$ and $y \in \mathrm{Emb}(\underline{\ell}, M)$, then $\mathrm{Mor}(x, y) = \{(j, \alpha)\}$ where $j : \underline{k} \to \underline{\ell}$ and α is a *reverse exit path* from x to $y \circ j$, meaning if points collided at some point of a path, they remain collided until the end. One has a natural functor from $\mathrm{Con}(M) \to \mathrm{Fin}$ that remembers only the set of points. Now the theorem is the following.

---

[5]There is still a question why the delooping $\overline{\mathrm{Emb}}_\partial(D^m, D^n) \cong \Omega^{m+1} \mathrm{hOper}(B_m, B_n)$ is compatible with Budney's B_{m+1}-action. (Obviously it is compatible when restricted on B_m.)

Theorem 11 *If $n - m \geq 3$, there is a homotopy Cartesian square*

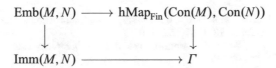

where Γ is the space of sections of $E \to M$ where $E = \{(m, n, \alpha)\}$ where m is in M, n is in N, and α is in $\mathrm{hMap}_{Fin}(\mathrm{Con}(T_m M), \mathrm{Con}(T_n N))$, which you'll see in a second is equivalent to $\mathrm{hOper}(B_m, B_n)$.

So what do they consider? They take the nerve of the category $\mathrm{Con}(M)$, this is a simplicial space, and the nerve of Fin. Then we need to consider the Rezk model category structure on simplicial spaces [19], a.k.a. homotopy theory of homotopy theories. The fibrant objects are complete Segal spaces. They work in the overcategory, the space of maps in this model category of objects over $N_\bullet$ Fin. There are two important statements.

Proposition 1 *When we apply the above construction to embeddings of discs $\mathrm{Emb}_\partial(D^m, D^n)$, we get the space $\mathrm{hMap}_{Fin_*}(\mathrm{Con}^\partial(D^m), \mathrm{Con}^\partial(D^n))$ which is contractible.[6]*

Notice that the map $\mathrm{Emb}_\partial(D^m, D^n) \to \mathrm{hMap}_{Fin_*}(\mathrm{Con}^\partial(D^m), \mathrm{Con}^\partial(D^n))$ factors through the space of topological embeddings, which is contractible by the Alexander trick. The statement of the proposition above is a "calculus version" of this trick.

Proposition 2 *One has $\mathrm{hMap}_{Fin}(\mathrm{Con}(\mathbb{R}^m), \mathrm{Con}(\mathbb{R}^n)) \cong \mathrm{hOper}(B_m, B_n)$.*

The nerve $N_\bullet \mathrm{Con}(\mathbb{R}^m)$ of the configuration category over $N_\bullet$ Fin is equivalent to a certain simplicial space C_{B_m} constructed from the operad B_m. If we have a sequence of maps of sets, we can assign to this a level tree. So once we have an operad $\mathcal{O}$, we can construct a simplicial space $C_\mathcal{O}$ over $N_\bullet$ Fin. That's essentially the idea of this construction that simplicial spaces over $N_\bullet$ Fin are some kind of leveled dendroidal spaces and thus are equivalent to operads. In particular they show that for any pair of operads $\mathcal{O}_1, \mathcal{O}_2$ with $\mathcal{O}_1(0) \cong \mathcal{O}_2(0) \cong *$, one gets $\mathrm{hMap}_{Fin}(C_{\mathcal{O}_1}, C_{\mathcal{O}_2}) \cong \mathrm{hOper}(\mathcal{O}_1, \mathcal{O}_2)$.

Acknowledgements The author/speaker is grateful to Gabriel C. Drummond-Cole for his amazing ability of simultaneous tex-typing during the lectures. The final version of this note is a slight improvement of his. The author/speaker is also grateful to P. Hackney and M. Robertson for organizing the conference and the MATRIX institute for providing support and base for this conference.

---

[6]The construction is slightly different in the case when we have boundary, that's why instead of Fin we get pointed sets Fin_*, the base point corresponding to the points escaping to the boundary.

References

1. Arone, G., Turchin, V.: On the rational homology of high dimensional analogues of spaces of long knots. Geom. Topol. **18**, 1261–1322 (2014)
2. Boavida de Brito, P., Weiss, M.: Manifold calculus and homotopy sheaves. Homol. Homot. Appl. **15**(2), 361–383 (2013)
3. Boavida de Brito, P., Weiss, M.: Spaces of smooth embeddings and configuration categories. To appear in J. Topology (2018)
4. Budney, R.: Little cubes and long knots. Topology **46**(1), 1–27 (2007)
5. Ducoulombier, J., Turchin, V.: Delooping manifold calculus tower on a closed disc (2017, preprint). arXiv:1708.02203
6. Dwyer, W., Hess, K.: Long knots and maps between operads. Geom. Topol. **16**(2), 919–955 (2012)
7. Dwyer, W., Hess, K.: Delooping the space of long embeddings (paper to appear)
8. Fiedorowicz, Z., Vogt, R.M.: An additivity theorem for the interchange of E_n structures. Adv. Math. **273**, 421–484 (2015)
9. Fresse, B., Willwacher, T.: The intrinsic formality of E_n operads (2015, preprint). arXiv:1503.08699
10. Fresee, B., Turchin, V., Willwacher, T.: The rational homotopy of mapping spaces of E_n operads (2017, preprint). arXiv:1703.06123
11. Goodwillie, T.G., Klein, J.: Multiple disjunction for spaces of smooth embeddings. J. Topol. **8**(3), 651–674 (2015)
12. Goodwillie, T.G., Weiss, M.: Embeddings from the point of view of immersion theory: part II. Geom. Topol. **3**, 103–118 (1999)
13. Haefliger, A.: Knotted $(4k - 1)$-spheres in $6k$-space. Ann. Math. (2) **75**, 452–466 (1962)
14. Haefliger, A.: Enlacements de sphères en codimension supérieure à 2. Comm. Math. Helv. **41**, 51–72 (1966–1967)
15. Kontsevich, M.: Operads and motives in deformation quantization. Lett. Math. Phys. **48**(1), 35–72 (1999). Moshé Flato (1937–1998)
16. Lambrechts, P., Volić, I.: Formality of the little N-disks operad. Mem. Am. Math. Soc. **230**(1079), viii+116 (2014)
17. McClure, J.E., Smith, J.H.: Cosimplicial objects and little n-cubes. I. Am. J. Math. **126**(5), 1109–1153 (2004)
18. Pirashvili, T.: Hodge decomposition for higher order Hochschild homology. Ann. Sci. Ecole Norm. Sup (4) **33**(2), 151–179 (2000)
19. Rezk, C.: A model for the homotopy theory of homotopy theory. Trans. Am. Math. Soc. **353**(3), 973–1007 (2001)
20. Robertson, M.: Spaces of Operad Structures. Preprint. arXiv:1111.3904
21. Sinha, D.: Operads and knot spaces. J. Am. Math. Soc. **19**(2), 461–486 (2006)
22. Tamarkin, D.E.: Formality of chain operad of little discs. Lett. Math. Phys. **66**(1–2), 65–72 (2003)
23. Turchin, V.: Context-free manifold calculus and the Fulton-MacPherson operad. Algebr. Geom. Topol. **13**(3), 1243–1271 (2013)
24. Turchin, V.: Delooping totalization of a multiplicative operad. J. Homotopy Relat. Struct. **9**(2), 349–418 (2014)
25. Turchin V., Willwacher, T.: Relative (non-)formality of the little cubes operads and the algebraic Cerf lemma. To appear in Amer. J. Math. (2018)
26. Weiss, M.: Embeddings from the point of view of immersion theory. I. Geom. Topol. **3**, 67–101 (1999)

Groups of Automorphisms and Almost Automorphisms of Trees: Subgroups and Dynamics

Adrien Le Boudec

Notes prepared by Stephan Tornier

Abstract These are notes of a lecture series delivered during the program *Winter of Disconnectedness* in Newcastle, Australia, 2016. The exposition is on several families of groups acting on trees by automorphisms or almost automorphisms, such as Neretin's groups, Thompson's groups, and groups acting on trees with almost prescribed local action. These include countable discrete groups as well as locally compact groups. The focus is on the study of certain subgroups, e.g. finite covolume subgroups, or subgroups satisfying certain normality conditions, such as commensurated subgroups or uniformly recurrent subgroups.

1 Introduction

The main $G^{\alpha}\mathfrak{a}$ theme on which these notes are based is the study of certain discrete and locally compact groups defined in terms of an action on a tree by automorphisms or almost automorphisms. Notorious examples of groups under consideration here include the finitely generated groups introduced by R. Thompson, as well as Neretin's groups.

This text is supposed to be accessible to people not familiar with the topic, and is organized as follows: Sect. 2 introduces basic results about groups acting on trees, and sketches the proof of Tits' simplicity theorem for groups satisfying Tits' independence property. In Sect. 3 we define the notion of almost automorphisms of trees and draw a brief survey about these groups. Section 4 concerns a family of groups acting on trees defined by prescribing the local action almost everywhere. It is shown that this construction provides locally compact groups with somehow unusual properties. Finally the focus in Sect. 5 is on the study of uniformly recurrent

A. Le Boudec (✉)
Université Catholique de Louvain, Chemin du Cyclotron 2, 1348 Louvain-la-Neuve, Belgium
e-mail: adrien.leboudec@uclouvain.be

© Springer International Publishing AG, part of Springer Nature 2018   501
D.R. Wood et al. (eds.), *2016 MATRIX Annals*, MATRIX Book Series 1,
https://doi.org/10.1007/978-3-319-72299-3_22

subgroups of countable groups having a so called micro-supported action on a Hausdorff topological space. Several classes of groups encountered in other sections of this text fall into that framework, but this class of groups is actually much larger.

We should warn the reader that these notes do not contain new results. Instead, they constitute an accessible introduction to the topic, and to recent developments around the groups under consideration here. Some of the theorems given here are proved in these notes, but most of them are stated without proofs, and we tried to indicate as much as possible references where the reader will be able to find complements and proofs of the corresponding results.

Finally the author would like to mention that Stephan Tornier should be credited with the existence of this text, for taking notes during the lectures and writing a substantial part of this text.

2 Groups Acting on Trees and Tits' Simplicity Theorem

In this section we recall classical results about groups acting on trees and outline a proof of Tits' simplicity theorem. For complements on groups acting on trees, the reader is invited to consult [36, 40, 44].

2.1 Classification of Automorphisms and Invariant Subtrees

Throughout, T denotes a simplicial tree and $\mathrm{Aut}(T)$ its automorphism group. To begin with, there is the following classical trichotomy for automorphisms of trees.

Proposition 1 Let $g \in \mathrm{Aut}(T)$. Then exactly one of the following holds:

(1) There is a vertex fixed by g.
(2) There are adjacent vertices permuted by g.
(3) There is a bi-infinite line along which g acts as a non-trivial translation.

Automorphisms of the first two kinds are called *elliptic*, and automorphisms of the third kind are called *hyperbolic*.

Proof Set $\|g\| := \min\{d(v, gv) \mid v \in V(T)\}$ and

$$\min(g) := \{v \in V(T) \mid d(v, gv) = \|g\|\}.$$

If $\|g\| = 0$ then (1) holds. Now assume $\|g\| > 0$. Let $s \in \min(g)$ and let $t \in V(T)$ be the vertex which is adjacent to s and contained in the geodesic segment $[s, gs]$. If $gt \in [s, gs]$ then either $gt = s$ and $gs = t$ and (2) holds. Otherwise, $\bigcup_{m \in \mathbb{Z}} g^m[s, gs]$ is a geodesic line and (3) holds. $\qquad\square$

Definition 1 Let $g \in \text{Aut}(T)$ be hyperbolic. The bi-infinite line along which g acts as a translation is called the *axis* of g, and is denoted L_g. The *endpoints* of g are the two ends of T defined by L_g.

With the classification of automorphisms of trees at hand we now turn to groups acting on trees. First, we record the following lemma.

Lemma 1 *Let* $g, h \in \text{Aut}(T)$ *by hyperbolic such that* L_g *and* L_h *are disjoint. Then* gh *is also hyperbolic and* L_{gh} *intersects both* L_g *and* L_h.

Proof The situation presents itself as in Fig. 1.

Consider $\alpha = [x, y]$. Then $(gh)\alpha \cap \alpha = \{y\}$. Thus gh is hyperbolic and $\alpha \subseteq L_{gh}$. $\square$

We derive the following proposition. For a group G acting on T by automorphisms, we denote by $\text{Hyp}(G)$ the set of hyperbolic elements of G.

Proposition 2 *Let G act on T and assume that* $\text{Hyp}(G) \neq \emptyset$. *Then there is a unique minimal G-invariant subtree, which is given by*

$$X = \bigcup_{g \in \text{Hyp}(G)} L_g.$$

Proof Let $g \in \text{Hyp}(G)$ and $h \in G$. Then $hgh^{-1} \in \text{Hyp}(G)$ and $L_{hgh^{-1}} = h(L_g)$. Hence X is G-invariant; it is a subtree by the previous lemma. As to minimality, let Y be a G-invariant subtree. Then for $y \in V(Y)$ and $g \in \text{Hyp}(G)$ we have $[y, gy] \subseteq Y$. Hence $Y \cap L_g \neq \emptyset$ and Y has to contain L_g. $\square$

Definition 2

(a) A subtree $X \subseteq T$ is called a *half-tree* if X is obtained as one of the components resulting from removing some edge of T (Fig. 2).

Fig. 1 Disjoint lines of hyperbolic elements

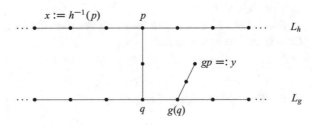

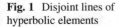

Fig. 2 Definition of half-tree

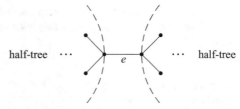

(b) An action of a group G on T is

   (1) *minimal* if there is no proper G-invariant subtree,
   (2) *lineal* if there is a G-invariant, bi-infinite line and $\mathrm{Hyp}(G) \neq \emptyset$.
   (3) *of general type* if there are $g_1, g_2 \in \mathrm{Hyp}(G)$ with no common endpoints.

2.2 Classification of Actions on Trees

In the following, *fixing* amounts to stabilizing point-wise and *stabilizing* amounts to stabilizing set-wise. Let G act on T. Then exactly one of the following happens.

(1) There is a vertex or an edge stabilized by G.
(2) The action is lineal.
(3) There is exactly one end fixed by G.
(4) The action is of general type.

An example of a lineal action is the action of $\mathbb{Z}$ on its standard Cayley graph. An example of case (3) is given by the action of a Baumslag-Solitar group $\mathrm{BS}(1, n)$, $n \geq 2$, on its Bass–Serre tree.

Proposition 3 *Let G act minimally and of general type on T. Then*

(1) for every half-tree X of T there is $g \in \mathrm{Hyp}(G)$ with $L_g \subseteq X$;
(2) every non-trivial normal subgroup $N \trianglelefteq G$ acts minimally of general type.

Proof For (1), let X be a half-tree in T. Then there is $g \in \mathrm{Hyp}(G)$ such that $L_g \cap X \neq \emptyset$, since otherwise there would be a proper invariant subtree in the complement of X by Proposition 2. Now if $h \in \mathrm{Hyp}(G)$ has no common endpoints with g, it is a simple verification that there must exist $n \in \mathbb{Z}$ such that $g^n h g^{-n}$ has its axis inside X. This shows (1).

Statement (2) is obtained by using the classification of group actions on trees. Details are left to the reader.   □

2.3 Tits' Simplicity Theorem

Let G act on T and let X be a (finite or infinite) geodesic in T. Further, let $\pi_X : T \to X$ denote the closest point projection on X, and $\mathrm{Fix}_G(X)$ the fixator of X in G (Fig. 3).

Fig. 3 Closest point projection

Given $x \in V(T)$, let $G^{(x)}$ be the permutation group induced by the action of $\mathrm{Fix}_G(X)$ on $\pi_X^{-1}(x)$. We have a morphism

$$\varphi_X : \mathrm{Fix}_G(X) \hookrightarrow \prod_{x \in X} G^{(x)}.$$

Definition 3 Retain the above notation. The group G satisfies Tits' independence property if φ_X is an isomorphism for all X.

The following lemma is the core of the proof of Tits' theorem. For the proof, we refer the reader to Tits' original article [44].

Lemma 2 (Commutator Lemma) *Let G act on T, $g \in \mathrm{Hyp}(G)$ and $X := L_g$. Assume that φ_X is an isomorphism. Then*

$$\mathrm{Fix}_G(X) = \{[g, h] \mid h \in \mathrm{Fix}_G(X)\}.$$

We now state Tits' simplicity theorem. Given a group G acting on T, we denote by G^+ the subgroup of G generated by fixators of edges: $G^+ := \langle \mathrm{Fix}_G(e) \mid e \in E(T) \rangle$. Clearly G^+ is a normal subgroup of G.

Theorem 1 *Let $G \leq \mathrm{Aut}(T)$ act minimally and of general type on T. If G satisfies Tits' independence property, then G^+ is either abstractly simple or trivial.*

Proof Assume that G^+ is non-trivial and let $N \trianglelefteq G^+$ be non-trivial. Two applications of Proposition 3 show that N acts minimally and of general type on T.

Let $e \in E(T)$. We show that $\mathrm{Fix}_G(e) \subseteq N$. By Tits' independence property and for symmetry reasons it suffices to show that $\mathrm{Fix}_G(X_1) \subseteq N$, where X_1 is one of the two half-trees defined by e. According to Proposition 3, there exists a hyperbolic element $g \in \mathrm{Hyp}(N)$ with $L_g \subseteq X_1$. Applying Lemma 2 to this element g, we obtain

$$\mathrm{Fix}_G(L_g) = [\mathrm{Fix}_G(L_g), g] \leq N,$$

where the last inclusion follows from the fact that N is normal in G. This finishes the proof as $\mathrm{Fix}_G(X_1) \subseteq \mathrm{Fix}_G(L_g)$.   □

Remark 1 This result has been generalized in various directions. See for instance Haglund–Paulin [20], Lazarovich [24] for cube complexes and Caprace [10] for buildings. Whereas the above generalizations vary the space, there is also work of Banks–Elder–Willis [4] and Möller–Vonk [31] for trees.

3   Almost-Automorphisms of Trees

In this section we draw a brief survey about groups of almost-automorphisms of trees, and discuss some recent results.

3.1 Definitions

For $d, k \geq 2$ let $T_{d,k}$ the rooted tree in which the root has degree k and the other vertices have degree $d + 1$. The *level* of a vertex is its distance from the root. For instance $T_{3,2}$ looks as in Fig. 4.

Fix a bijection between the vertices of $T_{d,k}$ and finite words that are either empty or of the form $xy_1y_2 \cdots y_j$, where $x \in \{0, \ldots, k-1\}$ and $y_i \in \{0, \ldots, d-1\}$. Further, let $X_{d,k} := \partial T_{d,k}$ denote the boundary of the tree $T_{d,k}$. Given $\xi, \xi' \in X_{d,k}$, we set $d(\xi, \xi') := d^{-N(\xi, \xi')}$ where $N(\xi, \xi') = \sup\{m \geq 1 \mid \xi_m = \xi'_m\}$ and ξ_m, ξ'_m denote the mth letter in the word ξ, ξ' respectively. This turns $(X_{d,k}, d)$ into a compact metric space homeomorphic to a Cantor set.

Remark 2 Note that there is a one-to-one correspondence between proper balls in $X_{d,k}$ and vertices of level at least one, in which a vertex $v \in V(T_{d,k})$ corresponds to set of ends of $T_{d,k}$ hanging below v.

Remark that any element of $\mathrm{Aut}(T_{d,k})$ induces a homeomorphism of $X_{d,k}$, and the action of $\mathrm{Aut}(T_{d,k})$ on $X_{d,k}$ is faithful and by isometries. The notion of almost automorphisms is a natural generalization of the one of automorphisms, and goes back to Neretin [32].

Definition 4 An element $g \in \mathrm{Homeo}(X_{d,k})$ is an *almost-automorphism* of $T_{d,k}$ if there is a partition $X_{d,k} = B_1 \sqcup \cdots \sqcup B_n$ of $X_{d,k}$ into balls, such that for every $i \in \{1, \ldots, n\}$ there is $\lambda_i > 0$ so that $\mathrm{dist}(gx, gy) = \lambda_i d(x, y)$ for all $x, y \in B_i$.

We denote by $\mathrm{AAut}(T_{d,k})$ the set of all almost-automorphisms of $T_{d,k}$, which is easily seen to be a subgroup of $\mathrm{Homeo}(X_{d,k})$. For an example of an almost-automorphism, consider Fig. 5

Remark 3 We record the following facts about $\mathrm{AAut}(T_{d,k})$.

(1) $\mathrm{Aut}(T_{d,k})$ is clearly a subgroup of $\mathrm{AAut}(T_{d,k})$.
(2) In the case $k = 2$, one may check that the group $\mathrm{AAut}(T_{d,2})$ coincides with the topological full group $[[\mathrm{Aut}(T_{d+1}), \partial T_{d+1}]]$, where T_{d+1} is a *non-rooted* regular tree of degree $(d + 1)$. This corresponds to Neretin's original definition [32] (although the terminology "topological full group" was not used there). Similarly, one can embed the group $\mathrm{Aut}(T_{d+1})$ into $\mathrm{AAut}(T_{d,k})$ for arbitrary k.

Fig. 4 The tree $T_{3,2}$

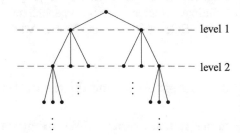

level 1

level 2

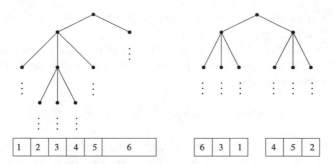

Fig. 5 An almost-automorphism

Remark that the group $\mathrm{Aut}(T_{d,k})$ is naturally a topological group, which is totally disconnected and compact. The proof of the following fact is left to the reader.

Proposition 4 *The group* $\mathrm{AAut}(T_{d,k})$ *admits a group topology which makes the inclusion of* $\mathrm{Aut}(T_{d,k})$ *continuous and open.*

Henceforth we implicitly consider $\mathrm{AAut}(T_{d,k})$ equipped with this topology.

Definition 5 (Higman–Thompson Group) Let $V_{d,k}$ be the set of $g \in \mathrm{Homeo}(X_{d,k})$ for which there is a partition $X_{d,k} = B_1 \sqcup \cdots \sqcup B_n$ such that $g|_{B_i}$ is a homothety and $g(w_i x) = w_{g(i)} x$ for every $w_i x \in B_i$ where w_i is the vertex defining B_i.

By a theorem of Higman (previously obtained by Thompson in the case $d = k = 2$), the group $V_{d,k}$ is finitely presented and has a simple subgroup of index at most two. One may check without difficulty that the group $\mathrm{AAut}(T_{d,k})$ is generated by its two subgroups $V_{d,k}$ and $\mathrm{Aut}(T_{d,k})$. Since $V_{d,k}$ is finitely generated and $\mathrm{Aut}(T_{d,k})$ is compact, this implies in particular that $\mathrm{AAut}(T_{d,k})$ is a compactly generated group.

Remark 4 It readily follows from the definitions that the group $V_{d,k}$ is dense in $\mathrm{AAut}(T_{d,k})$. Moreover $V_{d,k} \cap \mathrm{Aut}(T_{d,k})$ is exactly the group of finitary automorphisms of $T_{d,k}$, which is an infinite locally finite group. Since $\mathrm{Aut}(T_{d,k})$ is compact open in $\mathrm{AAut}(T_{d,k})$, this subgroup must be commensurated in $V_{d,k}$. See also Theorem 4(3) and the questions following it.

3.2   Some Results About $\mathrm{AAut}(T_{d,k})$

This paragraph further illustrates interesting properties satisfied by the groups $\mathrm{AAut}(T_{d,k})$.

(1) The group $\mathrm{AAut}(T_{d,k})$ is (abstractly) simple [22], and therefore belongs to the class of non-discrete, totally disconnected, compactly generated locally compact simple groups. The study of this class of groups recently received much attention, and we refer the reader to [12–14]. Note that the list of known examples of groups within this class is still quite restricted (see the introduction

of [14]). Note also that a stronger simplicity result for $AAut(T_{d,k})$ has recently been obtained in [43].

(2) The group $AAut(T_{d,k})$ coincides with the group of abstract commensurators of the profinite group $Aut(T_{d,k})$ [11]. Here, given a profinite group G, the group of commensurators of G is

$$Comm(G) = \{f : U \xrightarrow{\cong} V \mid U, V \leq_o G\}/\sim,$$

where $\sim$ identifies isomorphisms which agree on some open subgroup of G.

(3) The structure of subgroups of $AAut(T_{d,k})$ remains largely mysterious. On the one hand, the flexibility of the action of $AAut(T_{d,k})$ on $X_{d,k}$ readily implies that $AAut(T_{d,k})$ has "many" subgroups. On the other hand, it is very much unclear whether there are "large" discrete subgroups in $AAut(T_{d,k})$. A striking illustration of a restriction on discrete subgroups is given by the following result from [3].

Theorem 2 (Bader–Caprace–Gelander–Mozes) *The group* $AAut(T_{d,k})$ *does not admit lattices.*

For background and motivation for the problem of studying the existence of lattices in locally compact *simple* groups, we refer the reader to the introduction of [3]. Interestingly, the proof of Theorem 2 relies on finite group theoretic arguments, such as the study of subgroups of finite symmetric groups with a given upper bound on the index.

Other locally compact simple groups without lattices appear in Sect. 4 (Theorem 5). See also Remark 8.

(4) The group $AAut(T_{d,k})$ is compactly presented [25], and actually satisfies a stronger finiteness property, see Sauer–Thumann [39]. We mention that, although an upper bound has been obtained in [25], the Dehn function of the group $AAut(T_{d,k})$ is not known.

3.3 Commensurated Subgroups of Groups of Almost-Automorphisms of Trees

The goal of this section is to report on recent work concerning the study of commensurated subgroups of groups of almost automorphisms of trees, carried out in collaboration with Ph. Wesolek. For the proofs of the results mentioned in this section and for complements, we refer to the article [28].

Definition 6 Let G be a group. Two subgroups $H, K \leq G$ are *commensurable* if $H \cap K$ has finite index in both H and K. The subgroup H is *commensurated* if gHg^{-1} is commensurable with H for all $g \in G$.

Example 1

(1) Any normal subgroup is commensurated.
(2) Finite and finite index subgroups are commensurated.
(3) $SL(n, \mathbb{Z}) \leq SL(n, \mathbb{Q})$ is commensurated.
(4) Fundamental example: any compact open subgroup $U \leq G$ of a totally disconnected locally compact group G is commensurated.

Shalom-Willis classified the commensurated subgroups of S-arithmetic subgroups in certain simple algebraic groups [41]. For instance, in $SL(n, \mathbb{Z})$ $(n \geq 3)$ every commensurated subgroup is finite or of finite index.

Theorem 3 ([28]) *Let $H \leq \mathrm{AAut}(T_{d,k})$ be commensurated. Then either H is finite, $\overline{H}$ is compact open or $H = \mathrm{AAut}(T_{d,k})$.*

Remark 5 In Theorem 3, the conclusion cannot be strengthened to H itself being compact open in the second case, see [28, Ex. 4.4] for examples of non-closed commensurated subgroups.

One of the interests in studying commensurated subgroups is the fact that it provides information about possible embeddings into locally compact groups.

Corollary 1 *Any continuous embedding of $\mathrm{AAut}(T_{d,k})$ into a totally disconnected locally compact group has closed image.*

Remark 6 There are natural generalizations of $\mathrm{AAut}(T_{d,k})$ considered by Caprace–de Medts [11] for which Corollary 1 is not true, see [28, p. 25].

We now turn our attention to the family of Thompson's groups. For a pleasant introduction to these groups, we refer the reader to the notes [9].

Let $d = k = 2$. The group $V_{2,2}$ (see Definition 5) is known as Thompson's group V. Thompson's group T is the subgroup of $\mathrm{Homeo}(\mathbb{S}^1)$ consisting of those homeomorphisms which are piecewise linear, have slopes in $2^{\mathbb{Z}}$, all breakpoints at dyadics and only finitely many breakpoints in total. Finally, we let F denote the stabilizer of $0 \in \mathbb{S}^1$ in T. There are natural embeddings $F \leq T \leq V \leq \mathrm{AAut}(T_{2,2})$.

Theorem 4 ([28])

(1) Every commensurated subgroup of F is normal.
(2) Every commensurated subgroup of T is either finite or equal to T.
(3) Every commensurated subgroup of V is locally finite or equal to V.

We have seen in Remark 4 that there is an infinite and locally finite commensurated subgroup in Thompson's group V. The above theorem raises the question whether there exist other (non-commensurable) commensurated subgroups in V. Thanks to the process of Schlichting completion, a related question is the following: Are there locally compact groups other than $\mathrm{AAut}(T_{2,2})$ into which Thompson's group V embeds densely?

Corollary 2 *Every embedding of Thompson's groups F or T into a locally compact group has discrete image.*

Remark 7 If we denote respectively by $\mathrm{Homeo}^+([0, 1])$ and $\mathrm{Homeo}^+(\mathbb{S}^1)$ the groups of orientation-preserving homeomorphisms of the interval and the circle, endowed with their natural Polish topology, it is an exercise to check that Tompson's groups F and T are dense respectively in $\mathrm{Homeo}^+([0, 1])$ and $\mathrm{Homeo}^+(\mathbb{S}^1)$. Therefore F and T do appear as dense subgroups in some Polish groups, but cannot appear as dense subgroups in some locally compact groups by Corollary 2.

4 Groups Acting on Trees with Prescribed Local Actions

In this section we study examples of locally compact groups acting on trees, satisfying properties rather similar to groups of almost-automorphisms of trees from Sect. 3. We may think of them as analogues of groups of almost-automorphisms of trees, but more rigid and much smaller in a sense to be made precise. The reference for this section is [26].

4.1 Definitions

Let $d \geq 3$ and T_d denote the d-regular tree. Fix a set Ω of cardinality d. Fix a map $c : E(T_d) \to \Omega$ such that $c_v : E(v) \to \Omega$ is a bijection for every vertex $v \in V(T_d)$, where $E(v)$ is the set of edges around v. Given $g \in \mathrm{Aut}(T_d)$ and $v \in V(T_d)$ we obtain a permutation $\sigma(g, v) = c_{gv} \circ g \circ c_v^{-1} \in \mathrm{Sym}(\Omega)$.

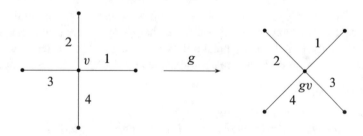

Definition 7 (Burger–Mozes [8]) Let $F \leq \mathrm{Sym}(\Omega)$. Define

$$\mathrm{U}(F) = \{g \in \mathrm{Aut}(T_d) \mid \forall v \in V(T_d) : \sigma(g, v) \in F\}.$$

We collect the following properties of $\mathrm{U}(F)$, the proof of which are left to the reader.

(1) $\mathrm{U}(F)$ is a closed subgroup of $\mathrm{Aut}(T_d)$, which is discrete if and only if the action $F \curvearrowright \Omega$ is free.
(2) $\mathrm{U}(\{1\})$ is vertex-transitive, and therefore a cocompact lattice in $\mathrm{U}(F)$.
(3) $\mathrm{U}(F)$ satisfies Tits' property (Definition 3).

(4) $U(F)^+$ (the subgroup generated by fixators of edges in $U(F)$) has index two in $U(F)$ if and only if the permutation group F is transitive and generated by its point stabilizers. In this case, $U(F)^+$ is transitive on geometric edges.

From now on, we assume $F \leq \mathrm{Sym}(\Omega)$ to be transitive.

Definition 8 (Bader–Caprace–Gelander–Mozes) For $F \leq \mathrm{Sym}(\Omega)$, set

$$G(F) := \{g \in \mathrm{Aut}(T_d) \mid \sigma(g, v) \in F \text{ for all but finitely many } v \in V(T_d)\}.$$

Note that $G(F)$ is a subgroup of $\mathrm{Aut}(T_d)$. In contrast to $U(F)$, the group $G(F)$ is *not* closed in $\mathrm{Aut}(T_d)$. One may actually check that $G(F)$ is a dense subgroup of $\mathrm{Aut}(T_d)$.

It can be shown that there is a unique group topology on $G(F)$ such that the inclusion of $U(F)$ into $G(F)$ is continuous and open (see for instance [26, p. 7]).

With respect to this topology, the action of $G(F)$ on the tree is continuous but not proper. More precisely, we have the following:

Proposition 5 *Let* $b \in V(T_d)$. *Then the stabilizer* $G(F)_b$ *is an increasing union of compact open subgroups.*

Proof Let $m \geq 1$.

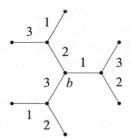

Set $K_m(b) = \{g \in G(F)_b \mid \forall v \notin B(b, m) : \sigma(g, v) \in F\}$. Then $K_m(b)$ is a subgroup of $G(F)$. Moreover it is a compact open subgroup as it contains the fixator of $B(b, m)$ in $U(F)$ as a finite index subgroup. Since $G(F)_b = \bigcup_{m \geq 1} K_m(b)$, the statement follows.   □

Definition 9 Let $F \leq F' \leq \mathrm{Sym}(\Omega)$. Set $G(F, F') := G(F) \cap U(F')$.

Note that $U(F) \leq G(F, F') \leq U(F')$ whence $G(F, F')$ is open in $G(F)$. For proofs of the following results, we refer the reader to [26].

(1) The group $G(F, F')$ satisfies a weak Tits' property.

 (a) There exist natural sufficient conditions on the permutation groups F and F' so that $G(F, F')$ is virtually simple.

(2) The group $G(F, F')$ is compactly generated but not compactly presented.
(3) The group $G(F, F')$ has asymptotic dimension one. This may be compared with the fact that $\mathrm{AAut}(T_{d,k})$ has infinite asymptotic dimension.

4.2 Lattices

We now turn to the study of lattices in the family of groups $G(F, F')$. Recall that if G is a locally compact group, a lattice Γ in G is a discrete subgroup of finite covolume, i.e. such that G/Γ carries a G-invariant finite measure.

We will state two different results, showing that the existence of lattices in the groups $G(F, F')$ strongly depends on the properties of permutation groups F and F'.

To this end, consider four permutation groups $F \leq F'$ and $H \leq H'$ such that $F \leq H$ and $F' \leq H'$. These conditions ensure the inclusion $G(F, F') \leq G(H, H')$.

Proposition 6 *Retain the above notation. Assume that $H \cap F' = F$ and $H' = HF'$. Then $G(F, F')$ is a closed cocompact subgroup of $G(H, H')$.*

If in addition the action $F \curvearrowright \Omega$ is free, then $G(F, F')$ is a cocompact lattice in $G(H, H')$.

For a proof of Proposition 6, see [26, Corollary 7.4].

Example 2 An example as in Proposition 6 is $d = 7$, $F = C_7$, $F' = \text{Alt}_7$, $H = D_7$ and $H' = \text{Sym}_7$, where C_7 and D_7 denote respectively the cyclic and dihedral group acting transitively on seven elements.

In another direction, we now provide sufficient conditions on F, F' which prevent the existence of lattices in the group $G(F, F')$.

Definition 10 Let G be a group. A subgroup $H \leq G$ is said to be *essential* in G if H intersects non-trivially every non-trivial subgroup of G.

The following criterion, the proof of which may be found in [26], provides sufficient conditions which prevent the existence of a lattice in a locally compact group.

Proposition 7 *Let G be a locally compact group with Haar measure μ. Suppose there are sequences of compact open subgroups $(U_m)_{m \in \mathbb{N}}$ and $(K_m)_{m \in \mathbb{N}}$ such that*

(1) $(U_m)_{m \in \mathbb{N}}$ is a neighbourhood basis of $1 \in G$.
(2) U_m is an essential subgroup of K_m for every $m \in \mathbb{N}$.
(3) $\mu(K_m) \xrightarrow{m \to \infty} \infty$.

Then G does not admit lattices.

Using this criterion we show that certain $G(F, F')$ do not contain lattices. For $F \leq \text{Sym}(\Omega)$ and $a \in \Omega$, we denote by F_a the stabilizer of a in F.

Theorem 5 ([26]) *Let $F \leq F' \leq \text{Sym}(\Omega)$ and $a \in \Omega$. Assume that*

(1) $F_a \leq F'_a$ is essential, and
(2) $|F'_a| < [F'_a : F_a]^{d-1}$.

Then $G(F, F')$ does not admit a lattice.

We point out that there are examples of groups $G(F, F')$ satisfying Theorem 5 and which are moreover (virtually) simple.

Example 3 Let $q \cong 1 \pmod 4$ be a prime power. Let $\Omega = \mathbb{P}^1(\mathbb{F}_q)$ be the projective line over the finite field $\mathbb{F}_q$, $F = \mathrm{PSL}(2, q)$ and $F' = \mathrm{PGL}(2, q)$. Set $a := \infty \in \mathbb{P}^1(\mathbb{F}_q)$. Then $F'_a = \mathbb{F}_q \rtimes \mathbb{F}_q^\times$ and $F_a = \mathbb{F}_q \rtimes \mathbb{F}_q^{\times,2}$, where we only take the squares in the multiplicative group. To see that F_a is essential in F'_a, consider the short exact sequence

$$1 \to \mathbb{F}_q^{\times,2} \to \mathbb{F}_q^\times \to C_2 \to 1.$$

The assumption $q \cong 1 \pmod 4$ implies that -1 is a square in $\mathbb{F}_q^\times$, and hence this short exact sequence does not split. Therefore, $\mathbb{F}_q^{\times,2}$ is essential in $\mathbb{F}_q^\times$ and hence so is $F_a \leq F'_a$. For the second condition, compute $|F'_a| = q(q-1) < 2^q = [F'_a, F_a]^{d-1}$ as $d = q + 1$.

Proof (Theorem 5) We construct $(U_m)_{m \in \mathbb{N}}$ and $(K_m)_{m \in \mathbb{N}}$ as in Proposition 7. For $m \geq 1$ and a fixed vertex $v_0 \in V(T_d)$ we set

$$U_m = \{ g \in \mathrm{U}(F) \mid g|_{B(v_0,m)} = \mathrm{id} \}$$

and

$$K_m = \left\{ g \in \mathrm{G}(F, F') \; \middle| \; \begin{array}{ll} g = \mathrm{id} & \text{on } B(v_0, m) \\ \sigma(g, v) \in F' & \text{for } v \in S(v_0, m) \\ \sigma(g, v) \in F & \text{for } d(v, v_0) \geq m + 1 \end{array} \right\}.$$

Note that by definition of the topology, (U_m) is a basis of neighbourhoods of the identity. It is easy to see that K_m is a subgroup of $G(F, F')$, which admits a semi-direct product decomposition $K_m = U_{m+1} \rtimes \prod_{S(v_0,m)} F'_a$. Moreover since F_a is essential in F'_a, and since being essential ascends to finite direct products, it follows that $U_m = U_{m+1} \rtimes \prod_{S(v_0,m)} F_a$ is essential in K_m. Furthermore,

$$\mu(K_m) = \mu(U_{m+1}) |F'_a|^{|S(v_0,m)|} = \mu(U_{m+1}) |F'_a|^{d(d-1)^{m-1}}$$

where, with the normalization $\mu(U_1) := 1$, we have

$$\mu(U_{m+1}) = \mu(U_1)[U_1 : U_{m+1}]^{-1} = [U_1 : U_{m+1}]^{-1}.$$

Furthermore, we have

$$[U_1 : U_{m+1}] = |F_a|^{|B(v_0,m)|} = |F_a|^{d \frac{(d-1)^m - 1}{d-2}}.$$

Combined with the assumption $|F'_a| < [F'_a : F_a]^{d-1}$ this implies, $\mu(K_m) \to \infty$. $\square$

Remark 8 Although the proofs of the absence of lattices in the groups $\mathrm{AAut}(T_{d,k})$ (Theorem 2) and in some of the groups $G(F, F')$ (Theorem 5) are very different, they share the same phenomenon that the absence of lattices is actually detected in some open locally elliptic subgroup of the ambient group. It would be interesting to know whether there exist compactly generated simple locally compact groups G not having lattices but such that all open locally elliptic subgroups $O \leq G$ do have lattices.

5 Micro-Supported Actions and Uniformly Recurrent Subgroups

In the previous sections we studied the structure of subgroups of particular families of groups acting on a tree by automorphisms or almost automorphisms, such as the groups $\text{AAut}(T_{d,k})$ and Thompson's groups (Sect. 3), or the groups $G(F, F')$ (Sect. 4). Yet another way to study the subgroups of a given group G is to view them as a whole by considering the Chabauty space of G, and to study the G-action on it. Here we will focus on the study of this action from the point of view of topological dynamics, through the notion of uniformly recurrent subgroups (URS).

The goal of this section is to give an account of joint work with N. Matte Bon [27]. The situation there is the study of URS's of a countable group G acting by homeomorphisms on a Hausdorff space X (with no further assumption). When all rigid stabilizers of this action are non-trivial (see Sect. 5.3 for the relevant terminology), many properties of rigid stabilizers are shown to be inherited by uniformly recurrent subgroups. This allows us to prove a C^*-simplicity criterion based on the non-amenability of rigid stabilizers. When the dynamics of the action of G on X is sufficiently rich, we obtain sufficient conditions ensuring that uniformly recurrent subgroups of G can be completely classified. This situation applies to several classes of groups which naturally come equipped with a micro-supported action; among which examples of groups encountered previously in our lectures such as Thompson's groups and the (countable) groups $G(F, F')$ of Sect. 4; as well as branch groups, groups of piecewise projective homeomorphisms of the real line [30] and topological full groups.

5.1 Uniformly Recurrent Subgroups

In this section G will always be a countable group. Let $\text{Sub}(G)$ be the Chabauty space of all subgroups of G, viewed as a subset of $\{0, 1\}^G$. When $\{0, 1\}^G$ is equipped with the product topology, the set $\text{Sub}(G)$ is a closed subset of $\{0, 1\}^G$, and hence is a compact space. Note that the conjugation action of G on $\text{Sub}(G)$ is an action by homeomorphisms.

The study of G-invariant (ergodic) probability measures on the space $\text{Sub}(G)$, called (ergodic) *Invariant Random Subgroups* (IRS) after [1], has recently received particular attention. In the next two lectures we deal with their topological counterparts:

Definition 11 (Glasner–Weiss [18]) A *Uniformly Recurrent Subgroup* (URS) of G is a minimal closed G-invariant subset of $\mathrm{Sub}(G)$.

Here *minimal* means that there is no proper non-empty G-invariant closed subset. This is obviously equivalent to the fact that every G-orbit is dense. We will denote by $\mathrm{URS}(G)$ the set of URS's of G.

Example 4

(1) If $N \in \mathrm{Sub}(G)$ is a normal subgroup of G, then $\{N\}$ is a URS of G. The URS associated to the trivial subgroup will be called the trivial URS.
(2) More generally if $H \in \mathrm{Sub}(G)$ has a finite conjugacy class, then $\{H^g \mid g \in G\}$ is a URS of G.

From the dynamical point of view, these examples of URS's present very few interest, and we will look after significantly different URS's.

Remark 9

(1) If $\mathcal{H} \in \mathrm{URS}(G)$ is countable, we claim that $\mathcal{H}$ must consist of a finite conjugacy class, i.e. $\mathcal{H}$ is of the form of Example 4. Indeed, being a countable compact space, $\mathcal{H}$ must have an isolated point by the Baire category theorem. Now the set of isolated points is an open G-invariant subset of $\mathcal{H}$, so it must be the entire $\mathcal{H}$ by minimality. Hence $\mathcal{H}$ is both compact and discrete, whereby $\mathcal{H}$ is finite. By minimality G must act transitively on $\mathcal{H}$, hence the claim.

   In particular if G has only countably many subgroups, then every URS is finite. This is for instance the case when every subgroup of G is finitely generated.
(2) Even "small" groups like the lamplighter group $\mathbb{Z}_2 \wr \mathbb{Z}$ may have many URS's [18].

Proposition 8 (Glasner–Weiss) *Let G be a countable group, and $G \curvearrowright X$ a minimal action of G by homeomorphisms on a compact space X. Then the closure of the image of the map*

$$\mathrm{Stab} : X \to \mathrm{Sub}(G), \ x \mapsto G_x$$

contains a unique URS. This URS is called the stabilizer URS of $G \curvearrowright X$, and is denoted $S_G(X)$.

For a proof of Proposition 8, see [18, Proposition 1.2].

We insist on the fact that the map $\mathrm{Stab} : X \to \mathrm{Sub}(G), x \mapsto G_x$ need not be continuous. This is for instance the case in the following example, which shows that a non-free action may plainly have a trivial stabilizer URS.

Example 5 Consider the action of the free group $\mathbb{F}_2$ on the boundary ∂T_4 of its standard Cayley graph. Then $S_{\mathbb{F}_2}(\partial T_4)$ is trivial. Equivalently, for every $g \in \mathbb{F}_2$, there is a sequence (g_n) of conjugates of g such that $(\langle g_n \rangle)$ converges to the trivial subgroup in $\mathrm{Sub}(\mathbb{F}_2)$.

The following example describes explicitly a URS of Thompson's group V (defined in Sect. 3).

Example 6 Consider Thompson's group V acting on the boundary of the rooted binary tree $T_{2,2}$, and set $\mathcal{H} = \{V_{\xi,0} \mid \xi \in \partial T_{2,2}\}$, where

$$V_{\xi,0} = \{g \in V \mid g \text{ fixes a neighbourhood of } \xi\}.$$

Then $\mathcal{H}$ is a URS of V. Actually $\mathcal{H}$ is the stabilizer URS $S_V(\partial T_{2,2})$ associated to the action $V \curvearrowright \partial T_{2,2}$.

5.2 C^*-*Simplicity*

One of the motivations for investigating URS's comes from the recently discovered connection with simplicity of reduced C^*-algebras, as we shall now explain. We shall mention that this may not be seen as the only motivation, and we believe that the notion of URS's is interesting in itself.

Let $\ell^2(G)$ be the Hilbert space of square summable complex valued functions on G. Then G acts on $\ell^2(G)$, giving rise to the left-regular representation $\lambda_G : G \to U(\ell^2(G))$. Recall that the *reduced* C^*-*algebra* $C^*_{red}(G)$ of G is by definition the closure in the operator norm of linear combinations of operators λ_g, $g \in G$.

Definition 12   A group G is C^*-simple if $C^*_{red}(G)$ is simple, i.e. $C^*_{red}(G)$ has no non-trivial 2-sided ideal.

For a pleasant introduction to the problem of C^*-simplicity and its historical development, we refer the reader to de la Harpe's survey [17].

Recall that every countable group admits an amenable normal subgroup $\mathrm{Rad}(G)$ containing all amenable normal subgroups, called the *amenable radical* of G.

Proposition 9 (Paschke–Salinas [35])   *Let G be a countable group. If $\mathrm{Rad}(G) \neq 1$, then G is not C^*-simple.*

The study of the C^*-simplicity of countable groups started with the result of Powers that the non-abelian free group $\mathbb{F}_2$ is C^*-simple [37]. The methods employed by Powers have been largely generalized and many classes of groups have been shown to be C^*-simple. In the following result we mention a few of these results, and refer to [17] for more references.

Theorem 6   *After modding out by the amenable radical, the following groups are C^*-simple:*

1. *Linear groups [5, 7, 38].*
2. *Acylindrically hyperbolic groups [15]. This generalizes the case of free products [35], Gromov-hyperbolic groups [16], relatively hyperbolic groups [2], mapping class groups and* $\mathrm{Out}(F_n)$ *[6].*
3. *free Burnside groups of sufficiently large odd exponents [34].*
4. *Tarski monsters [7, 21].*

In the sequel a URS $\mathcal{H} \in \mathrm{URS}(G)$ is said to be amenable if every $H \in \mathcal{H}$ is amenable.

Theorem 7 (Kalantar–Kennedy [21], Breuillard–Kalantar–Kennedy–Ozawa [7], Kennedy [23]) *For a countable group G, the following are equivalent:*

(1) G is C^*-simple.
(2) G acts freely on its Furstenberg boundary.
(3) G admits no non-trivial amenable URS.

5.3 From Rigid Stabilizers to Uniformly Recurrent Subgroups

Let G be a countable group, and let X be a Hausdorff space on which G acts faithfully by homeomorphisms.

Definition 13 Let $U \subseteq X$ be a non-empty open subset of X. The *rigid stabilizer* of U is the set of elements of G supported inside U:

$$G_U = \{g \in G \mid g = \mathrm{id} \text{ on } X \backslash U\}.$$

Definition 14 The action $G \curvearrowright X$ is *micro-supported* if the rigid stabilizer G_U is non-trivial for any non-empty open subset $U \subseteq X$.

Examples of countable groups admitting a micro-supported action are Thompson's groups (and many of their generalizations), branch groups (Sect. 5.5), or groups of piecewise projective homeomorphisms of the real line (Sect. 5.6). We refer to [27] for more examples.

For $H \in \mathrm{Sub}(G)$ we denote by $\mathscr{C}(H) \subset \mathrm{Sub}(G)$ the conjugacy class of H. If $H \in \mathrm{Sub}(G)$ belongs to a non-trivial URS of G, then the closure of $\mathscr{C}(H)$ does not contain the trivial subgroup in the Chabauty space $\mathrm{Sub}(G)$. In order to study URS's, it is then natural to study the subgroups of G whose conjugacy class closure does not contain the trivial subgroup.

Theorem 8 ([27]) *Let G be a countable group of homeomorphisms of a Hausdorff space X. Given $H \in \mathrm{Sub}(G)$, at least one of the following happens:*

(1) The closure of $\mathscr{C}(H)$ in $\mathrm{Sub}(G)$ contains the trivial subgroup.
(2) There exists $U \subseteq X$ open and non-empty such that H admits a subgroup $A \leq H$ which surjects onto a finite index subgroup of G_U.

Note that the first condition in Theorem 8 is intrinsic to G, in the sense that it does not depend on the space X, while the second condition is defined in terms of the action of G on X.

We deduce the following result, which says that many properties of rigid stabilizers associated to *one* action $G \curvearrowright X$ are inherited by *all* uniformly recurrent subgroups of G.

Corollary 3 *If for every non-empty open subset $U \subseteq X$ the rigid stabilizer G_U is non-amenable (resp. non-elementary amenable, contain $\mathbb{F}_2,\dots$) then the same is true for every non-trivial URS of G.*

Theorem 8 will follow from the following technical statement. In the sequel we fix G and X as in Theorem 8.

Proposition 10 *Fix $z \in X$ and $H \in \mathrm{Sub}(G)$. Then at least one of the following happens:*

(1) $\{1\}$ is contained in the closure of $\mathscr{C}(H)$.

(2) There is a neighbourhood $W \subseteq X$ of z in X such that for every $K \in \mathscr{C}(H)$, there exist an open subset $U \subseteq X$, a finite index subgroup $\Gamma \leq_{f.i.} G_U$ and a subgroup $A \subseteq K$ such that:

    (a) $A = \mathrm{id}$ on W.

    (b) A leaves U invariant and for every $\gamma \in \Gamma$ there is a $\in A$ so that $a = \gamma$ when restricted to U.

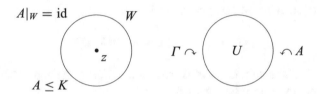

The above Proposition (for arbitrary z and W, and taking $K = H$) implies Theorem 8 because Γ is then a quotient of $A \leq H$. The rest of this paragraph is devoted to the proof of Proposition 10.

Lemma 3 *Let $H \in \mathrm{Sub}(G)$. Then the following are equivalent:*

(1) $\mathscr{C}(H)$ does not contain the trivial subgroup in its closure.

(2) There exist a finite set $P \in G\backslash\{1\}$ all of whose conjugates intersect H.

Proof The sets $\{H \leq G \mid H \cap P = \emptyset\}$ form a basis of the trivial subgroup in $\mathrm{Sub}(G)$, when P ranges over finite subsets of non-trivial elements. ☐

Lemma 4 *Suppose X has no isolated points and let $g_1, \dots, g_n \in \mathrm{Homeo}(X)$ be non-trivial. Then there are (non-empty) open $U_1, \dots, U_n \subseteq X$ such that $U_1, \dots, U_n,$ $g_1(U_1), \dots, g_n(U_n)$ are pairwise disjoint.*

For a proof of Lemma 4, see [27, Lemma 3.1].

Lemma 5 (B.H. Neumann) *Assume that a group Δ can be written $\Delta = \bigcup_{i=1}^{n} \Delta_i \gamma_i$ as a finite union of cosets of subgroups Δ_i, $i \in \{1, \dots, n\}$. Then at least one of the Δ_i has index at most n in Δ.*

For a proof of Lemma 5, see [33, Lemma 4.1].

As a consequence of Lemma 5, if $\Delta = \bigcup_{i=1}^{n} Y_i$ is a finite union of arbitrary subsets $Y_i \subseteq \Delta$, then putting $\Delta_i = \langle \gamma\delta^{-1}, \gamma, \delta \in Y_i \rangle$, we obtain that at least one of the Δ_i has index at most n in Δ.

We now go into the proof of Proposition 10. For the sake of simplicity we will only give the proof for $K = H$. As noted above, this is enough to obtain Theorem 8.

Proof (Proposition 10) Assume (1) does not hold. We prove (2). By Lemma 3 there is $P = \{g_1, \ldots, g_n\} \subseteq G \backslash \{1\}$ so that gPg^{-1} intersects H for all $g \in G$. Then Lemma 4 yields $U_1, \ldots, U_n \subseteq X$ open and non-empty so that $U_1, \ldots, U_n, g_1(U_1), \ldots, g_n(U_n)$ are disjoint. Let L be the subgroup generated by the subgroups G_{U_i} for $i \in \{1, \ldots, n\}$. Then $L = G_{U_1} \times \cdots \times G_{U_n}$. By definition of P we may write $L = \bigcup_{i=1}^{n} Y_i$, where $Y_i = \{g \in L \mid ggig^{-1} \in H\}$. By Lemma 5 there is $l \geq 1$ such that $\Delta_l = \langle \gamma \delta^{-1}, \gamma, \delta \in Y_l \rangle$ has finite index in L. Now consider for $\gamma, \delta \in Y_l$ the element $a_{\gamma, \delta} = (\gamma g_l \gamma^{-1})(\delta g_l \delta^{-1}) \in H$. Set $A = \langle a_{\gamma, \delta} \mid \gamma, \delta \in Y_l \rangle$. Let $\pi : L \to G_{U_l}$ be the canonical projection. Then $\Gamma = \pi(\Delta_l)$ has finite index in G_{U_l}.

Lemma 6 *For all $\gamma, \delta \in Y_l$ the element $a_{\gamma, \delta}$ leaves U_l invariant and coincides with $\gamma \delta^{-1}$ on U_l.*

Proof The statement follows from the definition of $U_1, \ldots, U_n$ and the fact that δ, γ are supported in $\bigcup_{i=1}^{n} U_i$. We leave the details to the reader. □
This lemma yields the conclusion because A will map onto Γ. □

Although we did not use it here, the existence of a neighbourhood W in Proposition 10 which is uniform for all conjugates of H is important for other applications. As we shall now briefly explain, we are able to say more on URS's of G if the action of G on X enjoys additional properties.

Definition 15 The action $G \curvearrowright X$ is *extremely proximal* if for every closed subset $C \subsetneq X$, there is $x \in X$ so that for every neighbourhood U of x, there is $g \in G$ such that $g(C) \subseteq U$.

Example 7

(1) Assume that G has an action on a locally finite tree T which is minimal and of general type. Then the action of $G \curvearrowright \partial T$ is extremely proximal.
(2) The action of Thompson's group F on $\mathbb{S}^1$ is extremely proximal.
(3) The action of Thompson's group V on $\partial T_{2,2}$ is extremely proximal.

The conclusion of the following result is much stronger than the one of Theorem 8 for the reason that we obtain some information about *subgroups* of non-trivial URS's, whereas Theorem 8 only deals with their subquotients.

Theorem 9 ([27]) *Suppose $G \curvearrowright X$ is extremely proximal. Let $\mathscr{H} \in \mathrm{URS}(G)$ be non-trivial, and let $H \in \mathscr{H}$. Then there is a non-empty open subset $U \subset X$ and a finite index subgroup Γ of G_U such that $[\Gamma, \Gamma] \leq H$.*

If we strengthen again the assumption on the dynamics of the action of G on X, we obtain sufficient conditions ensuring that the stabilizer URS of the action of G on X (see Proposition 8) is actually the only URS of G, apart from the points $\{1\}$ and $\{G\}$. This statement applies for example to Thompson's groups T and V, and to the groups $G(F, F')$ under appropriate assumptions on the permutation groups F and F'. See [27] for the proof.

Theorem 10 ([27]) *Let X be a compact space, and G $\curvearrowright$ X a minimal and extremely proximal action. Suppose that for every $U \subseteq X$ and $\Gamma \leq_{f.i.} G_U$, there is an open subset $V \subseteq X$ with $G_V \subseteq [\Gamma, \Gamma]$, and that point stabilizers G_x ($x \in X$) are maximal subgroups of G. Then the only URS's of G are $\{1\}$, $\{G\}$ and $S_G(X)$.*

We shall now explain the applications of these results to several classes of groups. We refer the reader to [27] for more applications.

5.4 Thompson's Groups

Recall that Thompson's group F is the group of piecewise $ax + b$ homeomorphisms of the interval $[0, 1]$, with finitely many pieces which are intervals with dyadic rationals endpoints, and where $a \in 2^{\mathbb{Z}}$ and $b \in \mathbb{Z}[1/2]$. Thompson's group T admits a similar description as group of homeomorphisms of the circle $\mathbb{S}^1$ [9], and Thompson's group V has been defined (as group of homeomorphisms of the Cantor set) in Sect. 3.

Theorem 11 (Classification of URS's of Thompson's Groups [27])

(1) The URS's of F are the normal subgroups of F. (Apart from the trivial subgroup, these are precisely the subgroups of F containing the commutator subgroup.)
(2) The URS's of T are $\{1\}$, $\{T\}$ and $S_T(\mathbb{S}^1)$.
(3) The URS's of V are $\{1\}$, $\{V\}$ and $S_V(\partial T_{2,2})$.

Since the stabilizer URS associated to the action $V \curvearrowright \partial T_{2,2}$ is non-amenable (see Example 6), by Theorem 7 we deduce the following result.

Corollary 4 *Thompson's group V is C^*-simple.*

It is a notorious open question to determine whether Thompson's group F is amenable. In 2014, Haagerup and Olesen [19] proved that in case Thompson's group T is C^*-simple, then Thompson's group F must be non-amenable. Theorem 11 shows that the problems of C^*-simplicity of T and non-amenability of F are actually equivalent, and that it is also equivalent to the C^*-simplicity of F. We refer to [27] for details (see also the references given there for some partial converse of the Haagerup–Olesen result previously obtained by Bleak–Juschenko and Breuillard–Kalantar–Kennedy–Ozawa).

5.5 Branch Groups

In this paragraph T will be a rooted tree, and $\mathrm{Aut}(T)$ will be the automorphism group of T. For a subgroup $G \leq \mathrm{Aut}(T)$ and a vertex $v \in V(T)$, we define

$$\mathrm{Rist}_G(v) = \{g \in G \mid g \text{ is supported inside the subtree below } v\}.$$

Furthermore, for $m \geq 1$, we set $\mathrm{Rist}_G(m) = \langle \mathrm{Rist}_G(v) \mid v \text{ is at level } m \rangle$.

Definition 16 A group G is a *branch group* if G acts transitively on each level of T and $\mathrm{Rist}_G(m)$ has finite index in G for all $m \geq 1$.

Many well-studied examples of branch groups are amenable, e.g. the Grigorchuk group and the Gupta-Sidki group. But non-amenable branch groups also exist:

Theorem 12 (Sidki–Wilson [42]) *There are finitely generated branch groups containing the free group* $\mathbb{F}_2$.

The following result shows that the class of branch groups satisfies the following strong dichotomy.

Theorem 13 ([27]) *A countable branch group is either amenable or C^*-simple.*

5.6   Piecewise Projective Homeomorphisms of $\mathbb{R}$

The group $\mathrm{PSL}(2, \mathbb{R})$ acts by Mobius transformations on the projective line $\mathbb{P}^1(\mathbb{R})$. Let $A \subseteq \mathbb{R}$ be a subring of $\mathbb{R}$, and define $H(A)$ to be the group of homeomorphisms of $\mathbb{R}$ which are piecewise $\mathrm{PSL}(2, A)$, with finitely many pieces, the endpoints of the pieces being endpoints of hyperbolic elements of $\mathrm{PSL}(2, A)$.

The recent interest in these groups comes from the work of Monod [30], who showed that they provide new examples answering the so-called von Neumann-Day problem:

Theorem 14 (Monod) *If A is a dense subring of $\mathbb{R}$ (e.g. $A = \mathbb{Z}[\sqrt{2}]$) then $H(A)$ is non-amenable and does not contain free subgroups.*

Lodha and Moore have then found a finitely presented subgroup $G_0 \leq H(\mathbb{R})$ which remains non-amenable [29].

Theorem 15 ([27]) *Retain the assumption of Theorem 14. Then the group $H(A)$ is C^*-simple. Moreover the Lodha-Moore group G_0 is C^*-simple.*

We shall mention that, although the conclusion of Theorem 15 on the group $H(A)$ is formally stronger than the one of Theorem 14, the non-amenability of the group $H(A)$ is used in an essential way in the proof of Theorem 15.

It was a question of de la Harpe [17] whether there exist countable C^*-simple groups without free subgroups. This question has been answered in the positive by Olshanskii and Osin [34]. The examples given there are finitely generated, but not finitely presented. Theorem 15 provides the first examples of finitely presented C^*-simple groups without free subgroups.

Acknowledgements I wish to thank the organizers of the program *Winter of Disconnectedness* which took place in Newcastle, Australia, in 2016, for the invitation to give this series of lectures, and also for the encouragement to make these notes available. I would particularly like to thank Colin Reid and George Willis for welcoming me so warmly. ERC grant #278469 partially supported my participation in this program, and this support is gratefully acknowledged. Finally thanks are due to Ben Brawn for pointing out typos and mistakes in a former version of these notes.

References

1. Abért, M., Glasner, Y., Virág, B.: Kesten's theorem for invariant random subgroups. Duke Math. J. **163**(3), 465–488 (2014)
2. Arzhantseva, G., Minasyan, A.: Relatively hyperbolic groups are C^*-simple. J. Funct. Anal. **243**(1), 345–351 (2007)
3. Bader, U., Caprace, P.-E., Gelander, T., Mozes, S.: Simple groups without lattices. Bull. Lond. Math. Soc. **44**(1), 55–67 (2012)
4. Banks, C., Elder, M., Willis, G.: Simple groups of automorphisms of trees determined by their actions on finite subtrees. J. Group Theory **18**(2), 235–261 (2015)
5. Bekka, M., Cowling, M., de la Harpe, P.: Some groups whose reduced C^*-algebra is simple. Publ. Math. l'IHÉS **80**, 117–134 (1994)
6. Bridson, M., de la Harpe, P.: Mapping class groups and outer automorphism groups of free groups are C^*-simple. J. Funct. Anal. **212**(1), 195–205 (2004)
7. Breuillard, E., Kalantar, M., Kennedy, M., Ozawa, N.: C^*-simplicity and the unique trace property for discrete groups (2014). Preprint. arXiv:1410.2518
8. Burger, M., Mozes, S.: Groups acting on trees: from local to global structure. Publ. Math. l'IHÉS **92**, 113–150 (2000)
9. Cannon, J.W., Floyd, W.J., Parry, W.R.: Introductory notes on Richard Thompson's groups. Enseign. Math. (2) **42**(3-4), 215–256 (1996)
10. Caprace, P.-E.: Automorphism groups of right-angled buildings: simplicity and local splittings (2012). Preprint. arXiv:1210.7549
11. Caprace, P.E., De Medts, T.: Simple locally compact groups acting on trees and their germs of automorphisms. Transformation Groups **16**(2), 375–411 (2011)
12. Caprace, P.-E., Monod, N.: Decomposing locally compact groups into simple pieces. Math. Proc. Camb. Philos. Soc. **150**, 97–128 (2011)
13. Caprace, P.-E., Reid, C., Willis, G.: Locally normal subgroups of totally disconnected groups. Part I: general theory (2013). arXiv:1304.5144v1
14. Caprace, P.-E., Reid, C., Willis, G.: Locally normal subgroups of totally disconnected groups. Part II: compactly generated simple groups (2014). arXiv:1401.3142v1
15. Dahmani, F., Guirardel, V., Osin, D.: Hyperbolically embedded subgroups and rotating families in groups acting on hyperbolic spaces (2011). Preprint. arXiv:1111.7048
16. de la Harpe, P.: Groupes hyperboliques, algebres d'opérateurs et un théoreme de jolissaint. CR Acad. Sci. Paris Sér. I Math. **307**(14), 771–774 (1988)
17. de la Harpe, P.: On simplicity of reduced C^*-algebras of groups. Bull. Lond. Math. Soc. **39**(1), 1–26 (2007)
18. Glasner, E., Weiss, B.: Uniformly recurrent subgroups. Recent Trends in Ergodic Theory and Dynamical Systems. Contemporary Mathematics, vol. 631, pp. 63–75. American Mathematical Society, Providence, RI, (2015)
19. Haagerup, U., Olesen, K.K.: The thompson groups t and v are not inner-amenable (2014). Preprint
20. Haglund, F., Paulin, F.: Simplicité de groupes d'automorphismes d'espaces a courbure négative. Geometry and Topology Monographs, vol. 1. International Press, Vienna (1998)
21. Kalantar, M., Kennedy, M.: Boundaries of reduced C^*-algebras of discrete groups. J. Reine Angew. Math. **727**, 247–267 (2017)
22. Kapoudjian, C.: Simplicity of Neretin's group of spheromorphisms. Ann. Inst. Fourier **49**, 1225–1240 (1999)
23. Kennedy, M.: Characterizations of C^*-simplicity (2015). arXiv:1509.01870v3
24. Lazarovich, N.: On regular cat (0) cube complexes (2014). Preprint. arXiv:1411.0178
25. Le Boudec, A.: Compact presentability of tree almost automorphism groups. Ann. Inst. Fourier (Grenoble) **67**(1), 329–365 (2017)
26. Le Boudec, A.: Groups acting on trees with almost prescribed local action. Comment. Math. Helv. **91**(2), 253–293 (2016)

27. Le Boudec, A., Matte Bon, N.: Subgroup dynamics and C^*-simplicity of groups of homeomorphisms (2016). arXiv:1605.01651
28. Le Boudec, A., Wesolek, P.: Commensurated subgroups in tree almost automorphism groups (2016). arXiv preprint arXiv:1604.04162
29. Lodha, Y., Moore, J.: A nonamenable finitely presented group of piecewise projective homeomorphisms. Groups Geom. Dyn. **10**(1), 177–200 (2016)
30. Monod, N.: Groups of piecewise projective homeomorphisms. Proc. Natl. Acad. Sci. **110**(12), 4524–4527 (2013)
31. Möller, R., Vonk, J.: Normal subgroups of groups acting on trees and automorphism groups of graphs. J. Group Theory **15**, 831–850 (2012)
32. Neretin, Yu.A.: Combinatorial analogues of the group of diffeomorphisms of the circle. Izv. Ross. Akad. Nauk Ser. Mat. **56**(5), 1072–1085 (1992)
33. Neumann, B.H.: Groups covered by permutable subsets. J. Lond. Math. Soc. **29**, 236–248 (1954)
34. Olshanskii, A., Osin, D.: C^*-simple groups without free subgroups (2014). arXiv:1401.7300v3
35. Paschke, W., Salinas, N.: C^*-algebras associated with free products of groups. Pac. J. Math. **82**(1), 211–221 (1979)
36. Pays, I., Valette, A.: Sous-groupes libres dans les groupes d'automorphismes d'arbres. Enseign. Math., Rev. Int., IIe Sér. 151–174 (1991)
37. Powers, R.T.: Simplicity of the C^*-algebra associated with the free group on two generators. Duke Math. J. **42**(1), 151–156 (1975)
38. Poznansky, T.: Characterization of linear groups whose reduced C^*-algebras are simple (2008). Preprint. arXiv:0812.2486
39. Sauer, R., Thumann, W.: Topological models of finite type for tree almost automorphism groups (2015). Preprint. arXiv:1510.05554
40. Serre, J.-P.: Trees, pp. ix+142. Springer, Berlin (1980)
41. Shalom, Y., Willis, G.: Commensurated subgroups of arithmetic groups, totally disconnected groups and adelic rigidity. Geom. Funct. Anal. **23**(5), 1631–1683 (2013)
42. Sidki, S., Wilson, J.S.: Free subgroups of branch groups. Arch. Math. **80**(5), 458–463 (2003)
43. Swiatoslaw, G., Gismatullin, J., with an appendix by N. Lazarovich, Uniform symplicity of groups with proximal action (2016). Preprint. arXiv:1602.08740
44. Tits, J.: Sur le groupe d'automorphismes d'un arbre. Essays on Topology and Related Topics, pp. 188–211. Springer, Berlin (1970)

Normal Subgroup Structure of Totally Disconnected Locally Compact Groups

Colin D. Reid

Abstract The present article is a summary of joint work of the author and Phillip Wesolek on the normal subgroup structure of totally disconnected locally compact second-countable (t.d.l.c.s.c.) groups. The general strategy is as follows: We obtain normal series for a t.d.l.c.s.c. group in which each factor is 'small' or a non-abelian chief factor; we show that up to a certain equivalence relation (called association), a given non-abelian chief factor can be inserted into any finite normal series; and we obtain restrictions on the structure of chief factors, such that the restrictions are invariant under association. Some limitations of this strategy and ideas for future work are also discussed.

1 Introduction

A common theme throughout group theory is the reduction of problems concerning a group G to those concerning the normal subgroup N and the quotient G/N, where both N and G/N have some better-understood structure; more generally, one can consider a decomposition of G via normal series. This approach has been especially successful for the following classes of groups: finite groups, profinite groups, algebraic groups, connected Lie groups and connected locally compact groups. To summarise the situation for these classes, let us recall the notion of chief factors and chief series.

Definition 1 Let G be a Hausdorff topological group. A **chief factor** K/L of G is a pair of closed normal subgroups $L < K$ such that there are no closed normal subgroups of G lying strictly between K and L. A **descending chief series** for G is a (finite or transfinite) series of closed normal subgroups $(G_\alpha)_{\alpha \leq \beta}$ such that $G = G_0$, $\{1\} = G_\beta$, $G_\lambda = \bigcap_{\alpha < \lambda} G_\alpha$ for each limit ordinal and each factor $G_\alpha/G_{\alpha+1}$ is chief.

C.D. Reid (✉)
University of Newcastle, University Drive, Callaghan, NSW 2308, Australia
e-mail: colin@reidit.net

© Springer International Publishing AG, part of Springer Nature 2018　　　　　525
D.R. Wood et al. (eds.), *2016 MATRIX Annals*, MATRIX Book Series 1,
https://doi.org/10.1007/978-3-319-72299-3_23

First, on the existence of chief series (or a good approximation thereof):

- Every finite group G has a finite chief series.
- Every profinite group has a descending chief series with finite chief factors.
- Every algebraic group has a finite normal series in which the factors are Zariski-closed and either abelian or a semisimple chief factor.
- Every connected Lie group has a finite normal series in which the factors are in the following list:

  connected centreless semisimple Lie groups; finite groups of prime order; $\mathbb{R}^n$, $\mathbb{Z}^n$ or $(\mathbb{R}/\mathbb{Z})^n$ for some n.

  We can also choose the series so that all factors are chief factors, except possibly for some occurrences of $\mathbb{Z}^n$ or $(\mathbb{R}/\mathbb{Z})^n$.
- Every connected locally compact group G has a descending series in which the factors come from connected Lie groups. G has a unique largest compact normal subgroup K, and all but finitely many factors of the series occur below K. (This can be generalised to the class of pro-Lie groups; see for example [7]. The fact that connected locally compact groups are pro-Lie is a consequence of the Gleason–Yamabe theorem.)

Second, on the structure of the factors occurring in such a series:

- A finite chief factor is a direct product of copies of a simple group.
- A chief factor that is a semisimple algebraic group is a direct product of finitely many copies of a simple algebraic group.
- A chief factor that is a semisimple Lie group is a direct product of finitely many copies of an abstractly simple connected Lie group.
- Finite simple groups, simple connected Lie groups and simple algebraic groups have been classified.

So given a group G in the above well-behaved classes, there exists a decomposition of G into 'known' groups. Moreover, it turns out that the non-abelian chief factors we see up to isomorphism are an invariant of G (not dependent on how we constructed the series).

Given the success of this approach to studying connected locally compact groups, one would hope to obtain analogous results for totally disconnected, locally compact (t.d.l.c.) groups. The ambition is expressed in the title of a paper of Pierre-Emmanuel Caprace and Nicolas Monod: 'Decomposing locally compact groups into simple pieces' [4]; similar approaches can also be seen in previous work of Marc Burger and Shahar Mozes [3] and of Vladimir Trofimov (see for instance [15]). We will not attempt to summarise these articles here; instead, we will note some key insights in [4] that are relevant to the project at hand.

1. It is advantageous to work with **compactly generated** t.d.l.c. groups, i.e. groups G such that $G = \langle X \rangle$ for some compact subset X. The advantage will be explained in Sect. 2 below. In this context, and more generally, it is no great loss to restrict attention to the second-countable (t.d.l.c.s.c.) case, that is, t.d.l.c. groups that have a countable base for the topology.

2. The class of t.d.l.c.s.c. groups includes all countable discrete groups. We cannot expect to develop a general theory of chief series for all such groups, and in any case, such a theory would lie beyond the tools of topological group theory. So instead, it is useful to have methods to ignore or exclude discrete factors.

3. Although compact groups are relatively well-behaved, in a given t.d.l.c.s.c. group there are likely to be many compact normal factors, and the tools for analysing them are of a different nature than those for studying the 'large-scale' structure of t.d.l.c. groups. Thus, as with the discrete factors, it is useful to find ways to ignore or exclude compact factors.

4. Given closed normal subgroups K and L of a locally compact group G, their product KL is not necessarily closed. In particular, $\overline{KL}/L$ need not be isomorphic to $K/(K \cap L)$.

5. To accommodate the previous point, the authors introduce a generalisation of the direct product, called a quasi-product (see Sect. 4.1 below). They show that compactly generated chief factors (as long as they are not compact, discrete or abelian) are quasi-products of finitely many copies of a topologically simple group.

6. A topologically simple group S can have dense normal subgroups; this fact turns out to be closely related to the existence of quasi-products of topologically simple groups that are not direct products.

Points (2) and (3) above immediately suggest a modification to the definition of chief series. We will restrict attention here to finite series; this will turn out to be sufficient for the analysis of compactly generated t.d.l.c.s.c. groups.

Definition 2 Let G be a t.d.l.c. group. An **essentially chief series** is a series

$$\{1\} = G_0 < G_1 < \cdots < G_n = G$$

of closed normal subgroups of G, such that for $1 \leq i \leq n$, the factor G_{i+1}/G_i is either compact, discrete, or a chief factor of G.

With point (5), there are two important caveats:

(a) A chief factor of a compactly generated t.d.l.c.s.c. group need not be itself compactly generated.

(b) Non-compactly generated chief factors can be quasi-products of finitely or infinitely many topologically simple groups, but they are *not necessarily* of this form.

These caveats are an important contrast with the situation of connected locally compact groups and account for much of the difficulty in developing a complete theory of normal subgroup structure for t.d.l.c. groups. In particular, we see that an essentially chief series does not by itself lead to a decomposition into simple factors, even if one is prepared to ignore all compact, discrete and abelian factors.

Based on the observations and results of Caprace–Monod, Burger–Mozes and Trofimov, the author and Phillip Wesolek have started a project to analyse the normal subgroup structure of t.d.l.c.s.c. groups by means of chief factors. Our proposed programme is as follows:

1. Obtain an essentially chief series for compactly generated t.d.l.c.s.c. groups.
2. Find a way to handle non-abelian chief factors that is independent of the choice of normal series, in other words, obtain 'uniqueness' or 'invariance' results.
3. Analyse (recursively) the chief factor structure of chief factors of t.d.l.c.s.c. groups. Try to 'reduce' to simple groups and low-complexity characteristically simple groups. Here 'low-complexity' means elementary with decomposition rank $\xi(G) \leq \alpha$, where α is some specified countable ordinal; it will turn out that a natural threshold to take here is $\alpha = \omega + 1$. (See Sect. 5 below for a brief discussion of decomposition rank.)
4. Develop a structure theory for the low-complexity characteristically simple t.d.l.c. groups and how these are built out of compactly generated and discrete groups. The most important case here appears to be the class of elementary t.d.l.c.s.c. groups of decomposition rank 2.
5. Find general properties of classes of topologically simple t.d.l.c. groups. Some general results have been obtained for compactly generated topologically simple groups: see [5]. In generalising from the compactly generated case, it is likely that some kind of non-degeneracy assumption must be made at the level of compactly generated subgroups to obtain useful structural results.

The goal of the rest of this article is to give an overview of progress made in this project to date. In this summary, some arguments will be sketched out for illustration, but for the full details it will be necessary to consult the articles [11, 12] and [13]. We focus for the most part on points (1)–(3) above; in the last section, some ideas for further work will be presented.

2 Compactly Generated Groups

2.1 The Cayley–Abels Graph

A finitely generated group G has a **Cayley graph**: this is a connected, locally finite graph Γ on which G acts vertex-transitively with trivial vertex stabilisers. Moreover, Γ is unique up to quasi-isometry.

Herbert Abels [1] showed that something similar is true for compactly generated t.d.l.c. groups G. Our strategy for obtaining an essentially chief series for G will be to use induction on the *degree* of the corresponding graph; to obtain the right notion of degree for this induction, we must be careful with the definition of graph we use (especially for the quotient graph; see Definition 4 below).

Definition 3 A **graph** Γ is a pair of sets $V\Gamma$ (vertices) and $E\Gamma$ (edges) together with functions $o : E\Gamma \rightarrow V\Gamma$ and $r : E\Gamma \rightarrow E\Gamma$ such that $r^2 = \mathrm{id}_{E\Gamma}$. (Given $e \in E\Gamma$, we do not require $r(e) \neq e$.) An **automorphism** α is a pair of bijections

α_V and α_E on $V\Gamma$ and $E\Gamma$ such that $o \circ \alpha_E = \alpha_V \circ o$ and $r \circ \alpha_E = \alpha_E \circ r$. (When clear from the context, we will omit the subscripts V and E.) Define $t(e) := o(r(e))$.

Given $v \in V\Gamma$, the **degree** $\deg(v)$ of v is defined to be $|o^{-1}(v)|$; Γ is **locally finite** if every vertex has finite degree. The **degree** of the graph Γ is $\deg(\Gamma) := \sup_{v \in V\Gamma} \deg(v)$.

Γ is **simple** if $t(e) \neq o(e)$ for all $e \in E\Gamma$ and the map $e \mapsto (o(e), t(e))$ is injective on $E\Gamma$. In this case, we can simply regard $E\Gamma$ as a symmetric binary relation on $V\Gamma$, identifying each edge with the pair $(o(e), t(e))$.

Let G be a compactly generated t.d.l.c. group. A **Cayley–Abels graph** for G is a graph Γ equipped with an action of G by automorphisms such that:

(i) Γ is connected and locally finite;
(ii) G acts transitively on $V\Gamma$;
(iii) For each $x \in V\Gamma \cup E\Gamma$, the stabiliser G_x is a compact open subgroup of G.

Theorem 1 (Abels [1]) *Let G be a compactly generated t.d.l.c. group.*

(i)   *For every compact open subgroup U of G, there is a simple Cayley–Abels graph with vertex set G/U;*
(ii) *Any two Cayley–Abels graphs are quasi-isometric.*

Recall that by Van Dantzig's theorem, every t.d.l.c. group has a base of identity neighbourhoods consisting of compact open subgroups, so Theorem 1(i) in particular ensures the existence of a Cayley–Abels graph for G.

The following lemma is a more detailed version of Theorem 1(i); we give a proof here as an illustration of the advantages of working with compact open subgroups. (The proof of Theorem 1(ii) is entirely analogous to that for Cayley graphs of finitely generated groups.)

Lemma 1 *Let G be a compactly generated t.d.l.c. group, let U be a compact open subgroup of G and let A be a compact symmetric subset of G such that $G = \langle U, A \rangle$.*

(i)   *There exists a finite symmetric subset B of G such that*

$$BU = UB = UBU = UAU.$$

(ii) *For any subset B satisfying part (i), then $G = \langle B \rangle U$ and the coset space G/U carries the structure of a simple locally finite connected graph, invariant under the natural G-action, where gU is adjacent to hU if and only if $(gU)^{-1}hU \subseteq UBU \setminus U$.*

Proof

(i)   The product of compact sets is compact, by continuity of multiplication. Thus UAU is a compact set. On the other hand, U is an open subgroup of G; thus G is covered by left cosets of U and finitely many suffice to cover UAU. That is, we have $UAU \subseteq \bigcup_{b \in B_1} bU$ such that B_1 is a finite subset of G. Moreover, we see that UAU is itself a union of left cosets of U; since the cosets partition G, we can in fact ensure $UAU = \bigcup_{b \in B_1} bU$. Now take $B = B_1 \cup B_1^{-1}$; it is easily verified that B satisfies the required equations.

(ii) Since $BU = UAU$, we have $A \in \langle U, B \rangle$; since $G = \langle U, A \rangle$, it follows that $G = \langle U, B \rangle$. Since $BU = UB$ and B is symmetric, we have $\langle U, B \rangle = \langle B \rangle U$. Now define a simple graph Γ with vertex set G/U and edges specified by the given adjacency relation. Note that gU is adjacent to hU if and only if $g^{-1}h \in UBU \setminus U$; in particular, we see that no vertex is adjacent to itself. Since $UBU \setminus U$ is a symmetric set, we have $g^{-1}h \in UBU \setminus U$ if and only if $h^{-1}g \in UBU \setminus U$, so the adjacency relation is symmetric.

We let G act on G/U by left translation. To show that G acts on the graph, it is enough to see that it preserves adjacency: given distinct vertices gU and hU, we note that $(xgU)^{-1}xhU = Ug^{-1}x^{-1}xhU = Ug^{-1}hU$, so (gU, hU) is an edge if and only if (xgU, xhU) is. The action of G is clearly also vertex-transitive. The graph is connected because xbU is either equal or adjacent to xU for all $b \in B$, and we have $G = \langle B \rangle U$. To show that Γ is locally finite, it suffices to see that $o^{-1}(U)$ is finite: specifically, we see that $o^{-1}(U) = \{(bU, U) \mid b \in B\}$, and hence $|o^{-1}(U)| \leq |B|$.

□

Define the **degree** $\deg(G)$ of a compactly generated t.d.l.c. group G to be the smallest degree of a Cayley–Abels graph of G. We can imagine the degree as analogous to 'dimension' or 'number of generators', depending on context.

The key difference between Cayley–Abels graphs and Cayley graphs is that vertex stabilisers are not necessarily trivial. In particular, it is useful to consider the action of a vertex stabiliser on the edges incident with that vertex.

Definition 4 Let G be a group acting on a graph Γ. Define the **local action** of G at v to be the permutation group induced by the action of G_v on $o^{-1}(v)$.

The **quotient graph** Γ/G is the graph with vertex set $\overline{V} = \{Gv \mid v \in V\Gamma\}$, edge set $\overline{E} = \{Ge \mid e \in E\Gamma\}$, such that $o(Ge) = G(o(e))$ and $r(Ge) = G(r(e))$.

If the action of G is vertex-transitive, we can refer to 'the' local action on Γ without reference to a specific vertex, since the action of G_v on $o^{-1}(v)$ will be permutation-isomorphic to the action of G_w on $o^{-1}(w)$.

Cayley–Abels graphs are well-behaved on passing to quotients. Moreover, we have good control of the degree.

Proposition 1 (See [12, Proposition 2.16]) *Let G be a compactly generated t.d.l.c. group, let Γ be a Cayley–Abels graph for G and let K be the kernel of the action of G on Γ. Let H be a closed normal subgroup of G.*

(i) *Γ/H is a Cayley–Abels graph for G/H.*

(ii) *We have $\deg(\Gamma/H) \leq \deg(\Gamma)$, with equality if and only if the local action of H is trivial. In particular, $\deg(G/H) \leq \deg(G)$.*

(iii) *Suppose that the local action of H on Γ is trivial. Then $H \cap K$ is a compact normal subgroup of G and $H/(H \cap K)$ is a discrete normal factor of G.*

Proof (sketch) For (i), one can show that the vertex Hv of Γ/H has stabiliser $G_v H/H$, which is a compact open subgroup of G, and that the graph is locally finite (see proof of part (ii)). The other conditions are clear.

For (ii), given $v \in V\Gamma$, we have a surjection ϕ from $o^{-1}(v)$ to $o^{-1}(Hv)$, since $o^{-1}(Hv) = Ho^{-1}(v)$. Thus $\deg(Hv) \leq \deg(v)$, with equality if and only if ϕ is injective. We see that ϕ is injective if and only if different edges incident with v lie in different H-orbits, which occurs if and only if H has trivial local action.

For (iii), we observe that for all $v \in V\Gamma$, then H_v fixes every edge incident with v, and hence every vertex adjacent to v. Since Γ is connected, it follows by induction on the distance from v that H_v fixes every $w \in V\Gamma$ and hence also every edge of Γ. Thus $H \cap G_v = H_v = H \cap K$. Clearly $H \cap K$ is normal; it is compact since K is compact; the equality $H \cap K = H \cap G_v$ shows that $H \cap K$ is open in H. Thus $H/(H \cap K)$ is discrete. $\qquad\square$

Remark 1 It remains an outstanding problem to classify non-discrete t.d.l.c. groups G with $\deg(G) = 3$, that is, non-discrete groups that act vertex-transitively with compact open stabilisers on a graph of degree 3. One can show (see for instance [6, Theorem 8.A.20]) that all such groups arise as $G = \widetilde{G}/D$, where $\widetilde{G}$ is a group acting on a regular tree T of degree 3 with the same local action, D is a discrete normal subgroup with trivial local action, and Γ arises as the quotient graph T/D. Moreover, it can be seen that there is a group $\widetilde{G} \leq H \leq \mathrm{Aut}(T)$, such that H has the same orbits on directed edges as $\widetilde{G}$ does and H is in the following list:

$$U(C_2), \ U(\mathrm{Sym}(3))_\delta, \ U(\mathrm{Sym}(3)),$$

where C_2 is a point stabiliser in $\mathrm{Sym}(3)$, $U(F)$ denotes the Burger-Mozes universal group with local action F (see [3]), and $U(F)_\delta$ is the stabiliser of an end in $U(F)$. (Note that $U(\mathrm{Sym}(3))_\delta$ has local action C_2.) So the structure of t.d.l.c. groups of degree 3 in principle reduces to understanding the subgroup structure of these three specific groups. At present, the least well-understood of these is $U(C_2)$.

2.2 Existence of Essentially Chief Series

We now reach our first goal, to show the existence of essentially chief series for compactly generated t.d.l.c.s.c. groups. In fact, given what is already known in the connected case, the result holds for all compactly generated locally compact groups.

Theorem 2 (See [12, Theorem 1.3]) *For every compactly generated locally compact group G, there is a finite series*

$$\{1\} = G_0 < G_1 < G_2 < \cdots < G_n = G$$

of closed normal subgroups of G, such that each G_{i+1}/G_i is compact, discrete or a chief factor of G.

Cayley–Abels graphs are used via the following lemma.

Lemma 2 (See [12, Lemma 3.1]) *Let G be a compactly generated t.d.l.c. group and Γ be a Cayley–Abels graph for G. Let $\mathscr{C}$ be a chain of closed normal subgroups of G.*

(i) Let $A = \overline{\bigcup_{H \in \mathscr{C}} H}$. Then $\deg(\Gamma/A) = \min\{\deg(\Gamma/H) \mid H \in \mathscr{C}\}$.
(ii) Let $D = \bigcap_{H \in \mathscr{C}} H$. Then $\deg(\Gamma/D) = \max\{\deg(\Gamma/H) \mid H \in \mathscr{C}\}$.

Proof For (i), it is enough to show that there exists $H \in \mathscr{C}$ such that A has trivial local action on Γ/H. This amounts to showing that there is some $H \in \mathscr{C}$ such that H_v and A_v have the same orbits on $o^{-1}(v)$, in other words $A_v = A_{v,1} H_v$, where $A_{v,1}$ is the subgroup of A fixing every edge in $o^{-1}(v)$. The existence of a suitable $H \in \mathscr{C}$ follows from the finiteness of the quotient $A/A_{v,1}$.

For (ii), given Proposition 1, we can assume $D = \{1\}$ without loss of generality. It is then enough to show that there exists $H \in \mathscr{C}$ that has trivial local action on Γ, in other words, such that $H \cap G_v \le G_{v,1}$. We see that $G_{v,1}$ is an open subgroup of the compact group G_v; since $\mathscr{C}$ is a chain of subgroups with trivial intersection, it follows by a compactness argument that indeed $H \cap G_v \le G_{v,1}$ for some $H \in \mathscr{C}$. $\quad\square$

Proof (Sketch Proof of Theorem 2) We will only consider the case when G is totally disconnected. Proceed by induction on $\deg(G)$; let Γ be a Cayley–Abels graph of smallest degree.

By Lemma 2(i) plus Zorn's lemma, there is a closed normal subgroup A that is maximal amongst closed normal subgroups such that $\deg(\Gamma/A) = \deg(\Gamma)$. By Proposition 1, there is a compact normal subgroup K of G such that $K \le A$ and A/K is discrete, and Γ/A is a Cayley–Abels graph for G/A.

By the maximality of A, we see that any closed normal subgroup of G that properly contains A will produce a quotient graph of Γ/A of smaller degree. By Lemma 2(ii), every chain of non-trivial closed normal subgroups of G/A has non-trivial intersection. By Zorn's lemma, there is a minimal closed normal subgroup D/A of G/A; in other words, D/A is a chief factor of G. We then have $\deg(\Gamma/D) < \deg(\Gamma/A)$, so $\deg(G/D) < \deg(G)$. By induction, G/D has an essentially chief series. We form an essentially chief series for G by combining the series for G/D with the G-invariant series $1 \le K \le A < D$ we have obtained for D. $\quad\square$

Lemma 2 and Proposition 1 also easily lead to chain conditions on closed normal subgroups, which are independently useful for understanding normal subgroup structure in t.d.l.c. groups.

Theorem 3 (See [12, Theorem 3.2]) *Let G be a compactly generated locally compact group and let $(G_i)_{i \in I}$ be a chain of closed normal subgroups of G.*

(i) For $K = \overline{\bigcup_i G_i}$, there exists i such that K/G_i has a compact open G-invariant subgroup.
(ii) For $L = \bigcap_i G_i$, there exists i such that G_i/L has a compact open G-invariant subgroup.

3 Equivalence Classes of Chief Factors

We have just seen that a compactly generated t.d.l.c.s.c. group G has an essentially chief series. However, the proof is non-constructive, and in general there could be many different essentially chief series without any natural choice of series. To obtain canonical structural properties of G, we wish to establish properties of essentially chief series that do not depend on the choices involved. In particular, we would like to say that the same factors always appear up to equivalence. In the process, we will obtain tools that are valid in a much more general setting; in particular, compact generation will not play a large role in this section.

In fact, many of the results in this section are naturally proved in the context of **Polish groups**, that is, topological groups G such that as a topological space, G is completely metrizable and has a countable dense set. A locally compact group is Polish if and only if it is second-countable; here we see the main technical motivation for our focus on t.d.l.c.s.c. groups as opposed to more general t.d.l.c. groups.

Let K and L be closed normal subgroups of a t.d.l.c.s.c. (more generally, Polish) group G. Consider the following normal series for G:

$$\{1\} \leq (K \cap L) \leq K \leq \overline{KL} \leq G;$$

$$\{1\} \leq (K \cap L) \leq L \leq \overline{KL} \leq G.$$

We want to think of these two series as having the same factors up to reordering. Specifically, $K/(K \cap L)$ corresponds to $\overline{KL}/L$ and $L/(K \cap L)$ to $\overline{KL}/K$.

In a discrete group, in fact $K/(K \cap L)$ is isomorphic to $\overline{KL}/L$ and $L/(K \cap L)$ is isomorphic to $\overline{KL}/K$, by the second isomorphism theorem. This is not true in the locally compact context.

Example 1 Let $G = \mathbb{Z}[\frac{1}{2}] \times \mathbb{Z}_2$, let $K = \{(x, 0) \mid x \in \mathbb{Z}[\frac{1}{2}]\}$ and let $L = \{(-y, y) \mid y \in \mathbb{Z}\}$. Then $K \cap L$ is trivial and $\overline{KL} = G$. We see that $K \cong \mathbb{Z}[\frac{1}{2}]$ and $L \cong \mathbb{Z}$, but $\overline{KL}/L \cong \mathbb{Q}_2$ and $\overline{KL}/K \cong \mathbb{Z}_2$.

We must therefore relax the notion of isomorphism to obtain a suitable equivalence relation on the chief factors.

On the other hand, there is a similarity between $K/(K \cap L)$ and $\overline{KL}/L$ that is not captured by group isomorphism, namely that the map

$$\varphi : K/(K \cap L) \to \overline{KL}/L; \ k(K \cap L) \mapsto kL$$

is a *G-equivariant* map with respect to the natural actions. In particular, we can exploit the fact that the image KL/L is a normal subgroup of G/L.

3.1   Normal Compressions

Definition 5 A **normal compression** of topological groups is a continuous homomorphism $\psi : A \to B$, such that ψ is injective and $\psi(A)$ is a dense normal subgroup of B. For example, there are natural normal compressions $\mathbb{Z} \to \mathbb{Z}_2$ and $\bigoplus \mathrm{Sym}(n) \to \prod \mathrm{Sym}(n)$.

An **internal compression** in a topological group G is a map

$$\varphi : K_1/L_1 \to K_2/L_2; \; kL_1 \mapsto kL_2,$$

where K_1/L_1 and K_2/L_2 are normal factors of G such that $K_2 = \overline{K_1 L_2}$ and $L_1 = K_1 \cap L_2$.

Given the ambient group G, we can also just say that K_2/L_2 is an internal compression of K_1/L_1, as the map φ is uniquely determined; given a normal compression $\psi : A \to B$, we will also simply say that B is a normal compression of A when the choice of ψ is clear from the context or not important.

The equations $K_2 = \overline{K_1 L_2}$ and $L_1 = K_1 \cap L_2$ are exactly what is needed to ensure φ is well-defined and injective with dense image; in other words, every internal compression is a normal compression. Conversely, in the class of t.d.l.c.s.c. groups (more generally, Polish groups), it turns out that every normal compression can be realised as an internal compression.

Let $\psi : A \to B$ be a normal compression. Then there is a natural action θ of B on A, which is specified by the equation

$$\psi(\theta(b)(a)) = b\psi(a)b^{-1}; \; a \in A, b \in B.$$

Write $A \rtimes_\psi B$ for the semidirect product formed by this action. It is easily seen that $A \rtimes_\psi B$ is a group; what is less clear is that the action of B on A is jointly continuous, so that the product topology on $A \rtimes_\psi B$ is a group topology. The joint continuity in this case follows from classical results on the continuity of maps between Polish spaces; see for example [8, (9.16)].

Proposition 2 ([11, Proposition 3.5]) *Let $\psi : A \to B$ be a normal compression where A and B are t.d.l.c.s.c. groups (Polish groups). Then $A \rtimes_\psi B$ with the product topology is a t.d.l.c.s.c. group (respectively, a Polish group).*

Here is an easy application.

Corollary 1 *Let $\psi : A \to B$ be a normal compression where A and B are Polish groups. Let K be a closed normal subgroup of A. Then $\psi(K)$ is normal in B.*

Proof We can identify K with the closed subgroup $K \times \{1\}$ of the semidirect product $G = A \rtimes_\psi B$. By Proposition 2, G is a Hausdorff topological group; in particular, the normaliser of any closed subgroup is closed. Thus $N_G(K)$ is closed in G. Moreover, $N_G(K)$ contains both A and $\psi(B)$, so $N_G(K)$ is dense in G and hence $N_G(K) = G$. In particular, K is preserved by the action of B on A, so that $\psi(K)$ is normal in B. □

We can use the semidirect product to factorise the normal compression map. We also see that the normal compression is realised as an internal compression of normal factors of the semidirect product.

Theorem 4 ([11, Theorem 3.6]) *Let* $\psi : A \rightarrow B$ *be a normal compression where* A *and* B *are Polish groups. Let* $\iota : A \rightarrow A \rtimes_\psi B$ *be given by* $a \mapsto (a, 1)$ *and* $\pi : A \rtimes_\psi B \rightarrow B$ *be given by* $(a, b) \mapsto \psi(a)b$.

- (i) $\psi = \pi \circ \iota$;
- (ii) ι *is a closed embedding;*
- (iii) π *is a quotient homomorphism and* $A \rightarrow \ker \pi$; $a \mapsto (a^{-1}, \psi(a))$ *is an isomorphism of topological groups.*

Corollary 2 *Let* $\psi : A \rightarrow B$ *be a normal compression where* A *and* B *are Polish groups. Then* ψ *is realised as an internal compression*

$$\gamma : A/\{1\} \rightarrow (A \rtimes_\psi B)/\ker \pi.$$

In the context of t.d.l.c.s.c. groups, instead of factorising the normal compression through $A \rtimes B$, we can factorise through $(A \rtimes U)/\Delta$, where U is a compact open subgroup of B and $\Delta = \{(w^{-1}, \psi(w)) \mid w \in W\}$, where W is a compact open subgroup of A such that $\psi(W)$ is normal in U. This allows us to be obtain tighter control over the relationship between A and B.

Theorem 5 ([13, Theorem 4.4]; see also [5, Proposition 5.17]) *Let* $\psi : A \rightarrow B$ *be a normal compression where* A *and* B *are t.d.l.c.s.c. groups. Let* U *be a compact open subgroup of* B. *Then there is a t.d.l.c. group* C *and continuous homomorphisms* $\alpha : A \rightarrow C$ *and* $\beta : C \rightarrow B$ *with the following properties:*

- (i) $\psi = \beta \circ \alpha$;
- (ii) α *is a closed embedding and* $C = \alpha(A)\widetilde{U}$ *with* $\widetilde{U} \cong U$;
- (iii) β *is a quotient homomorphism,* $\ker \beta$ *is discrete, and every element of* $\ker \beta$ *lies in a finite conjugacy class of* C.

As an example application, the following can be deduced from Theorem 5 together with standard properties of amenable groups.

Corollary 3 (See also [13, Proposition 5.6]) *Let* $\psi : A \rightarrow B$ *be a normal compression where* A *and* B *are t.d.l.c.s.c. groups. Then* A *is amenable if and only if* B *is amenable.*

3.2 The Association Relation and Chief Blocks

We now define a relation that will provide the promised equivalence relation on chief factors.

Definition 6 Say K_1/L_1 is **associated** to K_2/L_2 (write $K_1/L_1 \sim K_2/L_2$) if the following conditions are satisfied:

(i) $\overline{K_1 L_2} = \overline{K_2 L_1}$;
(ii) $K_i \cap \overline{L_1 L_2} = L_i$ for $i = 1, 2$.

Note that if K_1/L_1 and K_2/L_2 are associated, then K/L is an internal compression of both of them, where $K = \overline{K_1 K_2}$ and $L = \overline{L_1 L_2}$.

The **centraliser** $C_G(K/L)$ of a normal factor K/L is

$$C_G(K/L) := \{g \in G \mid \forall k \in K : [g, k] \in L\}.$$

In particular, $C_G(K/L)$ is a closed normal subgroup of G such that $L \leq C_G(K/L)$.

Using the fact that centralisers of (not necessarily closed) subsets of Hausdorff groups are closed, it is easy to see that the association relation preserves centralisers. For non-abelian chief factors, the converse holds.

Proposition 3 ([11, Proposition 6.8]) *Let K_1/L_1 and K_2/L_2 be normal factors of the topological group G.*

(i) If $K_1/L_1 \sim K_2/L_2$, then $C_G(K_1/L_1) = C_G(K_2/L_2)$.
(ii) If $C_G(K_1/L_1) = C_G(K_2/L_2)$ and if K_1/L_1 and K_2/L_2 are non-abelian chief factors of G, then they are associated.

Corollary 4 *Association defines an equivalence relation on the non-abelian chief factors of a topological group.*

Given a non-abelian chief factor K/L, define the **(chief) block** $\mathfrak{a} := [K/L]$ to be the class of non-abelian chief factors associated to K/L. Define also $C_G(\mathfrak{a}) = C_G(K/L)$.

At this point, the benefit of the additional abstraction of chief blocks is not clear. However, we will see in the rest of the article that chief blocks, and more generally sets of chief blocks, can usefully be manipulated in a way that would be awkward to do directly at the level of chief factors.

Association exactly characterises the *uniqueness* of occurrences of chief factors in normal series:

Theorem 6 ([11, Proposition 7.8]) *Let G be a Polish group, let*

$$\{1\} = G_0 \leq G_1 \leq \cdots \leq G_n = G$$

be a finite normal series for G, and let $\mathfrak{a}$ be a chief block of G. Then there is exactly one $i \in \{1, \ldots, n\}$ for which there exist $G_{i-1} \leq B < A \leq G_i$ with $A/B \in \mathfrak{a}$. Specifically, G_i is the lowest term in the series such that $G_i \not\leq C_G(\mathfrak{a})$.

We write $\mathfrak{B}_G$ for the set of chief blocks of G. Note that $\mathfrak{B}_G$ comes equipped with a partial order: we say $\mathfrak{a} \leq \mathfrak{b}$ if $C_G(\mathfrak{a}) \leq C_G(\mathfrak{b})$. Equivalently, we have $\mathfrak{a} < \mathfrak{b}$ if in every finite normal series (G_i) that includes representatives G_i/G_{i-1} and G_j/G_{j-1} of $\mathfrak{a}$ and $\mathfrak{b}$ respectively, then $G_j > G_i$.

3.3   Robust Blocks

Given a compactly generated t.d.l.c.s.c. group, it would be tempting to infer that every possible chief block is represented as a factor in every essentially chief series. However, this is not true: there can be infinitely many compact and discrete chief factors up to association, yet only finitely many of them will be represented in any given essentially chief series. We need to exclude compact and discrete factors in a way that is invariant under association.

Compactness and discreteness themselves are *not* invariant under association, even amongst non-abelian chief factors. However, there is a related property that is invariant.

Definition 7 The **quasi-centre** $QZ(G)$ of a topological group G is the set of all elements $x \in G$ such that $C_G(x)$ is open in G. A t.d.l.c.s.c. group G is **quasi-discrete** if its quasi-centre is dense.

Discrete factors of a t.d.l.c.s.c. group are certainly quasi-discrete. Profinite *chief* factors are direct products of finite simple groups, so they are also quasi-discrete (see for instance [14, Lemma 8.2.3]).

In a t.d.l.c.s.c. group (more generally, in any Polish group), a closed subgroup has countable index if and only if it is open; in particular, an element is quasi-central if and only if its conjugacy class is countable. It also follows from second-countability that every dense subgroup contains a countable dense subgroup. Consequently, a t.d.l.c.s.c. group is quasi-discrete if and only if it has a countable dense normal subgroup. Given a normal compression $\psi : A \to B$, if A has a countable dense normal subgroup D, then $\psi(D)$ is a countable dense *subnormal* subgroup of B, which does not allow us to conclude directly that B is quasi-discrete. However, quasi-discreteness is sufficiently well-behaved under normal compressions that the following holds.

Theorem 7 (See [13, Theorem 7.15]) *Let $\mathfrak{a}$ be a chief block of a t.d.l.c.s.c. group G. Then either all representatives of $\mathfrak{a}$ are quasi-discrete, or none of them are.*

It now makes sense to define a class of chief blocks that excludes quasi-discrete chief factors.

Definition 8 A chief factor K/L of a t.d.l.c.s.c. group is **robust** if it is not quasi-discrete; equivalently, $QZ(K/L) = \{1\}$. We say a chief block $\mathfrak{a}$ is **robust** if all (equivalently, some) of its representatives are robust.

Because robust chief factors cannot be associated to compact or discrete chief factors, we obtain the following corollary of Theorems 6 and 7.

Corollary 5 *Let G be a compactly generated t.d.l.c.s.c. group and let*

$$\{1\} = A_0 \leq A_1 \leq \cdots \leq A_m = G \text{ and } \{1\} = B_0 \leq B_1 \leq \cdots \leq B_n = G$$

be essentially chief series for G. Then the association relation induces a bijection between $\{A_i/A_{i-1} \text{ robust} \mid 1 \leq i \leq m\}$ and $\{B_j/B_{j-1} \text{ robust} \mid 1 \leq j \leq n\}$.

Consequently, the set $\mathfrak{B}_G^r$ of robust blocks of G is finite, and each robust block is represented exactly once in the factors of any given essentially chief series.

3.4 Canonical Representatives of Chief Blocks

We now obtain canonical representatives for the chief blocks. To discuss the relationship between normal subgroups and chief blocks, it will be useful to define what it means for a normal subgroup or factor to cover a block:

Definition 9 Let G be a t.d.l.c.s.c. group, let $\mathfrak{a}$ be a chief block and let $K \geq L$ be a closed normal subgroup of G. Say K/L **covers** $\mathfrak{a}$ if there exists $L \leq B < A \leq K$ such that $A/B \in \mathfrak{a}$. We say K covers $\mathfrak{a}$ if $K/\{1\}$ does.

Note that by Theorem 6, given any chief block $\mathfrak{a}$ and normal factor K/L, there are three mutually exclusive possibilities:

- L covers $\mathfrak{a}$, which occurs if and only if $L \not\leq C_G(\mathfrak{a})$;
- G/K covers $\mathfrak{a}$, which occurs if and only if $K \leq C_G(\mathfrak{a})$;
- K/L covers $\mathfrak{a}$, which occurs if and only if $L \leq C_G(\mathfrak{a})$ and $K \not\leq C_G(\mathfrak{a})$.

In particular, $C_G(\mathfrak{a})$ is the unique largest normal subgroup of G that does not cover $\mathfrak{a}$. Thus we obtain a canonical representative for $\mathfrak{a}$, the **uppermost representative**:

Proposition 4 ([11, Proposition 7.4]) *Let $\mathfrak{a}$ be a chief block of a Polish group G. Then $G/C_G(\mathfrak{a})$ has a unique smallest closed normal subgroup $G^{\mathfrak{a}}/C_G(\mathfrak{a})$. Given any $A/B \in \mathfrak{a}$, then $G^{\mathfrak{a}}/C_G(\mathfrak{a})$ is an internal compression of A/B.*

For the existence of an analogous lowermost representative, there would need to be a smallest closed normal subgroup K of G such that K covers $\mathfrak{a}$, in other words, $K \not\leq C_G(\mathfrak{a})$. An easy commutator argument shows that the set $\mathcal{K}$ of closed normal subgroups K such that $K \not\leq C_G(\mathfrak{a})$ is closed under *finite* intersections. However, in general we cannot expect $\mathcal{K}$ to be closed under arbitrary intersections. Consider for instance the situation when G is a finitely generated non-abelian discrete free group and G/N is an infinite simple group. Then $\mathfrak{a} = [G/N]$ is covered by every finite index normal subgroup and G is residually finite, so $\mathcal{K}$ has trivial intersection; yet the trivial group clearly does not cover $\mathfrak{a}$.

We say $\mathfrak{a}$ is **minimally covered** if there is in fact a least element $G_{\mathfrak{a}}$ of $\mathcal{K}$, in other words, $\mathcal{K}$ is closed under arbitrary intersections. The normal factor $G_{\mathfrak{a}}/C_{G_{\mathfrak{a}}}(\mathfrak{a})$ is then the **lowermost representative** of $\mathfrak{a}$.

Proposition 5 ([11, Proposition 7.13]) *Let $\mathfrak{a}$ be a minimally covered block of a Polish group G. Then $G_{\mathfrak{a}}$ has a unique largest closed G-invariant subgroup $C_{G_{\mathfrak{a}}}(\mathfrak{a})$. Given any $A/B \in \mathfrak{a}$, then A/B is an internal compression of $G_{\mathfrak{a}}/C_{G_{\mathfrak{a}}}(\mathfrak{a})$.*

One can picture a minimally covered block $\mathfrak{a}$ as a kind of bottleneck in the lattice $\mathscr{L}$ of closed normal subgroups of G. More precisely, $\mathscr{L}$ is partitioned into a principal filter and a principal ideal: every closed normal subgroup K of G satisfies exactly one of the inclusions $K \geq G_{\mathfrak{a}}$ or $K \leq C_G(\mathfrak{a})$.

In contrast to the situation for discrete groups, we find that as soon as we restrict to *robust* blocks of compactly generated groups, we do in fact obtain a lowermost representative. This is not so surprising when one considers that the minimally covered property is essentially a chain condition on closed normal subgroups, and that just such a chain condition is provided by Theorem 3.

Proposition 6 ([12, Proposition 4.10]) *Let $\mathfrak{a}$ be a robust block of a compactly generated t.d.l.c.s.c. group G. Then $\mathfrak{a}$ is minimally covered.*

To summarise the situation for compactly generated t.d.l.c.s.c. groups: to any t.d.l.c.s.c. group G we have associated two canonical finite sets of chief factors, namely the uppermost representatives and the lowermost representatives of the robust blocks. Moreover, given an *arbitrary* chief factor K/L of G, then either K/L is quasi-discrete, or else K/L interpolates between the lowermost and uppermost representatives of the corresponding block $\mathfrak{a} = [K/L]$, in the sense that we have internal compressions

$$G_{\mathfrak{a}}/C_{G_{\mathfrak{a}}}(\mathfrak{a}) \to K/L \to G^{\mathfrak{a}}/C_G(\mathfrak{a}).$$

The minimally covered property will also be important later, when studying blocks of characteristically simple groups (in particular, those groups that arise as chief factors of some larger group).

Normal compressions respect several of the properties of non-abelian chief factors discussed so far.

Theorem 8 (See [11, §8]) *Let $\psi : A \to B$ be a normal compression of t.d.l.c.s.c. groups. Then there is a canonical bijection $\widetilde{\psi} : \mathfrak{B}_A \to \mathfrak{B}_B$ such that, for $\mathfrak{a}, \mathfrak{b} \in \mathfrak{B}_A$:*

(i) $\mathfrak{a} \leq \mathfrak{b}$ if and only if $\widetilde{\psi}(\mathfrak{a}) \leq \widetilde{\psi}(\mathfrak{b})$;
(ii) $\mathfrak{a}$ is robust if and only if $\widetilde{\psi}(\mathfrak{a})$ is robust;
(iii) $\mathfrak{a}$ is minimally covered if and only if $\widetilde{\psi}(\mathfrak{a})$ is minimally covered.

Corollary 6 *Let K_1/L_1 and K_2/L_2 be associated non-abelian chief factors of a t.d.l.c.s.c. group G. Then $\mathfrak{B}_{K_1/L_1}$ and $\mathfrak{B}_{K_2/L_2}$ are canonically isomorphic as partially ordered sets, in a way that preserves the robust blocks and the minimally covered blocks.*

4 The Structure of Chief Factors

We now turn our attention from the existence and uniqueness of chief factors, to the structure of a chief factor $H = K/L$ as a topological group in its own right. Alternatively, we are interested in the structure of t.d.l.c.s.c. groups H that are **(topologically) characteristically simple**, meaning that a non-trivial subgroup N of H that is preserved by every automorphism of H as a topological group is necessarily dense in H.

Recall that our ambition in this article is "decomposing groups into simple pieces". Accordingly, we will not attempt to decompose further a t.d.l.c.s.c. group H that is topologically simple, that is, such that every *normal* subgroup is dense. If H is a chief factor of a t.d.l.c.s.c. group G that is not topologically simple, then H has a non-trivial lattice of closed normal subgroups and we can investigate the action of G on this lattice. (Analogously, if H is a characteristically simple group, we can investigate the action of $\mathrm{Aut}(H)$ on the lattice of closed normal subgroups.) Of course, we can take advantage of the fact that canonical structures arising from the collection of normal subgroups, such as the partially ordered set $\mathfrak{B}_H$ of chief blocks of H, or the subset $\mathfrak{B}_H^{\min}$ of minimally covered blocks, must also be preserved by automorphisms of H. However, here we run into the difficulty that the strong existence results we have so far for (minimally covered) chief factors only apply to *compactly generated* t.d.l.c.s.c. groups, and there is no reason for H to be compactly generated, even if G is.

In this section, we will focus attention on the situation where H has at least one minimally covered block. In Sect. 6 we will see that in fact, we can ensure the existence of minimally covered blocks of H quite generally, even without compact generation, as long as H has sufficient 'topological group complexity'.

4.1   Quasi-Products

Apart from being topologically simple, the tamest normal subgroup structure we can hope for in H is that H resembles a direct product of topologically simple groups, in that it has a (finite or countable) collection $\{S_i \mid i \in I\}$ of closed normal subgroups, each a copy of a topologically simple group S, such that H contains the direct sum of the S_i as a dense subgroup. However, even in this situation, the copies of S_i may be combined in a more complicated way than a direct product. (For one thing, the direct product of infinitely many non-compact groups is not even locally compact.) We now introduce a definition of quasi-product, generalising the definition of Caprace–Monod in order to account for possibly infinite sets of quasi-factors.

Definition 10   Let G be a topological group and let $\mathscr{S}$ be a set of non-trivial closed normal subgroups of G. Given $I \subseteq \mathscr{S}$, define $G_I := \overline{\langle N \in I \rangle}$.

$(G, \mathscr{S})$ is a **quasi-product** (or that G is a quasi-product of $\mathscr{S}$) if $G_{\mathscr{S}} = G$ and the map

$$d : G \mapsto \prod_{N \in \mathscr{S}} \frac{G}{G_{\mathscr{S} \setminus N}}; \ g \mapsto (g G_{\mathscr{S} \setminus N})_{N \in \mathscr{S}}$$

is injective. We then say $\mathscr{S}$ is a set of **quasi-factors** of G.

We have already seen a general situation in which quasi-products occur. The following is an easy consequence of the way normal compressions factor through the semidirect product:

Corollary 7 *Let $\psi : A \to B$ be a normal compression of Polish groups and let $G = A \rtimes_\psi B$. Then G is a quasi-product of two copies of A, namely $A_1 = \{(a, 1) \mid a \in A\}$ and $A_2 = \{(a^{-1}, \psi(a)) \mid a \in A\}$. We have $G = A_1 \times A_2$ abstractly if and only if ψ is surjective.*

Quasi-products are straightforward to identify in the case of centreless groups.

Lemma 3 (See [11, Proposition 4.4]) *Let G be a topological group and let $\mathscr{S}$ be a set of closed normal subgroups of G. Suppose the centre $Z(G)$ is trivial. Then G is a quasi-product if and only if $G_{\mathscr{S}} = G$ and any two distinct elements of $\mathscr{S}$ commute.*

We can now state the Caprace–Monod structure theorem for compactly generated characteristically simple t.d.l.c. groups.

Theorem 9 ([4, Corollary D]) *Let G be a topologically characteristically simple locally compact group. Suppose that G is compactly generated and neither compact, nor discrete, nor abelian. Then G is a quasi-product of finitely many copies of a compactly generated topologically simple group S.*

Remark 2 It is unknown if the conclusion of this theorem can be improved to say that G is a *direct* product of copies of S. It would be enough to show that there is no normal compression $\psi : S \to T$ into a t.d.l.c.s.c. group where $Z(T) = \{1\}$ and $\psi(S) \neq T$. Note that given any such normal compression, T would itself be compactly generated and topologically simple, but clearly not *abstractly* simple. So the question of whether such characteristically simple groups are necessarily direct products of simple groups is closely related to the open question of whether every compactly generated topologically simple t.d.l.c.s.c. group is abstractly simple.

Away from the case of compactly generated characteristically simple groups, there are many more possibilities for quasi-products; examples are given in [4, Appendix II]. If we allow infinitely many quasi-factors, there is a general construction. Notice that if $(G_i)_{i \in \mathbb{N}}$ is a sequence of non-compact t.d.l.c.s.c. groups, then $\prod_{i \in \mathbb{N}} G_i$ cannot be locally compact. However, given a choice of compact open subgroups of G_i, there is a natural way to obtain a locally compact quasi-product of $(G_i)_{i \in \mathbb{N}}$.

Definition 11 Let $(G_i)_{i \in \mathbb{N}}$ be a sequence of t.d.l.c.s.c. groups, and for each i let U_i be a compact open subgroup of G_i. The **local direct product** $P := \bigoplus_{i \in \mathbb{N}}(G_i, U_i)$ is the set of functions from $\mathbb{N}$ to $\sqcup G_i$ (with pointwise multiplication) such that $f(i) \in G_i$ for all i and $f(i) \in U_i$ for all but finitely many i. There is a natural inclusion $\iota : \prod_{i \in \mathbb{N}} U_i \to P$; we give P the unique group topology that makes ι continuous and open.

It is easily seen that the local direct product is a t.d.l.c.s.c. group, and that it is a quasi-product with the obvious factors. In general, the isomorphism type of $\bigoplus_{i \in \mathbb{N}}(G_i, U_i)$ is sensitive to the choice of U_i as well as G_i. So there will be many different local direct products of copies of a given group. Nevertheless, all local direct products of copies of a given t.d.l.c.s.c. group occur as chief factors:

Proposition 7 *Let $(S_i)_{i \in \mathbb{N}}$ be a sequence of copies of a fixed topologically simple t.d.l.c.s.c. group S, and for each i let U_i be a compact open subgroup of S_i*

(no consistency is required in the choice of U_i). Then $\bigoplus_{i \in \mathbb{N}}(S_i, U_i)$ *occurs as a chief factor of a t.d.l.c.s.c. group.*

Proof (sketch) Let F be the group of permutations of $\mathbb{N}$ of finite support, equipped with the discrete topology. It is easily verified that F admits an action on $P = \bigoplus_{i \in \mathbb{N}}(S_i, U_i)$ given by setting $(f.g)(i) = g(f^{-1}(i))$ for $f \in F$, $g \in P$ and $i \in \mathbb{N}$. (Here we exploit the fact that P is not sensitive to the choice of any finite subset of the compact open subgroups U_i.) Moreover, the semidirect product $G := P \rtimes F$ with this action of F is a t.d.l.c.s.c. group with the product topology. We see that the intersection $\bigcap_{i \in \mathbb{N}} C_G(S_i)$ is trivial, where S_i is regarded as a subgroup of P in the natural way. Thus given a non-trivial closed normal subgroup K of G, then $[K, S_i] \neq \{1\}$ for some i, which implies that $K \geq S_i$ for that i and hence $K \geq P$. Thus P is the smallest non-trivial closed normal subgroup of G; in particular, $P/\{1\}$ is a chief factor of G.   □

To some extent, the local direct product can also be used as a model of an arbitrary t.d.l.c.s.c. quasi-product.

Theorem 10 ([11, Proposition 4.8] and [13, Corollary 6.20]) *Let* $(G, \mathscr{S})$ *be a quasi-product such that G is a t.d.l.c.s.c. group and let U be a compact open subgroup of G. Then $\mathscr{S}$ is countable. Moreover, there is a canonical normal compression*

$$\psi : \bigoplus_{N \in \mathscr{S}}(N, N \cap U) \to G$$

such that ψ restricts to the identity on each $N \in \mathscr{S}$.

4.2 Extension of Chief Blocks

If H has a closed normal subgroup S that is non-abelian and topologically simple, then in particular H has a chief factor, namely $S/\{1\}$. Clearly $S/\{1\}$ is the lowermost representative of its block, so the corresponding block is minimally covered.

If H is a chief factor of some larger group G, say $H = K/L$, we can think of it as the chief factor of G 'generated' by a chief block of K (namely, the block of K corresponding to S). This situation can be generalised to talk about how chief blocks of a closed subgroup K of G form chief blocks of G.

Definition 12 Let G be a Polish group, let H be a closed subgroup of G and let $\mathfrak{a} \in \mathfrak{B}_H$. Say that $\mathfrak{b} \in \mathfrak{B}_G$ is the **extension** of $\mathfrak{a}$ to G, and write $\mathfrak{b} = \mathfrak{a}^G$, if for every normal factor K/L of G, then K/L covers $\mathfrak{b}$ if and only if $(K \cap H)/(L \cap H)$ covers $\mathfrak{a}$.

Extensions of blocks are unique, when they exist. Extensions are also transitive: given $H \leq R \leq G$, and $\mathfrak{a} \in \mathfrak{B}_H$, we have $\mathfrak{a}^G = (\mathfrak{a}^R)^G$ whenever either side of this equation makes sense. It is not clear in general which blocks extend from which subgroups. However, extensions of *minimally covered* blocks are better-behaved. Write $\mathfrak{B}_G^{\min}$ for the set of minimally covered blocks of G.

The following extendability criterion will be useful later.

Lemma 4 *Let G be a Polish group, let K be a closed subgroup of G and let $\mathfrak{a} \in \mathfrak{B}_K^{\min}$. Then $\mathfrak{a}$ extends to G if and only if there is $\mathfrak{b} = \mathfrak{a}^G \in \mathfrak{B}_G^{\min}$ such that $G_{\mathfrak{b}} \cap K$ covers $\mathfrak{a}$ and $C_G(\mathfrak{b}) \cap K$ does not cover $\mathfrak{a}$.*

Proof Suppose $\mathfrak{a}$ extends to G, with $\mathfrak{b} = \mathfrak{a}^G$. Let $\mathcal{H}$ be the set of closed normal subgroups of G that cover $\mathfrak{b}$. Then $L \cap K$ covers $\mathfrak{a}$ for all $L \in \mathcal{H}$; since $\mathfrak{a}$ is minimally covered, $\bigcap_{L \in \mathcal{H}} L \cap K$ covers $\mathfrak{a}$; hence $\bigcap_{L \in \mathcal{H}} L$ covers $\mathfrak{b}$. Thus $\mathfrak{b}$ is minimally covered. Certainly $G_{\mathfrak{b}} \cap K$ covers $\mathfrak{a}$ and $C_G(\mathfrak{b}) \cap K$ does not cover $\mathfrak{a}$.

Conversely, suppose there exists $\mathfrak{b} \in \mathfrak{B}_G^{\min}$ such that $G_{\mathfrak{b}} \cap K$ covers $\mathfrak{a}$ and $C_G(\mathfrak{b}) \cap K$ does not cover $\mathfrak{a}$. Let L be a closed normal subgroup of K. If L covers $\mathfrak{b}$, then $L \geq G_{\mathfrak{b}}$, so $L \cap K \geq G_{\mathfrak{b}} \cap K$, and hence $L \cap K$ covers $\mathfrak{a}$. If L does not cover $\mathfrak{b}$, then $L \leq C_G(\mathfrak{b})$, so $L \cap K \leq C_G(\mathfrak{b}) \cap K$, and hence $L \cap K$ does not cover $\mathfrak{a}$. Thus $\mathfrak{b}$ is the extension of $\mathfrak{a}$ to G. $\qquad\square$

If H is normal in G, the extendability criterion is always satisfied.

Proposition 8 ([11, Proposition 9.8]) *Let G be a Polish group, let K be a closed normal subgroup and let $\mathfrak{a} \in \mathfrak{B}_K^{\min}$. Then $\mathfrak{a}$ extends to a minimally covered block $\mathfrak{b} := \mathfrak{a}^G$ of G. The lowermost representative $G_{\mathfrak{b}}/C_{G_{\mathfrak{b}}}(\mathfrak{b})$ of $\mathfrak{b}$ is formed from the following subgroups of K:*

$$G_{\mathfrak{b}} = \overline{\langle gK_{\mathfrak{a}}g^{-1} \mid g \in G \rangle}; \quad C_{G_{\mathfrak{b}}}(\mathfrak{b}) = G_{\mathfrak{b}} \cap \bigcap_{g \in G} gC_K(\mathfrak{a})g^{-1}.$$

Corollary 8 *Given a Polish group G and a closed normal subgroup K, there is a well-defined map $\theta : \mathfrak{B}_K^{\min} \to \mathfrak{B}_G^{\min}$ given by $\mathfrak{a} \mapsto \mathfrak{a}^G$.*

Since K is normal in G, we have an action of G on $\mathfrak{B}_K^{\min}$ by conjugation. We can describe the structure of θ using the partial order on $\mathfrak{B}_K^{\min}$ together with conjugation action of G.

Theorem 11 ([13, Theorem 9.13]) *Let G be a Polish group, let K be a closed normal subgroup and let $\mathfrak{a}, \mathfrak{b} \in \mathfrak{B}_K^{\min}$. Then $\mathfrak{a}^G \leq \mathfrak{b}^G$ if and only if there exists $g \in G$ such that $g.\mathfrak{a} \leq \mathfrak{b}$.*

4.3 Three Types of Chief Factor

Let G be a Polish group with K a closed normal subgroup of G, let $\theta : \mathfrak{B}_K^{\min} \to \mathfrak{B}_G^{\min}$ be the extension map and fix $\mathfrak{c} \in \mathfrak{B}_G^{\min}$. There are three possibilities for $\theta^{-1}(\mathfrak{c})$:

1. $\theta^{-1}(\mathfrak{c})$ is empty;
2. $\theta^{-1}(\mathfrak{c})$ is a non-empty antichain (in other words $\mathfrak{a} \not< \mathfrak{b}$ for all $\mathfrak{a}, \mathfrak{b} \in \theta^{-1}(\mathfrak{c})$): then $\forall \mathfrak{a}, \mathfrak{b} \in \theta^{-1}(\mathfrak{c}) \ \exists g : g.\mathfrak{a} = \mathfrak{b}$.
3. $\theta^{-1}(\mathfrak{c})$ is non-empty and not an antichain: then $\forall \mathfrak{a}, \mathfrak{b} \in \theta^{-1}(\mathfrak{c}) \ \exists g : g.\mathfrak{a} < \mathfrak{b}$.

Now let K/L be a non-abelian chief factor of the Polish group G. We may as well pass to G/L, in other words we may assume $L = \{1\}$ and K is a minimal closed normal subgroup of G. Then $\mathfrak{c} = [K/\{1\}] \in \mathfrak{B}_G^{\min}$.

We still have a map $\theta : \mathfrak{B}_K^{\min} \to \mathfrak{B}_G^{\min}$. But now, since K is itself a chief factor, we have $\mathfrak{a}^G = \mathfrak{c}$ for *every* $\mathfrak{a} \in \mathfrak{B}_K^{\min}$. So $\theta^{-1}(\mathfrak{c}) = \mathfrak{B}_K^{\min}$, and hence $\mathfrak{B}_K^{\min}$ has one of the forms (1), (2), (3) described above.

Definition 13 Let H ($= K/L$) be a topologically characteristically simple Polish group (for instance, a chief factor of some Polish group). We say H is of:

1. **weak type** if $\mathfrak{B}_H^{\min} = \emptyset$;
2. **semisimple type** if $\mathfrak{B}_H^{\min}$ is a non-empty antichain;
3. **stacking type** if $\mathfrak{B}_H^{\min}$ has a non-trivial partial order.

Note that the types are completely determined by the internal structure of H: we no longer need to refer to the ambient group.

We recall moreover that if K_1/L_1 and K_2/L_2 are associated non-abelian chief factors, then $\mathfrak{B}_{K_1/L_1}^{\min} \cong \mathfrak{B}_{K_2/L_2}^{\min}$ as partially ordered sets. So all representatives of a chief factor are of the same type, and it makes sense to talk about the type of a chief block.

To justify the terminology, we note that 'semisimple type' chief factors do indeed break up into topologically simple pieces:

Proposition 9 *Let H be a Polish chief factor of semisimple type. Then H is a quasi-product of copies of a topologically simple group.*

Proof Without loss of generality we may suppose H is a minimal non-trivial closed normal subgroup of some ambient group G. Let $\mathfrak{a} \in \mathfrak{B}_H^{\min}$ and let $K = H_\mathfrak{a}$. Note that $\overline{[K,K]}$ also covers $\mathfrak{a}$, so we must have $K = \overline{[K,K]}$.

Let $g \in G$ and suppose that K covers $g.\mathfrak{a}$. Then the lowermost representative L of $g.\mathfrak{a}$ is a subgroup of K. It follows that every subgroup that covers $\mathfrak{a}$, also covers $g.\mathfrak{a}$; this is only possible if $g.\mathfrak{a} \leq \mathfrak{a}$. Since $\mathfrak{B}_H^{\min}$ is an antichain, we must have $\mathfrak{a} = g.\mathfrak{a}$. In particular, we see that $M = C_K(\mathfrak{a})$ does not cover $g.\mathfrak{a}$ for any $g \in G$, so $M \leq \bigcap_{g \in G} C_H(g.\mathfrak{a})$. On the other hand $\bigcap_{g \in G} C_H(g.\mathfrak{a})$ is a proper G-invariant subgroup of H; by minimality we conclude that M is trivial. Thus $K/\{1\} \in \mathfrak{a}$, in other words K is a minimal non-trivial closed normal subgroup of H.

The minimality of K ensures that, whenever $g \in G$ is such that $gKg^{-1} \neq K$, then $K \cap gKg^{-1} = \{1\}$. Since both K and gKg^{-1} are normal in H, it follows that in fact $[K, gKg^{-1}] = \{1\}$. Moreover, since H is a minimal non-trivial closed normal subgroup of G, we must have $H = \overline{\langle \mathscr{S} \rangle}$ where $\mathscr{S} = \{gKg^{-1} \mid g \in G\}$. Since H is non-abelian and characteristically simple, $Z(H) = \{1\}$. Since distinct elements of $\mathscr{S}$ commute, we conclude by Lemma 3 that $(H, \mathscr{S})$ is a quasi-product. In particular H is a quasi-product of copies of K and there is an internal compression from K to H/C, where $C = \overline{\langle \mathscr{S} \setminus K \rangle}$. We see that H/C is a representative of $\mathfrak{a}$, so H/C has no proper non-trivial closed normal (equivalently, H-invariant) subgroups and hence is topologically simple. By Corollary 1, every non-trivial closed normal subgroup of K has dense image in H/C. It can then be seen [11, Proposition 3.8] that every non-trivial closed normal subgroup of K contains the derived group of K; since K is topologically perfect, we conclude that K is topologically simple. $\qquad\square$

At this level of generality, weak type does not give us much to work with. As far as we know, a characteristically simple Polish group could be very complicated, but nevertheless not have any minimally covered blocks. The situation is different in the class of t.d.l.c.s.c. groups, as we will see in Sect. 6: here we have a precise notion of complexity, and we can control the structure of high-complexity chief factors via essentially chief series of compactly generated open subgroups.

The most interesting of the three types (and in some sense the generic type, at least in t.d.l.c.s.c. groups) is stacking type. If K is a minimal closed normal subgroup of G of stacking type, then K has a characteristic collection

$$\mathscr{N} = \{K_\mathfrak{a} \mid \mathfrak{a} \in \mathfrak{B}_K^{\min}\}$$

of closed normal subgroups, such that for all $A, B \in \mathscr{N}$ (including the case $A = B$), there exists $g \in G$ such that $A < gBg^{-1}$. To put this another way, we have a characteristic collection $\mathscr{C}$ of chief factors of K (specifically, the lowermost representatives of elements of $\mathfrak{B}_K^{\min}$), such that for every pair $A_1/B_1, A_2/B_2 \in \mathscr{C}$, then a G-conjugate of A_1/B_1 appears as a normal factor of the *outer automorphism group* of A_2/B_2 induced by K.

4.4 Examples of Chief Factors of Stacking Type

To see that stacking type chief factors occur naturally in the class of t.d.l.c. groups, we consider a construction of groups that act on trees, fixing an end. This construction and generalisations will be discussed in detail in a forthcoming article.

Let $T_\to$ be a tree (not necessarily locally finite) in which every vertex has degree at least 3, with a distinguished end δ. We define $\mathrm{Aut}(T_\to)$ to be the group of graph automorphisms that fix δ, equipped with the usual permutation topology (equivalently, the compact-open topology). Then there is a function f from $VT_\to$ to $\mathbb{Z}$ with the following properties:

(a) For every edge e of the tree, we have $|f(o(e)) - f(t(e))| = 1$;
(b) We have $f(t(e)) > f(o(e))$ if and only if e lies on a directed ray towards δ.

Thus $f(v)$ increases as we approach δ. The function f is unique up to an additive constant; its set $\{f^{-1}(i) \mid i \in \mathbb{Z}\}$ of fibres is therefore uniquely determined. The fibres are the **horospheres** centred at δ, and the sets $\{v \in VT_\to \mid f(v) \geq i\}$ for $i \in \mathbb{Z}$ are the **horoballs** centred at δ.

We also have an associated partial order on $VT_\to$: say $v \leq w$ if there is a path from v to w in the direction of δ, in other words, a path $v_0 v_1 \ldots v_n$, with $v = v_0$ and $w = v_n$, such that $f(v_{i-1}) < f(v_i)$ for $1 \leq i \leq n$.

Now let G be a topological group acting faithfully and continuously on $T_\rightarrow$, such that G fixes δ. For each vertex $v \in VT_\rightarrow$, define the **rigid stabiliser** $\mathrm{rist}_G(v)$ of v to be the subgroup of G that fixes every vertex w such that $w \not< v$ (including v itself). Let $G_i = \langle \mathrm{rist}_G(v) \mid v \in f^{-1}(i) \rangle$. Note that given $w \in VT_\rightarrow$ such that $f(w) < i$, then there exists $v \in f^{-1}(i)$ such that $w < v$ and hence $\mathrm{rist}_G(w) \le \mathrm{rist}_G(v)$. In particular, we have $G_i \le G_{i+1}$ for all $i \in \mathbb{Z}$. Since G preserves the set of horospheres, for every $g \in G$, there exists j such that $f(gv) = f(v) + j$ for all $v \in VT_\rightarrow$, and hence $gG_ig^{-1} = G_{i+j}$ for all $i \in \mathbb{Z}$. Thus $E = \overline{\bigcup_{i\in\mathbb{Z}} G_i}$ is a closed normal subgroup of G. Under some fairly mild assumptions, E is actually a minimal non-trivial normal subgroup of G; in particular, E is a chief factor of G.

Proposition 10 *Let G and E be as described above. Suppose that:*

(a) *For all $v \in f^{-1}(0)$, the group $\mathrm{rist}_G(v)$ is topologically perfect and does not fix any end of $T_\rightarrow$ other than δ;*
(b) *There exists $h \in G$ and $v \in VT_\rightarrow$ such that $f(hv) \ne f(v)$.*

Then E is a minimal non-trivial normal subgroup of G.

Proof Condition (a) ensures that E is non-trivial. Let K be a non-trivial closed subgroup of E, such that K is normal in G. We must show that $K = E$.

Condition (b) in fact ensures that h has hyperbolic action on T, with δ as one of the ends of the axis of h. Without loss of generality $f(hv) = f(v) + j$ for all $v \in VT_\rightarrow$, where $j > 0$. Consequently G_0 is not normal in G, and indeed $E = \overline{\bigcup_{n\ge 0} h^n G_i h^{-n}}$ for any given $i \in \mathbb{Z}$. Note also that $f(gv) = f(v)$ for all $g \in E$.

Let $v \in VT_\rightarrow$ be such that v is not fixed by K. There is then $n \in \mathbb{Z}$ such that $f(h^{n-1}v) < 0$ but $f(h^n v) \ge 0$. Let $w \in VT_\rightarrow$ be such that $h^{n-1}v < w \le h^n v$ and $f(w) = 0$. Then $h^n v$ is not fixed by K, say $kh^n v \ne h^n v$. We see that $kw \ne w$ but $f(kw) = f(w)$, and hence $\mathrm{rist}_G(w)$ and $\mathrm{rist}_G(kw)$ have disjoint support. In particular, y and kzk^{-1} commute for all $y, z \in \mathrm{rist}_G(w)$. Given $y, z \in \mathrm{rist}_G(w)$, we therefore have

$$[y, z] = [y, z(kz^{-1}k^{-1})] = [y, [z, k]] \in K.$$

Since $\mathrm{rist}_G(w)$ is topologically perfect, we conclude that $\mathrm{rist}_G(w) \le K$. Since $h^{n-1}v < w$, we see that $\mathrm{rist}_G(h^{n-1}v) \le \mathrm{rist}_G(w) \le K$; by conjugating by powers of h, it follows that $\mathrm{rist}_G(h^m v) \le K$ for all $m \in \mathbb{Z}$.

Since $\mathrm{rist}_G(w)$ does not fix any end of $T_\rightarrow$, we see that K does not preserve the axis of h. In particular, we could have chosen v to lie on the axis of h. Let us assume we have done so.

Now let $w' \in VT_\rightarrow$ be arbitrary. Then for n sufficiently large (depending on w') we have $w' \le h^n v$, and hence $\mathrm{rist}_G(w') \le \mathrm{rist}_G(h^n v) \le K$. So $\mathrm{rist}_G(w') \le K$ for every vertex w', and hence $K = E$. $\qquad\square$

It is clear that E is not of semisimple type, so to obtain a chief factor of stacking type, it suffices to impose conditions to ensure the existence of a minimally covered block of E. We leave the details to the interested reader.

Example 2 Let $T_\to$ be a regular locally finite tree of degree $d \geq 6$ with a distinguished end δ. Let G be the subgroup of $\mathrm{Aut}(\mathbf{T})$, such that the local action at every vertex is the alternating group $\mathrm{Alt}(d)$ of degree d. Then for each $v \in VT$, the rigid stabiliser $\mathrm{rist}_G(v)$ is an iterated wreath product of copies of $\mathrm{Alt}(d-1)$. The subgroup E described in Proposition 10, which in this case is actually the set of all elliptic elements of G, is then a chief factor of G of stacking type.

Considering the above example, one might still imagine that stacking type chief factors, by virtue of being characteristically simple, are 'built out of topologically simple groups' in an easily-understood way. The following much more general construction, which is inspired by the construction of Adrien Le Boudec in [9], should strike a cautionary note for any attempt to reduce the classification of chief factors to the topologically simple case.

Example 3 Let $T_\to$ be a tree with a distinguished end δ, such that every vertex has a countably infinite set of neighbours. We set a colouring function $\sigma : ET_\to \to \mathbb{N}$, such that $\sigma \circ r = \sigma$, there is a ray towards δ in which every edge has the colour 1, and at every vertex v, σ restricts to a bijection c_v between the set $t^{-1}(v)$ of in-edges and $\mathbb{N}$. Given $h \in \mathrm{Aut}(T_\to)$, the **local action** of h at v is a permutation of $\mathbb{N}$ given by $\sigma(h, v) = c_{hv}^{-1} \circ h \circ c_v$.

Let P be a transitive t.d.l.c.s.c. subgroup of $\mathrm{Sym}(\mathbb{N})$ in the permutation topology, and let U be a compact open subgroup of P. Define $E(P, U)$ to consist of all elements h of $\mathrm{Aut}(T_\to)$ such that $\sigma(h, v) \in G$ for all $v \in VT_\to$ and $\sigma(h, v) \in U$ for all but finitely many vertices. We see that $E(P, U)$ is a subgroup of $\mathrm{Aut}(T_\to)$. At the moment it is not locally compact, but we can rectify this by choosing a new topology.

Let $v \in VT$ and consider the stabiliser $E(U, U)_v$ of v in $E(U, U)$. It is straightforward to see that $E(U, U)_v$ is a closed profinite subgroup of $\mathrm{Aut}(T_\to)$. Moreover, there is a unique group topology on $E(P, U)$ so that the inclusion of $E(U, U)_v$ is continuous and open; this topology does not depend on the choice of v. We now equip $G = E(P, U)$ with this topology, and see that G is a t.d.l.c.s.c. group.

Regardless of the choice of P and U, the group G acts transitively on the vertices of v, and for each horosphere $f^{-1}(i)$, the fixator $f^{-1}(i)$ is a closed subgroup of G_i of G that is quasi-product of the rigid stabilisers of vertices in $f^{-1}(i)$. In particular, we see that for all $i \in \mathbb{Z}$, we have $G_i/G_{i-1} \cong \bigoplus_{j\in\mathbb{N}}(P, U)$. In turn, every element of G with elliptic action on $T_\to$ can be approximated in the topology of G by elements of $\bigcup_{i\in\mathbb{Z}} G_i$. Thus $E = \bigcup_{i\in\mathbb{Z}} G_i$ is a closed subgroup of G consisting of all elliptic elements of G. We see that $E(U, U)_v \leq E$, so E is open, and in fact $G \cong E \rtimes \mathbb{Z}$ as a topological group.

Suppose now that P is topologically perfect. It then follows that $\mathrm{rist}_G(v)$ is topologically perfect, and hence E is a chief factor of G by Proposition 10.

Given a t.d.l.c.s.c. group P, then P occurs as a transitive subgroup of $\mathrm{Sym}(\mathbb{N})$ provided that P does not have arbitrarily small compact normal subgroups. There are also many examples of topologically perfect t.d.l.c.s.c. groups; a general construction is to take the normal closure of $\mathrm{Alt}(5)$ in $G \wr \mathrm{Alt}(5)$, where G is some given t.d.l.c.s.c. group. So the conditions on P for Example 3 to produce a chief factor are quite weak, and by no means ensure that P has a well-understood (sub-)normal subgroup lattice. At the same time, a local direct product of copies of P appears as a normal factor of the chief factor E. So we have effectively buried the subnormal subgroup structure of P inside a chief factor E of another t.d.l.c.s.c. group. One can also take the resulting group E, set $P_2 = E$, and repeat the construction, iterating to produce increasingly complex chief factors.

5   Interlude: Elementary Groups

To introduce the right notion of topological group complexity for the next section, we briefly recall the class of elementary t.d.l.c.s.c. groups and their decomposition rank, as introduced by Wesolek. For a detailed account, see [16]; a more streamlined version is also given by Wesolek in these proceedings.

We will write $G = \varinjlim O_i$ as a shorthand to mean that G is a t.d.l.c.s.c. group, formed as an increasing union of compactly generated open subgroups O_i.

Definition 14 The class $\mathscr{E}$ of **elementary** t.d.l.c.s.c. groups is the smallest class of t.d.l.c.s.c. groups such that

  (i) $\mathscr{E}$ contains all countable discrete groups and second-countable profinite groups;
 (ii) Given a t.d.l.c.s.c. group G and $K \trianglelefteq G$ such that $K, G/K \in \mathscr{E}$, then $G \in \mathscr{E}$;
(iii) Given $G = \varinjlim O_i$ such that $O_i \in \mathscr{E}$, then $G \in \mathscr{E}$.

Notice that if G is in the class $\mathscr{S}$ of compactly generated, non-discrete, topologically simple t.d.l.c.s.c. groups, then G is *not* elementary. More generally, an elementary group cannot involve a group from $\mathscr{S}$, meaning that if G is elementary, then we cannot have closed subgroups $K \trianglelefteq H \leq G$ such that $H/K \in \mathscr{S}$. It is presently unknown if the converse holds. A candidate for a counterexample is the Burger–Mozes universal group $U(C_2)$ acting on the 3-regular tree mentioned in Remark 1; one can show that $U(C_2)$ is non-elementary, but it is not clear if it involves any groups in $\mathscr{S}$.

Elementary groups admit a canonical rank function, taking values in the countable successor ordinals, called the **decomposition rank** $\xi(G)$ of G. It will suffice for our purposes to recall some properties of how this rank behaves.

Write ω_1 for the set of countable ordinals; for convenience, if G is not elementary we will define $\xi(G) = \omega_1$. We also define the **discrete residual** $\mathrm{Res}(G)$ of a t.d.l.c. group G to be the intersection of all open normal subgroups of G.

Theorem 12 (See [16, §4.3]) *There is a unique mapping* $\xi : \mathscr{E} \to \omega_1$ *with the following properties:*

  (i) $\xi(1) = 1$;
 (ii) *If* $G \neq 1$ *and* $G = \varinjlim O_i$, *then* $\xi(G) = \sup\{\xi(\mathrm{Res}(O_i))\} + 1$.

Theorem 13 *Let* G *be a t.d.l.c.s.c. group.*

(i) *If $\psi : H \to G$ is a continuous injective homomorphism, then $\xi(H) \le \xi(G)$.*
   ([16, Corollary 4.10])
(ii) *If K is a closed normal subgroup of G, then $\xi(G/K) \le \xi(G) \le \xi(K) + \xi(G/K)$.*
   ([10, Lemma 6.4])
(iii) *If K is a closed normal cocompact subgroup of G, then $\xi(K) = \xi(G)$.*
   ([13, Lemma 3.8])

Of particular interest is the class $\{G \in \mathscr{E} \mid \xi(G) = 2\}$. These are the non-trivial t.d.l.c.s.c. groups G such that, for every compactly generated open subgroup O of G, then O is residually discrete. In fact, in this situation O is a SIN group, that is, O has a basis of identity neighbourhoods consisting of compact open normal subgroups; this was shown in [4, Corollary 4.1]. Clearly both profinite groups and discrete groups have rank 2; we also note that this class includes the quasi-discrete groups.

Lemma 5 *If G is a non-trivial t.d.l.c.s.c. group such that $QZ(G)$ is dense, then $\xi(G) = 2$.*

Proof Let O be a compactly generated open subgroup of G and let U be a compact open subgroup of O. Since $QZ(O) = QZ(G) \cap O$ is dense in O and O is compactly generated, we can choose a finite subset A of $QZ(O)$ such that $O = \langle A, U \rangle$. The group $V = \bigcap_{a \in A} C_U(a)$ is then an open subgroup of U. Since U is a profinite group, there is a base of identity neighbourhoods consisting of open normal subgroups W of U. Given $W \le V$ such that W is U-invariant, we see that W is centralised by $\langle A \rangle$ and hence W is normal in O. Thus O has a base of identity neighbourhoods consisting of open normal subgroups. In particular, $\text{Res}(O) = \{1\}$ and hence $\xi(\text{Res}(O)) = 1$. Since O was an arbitrary compactly generated open subgroup of G, it follows that $\xi(G) = 2$ as claimed. $\qquad\square$

It follows from Theorems 5 and 13 that normal compressions preserve the rank.

Proposition 11 ([13, Proposition 5.4]) *Let $\psi : A \to B$ be a normal compression where A and B are t.d.l.c.s.c. groups. Then $\xi(A) = \xi(B)$.*

Proof By Theorem 13, we have $\xi(A) \le \xi(B)$. On the other hand, by Theorem 5 we have a closed embedding $\alpha : A \to C$ and a quotient map $\beta : C \to B$, such that $\alpha(A)$ is a cocompact normal subgroup of C. It then follows by Theorem 13 that $\xi(B) \le \xi(C) = \xi(A)$, so in fact $\xi(A) = \xi(B)$. $\qquad\square$

In particular, given a chief block $\mathfrak{a}$, the rank of any representative of $\mathfrak{a}$ is the same as the rank of its uppermost representative. So given a block $\mathfrak{a} \in \mathfrak{B}_G$, one can define $\xi(\mathfrak{a}) := \xi(K/L)$ for some/any representative K/L of $\mathfrak{a}$.

Corollary 9 *Let G be a compactly generated t.d.l.c.s.c. group.*

(i) *Let $\mathfrak{a}$ be a chief block of G such that $\xi(\mathfrak{a}) > 2$. Then $\mathfrak{a}$ is robust, and hence minimally covered.*
(ii) *Suppose that $\xi(G)$ is infinite. Then there exist $n \in \mathbb{N}$ and robust blocks $\mathfrak{a}_1, \ldots, \mathfrak{a}_k$ of G satisfying*

$$\xi(G) \le \xi(\mathfrak{a}_1) + \xi(\mathfrak{a}_2) + \cdots + \xi(\mathfrak{a}_k) + n.$$

Proof Part (i) follows from Lemma 5; part (ii) follows from Theorem 13 together with the existence of essentially chief series, noting that every factor of such a series that is *not* a robust chief factor has rank 2, and also that $n + \alpha = \alpha$ whenever $n \in \mathbb{N}$ and α is an infinite ordinal.   □

6 Building Chief Factors from Compactly Generated Subgroups

6.1 Regional Properties

Unlike connected locally compact groups, t.d.l.c.s.c. groups are not necessarily compactly generated. However, we can always write a t.d.l.c.s.c. group G as $G = \varinjlim O_i$, where the groups O_i are open and compactly generated. In some situations we can hope to extract features of G from properties that hold for a *sufficiently large* compactly generated open subgroup. Such properties will then appear in any increasing exhaustion of G by compactly generated open subgroups O_i, independently of the choice of sequence (O_i), and we can potentially use the structure of compactly generated groups to describe that of non-compactly generated groups. In this section, our aim is to use this approach to obtain chief factors of G.

Definition 15 A property $\mathscr{P}$ of t.d.l.c.s.c. groups holds **locally** in G if every sufficiently small compact open subgroup of G has the property. The property is a **local property** if, whenever G has the property, then every open subgroup of G also has it. For example, compactness is a local property.

A property $\mathscr{P}$ of t.d.l.c.s.c. groups holds **regionally** in G if every sufficiently large compactly generated open subgroup has the property; that is, there is a compact subset X such that, whenever $X \subseteq O \leq G$ and O is a compactly generated open subgroup of G, then O has $\mathscr{P}$. The property is a **regional property** if the following happens: given G and H are compactly generated t.d.l.c.s.c. groups such that G is open in H, if G has the property, then so does H.

Some remarks on possibly controversial terminology are in order.

Remark 3 In topology, it is usual to use 'local' to refer to (small) open sets. In classical group theory, 'local' more often refers to finitely generated subgroups. Both notions are important in the theory of t.d.l.c. groups (with 'compactly generated' instead of 'finitely generated'). To avoid overloading the word 'local', we have chosen 'regional(ly)' to have the meaning 'pertaining to compactly generated (open) subgroups'. For example, the property that every compactly generated subgroup is compact, which is unfortunately rendered as 'locally elliptic' in the literature, would instead be 'regionally compact' or 'regionally elliptic'.

Remark 4 Many authors define local properties to be such that G has the property if and only if some open subgroup or subspace has it. However, it is useful here to distinguish between properties that are inherited 'downwards' (if G has it, then so does an open subgroup) from those that are inherited 'upwards' (regional properties).

The distinction between local and regional properties is neatly illustrated by elementary decomposition rank: given a countable ordinal α, then '$\xi(G) \leq \alpha$' is a local property whereas '$\xi(G) \geq \alpha$' is a regional property. We also see by Theorem 12 that for any ordinal α, we have $\xi(G) \leq \alpha+1$ if and only if $\xi(H) \leq \alpha+1$ for all compactly generated open subgroups H of G.

The following is a surprisingly powerful example of a regional property.

Definition 16 Say a compactly generated t.d.l.c.s.c. group G has property $\mathscr{RF}$ if there exists a Cayley–Abels graph Γ for G such that the action of G on Γ is faithful.

Lemma 6 $\mathscr{RF}$ is a regional property.

Proof We see that G has $\mathscr{RF}$ if and only if there is a compact open subgroup U of G such that $\bigcap_{g \in G} gUg^{-1} = \{1\}$. Suppose that this is the case and that G occurs as an open subgroup of the compactly generated t.d.l.c.s.c. group H. Then U is a compact open subgroup of H, and we have $\bigcap_{h \in H} hUh^{-1} \leq \bigcap_{g \in G} gUg^{-1} = \{1\}$. Thus H has $\mathscr{RF}$. □

We define a t.d.l.c.s.c. group G to be **regionally faithful** if some (and hence any sufficiently large) compactly generated open subgroup has $\mathscr{RF}$. Note that this allows, for example, G to be any discrete group, so the class of all regionally faithful groups is not so well-behaved. However, as long as the quasi-centre is not too large, we can use the regionally faithful property to obtain minimal non-trivial closed normal subgroups.

Definition 17 Say a t.d.l.c.s.c. group G has property $\mathscr{M}$ if $\mathrm{QZ}(G)$ is discrete (equivalently: G has a unique largest discrete normal subgroup) and every chain of non-trivial closed normal subgroups of $G/\mathrm{QZ}(G)$ has non-trivial intersection.

Lemma 7 $\mathscr{M}$ is a regional property, and regionally $\mathscr{M}$ groups have $\mathscr{M}$. In a group G with $\mathscr{M}$, every non-trivial closed normal subgroup of $G/\mathrm{QZ}(G)$ contains a minimal one.

Proof Let G be a t.d.l.c.s.c. group, such that some compactly generated open subgroup O of G has $\mathscr{M}$. We must show that G has $\mathscr{M}$.

We note first that $\mathrm{QZ}(G) \cap O = \mathrm{QZ}(O)$; since O is open, this ensures that $\mathrm{QZ}(G)$ is discrete. Moreover, the group $G/\mathrm{QZ}(G)$ has a compactly generated open subgroup isomorphic to $O/\mathrm{QZ}(O)$. So we may assume $\mathrm{QZ}(G) = \mathrm{QZ}(O) = \{1\}$.

Let $\mathscr{C}$ be a chain of non-trivial closed normal subgroups of G. For each $K \in \mathscr{C}$, we see that K is non-discrete, since any discrete normal subgroup of G would be contained in $\mathrm{QZ}(G)$. In particular, $K \cap O \neq \{1\}$. Thus $\{K \cap O \mid K \in \mathscr{C}\}$ is a chain of non-trivial closed normal subgroups of O; since O has $\mathscr{M}$, the intersection $\bigcap_{K \in \mathscr{C}} K \cap O$ is non-trivial, and hence $L = \bigcap_{K \in \mathscr{C}} K$ is non-trivial. Thus G has $\mathscr{M}$.

The last conclusion follows by Zorn's lemma. □

Proposition 12 *Let G be a t.d.l.c.s.c. group. Suppose that* $QZ(G) = \{1\}$ *and G is regionally faithful. Then G has $\mathcal{M}$; in particular, G has a minimal non-trivial closed normal subgroup.*

Proof Let O be a compactly generated open subgroup of G; choose O sufficiently large that O acts faithfully on some Cayley–Abels graph Γ. Then $QZ(O) = \{1\}$, so O has no non-trivial discrete normal subgroups. Let $\mathcal{C}$ be a chain of non-trivial closed normal subgroups of G. Then for each $K \in \mathcal{C}$, we see that K is not discrete, and therefore has non-trivial local action on Γ. By Lemma 2, the intersection $L = \bigcap_{K \in \mathcal{C}} K$ also has non-trivial local action on Γ; in particular, $L \neq \{1\}$. Thus O has $\mathcal{M}$, showing that G is a regionally $\mathcal{M}$ group. Hence G has $\mathcal{M}$ by Lemma 7.   □

Here we have a situation where we first obtain minimal normal subgroups *regionally*, and then conclude that we have minimal normal subgroups *globally*. More work is required to obtain an analogous result for chief factors that are not necessarily associated to minimal normal subgroups. The key ingredients are *robustness* (recall Sect. 3.3) and *extension of chief blocks* (recall Sect. 4.2), and the use of the *decomposition rank* (as described in Sect. 5) to ensure the existence of robust blocks of compactly generated open subgroups.

6.2   Regionally Robust Blocks

As we saw in Sect. 4.2, we can always extend minimally covered blocks from normal subgroups. Remarkably, many blocks extend from *open* subgroups, and moreover can be detected from compactly generated open subgroups.

Definition 18 Let G be a t.d.l.c.s.c. group and let $\mathfrak{a} \in \mathfrak{B}_G$. Say $\mathfrak{a}$ is a **regional block** if there exists $H \leq G$ and $\mathfrak{b} \in \mathfrak{B}_H$ such that H is compactly generated and open, and $\mathfrak{a} = \mathfrak{b}^G$. If $\mathfrak{b}$ is robust, we say $\mathfrak{a}$ is **regionally robust**. Write $\mathfrak{B}_G^{rr}$ for the set of regionally robust blocks of G.

Note that regional blocks manifest 'regionally', because if $\mathfrak{a} \in H$ extends to G, then it certainly extends to any $H \leq O \leq G$, including when O is compactly generated and open. If G itself is compactly generated, then every block is regional and 'regionally robust' just means 'robust'.

Here is the main theorem of this section.

Theorem 14 (See [13, §8]) *Let G be a t.d.l.c.s.c. group.*

   (i) *Every regionally robust block of G is minimally covered and robust, and there are at most countably many regionally robust blocks of G.*
  (ii) *Let $H \leq G$, such that H is either open in G or closed and normal in G, and let $\mathfrak{a} \in \mathfrak{B}_H^{rr}$. Then $\mathfrak{a}$ extends to a regionally robust block of G.*
 (iii) *Let N be a closed normal subgroup of G. Then every regionally robust block G/N lifts to a regionally robust block of G.*
  (iv) *Let K/L be a chief factor of G such that $\mathfrak{B}_{K/L}^{rr} \neq \emptyset$. Then $[K/L]$ is a regionally robust block of G.*

As a corollary, we observe that in any t.d.l.c.s.c. group G, a sufficiently complex normal factor (in the sense of elementary decomposition rank) covers a regionally robust block, and that sufficiently complex chief factors cannot be of weak type.

Corollary 10 *Let G be a t.d.l.c.s.c. group. Let K/L be a normal factor of G such that $\xi(K/L) > \omega + 1$.*

(i) *There exists $L \leq B < A \leq K$ such that A/B is a chief factor of G and $[A/B]$ is regionally robust. If K/L is non-elementary, then A/B can also be chosen to be non-elementary.*

(ii) *Suppose K/L is a chief factor of G. Then K/L is of semisimple or stacking type.*

Proof Since $\xi(K/L) > \omega + 1$, there must be a compactly generated open subgroup H of K/L such that $\xi(H)$ is infinite. It follows by Corollary 9 that H has a robust block $\mathfrak{a}$. By Theorem 14, $\mathfrak{a}$ extends to a regionally robust block of K/L and then to a regionally robust block of G/L; this block in turn lifts to a regionally robust block $\mathfrak{b}$ of G. We see that $\mathfrak{b}$ is covered by K/L, in other words, there exists $L \leq B < A \leq K$ such that A/B is a chief factor of G and $[A/B]$ is regionally robust.

If K/L is non-elementary, we can choose H to be non-elementary; by Corollary 9, $\mathfrak{a}$ can be chosen to be non-elementary; it then follows that $\xi(\mathfrak{b}) = \omega_1$, so A/B is non-elementary.

Now suppose K/L is a chief factor of G. We have seen that K/L has a regionally robust block, that is, $\mathfrak{B}^{rr}_{K/L}$ is non-empty; since regionally robust blocks are minimally covered, it follows that K/L is not of weak type. Thus K/L must be of one of the remaining two types, that is, semisimple type or stacking type. □

We will now sketch the core part of the proof of Theorem 14, which is to prove the following statement:

($*$) Let G be a t.d.l.c.s.c. group, let O be a compactly generated open subgroup of G and let $\mathfrak{a}$ be a robust block of O. Then $\mathfrak{a}$ extends to G.

Lemma 8 *Let H be a quasi-discrete t.d.l.c.s.c. group and let A/B be a non-trivial normal factor of H. Then $QZ(A/B) > 1$.*

Proof We see that H/B is quasi-discrete, so we may assume $B = \{1\}$. Suppose $QZ(A) = \{1\}$. We see that $QZ(H) \cap A$ is quasi-central in A, so $QZ(H) \cap A = \{1\}$. Thus $QZ(H)$ and A commute. But $QZ(H)$ is dense in H, so A is central in H. In particular A is abelian, so $QZ(A) = A$, a contradiction. □

Proof (Sketch Proof of ($*$)) For brevity we will write $H^O := H \cap O$.

Case 1: G is compactly generated.

Let $(G_i)_{i=0}^n$ be an essentially chief series for G. There must be some i such that G_{i+1}^O/G_i^O covers $\mathfrak{a}$. By Lemma 8, G_{i+1}^O/G_i^O cannot be quasi-discrete, so G_{i+1}/G_i cannot be quasi-discrete. Thus G_{i+1}/G_i is a robust, hence minimally covered, chief factor of G. Set $\mathfrak{b} = [G_{i+1}/G_i]$.

Let N/C be the uppermost representative of $\mathfrak{b}$. Since $N \geq G_{i+1}$, we see that N^O covers $\mathfrak{a}$. On the other hand C centralises G_{i+1}/G_i, so in particular C^O centralises G_{i+1}^O/G_i^O, and hence C^O cannot cover $\mathfrak{a}$.

Let I/J be the lowermost representative of $\mathfrak{b}$. Both I/J and N^O/C^O map injectively to N/C; moreover $\mathrm{QZ}(N/C) = \{1\} = \mathrm{QZ}(I/J)$ since $\mathfrak{b}$ is robust. In particular, I/J is not discrete.

The subgroup $I^O J/J$ is non-trivial, so $I^O C/C$ is non-trivial. Since $\mathrm{QZ}(N/C) = \{1\}$, it follows that $I^O C/C$ does not commute with the open subgroup $N^O C/C$ of N/C. One can deduce that $I^O \not\leq \mathrm{C}_O(\mathfrak{a})$. Apply Lemma 4 to conclude $\mathfrak{b} = \mathfrak{a}^G$.

Case 2: G is not compactly generated.

We can write G as $G = \varinjlim O_i$ where $O_1 = O$. By Case 1, $\mathfrak{a}_i$ extends to some block $\mathfrak{a}_i := \mathfrak{a}^{O_i}$ of O_i. Set $D := \bigcup_{n \geq 1} \bigcap_{i \geq n} \mathrm{C}_{O_i}(\mathfrak{a}_i)$. (In other words, D is the 'limit inferior' of the centralisers $\mathrm{C}_{O_i}(\mathfrak{a}_i)$.)

Observe that $D \cap O_n = \bigcap_{i \geq n} \mathrm{C}_{O_i}(\mathfrak{a}_i)$ for all n. It follows that D is a closed normal subgroup of G, and that D^O does not cover $\mathfrak{a}$. In fact, one sees that D is the *unique largest* closed normal subgroup of G such that D^O does not cover $\mathfrak{a}$.

Letting N range over the closed normal subgroups of G, the property 'N^O covers $\mathfrak{a}$' is closed under arbitrary intersections (since $\mathfrak{a}$ is minimally covered). So there is a smallest closed normal subgroup M such that M^O covers $\mathfrak{a}$.

We deduce that $\overline{MD}/D$ is the unique smallest non-trivial closed normal subgroup of G/D. Set $\mathfrak{b} := [\overline{MD}/D]$ and observe that M is the least closed normal subgroup that covers $\mathfrak{b}$, whilst $D = \mathrm{C}_G(\mathfrak{b})$. We conclude by Lemma 4 that $\mathfrak{b} = \mathfrak{a}^G$. $\square$

7 Some Ideas and Open Questions

In this last section, we discuss some possible further directions for research into the normal subgroup structure of t.d.l.c.s.c. groups, in particular focusing on the gaps left by the results presented in the previous sections.

7.1 *Elementary Groups of Small Rank*

We have seen that G is a t.d.l.c.s.c. group and K/L is a normal factor such that $\xi(K/L) > \omega + 1$, then K/L covers a (regionally robust) chief factor of G. As a complement to such a result, we would like to be able to say something about normal or characteristic subgroups of G when $\xi(G) \leq \omega + 1$.

For certain ranks $\xi(G)$, we can always produce a proper characteristic subgroup of G. We will use the following fact:

Lemma 9 ([13, Proposition 3.10]) *Let G be a t.d.l.c.s.c. group and let (R_i) be an increasing sequence of closed subgroups of G. Suppose $\mathrm{N}_G(R_i)$ is open for all i. Then*

$$\xi(\overline{\bigcup R_i}) = \sup \xi(R_i) + \varepsilon,$$

where $\varepsilon = 1$ if $\sup \xi(R_i)$ is a limit ordinal and $\varepsilon = 0$ otherwise.

Proposition 13 (See also [13, Proposition 3.18]) *Let G be an elementary t.d.l.c.s.c. group, $G = \varinjlim O_i$. Then exactly one of the following occurs:*

(i) $\xi(G) = \lambda + 1$ *where* λ *is a limit ordinal;*
(ii) $R = \bigcup \mathrm{Res}(O_i)$ *is a closed characteristic subgroup of G, which does not depend on the choice of (O_i), such that* $\xi(G) = \xi(R) + 1$ *and* $\xi(G/R) = 2$.

Proof Note that given $K \leq H \leq G$, then $\mathrm{Res}(K) \leq \mathrm{Res}(H)$. In particular, $\bigcup \mathrm{Res}(O_i)$ is an increasing union of subgroups, so R is a closed subgroup of G. Moreover, given a compactly generated open subgroup O of G, then $O_i \geq O$ eventually, so $\mathrm{Res}(O_i) \geq \mathrm{Res}(O)$. Thus R is the closure of the union of all discrete residuals of compactly generated open subgroups of G; in particular, R is characteristic and does not depend on the choice of (O_i).

We now have

$$\xi(G) = \sup\{\xi(\mathrm{Res}(O_i))\} + 1 \text{ and } \xi(R) = \sup\{\xi(\mathrm{Res}(O_i))\} + \varepsilon,$$

the latter by Lemma 9, where $\varepsilon = 0$ unless $\sup \xi(\mathrm{Res}(O_i))$ is a limit ordinal. If (i) holds, then $\sup\{\xi(\mathrm{Res}(O_i))\} = \lambda$ is a limit ordinal, so $\xi(R) \geq \xi(G)$ and (ii) does not hold. So from now on we may assume (i) fails, that is, $\xi(G) = \alpha + 2$ for some ordinal α, and aim to show that (ii) holds. In this case, we see that $\sup\{\xi(\mathrm{Res}(O_i))\} = \alpha + 1$ is not a limit ordinal, so $\xi(R) = \alpha + 1$, in other words, $\xi(G) = \xi(R) + 1$.

Certainly $R < G$, so $\xi(G/R) > 1$. To show $\xi(G/R) = 2$, it is enough to see that every compactly generated open subgroup of G/R is a SIN group. Let O/R be a compactly generated open subgroup of G/R. Then for i sufficiently large, $O \leq O_i R$, so O/R is isomorphic to a subgroup of a quotient of $O_i/\mathrm{Res}(O_i)$. By [4, Corollary 4.1], $O_i/\mathrm{Res}(O_i)$ is a SIN group; subgroups and quotients of SIN groups have SIN, so O/R is a SIN group. This completes the proof of (ii). □

The following corollary follows easily.

Corollary 11 *Let G be a non-trivial elementary t.d.l.c.s.c. group. Let*

$$\mathscr{L} = \{\alpha \in \omega_1 \mid \alpha = 2 \text{ or } \alpha \text{ is a limit ordinal}\}.$$

(i) There is a non-trivial closed characteristic subgroup R of G such that $\xi(R) \in \mathscr{L}$ and $\xi(G) < \xi(R) + \omega$.
(ii) Suppose that G is characteristically simple. Then $\xi(G) \in \mathscr{L}$.

For t.d.l.c.s.c. groups G with $\xi(G) \leq \omega + 1$, we can split into three cases: $\xi(G) = 2$, $\xi(G) = \omega + 1$ and $2 < \xi(G) < \omega$. (Recall that the rank is never a limit ordinal.)

- If $\xi(G) = 2$, then $G = \varinjlim O_i$ where O_i has arbitrarily small open normal subgroups. Many characteristically simple groups are of this form, and $\xi(G) = 2$ is implied by several natural conditions on t.d.l.c.s.c. groups.

- If $2 < \xi(G) < \omega$ then G has finite rank, and we obtain a finite characteristic series

$$G = R_0 > R_1 > \cdots > R_n = 1$$

such that $\xi(R_{i-1}/R_i) = 2$.
- If $\xi(G) = \omega + 1$, then $G = \varinjlim O_i$ where each O_i has finite rank (but $\xi(O_i) \to \omega$ as $i \to \infty$), and so O_i admits a characteristic decomposition as in the previous point. Perhaps G can be studied by comparing these characteristic series across different O_i.

Problem 1 Develop a theory of normal/characteristic subgroups for t.d.l.c.s.c. groups with $\xi(G) = 2$.

This class includes all profinite and discrete second countable groups, so what one hopes for are theorems that relate the more general situation to profinite/discrete groups in an interesting way. The discrete case is too wild to deal with directly, but at least in the profinite case, we know what the characteristically simple groups are.

Problem 2 Find examples of characteristically simple t.d.l.c.s.c. groups with $\xi(G) = \omega + 1$, *without* using a 'stacking' construction.

There are known examples of non-discrete topologically simple groups of rank 2, but reaching rank $\omega + 1$ is more difficult. By a 'stacking' construction, we mean a construction similar to that of Sect. 4.4; similar constructions can be used to produce weak type chief factors of rank $\omega + 1$, but only because one obtains a characteristically simple group in which every chief factor is abelian. More interesting would be to find an example of a characteristically simple group of rank $\omega + 1$ that has non-abelian chief factors, but such that none of those chief factors are minimally covered.

7.2 Well-Foundedness of Stacking Chief Factors

If we have a subnormal chain $K_0 \trianglelefteq K_1 \trianglelefteq \cdots \trianglelefteq K_n$ ($n \geq 1$) of closed subgroups of some ambient t.d.l.c.s.c. group, then any minimally covered block $\mathfrak{a} \in \mathfrak{B}_{K_0}^{\min}$ will extend to K_n. Let $\mathfrak{a}_i = \mathfrak{a}^{K_i}$ and let $\theta_i : \mathfrak{B}_{K_{i-1}}^{\min} \to \mathfrak{B}_{K_i}^{\min}$ be the extension map.

As the following proposition shows, we cannot produce essentially different semisimple type factors by extending chief blocks from subnormal subgroups; all we are doing is increasing the number of copies of the simple group and possibly normally compressing those copies.

Proposition 14 (See [11, Proposition 9.21]) *If $\mathfrak{a}_n$ is of semisimple type, then so is $\mathfrak{a}$, and $\theta_i^{-1}(\mathfrak{a}_i)$ is an antichain for all i.*

Once we are beyond rank $\omega + 1$, we also cannot produce a weak type chief factor. In other words, beyond this stage, the only way to increase the complexity of the chief factor via extensions from subnormal subgroups is to produce chief factors of stacking type, and moreover to 'stack' the blocks repeatedly (meaning that $\theta_i^{-1}(\mathfrak{a}_i)$ has a non-trivial partial order).

Given constructions like Sect. 4.4, we can certainly form n-fold stacking factors for every n. Perhaps this can be continued transfinitely. However, for any given stacking type chief factor, we might hope that we can reduce it to topologically simple groups and groups of rank at most $\omega + 1$. We thus have a well-foundedness question.

Question 1 Suppose that $G =: G_0$ is a topologically characteristically simple t.d.l.c.s.c. group. If G_0 is abelian, elementary with rank at most $\omega + 1$, or of semisimple type, we stop. Otherwise, we find a chief factor $G_1 := K/L$ of G_0 that is regionally robust. Continuing in this fashion produces a sequence $G_0, G_1, \ldots$ of l.c.s.c groups. Is it the case that any such sequence halts in finitely many steps? What about in the case that the group G is also elementary?

We do not know the answer even for elementary t.d.l.c.s.c. groups. In this case, to prove well-foundedness it would be enough (assuming $\xi(G_i) > \omega + 1$ and G_i is of stacking type) for every regionally robust chief factor G_{i+1} of G_i to be such that $\xi(G_{i+1}) < \xi(G_i)$. All elementary examples we know of have this property.

7.3 Contraction Groups

On 'large' stacking type chief factors K/L, the ambient group G has non-trivial local dynamics, which in particular imply the existence of a non-trivial contraction group.

Definition 19 For $\alpha \in \mathrm{Aut}(G)$, $\mathrm{con}(\alpha) := \{x \in G \mid \alpha^n(x) \to 1 \text{ as } n \to \infty\}$.
Given $g \in G$, $\mathrm{con}(g) := \{x \in G \mid g^n x g^{-n} \to 1 \text{ as } n \to \infty\}$.

Proposition 15 *Let G be a t.d.l.c.s.c. group and let K/L be a chief factor of stacking type, such that $\xi(K/L) > \omega + 1$. Then there exists $g \in G$ and $L < A \lhd K$ such that $gAg^{-1} < A$ and A/gAg^{-1} is non-discrete. Moreover, for any such g and A, we have $\mathrm{con}(g) \cap A \nleq L$.*

We appeal to the following observation due to George Willis:

Lemma 10 (Willis) *Let H be a t.d.l.c. group, let $\alpha \in \mathrm{Aut}(H)$ and let D be a closed subset of H such that $\alpha(D) \subseteq D$ and $\bigcap_{n \geq 0} \alpha(D) = \{1\}$. Then $D \cap \mathrm{con}(\alpha)$ is a neighbourhood of the identity in D.*

Proof Let U be a compact open subgroup of H, let $U_- = \bigcap_{n \geq 0} \alpha^{-n}(U)$ and let $U_{--} = \bigcup_{n \geq 0} \alpha^{-n}(U_-)$.

Let V be a compact open identity neighbourhood in H. We see that $\alpha^m(D) \cap U$ is a decreasing sequence of closed sets with intersection $\{1\}$. By the compactness of U, we have $\alpha^m(D) \cap U \subseteq V$ for m sufficiently large, showing that $\alpha^m(D) \cap U \subseteq \mathrm{con}(\alpha)$. On the other hand, given $x \in D \cap U_{--}$, then for n sufficiently large, $\alpha^n(x) \in U_- \leq U$, and hence $\alpha^{m+n} \in V$. Thus $D \cap U_{--} \subseteq \mathrm{con}(\alpha)$.

Now let $V = \alpha^{-1}(U)$, write $Y_i = \alpha^i(D) \cap U$ and let m be such that $Y_m \subseteq V$. Let $y \in Y_m$. Then $y \in V = \alpha^{-1}(U)$, so $\alpha(y) \in U$. But we also have $\alpha(y) \in \alpha^{m+1}(D) \subseteq \alpha^m(D)$, so in fact $\alpha(y) \in Y_m$. Thus $\alpha(Y_m)$ is a subset of Y_m; in particular, $\alpha^n(Y_m) \subseteq U$ for all $n \geq 0$. It follows that $Y_m \subseteq U_-$, and hence $\alpha^{-m}(Y_m) \subseteq U_{--}$. Now $\alpha^{-m}(Y_m) = D \cap \alpha^{-m}(U)$ is an identity neighbourhood in D contained in U_{--}, and hence in $D \cap \mathrm{con}(\alpha)$.   □

Remark 5 It can in fact be shown (Willis, private communication) that in the above lemma, $D \cap \mathrm{con}(\alpha)$ is compact and open in D and that $D \cap \mathrm{con}(\alpha) = D \cap U_{--}$ for any tidy subgroup U.

Proof (of Proposition 15) Since $\xi(K/L) > \omega + 1$, there exists $\mathfrak{a} \in \mathfrak{B}_{K/L}^{rr}$. Let $A/L = (K/L)_{\mathfrak{a}}$. Since K/L is of stacking type, there must exist $g \in G$ such that $gAg^{-1} < A$, and moreover A/gAg^{-1} covers $\mathfrak{a}$. So A/gAg^{-1} cannot be discrete.

Now suppose $g \in G$ and $L < A \lhd K$ are such that $gAg^{-1} < A$ and A/gAg^{-1} is non-discrete. Let $M = \bigcap_{n \in \mathbb{Z}} g^n A g^{-n}$. Then M is normal in K and A/M is not discrete. Let $D = A/M$, let $H = K/M$ and let α be the automorphism of H induced by conjugation by g. Then α acts on H in the manner of Lemma 10, so $\mathrm{con}_{K/M}(g)$ contains an open (in particular, non-trivial) subgroup of A/M.

By [2, Theorem 3.8], we have $\mathrm{con}_{K/M}(g) = \mathrm{con}_K(g)M$. So in fact $\mathrm{con}_K(g) \cap A \not\subseteq M$ and in particular $\mathrm{con}(g) \cap A \not\subseteq L$.   □

We observe from the proof that $M(\mathrm{con}(g) \cap A)$ is an open subgroup of A. Using the fact that $A/L = (K/L)_{\mathfrak{a}}$, we obtain the following.

Corollary 12 *Let A, g and M be as in the proof Proposition 15. Then*
$$A = M\langle k(\mathrm{con}(g) \cap A)k^{-1} \mid k \in K\rangle.$$

Note that we are not claiming that the group $\langle k(\mathrm{con}(g) \cap A)k^{-1} \mid k \in K\rangle$ is closed, but nevertheless the *abstract* product $M\langle k(\mathrm{con}(g) \cap A)k^{-1} \mid k \in K\rangle$ suffices to obtain every element of A.

It is tempting to speculate that the entirety of a stacking type chief factor is accounted for by contraction groups, as follows:

Question 2 In the situation of Proposition 15, do we in fact have

$$K = L\langle \mathrm{con}(h) \cap K \mid h \in G\rangle?$$

More generally, it would be useful to develop a dynamical approach to stacking type chief factors, analogous to the theory developed for compactly generated topologically simple groups with micro-supported action. Here is a sketch of how one might proceed:

Given a stacking type chief factor K/L, let $\mathscr{L}$ be the set of upward-closed subsets of the partially ordered set $\mathfrak{B}_{K/L}^{min}$; notice that $\mathscr{L}$ is a complete bounded distributive lattice under the operations of intersection and union. By Priestley duality, there is an associated ordered topological space X, which is a profinite space in which $\mathfrak{B}_{K/L}^{min}$ is embedded as a dense set of isolated points.

Problem 3 Recall for all $\mathfrak{a}, \mathfrak{b} \in \mathfrak{B}_{K/L}^{\min}$, there exists $g \in G$ such that $g.\mathfrak{a} < \mathfrak{b}$. Reinterpret this as a property of (ordered) topological dynamics of the action of G/K on $X \setminus \mathfrak{B}_{K/L}^{\min}$, and use these dynamics to obtain further restrictions on the structure of G or its contraction groups as topological groups.

A t.d.l.c. group G is **anisotropic** (or **pointwise distal**) if $\mathrm{con}(g) = \{1\}$ for all $g \in G$. This is a property that is clearly inherited by closed subgroups; by [2, Theorem 3.8] it is also inherited by quotients. However, as it is a 'pointwise' property, it in no way prevents G from having complicated dynamics globally.

The class of anisotropic t.d.l.c.s.c. groups is mysterious at present. For example, there are topologically simple anisotropic groups G, but it is unknown if a topologically simple anisotropic group G can be in $\mathscr{S}$, or more generally, whether G can be non-elementary.

Nevertheless, the essential role of contraction groups in stacking type shows that if anisotropic groups can be non-elementary or achieve large decomposition ranks, then topologically simple groups are the major source of complexity. We have found a potentially non-trivial situation where we really can break a t.d.l.c.s.c. group into topologically simple pieces (plus low rank pieces).

Proposition 16 *Let G be an anisotropic t.d.l.c.s.c. group.*

 (i) *Every chief factor of G of rank greater than $\omega + 1$ is of semisimple type.*
 (ii) *Let K/L be a normal factor of G such that $\xi(K/L) > \omega^2 + 1$ (or K/L non-elementary).*
   Then there is $L \leq B < A \leq K$ such that A/B is a chief factor of G, $\xi(A/B) \geq \omega^2 + 1$ (respectively, A/B is non-elementary) and A/B is a quasi-product of copies of a topologically simple group.

Proof

 (i) Let K/L be a chief factor of G of rank greater than $\omega + 1$. Then K/L is not of weak type by Theorem 14, and it is not of stacking type by Proposition 15. Thus K/L must be of semisimple type.
 (ii) Now let K/L be a normal factor of G such that $\xi(K/L) > \omega^2 + 1$. Then there is a compactly generated open subgroup H of K/L of rank at least $\omega^2 + 1$; if K/L is non-elementary, we can choose H to be non-elementary. By Corollary 9, H has a chief factor R/S such that $\xi(R/S) \geq \omega^2 + 1$ (respectively, R/S is non-elementary). The corresponding block of H then lifts via Theorem 14 to a block $\mathfrak{a}$ of G, with $\xi(\mathfrak{a}) \geq \xi(R/S) \geq \omega^2 + 1$. We see that K/L covers $\mathfrak{a}$, so there exists $L \leq B < A \leq K$ such that A/B such that $A/B \in \mathfrak{a}$. By part (i), A/B is of semisimple type, that is, A/B a quasi-product of copies of a topologically simple group. $\square$

Acknowledgements The author is an ARC DECRA fellow; this article is based on research supported in part by ARC Discovery Project DP120100996. I thank Pierre-Emmanuel Caprace and Phillip Wesolek for their very helpful feedback on the presentation of this material. I also thank George Willis for his insights on contraction groups.

References

1. Abels, H.: Specker-Kompaktifizierungen von lokal kompakten topologischen Gruppen. Math. Z. **135**, 325–361 (1974)
2. Baumgarnter, U., Willis, G.A.: Contraction groups and scales of automorphisms of totally disconnected locally compact groups. Isr. J. Math. **142**, 221–248 (2004)
3. Burger, M., Mozes, S.: Groups acting on trees: from local to global structure. Publ. Math. IHÉS **92**, 113–150 (2000)
4. Caprace, P.-E., Monod, N.: Decomposing locally compact groups into simple pieces. Math. Proc. Cambridge Philos. Soc. **150**(1), 97–128 (2011)
5. Caprace, P.-E., Reid, C.D., Willis, G.A.: Locally normal subgroups of totally disconnected groups; Part II: Compactly generated simple groups (2016, preprint). arXiv:1401.3142v2
6. Cornulier, Y., de la Harpe, P.: Metric geometry of locally compact groups (2016, preprint of book in preparation). arXiv:1403.3796v4
7. Hofmann, K., Morris, S.: The Lie Theory of Connected Pro-Lie Groups: A Structure Theory for Pro-Lie Algebras, Pro-Lie Groups, and Connected Locally Compact Groups. European Mathematical Society, Zürich (2007)
8. Kechris, A.S.: Classical Descriptive Set Theory. Graduate Texts in Mathematics, vol. 156. Springer, New York (1995)
9. Le Boudec, A.: Groups acting on trees with almost prescribed local action (2016, preprint). arXiv:1505.01363v2
10. Reid, C.D., Wesolek, P.R.: Homomorphisms into totally disconnected, locally compact groups with dense image (2015, preprint). arXiv:1509.00156v1
11. Reid, C.D., Wesolek, P.R.: Chief factors in Polish groups (with an appendix by François Le Maître) (2016, preprint). arXiv:1509.00719v3
12. Reid, C.D., Wesolek, P.R.: The essentially chief series of a compactly generated locally compact groups. Math. Ann. (2017). https://doi.org/10.1007/s00208-017-1597-0
13. Reid, C.D., Wesolek, P.R.: Dense normal subgroups and chief factors in locally compact groups. Proc. Lond. Math. Soc. https://doi.org/10.1112/plms.12088
14. Ribes, L., Zalesskii, P.: Profinite groups. In: Ergebnisse der Mathematik und ihrer Grenzgebiete, 3. Folge/A Series of Modern Surveys in Mathematics, vol. 40. Springer, Berlin (2010)
15. Trofimov, V.I.: On the action of a group on a graph. Acta Appl. Math. **29**(1), 161–170 (1992)
16. Wesolek, P.R.: Elementary totally disconnected locally compact groups. Proc. Lond. Math. Soc. **110**(6), 1387–1434 (2015)

Automorphism Groups of Combinatorial Structures

Anne Thomas

Abstract This is a series of lecture notes taken by students during a five lecture series presented by Anne Thomas in 2016 at the MATRIX workshop: The Winter of Disconnectedness.

1 Introduction

These lectures are based on the survey paper [21] of Farb, Hruska and Thomas. We give additional background in these lectures and update some results, including some answers to questions posed in [21]. Both the survey [21] and these lectures were inspired by the paper [31], which explores the theory of tree lattices and relates results to the classical case. The primary goal of these lectures was to present the main examples of polyhedral complexes and then use them to generate interesting examples of locally compact groups. Results are stated about these groups with a focus on lattices and comparisons with the well developed theory of Lie groups. Some results will be identical to results from Lie groups with identical proofs, some will require different techniques to prove and some will directly contrast. Using this point of view, it is possible to pull insight from one case to another, thus it is possible to gain insight into Lie groups by studying groups acting on polyhedral complexes. When possible, at the start of each section we give references which provide the reader with a deeper background into the topic which is beneficial but not required. The interested reader may want to read further than what is presented here. In that case we recommend the survey [21] which presents a long, but motivated, list of research problems.

Notes prepared by Ben Brawn, Tim Bywaters and Thomas Taylor.

A. Thomas (✉)
School of Mathematics and Statistics, The University of Sydney, Carslaw Building, Camperdown, NSW 2006, Australia
e-mail: anne.thomas@sydney.edu.au

© Springer International Publishing AG, part of Springer Nature 2018
D.R. Wood et al. (eds.), *2016 MATRIX Annals*, MATRIX Book Series 1,
https://doi.org/10.1007/978-3-319-72299-3_24

For the remainder of this section we define a notion of curvature on metric spaces and consider the basics of isometries on these space. We then give a short introduction to lattices and provide elementary examples. We finish with a rough intuition for symmetric spaces. This intuition is not meant to be a definition, but instead places future results in context.

In Sect. 2 we standardise the notation we will be using for graphs and specifically trees. We then provide some results on tree lattices. In Sect. 3 we define a general polyhedral complex and see how previous examples fit this mould. We then explore results related to buildings which are an important example. Finally in Sect. 4 we present examples of polyhedral complexes which have received recent attention in the literature.

1.1 Models of Metric Spaces

In this section we introduce the concept of metric spaces. We then define three fundamental examples of metric spaces. These are the unique n-dimensional Riemannian manifolds with constant sectional curvature 1, 0 and -1. In later sections, we will compare arbitrary metric spaces with these to define a notion of curvature in a general geodesic metric space. We suggest [10] as a very complete reference for this section.

Definition 1 Suppose (X, d) is a metric space. A geodesic segment is a function $\gamma : [a, b] \subset \mathbb{R} \to X$ such that for all $s, t \in [a, b]$ we have

$$d(\gamma(s), \gamma(t)) = |s - t|. \tag{1}$$

A geodesic ray is a function $\gamma : [a, \infty) \to X$ such that for all $s, t \in [a, \infty)$ Eq. (1) holds. A geodesic line is a function $\gamma : (-\infty, \infty) \to X$ such that for all $s, t \in (-\infty, \infty)$ Eq. (1) holds.

Remark 1 We will often refer to a geodesic segment, ray or line by just geodesic when the context is clear. We also identify a geodesic γ with its image in X.

Example 1 Here we provide the examples of the three unique n-dimensional Riemannian manifolds with constant curvature 1, 0 and -1 respectively. Instead of defining the metric explicitly, it is equivalent to describe the geodesics in the space, as they determine the metric.

- Let $\mathbb{S}^n$ denote the unit sphere in $\mathbb{R}^n$ with the induced Euclidean metric. Then all geodesics are arcs of great circles. If two points are not antipodal then there is a unique geodesic connecting them. Alternatively there are infinitely many geodesics between two antipodal points.
- Let $\mathbb{E}^n$ denote n-dimensional Euclidean space. We make the distinction between $\mathbb{E}^n$ and $\mathbb{R}^n$; $\mathbb{E}^n$ is not equipped with a vector space structure, nor is there a fixed origin. This allows us to work in a coordinate free manner. The geodesics in Euclidean space are straight lines.
- Let $\mathbb{H}^n$ denote n-dimensional real hyperbolic space. There are two models of hyperbolic space that we will use when convenient.

  - Consider the half plane model

$$\mathcal{U} = \{z \in \mathbb{C} : \Im(z) > 0\}$$

  for $\mathbb{H}^2$. Then geodesics take the form of vertical lines or segments of semicircles perpendicular to the Real axis. These are shown in Fig. 1.
  - An alternative model $\mathbb{H}^2$ is the Poincaré disk $\mathcal{D}$. This is given by

$$\mathcal{D} = \{x \in \mathbb{C} : |x| < 1\}.$$

  Geodesics in the disk are either diameters or arcs of circles perpendicular to the boundary of the unit disk. These are shown in Fig. 2. There is an isometry $\mathcal{U} \to \mathcal{D}$ which wraps the Real axis into a circle with endpoints meeting at the point at infinity.

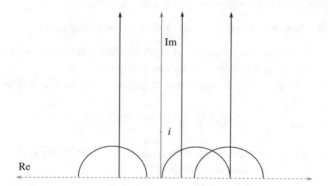

Fig. 1 Geodesics in the upper half plane model of hyperbolic space

Fig. 2 Geodesics in the
Poincaré disk model of
hyperbolic space

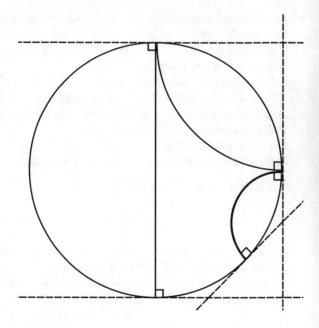

1.2 Curvature Condition and Isometries

Given the spaces with constant sectional curvature described above we describe
a notion of curvature in a general geodesic space. We focus on spaces with non-
positive curvature. As we will see later, these spaces appear naturally in many
settings. Again, [10] is a good reference for this section.

Definition 2 For a metric space (X, d) we define the following:

- We say (X, d) is geodesic if any two points can be connected by a geodesic. Note
  that we do not require this geodesic to be unique.
- A geodesic triangle $\Delta(x_1, x_2, x_3)$ between points $x_1, x_2, x_3 \in X$ is a union

$$\bigcup \{[x_i, x_j] : i, j \in \{1, 2, 3\}\}$$

where $[x_i, x_j]$ is a geodesic from x_i to x_j. A comparison triangle $\Delta(\bar{x}_1, \bar{x}_2, \bar{x}_3)$ for
$\Delta(x_1, x_2, x_3)$ in $\mathbb{E}^2$ is a union of geodesics

$$\bigcup \{[\bar{x}_i, \bar{x}_j] : i, j \in \{1, 2, 3\}\}$$

where $[\bar{x}_i, \bar{x}_j]$ is the unique geodesic between points $\bar{x}_i, \bar{x}_j \in \mathbb{E}^2$ which are chosen
to satisfy $d(x_i, x_j) = d(\bar{x}_i, \bar{x}_j)$. If $p \in [x_i, x_j] \subset \Delta(x_1, x_2, x_3)$, then a comparison
point for p is the unique $\bar{p} \in [\bar{x}_i, \bar{x}_j]$ with $d(\bar{p}, \bar{x}_i) = d(p, x_i)$.

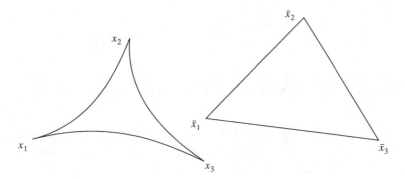

Fig. 3 On the left we have a triangle in a CAT(0) space and on the right a comparison triangle in Euclidean space

- We say $\Delta(x_1, x_2, x_3)$ satisfies the CAT(0) inequality if for any $p, q \in \Delta(x_1, x_2, x_3)$ we have

$$d(p, q) \le d(\bar{p}, \bar{q}).$$

Note that satisfying the CAT(0) inequality is independent of choice of comparison triangle (see Fig. 3).
- We call X a CAT(0) space if every geodesic triangle in X satisfies the CAT(0) inequality.

Remark 2 Similarly we can define CAT(-1) and CAT(1) by comparing with $\mathbb{H}^2$ and $\mathbb{S}^2$ respectively. Because the sphere has finite diameter and so geodesics have finite length, to show a space is CAT(1) it only makes sense to compare triangles with diameters less than 2π [10, Part II]. It can also be seen that CAT(-1) implies CAT(0) and CAT(0) implies CAT(1) [10, Theorem 1.12].

Example 2 It is easy to see that $\mathbb{E}^n$ is CAT(0) for all $n \in \mathbb{N}$. More generally a normed vector space is CAT(0) if and only if it is an inner product space, for a proof see [10, Proposition 1.14]. This shows that any Banach space that is not a Hilbert space is not CAT(0).

Proposition 1 *Suppose X is a CAT(0) metric space. Then there exists a unique geodesic between any two distinct points.*

Proof The following argument is presented pictorially in Fig. 4. Suppose x_1 and x_2 are two points and γ_1 and γ_2 are two geodesics from x_1 to x_2. Then for any $p_1 \in \gamma_1$ there exists a unique $p_2 \in \gamma_2$ such that $d(x_1, p_1) = d(x_1, p_2)$. We will show $p_1 = p_2$.

Consider the triangle $\Delta(x_1, p_1, x_2)$ which is given by $\gamma_1 \cup \gamma_2$. Taking a comparison triangle $\Delta(\bar{x}_1, \bar{p}_1, \bar{x}_2)$ we must have

$$d(\bar{x}_1, \bar{p}_1) + d(\bar{x}_2, \bar{p}_1) = d(\bar{x}_1, \bar{x}_2).$$

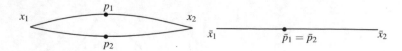

Fig. 4 Two geodesics between two distinct points form a geodesic triangle which we can compare with a comparison triangle in Euclidean space. We see that the two geodesics must be equal

This can only happen if $\bar{p}_1$ is on the unique geodesic from $\bar{x}_1$ to $\bar{x}_2$. This shows $\Delta(\bar{x}_1, \bar{p}_1, \bar{x}_2)$ is in fact a line. Taking a comparison point $\bar{p}_2$ for p_2, we must have $\bar{p}_2 = \bar{p}_1$. Applying the CAT(0) inequality we have

$$d(p_1, p_2) \leq d(\bar{p}_1, \bar{p}_2) = 0.$$

This shows $\gamma_1 \subset \gamma_2$. A symmetric argument gives the reverse containment and so we must have equality.

We now define a natural way to compare metric spaces. From this we will be able to generate automorphism groups of a metric space. We consider examples and state some elementary results.

Definition 3 An isometry $\varphi : X_1 \to X_2$ is a surjective map between metric space (X_1, d_1) and (X_2, d_2) such that

$$d_1(x, y) = d_2(\varphi(x), \varphi(y)).$$

It is easy to see that the set of isometries $X \to X$ forms a group under composition. Denote the group of isometries of X by $\mathrm{Isom}(X)$.

It is not hard to see that an isometry is in fact a homeomorphism and that the existence of an isometry between two spaces is an equivalence relation which we call isometric.

Example 3 In all of the following examples $\mathrm{Isom}(X)$ is in fact a Lie group. More details can be found in [10, Theorem 2.4].

- For $X = S^n$ we have $\mathrm{Isom}(S^n) = O(n)$ where

$$O(n) = \{A \in \mathrm{GL}(n, \mathbb{R}) : AA^t = \mathrm{Id}\}$$

  is the orthogonal group.
- Since all isometries of Euclidean space are compositions of translations and rotations, we have $\mathrm{Isom}(\mathbb{E}^n) = O(n) \ltimes \mathbb{R}^n$.
- For $X = \mathscr{U}$, denote the orientation preserving isometries of $\mathscr{U}$ by $\mathrm{Isom}^+(\mathscr{U})$. Then

$$\mathrm{Isom}^+(\mathscr{U}) = \mathrm{PSL}(2, \mathbb{R}),$$

which is the projective special linear group. The action is by Möbius transforms. That is, if

$$\left[\begin{pmatrix} a & b \\ c & d \end{pmatrix}\right] \in \mathrm{PSL}(2, \mathbb{R})$$

and $z \in \mathscr{U}$, then

$$\left[\begin{pmatrix} a & b \\ c & d \end{pmatrix}\right] z = \frac{az + b}{cz + d}.$$

This action is transitive on the unit tangent bundle.

Example 4 Examples of elements in $\mathrm{Isom}^+(\mathscr{U})$. These are also examples of elements which are in the distinct classes outlined in Theorem 1 below. We recommend [27] as a reference for hyperbolic space and its isometries.

- $z \mapsto -\dfrac{\bar{z}}{|z|^2} = \dfrac{-1}{z} = \left[\begin{pmatrix} 0 & 1 \\ -1 & 0 \end{pmatrix}\right] z.$

  Note that this isometry fixes the point i.
- $z \mapsto z + 1 = \left[\begin{pmatrix} 1 & 1 \\ 0 & 1 \end{pmatrix}\right] z.$

  This isometry shifts the whole upper half plan to the right. It fixes one point in the boundary which is the point at infinity.
- $z \mapsto 2z = \left[\begin{pmatrix} \sqrt{2} & 0 \\ 0 & \frac{1}{\sqrt{2}} \end{pmatrix}\right] z.$ This isometry is a dilation away from the origin. It

  fixes two points on the boundary; namely the point at infinity and 0. It acts as a translation along the imaginary axis.

The next theorem shows that the properties exhibited in the previous example classify isometries of $\mathscr{U}$. We will soon see that this classification extends beyond hyperbolic space.

Theorem 1 ([10]) *Let φ be an orientation preserving isometry of the upper half plane. Then exactly one of the following holds:*

- *φ fixes at least one point in $\mathscr{U}$, in this case we call φ elliptic;*
- *φ is not elliptic and fixes precisely one point on the boundary, in this case we call φ parabolic,*
- *φ is not elliptic and fixes precisely two points on the boundary, in this case we call φ hyperbolic.*

Furthermore φ is either elliptic, parabolic or hyperbolic if its trace is less than 2, equal to 2 or greater than 2 respectively.

To extend the previous result to more general spaces we need an appropriate definition of boundary.

Definition 4 Let X be a complete CAT(0) space. Call two geodesic rays γ_1 and γ_2 equivalent if there exists a constant $K \geq 0$ such that $d(\gamma_1(t), \gamma_2(t)) \leq K$. We define the visual boundary ∂X of X to be the collection of equivalence classes of rays.

Example 5 Standard examples of visual boundaries.

- Since there are no geodesic rays in S^2 we have $\partial S^2 = \varnothing$;
- Two rays in $\mathbb{E}^2$ are equivalent if and only if they are parallel. It follows that the visual boundary $\partial \mathbb{E}^2 = S^1$. Note that there is a natural topology on the visual boundary of a CAT(0) space that makes this identification with the circle a homeomorphism.
- In $\mathscr{U}$ we have two rays γ_1 and γ_2 equivalent if and only if they are asymptotic. Therefore the visual boundary of $\mathscr{U}$ is the real axis union a point at infinity.

Theorem 2 ([10]) *Suppose X is a complete CAT(0) space with φ an isometry of X. Defining elliptic, hyperbolic and parabolic as in Theorem 1, we have one of the following:*

- *φ is elliptic;*
- *φ is parabolic;*
- *φ is hyperbolic.*

The following result is a useful consequence of the CAT(0) condition which we will refer to later.

Proposition 2 *Suppose a group G acts by isometries on a CAT(0) space X. Then:*

1. if G has a bounded orbit, for example if G is finite, then G fixes a point in X.
2. the fixed set of G in X, denoted X^G, is convex.

Proof (Proof of 2) Let $g \in G$, $x, y \in X^G$ and γ be the geodesic between x and y. If $\gamma \neq g \cdot \gamma$ pointwise, then we contradict uniqueness of geodesics in CAT(0) spaces. Therefore $\gamma \subset X^G$ and X^G is convex. This argument can be seen as a diagram in Fig. 5.

Fig. 5 If the fixed point set of a group element is not convex, then X does not have unique geodesics

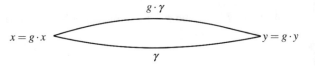

1.3   Lattices

In this section we introduce the concept of a lattice. This will be important for later sections and is the focus of many results. Lattices can be thought of as discrete approximations to non-discrete groups and so many results mentioned will be exploring how close this approximation can be. Results on the generalities of lattices can be found in [34].

Definition 5 Let G be a locally compact group with Haar measure μ. A lattice in G is a subgroup $\Gamma \le G$ so that

- Γ is a discrete subgroup of G.
- Γ has finite covolume, that is $\mu(G/\Gamma) < \infty$.

A lattice Γ is cocompact (or uniform) if G/Γ is compact, and otherwise is non-cocompact (or nonuniform).

Lattices were originally studied in the setting of Lie groups. As tools, they have provided a large number of rigidity results. It is in recent work, which we will discuss later, that the study of lattices in groups which are not necessarily Lie groups has been explored. This is still an active area of research with many open problems. For a survey which includes many open problems and questions see [21].

For the following examples a vague notion of fundamental domain is used. If a group G acts on a space X, then the fundamental domain of this action is a set whose interior contains precisely one point of each orbit, usually with some other nice topological conditions.

Example 6 The following are examples of lattices.

- Let $\Gamma = \mathbb{Z}^n \le \mathbb{R}^n = G$ with group structure given by addition. The Haar measure μ on G is the usual Lebesgue measure. Then Γ is clearly discrete and G/Γ is the n-torus. Thus Γ is a cocompact lattice in G. It can be shown that any lattice in $\mathbb{R}^n$ is isomorphic to $\mathbb{Z}^n$ and hence is cocompact. Figure 6 gives a diagram of the fundamental domain for this action in the case of $n = 2$.

Fig. 6 The fundamental domain for $\mathbb{Z}^2$ acting on $\mathbb{R}^2$

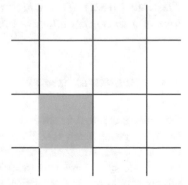

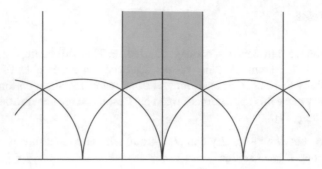

Fig. 7 The fundamental domain for $\Gamma = \mathrm{SL}(2, \mathbb{Z})$ acting on $\mathscr{U}$. The domain is a triangle with a corner at infinity

- Let $\Gamma = \mathrm{SL}(2, \mathbb{Z})$, $G = \mathrm{SL}(2, \mathbb{R})$. The following also applies to projective special linear groups. For the sake of avoiding cosets we restrict our attention to the special linear groups.It is clear that Γ is a discrete subgroup of G.

  Theorem 3 Γ *is a non-cocompact lattice in* G.
  Set $K = \mathrm{SO}(2, \mathbb{R}) = \mathrm{Stab}_G(i)$. Then G/K can be identified with $\mathscr{U}$ by the map $g \mapsto g(i)$ and action of Γ on $\mathscr{U}$ induces a tessellation. Now Γ is generated by

  $$u = \begin{pmatrix} 1 & 1 \\ 0 & 1 \end{pmatrix} \text{ and } v = \begin{pmatrix} 0 & 1 \\ -1 & 0 \end{pmatrix}.$$

  The fundamental domain of Γ is a triangle with one vertex at ∞, see Fig. 7. It can be shown that Γ is a nonuniform lattice as this triangle has finite area. It can also be shown $\mathscr{U}/\Gamma = (G/K)/\Gamma$ is a modular surface.

The group $\Gamma = \mathrm{SL}(2, \mathbb{Z})$ in the previous example is a first example of an arithmetic group. Roughly speaking, an arithmetic group is commensurable to integer points. The next result gives a restriction on which groups can appear as lattices in Lie groups.

Theorem 4 ([32]) *If G is a higher rank semisimple Lie group, for example* $\mathrm{SL}(n, \mathbb{R})$ *for* $n \geq 3$, *then every lattice in G is arithmetic.*

1.4 Symmetric Spaces

Roughly, a symmetric space is a Riemannian Manifold with a "highly transitive" isometry group such that the stabiliser at each point contains an element with derivative -1. The spaces $\mathbb{S}^n$, $\mathbb{E}^n$ and $\mathbb{H}^n$ are all examples of symmetric spaces. Symmetric spaces are used to study Lie groups and their lattices. There is a current research focus to use trees, buildings, and other polyhedral complexes to study other locally compact groups and their lattices, for example $SL_n(\mathbb{Q}_p)$, $SL_n(\mathbb{F}_q((t)))$ and Kac-Moody groups. These two methods of study can be seen as analogues. By comparing the two situations, we gain insights into classical Lie groups from the non-Lie group setting. In the other direction, insights from the Lie group setting can be applied to locally compact groups, even if the techniques used to prove results are vastly different. We give two references for more information concerning symmetric spaces. We give [26] as a standard reference and [19] as an almost self-contained treatment of the nonpositive curved case.

The following result was proved by Weil [42, 43] for Lie Groups and has recently been generalised. The proof uses the Rips complex whereas the proof for Lie groups relied on the differential structure of a Lie group.

Theorem 5 ([22, Theorem 1.1]) *If G is a compactly generated locally compact group and Γ is a cocompact lattice in G, then Γ is topologically locally rigid.*

2 Graphs and Trees

The automorphism group of a regular tree is a standard example of a locally compact group outside the realm of Lie groups. They are quite simple to define, yet there are many interesting results for which they are the focus. In this section we define these groups and consider some results concerning lattices.

2.1 Graphs and Notation

Graphs are versatile tools for generating interesting examples of groups and then studying them as well as being of interest on their own. For example, automorphism groups of trees and their subgroups will the be the subject of many results concerning lattices in locally compact groups. We will also consider right-angled buildings which are encoded by graphs. The following definition is to ensure notation is standard. There are many introductory texts on graphs.

Definition 6 A graph X consists of a vertex set VX, an edge set EX with an involution $e \mapsto \bar{e}$ such that $e \neq \bar{e}$, and maps $i, t : EX \to VX$ such that for all $e \in EX$, $i(\bar{e}) = t(e)$ and $t(\bar{e}) = i(e)$. We call $\bar{e}$ the ghost edge of e and often omit it in diagrams (Fig. 8). For a given graph X we define the following:

- A path is a sequence of edges $e_1, \ldots, e_n$ so that $t(e_i) = i(e_{i+1})$ (see Fig. 9).
- A reduced path is a path where $e_{i+1} \neq \bar{e_i}$.
- We say X is connected if any two vertices are connected by a path.
- We say X is a tree if X is a connected and any two vertices are connected by a unique path (Fig. 10).

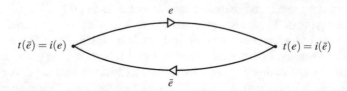

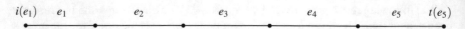

Fig. 8 An edge with its ghost edge

Fig. 9 A path in a graph

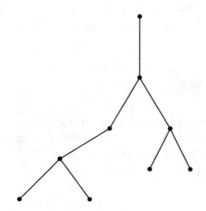

Fig. 10 A finite tree with ghost edges omitted

- If X is a tree, then we call X regular with degree n if $|i^{-1}(v)| = n$ for all $v \in VX$. In general we call $|i^{-1}(v)|$ the degree of v. We denote by T_n the unique infinite regular tree of degree n.
- We can view X as a metric space with the path metric. We identify edges with the unit interval and have length 1. The distance between two vertices is the length of a shortest path between them.
- A graph automorphism ϕ is a pair of bijections $\phi_V : VX \to VX$ and $\phi_E : EX \to EX$ such that ϕ_V commutes with t and i and ϕ_E commutes with the involution $e \to \bar{e}$. If X is a tree, ϕ_V completely determines ϕ_E and so we identify ϕ with ϕ_V.
- The set of ends of a tree is the set of geodesic rays originating from a fixed vertex v. It is an exercise to show that this definition is independent of choice of v and corresponds to the visual boundary.

It is an exercise to show that a graph is CAT(0) if and only if it is a tree. This can be done by recalling that CAT(0) spaces have unique geodesics. We can classify tree automorphisms using Theorem 2. It is an exercise to show that trees have no parabolic isometries.

Fig. 11 A depiction of the action of an automorphism of a tree which fixes a vertex r and but swaps the vertices v_1 and v_2

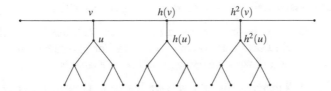

Fig. 12 A depiction of the action of a hyperbolic automorphism h of T_3

Elliptic automorphisms can be viewed as automorphisms of rooted trees. If ϕ fixes a vertex r, then we can view ϕ as an automorphism of the tree with root r. See Fig. 11 for a depiction of this action. If ϕ flips an edge, that is $\phi(e) = \bar{e}$, then we can subdivide e into two edges. Now ϕ fixes a new vertex which was originally the midpoint of e. A group Γ acts without inversions on a tree T if for all $e \in ET$ and for all $g \in \Gamma$, $ge \neq \bar{e}$. This can always be achieved by subdividing edges if necessary. This is the same as taking a barycentric subdivision of each edge as described in Fig. 25.

The other type of automorphism of a tree is hyperbolic. If an automorphism h is hyperbolic, then h translates along a bi-infinite geodesic between the two distinct points in the boundary which are fixed by the automorphism. See Fig. 12 for more information.

2.2 Tree Lattices

If T is a locally finite tree, that is the degree of each vertex is finite, then Aut(T) is totally disconnected and locally compact. Assume that Aut(T) acts without inversions, then vertex stabilisers in Aut(T) are precisely maximal compact open subgroups. They can be realised as the projective limit of finite groups by considering the action of the stabiliser on balls of increasing radius centred at the fixed vertex. For a complete reference on lattices in trees we recommend [4].

Proposition 3 *We have the following results concerning lattices in* Aut(T). *We use* T/Γ *to the denote the quotient of T by the action of Γ.*

- A subgroup $\Gamma \leq \mathrm{Aut}(T)$ is discrete if and only if Γ acts on T with finite stabilisers.
- If Γ is discrete, then:

  1. Γ is a cocompact lattice if and only if the quotient T/Γ is finite.
  2. Γ is a non-cocompact lattice if and only if T/Γ is infinite and

  $$\sum \frac{1}{|\mathrm{Stab}_\Gamma(v)|} < \infty,$$

  where this series is the sum over representative vertices v from every Γ-orbit on T. This assumes that $\mathrm{Aut}(T)$ acts cocompactly on T.

Theorems 6 and 7 highlight the different behaviours exhibited by lattices in Lie groups and lattices in $\mathrm{Aut}(T)$. Theorem 6 shows that for a given measure on $\mathrm{SL}_2(\mathbb{R})$, we are limited by how closely we can approximate by lattices.

Theorem 6 ([37, Theorem 5]) *If Γ is any lattice in $G = \mathrm{SL}_2(\mathbb{R})$ and μ is the standard Haar measure on G, then $\mu(G/\Gamma) \geq \frac{\pi}{21}$. More generally for any Haar measure μ on G, there exists $\varepsilon > 0$, dependent on μ, such that for all lattices $\Gamma \leq G$, $\mu(G/\Gamma) \geq \varepsilon$.*

Theorem 7 is in direct contrast with Theorem 6 by not only saying that there exist lattices with arbitrarily small covolume in $\mathrm{Aut}(T)$, but they can also be chosen to be in different commensurability classes. We say that two subgroups H_1 and H_2 of a group G are commensurable if there exists $g \in G$ such that $gH_1g^{-1} \cap H_2$ has finite index in gH_1g^{-1} and H_2. In particular conjugate lattices are commensurable. Being commensurable is an equivalence relation on subgroups of G and hence we can define commensurability classes.

Theorem 7 ([20, Corollary 1.2]) *Let $G = \mathrm{Aut}(T_m)$, $m \geq 3$, then for every $r > 0$, there exists uncountably lattices in G of covolume r, all in different commensurability classes.*

A key tool for studying tree lattices is Bass-Serre Theory. The fundamental theorem of Bass-Serre theory gives a correspondence between groups acting on trees without inversion and graphs of groups. See [36] for more information.

Example 7 Using Bass-Serre theory, we can describe certain lattices in $\mathrm{Aut}(T_3)$ using graphs of groups.

- Let Γ be the fundamental group of the edge of groups in Fig. 13. Then Γ is equal to $C_3 * C_3$, where $*$ denotes the free product, and Γ is a cocompact lattice in $\mathrm{Aut}(T_3)$, acting with quotient a single edge.

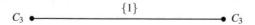

Fig. 13 An edge of groups which gives a lattice in $\mathrm{Aut}(T_3)$

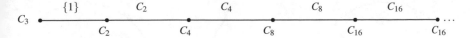

Fig. 14 A graph of groups which gives a non-cocompact lattice in Aut(T_3)

- The subgroup of Aut(T_3) determined by the graph of groups shown in Fig. 14 is
  a lattice which is not cocompact. Checking the condition of Proposition 3 part 2,
  we have

$$\sum \frac{1}{|\Gamma_v|} = \frac{1}{3} + \frac{1}{2} + \frac{1}{4} + \frac{1}{8} + \frac{1}{16} + \cdots < \infty.$$

3   Polyhedral Complexes

Polyhedral complexes are a geometric construction which can be used to study
classes of groups. There are many ways this can be done. In this section we
will focus on buildings which are geometric objects that can be associated to
certain groups. Results show that they essentially determine the group that they
are associated to. We will see that infinite trees where each vertex degree at least 2,
which are precisely the buildings of dimension 1, behave differently to buildings of
higher dimension. Along the way we will define some geometric tools that can be
used to study more general polyhedral complexes and state how these tools can be
applied to the study of buildings. We finish the section by comparing results in the
different classes of groups we have seen thus far.

3.1   The Definition of a Building

We define buildings from the view of polyhedral complexes. This is one of many
ways one can define a building. Once we have reached the definition and given
some elementary examples we will provide a short history of their development.
The theory of buildings is very developed and is deeper than presented here. For
more information we suggest [29].

For this section let X^n denote either $\mathbb{S}^n$, $\mathbb{E}^n$, or $\mathbb{H}^n$. For the following definition a
polytope is a finite intersection of half spaces. We say a polytope is simple if each
vertex is adjacent to precisely $n - 1$ edges. This is equivalent to the link of each
vertex, as seen in Fig. 17 and Definition 10, being an $n - 1$ simplex.

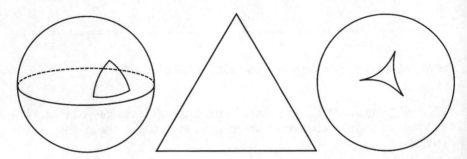

Fig. 15 Triangles for the triples of integers $(2, 2, 2)$, $(3, 3, 3)$ and $(4, 4, 4)$ in $\mathbb{S}^2$, $\mathbb{E}^2$, $\mathbb{H}^2$ respectively Here we are using the Poincaré disk model of hyperbolic space

Definition 7 A Coxeter polytope P in $\mathbb{X}^n$ is a convex simple compact polytope contained in $\mathbb{X}^n$ such that for any two codimension 1 faces F_i, F_j of P, either F_i and F_j are disjoint or they meet at dihedral angle $\frac{\pi}{m_{ij}}$, where $m_{ij} \geq 2$ is an integer.

Example 8 Suppose p, q, r are integers at least 2. If $\frac{\pi}{p} + \frac{\pi}{q} + \frac{\pi}{r}$ is greater than π, equal to π, or less than π, then there exists a triangle in $\mathbb{S}^2$, $\mathbb{E}^2$ or $\mathbb{H}^2$ respectively with interior angles $\frac{\pi}{p}, \frac{\pi}{q}, \frac{\pi}{r}$. For explicit examples see Fig. 15.

Theorem 8 ([17]) *Let P be a Coxeter polytope. For each codimension 1 face F_i let s_i be the reflection of $\mathbb{X}^n$ in the hyperplane supporting F_i. Then the reflection group $W := \langle s_i \rangle \leq \mathrm{Isom}(\mathbb{X}^n)$ is a Coxeter group. Moreover, W is a discrete subgroup of $\mathrm{Isom}(\mathbb{X}^n)$ and the action of W tessellates $\mathbb{X}^n$ by copies of P.*

Definition 8 A spherical, Euclidean or hyperbolic polyhedral complex X is a CW-complex where each n-cell is metrised as a convex, compact polytope in $\mathbb{S}^n$, $\mathbb{E}^n$ or $\mathbb{H}^n$ respectively, such that the metrics agree on intersections of closed cells.

Example 9 The following are examples of polyhedral complexes:

- The geometric realisation of a graph with the path metric has each 1-cell metrised as $[0, 1] \subset \mathbb{E}^1$.
- Tessellations of $\mathbb{X}^n$ by regular polygons.

Consider a path, that is a continuous image of the closed interval, between two points in a polyhedral complex. This path must intersect with finitely many polytopes. Call the path a string if the intersection with each polytope is a geodesic in that polytope. We can define a length for each string by summing up these the lengths of the geodesics which union to the whole string. The taut string metric can be defined by taking the infimum of lengths of strings between any two points.

Theorem 9 ([8, Theorem 1.1]) *If a polyhedral complex X has finitely many isometry types of cells, then X is a geodesic metric space when equipped with the taut string metric.*

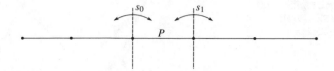

Fig. 16 An apartment in a rank 1 Euclidean building

Definition 9 Let $P \subset \mathbb{X}^n$ be a Coxeter polytope and $W = W(p)$ be the group generated by reflections in its codimension 1 faces. A building of type W, Δ, is a polyhedral complex which is a union of subcomplexes, called apartments, each isometric to the tessellation of $\mathbb{X}^n$ induced by the action of W. Each copy of P in Δ is called a chamber. We require that the following axioms hold:

- Any two of the chambers are contained in a common apartment.
- Given any two apartments A_1, A_2, there is an isometry $A_1 \to A_2$ which fixes $A_1 \cap A_2$ pointwise.

A building is spherical, Euclidean or hyperbolic as $\mathbb{X}^n$ is $\mathbb{S}^n$, $\mathbb{E}^n$ or $\mathbb{H}^n$ respectively. It is common for a Euclidean building to be referred to as an affine building. The rank, or dimension, of the building is defined to be n.

Example 10 We have the following examples of buildings:

- Take $\mathbb{X} = \mathbb{E}$, $P = [0, 1]$ and $W = \langle s_0, s_1 \rangle = \langle s_0, s_1 | s_0^2 = s_1^2 = 1 \rangle \cong D_\infty$. Then building of type W is a tree without leaves, such as T_3. Apartments are copies of the line tessellated by unit intervals, and chambers are edges in the tree. An apartment with the action of W is given in Fig. 16.
- Take $\mathbb{X}^2 = \mathbb{E}^2$, $P = [0, 1] \times [0, 1]$ and

$$W = \langle s_1, s_2, s_3, s_4 \rangle = \langle s_1, s_3 \rangle \times \langle s_2, s_4 \rangle \cong D_\infty \times D_\infty.$$

A building of type W is a product of trees, and the chambers are Euclidean squares. For a given vertex in the product we can define a graph which captures the local structure at that vertex. We call this graph the link of v. It has a vertex for each 1 dimensional face incident to v. Two vertices are connected by an edge if the are part of the same 2 dimensional face. For example, take $v \in T_3 \times T_3$ which is shown in Fig. 17. Then v is incident to precisely six edges. Forming vertices for each of these edges and connecting them by an edge if they form part of a square, we see that the link of v is precisely the complete bipartite graph with six vertices $K_{3,3}$, see Fig. 22. Conversely, any simply connected square complex such that the link of each vertex is a complete bipartite graph is a product of trees, see [11].
- For $p \geq 5$ and $q \geq 2$ define the building $I_{p,q}$ to be the unique simply connected hyperbolic polygonal complex (2-dimensional polyhedral complex) in which all faces (2-cells) are regular right angled hyperbolic p-gons, and the link of every vertex is $K_{p,q}$. $I_{p,q}$ is a building of type W where $W = \langle$ reflections in faces of $P \rangle$

Fig. 17 A vertex with v the product $T_3 \times T_3$ with all chambers incident to v

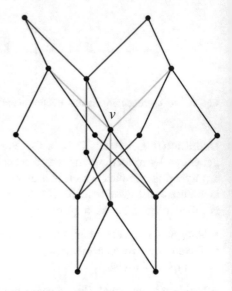

and P is the regular right-angled hyperbolic p-gon. One apartment is a tessellation of $\mathbb{H}^2$. This example is referred to as Bourdon's building and was first studied in [6]. It is the first example of a hyperbolic building. It is locally a product space, but not globally a product.

3.2 Original Motivation for Buildings and Bass-Serre Theory

Buildings were originally developed to study reductive algebraic groups over non-archimedean local fields. Examples include $SL_n(\mathbb{Q}_p)$ and $SL_n(\mathbb{F}_q((t)))$. Given such a group $G(F)$, the affine building for $G(F)$ is analogous to the symmetric space for a Lie group. The building has a set of chambers $G(F)/I$, where I is an Iwahori subgroup of $G(F)$. The subgroup I is a compact open subgroup of $G(F)$ and is analogous to a Borel subgroup of an algebraic group. If $G = SL_n(\mathbb{Q}_p)$, then

$$
I = \left\{ \begin{pmatrix} \mathbb{Z}_p^\times & \mathbb{Z}_p & \cdots & \cdots & \mathbb{Z}_p \\ p\mathbb{Z}_p & \mathbb{Z}_p^\times & & & \vdots \\ \vdots & & \ddots & & \vdots \\ \vdots & & & \mathbb{Z}_p^\times & \mathbb{Z}_p \\ p\mathbb{Z}_p & \cdots & \cdots & p\mathbb{Z}_p & \mathbb{Z}_p^\times \end{pmatrix} \in G \right\}.
$$

Apartments of the affine building associated to $G(F)$ are tessellations of Euclidean space induced by the action of the associated affine Weyl group W. In the case when

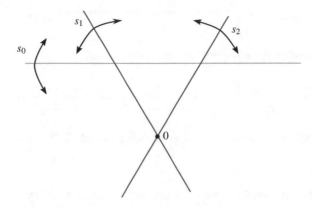

Fig. 18 The action of $S_3 \ltimes \mathbb{Z}^2$ on $\mathbb{R}^2$

$G = \mathrm{SL}_n(\mathbb{Q}_p)$ and $W = S_3 \ltimes \mathbb{Z}^2$, W is generated by three reflections in Euclidean space. Their action is demonstrated pictorially Fig. 18.

Vertices of the building for $G(F)$ are cosets in $G(F)$ of maximal compact subgroups of $G(F)$. The group $G(F)$ acts chamber transitively on its building with compact stabilisers.

It is not the case that every building is associated to a group in the way we have described. When the building is associated to a group, Tits showed that in this setting, the group is essentially determined by the building.

Theorem 10 ([41]) *If Δ is an irreducible (not a product) spherical building of dimension at least 2, or an irreducible Euclidean building of dimension at least 3, then Δ is the building for some $G(F)$. Moreover,* $\mathrm{Aut}(\Delta) = G(F) \rtimes \mathrm{Aut}(F)$.

There are many differences between affine buildings of rank 1 and buildings of higher rank. This is especially prominent when studying the groups for which the building is associated to. Part of the motivation for Bass-Serre theory was the study of this special case. This can be done since the affine building for a rank 1 group is by definition a tree.

Example 11 The following are properties of rank 1 groups that can be shown using Bass-Serre theory. More details can be found in [36].

- Ihara's Theorem. We can decompose $\mathrm{SL}_2(\mathbb{Q}_p)$ into the amalgamated free product

$$\mathrm{SL}_2(\mathbb{Q}_p) \cong \mathrm{SL}_2(\mathbb{Z}_p) *_I \mathrm{SL}_2(\mathbb{Z}_p).$$

If $\Delta = T_p$ is the building associated to $\mathrm{SL}_2(\mathbb{Q}_p)$, then the result can be seen by noting that $\Delta/\mathrm{SL}_2(\mathbb{Q}_p)$ is the edge of groups Fig. 19.
- Set

$$\Gamma = \mathrm{SL}_2(\mathbb{F}_q[t]) \leq G = \mathrm{SL}_2(\mathbb{F}_q((t^{-1}))).$$

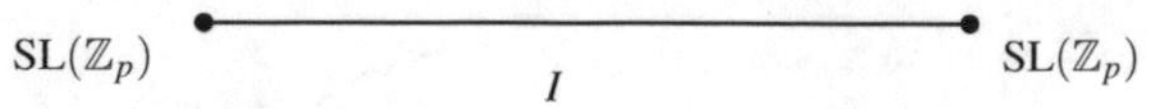

Fig. 19 An edge of groups which represents the action of $SL_2(\mathbb{Q}_p)$ on T_p

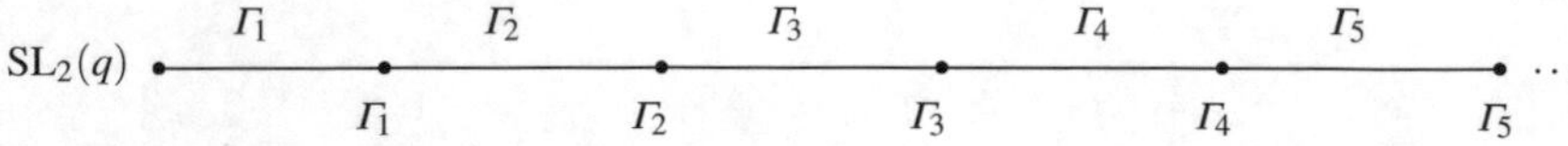

Fig. 20 The graph of groups for a non-cocompact lattice in $SL_2(\mathbb{F}_q((t^{-1})))$, where Γ_a is given by Eq. (2)

The graph of groups for Γ has edge and vertex groups given by

$$\Gamma_a = \left\{ \begin{pmatrix} a & b \\ 0 & a^{-1} \end{pmatrix} \middle| a \in \mathbb{F}_q^*, b \in \mathbb{F}_q[t], \deg(b) \leq n \right\}, \tag{2}$$

and is shown in Fig. 20. It can show that Γ is a non-cocompact lattice in G.

3.3 Gromov's Link Condition

We briefly explored the link of a vertex when considering a product of trees. In that setting we gave a sufficient condition for a polyhedral complex to be a product of trees. The sufficient condition was that the link of each vertex was a complete bipartite graph. We now formalise this notion of link and state more results that show how global structure can be inferred from local structure.

Definition 10 Let X be a polyhedral complex and $v \in VX$. The link of v, denoted by $Lk(v, X)$, is the spherical polyhedral complex obtained by intersecting X with a small sphere centred at v.

Although the link is a local structure, Theorem 11, also known as Gromov's link condition, shows how one can imply global structure from local structure.

Theorem 11 ([10]) *Let X be a simply connected polyhedral complex.*

1. *If X is Euclidean, then X is CAT(0) if and only if $Lk(v, X)$ is CAT(1) for each $v \in VX$.*
2. *If X is hyperbolic, then X is CAT(-1) if and only if $Lk(v, X)$ is CAT(1) for each $v \in VX$.*

There are two cases where it is easy to check whether vertex links are CAT(0):

* When X has dimension 2, $Lk(v, X)$ is a graph with edge lengths given by angles in X. An example where X is the built from squares in $\mathbb{R}^2$ is given in Fig. 21.

    In this setting $Lk(v, X)$ is CAT(1) if and only if each embedded cycle has length at least 2π. Similarly, if X is a product of trees or Bourdon's building, then

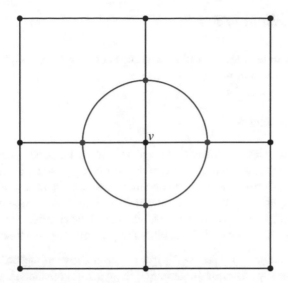

Fig. 21 We have a link of a vertex $v \in \mathbb{Z}^2$ which is the 1-skeleton of a polyhedral complex which is a tessellation of $\mathbb{R}^2$ by the unit squares. The link is the graph given by the circle. We can assign lengths of $\frac{\pi}{2}$ to each edge

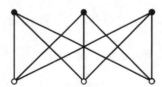

Fig. 22 The link of a vertex in the product of two regular trees of degree 3 is a complete bipartite graph with six vertices. All the edges have length $\frac{\pi}{2}$ and so the length of the shortest cycle is 2π

$Lk(v, X)$ is the complete bipartite graph with edge lengths $\frac{\pi}{2}$. Using Theorem 11 we can show that products of trees and Bourdon's building are CAT(0).

- X is a cube complex, that is, each n-cell is metrised as a unit cube in $\mathbb{E}^n$. Here, links are spherical simplicial complexes with all angles $\frac{\pi}{2}$. Figure 21 demonstrates this.

A simplicial complex is flag if whenever it contains the 1-skeleton of a simplex, it contains that simplex. The following results can be shown using Theorem 11.

Theorem 12 ([23]) *An all-right spherical simplicial complex is* CAT(1) *if and only if it is flag.*

Theorem 13 ([16, 17, 33]) *Euclidean buildings are* CAT(0), *hyperbolic buildings are* CAT(−1).

Theorem 14 ([9]) *Let X be a* CAT(0) *polyhedral complex. Then* Aut(X) *contains no parabolic isometries.*

3.4   Comparisons of Results

Here we compare analogous results for groups acting on trees, higher-dimensional complexes and Lie groups.

3.4.1   Quasi-Isometries

Quasi-isometries play a fundamental role in geometric group theory. Often the notion of isometry is too exact, for example a group may have two different Cayley graphs, generated from different generating sets, which may not be isometric. However, the graphs will be quasi-isometric and so results shown about groups from their Cayley graphs will be quasi-isometric invariants. Properties such as hyperbolicity, growth rate and amenability are preserved by quasi-isometry.

Definition 11 Suppose $f : X_1 \rightarrow X_2$ is a function between metric spaces (X_1, d_1) and (X_2, d_2). We say f is a quasi-isometry if there exists constants $A \geq 1$ and $B, C \geq 0$ such that:

- For all $x, y \in X_1$

$$\frac{1}{A} d_1(x, y) - B \leq d_2(f(x), f(y)) \leq A d_1(x, y) + B;$$

- For each $y \in X_2$ there exists $x \in X_1$ such that

$$d_2(f(x), y) \leq C.$$

It is an exercise to show that the existence of a quasi-isometry between two spaces is an equivalence relation and so it makes sense to talk about metric spaces as being quasi-isometric.

Quasi-isometries can be seen as preserving the coarse metric structure as the following examples suggest.

Example 12 Here we give examples of quasi-isometric spaces. The different examples we give are from different quasi-isometry classes.

- The metric spaces $\mathbb{R}^2$ and $\mathbb{Z}^2$ are quasi-isometric, with $(x, y) \mapsto (\lfloor x \rfloor, \lfloor y \rfloor)$ a quasi-isometry;
- If T and T' are infinite regular trees with degree at least 3, then T and T' are quasi-isometric;
- All compact metric spaces are quasi-isometric, in particular they are quasi-isometric to a single point.

The previous example demonstrates that there are many quasi-isometric rank 1 buildings which are not isometric, namely trees. The following result shows that this property is not shared with Euclidean buildings of higher rank.

Theorem 15 ([28]) *Suppose Δ_1 and Δ_2 are two quasi-isometric Euclidean buildings with dimension at least 2. Then Δ_1 and Δ_2 are isometric.*

It was also shown in [28] that the above result holds for symmetric spaces of rank at least 2 and for nilpotent simply connected Lie groups. The two results are shown using a single proof. This is in contrast with the next result which requires a proof which differs greatly from the proof for symmetric spaces. The proof utilises conformal analysis on the visual boundary

Theorem 16 ([7, 45]) *If X and Y are any two hyperbolic buildings of dimension 2 such that X and Y are quasi-isometric, then X and Y are isometric. For example, $I_{p,q}$ is quasi-isometric to $I_{p',q'}$ if and only if $p = p'$ and $q = q'$.*

3.4.2 Lattices

The structure of lattices contained within a group has been an active research area. In the case of Lie groups, these lattices are residually finite [46]. Residually finite groups have ample normal subgroups. This is a strong contrast to the following result.

Theorem 17 ([11]) *There exist simple uniform lattices in the automorphism group of a product of trees.*

The proof of the theorem uses some ideas and techniques from the study of lattices in Lie groups, for example the normal subgroup theorem, via Property (T), see [5]. It can also be shown that, unlike in other cases, a uniform tree lattice is virtually free.

The original motivation for Kazhdan's Property (T) was to show that lattices in higher rank Lie groups are all finitely generated. If we instead consider the case of lattices in other groups the situation is quite different. We have the following result for the rank 1 case.

Theorem 18 ([4]) *Non-uniform tree lattices (including lattices in groups whose building is a tree) are never finitely generated.*

If $\dim(X) \geq 2$, the situation for lattices in Aut(X) is mixed. In some situations, for example an $\tilde{A}_2$ building or a triangular hyperbolic building (that is, chambers are hyperbolic triangles), then Aut(X) has Property (T), hence lattices in Aut(X) are finitely generated, see ([3, 15] and [47]). On the other hand, in [40] it is shown that if a non-uniform lattice $\Gamma \leq$ Aut$(I_{p,q})$ has a strict fundamental domain (that is, there exists a subcomplex $Y \subset I_{p,q}$ containing exactly one point from each Γ-orbit), then Γ is not finitely generated.

The case of products of trees is still unknown. If Γ is an irreducible non-uniform lattice on a product of trees is Γ finitely generated? Any known examples of such a lattice is either an arithmetic group or a Kac-Moody group. All of these are finitely generated.

4   More Examples of Polyhedral Complexes

In this section we consider specific examples of combinatorial objects which have yielded interesting group actions. Some of these examples will be specific types of buildings while another will be a polyhedral complex with a slightly weaker homogeneity condition than that of a building.

4.1   Right-Angled Coxeter Groups, Their Davis Complexes and Associated Buildings

The automorphism group of a right-angled building has recently been shown to have interesting properties which contrast with the results we have seen so far.

4.1.1   Right-Angled Coxeter Groups and Davis Complexes

Let Γ be a finite simplicial graph with vertex set S. The associated right-angled Coxeter group W_Γ is

$$W_\Gamma = \langle S \mid s^2 = 1 \; \forall s \in S, \, st = ts \text{ if } s \text{ and } t \text{ are adjacent in } \Gamma \rangle.$$

Example 13   Examples of right angled Coxeter groups associated with graphs:

- Suppose Γ consists of two disconnected vertices. Then

$$W_\Gamma = \langle s, t \mid s^2 = t^2 = 1 \rangle \cong D_\infty.$$

  Note that this group is infinite.
- If Γ is the complete graph on n vertices, then

$$W_\Gamma = \langle s_1, \ldots, s_n \mid s_i^2 = 1, \, s_i s_j = s_j s_i \; \forall i, j \rangle \cong (C_2)^n.$$

  In contrast with the case when Γ is given by two disconnected vertices, this group is finite.

We can consider the Cayley graph $\mathrm{Cay}(W_\Gamma, S)$. Recall that this is the graph with vertex set W_Γ and directed edges of the form (w, ws) for $w \in W_\Gamma$ and $s \in S$. For each edge (s, t) in Γ we get a square in $\mathrm{Cay}(W_\Gamma, S)$. For each triangle in Γ we get a cube in $\mathrm{Cay}(W_\Gamma, S)$. For each K_n in Γ we get a 1-skeleton of n-cube in $\mathrm{Cay}(W_\Gamma, S)$. These ideas are shown in Figs. 23 and 24. In general, for each K_n in Γ we get a 1-skeleton of an n-cube in $\mathrm{Cay}(W_\Gamma, S)$.

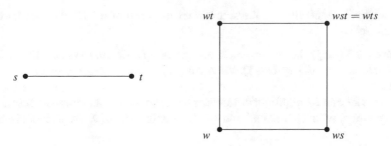

Fig. 23 If we take an edge in Γ, we get a square in the Cayley graph of the right-angled Coxeter group associated to Γ

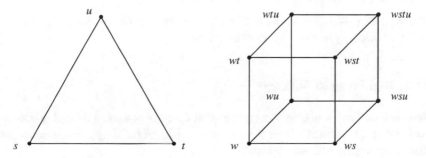

Fig. 24 If we take a triangle in Γ, we get a cube in the Cayley graph of the right-angled Coxeter group associated to Γ

Recall that a graph is CAT(0) if and only if it is a tree. Since trees have no cycles we see that the Cayley graph $\mathrm{Cay}(W_\Gamma, S)$ is CAT(0) if and only if Γ has no edges. This is equivalent to W_Γ begin isomorphic to a free product of copies of C_2.

By identifying each n-cube with $[0, 1]^n$ in $\mathbb{R}^n$, we can fill in the cubes of $\mathrm{Cay}(W_\Gamma, S)$ and realise $\mathrm{Cay}(W_\Gamma, S)$ as a CAT(0) space. The fact that the resultant space is CAT(0) follows from considering the link of each vertex. It is not hard to see that these are all flag. Applying Theorem 11, one can prove Theorem 19.

Theorem 19 ([17]) *If Γ a finite simplicial graph with vertex set X, then* $\mathrm{Cay}(W_\Gamma, S)$ *is the* 1-*skeleton of a* CAT(0) *cubical complex.*

The resulting space is the Davis complex for W_Γ. For any Coxeter system (W, S) the Davis complex Σ is a piecewise Euclidean, CAT(0), finite dimensional, locally finite, contractible, polyhedral complex on which W acts cocompactly with finite stabilisers. In particular, W is a cocompact lattice in $\mathrm{Aut}(\Sigma)$. For more information we suggest [17] as a reference.

For a general Coxeter system (W, S), let Γ be the graph with vertex set S, an edge labelled m_{ij} between each pair of vertices s_i, s_j, where $(s_i s_j)^{m_{ij}} = 1$. We call Γ the Coxeter graph of (W, S) and say that Γ is flexible if it has a non-trivial label preserving automorphism ϕ that fixes the star of some vertex. Theorem 20 shows that flexible is equivalent to the automorphism group of the Davis complex being non-discrete.

Theorem 20 ([25]) *Let Γ be the Coxeter graph of a Coxeter system (W, S). Let Σ be the Davis complex of (W, S). Then $\mathrm{Aut}(\Sigma)$ is non-discrete if and only if Γ is flexible.*

Theorems 21 and 22 highlight differences between the Lie and non-Lie cases. They are in contrast with Theorems 6 and 18. The conditions on Σ are technical and left out.

Theorem 21 ([39]) *For certain Σ with non-discrete $G = \mathrm{Aut}(\Sigma)$, there exists a sequence of uniform lattices (Γ_n) with $\mu(G/\Gamma_n) \longrightarrow \mu(G/\Gamma_\infty)$ for Γ_∞ a non-uniform lattice. Moreover, Γ_∞ is not finitely generated.*

Theorem 22 ([44]) *For certain Σ with non-discrete $G = \mathrm{Aut}(\Sigma)$, G admits uniform lattices of arbitrarily small covolume.*

4.1.2 Right-Angled Buildings

Now that we have a notion of a right-angled Coxeter system, a natural question to ask is what this extra structure brings to the setting of buildings. In this section we state some results with this in mind.

Theorem 23 ([2, 24]) *Given any right angled Coxeter system (W_Γ, S) and a collection of cardinalities $\{q_s : s \in S\}$ each of at least 2, there exists a unique building of type (W_Γ, S) such that each panel of type s has q_s chambers incident.*

Remark 3 The apartments in the above building are copies of the barycentric subdivision of the Davis complex for W_Γ. An overview of a barycentric subdivision is given in Fig. 25. The automorphism group of this building is locally compact if each q_s is finite, it is non-discrete if W_Γ is infinite and there exists $q_s, q_t \geq 3$ with $st \neq ts$. They are also a source of simple groups as shown in Theorem 24.

Theorem 24 ([12]) *The automorphism group of a right-angled building is virtually simple.*

Theorem 25 shows that a right-angled building is a CAT(0) space.

Theorem 25 ([16, 17, 33]) *A right-angled building Δ of type W_Γ is a CAT(0) space. The following are equivalent:*

- *Δ can be equipped with a piecewise metric so it is CAT(-1) and hence Gromov hyperbolic.*
- *W_Γ is word hyperbolic;*
- *Γ contains no non-empty squares; that is, if Γ contains a 4-cycle, then it contains a diagonal.*

Corollary 1 *Suppose W_Γ is word hyperbolic and Λ is a uniform lattice in a right angle building of type W_Γ. Then Λ is finitely-generated linear group.*

Fig. 25 Here we have the barycentric subdivision of a square. The original square is shown in blue. For each face of the square there exists a unique point, called the barycentre, which is fixed by every isometry of the face. This can easily be seen as 'the middle' of each face. For each chain of faces $F_1 \subset F_2 \subset \cdots$, we take the convex hull of the barycentres of each face. We then split our square into the union of the hulls to obtain a simplicial complex

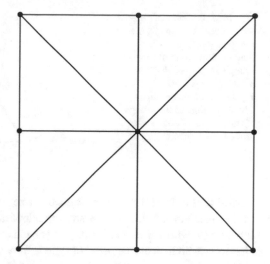

Proof (Proof Idea) The proof uses Agol's Theorem, see [1], which states that if a group acts properly, discontinuously and cocompactly on a Gromov hyperbolic cube complex, then it is linear.

Compare Burger-Mozes groups acting on products of trees which are certainly linear, since finitely generated linear groups are residually finite.

Recall that a tree is regular if each vertex has degree n. A tree is bi-regular of degrees n_1 and n_2 if each vertex has degree n_1 or n_2 and if u and v are adjacent vertices with degree of v is n_1, then the degree of u is n_2. For $n_1, n_2 \geq 2$, Denote the unique infinite bi-regular tree of degrees n_1 and n_2 by T_{n_1, n_2}.

Theorem 26 ([38]) *If* $\mathrm{Aut}(\Delta)$ *is non-discrete where* Δ *is a right-angled building, then for each* $r > 0$ *there exists uncountable many commensurability classes of lattices in* $\mathrm{Aut}(\Delta)$ *with covolume r.*

Proof (Proof Idea) Promote groups acting on T_{q_s, q_t} to complexes of groups acting on Δ. This takes uniform (respectively, non-uniform) lattices in $\mathrm{Aut}(T_{q_s, q_t})$ to uniform (respectively, non-uniform) lattices in $\mathrm{Aut}(\Delta)$ whilst preserving covolume.

Remark 4 Caution should be used with this technique. Unlike the case of graphs of groups, not every complex of groups comes from a group action. This is demonstrated in Fig. 26.

There is a sufficient condition for complexes of groups to be developable, that is local groups embed into fundamental groups. A complex of groups is developable if and only if the complex of groups has a universal cover on which its fundamental group acts inducing the complex of groups.

For more information on complexes of groups we suggest [10].

Recently in [18] the concept of universal groups acting on trees with prescribed local action has been extended to groups acting on right-angled buildings. Then the authors find examples of simple totally disconnected locally compact groups.

Fig. 26 A complex of groups which does not come from a group action. The fundamental group of this triangle is $\langle a, b, c | aba^{-1} = b^2, cac^{-1} = a^2, bcb^{-1} = c^2 \rangle$. However, this group is trivial. So none of the groups at each vertex or edge embed into the fundamental group

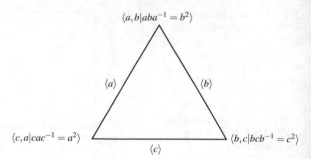

$\langle a, b | aba^{-1} = b^2 \rangle$

$\langle a \rangle$ $\langle b \rangle$

$\langle c, a | cac^{-1} = a^2 \rangle$ $\langle b, c | bcb^{-1} = c^2 \rangle$

$\langle c \rangle$

To define a universal group acting on a tree, one starts by labelling the edge set of the tree. For a building, the authors adopt a combinatorial point of view of Δ. It's a graph with vertex set the chambers and an edge of colour s between any two chambers which meet in a panel of type s. Now an apartment is a copy of the Cayley graph for W_Γ with generators S. An s-panel is a complete subgraph on q_s vertices. The labelling they choose for each s-panel is a finite subgroup $F_s \leq \mathrm{Sym}(q_s)$.

4.2 Kac-Moody Buildings

Here we give a short mention of Kac-Moody groups and associated buildings. These groups can be thought of as infinite dimensional Lie groups. Alternatively, they also behave like arithmetic groups. Suppose Λ is a minimal Kac-Moody group over $\mathbb{F}_q$. That is, Λ has a presentation with generating set which are the root subgroups

$$U_\alpha \cong (\mathbb{F}_q, +)$$

and commutator relations. Then Λ has a Weyl group W, which is a Coxeter group with presentation

$$\langle S : s_i^2 = 1, (s_i s_j)^{m_{ij}} = 1 \rangle,$$

where $m_{ij} \in \{2, 3, 4, 6\} \cup \{\infty\}$. Note that by convention $m_{ij} = \infty$ means the $s_i s_j$ has infinite order.

Now Λ has twin Iwahori subgroups I^+ and I^-. We can define two buildings Δ^+ and Δ^- of type W such that $\Delta^\pm$ has chambers $\Lambda/I^\pm$ and apartments isomorphic to copies of the Davis complex of W. Then Λ acts on $\Delta^+ \times \Delta^-$.

Theorem 27 ([14, 35]) *With Λ as above we have; for q large enough:*

- *Λ is a non-uniform lattice in $\mathrm{Aut}(\Delta^+ \times \Delta^-)$. This generalises $\mathrm{SL}_n(\mathbb{F}_q[t, t^{-1}])$, which is an irreducible lattice in $\mathrm{SL}_n(\mathbb{F}_q((t))) \times \mathrm{SL}_n(\mathbb{F}_q((t^{-1})))$.*

- *The stabiliser of $v \in \Delta^+$ in Λ is a non-uniform lattice in* $\mathrm{Aut}(\Delta^-)$. *This generalises the Nagao Lattices* $\mathrm{SL}_n(\mathbb{F}_q[t])$ *in* $\mathrm{SL}_n(\mathbb{F}_q((t^{-1})))$.

Theorem 28 ([13]) *If the Weyl group W is 2-spherical, that is if m_{ij} is finite for all i and j, infinite, non-affine and q is sufficiently large depending on W, then Λ mod out by its finite centre is abstractly simple.*

Remark 5 Compare Theorem 28 with the example given by the universal Burger Mozes lattices.

4.3 (k, L)-Complexes

Here we define a complex which is a generalisation of many of the geometric structures we have seen previously. Preliminary results hint that this is a promising setting for results. A (k, L)-complex is defined by choosing a k-gon for which all 2-dimensional faces will be isometric to and graph L which describes how the faces are glued together at each vertex. The resulting structure are still homogeneous in the sense that the link at each vertex is the same.

Definition 12 A (k, L)-complex is a simply connected polyhedral complex where each 2-dimensional cell is a regular k-gon and the link of each vertex is the graph L.

Example 14 The product of trees $T_m \times T_n$ is a $(4, K_{m,n})$-complex.
It is interesting to ask for which pairs (k, L) do there exist (k, L)-complexes and, given existence, is any such complex unique. The following results are progress towards an answer.

Theorem 29 ([2]) *Provided (k, L) satisfy the Gromov link condition, there exists at least one (k, L)-complex.*

Theorem 30 ([2, 24]) *There exist uncountably many $(6, K_4)$-complexes which are not pairwise isomorphic.*

Definition 13 For a vertex v in a graph L, define $\mathrm{St}(v)$ to be the subgraph containing v and all vertices linked to v by an edge. For an edge e in L, define $\mathrm{St}(e)$ to be the collection of edges which share an endpoint with e.

A graph L is vertex (respectively edge) star transitive if for all $u, v \in VL$ (respectively EL), every isomorphism $\mathrm{St}(u) \to \mathrm{St}(v)$ extends to an automorphism of L.

Theorem 31 ([30]) *If (k, L) satisfies the Gromov link condition, $k \geq 4$ and L is vertex star transitive and edge star transitive, then there exists a unique (k, L)-complex.*

Example 15 There are examples of pairs (k, L), where L is vertex star transitive and edge star transitive but the associated (k, L)-complex is not a building. Possible

Fig. 27 A well known
example of an odd graph,
namely the Peterson graph

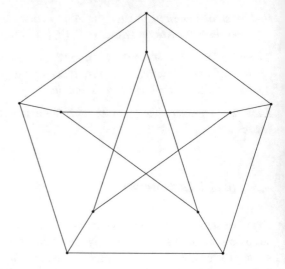

choices for L include complete bipartite graphs and odd graphs. For a well known
example of an odd graph see Fig. 27.

Acknowledgements We would like to thank John J. Harrison for kindly sharing his notes. We
would also like to thank MATRIX for funding and hosting the Winter of Disconnectedness
workshop where these lectures were presented.

References

1. Agol, I.: The virtual Haken conjecture. Doc. Math. **18**, 1045–1087 (2013). With an appendix
   by Agol, Daniel Groves, and Jason Manning
2. Ballmann, W., Brin, M.: Polygonal complexes and combinatorial group theory. Geom.
   Dedicata. **50**(2), 165–191 (1994)
3. Ballmann, W., Świątkowski, J.: On L^2-cohomology and property (T) for automorphism groups
   of polyhedral cell complexes. Geom. Funct. Anal. **7**(4), 615–645 (1997)
4. Bass, H., Lubotzky, A.: Tree Lattices. Progress in Mathematics, vol. 176. Birkhäuser, Boston
   (2001). With appendices by Bass, Carbone, L., Lubotzky, Rosenberg, G., Tits, J.
5. Bekka, B., de la Harpe, P., Valette, A.: Kazhdan's Property (T). New Mathematical Monographs, vol. 11. Cambridge University Press, Cambridge (2008)
6. Bourdon, M.: Immeubles hyperboliques, dimension conforme et rigidité de Mostow. Geom.
   Funct. Anal. **7**(2), 245–268 (1997)
7. Bourdon, M., Pajot, H.: Rigidity of quasi-isometries for some hyperbolic buildings. Comment.
   Math. Helv. **75**(4), 701–736 (2000)
8. Bridson, M.R.: Geodesics and curvature in metric simplicial complexes. Ph.D. thesis, Cornell
   University. ProQuest LLC, Ann Arbor (1991)
9. Bridson, M.R.: On the semisimplicity of polyhedral isometries. Proc. Amer. Math. Soc. **127**(7),
   2143–2146 (1999)
10. Bridson, M.R., Haefliger, A.: Metric Spaces of Non-positive Curvature, vol. 319. Springer,
    New York (2011)

11. Burger, M., Mozes, S.: Lattices in product of trees. Inst. Hautes Études Sci. Publ. Math. **92**, 151–194 (2000/2001)
12. Caprace, P.-E.: Automorphism groups of right-angled buildings: simplicity and local splittings. Fund. Math. **224**(1), 17–51 (2014)
13. Caprace, P.-E., Remy, B.: Simplicity and superrigidity of twin building lattices. ArXiv Mathematics e-prints (July, 2006)
14. Carbone, L., Garland, H.: Existence of lattices in Kac-Moody groups over finite fields. Commun. Contemp. Math. **05**(05), 813–867 (2003)
15. Cartwright, D.I., Mantero, A.M., Steger, T., Zappa, A.: Groups acting simply transitively on the vertices of a building of type $\tilde{A}_2$. I. Geom. Dedicata. **47**(2), 143–166 (1993)
16. Davis, M.W.: Buildings are CAT(0). In: Geometry and Cohomology in Group Theory (Durham, 1994). London Mathematical Society Lecture Note Series, vol. 252, pp. 108–123. Cambridge University Press, Cambridge (1998)
17. Davis, M.W.: The geometry and topology of Coxeter groups. In: Introduction to Modern Mathematics. Advanced Lectures in Mathematics (ALM), vol. 33, pp. 129–142. International Press, Somerville (2015)
18. De Medts, T., Silva, A.C., Struyve, K.: Universal groups for right-angled buildings. ArXiv e-prints (March, 2016)
19. Eberlein, P.B.: Geometry of Nonpositively Curved Manifolds. Chicago Lectures in Mathematics. University of Chicago Press, Chicago (1996)
20. Farb, B., Hruska, G.C.: Commensurability invariants for nonuniform tree lattices. Isr. J. Math. **152**, 125–142 (2006)
21. Farb, B., Hruska, C., Thomas, A.: Problems on automorphism groups of nonpositively curved polyhedral complexes and their lattices. In: Geometry, Rigidity, and Group Actions. Chicago Lectures in Mathematics, pp. 515–560. University of Chicago Press, Chicago (2011)
22. Gelander, T., Levit, A.: Local rigidity of uniform lattices. ArXiv e-prints (May 2016)
23. Gromov, M.: Hyperbolic groups. In: Essays in Group Theory. Publications of the Research Institute for Mathematical Sciences, vol. 8, pp. 75–263. Springer, New York (1987)
24. Haglund, F.: Les polyèdres de Gromov. C. R. Acad. Sci. Paris Sér. I Math. **313**(9), 603–606 (1991)
25. Haglund, F., Paulin, F.: Simplicité de groupes d'automorphismes d'espaces à courbure négative. In: The Epstein Birthday Schrift. Geometry & Topology Monographs, vol. 1, pp. 181–248 (electronic). Geometry & Topology Publications, Coventry (1998)
26. Helgason, S.: Differential Geometry, Lie Groups, and Symmetric Spaces. Graduate Studies in Mathematics, vol. 34. American Mathematical Society, Providence (2001). Corrected reprint of the 1978 original
27. Katok, S.: Fuchsian Groups. University of Chicago Press, Chicago (1992)
28. Kleiner, B., Leeb, B.: Rigidity of quasi-isometries for symmetric spaces and Euclidean buildings. C. R. Acad. Sci. Paris Sér. I Math. **324**(6), 639–643 (1997)
29. Krifka, Y.: Geometric and topological aspects of Coxeter groups and buildings (June, 2016). https://people.math.ethz.ch/~ykrifka/docs/lecturenotes.pdf
30. Lazarovich, N.: Uniqueness of homogeneous CAT(0) polygonal complexes. Geom. Dedicata. **168**(1), 397–414 (2014)
31. Lubotzky, A.: Tree-lattices and lattices in Lie groups. In: Combinatorial and Geometric Group Theory (Edinburgh, 1993). London Mathematical Society Lecture Note Series, vol. 204, pp. 217–232. Cambridge University Press, Cambridge (1995)
32. Margulis, G.A.: Discrete Subgroups of Semisimple Lie Groups. Ergebnisse der Mathematik und ihrer Grenzgebiete (3) [Results in Mathematics and Related Areas (3)], vol. 17. Springer, Berlin (1991)
33. Moussong, G.: Hyperbolic coxeter groups. Ph.D. thesis, The Ohio State University. ProQuest LLC, Ann Arbor (1988)
34. Raghunathan, M.S.: Discrete subgroups of Lie groups. Math. Student **Special Centenary Volume** (2007), 59–70 (2008)
35. Rémy, B.: Construction de réseaux en théorie de Kac-Moody. C. R. Acad. Sci. Paris Sér. I Math. **329**(6), 475–478 (1999)

36. Serre, J.P.: Trees Translated from the French by John Stillwell. Springer, Berlin (1980)
37. Siegel, C.L.: Some remarks on discontinuous groups. Ann. Math. **46**(4), 708–718 (1945)
38. Thomas, A.: Lattices acting on right-angled buildings. Algebr. Geom. Topol. **6**, 1215–1238 (2006)
39. Thomas, A.: Existence, covolumes and infinite generation of lattices for Davis complexes. Groups Geom. Dyn. **6**(4), 765–801 (2012)
40. Thomas, A., Wortman, K.: Infinite generation of non-cocompact lattices on right-angled buildings. Algebr. Geom. Topol. **11**(2), 929–938 (2011)
41. Tits, J.: Buildings of Spherical Type and Finite BN-Pairs. Lecture Notes in Mathematics, vol. 386. Springer, Berlin/New York (1974)
42. Weil, A.: On discrete subgroups of Lie groups. Ann. Math. (2) **72**, 369–384 (1960)
43. Weil, A.: On discrete subgroups of Lie groups. II. Ann. Math. (2) **75**, 578–602 (1962)
44. White, G.: Covolumes of latticies in automorphism groups of trees and Davis complexes (2012). MSc thesis
45. Xie X.: Quasi-isometric rigidity of Fuchsian buildings. Topology **45**(1), 101–169 (2006)
46. Zimmer, R.J.: Ergodic Theory and Semisimple Groups. Monographs in Mathematics, vol. 81. Birkhäuser, Basel (1984)
47. Żuk, A.: La propriété (T) de Kazhdan pour les groupes agissant sur les polyèdres. C. R. Acad. Sci. Paris Sér. I Math. **323**(5), 453–458 (1996)

A Survey of Elementary Totally Disconnected Locally Compact Groups

Phillip Wesolek

Abstract The class of elementary totally disconnected locally compact (t.d.l.c.) groups is the smallest class of t.d.l.c. second countable (s.c.) groups which contains the second countable profinite groups and the countable discrete groups and is closed under taking closed subgroups, Hausdorff quotients, group extensions, and countable directed unions of open subgroups. This class appears to be fundamental to the study of t.d.l.c. groups. In these notes, we give a complete account of the basic properties of the class of elementary groups. The approach taken here is more streamlined than previous works, and new examples are sketched.

1 Introduction

In the general study of totally disconnected locally compact (t.d.l.c.) groups, one often wishes to avoid discrete groups and compact t.d.l.c., equivalently profinite, groups. For example, considering finitely generated groups as lattices in themselves is unenlightening, and the scale function on a profinite group is trivial. However, non-discreteness and non-compactness are often not enough by themselves. For example, every finitely generated group is a lattice in a non-discrete t.d.l.c. group simply by taking a direct product with an infinite profinite group. We thus wish to study t.d.l.c. groups that are 'sufficiently non-discrete.'

What we mean by 'sufficiently non-discrete' is that there is a suitably rich interaction between the topological structure and the large-scale structure of the group in question. With this in mind, let us consider examples. Certainly discrete groups have weak interaction between topological and large-scale structure, since they have trivial topological structure. The profinite groups have the opposite problem: they have local structure but trivial large-scale structure. On the other hand, compactly generated t.d.l.c. groups which are non-discrete and topologically simple

---

P. Wesolek (✉)

Department of Mathematical Sciences, Binghamton University, PO Box 6000, Binghamton, NY 13902, USA

e-mail: pwesolek@binghamton.edu

© Springer International Publishing AG, part of Springer Nature 2018

D.R. Wood et al. (eds.), *2016 MATRIX Annals*, MATRIX Book Series 1,

https://doi.org/10.1007/978-3-319-72299-3_25

Fig. 1 Interaction between
large-scale structure and
topological structure

Weak interaction	Strong interaction
Profinite groups, discrete groups	Aut(T_n)
abelian groups	The Neretin groups
profinite-by-discrete groups, e.g. $A_5^{\mathbb{Z}} \rtimes \mathbb{Z}$	$SL_n(\mathbb{Q}_p)$

have a rich interaction between topological and large-scale structure; examples of these include the Neretin groups, $\mathrm{Aut}(T_n)^+$ for T_n the n-regular tree, the simple algebraic groups, and many others. Thinking further, it is clear that abelian groups and compact-by-discrete groups have much weaker interaction between topological and large-scale structure than that of the aforementioned simple groups, so these groups should be collected with the profinite groups and discrete groups (Fig. 1). At this point, it seems natural to conclude that any 'elementary' combination of groups with weak interaction should again have weak interaction. We thus arrive to the central definition of these notes:

Definition 1 The class of **elementary groups** is the smallest class $\mathscr{E}$ of t.d.l.c.s.c. groups such that

   (i) $\mathscr{E}$ contains all second countable profinite groups and countable discrete groups.
  (ii) $\mathscr{E}$ is closed under taking closed subgroups.
 (iii) $\mathscr{E}$ is closed under taking Hausdorff quotients.
 (iv) $\mathscr{E}$ is closed under taking group extensions.
  (v) If G is a t.d.l.c.s.c. group and $G = \bigcup_{i \in \mathbb{N}} O_i$ where $(O_i)_{i \in \mathbb{N}}$ is an $\subseteq$-increasing sequence of open subgroups of G with $O_i \in \mathscr{E}$ for each i, then $G \in \mathscr{E}$. We say that $\mathscr{E}$ is **closed under countable increasing unions**.

The operations (ii)–(v) are often called the **elementary operations**.

Remark 1 We restrict to the second countable t.d.l.c. groups. This is a mild and natural assumption which makes our discussion much easier. Any notion of being 'elementary' must be 'regional' in the sense that it reduces to compactly generated subgroups, and compactly generated groups are second countable modulo a compact normal subgroup. Generalizing our notion of elementary groups to the non-second countable setting thus adds little to the theory.

In these notes, we explore the class of elementary groups. In particular, the class is shown to enjoy strong permanence properties and to admit a well-behaved, ordinal valued rank function. This rank function, aside from being an important tool to study elementary groups, gives a quantitative measure of the level of interaction between topological and large scale structure in a given elementary group.

Remark 2 The primary reference for these notes is [15]; the reader may also wish to consult the nice survey of Cesa and Le Maître [5]. The general approach developed in these notes is different from that of [15]. Our approach follows that of [16]; in loc. cit., the class of elementary amenable *discrete* groups is studied, but the parallels are obvious.

2 Preliminaries

For G a t.d.l.c. group, we shall use $\mathcal{U}(G)$ to denote the set of compact open subgroups of G. For K a subgroup of a group G, we use $\langle\langle K \rangle\rangle_G$ to denote the normal subgroup generated by K in G. When clear from context, we drop the subscript. If H is an open subgroup of G, we write $H \leq_o G$. If H is a closed subgroup of G such that G/H is compact in the quotient topology, we say that H is a **cocompact** subgroup and write $H \leq_{cc} G$. It is a classical result that cocompact closed subgroups of compactly generated t.d.l.c. groups are themselves compactly generated.

2.1 Ordinals

Ordinal numbers are used frequently in these notes; Kunen [8] contains a nice introduction to ordinal numbers and ordinal arithmetic. Recalling that a **well-order** is a total order with no infinite descending chains, the easiest definition of an ordinal number is due to J. von Neumann: Each ordinal is the well-ordered set of all smaller ordinals with $0 := \emptyset$. For example, $2 = \{0, 1\}$ and $3 = \{0, 1, 2\}$. Ordinal numbers are in particular well-orders themselves. For example, 2 is the two element well-order, and the first transfinite ordinal is $\omega := \mathbb{N}$. The second transfinite ordinal, $\omega + 1$, is the well-order given by a copy of $\mathbb{N}$ followed by one point. The first uncountable ordinal is denoted by ω_1. An important feature of ω_1, which is often used implicitly, is that there is no countable cofinal subset. That is to say, there is no countable sequence of countable ordinals $(\alpha_i)_{i \in \mathbb{N}}$ such that $\sup_{i \in \mathbb{N}} \alpha_i = \omega_1$. We stress that ω_1 is much larger than any countable ordinal. Ordinals such as ω^ω or ω^{ω^ω} are still strictly smaller than ω_1. Indeed, one can never reach ω_1 via arithmetic combinations of countable ordinals.

Given ordinals α and β, the ordinal $\alpha + \beta$ is the well-order given by a copy of α followed by a copy of β. Observe that the well-orders $1 + \omega$ and $\omega + 1$ are thus not equal, since the former is order isomorphic to ω while the latter is not, hence addition is *non-commutative*. Multiplication and exponentiation can be defined similarly. We shall not use ordinal arithmetic in a complicated way. The reader is free to think of ordinal arithmetic as usual arithmetic keeping in mind that *it is non-commutative*.

Ordinals of the form $\alpha + 1$ for some ordinal α are called **successor ordinals**. A **limit ordinal** is an ordinal which is not of the form $\alpha + 1$ for some ordinal α. The ordinals ω, $\omega + \omega$, and ω_1 are examples of limit ordinals. We stress that our definition implies 0 is a limit ordinal.

An important feature of ordinals is that they allow us to extend induction arguments transfinitely. Transfinite induction proceeds just as the familiar induction with one additional step: One must check the inductive claim holds for limit ordinals λ given that the claim holds for all ordinals $\alpha < \lambda$. In the induction arguments in these notes, the limit case of the argument will often be trivial.

2.2 Descriptive-Set-Theoretic Trees

We will require the notion of a descriptive-set-theoretic tree. This notion of a tree differs from the usual graph-theoretic definition; it is similar to the notion of a rooted tree used in the study of branch groups. The definitions given here are restricted to the collection of finite sequences of natural numbers; see [7, 2.A] for a general account.

Denote the collection of finite sequences of natural numbers by $\mathbb{N}^{<\mathbb{N}}$. For sequences $s := (s_0, \ldots, s_n) \in \mathbb{N}^{<\mathbb{N}}$ and $r := (r_0, \ldots, r_m) \in \mathbb{N}^{<\mathbb{N}}$, we write $s \sqsubseteq r$ if s is an **initial segment** of r. That is to say, $n \leq m$ and $s_i = r_i$ for $0 \leq i \leq n$. The empty sequence, denoted by $\emptyset$, is considered to be an element of $\mathbb{N}^{<\mathbb{N}}$ and is an initial segment of any $t \in \mathbb{N}^{<\mathbb{N}}$. We define the **concatenation** of s with r to be

$$s^\frown r := (s_0, \ldots, s_n, r_0, \ldots, r_m).$$

For $t = (t_0, \ldots, t_k) \in \mathbb{N}^{<\mathbb{N}}$, the **length** of t, denoted by $|t|$, is the number of coordinates; i.e. $|t| := k + 1$. If $|t| = 1$, we write t as a natural number, as opposed to a sequence of length one. For $0 \leq i \leq |t| - 1$, we define $t(i) := t_i$. For an infinite sequence $\alpha \in \mathbb{N}^{\mathbb{N}}$, we set $\alpha \restriction_n := (\alpha(0), \ldots, \alpha(n-1))$, so $\alpha \restriction_n \in \mathbb{N}^{<\mathbb{N}}$ for any $n \geq 0$.

Definition 2 A set $T \subseteq \mathbb{N}^{<\mathbb{N}}$ is a **tree** if it is closed under taking initial segments. We call the elements of T the **nodes** of T. If $s \in T$ and there is no $n \in \mathbb{N}$ such that $s^\frown n \in T$, we say s is a **leaf** or **terminal node** of T. An **infinite branch** of T is a sequence $\alpha \in \mathbb{N}^{\mathbb{N}}$ such that $\alpha \restriction_n \in T$ for all n. If T has no infinite branches, we say that T is **well-founded**.

For T a well-founded tree, there is an ordinal valued rank, denoted by ρ_T, on the nodes of T defined inductively as follows: If $s \in T$ is terminal, $\rho_T(s) := 0$. For a non-terminal node s,

$$\rho_T(s) := \sup\{\rho_T(r) + 1 \mid s \sqsubset r \in T\}.$$

The reader is encouraged to verify that this function is defined on all nodes of a well-founded tree. The **rank** of a well-founded tree T is defined to be

$$\rho(T) := \sup\{\rho_T(s) + 1 \mid s \in T\}.$$

When T is the empty tree, $\rho(T) = 0$, and for all other well-founded trees, it is easy to verify that $\rho(T) = \rho_T(\emptyset) + 1$. We thus see that $\rho(T)$ is always either a successor ordinal or zero. We extend ρ to ill-founded trees by declaring $\rho(T) = \omega_1$ for T an ill-founded tree.

There is an important, well-known relationship between the rank ρ_T on the nodes of T and the rank ρ on well-founded trees; we give a proof for completeness. For T a tree and $s \in T$, we put $T_s := \{r \in \mathbb{N}^{<\mathbb{N}} \mid s^\frown r \in T\}$. The set T_s is the tree obtained by taking the elements in T that extend s and deleting the initial segment s from each.

Lemma 1 *Suppose that $T \subseteq \mathbb{N}^{<\mathbb{N}}$ is a well-founded tree and $s \in T$. Then*

(1) $\rho_T(s) + 1 = \rho(T_s)$ and
(2) $\rho(T) = \sup\{\rho(T_i) \mid i \in T\} + 1$.

Proof Fixing $s \in T$, we first argue by induction on $\rho_{T_s}(r)$ that $\rho_{T_s}(r) = \rho_T(s^\frown r)$. For the base case, $\rho_{T_s}(r) = 0$, the node r is terminal in T_s. The node $s^\frown r$ is thus terminal in T, hence $\rho_T(s^\frown r) = 0$.

Suppose that the inductive claim holds for all $r \in T_s$ with $\rho_{T_s}(r) < \beta$ and say that $\rho_{T_s}(r) = \beta$. We now deduce that

$$\begin{aligned}
\rho_{T_s}(r) &= \sup\{\rho_{T_s}(t) + 1 \mid r \sqsubset t \in T_s\} \\
&= \sup\{\rho_T(s^\frown t) + 1 \mid s^\frown r \sqsubset s^\frown t \in T\} \\
&= \sup\{\rho_T(t) + 1 \mid s^\frown r \sqsubset t \in T\} \\
&= \rho_T(s^\frown r)
\end{aligned}$$

where the second equality follows from the inductive hypothesis. Our induction is complete.

Taking $r = \emptyset$, we deduce that $\rho_{T_s}(\emptyset) = \rho_T(s)$. Therefore, $\rho(T_s) = \rho_{T_s}(\emptyset) + 1 = \rho_T(s) + 1$, which verifies (1). Claim (2) follows from (1). $\blacksquare$

3 Elementary Groups and Well-Founded Trees

Classes defined by axioms, such as $\mathscr{E}$, are often studied via induction on the class formation axioms. In the case of $\mathscr{E}$, this approach has the unfortunate side-effect of cumbersome and technical proofs. We thus begin by characterizing $\mathscr{E}$ in terms of well-founded descriptive-set-theoretic trees. This gives an elegant and natural approach to the class of elementary groups.

To motivate our characterization, consider a game in which a friend builds a t.d.l.c.s.c. group and asks you to determine if it is or is not elementary. Since your friend built the group, there must be some way in which the group can be disassembled. You could thus, in principle, devise a general strategy to disassemble the group which halts exactly when the group is elementary.

Our characterization will be exactly such a strategy. Our decomposition strategy will alternate between eliminating discrete quotients and passing to compactly generated open subgroups. These operations will "undo" the closure properties (iv) and (v). A priori, there are other elementary operations that must also be "undone." It will turn out that it is indeed enough to only consider (iv) and (v). (This is unsurprising in view of [10].)

3.1   Decomposition Trees

Eliminating discrete quotients is accomplished by taking the discrete residual.

Definition 3 For a t.d.l.c. group H, the **discrete residual** of H is

$$\mathrm{Res}(H) := \bigcap \{O \mid O \trianglelefteq_o H\}.$$

The discrete residual is a closed characteristic subgroup of H. The quotient $H/\mathrm{Res}(H)$ also has a special structure. A t.d.l.c. **SIN group** is a t.d.l.c. group which admits a basis at 1 of compact open *normal* subgroups; note that t.d.l.c. SIN groups are elementary.

Proposition 1 ([3, Corollary 4.1]) *For G a compactly generated t.d.l.c. group, the quotient $G/\mathrm{Res}(G)$ is a SIN group.*

   To reduce to compactly generated open subgroups, we define a second operation. Let G be a t.d.l.c.s.c. group and $U \in \mathscr{U}(G)$. Fix γ a choice of a countable dense subset of *every* closed subgroup of G; we call γ a **choice function** for G. Formally, γ is a map that sends a closed subgroup $H \leq G$ to a countable dense subset $\{h_i\}_{i \in \mathbb{N}}$ of H; the axiom of choice ensures such a γ exists. If L is a closed subgroup of G, then the restriction of γ to closed subgroups of L obviously induces a choice function for L. We will abuse notation and say that γ is a choice function for L.

   For $H \leq G$ closed and $n \in \mathbb{N}$, we now define

$$R_n^{(U,\gamma)}(H) := \langle U \cap H, h_0, \dots, h_n \rangle$$

where the $h_0, \dots, h_n$ are the first $n + 1$ elements of the countable dense set of H picked out by γ. For each $n \in \mathbb{N}$, the subgroup $R_n^{(U,\gamma)}(H)$ is a compactly generated open subgroup of H. Furthermore, $R_n^{(U,\gamma)}(H) \leq R_{n+1}^{(U,\gamma)}(H)$ for all n, and

$$H = \bigcup_{i \in \mathbb{N}} R_i^{(U,\gamma)}(H).$$

The subgroups $R_n^{(U,\gamma)}(H)$ thus give a canonical increasing exhaustion of H by compactly generated open subgroups.

   We now define a tree $T_{(U,\gamma)}(G)$ and associated closed subgroups G_s of G for each $s \in T_{(U,\gamma)}(G)$. Put

- $\emptyset \in T_{(U,\gamma)}(G)$ and $G_\emptyset := G$.
- Suppose we have defined $s \in T_{(U,\gamma)}(G)$ and $G_s \leq G$. If $G_s \neq \{1\}$ and $n \in \mathbb{N}$, then put $s^\frown n \in T_{(U,\gamma)}(G)$ and set

$$G_{s^\frown n} := \mathrm{Res}\left(R_n^{(U,\gamma)}(G_s)\right).$$

Definition 4 For G a t.d.l.c.s.c. group, $U \in \mathscr{U}(G)$, and γ a choice function for G, we call $T_{(U,\gamma)}(G)$ the **decomposition tree** of G with respect to U and γ.

The decomposition tree is always non-empty, and the subgroup associated to any terminal node is the trivial group. We make one further observation; the proof is straightforward and therefore left to the reader. Recall that T_s is the tree below the node s in the tree T; precisely, $T_s = \{r \in \mathbb{N}^{<\mathbb{N}} \mid s^\frown r \in T\}$. For a decomposition tree $T_{(U,\gamma)}(G)$, we shall write $T_{(U,\gamma)}(G)_s$, instead of the more precise $(T_{(U,\gamma)}(G))_s$, for the tree below s.

Observation 1 *For any $s \in T_{(U,\gamma)}(G)$, $T_{(U,\gamma)}(G)_s = T_{(G_s \cap U,\gamma)}(G_s)$. Further, for $r \in T_{(G_s \cap U,\gamma)}(G_s)$, the associated subgroup $(G_s)_r$ is the same as the subgroup $G_{s^\frown r}$ associated to $s^\frown r \in T_{(U,\gamma)}(G)$.*

Remark 3 By classical results in descriptive set theory, the choice function γ can indeed be constructed in a Borel manner using selector functions; see [7, (12.13)]. The advantage of using selector functions to produce γ is that the assignment $G \mapsto T_{(U,\gamma)}(G)$ is Borel, when considered as a function between suitable parameter spaces. This allows for further descriptive-set-theoretic analysis of the class of elementary groups. See [16] for an example of such an analysis in the space of marked groups.

The decomposition tree plainly depends on the choices of compact open subgroup U and choice function γ, so there is no hope the decomposition tree outright is an invariant of the group. However, a decomposition tree comes with an ordinal rank, and this rank is a group invariant. That is to say, the rank of a decomposition tree does not depend on the choices of compact open subgroup and choice function.

Proposition 2 *Suppose that G is a t.d.l.c.s.c. group, $U \in \mathscr{U}(G)$, and γ is a choice function for G. Suppose additionally that H is a t.d.l.c.s.c. group, $W \in \mathscr{U}(H)$, and δ is a choice function for H. If $\psi : H \to G$ is a continuous, injective homomorphism, then*

$$\rho(T_{(W,\delta)}(H)) \leq \rho(T_{(U,\gamma)}(G)).$$

Proof We induct on $\rho(T_{(U,\gamma)}(G))$ simultaneously for all G, $U \in \mathscr{U}(G)$, and γ a choice function for G. The base case is obvious since $\rho(T_{(U,\gamma)}(G)) = 1$ implies $G = \{1\}$. We may also ignore the case of $\rho(T_{(U,\gamma)}(G)) = \omega_1$, since the proposition obviously holds here.

Suppose $\rho(T_{(U,\gamma)}(G)) = \beta + 1$. For each i, the subgroup $R_i^{(W,\delta)}(H)$ is compactly generated, so there is $n(i)$ with $\psi\left(R_i^{(W,\delta)}(H)\right) \leq R_{n(i)}^{(U,\gamma)}(G)$. We thus have that

$$\psi(H_i) = \psi\left(\mathrm{Res}\left(R_i^{(W,\delta)}(H)\right)\right) \leq \mathrm{Res}\left(R_{n(i)}^{(U,\gamma)}(G)\right) = G_{n(i)}.$$

The map ψ thereby restricts to $\psi : H_i \to G_{n(i)}$. Lemma 1 and Observation 1 imply

$$\rho\left(T_{(G_{n(i)} \cap U,\gamma)}(G_{n(i)})\right) = \rho\left(T_{(U,\gamma)}(G)_{n(i)}\right) \leq \beta.$$

Applying the inductive hypothesis, we deduce that

$$\rho\left(T_{(H_i \cap W, \delta)}(H_i)\right) \leq \rho\left(T_{(G_{n(i)} \cap U, \gamma)}(G_{n(i)})\right).$$

Therefore,

$$\begin{aligned}
\rho(T_{(W,\delta)}(H)) &= \sup_{i \in \mathbb{N}} \rho\left(T_{(H_i \cap W, \delta)}(H_i)\right) + 1 \\
&\leq \sup_{i \in \mathbb{N}} \rho\left(T_{(G_{n(i)} \cap U, \gamma)}(G_{n(i)})\right) + 1 \\
&\leq \rho(T_{(U,\gamma)}(G)),
\end{aligned}$$

so $\rho(T_{(W,\delta)}(H)) \leq \beta + 1$. This finishes the induction, and we conclude the proposition.

Proposition 2 ensures that the rank of a decomposition tree is indeed a group-theoretic property.

Corollary 1 *For G a t.d.l.c.s.c. group, $U, W \in \mathcal{U}(G)$, and γ and δ choice functions for G, $\rho(T_{(U,\gamma)}(G)) = \rho(T_{(W,\delta)}(G))$. In particular, $T_{(U,\gamma)}(G)$ is well-founded for some U and γ if and only if $T_{(U,\gamma)}(G)$ is well-founded for all U and γ.*

In view of Corollary 1, we make a definition.

Definition 5 For a t.d.l.c.s.c. group G, the **decomposition rank** of G is

$$\xi(G) := \rho(T_{(U,\gamma)}(G))$$

for some (any) $U \in \mathcal{U}(G)$ and γ a choice function for G.

Decomposition trees are a strategy to disassemble t.d.l.c.s.c. groups. Requiring the resulting decomposition tree to be well-founded is the obvious halting condition for this decomposition strategy. With this in mind, we define the following class:

Definition 6 The class $\mathscr{WF}$ is defined to be the class of t.d.l.c.s.c. groups G with $\xi(G) < \omega_1$. Equivalently, $\mathscr{WF}$ is the collection of t.d.l.c.s.c. groups with some (equivalently every) decomposition tree well-founded.

Our goal is to show that indeed $\mathscr{WF} = \mathscr{E}$, verifying that well-founded decomposition trees exactly isolate the elementary groups; the notation "$\mathscr{WF}$" will be discarded after establishing $\mathscr{WF} = \mathscr{E}$. We shall argue for $\mathscr{E} \subseteq \mathscr{WF}$ by verifying that $\mathscr{WF}$ satisfies the same closure properties; the next section will make these verifications. The converse inclusion will be an easy induction argument.

3.2   The Class $\mathscr{WF}$

Our analysis of the class $\mathscr{WF}$ is via induction on the decomposition rank, so we first establish a computation technique for the rank. This technique allows us to avoid discussing decomposition trees. To establish this technique, let us first recast Proposition 2; our restatement also gives a first closure property of $\mathscr{WF}$.

Proposition 3 *Suppose that G and H are t.d.l.c.s.c. groups. If $\psi : H \to G$ is a continuous, injective homomorphism, then $\xi(H) \leq \xi(G)$. In particular, if $H \leq G$, then $\xi(H) \leq \xi(G)$, so $\mathscr{WF}$ is closed under taking closed subgroups.*

Proposition 4 *Suppose $G \in \mathscr{WF}$ is non-trivial.*

(1) If $G = \bigcup_{i \in \mathbb{N}} O_i$ with $(O_i)_{i \in \mathbb{N}}$ an $\subseteq$-increasing sequence of compactly generated open subgroups of G, then $\xi(G) = \sup_{i \in \mathbb{N}} \xi(\mathrm{Res}(O_i)) + 1$.
(2) If G is compactly generated, then $\xi(G) = \xi(\mathrm{Res}(G)) + 1$.

Proof For (1), fix $U \in \mathscr{U}(G)$ and a choice function γ for G. For each i, there is $n(i)$ such that $O_i \leq R_{n(i)}^{(U,\gamma)}(G)$, since O_i is compactly generated. Therefore,

$$\mathrm{Res}(O_i) \leq \mathrm{Res}\left(R_{n(i)}^{(U,\gamma)}(G)\right) = G_{n(i)},$$

and Proposition 2 implies $\xi(\mathrm{Res}(O_i)) \leq \xi(G_{n(i)})$. We conclude that

$$\sup_{i \in \mathbb{N}} \xi(\mathrm{Res}(O_i)) + 1 \leq \sup_{j \in \mathbb{N}} \xi(G_j) + 1 = \xi(G).$$

On the other hand, $(O_i)_{i \in \mathbb{N}}$ is an exhaustion of G by open subgroups, so for each j, there is $n(j)$ with $R_j^{(U,\gamma)}(G) \leq O_{n(j)}$. Therefore, $G_j \leq \mathrm{Res}(O_{n(j)})$, and applying Proposition 2 again,

$$\xi(G) = \sup_{j \in \mathbb{N}} \xi(G_j) + 1 \leq \sup_{i \in \mathbb{N}} \xi(\mathrm{Res}(O_i)) + 1.$$

Hence, $\xi(G) = \sup_{i \in \mathbb{N}} \xi(\mathrm{Res}(O_i)) + 1$, as required.

Claim (2) now follows immediately from (1) by taking the sequence $(O_i)_{i \in \mathbb{N}}$ with $O_i = G$ for all i.

We now begin in earnest to verify that $\mathscr{WF}$ satisfies the same closure properties as $\mathscr{E}$. A t.d.l.c. group G is **residually discrete** if $\mathrm{Res}(G) = \{1\}$. From the definition of a decomposition tree, we see that any decomposition tree for such a group has rank at most 2. We thus deduce the following proposition:

Proposition 5 *All residually discrete groups are elements of $\mathscr{WF}$. In particular, all second countable profinite groups and countable discrete groups are elements of $\mathscr{WF}$.*

We next consider countable unions; we prove a slightly more general result for later use.

Proposition 6 *Suppose G is a t.d.l.c.s.c. group and $(O_i)_{i \in \mathbb{N}}$ is an $\subseteq$-increasing exhaustion of G by compactly generated open subgroups. If $\xi(\mathrm{Res}(O_i)) < \omega_1$ for all i, then $G \in \mathscr{WF}$. In particular, $\mathscr{WF}$ is closed taking countable increasing unions.*

Proof Fix $U \in \mathscr{U}(G)$ and γ a choice function for G. Via Observation 1, the tree $T_{(U,\gamma)}(G)$ is well-founded exactly when $T_{(G_j \cap U,\gamma)}(G_j)$ is well-founded for all $j \in \mathbb{N}$.

For each $j \in \mathbb{N}$, there is $i \in \mathbb{N}$ such that $R_j^{(U,\gamma)}(G) \leq O_i$, since $R_j^{(U,\gamma)}(G)$ is compactly generated. We deduce that $G_j \leq \mathrm{Res}(O_i)$. Proposition 3 now ensures that $\xi(G_j) < \omega_1$, and thus, $T_{(G_j \cap U,\gamma)}(G_j)$ is well-founded. We conclude that $G \in \mathscr{WF}$.

We now turn our attention to quotients and group extensions. Our arguments here require several preliminary results. The first observation is immediate from the relevant definitions.

Observation 2 *If G is a t.d.l.c. SIN group and $L \trianglelefteq G$, then G/L is a t.d.l.c. SIN group.*

Let us also note an easy fact about the discrete residual.

Lemma 2 *If G is a compactly generated t.d.l.c.s.c. group and $L \trianglelefteq G$, then $\mathrm{Res}(G/L) = \overline{\mathrm{Res}(G)L/L}$.*

Proof Let $\pi : G \to G/L$ be the usual projection map. For every open normal $O \trianglelefteq G/L$, the subgroup $\pi^{-1}(O)$ is an open normal subgroup of O. Hence, $\mathrm{Res}(G) \leq \pi^{-1}(\mathrm{Res}(G/L))$, and we deduce that $\overline{\mathrm{Res}(G)L/L} \leq \mathrm{Res}(G/L)$.

Conversely, the group $(G/L)/(\overline{\mathrm{Res}(G)L/L})$ is a quotient of the SIN group $G/\mathrm{Res}(G)$. Observation 2 ensures $(G/L)/(\overline{\mathrm{Res}(G)L/L})$ is a SIN group and therefore residually discrete. We conclude that $\mathrm{Res}(G/L) \leq \overline{\mathrm{Res}(G)L/L}$, verifying the proposition.

A non-trivial permanence property of t.d.l.c. SIN groups will be needed. The argument requires the following easy application of the Baire category theorem, which we leave as an exercise: *Every element of a discrete normal subgroup of a t.d.l.c.s.c. group has an open centralizer.*

Lemma 3 *If G is a compactly generated t.d.l.c. group and $N \trianglelefteq_{cc} G$ is a SIN group, then G is a SIN group.*

Proof Fix $U \in \mathscr{U}(G)$ and form the subgroup UN. Since N is a SIN group, we may find $W \in \mathscr{U}(N)$ with $W \leq U$ and $W \trianglelefteq N$. The normal closure $J := \langle\langle W \rangle\rangle$ of W in UN is generated by U-conjugates of W, and thus $J \leq U$. Since N is cocompact in G, UN has finite index in G, so $N_G(J)$ has finite index in G. Letting $g_1, \ldots, g_n$ list left coset representatives for $N_G(J)$ in G, we see that

$$\bigcap_{g \in G} gJg^{-1} = \bigcap_{i=1}^{n} g_i J g_i^{-1}.$$

Defining $K := \bigcap_{g \in G} gJg^{-1}$, it follows that $K \in \mathscr{U}(N)$ and that $K \trianglelefteq G$.

Passing to G/K, the image $\pi(N)$ is normal and discrete in G/K where $\pi : G \to G/K$ is the usual projection. The subgroup N is compactly generated, since cocompact in a compactly generated group, hence the subgroup $\pi(N)$ is finitely generated. Moreover, since each generator of $\pi(N)$ has an open centralizer, $\pi(N)$ has an open centralizer. Say that $Q \leq_o \pi(U)$ centralizes $\pi(N)$. Clearly, $Q \trianglelefteq Q\pi(N)$, and using that $\pi(N)$ is cocompact in G/K, we additionally see that $Q\pi(N)$ has finite index in G/K. Just as in the previous paragraph, there is $L \leq_o Q$ with $L \trianglelefteq G/K$. It now follows that $\pi^{-1}(L)$ is an open normal subgroup of G contained in U.

We conclude that inside every compact open subgroup U of G, we may find a compact open normal subgroup of G. That is to say, G is a SIN group.

Our final subsidiary result is important outside the immediate application, because it allows one to go from a closed normal subgroup to an *open* subgroup with the same rank.

Proposition 7 ([11, Lemma 2.9]) *If $G \in \mathscr{WF}$ with $N \trianglelefteq_{cc} G$ closed and non-trivial, then $\xi(G) = \xi(N)$.*

Proof Fix $(O_i)_{i \in \mathbb{N}}$ a countable $\subseteq$-increasing exhaustion of G by compactly generated open subgroups of G and put $N_i := N \cap O_i$. Each N_i is open in N, and since $N_i \trianglelefteq_{cc} O_i$, it is also compactly generated. Proposition 3 ensures $N \in \mathscr{WF}$, and in view of Proposition 4, we infer that

$$\xi(N) = \sup_{i \in \mathbb{N}} \xi(\text{Res}(N_i)) + 1.$$

We now consider the group $O_i/\text{Res}(N_i)$. The subgroup $N_i/\text{Res}(N_i)$ is a SIN group via Proposition 1, and it is cocompact in $O_i/\text{Res}(N_i)$. Lemma 3 thus implies that $O_i/\text{Res}(N_i)$ is also a SIN group, hence $O_i/\text{Res}(N_i)$ is residually discrete. It now follows that $\text{Res}(O_i) = \text{Res}(N_i)$. Applying Proposition 4 again, we conclude that

$$\xi(G) = \sup_{i \in \mathbb{N}} \xi(\text{Res}(O_i)) + 1 = \sup_{i \in \mathbb{N}} \xi(\text{Res}(N_i)) + 1 = \xi(N),$$

verifying the lemma.

We are now prepared to show $\mathscr{WF}$ is closed under taking quotients; the proof is an instructive illustration of the utility of Proposition 7.

Proposition 8 *If $G \in \mathscr{WF}$ and $L \trianglelefteq G$ is closed, then $G/L \in \mathscr{WF}$ with $\xi(G/L) \leq \xi(G)$.*

Proof Fix $U \in \mathscr{U}(G)$ and fix $(O_i)_{i \in \mathbb{N}}$ an $\subseteq$-increasing exhaustion of G by compactly generated open subgroups such that $U \leq O_0$.

We induct on $\xi(G)$ for the proposition. The case of $\xi(G) = 1$ is obvious, and it will be convenient to take $\xi(G) = 2$ as the base case. Proposition 4 ensures that $\text{Res}(O_i) = \{1\}$ for all i, and in view of Proposition 1, we deduce that each O_i is a SIN group. Since the class of SIN groups is stable under taking Hausdorff quotients, $O_i/O_i \cap L$ is also a SIN group for all $i \in \mathbb{N}$. On the other hand, G/L is the union of the increasing sequence $(O_i L/L)_{i \in \mathbb{N}}$, and since $O_i L/L \simeq O_i/O_i \cap L$, each term of the sequence is a SIN group. Proposition 5 now ensures each $O_i L/L$ is in $\mathscr{WF}$, so we conclude that $G/L \in \mathscr{WF}$ via Proposition 6. From Proposition 4, we deduce further that $\xi(G/L) \leq 2$.

Let us now suppose that $\xi(G) = \beta + 1$ with $\beta > 1$. In view of Proposition 4, each $R_i := \text{Res}(O_i)$ has rank at most β. Furthermore, it cannot be the case that $R_i = \{1\}$ for all i, since then G has rank two. Throwing out finitely many O_i if needed, we may assume each R_i is non-trivial. Each R_i is then a non-trivial cocompact normal subgroup of UR_i, so Proposition 7 implies $\xi(UR_i) = \xi(R_i)$. Applying the inductive hypothesis, we infer that

$$UR_i/UR_i \cap L \simeq UR_iL/L$$

has rank at most β for each i. As $\overline{R_iL}/L$ is a closed subgroup of UR_iL/L, we deduce further that $\xi(\overline{R_iL}/L) \leq \beta$, via Proposition 3.

The quotient G/L is the increasing union of the compactly generated open subgroups $W_i := O_iL/L$. Lemma 2 shows that $\mathrm{Res}(W_i) = \overline{R_iL}/L$, so our work above implies $\xi(\mathrm{Res}(W_i)) \leq \beta$. Applying Proposition 6, we deduce that $G/L \in \mathscr{WF}$, and via Proposition 4, $\xi(G/L) \leq \beta + 1$, completing the induction.

We next show $\mathscr{WF}$ is closed under forming group extensions; our proof is inspired by a similar argument in [10].

Proposition 9 ([12, Lemma 7.4]) *Suppose*

$$\{1\} \to N \to G \to Q \to \{1\}$$

is a short exact sequence of t.d.l.c.s.c. groups. If N and Q are members of $\mathscr{WF}$, then $G \in \mathscr{WF}$ with

$$\xi(G) \leq \xi(N) + \xi(Q).$$

In particular, $\mathscr{WF}$ is closed under group extensions.

Proof We induct on $\xi(Q)$ for the proposition. The base case, $\xi(Q) = 1$, is obvious, so we suppose $\xi(Q) = \beta + 1$.

Let $\pi : G \to Q$ be the projection given in the short exact sequence, fix $(O_i)_{i \in \mathbb{N}}$ an $\subseteq$-increasing exhaustion of G by compactly generated open subgroups, and put $W_i := \pi(O_i)$. The sequence $(W_i)_{i \in \mathbb{N}}$ is an exhaustion of Q by compactly generated open subgroups. Fix $i \in \mathbb{N}$, form $R := \mathrm{Res}(O_i)$, and put $M := \overline{RN}$. The group M/N is a closed subgroup of $\mathrm{Res}(W_i)$, hence $\xi(M/N) \leq \beta$ via Proposition 3. The inductive hypothesis implies $M \in \mathscr{WF}$ with $\xi(M) \leq \xi(N) + \beta$, and since $R \leq M$, a second application of Proposition 3 ensures that $\xi(R) \leq \xi(N) + \beta$. In view of Propositions 4 and 6, we conclude that $G \in \mathscr{WF}$ with $\xi(G) \leq \xi(N) + \beta + 1$, verifying the inductive claim.

3.3   The Return of Elementary Groups

We now argue that $\mathscr{WF}$ is exactly the class of elementary groups. Our argument will have the added benefit of showing that some of the elementary operations used to define $\mathscr{E}$ are redundant.

Definition 7 The class $\mathscr{E}^*$ is the smallest class of t.d.l.c.s.c. groups such that the following hold:

(i) $\mathscr{E}^*$ contains all second countable profinite groups and countable discrete groups.

(ii) $\mathscr{E}^*$ is closed under taking group extensions of second countable profinite or countable discrete groups. That is, if G is a t.d.l.c.s.c. group and $H \trianglelefteq G$ is a closed normal subgroup with $H \in \mathscr{E}^*$ and G/H profinite or discrete, then $G \in \mathscr{E}^*$.

(iii) If G is a t.d.l.c.s.c. group and $G = \bigcup_{i \in \mathbb{N}} O_i$ where $(O_i)_{i \in \mathbb{N}}$ is an $\subseteq$-increasing sequence of open subgroups of G with $O_i \in \mathscr{E}^*$ for each i, then $G \in \mathscr{E}^*$.

Obviously $\mathscr{E}^*$ is contained in $\mathscr{E}$. It turns out this containment is indeed an equality.

Theorem 1 $\mathscr{E} = \mathscr{W}\mathscr{F} = \mathscr{E}^*$.

Proof Since $\mathscr{E}^* \subseteq \mathscr{E}$, it suffices to show the inclusions $\mathscr{E} \subseteq \mathscr{W}\mathscr{F} \subseteq \mathscr{E}^*$. For the first inclusion, since $\mathscr{E}$ is defined to be the smallest class such that certain closure properties hold, it is enough to show that $\mathscr{W}\mathscr{F}$ satisfies the same properties. That $\mathscr{W}\mathscr{F}$ contains the profinite groups and discrete groups is given by Proposition 5. The class $\mathscr{W}\mathscr{F}$ is closed under taking closed subgroups, Hausdorff quotients, and countable increasing unions via Propositions 3, 8, and 6, respectively. Proposition 9 ensures $\mathscr{W}\mathscr{F}$ is closed under forming group extensions.

For the second inclusion, we argue by induction on $\xi(G)$. If $\xi(G) = 1$, then $G = \{1\}$ is plainly in $\mathscr{E}^*$. Suppose $H \in \mathscr{E}^*$ for all $H \in \mathscr{W}\mathscr{F}$ with $\xi(H) \leq \beta$ and consider $G \in \mathscr{W}\mathscr{F}$ with $\xi(G) = \beta + 1$. Fix $(O_i)_{i \in \mathbb{N}}$ an $\subseteq$-increasing exhaustion of G by compactly generated open subgroups. In view of Proposition 4, each O_i is such that $\xi(\mathrm{Res}(O_i)) \leq \beta$, so the inductive hypothesis implies $\mathrm{Res}(O_i) \in \mathscr{E}^*$. The quotient $O_i/\mathrm{Res}(O_i)$ is a SIN group via Proposition 1. We may then fix $W \trianglelefteq O_i/\mathrm{Res}(O_i)$ a compact open normal subgroup. Letting $\pi : O_i \to O_i/\mathrm{Res}(O_i)$ be the usual projection, $\mathrm{Res}(O_i)$ is a cocompact normal subgroup of $\pi^{-1}(W)$, and as $\mathscr{E}^*$ is closed under extensions of profinite groups, we deduce that $\pi^{-1}(W) \in \mathscr{E}^*$. On the other hand, the quotient $O_i/\pi^{-1}(W)$ is discrete. As $\mathscr{E}^*$ is closed under extensions of discrete groups, we can conclude that $O_i \in \mathscr{E}^*$. It now follows that $G \in \mathscr{E}^*$, completing the induction. $\square$

As an immediate consequence, we obtain a simpler characterization of elementary groups.

Corollary 2 *The class of elementary groups is the smallest class $\mathscr{E}$ of t.d.l.c.s.c. groups such that the following hold:*

(i) *$\mathscr{E}$ contains all second countable profinite groups and countable discrete groups.*

(ii) *$\mathscr{E}$ closed under taking group extensions of second countable profinite or countable discrete groups; that is, if G is a t.d.l.c.s.c. group and $H \trianglelefteq G$ is a closed normal subgroup with $H \in \mathscr{E}$ and G/H profinite or discrete, then $G \in \mathscr{E}$.*

(iii) *If G is a t.d.l.c.s.c. group and $G = \bigcup_{i \in \mathbb{N}} O_i$ where $(O_i)_{i \in \mathbb{N}}$ is an $\subseteq$-increasing sequence of open subgroups of G with $O_i \in \mathscr{C}$ for each i, then $G \in \mathscr{E}$.*

We note a second consequence, which is quite useful in the study of elementary groups.

Corollary 3 *If G is a non-trivial compactly generated elementary group, then G has a non-trivial discrete quotient.*

Proof Via Theorem 1, G is a member of $\mathscr{WF}$, and Proposition 4 implies $\xi(G) = \xi(\mathrm{Res}(G)) + 1$. We conclude that $\mathrm{Res}(G) \lneq G$, and thus, G has a non-trivial discrete quotient. $\square$

4 Examples and Non-examples of Elementary Groups

We conclude with a discussion of examples and non-examples. In particular, we will exhibit a family of examples with unboundedly large finite rank and compactly generated examples with transfinite rank.

4.1 Non-examples

Our motivation to form the class of elementary groups is to make precise the class of groups with weak interaction between topological and large-scale structure. The groups which surely have strong interaction between topological and large-scale structure are the compactly generated t.d.l.c.s.c. groups which are non-discrete and simple. Our notion of an elementary group excludes these simple groups.

Proposition 10 *If G is a compactly generated t.d.l.c.s.c. group that is non-discrete and topologically simple, then G is not elementary.*

Proof Since G is topologically simple and non-discrete, it has no non-trivial discrete quotients. In view of Corollary 3, that G is compactly generated ensures that it is non-elementary. $\square$

We note that there are many compactly generated t.d.l.c.s.c. groups that are topologically simple and non-discrete. For the n-regular tree T_n with $n \geq 3$, work of Tits [14] shows that there is an index two closed subgroup of $\mathrm{Aut}(T_n)$, denoted by $\mathrm{Aut}^+(T_n)$, that is topologically simple, compactly generated, and non-discrete. The projective special linear groups $PSL_n(\mathbb{Q}_p)$ where $\mathbb{Q}_p$ is the p-adic numbers and $n \geq 2$ are further examples; cf. [1, 6]. There are in fact continuum many compactly generated t.d.l.c.s.c. groups that are topologically simple and non-discrete by work of Smith [13].

4.2 Finite Rank Examples

Our construction requires a couple of general notions. A group is called **perfect** if it is generated by commutators; a **commutator** is an element of the form $[g, h] := ghg^{-1}h^{-1}$ for group elements g and h. We also require the notion of a local direct product.

Definition 8 Suppose that $(G_i)_{i\in\mathbb{N}}$ is a sequence of t.d.l.c. groups and suppose that there is a distinguished compact open subgroup $U_i \leq G_i$ for each $i \in \mathbb{N}$. The **local direct product** of $(G_i)_{i\in\mathbb{N}}$ over $(U_i)_{i\in\mathbb{N}}$ is defined to be

$$\left\{ f : \mathbb{N} \to \bigsqcup_{i\in\mathbb{N}} G_i \mid f(i) \in G_i, \text{ and } f(i) \in U_i \text{ for all but finitely many } i \in \mathbb{N} \right\}$$

with the group topology such that $\prod_{i\in\mathbb{N}} U_i$ continuously embeds as an open subgroup. We denote the local direct product by $\bigoplus_{i\in\mathbb{N}} (G_i, U_i)$.

The following property of local direct products is an easy consequence of the definitions; we leave the proof to the reader.

Proposition 11 *If $(G_i)_{i\in\mathbb{N}}$ is a sequence of elementary groups with U_i a distinguished compact open subgroup for each i, then $\bigoplus_{\mathbb{N}}(G_i, U_i)$ is an elementary group.*

We are now ready to construct our groups. Let A_5 be the alternating group on five letters and let S be an infinite finitely generated perfect group. Form $H := S^{[5]} \rtimes A_5$ where $A_5 \curvearrowright S^{[5]}$ by shift and fix a transitive, free action of H on $\mathbb{N}$.

Lemma 4 *The normal subgroup of H generated by A_5 equals H.*

Proof Identify S with the copy of S in $S^{[5]}$ supported on 0 and take $a \in A_5$ so that $a(0) \neq 0$. For $g, h \in S \leq S^{[5]}$, the element aga^{-1} has support disjoint from both g and h, hence aga^{-1} commutes with both g and h. An easy calculation now shows that $[h, [g, a]] = [h, g]$. Since $[g, a] \in \langle\langle A_5 \rangle\rangle$, we deduce that $[h, [g, a]] \in \langle\langle A_5 \rangle\rangle$. The group $\langle\langle A_5 \rangle\rangle$ thus contains all commutators of S, and since S is perfect, $S \leq \langle\langle A_5 \rangle\rangle$. It now follows that $\langle\langle A_5 \rangle\rangle = H$.

Starting from the group H, we inductively define compactly generated elementary groups L_n with a distinguished $K_n \in \mathcal{U}(L_n)$ such that $\langle\langle K_n \rangle\rangle = L_n$. For the base case, $n = 1$, define $L_1 := H$ and $K_1 := A_5$. The group L_1 is compactly generated, K_1 is a compact open subgroup of L_1, and $\langle\langle K_1 \rangle\rangle = L_1$, via Lemma 4.

Suppose we have defined a compactly generated group L_n with a compact open subgroup K_n such that $\langle\langle K_n \rangle\rangle = L_n$. Let $(L_n^i)_{i\in\mathbb{N}}$ and $(K_n^i)_{i\in\mathbb{N}}$ list countably many copies of L_n and K_n and form the local direct product $\bigoplus_{i\in\mathbb{N}}(L_n^i, K_n^i)$. Taking the previously fixed action of H on $\mathbb{N}$, we have that $H \curvearrowright \bigoplus_{i\in\mathbb{N}}(L_n^i, K_n^i)$ by shift, so we may form

$$L_{n+1} := \bigoplus_{i\in\mathbb{N}}(L_n^i, K_n^i) \rtimes H.$$

The group L_{n+1} is a t.d.l.c. group under the product topology, and $K_{n+1} := K_n^{\mathbb{N}} \rtimes A_5$ is a compact open subgroup. Letting X be a compact generating set for L_n^0 and F be a finite generating set for H in L_{n+1}, one verifies that $X \times \prod_{i>0} K_n^i \cup F$ is a compact generating set for L_{n+1}. It is easy to further verify that $\langle\langle K_{n+1}\rangle\rangle_{L_{n+1}} = L_{n+1}$. This completes our inductive construction.

Proposition 12 *For each $n \geq 1$, $L_n \in \mathscr{E}$ with $\xi(L_n) \geq n + 1$.*

Proof In view of Proposition 11, an easy induction argument verifies that $L_n \in \mathscr{E}$ for all $n \geq 1$. For the lower bound on the rank, we argue by induction on n. For the base case, $L_1 = H$ is non-trivial and discrete. Since the trivial group has rank 1, Proposition 4 implies that $\xi(L_1) = 2$.

Suppose the inductive hypothesis holds up to n and consider L_{n+1}. We first compute $\operatorname{Res}(L_{n+1})$. Consider $O \trianglelefteq_o L_{n+1}$. Since $K_n^{\mathbb{N}}$ is a compact open subgroup of L_{n+1}, the subgroup O must contain

$$K_n^{[k,\infty]} := \{f : \mathbb{N} \to K_n \mid f(0) = \cdots = f(k) = 1\}$$

for some $k \in \mathbb{N}$. Since H acts transitively on $\mathbb{N}$ and O is normal, O indeed contains $K_n^{\mathbb{N}}$. Recalling that $\langle\langle K_n\rangle\rangle_{L_n} = L_n$, we conclude that

$$\bigoplus_{i\in\mathbb{N}}(L_n^i, K_n^i) = \langle\langle K_n^{\mathbb{N}}\rangle\rangle_{L_{n+1}} \leq O.$$

It now follows that $\operatorname{Res}(L_{n+1}) = \bigoplus_{i\in\mathbb{N}}(L_n^i, K_n^i)$.

In view of Proposition 4, $\xi(L_{n+1}) = \xi(\operatorname{Res}(L_{n+1})) + 1$, because L_{n+1} is compactly generated. The group L_n admits a continuous injection into $\operatorname{Res}(L_{n+1})$, so

$$\xi(\operatorname{Res}(L_{n+1})) \geq \xi(L_n) \geq n + 1$$

via Proposition 3 and the inductive hypothesis. We conclude that $\xi(L_{n+1}) \geq n + 2$, and the induction is complete.

It is indeed the case that $\xi(L_n) = n + 1$ for all $n \geq 1$; one can devise a proof of this using the computation of the rank of a quasi-product given in [11]. The set $\{L_n \mid n \geq 1\}$ is thus a family of elementary groups with members of arbitrarily large finite decomposition rank. From this family, we obtain a first example of an elementary group with transfinite rank.

Corollary 4 *The group $G := \bigoplus_{n\geq 1}(L_n, K_n)$ is elementary with $\xi(G) \geq \omega + 1$.*

Proof For each $n \geq 1$, there is a continuous injection $L_n \hookrightarrow G$. Via Proposition 3, $n + 1 \leq \xi(G)$ for all $n \geq 1$, so $\omega \leq \xi(G)$. Since the decomposition rank is always a successor ordinal, we conclude that $\omega + 1 \leq \xi(G)$.

Remark 4 The examples above demonstrate a strategy for finding examples of higher rank. Suppose that we have $H \in \mathscr{E}$ with rank α and suppose that we construct a compactly generated $G \in \mathcal{E}$ for which $H \hookrightarrow \text{Res}(G)$. Applying Proposition 4, we then have that $\xi(G) \geq \alpha + 1$. The problem, of course, is finding the group G. We stress that one should not expect general embedding theorems which produce such a G. Indeed, there are groups in $\mathcal{E}$ which do not embed into *any* compactly generated t.d.l.c.s.c. group; see [4].

4.3 Compactly Generated Elementary Groups with Transfinite Rank

We here describe a technique which produces compactly generated elementary groups with transfinite rank. We omit proofs as they are somewhat technical; the full details of the construction will appear in a later article. The construction is inspired by ideas from [2, 9, 13], and the reader familiar with [9] and the theory of elementary groups can likely fill in the proofs.

Let T be the countable regular tree and fix δ an end of T. We orient the edges of T such that all edges point toward the end δ. The resulting directed graph is denoted by $\mathcal{T}$, and we call δ the **distinguished end** of $\mathcal{T}$. Given a countable set X, a **coloring** of $\mathcal{T}$ is a function $c : E\mathcal{T} \to X$ such that for each $v \in V\mathcal{T}$,

$$c_v := c \upharpoonright_{\text{inn}(v)} : \text{inn}(v) \to X$$

is a bijection. The set $\text{inn}(v)$ is the collection of directed edges with terminal vertex v. We call the coloring **ended** if there is a monochromatic directed ray which is a representative of the distinguished end δ; *we shall always assume our colorings are ended*. The coloring allows us to define the **local action** of $g \in \text{Aut}(\mathcal{T})$ at $v \in V\mathcal{T}$:

$$\sigma(g, v) := c_{g(v)} \circ g \circ c_v^{-1} \in \text{Sym}(X).$$

The local action allows us to isolate the groups we wish to consider. It shall be convenient to make a definition: A **t.d.l.c.s.c. permutation group** is a pair (G, X) where G is a t.d.l.c.s.c. group and X is a countable set on which G acts faithfully with compact open point stabilizers. We stress that X is assumed to be infinite.

Definition 9 Suppose that (G, X) is a t.d.l.c.s.c. permutation group with $U \in \mathcal{U}(G)$ and color the tree $\mathcal{T}$ by X. We define the group $E_X(G, U) \leq \text{Aut}(\mathcal{T})$ as follows: $E_X(G, U)$ is the set of $g \in \text{Aut}(\mathcal{T})$ such that $\sigma(g, v) \in G$ for all $v \in V\mathcal{T}$ and that $\sigma(g, v) \in U$ for all but finitely many $v \in V\mathcal{T}$.

It is easy to verify that $E_X(G, U)$ is an abstract group. With more care, one can also identify a natural t.d.l.c.s.c. group topology on $E_X(G, U)$. One first verifies that the vertex stabilizer $E_X(U, U)_{(v)}$ is compact in the topology on $\text{Aut}(\mathcal{T})$. The group $\text{Aut}(\mathcal{T})$ is given the topology of pointwise convergence; this topology is not locally compact, since the tree is locally infinite. One then argues for the following proposition:

Proposition 13 *For (G, X) a t.d.l.c.s.c. permutation group and $U \in \mathscr{U}(G)$, there is a t.d.l.c.s.c. group topology on $E_X(G, U)$ such that the inclusion $E_X(U, U)_{(v)} \hookrightarrow E_X(G, U)$ is continuous with a compact open image for any $v \in V\mathscr{T}$.*

The resulting t.d.l.c.s.c. group $E_X(G, U)$ yields the desired examples.

Theorem 2 *Suppose that (G, X) is a transitive t.d.l.c.s.c. permutation group. If G is compactly generated and elementary, then $E_X(G, U)$ is compactly generated and elementary with*

$$\xi(E_X(G, U)) \geq \xi(G) + \omega + 2$$

for any non-trivial $U \in \mathscr{U}(G)$.

Take G any infinite finitely generated group with a non-trivial finite subgroup U such that U has a trivial normal core in U. Letting $X := G/U$ and $G \curvearrowright X$ by left multiplication, the pair (G, X) is a t.d.l.c.s.c. permutation group. Theorem 2 now implies that $E_X(G, U)$ is elementary with rank at least $\omega + 2$. (It is indeed the case that $E_X(G, U)$ has rank exactly $\omega + 2$.)

Applying Theorem 2 repeatedly allows us to build elementary groups with even larger rank.

Corollary 5 *For each $0 \leq n < \omega$, there is a compactly generated elementary group L_n with $\xi(L_n) \geq \omega \cdot n + 2$.*

Acknowledgements This survey originated from a mini-course given at the Mathematical Research Institute MATRIX. The author thanks the institute for its hospitality. He also thanks Colin Reid and Simon M. Smith for their many suggestions for improvements to these notes and for reading an initial draft. The author is supported by ERC grant #278469.

References

1. Bekka, B., de la Harpe, P., Valette, A.: Kazhdan's property (T), New Mathematical Monographs, vol. 11. Cambridge University Press, Cambridge (2008). https://doi.org/10.1017/CBO9780511542749
2. Brin, M.G.: Elementary amenable subgroups of R. Thompson's group F. Int. J. Algebra Comput. **15**(4), 619–642 (2005). https://doi.org/10.1142/S0218196705002517
3. Caprace, P.E., Monod, N.: Decomposing locally compact groups into simple pieces. Math. Proc. Camb. Philos. Soc. **150**(1), 97–128 (2011). https://doi.org/10.1017/S0305004110000368
4. Caprace, P.E., Cornulier, Y.: On embeddings into compactly generated groups. Pac. J. Math. **269**(2), 305–321 (2014). https://doi.org/10.2140/pjm.2014.269.305
5. Cesa, M., Le Maître, F.: Elementary totally disconnected locally compact groups, after Wesolek (2015). https://webusers.imj-prg.fr/~francois.le-maitre/articles/surveyETDLC.pdf
6. Dieudonné, J.: La géométrie des groupes classiques. Springer, Berlin (1971). Troisième édition, Ergebnisse der Mathematik und ihrer Grenzgebiete, Band 5
7. Kechris, A.: Classical Descriptive Set Theory. Graduate Texts in Mathematics, vol. 156. Springer, New York (1995). https://doi.org/10.1007/978-1-4612-4190-4
8. Kunen, K.: Set Theory: An Introduction to Independence Proofs. Studies in Logic and the Foundations of Mathematics, vol. 102. North-Holland Publishing Co., Amsterdam (1980)

9. Le Boudec, A.: Groups acting on trees with almost prescribed local action. Comment. Math. Helv. **91**(2), 253–293 (2016). https://doi.org/10.4171/CMH/385
10. Osin, D.V.: Elementary classes of groups. Mat. Zametki **72**(1), 84–93 (2002). https://doi.org/10.1023/A:1019869105364
11. Reid, C.D., Wesolek, P.R.: Dense normal subgroups and chief factors in locally compact groups. Proc. London Math. Soc. (2017). https://doi.org/10.1112/plms.12088
12. Reid, C.D., Wesolek, P.R.: Homomorphisms into totally disconnected, locally compact groups with dense image (2015). http://arxiv.org/abs/1509.00156
13. Smith, S.M.: A product for permutation groups and topological groups. Duke Math. J. **166**(15), 2965–2999 (2017)
14. Tits, J.: Sur le groupe des automorphismes d'un arbre. In: Essays on topology and related topics (Mémoires dédiés à Georges de Rham), pp. 188–211. Springer, New York (1970)
15. Wesolek, P.: Elementary totally disconnected locally compact groups. Proc. Lond. Math. Soc. (3) **110**(6), 1387–1434 (2015). https://doi.org/10.1112/plms/pdv013
16. Wesolek, P., Williams, J.: Chain conditions, elementary amenable groups, and descriptive set theory. Groups Geom. Dyn. (2015, accepted for publication). http://arxiv.org/abs/1410.0975

The Scale, Tidy Subgroups and Flat Groups

George Willis

Abstract These notes discuss the scale, tidy subgroups, subgroups associated with endomorphisms and flat groups on totally disconnected locally compact (t.d.l.c) groups. The first section discusses the structure theory of subgroups which are *minimizing* for an endomorphism and introduces the *scale* of an endomorphism. The second section discusses the applications and properties of the scale function. Section 3 discusses other subgroups which may be associated with endomorphisms in a unique way. Section 4 discusses flat groups of automorphisms, the flat rank and various results about flat groups. The final section discusses the geometry of t.d.l.c groups.

1 Subgroups Tidy for an Endomorphism

1.1 Introduction

A group G is *locally compact* if it has a locally compact topology such that the group operations are continuous. Locally compact groups have a structure theory; there exists a short exact sequence

$$G_0 \hookrightarrow G \twoheadrightarrow G/G_0$$

where G_0 is the connected component of the identity and G/G_0 is totally disconnected [21].

Notes prepared by John J. Harrison.

G. Willis (✉)
University of Newcastle, University Drive, Callaghan, NSW 2308, Australia
e-mail: george.willis@newcastle.edu.au

If G is a connected locally compact group and $\mathcal{N}$ is a neighbourhood of the identity, then there exists a compact normal subgroup U contained in $\mathcal{N}$ such that G/U is a Lie group [20]. In other words, connected groups are approximated by Lie groups. This was the solution to Hilbert's fifth problem.

Matrix groups over a locally compact field are important examples of locally compact groups. Every topological field is either connected or totally disconnected, and the group $SL(n, \mathbb{F})$ is connected or totally disconnected depending on whether $\mathbb{F}$ is. The connected locally compact fields are $\mathbb{R}$ and $\mathbb{C}$. Every totally disconnected topological field is either discrete or non-discrete. The non-discrete totally disconnected locally compact fields are the p-adics and their finite extensions, which have characteristic zero, and the formal Laurent series over finite fields, which have positive characteristic [8].

The focus of these notes is non-discrete totally disconnected locally compact groups.

1.2 Totally Disconnected Groups

From now on, suppose that G is a totally disconnected locally compact group and that $\mathcal{N}$ is a neighbourhood of the identity. Then there exists a compact open subgroup O contained in $\mathcal{N}$ [12]. This subgroup is not necessarily normal. We denote by COS (G) the set of compact open subgroups of G. The compact open subgroups are also called *0-dimensional groups* because they have inductive dimension equal to zero.

All compact metrizable totally disconnected spaces are homeomorphic to a Cantor set [14]. There are therefore no topological invariants–such as dimension– which are useful for distinguishing them.

Definition 1 Let G be a totally disconnected locally compact (*t.d.l.c.*) group. An *endomorphism* is a continuous homomorphism on G. The set of endomorphisms on G form a semi-group under composition, denoted by $\mathsf{End}(G)$. An *automorphism* on G is an endomorphism which is a bijection with a continuous inverse. The group of automorphisms on G is denoted by $\mathsf{Aut}(G)$.

1.3 Endomorphisms and Minimizing Subgroups

Suppose that α is an endomorphism of G and that U is a compact open subgroup. Then the set $\alpha(U) \cap U$ is open in the subspace topology on $\alpha(U)$ and $\alpha(U)$ is compact. Hence $[\alpha(U) : \alpha(U) \cap U]$ is finite. The following definition was made for automorphisms in [23, 24] and for endomorphisms in [27].

Definition 2 Suppose that α is an endomorphism on a t.d.l.c. group G. The scale of α is

$$s(\alpha) = \min\{[\alpha(U): U \cap \alpha(U)] : U \in \mathrm{COS}(G)\}.$$

A compact open subgroup U is said to be *minimizing for* α if

$$s(\alpha) = [\alpha(U) : U \cap \alpha(U)].$$

Suppose that α is an endomorphism of G. For each compact open subgroup U, let

$$U_+ = \{x \in U : \exists \{x_n\}_{n \in \mathbb{N}} \subset U \text{ with } x_0 = x \text{ and } \alpha(x_{n+1}) = x_n \ \forall n \in \mathbb{N}\}$$

and let

$$U_- = \{x \in U : \alpha(x) \in U \ \forall n \in \mathbb{N}\}.$$

Lemma 1 *Suppose that α is an automorphism of G and that U is a compact open subgroup of G. Then*

$$U_+ = \bigcap_{k \geq 0} \alpha^k(U)$$

and

$$U_- = \bigcap_{k \geq 0} \alpha^{-k}(U)$$

Remark 1 These expressions for U_- and U_+ are usually given as the definition in the case of automorphisms. The fact that endomorphisms may lack an inverse is why the definitions must be changed to accommodate endomorphisms.

In the above context, note that α expands U_+ and contracts U_-. See Fig. 1 for an illustration of an automorphism α with $s(\alpha) = 3$ and $s(\alpha^{-1}) = 2$. The figure is not accurate for endomorphisms which may have range much thinner than U.

The following characterisation of minimising subgroups in terms of their structure is given in [27].

Theorem 1 (The Structure of Minimising Subgroups) *A compact open subgroup U of a locally compact totally disconnected group is minimizing for an endomorphism α if and only if it satisfies:*

(TA) $U = U_+ U_-$.
(TB1) $U_{++} = \bigcup_{n \geq 0} \alpha^n(U_+)$ *is closed.*
(TB2) *The sequence of integers $[\alpha^{n+1}(U_+) : \alpha^n(U_+)]$ is constant.*

When these conditions are satisfied $s(\alpha) = [\alpha(U_+) : U_+]$.

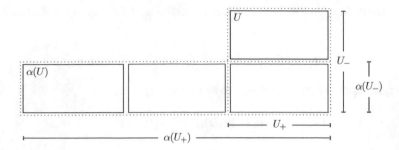

Fig. 1 Illustration of an automorphism α with $s(\alpha) = 3$ and $s(\alpha^{-1}) = 2$

Remark 2 The property (TB2) is not needed if α is an automorphism.

Remark 3 It is immediate from the definition that $\alpha(U_+) \geq U_+$. It follows that U_{++} is a subgroup for this reason.

Remark 4 One problem with working with endomorphisms instead of automorphisms is the fact that if α is an endomorphism, then $\alpha(A \cap B)$ is in general not necessarily equal to $\alpha(A) \cap \alpha(B)$.

It is immediate that U_+, U_- and U_{++} are subgroups, that $\alpha(U_+) \geq U_+$, and $\alpha(U_-) \leq U_-$, U_+ and U_- are all closed.

Definition 3 Let G be a totally disconnected locally compact group and let α be an endomorphism on G. A compact open subgroup U is called *tidy above for* α if it satisfies (TA) and *tidy below for* α if it satisfies both (TB1) and (TB2). If U is both tidy above and tidy below for α, then it is simply called *tidy*.

The motivation for the names 'tidy above' and 'tidy below' comes from a *tidying procedure*. Given any compact open subgroup U, the tidying procedure produces a tidy compact open subgroup. There are two steps to the procedure. The first step produces a compact open subgroup, V, which is tidy above. The second step takes this tidy above subgroup and produces a new compact open subgroup, W, which is both tidy above and tidy below. Each step of the tidying process reduces the index, so that

$$[\alpha(U) : \alpha(U) \cap U] \geq [\alpha(V) : \alpha(V) \cap V] \geq [\alpha(W) : \alpha(W) \cap W].$$

If U and V are tidy subgroups for an endomorphism α, then they have the same index,

$$[\alpha(U) : \alpha(U) \cap U] = [\alpha(V) : \alpha(V) \cap V].$$

Lemma 2 (Tidying Procedure Part One) *Let α be an endomorphism on G and let U be a compact open subgroup in G. Then there exists a natural number n such* *that*

$$V = U_n = \bigcap_{k=0}^{n} \alpha^n(U) = \{u \in U : \alpha^{-k}(u) \in U \text{ for } 0 \le k \le n\}$$

is tidy above.

Proof We suppose that α is an automorphism, and only give the proof in that case. We first note that each U_{-n} is open, because it is a finite intersection of open subgroups. Recall that

$$U_+ = \bigcap_{k \ge 0} \alpha^k(U)$$

and

$$U_- = \bigcap_{k \ge 0} \alpha^{-k}(U).$$

Let

$$U_k = \bigcap_{0 \le j \le k} \alpha^j(U)$$

and see Fig. 2 for an illustration of these sets. Then

$$\alpha(U_+) = \bigcap_{0 \le j \le k} \alpha(U_j).$$

(That this modest claim fails for endomorphisms is one reason that definitions and arguments must be modified.) Furthermore, $\{\alpha(U_j)\}_{j=0}^{\infty}$ is a decreasing sequence of compact open subgroups. Since $\alpha(U_+)U$ is an open neighbourhood of $\alpha(U_+)$, there must be a natural number n such that $\alpha(U_n) \subseteq U_+U$.

Since $\alpha(U_n) \subseteq \alpha(U_+)U$, if y is in $\alpha(U_n)$, then $y = zu$, for some z in $\alpha(U_+)$ and u in U. Then $u = z^{-1}y$ is in $\alpha(U_n) = \bigcap_{j \ge 1}^{n+1} \alpha^j(U)$, which implies that u is in U_{n+1}. In fact, $\alpha(U_n) \subseteq \alpha(U_+)U_{n+1}$.

In order to complete the proof, we will prove the claim that

$$\alpha^l(U_n) = \alpha^l(U_+)U_{n+l}$$

for all non-negative integers l. We do so by induction.

$$\alpha^{l+1}(U_n) = \alpha(\alpha^l(U_+)U_{n+l})$$
$$= \alpha^{l+1}(U_+)\alpha(U_{n+l})$$

George Willis

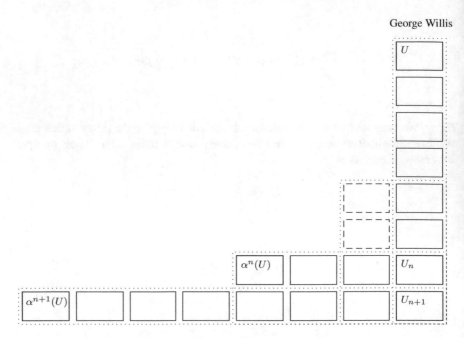

Fig. 2 Illustration of the subsets $U_n = \bigcap_{0 \leq j \leq n} \alpha^j(U)$

$$= \alpha^{l+1}(U_+)\alpha(U_+)U_{n+l+1}$$
$$= \alpha^{l+1}(U_+)U_{n+l+1}.$$

Let y be an element of our compact open subgroup $V = U_n$ and let

$$C_j = \{z \in V_+ : \alpha^j(y) \in \alpha^j(z)V_j\} \neq \emptyset.$$

Note that C_j is compact, and that C_{j+1} is contained within C_j for all natural numbers j. Now choose some z contained in the intersection $\bigcap_{j \geq 0} C_j$. Observe that z is in V_+, and that $z^{-1}y$ is in V, for all $j \geq 0$. This implies that $z^{-1}y$ is in V_-.

2   Scale of an Endomorphism

Recall Definition 2, which states that if G is a t.d.l.c. group, then the scale of α is

$$s(\alpha) = \min\{[\alpha(U): U \cap \alpha(U)] : U \in \mathrm{COS}(G)\}.$$

The scale defines a function from the endomorphisms of the group to the natural numbers. The scale of an endomorphism is 1 if and only if there exists a compact open subgroup U such that $\alpha(U)$ is a subgroup of U.

The scale of an endomorphism α satisfies

$$s(\alpha^n) = s(\alpha)^n \tag{1}$$

for all integers n. The proof of this fact is a consequence of the following lemma, which is [27, Proposition 16].

Lemma 3 *Suppose that U is a compact open subgroup which is tidy for an endomorphism α on a t.d.l.c. group G. Then U is tidy for α^n for every natural number n. Furthermore,*

$$s(\alpha^n) = [\alpha^n(U_+) : U_+].$$

To deduce (1) observe that

$$
\begin{aligned}
s(\alpha^n) &= [\alpha^n(U_+) : U_+] \\
&= \prod_{k=0}^{n-1} [\alpha^{k+1}(U_+) : \alpha^k(U_+)] \\
&= [\alpha(U_+) : U_+]^n,
\end{aligned}
$$

because $\{[\alpha^{k+1}(U_+) : \alpha^k(U_+)]\}$ is constant.

Another characterisation of the scale, known as *Møller's spectral radius formula*, may be derived from (1). This formula asserts that for any endomorphism α

$$s(\alpha) = \lim_{n\to\infty} [\alpha^n(U) : \alpha^n(U) \cap U]^{\frac{1}{n}} \tag{2}$$

where U is any compact open subgroup, not necessarily minimizing. (This formula is analogous to the spectral radius of a bounded linear operator T

$$\rho(T) = \lim_{n\to\infty} \|T^n\|^{\frac{1}{n}}.)$$

Proof (Møller's Formula) Since (2) holds when U is tidy, by (1), it suffices to show that the $\lim_{n\to\infty} [\alpha^n(U) : \alpha^n(U) \cap U]^{\frac{1}{n}}$ is independent of U. For this, it suffices to show that the limit is the same for any compact open subgroup V containing U. To do this, consider

$$[\alpha^n(V) : \alpha^n(U) \cap U] = [\alpha^n(V) : \alpha^n(U)][\alpha^n(U) : \alpha^n(U) \cap U] \tag{3}$$

$$= [\alpha^n(V) : \alpha^n(V) \cap V][\alpha^n(V) \cap V : \alpha^n(U) \cap U]. \tag{4}$$

Note that, since α is an endomorphism on G,

$$[\alpha^n(V) : \alpha^n(U)] \le [V : U]$$

and

$$[\alpha^n(V) \cap V : \alpha^n(U) \cap U] = [\alpha^n(V) \cap V : \alpha^n(V) \cap U][\alpha^n(V) \cap U : \alpha^n(U) \cap U]$$
$$\le [V : U]^2.$$

Note further that all indices are greater than or equal to one. Hence (3) and (4) imply that

$$[V : U]^{-1}[\alpha^n(V) : \alpha^n(V) \cap V] \le [\alpha^n(U) : \alpha^n(U) \cap U]$$
$$\le [V : U]^2[\alpha^n(V) : \alpha^n(V) \cap V]$$

Therefore,

$$\lim_{n \to \infty} [\alpha^n(V) : \alpha^n(V) \cap V]^{\frac{1}{n}} = \lim_{n \to \infty} [\alpha^n(U) : \alpha^n(U) \cap U]^{\frac{1}{n}}.$$

$\square$

Rögnvaldur Möller originally derived this formula from an alternative graph-theoretic characterisation of tidy subgroups and the scale he established in [16].

Automorphisms No further multiplicativity or submultiplicativity properties hold for the scale in general. More can be said for automorphisms.

Theorem 2 *Suppose that α is an automorphism on G. Then U is minimizing for α if and only if it is minimizing for α^{-1} and $\Delta(\alpha) = s(\alpha)/s(\alpha^{-1})$ where $\Delta : \mathsf{Aut}(G) \to (\mathbb{R}, \times)$ is the modular function.*

Proof Let U be any compact open subgroup of G and m be the Haar measure on G. Then by definition of the modular function

$$\Delta(\alpha) = \frac{m(\alpha(U))}{m(U)}$$
$$= \frac{m(\alpha(U))}{m(\alpha(U) \cap U)} \times \frac{m(\alpha(U) \cap U)}{m(U)}$$
$$= \frac{[\alpha(U) : \alpha(U) \cap U]}{[U : \alpha(U) \cap U]}$$
$$= \frac{[\alpha(U) : \alpha(U) \cap U]}{[\alpha^{-1}(U) : \alpha^{-1}(U) \cap U]}.$$

Choosing U minimizing for α gives $\Delta(\alpha) \leq \frac{s(\alpha)}{s(\alpha^{-1})}$, and choosing U minimizing for α^{-1} gives that $\frac{s(\alpha)}{s(\alpha^{-1})} \leq \Delta(\alpha)$. Hence U is minimizing for α if and only if it is minimizing for α^{-1} and

$$\Delta(\alpha) = \frac{s(\alpha)}{s(\alpha^{-1})}.$$

$\square$

The *Braconnier topology* on $\mathsf{Aut}(G)$ is the topology with base elements

$$N_1(K, O) = \{\alpha \in \mathsf{Aut}(G) : \alpha(K) \leq O\}, \text{ and}$$

$$N_2(K, O) = \{\alpha \in \mathsf{Aut}(G) : \alpha^{-1}(K) \leq O\}$$

ranging over all compact subsets K in G and open subsets O in G. This topology is formally stronger than the *compact open topology*, which is the topology with base formed from the sets $N_1(K, O)$, but these two topologies are equal in many cases. The compact open topology is not in general a group topology, because the inverse mapping may fail to be continuous for it, and the Braconnier topology remedies that difficulty.

Examples show that the scale function $s : \mathsf{Aut}(G) \to \mathbb{N}$ need not be continuous with respect to the Bracconier topology on $\mathsf{Aut}(G)$ and discrete topology on $\mathbb{N}$.

Question 1 Is there a topology on $\mathsf{Aut}(G)$, or possibly $\mathsf{End}(G)$, such that the scale is continuous? Is there a topology on $\mathsf{End}(G)$ with $\mathsf{Aut}(G)$ an open subgroup? The second of these questions is motivated by the fact that the group of invertible operators on a normed space is open in the semigroup of all endomorphisms.

Inner Automorphisms

Each $x \in G$ determines an inner automorphism $\alpha_x : y \mapsto xyx^{-1}$. The homomorphism $G \to \mathsf{Aut}(G)$, $x \mapsto \alpha_x$ induces a function on G called the scale on G, which is also denoted by s.

Theorem 3 *The scale $s : G \to \mathbb{N}$ is continuous with respect to the given topology in G and the discrete topology on $\mathbb{N}$.*

The next theorem gives much more precise information.

Theorem 4 *Suppose that $x \in G$ and let U be a compact open subgroup tidy for x. Then U is tidy for every $y \in UxU$ and $s(y) = s(x)$.*

Proof In the first instance, let u be an element of U. Consider $y = xu$ and let $u = u_- u_+$ for some $u_- \in U_-$ and $u_+ \in U_+$. Then

$$(xu)^2 = xuxu$$

$$= xu_- u_+ xu_- u_+$$

$$= (xu_- x^{-1})xu_+ (xu_- x^{-1})xu_+ .$$

But xu_-x^{-1} is in U_-, since conjugation by x shrinks u_-. Hence,

$$(xu)^2 = u'_-xvxu'_+$$
$$= u'_-(xv_-x^{-1})x^2(x^{-1}v_+x)u'_+$$
$$= u''_-x^2u''_+$$

for $u'_-, u''_- \in U_-$ and $u'_+, u''_+ \in U_+$ and some $v = v_-v_+$, where $v_- \in U_-$ and $v_+ \in U_+$. A similar calculation shows that, more generally, if $y = u_1xu_2 \in UxU$ and $n \geq 0$, then $(u_1xu_2)^n = u''_-x^nu''_+$ for some $u''_\pm \in U_\pm$. Hence,

$$[y^nUy^{-n} : y^nUy^{-n} \cap U]$$
$$= [(u_1xu_2)^nU(u_1xu_2)^{-n} : (u_1xu_2)^nU(u_1xu_2)^{-n} \cap U]$$
$$= [(u_-x^nu_+)U(u_-x^nu_+)^{-1} : (u_-x^nu_+)U(u_-x^nu_+)^{-1} \cap U]$$
$$= [u_-x^n(u_+Uu_+^{-1})x^nu_-^{-1} : u_-x^n(u_+Uu_+^{-1})x^nu_-^{-1} \cap U]$$
$$= [u_-x^nUx^{-n}u_-^{-1} : u_-x^nUx^{-n}u_-^{-1} \cap U]$$
$$= [x^nUx^{-n} : x^nUx^{-n} \cap U]$$
$$= s(x^n).$$

Hence, by Möller's spectral radius formula,

$$s(y) = \lim_{n \to \infty} [y^nUy^{-n} : y^nUy^{-n} \cap U]^{1/n} = s(x).$$

The $n = 1$ case of the calculation then shows that

$$[yUy^{-1} : yUy^{-1} \cap U] = s(x) = s(y)$$

and so U is minimizing for y. □

2.1 An Application of the Scale Function

Define, for G a non-discrete totally disconnected locally compact group,

$$\mathrm{Per}(G) = \{x \in G : \overline{\langle x \rangle} \text{ is compact}\}.$$

Theorem 5 $\mathrm{Per}(G)$ *is a closed subset of* G.
The theorem answers a question posed by Karl Heinrich Hofmann that was motivated by the following considerations. When G is discrete $\mathrm{Per}(G)$ is clearly

closed. However Per(G) is not always closed when G is connected. For example, translations can be approximated by rotations in the affine group of the plane.

Lemma 4 *Suppose that G is a totally disconnected locally compact group. Let x be an element of* Per(G). *Then,*

$$s(x) = 1 = s(x^{-1}).$$

Proof The quantity $s(\overline{\langle x \rangle})$ is finite because the scale is continuous and the image of a compact set must therefore be finite. But since

$$s(x^n) = s(x)^n,$$

the finiteness of the image implies that $s(x) = 1$. A similar argument shows that $s(x^{-1}) = 1$. □

Now suppose that x is in $\overline{\text{Per}(G)}$. Choose U tidy for x. Then $xUx^{-1} = U$ and $x = yu$ for some y in Per(G) and u in U. Hence $yUy^{-1} = U$ and it follows that

$$\langle x \rangle \subseteq \langle y \rangle U \subseteq \overline{\langle y \rangle} U, \text{ which is compact.}$$

□

2.2 The Scale and Tidy Subgroups for Homomorphisms

Question 2 Do the concepts of scale and tidy subgroup extend to homomorphisms $\tau : G \to H$?

The answer is 'probably not'—the scale is analogous to the eigenvalues of a linear transformation $T : V \to V$. There is no concept of eigenvalue for linear maps between different vector spaces. Here is a related question.

Question 3 Suppose that $\tau : G \to H$ and $\sigma : H \to G$ are homomorphisms. Is $s(\tau \circ \sigma) = s(\sigma \circ \tau)$?

The following special case asks for an analogue of singular values. Let G and H be self-dual abelian t.d.l.c. groups, with $\iota_G : \hat{G} \to G$ and $\iota_H : \hat{H} \to H$ isomorphisms. Let $\tau : G \to H$ be a homomorphism and put $\sigma = \iota_G \circ \hat{\tau} \circ \iota_H^{-1}$. Is $s(\tau \circ \sigma) = s(\sigma \circ \tau)$?

3 Subgroups Associated with Endomorphisms

3.1 Minimising Subgroups and Their Associates

We begin by recalling the following from Sect. 1.

Theorem 6 (The Structure of Minimising Subgroups) *A compact open subgroup U of a locally compact totally disconnected group is minimizing for an endomorphism α if and only if it is tidy.*

Definition 4 A compact open subgroup U of a locally compact totally disconnected group is *tidy* for an endomorphism α if it satisfies:

(TA) $U = U_+U_-$.
(TB1) $U_{++} = \bigcup_{n\geq 0} \alpha^n(U_+)$ is closed.
(TB2) The sequence of integers $[\alpha^{n+1}(U_+) : \alpha^n(U_+)]$ is constant. (This property is only needed for endomorphisms)

where

$$U_+ = \{x \in U : \exists\{x_n\}_{n\in\mathbb{N}} \subset U \text{ with } x_0 = x \text{ and } \alpha(x_{n+1}) = x_n \; \forall n \in \mathbb{N}$$

and

$$U_- = \{x \in U : \alpha^n(x) \in U \; \forall n \in \mathbb{N}\}.$$

There may be many subgroups that are tidy for a given endomorphism α. For example, if $\alpha \in \mathsf{Aut}(G)$ and U is tidy for α, then so are $\alpha^n(U)$ and $\bigcap_{k=0}^n \alpha^k(U)$ for every integer n. The associated subgroups U_+, U_-, U_{++}, and U_{--} may depend on the choice of tidy subgroup U.

Other subgroups of G may be associated with a given endomorphism α in a unique way.

Definition 5 (The Parabolic and Levi Subgroups) Suppose that G is a totally disconnected locally compact group and let α be an endomorphism on G. Define

- the *parabolic subgroup* to be

$$\overrightarrow{\mathsf{par}}(\alpha) = \{x \in G \mid \{\alpha^n(x)\}_{n\in\mathbb{N}} \text{ is pre-compact}\},$$

- the *anti-parabolic subgroup* to be the subgroup

$$\overleftarrow{\mathsf{par}}(\alpha) = \{x \in G \mid \exists\{x_n\}_{n=0}^{\infty} \text{ pre-compact with } x_0 = x \text{ and } \alpha(x_{n+1}) = x_n\}, \text{ and}$$

- the *Levi subgroup* to be the intersection of the parabolic subgroup and the anti-parabolic subgroup:

$$\mathsf{lev}(\alpha) = \overrightarrow{\mathsf{par}}(\alpha) \cap \overleftarrow{\mathsf{par}}(\alpha).$$

It may be checked that these are subgroups of G but an argument using a subgroup tidy for α shows more.

Theorem 7 $\overrightarrow{\mathrm{par}}(\alpha)$, $\overleftarrow{\mathrm{par}}(\alpha)$ *and* $\mathrm{lev}(\alpha)$ *are closed subgroups of G.*

Proof (Sketch of the Proof That $\overrightarrow{\mathrm{par}}(\alpha)$ *is Closed)* Show that $\overrightarrow{\mathrm{par}}(\alpha) \cap U = U_-$, which is closed. A classical lemma of Bourbaki implies that U is closed. ☐

Definition 6 Suppose that $\alpha \in \mathrm{End}(G)$. Define

- the *contraction subgroup* to be

$$\overrightarrow{\mathrm{con}}(\alpha) = \{x \in G \mid \alpha^n(x) \to 1 \text{ as } n \to \infty\},$$

- the *iterated kernel* to be

$$\ker^\infty(\alpha) = \{x \in G \mid \exists n \geq 0 \text{ with } \alpha^n(x) = 1\}, \text{ and}$$

- the *anti-contraction subgroup* to be

$$\overleftarrow{\mathrm{con}}(\alpha) = \{x \in G \mid \exists \{x_n\}_{n=0}^\infty \text{ such that } x_n \to 1 \text{ and } \alpha(x_{n+1}) = x_n\}.$$

Clearly, $\ker^\infty(\alpha) \leq \overrightarrow{\mathrm{con}}(\alpha)$ and is a normal subgroup of G. Furthermore, $\overrightarrow{\mathrm{con}}(\alpha) \leq V_{--}$ and $\overleftarrow{\mathrm{con}}(\alpha) \leq V_{++}$ for every subgroup V tidy for α. When α is an automorphism, we have that $\overleftarrow{\mathrm{con}}(\alpha) = \overrightarrow{\mathrm{con}}(\alpha^{-1})$. The contraction subgroup for α will be denoted by $\mathrm{con}(\alpha)$ in this case. It is related to the scale of α^{-1}.

Theorem 8 (Baumgartner and Willis, Jaworski) *Suppose that α is an automorphism on G. Then*

$$\overline{\mathrm{con}(\alpha)} = \bigcap\{U_{--} \in \mathrm{COS}\,(G) \mid U \text{ is tidy for } \alpha\}, \text{ and}$$

$$s(\alpha^{-1}|_{\overline{\mathrm{con}(\alpha)}}) = s(\alpha^{-1}).$$

Remark 5 Theorem 8 was established for metrisable groups in [3] and the metrisability condition removed in [13]. The second part of the theorem implies in particular that, if $s(\alpha^{-1}) > 1$, then the contraction subgroup for α is not trivial.

Question 4 Extend this result to endomorphisms. There will need to be two theorems. One for $\overrightarrow{\mathrm{con}}(\alpha)$ and one for $\overleftarrow{\mathrm{con}}(\alpha)$, e.g.

$$s(\alpha|_{\overrightarrow{\mathrm{con}(\alpha)}}) = s_G(\alpha).$$

Remark 6 Since this lecture was delivered, results about the contraction and anti-contraction subgroups that extend Theorem 8 to endomorphisms have been established by T. Bywaters, H. Glöckner and S. Tillman.

Definition 7 The *nub* subgroup for the endomorphism α on a totally disconnected locally compact group G is

$$\mathsf{nub}(\alpha) = \bigcap \{U \in \mathrm{COS}\,(G) \mid U \text{ is tidy for } \alpha\}\,.$$

The following was established in [3], although the nub terminology was not used.

Theorem 9 (Baumgartner and Willis) *Let $\alpha \in \mathsf{Aut}(G)$. Then, the following are equivalent:*

- $\mathsf{nub}(\alpha) = \{1\}$;
- $\mathsf{con}(\alpha)$ *is closed; and*
- *if U is tidy above for α, then U is tidy for α.*

Since $\mathsf{nub}(\alpha) = \mathsf{nub}(\alpha^{-1})$, it follows as well from the theorem that $\mathsf{con}(\alpha^{-1})$ is closed whenever $\mathsf{nub}(\alpha)$ is trivial.

Example 1 Let F be a finite group and put $G = F^{\mathbb{Z}}$. Then G is a compact t.d.l.c. group. Define $\alpha \in \mathsf{Aut}(G)$ by

$$\alpha(f)_n = f_{n+1}, \quad (f \in F^{\mathbb{Z}}).$$

Then $\mathsf{nub}(\alpha) = G$.

Note that, in the example,

$$\mathsf{con}(\alpha) = \{f \in G \mid \exists N \in \mathbb{Z} \text{ such that } f_n = 1 \text{ if } n \geq N\}$$

and is dense in G. Moreover, $\mathsf{con}(\alpha) \cap \mathsf{con}(\alpha^{-1})$ is equal to the subgroup of functions with finite support, which is also dense.

It may be shown that $\mathsf{nub}(\alpha)$ is also the largest closed subgroup of G on which the restriction of α is ergodic. This fact extends a result due to Aoki in the 1980's who proved for t.d.l.c. groups a conjecture of Halmos that any locally compact group for which there is an ergodic automorphism must be compact. The method of tidy subgroups allows this to be proved in a few lines.

Another characterisation of the nub is that

$$\mathsf{nub}(\alpha) = \overline{(\mathsf{con}(\alpha) \cap \mathsf{par}(\alpha^{-1}))}$$

$$= \left\{x \in G : \alpha^n(x) \to 1 \text{ as } n \to \infty, \{\alpha^{-n}(x)\}_{n=0}^{\infty} \text{ is precompact}\right\}^{-}.$$

The structure of nub subgroups may be described in some detail, see [26, 27]. Among the results is that $\mathsf{con}(\alpha|_{\mathsf{nub}(\alpha)})$ is dense in $\mathsf{nub}(\alpha)$, and $\mathsf{con}(\alpha|_{\mathsf{nub}(\alpha^{-1})})$ is dense in $\mathsf{nub}(\alpha^{-1})$. Their intersection may fail to be dense however.

It may happen that $\mathsf{con}(\alpha)$ is closed. That is the case when G is a p-adic Lie group for example, see [3]. A key example is a *restricted product with shift* which, for some given finite group F, is the group

$$G = \{f = (f_n)_{n \in \mathbb{Z}} : f_n \in F, \ \exists N \text{ such that } f_n = 1 \text{ for all } n \geq N\}$$

with the *shift automorphism* on G defined by

$$\alpha(f)_n = f_{n+1}, \qquad (n \in \mathbb{Z}).$$

The shift automorphisms satisfies $\mathsf{con}(\alpha) = G$, which is closed. The structure of general closed contraction groups may be described, see [10].

Theorem 10 (Glöckner and Willis) *Let G be a t.d.l.c. group and $\alpha \in \mathsf{Aut}(G)$. Suppose that $\mathsf{con}(\alpha) = G$. Then*

- $G = N \times T$, *where N and T are α-invariant, N is a divisible subgroup of G and T a torsion subgroup;*
- N *is isomorphic to the direct sum of a finite number of nilpotent p-adic Lie groups for primes p dividing $s(\alpha^{-1})$; and*
- T *has a composition series*

$$T_0 = \{1\} \lhd T_1 \lhd \cdots \lhd T_j \lhd \cdots \lhd T_{r-1} \lhd T_r = T$$

of closed α-invariant subgroups such that T_{j+1}/T_j is isomorphic to a restricted product with shift.

4 Flat Groups of Automorphisms

Analogies with linear algebra are suggested by, or have motivated, several of the ideas seen so far. The 'spectral radius' formula is one of the ideas suggesting an analogy between the scale and eigenvalues of a linear transformation; and the fact that the methods of linear algebra apply to all linear transformations and not just invertible ones was one reason for thinking that the characterisation of subgroups minimising for automorphisms would extend to endomorphisms.

How the method of tidy subgroups might extend to more than one automorphism simultaneously is also suggested by this analogy. Finding a subgroup tidy for an endomorphism is the analogue of finding a Jordan basis for a linear transformation, which essentially can only be done when the linear transformations commute. On the other hand, when two linear transformations do share a common Jordan basis, they commute modulo upper triangular matrices. The following results were suggested by these observations and established in [25].

Theorem 11 *Let $\{\alpha_1, \ldots, \alpha_k\} \subset \mathsf{Aut}(G)$ be a commuting set of automorphisms of the t.d.l.c. group G. Then there is $U \in \mathsf{COS}(G)$ that is tidy for every α in $\langle \alpha_1, \ldots, \alpha_k \rangle$.*

Theorem 12 *Suppose, for some* $\alpha, \beta \in \mathsf{Aut}(G)$, *that there is* $U \in \mathsf{COS}\,(G)$ *that is tidy for every* $\gamma \in \langle \alpha, \beta \rangle$. *Then* $s([\alpha, \beta]) = 1$, *that is,* α *and* β *commute modulo the uniscalar elements in* $\langle \alpha, \beta \rangle$.

Remark 7 It is important that these results refer to *groups* of automorphisms. It is not automatically the case that, if α and β share a common tidy subgroup U, then U is tidy for every $\gamma \in \langle \alpha, \beta \rangle$. This complicates the proof of the first theorem and means that the hypothesis of the second theorem needs to be strictly stronger than that α and β should share a common tidy subgroup.

Example 2 Let $G = \mathbb{Q}_p^d$ and define $\alpha, \beta \in \mathsf{Aut}(G)$ by

$$\alpha(x_1, \ldots, x_d) = p(x_1, \ldots, x_d)$$

$$\text{and } \beta(x_1, x_2, \ldots, x_d) = (px_1, p^2 x_2, \ldots, p^d x_d).$$

Then $U = \mathbb{Z}_p^d$ is tidy for every $\gamma \in \langle \alpha, \beta \rangle$. However,

$$V = \{(z_1, \ldots, z_d) \in U \mid z_i \equiv z_j \pmod{p}, \ i, j \in \{1, \ldots, d\}\}$$

is tidy for α and β, and indeed for every γ in the semigroup generated by α and β, but not for every $\gamma \in \langle \alpha, \beta \rangle$.

Proof (d=2) In this case,

$$V = \{(z_1, z_2) \in \mathbb{Z}_p^2 : z_1 \equiv z_2 \pmod{p}\}$$

and

$$\alpha(V) = \{(pz_1, pz_2) \in \mathbb{Z}_p^2 : z_1 \equiv z_2 \pmod{p}\} \le V$$

because $pz_1 \equiv pz_2 \equiv 0 \pmod{p}$. We also have

$$\beta(v) = \{(pz_1, p^2 z_2) : z_1 \equiv z_2 \pmod{p})\} \le V.$$

Hence, V is tidy for α and β, $V_+ = \{0\}$, $V_- = V$ and $V_{--} = G$ for both α and β. We also have that $s(\alpha) = 1 = s(\beta)$, $s(\alpha^{-1}) = p^2$ and $s(\beta^{-1}) = p^3$. Hence V is tidy for α and β. Moreover, if $\gamma = \alpha^m \beta^n$, $m, n \ge 0$, then $\gamma(V) \subseteq V$, so that V is tidy.

The subgroup V is not minimizing for $\alpha \beta^{-1}$ however. Calculation shows that

$$V \cap \alpha \beta^{-1}(V) = \{(w_1, w_2) \in \mathbb{Z}_p^2 : w_i \equiv 0 \pmod{p}\} = p\mathbb{Z}_p^2.$$

$$[\alpha \beta^{-1}(U) : p\mathbb{Z}^2] = p^2,$$

which is larger than the corresponding index found for U,

$$[\alpha\beta^{-1}(\mathbb{Z}_p^2) : \alpha\beta^{-1}(\mathbb{Z}_p^2) \cap \mathbb{Z}_p^2] = [\alpha\beta^{-1}(\mathbb{Z}_p^2) : \mathbb{Z}_p^2] = p.$$

It may also be seen that V is not tidy for $\alpha\beta^{-1}$, for

$$V_+ = \bigcap_{k \geq 0}^{\infty} (\alpha\beta^{-1})^k(V) = p\mathbb{Z}_p^2 \text{ and}$$

$$V_- = \bigcap_{k \geq 0}^{\infty} (\alpha\beta^{-1})^{-k}(V) = p\mathbb{Z}_p \oplus \{0\}, \text{ so that}$$

$$V_+ V_- = p\mathbb{Z}_p^2 \neq V \text{ and } V \text{ is not tidy above.}$$

$\square$

4.1   Flat Groups and the Flat-Rank

Definition 8 A subgroup $\mathcal{H} \leq \text{Aut}(G)$ is *flat* if there is $U \in \text{COS}(G)$ that is tidy for every $\alpha \in \mathcal{H}$. The *uniscalar subgroup* of the flat group $\mathcal{H}$ is

$$\mathcal{H}_1 = \{\alpha \in \mathcal{H} \mid s(\alpha) = 1 = s(\alpha^{-1})\}.$$

$\mathcal{H}_1$ is a subgroup because $\alpha \in \mathcal{H}_1$ if and only if $\alpha(U) = U$ for any, and hence all, subgroups tidy for $\mathcal{H}$.

Theorem 13 *Suppose that $\mathcal{H} \leq \text{Aut}(G)$ is finitely generated and flat, and let U be tidy for $\mathcal{H}$. Then $\mathcal{H}_1 \lhd \mathcal{H}$ and there is $r \in \mathbb{N}$ such that*

$$\mathcal{H}/\mathcal{H}_1 \cong \mathbb{Z}^r.$$

1. There is $d \in \mathbb{N}$ such that

$$U = U_0 U_1 \ldots U_d,$$

where for every $\alpha \in \mathcal{H}$: $\alpha(U_0) = U_0$ and,
for every $j \in \{1, 2, \ldots, d\}$, either $\alpha(U_j) \leq U_j$ or $\alpha(U_j) \geq U_j$.

2. *For each $j \in \{1, 2, \ldots, d\}$ there is a homomorphism $\rho_j : \mathcal{H} \to \mathbb{Z}$ and a positive integer s_j such that*

$$[\alpha(U_j) : U_j] = s_j^{\rho_j(\alpha)},$$

where

$$[\alpha(U_j) : U_j] = \begin{cases} [\alpha(U_j) : U_j], & \text{if } U_j \leq \alpha(U_j), \\ [U_j : \alpha(U_j)]^{-1}, & \text{if } U_j \geq \alpha(U_j). \end{cases}$$

3. For each $j \in \{1, 2, \ldots, d\}$,

$$\widetilde{U}_j := \bigcup_{\alpha \in \mathscr{H}} \alpha(U_j)$$

is a closed subgroup of G.
4. The natural numbers r and d, the homomorphisms $\rho_j : \mathscr{H} \to \mathbb{Z}$ and positive integers s_j are independent of the subgroup U tidy for α.

Remark 8 The numbers $s_j^{\rho_j(\alpha)}$ are analogues of absolute values of eigenvalues for α and the subgroups $\widetilde{U}_j = \bigcup_{\alpha \in \mathscr{H}} \alpha(U_j)$ are the analogues of common eigenspaces for the automorphisms in $\mathscr{H}$.

Example 3 (A) Take $G = \mathbb{Q}_p^d$, and α, β as before. Take $U = \mathbb{Z}_p^d$ as a tidy subgroup. The number of factors will be d. How do we obtain the factors? Note that

$$U_{\alpha+}, U_{\beta+} = \{0\}$$
$$U_{\alpha-}, U_{\beta-} = U_-.$$

Choose $\alpha\beta^{-1}$, then

$$\alpha\beta^{-1}(z_1, \cdots, z_d) = (z_1, p^{-1}z_2, \ldots, p^{1-d}z_d).$$

Calculating the factoring of U determined by $\alpha\beta^{-1}$ we obtain

$$U_{\alpha\beta^{-1}+} = U, \quad U_{\alpha\beta^{-1}-} = \mathbb{Z}_p \oplus \{0\},$$

which identifies one factor but not the others. To separate out these factors choose, for each $i \in \{1, \ldots, d\}$ the element $\gamma = \alpha^i\beta^{-1}$. Then

$$U_{\gamma+} = \{0\} \oplus (\mathbb{Z}_p)^{d-i+1}, \quad U_{\gamma-} = (\mathbb{Z}_p)^i \oplus \{0\} \text{ and } U_{\gamma+} \cap U_{\gamma-} = \{0\} \oplus \underbrace{\mathbb{Z}_p}_{i} \oplus \{0\}.$$

We see that generators alone are insufficient to separate all the factors in Theorem 13.1 but they can be separated by using additional elements. That is the strategy of the proof of Theorem 13.

Example 4 Let $G = SL(n, \mathbb{Q}_p)$ and let H be subgroup of the diagonal matrices in G. Let $\alpha_h(x) = hxh^{-1}$. Then:

- $r = n - 1$;
- $d = n(n - 1)$;
- ρ_j are roots of H; and
- U_j are root subgroups of G.

Definition 9 The number r appearing in the theorem is the *rank* of the flat group $\mathcal{H}$. The maximum rank of any flat group of inner automorphisms of the t.d.l.c. group G is the *flat-rank* of G.

Example 5

- Let $G = \mathbb{Q}_p^d \rtimes \langle \alpha, \beta \rangle$. Then G has flat-rank 2.
- Let $G = SL(n, \mathbb{Q}_p)$. Then G has flat-rank $n - 1$.
- Let $G = \text{AAut}(T)$ be the group of almost automorphisms of the regular tree T. Then G has infinite flat-rank.

4.2 Further Results About Flat Groups

The theorem on finitely generated flat groups can be applied to show that a flat group of automorphisms contains uniscalar elements when the group is not abelian. The following is established in [19].

Theorem 14 *Let G be a t.d.l.c. group. Then,*

- *Every finitely generated nilpotent subgroup of $\text{Aut}(G)$ is flat.*
- *Every polycyclic group subgroup of $\text{Aut}(G)$ is virtually flat, that is, has a flat subgroup of finite index.*

The subgroup of upper triangular matrices in $SL(n, \mathbb{Z})$ is nilpotent and is non-abelian when $n \geq 3$. Hence, if $n \geq 3$ and $\rho : SL(n, \mathbb{Z}) \rightarrow G$ is a homomorphism with G a t.d.l.c. group, then there is an upper triangular T such that $\rho(T)$ is uniscalar. Further work using deep theorems about $SL(n, \mathbb{Z})$ and a theorem about groups that commensurate in bounded fashion deduce from this that $\rho(SL(n, \mathbb{Z}))$ normalises a compact open subgroup of G, see Remark 10 below. For details and additional references see [19].

5 T.d.l.c. Groups and Geometry

In this section we only consider automorphisms of t.d.l.c. groups. The aim is to survey actions of t.d.l.c. groups that may be viewed as geometric.

5.1 Symmetric Spaces Modulo a Compact Open Subgroup

Many groups have geometric representations that aid understanding of the group. Semi-simple real Lie groups, for example, act on a real symmetric space [11] and semi-simple Lie groups over a totally disconnected locally compact field may be represented as acting on a simplicial complex called an *affine building* and also on a related simplicial complex called a *spherical building* [2, 9]. In the case when the group has rank 1, e.g., $SL(2, \mathbb{Q}_p)$, the affine building is a tree and the spherical building is the set of ends of the tree. Other examples of t.d.l.c. groups, such as Kac-Moody groups [22], also act on buildings and on the boundary of the building. Automorphism groups of buildings are themselves t.d.l.c. groups and which then come with their own natural geometric representation. They, and their closed subgroups, are a rich source of examples of t.d.l.c. groups.

The so-called 1-skeleton of an affine building is a graph and path length then defines a metric on the set of vertices of this graph. As a metric space, it contains geometric 'flats', which are subsets quasi-isometric to $\mathbb{Z}^r$ for some r. This number r is the *geometric rank* of the building. In many cases of groups acting on a building, such as Kac-Moody groups, the group also has an algebraic rank. Under certain hypotheses, it may be shown that the geometric rank of the building, the algebraic rank of the group and the flat-rank are all equal, see [5].

Example 6 The group $SL(2, \mathbb{Q}_p)$ has flat-rank 1 and acts on the regular tree with valency $p + 1$. Trees have geometric rank equal to 1.

Vertex stabilisers for this action are maximal compact subgroups of $SL(2, \mathbb{Q}_p)$. Indeed, there is $v \in V(T)$ such that $\text{stab}_G(v) = SL(2, \mathbb{Z}_p)$, which is one of the maximal compact subgroups of $SL(2, \mathbb{Z}_p)$. The homogeneous space $SL(2, \mathbb{Q}_p)/SL(2, \mathbb{Z}_p)$ may thus be identified with the $SL(2, \mathbb{Q}_p)$-orbit of v, which is one of two such orbits in $V(T)$. Note that any compact subgroup of a t.d.l.c. group is contained in an open compact subgroup, so that these maximal compact subgroups are open and the quotient topology on $SL(2, \mathbb{Q}_p)/SL(2, \mathbb{Z}_p)$ is discrete.

5.2 Cayley-Abels Graphs

Suppose that G is a compactly generated t.d.l.c. group and let $U \in \text{COS}(G)$. A graph, $\Gamma(K, U)$, may be defined by choosing a compact, symmetric generating set, K, for G and setting

$$V(\Gamma) = G/U \text{ and } E(\Gamma) = \left\{ (gU, hU) \in V(\Gamma)^2 \mid h^{-1}g \in UKU \right\}.$$

Then $\Gamma(K, U)$ is a locally finite graph and the translation action of G on Γ is by graph automorphisms. This action is transitive and vertex stabilisers are all conjugates of U.

Any graph on which G acts vertex-transitively and with compact open vertex stabilisers is called a *Cayley-Abels graph* for G. Hence $\Gamma(K, U)$ is a Cayley-Abels graph. The graphs $\Gamma(K, U)$ are not unique and depend on the choices of K and U. All Cayley-Abels graphs for G are quasi-isometric however, see [1, 15, 17].

The Cayley-Abels graph guarantees that every compactly generated t.d.l.c. group acts on a locally finite connected graph. This graph is not canonical however because there may be many non-isomorphic Cayley-Abels graphs. On the other hand, the automorphism group of any locally finite connected graph Γ is totally disconnected when equipped with the topology of uniform convergence on compact sets. The vertex stabilisers will be compact open subgroups of the automorphism group and so Γ is a Cayley-Abels graph for its automorphism group. There is thus an equivalence between compactly generated t.d.l.c. groups and closed subgroups of automorphism groups of connected locally finite graphs.

Example 7

- Let Γ be a regular tree. Then $G = \mathsf{Aut}(\Gamma)$ is a t.d.l.c. group and Γ is a Cayley-Abels graph for G.
- The group $PSL(n, \mathbb{Q}_p)$ acts on a Bruhat-Tits building of rank $n-1$. The 1-skeleton of this building is not a Cayley-Abels graph for the group because the action is not transitive. However, the building is quasi-isometric to a Cayley-Abels graph because vertex stabilisers are compact and there are only finitely many orbits for the G-action.

Remark 9 In many examples, the orbit $H.v \subset \Gamma(K, U)$ of a flat subgroup $H \leq G$ is quasi-isometric to $\mathbb{Z}^r$, where r is the flat-rank of H. However, this only holds when H_1, the uniscalar subgroup in H, is compact.

5.3 Actions on Sets of Subgroups of G

The set COS (G) is a discrete metric space with the metric defined by

$$d(U, V) = \log([U : U \cap V][V : U \cap V]).$$

For each $\alpha \in \mathsf{Aut}(G)$ the map $U \mapsto \alpha(U)$ is an isometry of COS (G) and the map $\mathsf{Aut}(G) \to \mathrm{Isom}(\mathrm{COS}\,(G))$ is a homomorphism.

Subgroups tidy for α may be characterised as those whose α-orbit is a straight line in COS (G), see [4].

Proposition 1

> U *is tidy for* $\alpha \in \mathsf{Aut}(G)$
>
> $\iff d(\alpha^m(U), \alpha^n(U)) = |m - n|\, d(U, \alpha(U))$ *for every* $m \geq 0$.

The relationship between Cayley-Abels graphs and the metric G-space COS (G) is seen in the following.

Proposition 2 *The function* $\psi : \Gamma(K, U) \rightarrow \mathrm{COS}\,(G)$ *defined by*

$$\psi(xU) = xUx^{-1}, \quad xU \in V(\Gamma(K, U)),$$

is bounded with respect to the geodesic distance on $\Gamma(K, U)$ *and is injective if and only if* $N_G(U) = U$.

Remark 9 points out that if $H \leq G$ is flat with rank r, then there is an H-orbit in $\Gamma(K, U)$ that is quasi-isometric to $\mathbb{Z}^r$ if and only if the uniscalar subgroup of H is compact. On the other hand, H-orbits in COS $((G))$ are always quasi-isometric to $\mathbb{Z}^r$ if H is flank with rank r. The following, proved in [6], goes in the opposite direction.

Theorem 15 (Baumgartner, Schlichting, Willis) *Suppose that all balls in the metric space* COS (G) *are finite. Let* $\mathcal{H} \leq \mathrm{Aut}(G)$ *be such that the* $\mathcal{H}$-*orbit* $\{\alpha(U) \mid \alpha \in \mathcal{H}\}$ *is quasi-isometric to* $\mathbb{Z}^r$. *Then* $\mathcal{H}$ *is virtually flat.*

Question 5 Does the conclusion of Theorem 15 hold for all t.d.l.c. groups, rather than just those for which all balls in COS (G) are finite?

Remark 10 The answer to this question is 'yes' in the flat-rank 0 case, see [7, 18]. In other words, if $\{d(U, hUh^{-1}) : h \in H\}$ is bounded for some $U \in \mathrm{COS}\,(G)$, then there is $V \in \mathrm{COS}\,(G)$ such that $hVh^{-1} = V$ for every $h \in H$. This is one of the additional theorems used in [19] that was referred to in the comments following Theorem 14.

The Space of Directions

Definition 10 The *ray* generated by $\alpha \in \mathrm{Aut}(G)$ and based at $U \in \mathrm{COS}\,(G)$ is the sequence $\{\alpha^n(U)\}_{n=0}^{\infty}$. An automorphism α on G *moves towards infinity* if for any pair $V \leq W \in \mathrm{COS}\,(G)$ there is $n \geq 1$ such that $\alpha^n(V) \not\leq W$.

It may be seen that α moves towards infinity if and only if $s(\alpha) > 1$. A pseudometric may be defined on all rays $\{\alpha^n(U)\}_{n=0}^{\infty}$ such that α moves towards infinity. Identifying rays that are distance 0 apart and completing with respect to the metric yields the *space of directions*, see [4].

Example 8 Let $G = \mathrm{Aut}(T)$, with T a regular tree. Then the space of directions is the set of ends of the tree with all distinct points being distance 2 apart.

The space of directions is computed in a number of cases in [4] and is seen to be a familiar space in many well-known examples. However, it is also seen that it can be quite complicated. An example is given in [4] where the space of directions is isometric to the set of Borel subsets on $[0, 1]$ with the metric $d(A, B) = m(A \triangle B)$, where m is the Lebesgue measure and two sets are identified if they differ on a set of measure 0. It might be of interest to know the space of directions for groups whose geometric structure is not well understood.

Question 6 What is the space of directions for Neretin's group?

The Chabauty Space

Definition 11 Let G be a locally compact group. The *Chabauty space*, $\mathrm{SUB}(G)$, is the set of all closed subgroups of G equipped with the topology generated by the subsets

$$\mathcal{N}_{K,O}(C) = \{D \in \mathrm{SUB}(G) \mid D \cap K \subset CO, \ C \cap K \subset DO\},$$

where $K \subset G$ is compact and $O \subset G$ is a neighbourhood of 1.

The set $\mathrm{SUB}(G)$ is a compact topological space and the action of $\mathsf{Aut}(G)$ on G induces a natural action on $\mathrm{SUB}(G)$ by homeomorphisms.

We use the following lemma in the proof of the next theorem.

Lemma 5 *Let O be a neighbourhood of 1. Let $\alpha \in \mathsf{Aut}(G)$, suppose that U is tidy for α and let $U_0 = U_+ \cap U_- = \bigcap_{j \in \mathbb{Z}} \alpha^j(U)$ be the largest α-invariant subgroup of U. Then there is a non-negative integer N such that $\alpha^n(U_-) \subset U_0 O$ for all natural numbers $n \geq N$.*

Proof The sequence $\{\alpha^n(U_-)\}_{n \geq 0}$ of compact subgroups of U_- decreases U_0. Since $U_0 O$ is an open neighbourhood of U_0, there is N such that $\alpha^N(U_-) \subset U_0 O$. $\quad\square$

Proposition 3 *Let $\alpha \in \mathsf{Aut}(G)$ and suppose that $U \in \mathrm{COS}\,(G) \subset \mathrm{SUB}(G)$ is tidy for α. Then $\alpha^n(U) \to U_{++}$ with respect to the Chabauty topology in $\mathrm{SUB}(G)$.*

Proof Let $K \subset G$ be compact and let $O \subset G$ be an open neighbourhood of 1. Consider $\mathcal{N}_{K,O}(U_{++})$. We will find a natural number M such that $\alpha^n(U) \in \mathcal{N}_{K,O}(U_{++})$ whenever $n \geq M$.

Now, $K \cap U_{++}$ is a compact subgroup of U_{++}. We have $U_{++} = \bigcup_{k \geq 0} \alpha^k(U_+)$ and U_+ is relatively open in U_{++}. Hence there is a k such that $\alpha^k(U_+) \geq K \cap U_{++}$. Hence,

$$K \cap U_{++} \subset \alpha^k(U_+) \subset \alpha^n(U) \text{ for all } n \geq k. \tag{5}$$

Choose N as in Lemma 5, so that $\alpha^n(U_-) \subset U_0 O$ for every $n \geq N$. Then we have

$$n \geq N \implies \alpha^n(U) = \alpha^n(U_+)\alpha^n(U_-) \tag{6}$$
$$\subset \alpha^n(U_+)O \subset U_{++}O.$$

Equations (5) and (6) imply that, if $n \geq \max(k, M)$, then

$$K \cap U_{++} \subset \alpha^n(U)O \text{ and } K \cap \alpha^n(U) \subset U_{++}O,$$

i.e. $\alpha^n(U) \in \mathcal{N}_{K,O}(U_{++})$ and so $\alpha^n(U) \to U_{++}$ as $n \to \infty$. $\quad\square$

Acknowledgements We would like to thank Stephan Tornier, who prepared some of the figures used in these lecture notes.

References

1. Abels, H.: Kompakt definierbare topologische Gruppen. Math. Ann. **197**, 221–233 (1972)
2. Abramenko, P., Brown, K.S.: Buildings: Theory and Applications. Springer Graduate Texts in Mathematics, vol. 248. Springer, New York (2008)
3. Baumgartner, U., Willis, G.A.: Contraction groups for automorphisms of totally disconnected groups. Isr. J. Math. **142**, 221–248 (2004)
4. Baumgartner, U., Willis, G.A.: The direction of an automorphism of a totally disconnected locally compact group. Math. Z. **252**, 393–428 (2006)
5. Baumgartner, U., Rémy, B., Willis, G.A.: Flat rank of automorphism groups of buildings. Transformation Groups **12**, 413–436 (2007)
6. Baumgartner, U., Schlichting, G., Willis, G.A.: Geometric characterization of flat groups of automorphisms. Groups Geom. Dyn. **4**, 1–13 (2010)
7. Bergman, G.M., Lenstra, Jr., H.W.: Subgroups close to normal subgroups. J. Algebra **127**(1), 80–97 (1989)
8. Bourbaki, N.: Commutative Algebra. Springer, New York (1989)
9. Bruhat, F., Tits, J.: Groupes réductifs sur un corps local, I. Données radicielles values. Publ. Math. IHES **41**, 5–251 (1972)
10. Glöckner, H., Willis, G.A.: Classification of the simple factors appearing in composition series of totally disconnected contraction groups. J. Reine Angew. Math. **643**, 141–169 (2010)
11. Helgason, S.: Differential Geometry, Lie Groups and Symmetric Spaces. Academic, New York (1978)
12. Hewitt, E., Ross, K.A.: Abstract Harmonic Analysis. Springer Science and Business Media, Alemania (1994)
13. Jaworski, W.: On contraction groups of automorphisms of totally dis-connected locally compact groups. Isr. J. Math. **172**, 1–8 (2009)
14. Kechris, A.S.: Classical Descriptive Set Theory. Springer, New York (1995)
15. Krön, B., Möller, R.G.: Analogues of Cayley graphs for topological groups. Math. Z. **258**, 637–675 (2008)
16. Möller, R.G.: Structure theory of totally disconnected locally compact groups via graphs and permutations. Can. J. Math. **54**, 795–827 (2002)
17. Monod, N.: Continuous bounded cohomology of locally compact groups. Lecture Notes in Mathematics, vol. 1758. Springer, Berlin (2001)
18. Schlichting, G.: Operationen mit periodischen Stabilisatoren. Arch. Math. **34**(2), 97–99 (1980)
19. Shalom, Y., Willis, G.A.: Commensurated subgroups of arithmetic groups, totally disconnected groups and adelic rigidity. Geom. Funct. Anal. **23**, 1631–1683 (2013)
20. Stroppel, M.: Locally Compact Groups. European Mathematical Society, Zürich (2006)
21. Tao, T.: Hilbert's Fifth Problem and Related Topics. American Mathematical Society, Providence (2014)
22. Tits, J.: Groupes associés aux algèbres de Kac-Moody, Astérisque, no. 177–178, Exp. No. 700, 7–31; Séminaire Bourbaki, vol. 1988/89
23. Willis, G.A.: The structure of totally disconnected, locally compact groups. Math. Ann. **300**, 341–363 (1994)
24. Willis, G.A.: Further properties of the scale function on totally disconnected groups. J. Algebra **237**, 142–164 (2001)
25. Willis, G.A.: Tidy subgroups for commuting automorphisms of totally disconnected groups: an analogue of simultaneous triangularisation of matrices. N. Y. J. Math. **10**, 1–35 (2004). Available at http://nyjm.albany.edu:8000/j/2004/Vol10.htm
26. Willis, G.A.: The nub of an automorphism of a totally disconnected, locally compact group. Ergodic Theory Dyn. Syst. **34**, 1365–1394 (2014)
27. Willis, G.A.: The scale and tidy subgroups for endomorphisms of totally disconnected locally compact groups. Math. Ann. **361**, 403–442 (2015)

Introduction to Quantum Invariants of Knots

Roland van der Veen

Abstract By introducing a generalized notion of tangles we show how the algebra behind quantum knot invariants comes out naturally. Concrete examples involving finite groups and Jones polynomials are treated, as well as some of the most challenging conjectures in the area. Finally the reader is invited to design his own invariants using the Drinfeld double construction.

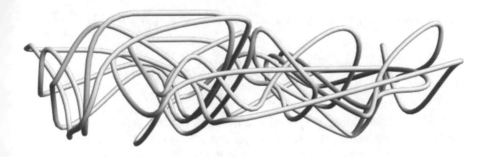

1   Introduction

The purpose of these three lectures is to explain some of the topological motivation behind quantum invariants such as the colored Jones polynomial. In the first lecture we introduce a generalized notion of knots whose topology captures the ribbon Hopf algebra structure that is central to quantum invariants. The second lecture actually defines such algebras and shows how to obtain quantum invariants from them. As an

R. van der Veen (✉)
Leiden University, Leiden, Netherlands
e-mail: r.i.van.der.veen@math.leidenuniv.nl

© Springer International Publishing AG, part of Springer Nature 2018   637
D.R. Wood et al. (eds.), *2016 MATRIX Annals*, MATRIX Book Series 1,
https://doi.org/10.1007/978-3-319-72299-3_27

example we discuss the colored Jones polynomial and present a couple of intriguing conjectures related to it. The final lecture is about constructing new examples of such algebras and invariants using the Drinfeld double.

Although the prerequisites for these lectures are low the reader will probably appreciate the lectures most after having studied some elementary knot theory. Knot diagrams and Reidemeister moves up to the skein-relation definition of the Jones polynomial should be sufficient. Beyond that a basic understanding of the tensor product is useful. Finally if you ever wondered why and how things like the quantum group $U_q\mathfrak{sl}_2$ arise in knot theory then these lectures may be helpful.

To illustrate the nature of the algebras at hand recall that the colored Jones polynomial is closely related to $U_q\mathfrak{sl}_2$, the quantized enveloping algebra of $\mathfrak{sl}_2$. Following [8] $U_q\mathfrak{sl}_2$ is the algebra generated by $1, E, F, H$ defined by the following operations and relations. Our purpose is to demystify and motivate such constructions from a topological/knot theoretical viewpoint.

$$HE - EH = 2E \quad \Delta(E) = E \otimes q^{\frac{H}{2}} + 1 \otimes E \quad S(E) = -Eq^{-\frac{H}{2}} \quad \epsilon(E) = 0$$

$$HF - FH = -2F \quad \Delta(F) = F \otimes 1 + q^{-\frac{H}{2}} \otimes F \quad S(F) = -q^{\frac{H}{2}}F \quad \epsilon(F) = 0$$

$$EF - FE = [H] \quad \Delta(H) = H \otimes 1 + 1 \otimes H \quad S(H) = -H \quad \epsilon(H) = 0 \quad [x] = \frac{q^{\frac{x}{2}} - q^{-\frac{x}{2}}}{q^{\frac{1}{2}} - q^{-\frac{1}{2}}}$$

$$R = q^{\frac{H \otimes H}{4}} \sum_{n=0}^{\infty} \frac{q^{n(n-1)/4}}{[n]!} (q^{\frac{1}{2}} - q^{-\frac{1}{2}})^n (E^n \otimes F^n) \quad \alpha = q^{\frac{H}{2}}$$

The N-colored Jones polynomial arises out of this algebra using the N-dimensional irreducible representation $\rho_N : U_q\mathfrak{sl}_2 \to Mat(N \times N)$ defined by the matrices. For a more detailed description see lecture 2.

$$\rho_N(E) = \begin{pmatrix} 0 & [N-1] & 0 & 0 & 0 & \dots \\ 0 & 0 & [N-2] & 0 & \dots & 0 \\ 0 & 0 & 0 & \dots & 0 & 0 \\ 0 & 0 & \dots & & 0 & [2] & 0 \\ 0 & \dots & & 0 & 0 & 0 & [1] \\ \dots & 0 & & 0 & 0 & 0 & 0 \end{pmatrix}$$

$$\rho_N(F) = \begin{pmatrix} 0 & 0 & 0 & 0 & 0 & \dots \\ [1] & 0 & 0 & 0 & \dots & 0 \\ 0 & [2] & 0 & \dots & 0 & 0 \\ 0 & 0 & \dots & 0 & 0 & 0 \\ 0 & \dots & 0 & [N-2] & 0 & 0 \\ \dots & 0 & 0 & 0 & [N-1] & 0 \end{pmatrix}$$

$$\rho_N(H) = \begin{pmatrix} (n-1)/2 & 0 & 0 & 0 & 0 & \cdots \\ 0 & (n-3)/2 & 0 & 0 & \cdots & 0 \\ 0 & 0 & (n-5)/2 & \cdots & 0 & 0 \\ 0 & 0 & \cdots & -(n-5)/2 & 0 & 0 \\ 0 & \cdots & 0 & 0 & -(n-3)/2 & 0 \\ \cdots & 0 & 0 & 0 & 0 & -(n-1)/2 \end{pmatrix}$$

We will explain how the topology of knots or rather tangles leads one to consider such algebraic structures, known as ribbon Hopf algebras. Next we will show how one can design ribbon Hopf algebras at will using Drinfeld's double construction. Again emphasizing that the double construction too is directly forced on us by the topological problems we want to solve: i.e. finding invariants of knots.

Our exposition is meant to complement the literature rather than being exhaustive. We chose not to say anything about quantum field theory, which of course is one of the main driving forces of the subject. Our main sources are Ohtsuki [8], Etingof and Schiffmann [1] and Kauffman [6]. Much of this work is inspired by conversations with Dror Bar-Natan.

2 Lecture 1: Tangles as a Ribbon Hopf Algebra

The goal of this lecture is to present tangles in a rather non-standard way that will allow us to define many algebraic operations on them. These operations turn out to be the operations that one can do on the tensor algebra of a ribbon Hopf algebra such as $U_q\mathfrak{sl}_2$. To find out what that algebra should do we turn to topology to watch and listen what the tangles have to tell us.

2.1 rv-Tangles

Loosley following Kauffman [6] and Bar-Natan we work with a version of rotational virtual tangles, abbreviation rv-tangles. To make sure the algebra comes out unmangled our set up is rather abstract. Instead of relying on diagrams in the plane[1] we prefer working with ribbon graphs with some decorations modulo the usual Reidemeister relations viewed locally. Recall that a ribbon graph is a graph together with a cyclic orientation on the half-edges around each vertex.

Definition 1 An rv-ribbon graph is a ribbon graph G with a labelling of both vertices and edges by integers satisfying the following requirements. The degree one vertices (called ends) are required to come in pairs one labelled $+i$ and one

---

[1]This is the main difference with Kauffman's approach.

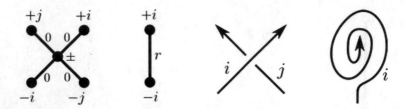

Fig. 1 (Left) The two fundamental rv-ribbon graphs from which all rv-tangles are built by disjoint union and multiplication. The first is the $\pm$ crossing $X_{ij}^{\pm}$ and the second is an edge α_i^r with rotation number r. (Right) the usual way of depicting the fundamental graphs in the plane, here the rotation number r is 2

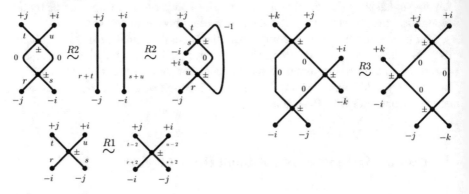

Fig. 2 The equivalence relations (Reidemeister moves)

$-i < 0$, the set of absolute values of end labels is denoted I_G. Each internal vertex is labelled ± 1.

Two key examples of rv-ribbon graphs are the $\pm$ crossing and the edge shown in Fig. 1 (left). The two figures on the right show the interpretation in terms of planar diagrams of knots we have in mind. We think of the edge labels as rotation numbers, often arising from taking a braid closure (rotation number ± 1). Edge label 0 will often be omitted for clarity. We define two operations on rv-ribbon graphs, disjoint union and multiplication. Together they suffice to build any graph we need. Disjoint union is simply disjoint union of graphs, where we assume the labels of the ends are all distinct. Multiplication is more interesting (Fig. 2).

Definition 2 For $i, j \in I_G$ and $k \notin I_G$ define $m_k^{ij}(G)$ to be the rv-ribbon graph obtained from rv-ribbon graph G by merging the edge that ends in $+i$ with the edge that ends on $-j$. The edge label for the new edge is the sum of the labels of the merged edges and the remaining ends $-i, +j$ are renamed $-k, +k$ respectively.

With these definitions in place we can turn to the tangles we are interested in. See Fig. 3 below to see the multiplication in action to build a diagram of a knot.

Definition 3 An rv-tangle is an rv-ribbon graph obtained from multiplying finitely many crossings and edges as in Fig. 1. rv-tangles are considered up to the equiva-

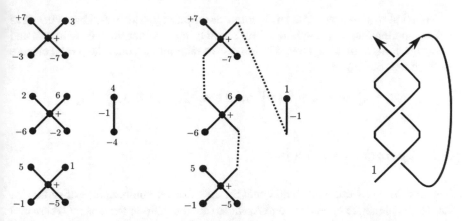

Fig. 3 (Right) The trefoil knot as a long knot (one component usual tangle). (Left) The seven fundamental rv tangles that can be assembled to produce the rv-tangle corresponding to the usual trefoil on the right. (Middle) An intermediate stage where we already multiplied ends $+1$ with -2 calling everything 1, then multiplied with component 3 and then with 4. The newly made connections are dotted and are abstract (not in the plane!)

lences $R0, R1, R2, R3$ as shown below. $R0$ is relabeling of the ends (multiplication by an edge labelled 0).

Larger tangles are understood to be equivalent if they contain equivalent factors. rv-tangles are meant as a language for dealing with diagrams of ordinary knots and tangles more efficiently. To interpret rv-tangles we should view the four-valent vertices as crossings, with sign as indicated. The edges are a disjoint union of straight paths in the graph that go from end $-i$ to $+i$ and are directed this way. These straight paths are the components of the tangle, by straight we mean that it takes the second right (straight on a roundabout) at every crossing. The integers on the edges are supposed to represent the rotation number of a tangent vector as it runs from one vertex to the next. Not all rv-tangles correspond to usual tangle diagrams as the crossings may be connected in ways that are impossible in the plane. However all usual tangle diagrams without closed components are included. In particular knots can be studied as one component tangles. Also, two usual tangles are equivalent (regular isotopic) if and only if the corresponding rv-tangles are.

Theorem 1 (Kauffman, Bar-Natan) *Two usual oriented tangles without closed components are regular isotopic if and only if the corresponding rv-tangles are equivalent.*

We emphasize that in our set up the crossings in an rv-tangle are connected *abstractly*, not necessarily in the plane. This is like Kauffman's rotational virtual tangles [6] but without the need to explicitly discuss 'virtual crossings'. Even if the reader is only interested in usual tangles in the plane, the language of rv-tangles is still an elegant and effective way to encode such. As an additional bonus we will see that it brings out the algebra very naturally.

As a first example we write an algebraic description for the trefoil knot viewed as a 1-component tangle (long knot) T by multiplying together three crossings X and one edge containing a negative rotation α^{-1} to take into account the partial closure of the braid. See Fig. 3.

$$T = m_1^{17} \circ m_1^{16} \circ m_1^{15} \circ m_1^{14} \circ m_1^{13} \circ m_1^{12}(X_{15}^+ \sqcup X_{62}^+ \sqcup X_{37}^+ \sqcup \alpha_4^{-1})$$

2.2 Operations on Tangles

The goal of this section is to show that the set of linear combinations of rv-tangles has all the algebraic operations and relations that are valid in the tensor algebra of a ribbon Hopf algebra. Instead of defining ribbon Hopf algebras we will dive right in and list two operations and some natural relations between them. The names of the relations reflect the algebraic structure intended.

The easiest one is tensor product of two rv-tangles, it is just another name for disjoint union considered above. We also have already seen multiplication $m_k^{i,j}$. Both these operations satisfy a form of associativity, let's write out what that means for m. Given three components labelled i, j, k it does not matter whether we first connect i to j and the result to k or first connect the j to k and then connect i to the result. In formulas: $m_x^{r,k} \circ m_r^{i,j} = m_x^{i,r} \circ m_r^{j,k}$. Here we called the intermediate result r and the end result x.

Next there is also the unit operation η_i which is disjoint union with a new edge labelled 0 with ends labelled $\pm i$ assuming the label i had not been used before. Dually there is a co-unit operation ϵ_i that deletes the component i. More interestingly there is the co-multiplication $\Delta_{j,k}^i$ that takes component i and doubles it. By this we mean it replaces component i with two new components j, k running parallel to i (with the same rotation numbers on parallel edges). Finally there is the antipode operation S_i which roughly speaking reverses the orientation of all the arrows on component i.

To give precise definitions of the operations mentioned we show explicitly what they do to the generators and extend them multiplicatively, see Fig. 4.

By multiplicativity we mean the following. For the co-unit it means $\epsilon_i(m_i^{ab}) = \epsilon_b \sqcup \epsilon_a$. For the co-product it means that first multiplying two components i, j calling the result k and then doubling that component calling the results x, y is the same as first doubling i and j calling the results i', i'' and j', j'' and then multiplying i', j' and i'', j'' calling the results x and y. In formulas $\Delta_{x,y}^k \circ m_k^{i,j} = m_y^{i'',j''} \circ m_x^{i',j'} \circ \Delta_{j',j''}^j \circ \Delta_{i',i''}^i$. The algebra looks complicated but the pictures are really simple!

Multiplicativity for the antipode S_i is actually anti-multiplicativity, because if we reverse a component built out of many segments, the order of the segments gets reversed! $S_k \circ m_k^{i,j} = m_k^{j,i} \circ S_j \circ S_i$.

The operations listed above are precisely those of a Hopf algebra and the following relations hold between them. First there is co-associativity: it does not matter if you split component x calling the results i, r and then split r into

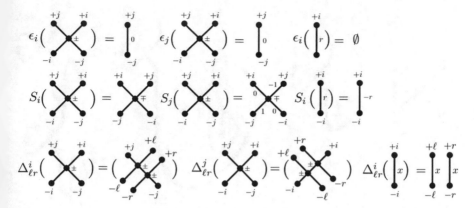

Fig. 4 The operations on the generators

components j, k or do it the other way around: first split x into r, k and then split r into i, j. In formulas $\Delta^r_{j,k} \circ \Delta^x_{i,r} = \Delta^r_{i,j} \circ \Delta^x_{r,k}$. Be careful that we really keep track of the order of the two components coming out of Δ.

Even more striking is the following relation between all the operations we have: Take a component i double it to get components j, k, reverse k and multiply j with k, what do you get? The band spanned by j, k may be retracted and all that is left is a little component, called x, without any internal vertices! This is the same as deleting component i and putting back a single edge called x with label 0. In formulas $m^{j,k}_x \circ S_k \circ \Delta^i_{j,k} = \eta_x \epsilon_i$.

We should also check that the operations described actually work on equivalence classes of rv-tangles. Doing the operation on two rv-tangles should always yield the same result (Exercise!).

The following relation is closely related to the $R3$ move. For any tangle T with component i:

$$m^{yj}_w \circ m^{xk}_z X^{\pm}_{xy} \sqcup \Delta^i_{jk}(T) = m^{jy}_w \circ m^{kx}_z X^{\pm}_{xy} \sqcup \Delta^i_{kj}(T)$$

It looks complicated but a picture makes it obvious (Exercise!).

The famous Drinfeld element is $U_k = m^{ji}_k \circ S_j(X^+_{ij})$, see Fig. 5 for a picture. For any tangle T_j it satisfies $m^{ji}_k U_j \sqcup T_i = m^{ij}_k U_j \sqcup S^2_i(T_i)$. The inverse of U_k is $U^{-1}_k = m^{ji}_k \circ S^{-1}_j(X^{-1}_{ij})$. Here inverse means that $m^{ij}_k(U^{-1}_i \sqcup U_j) = \alpha^0_k = m^{ji}_k(U^{-1}_i \sqcup U_j)$

Some more relations are $m^{jr}_x m^{ks}_y \Delta^i_{jk}(U_i) \sqcup m^{bc}_r m^{ad}_s X^+_{ab} \sqcup X^+_{cd} = U_x \sqcup U_y$ The same relation holds when we replace each U_h by $S_h(U_h)$. Also $\epsilon_j(U_j) = \alpha^0_j$. The element $W_k = m^{ij}_k S(U_i) \sqcup U_j$ commutes with everything and satisfies $S_k(W_k) = W_k$. There exists a square root V_k of W_k, this is called the ribbon element satisfying the same equations as W_k does. How does the ribbon element relate to α_k?

At this point at least some of the symbols in $U_q \mathfrak{sl}_2$ should look more familiar. In the next lectures we will focus more on the algebras and how they lead to invariants

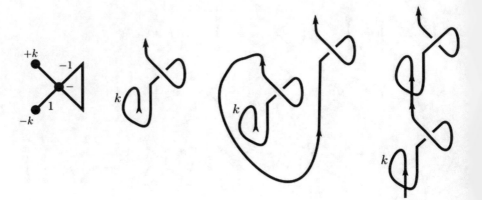

Fig. 5 The rv-graph corresponding to the Drinfeld element U_k (left). The second picture shows the same Drinfeld element as a usual tangle drawn in the plane. The third picture shows the square of the Drinfeld element $m_k^{ij} U_i \sqcup U_j$ again drawn as a usual tangle in the plane. Finally the last picture shows a more abstract version of this square where we allow ourselves a more schematic representation of the rotation numbers involved using abstract curls (not crossings!). One of the main points of rv-tangles is to not let the plane hold us back and let the algebra and topology mix freely

of tangles and knots. Looking back we emphasize that although strange looking, our presentation of knot theory is cleaner and more precise than the standard one.

2.3 Exercises

Exercise 1 Draw an rv-tangle diagram for a figure eight knot 4_1 (viewed as a long knot).

Exercise 2 Apply S_i to the top-left diagram in Fig. 2 and show you get the same as the third diagram on the same row of the figure.

Exercise 3 Draw diagrams to interpret the relations at the end of the lecture topologically.

Exercise 4 What happens to the R-matrix of $U_q \mathfrak{sl}_2$ when we set $q = 1, h = 0$? Look up what the universal enveloping algebra of a Lie algebra is. Do you recognise anything?

Exercise 5 (Skein Relation) First show that the matrix for $\rho_2(R)$ with respect to the basis $x \otimes x, x \otimes y, y \otimes x, y \otimes y$ of $\mathbb{C}^2 \otimes \mathbb{C}^2$ is

$$\rho_2(R) = \begin{pmatrix} q^{\frac{1}{4}} & 0 & 0 & 0 \\ 0 & q^{-\frac{1}{4}} & q^{\frac{1}{4}} - q^{-\frac{3}{4}} & 0 \\ 0 & 0 & q^{-\frac{1}{4}} & 0 \\ 0 & 0 & 0 & q^{\frac{1}{4}} \end{pmatrix}$$

Let P be the matrix for the linear transformation that sends $a \otimes b$ to $b \otimes a$. The skein relation is:

$$q^{\frac{1}{4}} PR - q^{-\frac{1}{4}} (PR)^{-1} = (q^{\frac{1}{2}} - q^{-\frac{1}{2}})I$$

3 Lecture 2: Quantum Invariants and Hopf Algebras

In this lecture we define the algebraic counterpart of the rv-tangles and their operations. The rough idea is to have a copy of some algebra correspond to each component of our tangle. This leads us to define quantum invariants of tangles. In particular $U_q\mathfrak{sl}_2$ is such an algebra and we will show how to obtain the colored Jones polynomial from it. Finally some of the most beautiful and challenging conjectures involving the Jones polynomial are mentioned.

Recall that an algebra is a vector space A together with a bilinear, associative multiplication map $m : A \times A \to A$. Good examples of algebras to keep in mind are the group algebra of a finite group $\mathbb{C}G$ and the universal enveloping algebra of a Lie algebra $U(\mathfrak{g})$. Elements in both algebras are defined to be formal linear combinations of products.

3.1 Quantum Knot Invariants

Given an algebra A and a set I define a bigger algebra A_I to be the algebra generated by elements a_i for $a \in A, i \in I$ such that $a_i a'_{i'} = a'_{i'} a_i$ if $i \neq i'$ and satisfy the same relations as $a, a' \in A$ would if $i = i'$. Really A_I is just a tensor product $\bigotimes_{i \in I} A_i$ where all A_i are isomorphic to A. We prefer the subscripts because they are more flexible about the ordering of the tensor factors and we can write the tensor product as a formal product. One should think of the set I as the index set I_G of some rv-ribbon graph G. For $I = \emptyset$ we define $A_I = \mathbb{C}$.

To really make the connection to the topology of the last lecture we need to define a multiplication map on A_I. For $i, j, k \notin I$ define $m_k^{i,j} : A_{I \cup \{i,j\}} \to A_{I \cup \{k\}}$ as follows. $m_k^{i,j}(x)$ is the result of moving all factors a_i in x to the left and then replacing all subscripts i, j by k. Notice that by making both subscripts i, j equal to k we are effectively multiplying the elements with subscript i with those of subscript j.

Definition 4 Suppose A is an algebra. By a quantum knot invariant Z we mean a way of assigning to each rv-tangle T an element $Z(T) \in A_{I_T}$ where I_T is the set of end-labels of T, in such a way that

$$m_k^{ij} Z(T) = Z(m_k^{ij} T) \quad Z(T \sqcup T') = Z(T)Z(T') \quad T \sim T' \Rightarrow Z(T) = Z(T')$$

Since multiplication can be done either algebraically or topologically, finding a quantum invariant comes down to finding suitable values for the fundamental rv-tangles: crossing $Z(X_{ij}^{\pm})$ and edge with rotation $Z(\alpha_i)$. By suitable we mean all the equivalences $R0 - R3$ from Fig. 2 should be satisfied. Notice that each of these becomes an explicit equation in terms of the values of the fundamental tangles. For example on of the equations implied by $R2$ is

$$m_i^{ik} \circ m_j^{jl}(X_{ij}^+ \sqcup X_{kl}^-) = \alpha_i^0 \sqcup \alpha_j^0$$

It should be noted that the usual quantum Reshetikhin-Turaev quantum invariants such as Jones, HOMFLY, Kauffman etc come from taking our notion of quantum knot invariant and composing it with a representation. However separating the representation from the invariant itself may clarify some issues. For future reference or perhaps as a definition (!) of ribbon Hopf algebra we state the following theorem [8]:

Theorem 2 *Any ribbon Hopf algebra A with R-matrix R and ribbon element α gives rise to a quantum knot invariant sending the crossing to R and the edge with a single rotation to α.*

One goal of these lectures is to introduce the Drinfeld double construction. This is a recipe for constructing a ribbon Hopf algebra and hence a knot invariant starting with a much simpler algebra, a Hopf algebra. In this way we can construct our own invariants instead of only focusing on the well known ones like the Jones polynomials.

For now let's focus on the particular algebra $A = U_q(\mathfrak{sl}_2)$, which happens to be a ribbon Hopf algebra. It yields the Jones polynomial as follows. Define Z by $Z(X_{12}^+) = R$ and $Z(\alpha) = \alpha$ referring to the formulas at the very beginning of the first lecture. Here we interpret $E^n \otimes F^n$ as $E_1^n F_2^n$ and do not worry about convergence issues.

Definition 5 If Z is the quantum invariant corresponding to the algebra $A = U_q(\mathfrak{sl}_2)$ as above then the N-colored Jones polynomial of knot K, notation $J_N(K;q)$ is defined as $\frac{1}{N}\text{Tr}\rho_N(Z(K'))$. Here ρ_N is its N-dimensional representation given in the first lecture and K' is 1-component rv-tangle whose closure is K.

To get a feel for this construction let's get our hands dirty and make an attempt to compute the 2-colored Jones polynomial of the trefoil knot T using the tangle description from the last lecture:

$$T' = m_1^{15} \circ m_5^{57} \circ m_5^{56} \circ m_1^{14} \circ m_1^{13} \circ m_1^{12}(X_{15}^+ \sqcup X_{62}^+ \sqcup X_{37}^+ \sqcup \alpha_4^{-1})$$

The formula should thus be $J_2(T;q) =$

$$\frac{1}{2}\text{Tr}\rho_2 Z(T') = \text{Tr}\rho_2\, m_1^{17} \circ m_1^{16} \circ m_1^{15} \circ m_1^{14} \circ m_1^{13} \circ m_1^{12}(Z(X_{15})Z(X_{26})Z(X_{37})Z(\alpha_4)^{-1})$$

Now the representation ρ_2 extends to tensor products as $\rho_2(a \otimes b) = \rho_2(a) \otimes \rho_2(b)$ or in other words $\rho_2(a_i b_j) = \rho_2(a)_i \rho_2(b)_j$. Recall that

$$\rho_2(E) = \begin{pmatrix} 0 & 1 \\ 0 & 0 \end{pmatrix} \quad \rho_2(F) = \begin{pmatrix} 0 & 0 \\ 1 & 0 \end{pmatrix} \quad \rho_2(H) = \begin{pmatrix} 1 & 0 \\ 0 & -1 \end{pmatrix}$$

Therefore $\rho_2(\alpha) = \rho_2(q^{H/2}) = \begin{pmatrix} q^{\frac{1}{2}} & 0 \\ 0 & q^{-\frac{1}{2}} \end{pmatrix}$ Also since $\rho_2(E)^2 = \rho_2(F)^2 = 0$ it suffices to only keep the first two terms of the complicated series for $R_{ij} = Z(X_{ij}^+)$. What remains is $R_{ij} = q^{\frac{H_i H_j}{4}}(1 + (q^{\frac{1}{2}} - q^{-\frac{1}{2}})E_i F_j)$ Before applying the multiplications we find the invariant for the four disjoint tangles to be: $Z(X_{15}^+)Z(X_{62}^+)Z(X_{37}^+)Z(\alpha_4^{-1}) =$

$$q^{\frac{H_1 H_5}{4}}(1 + vE_1 F_5)q^{\frac{H_6 H_2}{4}}(1 + vE_6 F_2)q^{\frac{H_3 H_7}{4}}(1 + vE_3 F_7)q^{-\frac{H_4}{2}} = D$$

Here we set $v = q^{\frac{1}{2}} - q^{-\frac{1}{2}}$ and only include the terms that are non-zero when applying ρ_2. The variables mostly commute because they have different subscripts (are on different components), this will change once we start multiplying (joining components). Already now we should be careful that $H_1 E_1 = E_1 H_1 = 2E_1$. All but the last multiplication are really easy. For example m_1^{12} means we should move all subscripts 1 to be left of the subscripts 2 and then change all 1 or 2 subscripts to 1. The same works for the next two m_1^{13} and m_1^{14}, we obtain:

$$W = m_5^{57} \circ m_5^{56} \circ m_1^{14} \circ m_1^{13} \circ m_1^{12}(D) =$$

$$q^{\frac{H_1 H_5}{4}}(1 + vE_1 F_5)q^{\frac{H_5 H_1}{4}}(1 + vE_5 F_1)q^{\frac{H_1 H_5}{4}}(1 + vE_1 F_5)q^{-\frac{H_1}{2}}$$

To be able to carry out the last step m_1^{15} we have to move all subscripts 5 to the right of the subscripts 1. The powers of q can be moved using the relations $q^{cH}E = q^{2c}Eq^{cH}$ and $q^{cH}F = q^{-2c}Fq^{cH}$, (Exercise!). For example $E_5 F_1 q^{\frac{H_5 H_1}{4}} = q^{\frac{H_5 H_1}{4} - \frac{H_5}{2} + \frac{H_1}{2} + 1}E_5 F_1$.

For clarity let us write out $W = \sum_{j=1}^{8} W^j$ into eight terms W^j and compute m_1^{15} for each term individually. The first term is $W^1 = q^{\frac{H_1 H_5}{4}}q^{\frac{H_5 H_1}{4}}q^{\frac{H_1 H_5}{4}}q^{-\frac{H_1}{2}}$. Since only H is involved we may move the first term to the far right without cost and then set H_5 to H_1 to get $m_1^{15}W^1 = q^{\frac{3H_1^2}{4} - \frac{H_1}{2}}$. Taking the trace in the ρ_2 representation our term gives

$$\text{Tr}\rho_2(m_1^{15}W^1) = (1 + q)q^{\frac{1}{4}}$$

Next we work with $W^2 = q^{\frac{H_1 H_5}{4}} v E_1 F_5 q^{\frac{H_5 H_1}{4}} q^{\frac{H_1 H_5}{4}} q^{-\frac{H_1}{2}}$. It suffices to bring the q-powers to the middle, the E_1, F_1 to the left and the E_5, F_5 to the right. We find

$$\mathrm{Tr}\rho_2 m_1^{15} W^2 = \mathrm{Tr}\rho_2 v E_1 q^{\frac{H_1 H_5}{4} + \frac{H_5}{2}} q^{\frac{H_5 H_1}{2} + H_1} q^{-\frac{H_1}{2}} F_5|_{1=5}$$

$$= \mathrm{Tr}\rho_2 v E_1 q^{\frac{H_1^2}{4} + H_1} F_1 = v q^{\frac{3}{4}} (q + q^{-1})$$

Applying $\mathrm{Tr}\rho_2 m_1^{15}$ to the remaining six terms $W^3, \ldots W^8$ and summing should yield the 2-colored Jones polynomial of the trefoil knot, i.e the ordinary Jones polynomial. There are easier ways of getting the same result but those tend to hide what is going on, making them of less use for serious applications such as the ones below. The technique illustrated here can be carried out even for Z itself without applying any representation.

3.2   Conjectures on the Colored Jones Polynomial

We pause our account of quantum invariants to illustrate the depth and lure of the subject by stating a few famous conjectures on the asymptotics of the colored Jones polynomial: The modularity conjecture, the AJ-conjecture and the slope conjecture. Each connects the Jones polynomial to an apparently completely different field. There may be more natural perspectives on these conjectures from quantum field theory but our purpose here is mainly to state some challenging problems in a concise way.

Recall our notation for the N-colored Jones polynomial of knot K is $J_N(K; q)$. It is always a Laurent polynomial in $q^{1/2}$.

The Modularity Conjecture is a radical generalization of the volume conjecture [5] connecting the Jones polynomial to hyperbolic geometry. Or perhaps more fundamentally to $SL(2, \mathbb{C})$ Chern-Simons theory. We take for granted the amazing facts that many knots allow a unique hyperbolic (finite volume complete) metric on their complement in the three-sphere. Such knots are called hyperbolic. By uniqueness (Mostow Rigidity) any property of the metric is a topological invariant of our knot K. Denote by Vol_K the volume of the complement with respect to that metric. Also denote the field generated by the traces of the holonomy representation of the knot group into $PSL(2, \mathbb{C}) = Isom^+(\mathbb{H}^3)$ by F_K. It is known as the trace field of K and is always a number field.

Modularity Conjecture [9]   Consider a hyperbolic knot K. Define $J : \mathbb{Q}/\mathbb{Z} \to \mathbb{C}$ by $J(\frac{r}{s}) = J_s(K; e^{\frac{2\pi}{s}})$. For any $a, b, c, d \in \mathbb{Z}$ with $ad - bc = 1$ and any sequence $(X_n) \in \mathbb{Q}$ going to infinity with bounded denominators, there exist $\Delta(\frac{a}{c}), A_j(\frac{a}{c}) \in \mathbb{C}$

such that

$$\frac{J(\frac{aX_n+b}{cX_n+d})}{J(X_n)} \sim_{n\to\infty} \left(\frac{2\pi}{h_n}\right)^{\frac{3}{2}} e^{\frac{\text{Vol}_K}{h_n}} \sum_{j=0}^{\infty} A_j \left(\frac{a}{c}\right) h_n^j$$

where $h_n = \frac{2\pi}{X_n + \frac{d}{c}}$ and $A_j(\frac{a}{c})$, $\Delta^{2c}(\frac{a}{c}) \in F_K(e^{\frac{2\pi a}{c}})$.

Another interesting conjecture is the AJ-conjecture. Define the function $J.(K;q)$: $\mathbb{N} \to \mathbb{Z}[q^{\frac{1}{2}}, q^{-\frac{1}{2}}]$ by sending N to $J_N(K;q)$. It was shown that $J.(K;q)$ satisfies a q-difference equation (recursion) in the following sense. Define operators $\hat{M}$ and $\hat{L}$ on functions on $\mathbb{N}$ by $(\hat{L}f)(N) = f(N+1)$ and $(\hat{M}f)(N) = q^N f(N)$. Then there is a non-commutative polynomial $\hat{A}(\hat{M}, \hat{L}, q)$ in $\hat{M}, \hat{L}$ with coefficients in $\mathbb{Z}[q^{\frac{1}{2}}, q^{-\frac{1}{2}}]$ such that $\hat{A}J.(K;q) = 0$ as a function. Up to some unimportant factors this polynomial $\hat{A}$ is unique so the following conjecture makes sense:

AJ Conjecture [2, 3] Setting $q = 1$ in $\hat{A}$ yields the $SL(2, \mathbb{C})$ A-polynomial of the knot K.

Roughly speaking the A-polynomial is a plane curve in $\mathbb{C}^2$ specifying which values M, L for eigenvalues of the peripheral subgroup of the knot group can be extended to a representation of the knot group into $SL(2, \mathbb{C})$.

Finally the slope conjecture makes a connection with an apparently different field of low-dimensional topology: essential orientable surfaces in the knot complement. Viewed in the knot exterior an essential surface Σ may end on the boundary of the knot in a certain homology class $a\mu + b\lambda$. Note that the surface may intersect the torus boundary in several disjoint components, it does not have to be a Seifert surface. In that case we say that Σ has slope $\frac{a}{b}$. According to the slope conjecture some slopes of surfaces are detected by the degree of the colored Jones polynomial. More precisely it is known that for sufficiently large N there is a $p \in \mathbb{N}$ and quadratic polynomials $Q_0, \ldots, Q_{p-1}$ (all dependent on the knot) such that for all $0 \le r \le p-1$ we have $\deg J_N(K;q) = Q_r(N)$ whenever $N = r \mod p$.

Slope Conjecture [4]
For any knot K the leading coefficient of Q_r is the slope of an essential surface in the complement of K.

3.3 Exercises

Exercise 1 Let A be the algebra of complex valued 2×2 matrices, $\text{End}(C^2)$. Write out all the equations one needs to solve for a 4×4 matrix $Z(X_{ij})$ and a 2×2 matrix $Z(\alpha_i)$ to obtain a quantum invariant in A.

Exercise 2 Compute the 2-colored Jones polynomial of the trefoil using the formulas at the beginning of lecture 1 and the topological description of the trefoil.

Exercise 3 The universal enveloping algebra of a Lie algebra $\mathfrak{g}$ is defined to be the vector space spanned by formal non-commutative products of Lie algebra elements modulo the relations $[X, Y] = XY - YX$ for any $X, Y \in \mathfrak{g}$. What operations make the universal enveloping algebra into a Hopf algebra?

Exercise 4 Prove the identities $q^{cH}E = q^{2c}Eq^{cH}$ and $q^{cH}F = q^{-2c}Fq^{cH}$.

Exercise 5 Find out how the volume conjecture is a special case of the modularity conjecture.

4 Lecture 3: Drinfeld Double

The goal of the last lecture is to show how new quantum invariants may be constructed from Hopf algebras. Hopefully this will inspire the reader to look for interesting and useful invariants beyond the usual ones. We will illustrate the technique by working out the case of the group algebra of a finite group $\mathbb{C}G$ carefully. The construction is known as the Drinfeld double construction. It may seem foreign at first but is actually very natural in the sense that its algebraic structure is forced on us by topology. Not the other way around as it often appears.

4.1 Hopf Algebras

An important example of a Hopf algebra is the group algebra of a finite group $\mathbb{C}G$ and another example is its dual, the functions on G, say $Fun(G)$. One instance of the Drinfeld double is $D(G) = \mathbb{C}C \rtimes Fun(G)$. In the final lecture we will see that the resulting quantum invariant counts the number of representations of the fundamental group of the knot complement into G. If one hopes to understand invariants like colored Jones that relate to representations of the fundamental group into $G = SL(2, \mathbb{C})$ it is a good idea to first understand similar invariants for G finite.

A definition of a Hopf algebra is given below. Notice that in the context of tangles we also had multiplication m, a unit component η, doubling of a component Δ, reversal S of a component and deletion of a component ϵ. They satisfied certain natural relations and those are precisely the axioms for Hopf algebras. Notice however that no notion of crossing is present here.

Definition 6 A Hopf algebra is an algebra H together with for any set I algebra morphisms $\epsilon_i : H_{I \sqcup \{i\}} \to H_I$ and $\Delta^i_{jk} : H_I \to H_{I \sqcup \{j,k\}}$, and an anti-algebra morphism $S : H_I \to H_I$. Satisfying the following axioms:

a. $\Delta^i_{ij} \circ \Delta^i_{ik} = \Delta^k_{jk} \circ \Delta^i_{ik}$
b. $\epsilon_i \circ \Delta^i_{ij} = \epsilon_j \circ \Delta^i_{ij} = \mathrm{id}_i$
c. $m^{ij}_i \circ S_i \circ \Delta^i_{ij} = m^{ij}_i \circ S_j \circ \Delta^i_{ij} = 1_i \epsilon_i$

In our running example of the group algebra (which is the vector space with basis the group elements), the multiplication is multiplication in the group extended linearly to the whole space. The unit is the unit in the group and $\Delta^i_{jk}(g_i) = g_j g_k$ for all $g \in G$. Again this definition is extended linearly to the whole of $\mathbb{C}G$. The co-unit is defined for $g \in G$ by $\epsilon_i(g_i) = 0$ if $g \neq 1$ and $g_i(1_i) = 1$. The antipode S_i is defined by $S_i(g_i) = g_i^{-1}$. When $i \neq z$ we set $\Delta^i_{jk}(g_z) = g_z$ and $\epsilon_i(g_z) = g_z$ and $S_i(g_z) = g_z$. In this case all the axioms listed are easy to check (Exercise!).

The complex valued functions on G, with pointwise multiplication also form a Hopf algebra called Fun(G) (Exercise!). The Hopf algebra structure can be described conveniently in terms of the basis of delta functions. For each $g \in G$ define the delta function $\delta^g \in Fun(G)$ by $\delta^g(h) = 0$ if $g \neq h$ and 1 if $g = h$. The co-unit is defined by $\epsilon_i(\delta^g_i) = \delta^g(1)$. The co-product is defined by $\Delta^i_{jk}(\delta^g_i) = \sum_{g=ab} \delta^a_j \delta^b_k$ and $S_i(\delta^g_i) = \delta^{g^{-1}}_i$. As before, when $i \neq z$ the functions Δ^i_{jk} and ϵ_i and S_i send δ^g_z to itself and are extended linearly.

Alternatively we can describe Fun(G) as the dual space $\mathbb{C}G^*$. The basis dual to the basis $\{g\}_{g \in G}$ of $\mathbb{C}G$ is the basis $\{\delta^g\}_{g \in G}$ of delta functions. The co-multiplication in Fun(G) is just the transpose of the multiplication in $\mathbb{C}G$ and the dual of the multiplication in Fun(G) is the comultiplication in $\mathbb{C}G$.

The group algebra itself is a little too simple to accommodate our rv-tangle language. In particular there is no natural candidate element in $\mathbb{C}G_{\{12\}}$ for the crossing X_{12} to map to. In the final lecture we will combine $\mathbb{C}G$ with its dual Fun(G) to make a bigger algebra where we can represent crossings and all the other properties of rv-tangles. This construction works for any Hopf algebra and is known as the Drinfeld double construction. The more involved quantum group $U_q\mathfrak{sl}_2$ also comes out of this construction in a natural way. Perhaps more importantly it allows you to design your own knot invariant!

4.2 Drinfeld Double

Before introducing the Drinfeld double construction let us recall two crucial properties of rv-tangles. First and foremost there is a notion of crossing, the fundamental tangle $X^{\pm}_{ij}$ satisfying a couple of natural algebraic properties. First we know what happens when we double one of the components, this was included in our definition of Δ^i_{jk}. In formulas (draw the pictures!)

$$\Delta^i_{xy} X^+_{ij} = m^{jz}_j X^+_{y,j} \sqcup X^+_{xz}$$

$$\Delta^j_{xy} X^+_{ij} = m^{iz}_i X^+_{i,x} \sqcup X^+_{zy} \tag{1}$$

$$m^{yj}_w \circ m^{xk}_z X^{\pm}_{xy} \sqcup \Delta^i_{jk}(T) = m^{jy}_w \circ m^{kx}_z X^{\pm}_{xy} \sqcup \Delta^i_{kj}(T)$$

The last line is not included in the definition of Δ but is a rather simple compatibility between the crossing and the doubling of a component. A direct algebraic consequence of these three relations is the Reidemeister $R3$ relation or Yang-Baxter equation (Exercise!)

$$m_i^{i,x}m_j^{j,y}m_k^{k,z}X_{ij}^+ \sqcup X_{xk}^+ \sqcup X_{yz}^+ = m_k^{k,x}m_j^{j,z}m_i^{i,y}X_{jk}^+ \sqcup X_{ix}^+ \sqcup X_{yz}^+ \tag{2}$$

To find a knot invariant Z we start with an unknown algebra A and assume Z already intertwines $\sqcup$ and m_k^{ij}. Applying Z to the above equations then yields algebraic equations we would like to solve.

The main idea of the Drinfeld double construction is to start with Eqs. (1) and (2) and a candidate solution and *find the algebra in which that candidate solution actually solves the equation*. In this way we really let the topology decide what the algebra should be and make sure the answer $R_{ij} = Z(X_{ij}^+)$ is nice to begin with.

Drinfeld's idea is to start with any Hopf algebra H and form $D(H) = H^* \otimes H$.

Definition 7 Define the Drinfeld Double of a Hopf algebra H to be the vector space $H^* \otimes H$ with the following properties: Writing elements $\phi \otimes h \in D(H)$ as ϕh we assume that the Hopf algebra rules from H or H^* are still valid when either $\psi = 1$ or $h = 1$. Define a coproduct and counit as follows:

$$\Delta_{jk}^i(\psi_i h_i) = \Delta_{jk}^i(\psi_i)\Delta_{jk}^i(h_i) \quad \epsilon_i(\psi_i h_i) = \epsilon_i(\psi_i)\epsilon_i(h_i)$$

The book [1] is a useful reference for the following theorem.

Theorem 3 Let $\{h^n\}$ be a basis for H and $\{\phi^n\}$ the dual basis of H^*. If $R_{ij} = \sum_n \phi_i^n h_j^n \in D(H)_{\{ij\}}$ satisfies Eq. (1) then the multiplication in $D(H)$ **must** be defined as follows:

$$\phi h \psi g = \sum_{n,m}(\psi_1^{1,n}(S_1 h_1^{1,m})\psi_3^{3,n}(h_3^{3,m}))\phi\psi^{2,n}h^{2,m}g$$

where $\Delta_{23}^2\Delta_{12}^1(x_1) = \sum_n x_1^{1,n}x_2^{2,n}x_3^{3,n}$ for any x. Also the antipode must be $S(\phi h) = S(h)S^{-1}(\phi)$

Let us see what the Drinfeld double $D(\mathbb{C}G)$ of the group algebra is. It is the vector space $\text{Fun}(G) \otimes \mathbb{C}G$. Elements in this space will be written as sums of formal products ϕg where $\phi \in \text{Fun}(G)$ and $g \in G$. To write down the product rule explicitly we first compute $\Delta_{23}^2\Delta_{12}^1(g_1) = g_1 g_2 g_3$ and $\Delta_{23}^2\Delta_{12}^1(\delta_1^a) = \sum_{rst=a}\delta_1^r\delta_2^s\delta_3^t$. Since $S(g) = g^{-1}$ the product rule becomes

$$\delta^a h \delta^b g = \sum_{rst=b}(\delta^r(h^{-1})\delta^t(h))\delta^a\delta^s hg = \delta^a\delta^{hbh^{-1}}hg$$

because we can solve $s = r^{-1}bt^{-1}$ and the delta functions tell us that $r = h^{-1}$ and $t = h$.

As an illustration of the theorem we derive the above product rule directly from imposing the Yang-Baxter relation. Write $R_{ij} = \sum_{g \in G} \delta_i^g g_j$ then Eq. (2) reads:

$$\sum_{f,g,h} \delta_1^f \delta_1^g f_2 \delta_2^h g_3 h_3 = \sum_{a,b,c} \delta_1^b \delta_1^c \delta_2^a c_2 a_3 b_3$$

Since $\delta^x \delta^y = \delta^x(y)\delta^x$ the equation simplifies to

$$\sum_{f,h} \delta_1^f f_2 \delta_2^h f_3 h_3 = \sum_{a,c} \delta_1^c \delta_2^a c_2 a_3 c_3$$

Comparing terms in the first and third component we must have $f = c$ and $fh = ac$ and hence $a = fhf^{-1}$. From the second component we find $f\delta^h = \delta^{fhf^{-1}} f$ exactly the product rule prescribed by the Drinfeld double.

To get a full quantum invariant we must also find the value of the rotation α. Referring to the first lecture we start by computing the Drinfeld element $U = \sum_g g^{-1} \delta^g = \sum_g \delta^g g^{-1}$. We also need to compute $S(U) = \sum_g \delta^{g^{-1}} g = U$. We have seen that $\alpha^2 = US(U)$ so we may take $\alpha = U$.

We now consider a representation $\rho : D(\mathbb{C}G) \to \text{End}(\mathbb{C}G)$ defined by $\rho(\phi h)(a) = \phi(hah^{-1})hah^{-1}$ for any $a \in \mathbb{C}G$. The reader should check that $\rho(xy) = \rho(x)\rho(y)$ (Exercise!). In this representation the crossing is sent to the following map $\rho(R_{ij}) \in \text{End}(\mathbb{C}G_{\{ij\}})$.

$$\rho(R_{ij})(a_i b_j) = \sum_g \rho(\delta^g)_i \rho(g)_j (a_i b_j) = \sum_g \delta^g(a)a_i(gbg^{-1})_j = a_i g_j b_j g_j^{-1}$$

Presenting a knot as the closure of a braid $\beta = \prod \sigma$ we get a map $\rho(\beta) \in \text{End}(\mathbb{C}G_{\{1,2,\dots,n\}})$. Using the standard basis $\{g^i\}$ any basis element $g_1^1 g_2^2 \dots g_n^n$ of $\mathbb{C}G_{\{1,2,\dots,n\}}$ gets sent by $\rho(\beta)$ to some other element in such a way that we can mark the arcs of the braid diagram with group elements such that the basis element is below and the image is on top and at each crossing the incoming under-arc gets conjugated by the upper arc to produce the outgoing under-arc. Taking the trace of $\rho(\beta)$ sums over the initial basis elements and forces the output at the top to be equal to the input. This is precisely the setup for the Wirtinger presentation of the knot group. Hence we see that $\text{Tr}\rho(\beta) = \#\{\text{representations of } \pi_1(S^3 - K) \text{ into } G\}$.

4.3 $U_q \mathfrak{sl}_2$ as a Drinfeld Double

Coming back to our initial object of interest $U_q \mathfrak{sl}_2$ we would like to show how what we learned so far helps to demystify the formulas we started with. So far we studied finite groups, to make the connection to Lie groups and their algebras we should replace the group ring $\mathbb{C}G$ by the universal enveloping algebra $U(\mathfrak{g})$. This is the algebra of formal monomials of Lie algebra elements modulo the relations

$[X, Y] = XY - YX$. Roughly the idea is that every element of a Lie group is generated by elements $\exp(X)$, where $X \in \mathfrak{g}$. Such exponentials naturally produce sums of Lie algebra elements that we may interpret in $U(\mathfrak{g})$.

Focusing on $\mathfrak{g} = \mathfrak{sl}_2$ with generators E, F, H its universal enveloping algebra $U(\mathfrak{sl}_2)$ is the algebra with 1 generated by E, F, H subject to the relations $HE - EH = 2E$ and $HF - FH = -2F$ and $EF - FE = H$ (compare to the formulas in lecture 1). The natural way to turn this into a Hopf algebra is to set the coproduct to be $\Delta^i jk(X_i) = X_j + X_k$ and $\epsilon(X) = 0$ and $S(X) = -X$ for $X = E, F$ or H.

To understand how $U_q \mathfrak{sl}_2$ arises from this simple setup we restrict ourselves to the Lie subalgebra $\mathfrak{b}$ generated by E, H only. By the same formulas the universal enveloping algebra $U(\mathfrak{b})$ is still a Hopf algebra. We claim its Drinfeld double $D(U(\mathfrak{b}))$ is almost isomorphic to $U(\mathfrak{sl}_2)$. The only difference is that in $D(U(\mathfrak{b}))$ one gets generators H, E and H^* and E^*. It is natural to identify F with E^* but to get $U(\mathfrak{sl}_2)$ one has to quotient out by the additional relation $H^* = H$.

Already this Drinfeld double gives an interesting knot invariant, it produces the Alexander polynomial of a knot (Exercise!).

To get the quantized enveloping algebra $U_q \mathfrak{sl}_2$ we follow the same procedure, but we use a modified version of $U(\mathfrak{b})$ called $U_q(\mathfrak{b})$. Its relations are $HE - EH = 2E$ as before but with a modified coproduct $\Delta(H) = H_1 + H_2$ and $\Delta(E) = E_1 q^{\frac{H_2}{2}} + E_2$, with $q = e^h$. This may seem arbitrary but there are not many possibilities if one wants a co-associative Δ that equals the usual one setting $q = 1$.

Applying the double construction to this modified $U_q(\mathfrak{b})$ yields $U_q \mathfrak{sl}_2$ after setting $H^* = H$. The form of the R-matrix R should now be recognizable as consisting of dual basis F^n and basis E^n. The coefficients are there to normalize properly [1].

4.4 The Dual of $U_q \mathfrak{sl}_2$

In our finite group examples the dual of the group algebra $\mathbb{C}G$ was Fun(G). In the context of Lie groups and algebras it still makes sense to talk about the functions on the group but it is easier if one only allows nice functions. For $G = SL(2)$ we consider the polynomial functions on G. These are all generated by the matrix elements. In this way we see that $\text{Fun}(SL(2)) = \mathbb{C}[a, b, c, d]/(ad - bc = 1)$.

$\mathbb{C}G^*$ is supposed to be isomorphic to Fun(G) and in our present example there is at least a way to pair the equivalent $U(\mathfrak{sl}_2)$ in a non-degenerate way with Fun($SL(2)$). Given $X \in \mathfrak{sl}_2$ and $f \in \text{Fun}(G)$ we consider $f \rho_2(X)$, i.e. evaluation of f on the representation of X as a 2 by 2 matrix.

Instead of deforming the Lie algebra we consider deforming the dual matrix group. This gives a different perspective on $U_q \mathfrak{sl}_2$ or rather it dual. To find a natural way of deforming Fun($SL(2)$) we call on one of the essential properties of $SL(2)$ namely its action on the plane. This sends a vector with coordinates (x, y) to a new vector in the plane with coordinates $(ax + by, cx + dy)$. Manin [7] proposed to first deform the functions on the plane, that is $\mathbb{C}[x, y]$. His idea was that any such deformation will naturally lead to a deformation of the symmetries of the plane

$SL(2)$. All we need to do is insist that the deformed $SL(2)$ still acts as symmetries of the deformed plane.

Calling on quantum mechanics a natural way to view a pair of coordinates is as position z and momentum p of a single particle on a line. These coordinates naturally deform to a pair of non-commuting variables Z, P satisfying the Heisenberg commutation relation $ZP - PZ = -ih$. Setting $x = e^Z$ and $y = e^P$ we find (Exercise!) that $yx = qxy$ with $q = e^{-ih}$ and absorbing the factor $-i$ into h we find our desired quantum plane, or rather its functions on it form the following algebra: $\mathrm{Fun}_q(\mathbb{C}^2) = \mathbb{C}\langle x, y\rangle/(yx = qxy)$. Here the angled brackets mean non-commuting polynomials in x, y.

With this preparation we can carry out Manin's proposal and ask what commutation relations a, b, c, d need to satisfy so that the quantum plane is preserved. Since (x, y) satisfy $yx = qxy$ we require $(ax + by, cx + dy)$ to satisfy the same relation: $(cx + dy)(ax + by) = q(ax + by)(cx + dy)$. Assuming a, b, c, d commute with x, y and q is a scalar we can compare coefficients of x^2 on the left and right hand side to find $ca = qac$.

To get more commutation relations we also require the transposed action (on row vectors) to be preserved so (x, y) gets sent to $(ax + cy, cx + dy)$. This yields the additional relation $(bx + dy)(ax + cy) = q(ax + cy)(bx + dy)$. The reader should check that these requirements yield the following six relations: $ba = qab, ca = qac$, $dc = qcd, db = qbd$ and $cb + qda = qad + q^2bc$ and $bc + qda = qad + q^2cb$. Assuming $q^2 \neq 1$ the last two are equivalent to $bc = cb$ and $ad - q^{-1}bc = da - qbc$. We recognize this last equation as statement about the deformed determinant. It expresses the fact that $ad - q^{-1}bc$ is a central element so we may quotient out by the relation $ad - q^{-1}bc = 1$ to obtain $\mathrm{Fun}_q(SL(2)) =$

$$\langle a, b, c, d\rangle/(ba = qab, dc = qcd, db = qbd, dc = qcd, bc = cb, ad - q^{-1}bc = 1)$$

Viewing the a, b, c, d as deformed matrix elements we may still use the same duality pairing with ρ_2 of $U_q\mathfrak{sl}_2$ to see that the two deformations are compatible and actually dual. Our arguments with the quantum plane are just another way of describing and motivating the Drinfeld double construction. In the end it all comes down to the same non-commutative geometry, whose applications to low-dimensional topology and other fields are endless.

4.5 Exercises

Exercise 1 Prove that both the group algebra of a finite group and its dual are Hopf algebras.

Exercise 2 Compute the Alexander polynomial of the trefoil by looking at the quantum invariant coming from the Drinfeld double of $U(\mathfrak{b})$ with the usual Hopf algebra structure.

Exercise 3 Show that the q-determinant $ad - q^{-1}bc$ in the last section is indeed central. What is the right notion of q-trace here?

Exercise 4 Find an interesting Hopf algebra and use the Drinfeld double construction to produce your own quantum invariant!

5 Epilogue

The purpose of these notes was to introduce quantum invariants, show that they connect to interesting parts of mathematics and convince the reader to construct their own invariants. Hopefully our discussion of $U_q\mathfrak{sl}_2$ makes the generators and relations we started with look less arbitrary. Much more can be said for example in terms of deformation quantization of Poisson-Lie groups but it is also important to note that the field of quantum invariants is still young. The simple things always come last and we're not there yet.

References

1. Etingof, P., Schiffmann, O.: Lectures on Quantum Groups. International Press, Somerville (2002)
2. Frohman, C., Gelca, R., LoFaro, W.: The A-polynomial from the noncommutative viewpoint. Trans. Am. Math. Soc. **354**, 735–747 (2002)
3. Garoufalidis, S.: On the characteristic and deformation varieties of a knot. In: Proceedings of the Casson Fest, Geometry and Topology Monographs, vol. 7 (2004)
4. Garoufalidis, S.: The Jones slopes of a knot. Quantum Topol. **2**, 43–69 (2011)
5. Kashaev, R.: A link invariant from quantum dilogarithm. Mod. Phys. Lett. A. **10**, 1409–1418 (1995)
6. Kauffman, L.: Rotational virtual knots and quantum link invariants (2015). Preprint. arXiv:1509.00578
7. Manin, Y.: Quantum Groups and Non Commutative Geometry. Centre De Recherches Mathematiques, Montréal (1988)
8. Ohtsuki, T.: Quantum Invariants. World Scientific, Singapore (2001)
9. Zagier, D.: Quantum modular forms. In: Quanta of Maths, Clay Mathematics Proceedings, vol. 11 (2010)

Printed in the United States
By Bookmasters